THE EFFECTS OF DRUG ABUSE ON THE HUMAN NERVOUS SYSTEM

THE EFFECTS OF DRUG ABUSE ON THE HUMAN NERVOUS SYSTEM

Edited by

BERTHA MADRAS AND MICHAEL KUHAR

Amsterdam • Boston • Heidelberg • London
New York • Oxford • Paris • San Diego
San Francisco • Sydney • Tokyo

Academic Press is an imprint of Elsevier

Academic Press is an imprint of Elsevier

The Boulevard, Langford Lane, Kidlington, Oxford OX5 1GB, UK

Radarweg 29, PO Box 211, 1000 AE Amsterdam, The Netherlands

225 Wyman Street, Waltham, MA 02451, USA

525 B Street, Suite 1800, San Diego, CA 92101-4495, USA

First edition 2014

Library of Congress Cataloging-in-Publication Data

The effects of drug abuse on the human nervous system / edited by Bertha Madras, Michael Kuhar. -- First edition.
   p. ; cm.
  Includes bibliographical references.
  ISBN 978-0-12-418679-8
I. Madras, Bertha, editor of compilation. II. Kuhar, Michael J., editor of compilation.
  [DNLM: 1. Substance-Related Disorders--physiopathology. 2. Nervous System--drug effects. 3. Risk Factors. 4. Substance-Related Disorders--epidemiology. WM 270]
  RC564
  362.29--dc23
            2013039438

British Library Cataloguing in Publication Data

A catalogue record for this book is available from the British Library

ISBN: 978-0-12-418679-8

The charcoal drawing on the cover is by Vivian Felsen and is from the collection of Bertha Madras.

For information on all Academic Press publications visit
our web site at store.elsevier.com

Printed and bound in USA
14 15 16 17 18   10 9 8 7 6 5 4 3 2 1

CONTENTS

LIST OF CONTRIBUTORS

Peter H. Addy
Psychiatry Service, VA Connecticut Healthcare System, West Haven, CT

Scott Bowen
Department of Psychology, Wayne State University, Detroit, MI, USA

Kathleen T. Brady
Mental Health Service, Ralph H. Johnson VA Medical Center, Charleston, SC, USA; Department of Psychiatry and Behavioral Sciences, Medical University of South Carolina, Charleston, SC, USA

James Robert Brašić
Section of High Resolution Brain Positron Emission Tomography Imaging, Division of Nuclear Medicine, The Russell H. Morgan Department of Radiology and Radiological Science, The Johns Hopkins University School of Medicine, Johns Hopkins Outpatient Center, Baltimore, MD, USA

Andreas Büttner
Institute of Forensic Medicine, University of Rostock, Rostock, Germany

Ryan HA. Chan
Addiction and Pharmacology Research Laboratory, California Pacific Medical Center Research Institute, CA, USA

Domenic A. Ciraulo
Department of Psychiatry, Boston Medical Center, Boston University School of Medicine, Boston, MA, USA

Wilson M. Compton
Division of Epidemiology, Services and Prevention Research, National Institute on Drug Abuse, Bethesda, MD, USA

Caryne P. Craige
Department of Pharmacology and Center for Substance Abuse Research, Temple University School of Medicine, Philadelphia, PA, USA

Silvia L. Cruz
Departamento de Farmacobiología, Cinvestav, Sede Sur, Mexico, Federal District, Mexico

Tomas Drgon
Molecular Neurobiology Branch, NIH-IRP (NIDA), Baltimore, MD, USA

Deepak Cyril D'Souza
Psychiatry Service, VA Connecticut Healthcare System, West Haven, CT; Abraham Ribicoff Research Facilities, Connecticut Mental Health Center, New Haven, CT; Department of Psychiatry, Yale University School of Medicine, New Haven, CT

Nicole M. Enman
Department of Pharmacology and Center for Substance Abuse Research, Temple University School of Medicine, Philadelphia, PA, USA

A. Eden Evins
Department of Psychiatry, Massachusetts General Hospital, Boston, MA; Harvard Medical School, Boston, MA

William E. Fantegrossi
Department of Pharmacology and Toxicology, University of Arkansas for Medical Sciences, Little Rock, AR, USA

Larry Gentilello
Department of Surgery, University of Texas, Dallas, TX, USA

Aryeh I. Herman
Department of Psychiatry and VA Connecticut Healthcare System, School of Medicine, Yale University, West Haven, CT, USA

Harold Kalant
Department of Pharmacology & Toxicology, University of Toronto, ON, Canada; Centre for Addiction and Mental Health, Toronto, ON, Canada

Jongho Kim
Department of Diagnostic Radiology and Nuclear Medicine University of Maryland Medical Center, Baltimore, MD, USA

Stephen J. Kish
Human Brain Laboratory, Centre for Addiction and Mental Health, Departments of Psychiatry and Pharmacology, University of Toronto, Toronto ON, Canada

Marsha Lopez
Division of Epidemiology, Services and Prevention Research, National Institute on Drug Abuse, Bethesda, MD, USA

Bertha Madras
Department of Psychiatry, Harvard Medical School and Massachusetts General Hospital, New England Primate Research Center, Southborough, MA, USA

Diana Martinez
Department of Biological Sciences, Rutgers University, Newark, NJ, USA

Una D. McCann
Department of Psychiatry, The Johns Hopkins University School of Medicine, Baltimore, MD, USA

John E. Mendelson
Addiction and Pharmacology Research Laboratory, California Pacific Medical Center Research Institute, CA, USA

David E. Nichols
Division of Chemical Biology and Medicinal Chemistry, Eshelman School of Pharmacy, University of North Carolina, Chapel Hill, NC, USA

Mark Oldham
Department of Psychiatry, Boston Medical Center, Boston University School of Medicine, Boston, MA, USA

Zev Schuman-Olivier
Department of Psychiatry, Cambridge Hospital, Cambridge, MA; Cambridge Health Alliance, Somerville, MA; Harvard Medical School, Boston, MA

Rajiv Radhakrishnan
Psychiatry Service, VA Connecticut Healthcare System, West Haven, CT; Department of
Psychiatry, Yale University School of Medicine, New Haven, CT

Mohini Ranganathan
Psychiatry Service, VA Connecticut Healthcare System, West Haven, CT; Abraham Ribicoff
Research Facilities, Connecticut Mental Health Center, New Haven, CT; Department of
Psychiatry, Yale University School of Medicine, New Haven, CT

George A. Ricaurte
Department of Neurology, The Johns Hopkins University School of Medicine, Baltimore, MD, USA

Cendrine Robinson
Department of Medical and Clinical Psychology, Uniformed Services University of the Health
Science, Bethesda, MD, USA

R. Andrew Sewell
Psychiatry Service, VA Connecticut Healthcare System, West Haven, CT; Abraham Ribicoff
Research Facilities, Connecticut Mental Health Center, New Haven, CT; Department of
Psychiatry, Yale University School of Medicine, New Haven, CT

Patrick D. Skosnik
Psychiatry Service, VA Connecticut Healthcare System, West Haven, CT; Abraham Ribicoff
Research Facilities, Connecticut Mental Health Center, New Haven, CT; Department of
Psychiatry, Yale University School of Medicine, New Haven, CT

Mehmet Sofuoglu
Department of Psychiatry and VA Connecticut Healthcare System, School of Medicine, Yale
University, West Haven, CT, USA

Luke E. Stoeckel
Department of Psychiatry, Massachusetts General Hospital, Boston, MA; Harvard Medical
School, Boston, MA

Pierre Trifilieff
School of Public Health, Columbia University, New York, NY, USA

George R. Uhl
Molecular Neurobiology Branch, NIH-IRP (NIDA), Baltimore, MD, USA

Ellen M. Unterwald
Department of Pharmacology and Center for Substance Abuse Research, Temple University
School of Medicine, Philadelphia, PA, USA

Donna Walther
Molecular Neurobiology Branch, NIH-IRP (NIDA), Baltimore, MD, USA

Andrew J. Waters
Department of Medical and Clinical Psychology, Uniformed Services University of the Health
Science, Bethesda, MD, USA

Naimah Weinberg
Division of Epidemiology, Services and Prevention Research, National Institute on Drug
Abuse, Bethesda, MD, USA

Erika Weisz
Department of Psychiatry, Massachusetts General Hospital, Boston, MA

CHAPTER ONE

Drug Use and Its Consequences

Bertha Madras
Department of Psychiatry, Harvard Medical School and Massachusetts General Hospital,
New England Primate Research Center, Southborough, MA, USA

1. INTRODUCTION

Humans are explorers of territory, new ideas, social contacts, mates, and sources of food. Successful exploration produced rewards, reinforced behaviors, and enhanced survival. Over millennia, our ancestors explored plants as food sources and serendipitously discovered that certain plants engendered unique rewarding stimuli. Some ingested phytochemicals were mildly arousing (e.g. nicotine, caffeine), others enhanced mood or altered perception, reduced dysphoria and pain, or intoxicated with mild or intense euphoria (alcohol, marijuana, hallucinogens, opiates, cocaine). Over the past two centuries, consumption of these substances expanded exponentially. Isolation from source materials, purification, chemical modification, delivery by chemical mechanisms or devices for maximum effect, and global marketing contributed to this expansion. Modern chemistry, production, and marketing methods produced an array of consumed drugs capable of generating hedonic signals that usurped motivational and volitional control of behaviors essential for survival. Drug use (tobacco, alcohol, other drugs) now accounts for nearly 25% of deaths annually in the United States. Death is not the sole peril. We have witnessed an unprecedented level of adverse biological, behavioral, medical, and social consequences.

1.1. Early Origins

The use of psychoactive drugs for religious, ritualistic, and medical purposes is an ancient practice, documented in texts, evidenced in artifacts (e.g. seeds, pipes), in trace chemical signatures, and artistic and sculptural images. References to excessive alcohol consumption are found in ancient, historical documents and literary prose (e.g. the historian Josephus, and William Shakespeare). Opiates are implicated in this quote from Homer (ninth century B.C.): "presently she cast a drug into the wine of which they drank to lull all pain and anger and bring forgetfulness of every sorrow…" Opiates were used for medicinal or psychoactive purposes, as they migrated from Sumeria to India, China and Western Europe (Brownstein, 1993). By the sixteenth century, manuscripts describing opioid drug abuse and tolerance were published in various countries, a consequence greatly

The Effects of Drug Abuse on the Human Nervous System
http://dx.doi.org/10.1016/B978-0-12-418679-8.00001-0
1

accelerated by the isolation of morphine from the opium poppy in 1803 (Brownstein, 1993). Marijuana is another ancient drug, used by Eastern cultures for medical and psychoactive purposes. Physical evidence of its use was found in ashes beside a skeleton of a 14-year-old girl, apparently in the midst of a failed breech birth (Zias et al., 1993). Cryptic mentions of mystifying drug effects in ancient texts (Dannaway et al., 2006; Dannaway, 2010), or religious prohibitions are scattered in various sources, but there is scant evidence that ancient drug use was as extensive or propagated the same public health, welfare, safety concerns, and consequences or responses, as in modern times.

1.2. Modern Era

The past two centuries have witnessed an exponential rise in drug use and a corresponding increase in associated consequences. The increase has been fueled by modernization: (1) the discovery and cross-cultural propagation of psychoactive drugs by explorers of new continents; (2) the advent of organic chemistry, which enabled isolation of pure, potent drugs from plants (e.g. cocaine, morphine) and de novo synthesis of new drugs (e.g. oxycodone, methamphetamine, amphetamine, cannabicyclohexanol) guided by structures of isolated phytochemicals (morphine, ephedrine, THC or Δ-9-tetrahydrocannbinol); (3) the development of modern drug delivery systems, the needle/syringe, the cigarette-rolling machine, and synthetic salt forms of drugs that enable efficient drug delivery systems; (4) the advent of sophisticated agricultural and purification methods increased drug concentrations in plants and improved crop yields; (5) modern capitalism increased prosperity and expendable income across classes. Expansion of user markets raised the profitability for manufacturing and sales of drugs; (6) sophisticated global marketing exploited modern, efficient communication and transportation systems; (7) cultural shifts eroded parental/family oversight at earlier stages of development; (8) drug use was normalized by cultural icons, media, and internet sites; (9) drug use was promoted by wealthy individuals for cryptic reasons, by underwriting state ballot and legislative initiatives to promote drug normalization, and by profit-seeking industries using advertising targeted to youth. The net effect was to make highly potent drugs widely available. A new enterprise, distribution of simple chemicals isolated and purified from plants, was born (Figure 1). The nineteenth century came to a close, with a cocaine and morphine epidemic in the United States, and a severe opium epidemic, especially in Asia. The twentieth century closed with global marketing and consumption of an array of phytochemicals and synthetic, "designer" drugs. The trajectory of the twenty-first century will be driven by national and international laws, regulations, shaped by biomedical science and informed public opinion.

In the late nineteenth century, drugs were advertised, freely available, unregulated in patent medicines, sold freely in drugstores, dissolved in popular drinks (colas and wines), as tonics, elixirs, and remedies. The major drugs at that time, heroin, morphine, cocaine,

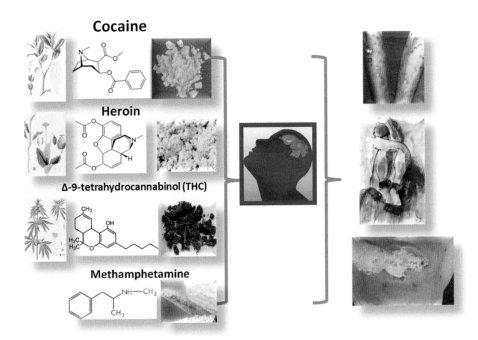

Figure 1 The biology of addiction. A simple ingested chemical, isolated from a plant and of molecular weight less than 1000, can profoundly affect the brain and body. On top right is a photo of "skin popping", a method of injection of cocaine under the skin that leaves lesions. The bottom right is a photo of a person with respiratory depression, resulting from a heroin overdose.

and marijuana, were marketed without restraint and had vocal or covert supporters, including high-profile physicians, Sigmund Freud and William Halsted, who succumbed to severe addiction (Musto, 1968, 2002; White, 1998; Musto et al., 2002; Gay et al., 1975; Cohen, 1975).

Problems with cocaine were evident from the beginning. By the turn of the twentieth century, 200,000 people are estimated to have been addicted to drugs in the United States alone. Increased availability, rapid rates of brain entry, distribution of multiple drugs, and initiation by younger populations more susceptible to addiction created an unfettered market for drugs.

1.3. Advent of Regulations and Laws

The adverse consequences aroused attention and legislative responses from physicians, national governments, and international organizations. As the medical historian David Musto stated "from repeated observation of the damage to acquaintances and society", awakened national and international governments to counteract these trends with regulatory, taxation and laws. In 1875 opium dens were outlawed in San Francisco. In 1906, the federal government passed the Pure Food and Drugs Act, a law a quarter-century in the making, that prohibited interstate commerce in adulterated and misbranded food and drugs and required accurate labeling of patent medicines containing opium and other drugs. The modern regulatory functions of the Food and Drug Administration (FDA) began with the passage of the 1906 Pure Food and Drugs Act, which provided basic elements of protection that consumers had not known before that time. Despite rapid metamorphic changes in our medical, cultural, economic, and political institutions over the past century, the core public health mission of the FDA retains a protective barrier against unsound claims and unsafe, ineffective drugs.

The 1906 legislation was extended by passage of the Harrison Act in 1914 forbidding the sale of narcotics or cocaine, except by licensed physicians. Regulatory mechanisms marched in tandem with newly emerging drugs, restricting harm to individuals by restricting access to drugs. Prior to the 1960s, Americans did not see drug use as an acceptable behavior, or an inevitable fact of life. Tolerance of drug use led to a dramatic rise in crime between 1960s and early 1990s, and the landscape of America was altered forever, (DEA). Consequently, the Drug Enforcement Administration (DEA) was created in 1973 by Executive Order to establish a single unified command over legal control of drugs and address America's growing drug problem. Congress passed the Controlled Substances Act to consolidate and replace, by then, more than 50 pieces of drug legislation. It established a single system of control for both narcotic and psychotropic drugs for the first time in the U.S. history (DEA).

Since the creation of the DEA, drug policy has been debated as choices between activists for free access to drugs and advocates for restrictive policies (Dupont et al., 2011). Activists view regulations as a restraint on their right to freely pursue "victimless" drug-induced pleasure, expansion of consciousness and of potential, self-medication, and profit. They are buoyed by narrow views that claim few people become addicted and that some addicts are productive. For example, Nikki Sixx documents in his book "The Heroin Diaries: a Year in the Life of a Shattered Rock Star", his ability, albeit limited, to perform during a severe addiction. Significantly, drug use is highest among 15–25 year olds, the "age of invulnerability". Advocates of stringent policies view drug use issues through a prism of human health, welfare, social, and safety concerns. The resistance to drugs and a shift in perceptions takes years to penetrate the public opinion, when drug use becomes viewed as reducing natural potential, and the consequences of drugs in family members, schools, and the workplace begin to take a toll (Musto, 1995). The counterclaims to restrictive legal and social containment of drug commerce and consumption are based on drug-seeking and use as historical, normative, acceptable, inevitable, a rite of passage, an expression of personal liberty, an extension of natural potential, and a victimless social activity. Some advocates acknowledge the evidence that drugs can produce adverse consequences to individual users (overdose and death, HIV-AIDS, dehydration), and focus on reducing "drug-associated harm". Needle exchange programs (designated syringe exchange program by advocates to substitute the pejorative delivery system "needle" with a container designation "syringe"), provision of water bottles at ecstasy-infused "rave" parties, and advocating for over-the-counter naloxone for opioid overdose crises, are practical solutions to "harm reduction". Reducing supply or demand for drugs and prevention and intervention program are not emphasized in this movement.

From my perspective, "harm reduction" is incompatible with strong evidence from addiction biology and medicine that drug use is associated with unacceptable, elevated risks in multiple domains: physical, mental, cognitive, behavioral, safety, education, and employment. Fundamental questions are rarely addressed in the case made for legalization: "will addiction rates rise and will people who initiate drug use intend to become addicted?"; "Do recovering addicts regret their recovery and desire the addicted state?" "Do addicted people benefit more, physically, personally, socially, emotionally and psychologically, from programs/services that accept and facilitate continued, uncontrollable drug use or from treatment programs and recovery services that emphasize abstention?"; "Is drug use a victimless activity?" Acceptance of the inevitability of use and mitigation of potential adverse consequences, without advocacy for prevention and treatment per se, is poor medical, public health, and national policy.

1.4. Current Legislative Initiatives

The front lines of this debate reside in the status of marijuana. In a 1980 Gallup Poll, 53% of the population favored legalization. Within 6 years, the number fell to 27%, and by 2011, it rose to 50%. Currently, there is a concerted political and media campaign to erode or eliminate many of the federally driven legal constraints, implemented over the past century, with the goal of legalizing marijuana, initially as a medicine. At the federal level, the FDA reaffirmed in 2006, that "there are no sound scientific studies supporting the medical use of marijuana. Bills introduced to legalize marijuana and to restrict the reach of the FDA in states that approved marijuana as a medicine have not progressed through the legislative process. The DEA reaffirmed that it would not shift marijuana to Schedule II, the IRS issued a ruling that prohibits business-related tax deductions for businesses selling or cultivating marijuana. Most of the legislative actions are occurring at the state level. In 2011, no fewer than 130 pro-legalization legislative bills were introduced in states, more than double the rate in 2008 (SOS Annual Report, 2011). In the same year, 49 states introduced 299 pieces of antidrug legislation, and of these 77 were signed into law. Initiatives defeated introduction of medical marijuana in 16 states, legalization of taxation and regulation bills in five states, and decriminalization in six states. Some focused on the status of marijuana, others restricted the sale and possession of a synthetic cannabinoid (K2), "spice and bath salts", or promoted veteran's drug treatment courts, enhanced prescription drug monitoring laws, and established Good Samaritan laws.

2.1. DRUGS AND CONSEQUENCES

2.1.1. Drug Classes and Types

The Drug Enforcement Agency (DEA) has drawn legal distinctions among substances, but these boundaries are frequently infringed by polysubstance abusers. Drugs in various chemical classes rise to the level of national concern regardless of legal classification: (1) nicotine, a legal nonintoxicating, but addictive drug with significant health risks; (2) alcohol, a legal drug, which in small, infrequent doses, is nonintoxicating and has health benefits but, at high and/or frequent doses, is an intoxicating, addictive drug with significant health risks; (3) inhalants, legal substances such as toluene or gasoline, which engender psychoactive, intoxicating effects, can be severely addictive and cause serious neurological and organ system damage; (4) Schedule I drugs are illegal, have high potential for abuse, and no currently accepted medical use in treatment. Examples of these include heroin and other specific synthetic opioids, cathinone, LSD (lysergic acid diethylamide) and other hallucinogens, marijuana, PCP, mescaline, mephedrone, GHB (gamma-hydroxybutyrate), MDMA or ecstasy (3,4-methylenedioxymethamphetamine); (5) Legal psychoactive drugs in Schedule II that have accepted medical use but a high potential for abuse, which may lead to severe psychological or physical dependence. Examples of these drugs are opioids (methadone, oxycodone, hydromorphone, meperidine, fentanyl), stimulants (cocaine, amphetamine, methamphetamine, methylphenidate), and barbiturates; (6) Drugs in Schedule III (hydrocodone, buprenorphine, codeine, ketamine and anabolic steroids), IV (propoxyphene, benzodiazepines), and V (low dose narcotics for cough, diarrhea, or pain) have decreasing abuse liability, but accepted medical uses.

Beyond these classifications are other forms of encroachment of drug schedules and regulations. Prescription medications, largely opioid analgesics, may be legally obtained by a single physician or by doctor-shopping and used inappropriately by intended or unintended persons for psychoactive purposes. They have high abuse, addictive, and overdose potential. Some prescription drugs are obtained without an appropriate diagnosis with/without prescription are used by intended/unintended persons, for "chemical coping", "self-medication", and functional improvement (e.g. antihyperactivity drugs for cognitive enhancement, opioid analgesics for sleep, anxiety problems, anabolic steroids for performance enhancement). Some over-the-counter psychoactive drugs (e.g. cough medicines) with overdose and addictive potential are used for psychoactive purposes. In the past, some over-the-counter drugs were purchased as precursors for production of illegal drugs (e.g. pseudoephedrine converted to methamphetamine). Marijuana can be "recommended", but not prescribed by a physician in states that have approved its use, is not approved by the FDA. There is a scant evidence for a number of medical conditions (e.g. Alzheimer's disease) embedded in state ballot initiatives and claimed to be alleviated by the inhaled smoke. The legal and powerful hallucinogen Salvinorin A has inadequate

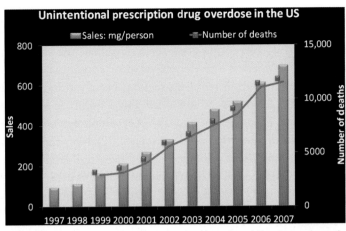

Figure 2 The past decade has seen a steep rise in unintentional drug overdose deaths in the US, primarily prescription opioids, that paralleled an unprecedented rise in prescriptions for opioid pain-relievers. *Paulozzi et al. (2011).*

evidence for regulating its disctribution. Naloxone, a prescription drug and an opioid receptor antagonist that effectively reverses opioid overdose, is currently dispersed to heroin addicts, friends, family for at home resuscitation. Emerging designer drugs, including K2, mephedrone, MDPV, or methylenedioxypyrovalerone, some distributed as "bath salts" (Murray et al., 2012; Prosser et al., 2012), have resulted in emergencies related to their pharmacological effects.

Escalating use of prescription drugs for nonmedical purposes, primarily opioid analgesics in Schedule II and III, is a new public health challenge: nonmedical use of opioid prescription drugs currently ranks second, after marijuana in number of users, number of new initiates of drug use (NSDUH, 2011), second among those dependent on illicit drugs (TEDS-prescription pain-relievers), and first among drug-related deaths, which currently exceed heroin- or cocaine-related deaths (Paulozzi et al., 2011). Overdose deaths involving opioid analgesics are correlated with per capita sales of opioid analgesics and now are at the highest levels seen in decades (Figure 2). The phenomenon of nonmedical use of prescription drugs weakens the case for legalization of currently illegal drugs. It demonstrates unequivocally that legal drugs can be diverted and abused for nonmedical purposes and increased access has driven increased addiction and deaths.

2.1.2. Definitions of Use, Abuse, and Addiction

Drugs can be consumed through a variety of routes of administration: oral, insufflation, huffing, inhalation, intravenous, intramuscular, subcutaneous, and others. It is generally accepted that the faster the rate of drug entry into the brain, the higher the addictive potential (Samaha and Robinson, 2005; Spencer et al., 2006). Substance use, abuse, misuse, problematic use risky use, nonmedical use, dependence, addiction, and similar terms are widely used to describe the use patterns. The DSM-IV (American Psychiatric Association, Diagnostic and Statistical Manual of Mental Disorders, 2000) divides abuse and addiction into separate categories; abuse criteria reflect use despite adverse consequences; addiction is defined as use despite adverse consequences combined with evidence of compulsive, uncontrollable drug-seeking behavior and adaptive responses. Specifically, the DSM-IV definition of substance abuse fulfills at least one of the following four criteria in a 12-month period: (1) Recurrent use resulting in failure to fulfill major role obligation at work, home, or school; (2) Recurrent use in physically hazardous situations; (3) Recurrent substance related legal problems; (4) Continued use despite persistent or recurrent social or interpersonal problems caused or exacerbated by substance. Addiction is defined by the following criteria and requires at least three positive responses in a 12 month period: (1) Tolerance (marked increase in amount; marked decrease in effect); (2) Characteristic withdrawal symptoms; substance taken to relieve withdrawal; (3) Substance taken in larger amount and for longer period than intended; (4) Persistent desire or repeated unsuccessful attempt to quit; (5) Much time/activity to obtain, use, recover; (6) Important social, occupational, or recreational activities given up or reduced; (7) Use continues despite knowledge of adverse consequences (e.g. failure to fulfill role obligation, use when physically hazardous). Based on these definitions, over 20 million people in the United States harbor a medical diagnosis of abuse/addiction, (Figure 3) yet do not seek treatment and remain unidentified (NSDUH, 2011).

Previous definitions focused on symptoms of biological adaptation, drug tolerance and withdrawal, but these can occur with drugs that are minimally or nonaddictive. Repeated use of alcohol and heroin, historical prototypes of addiction, engenders

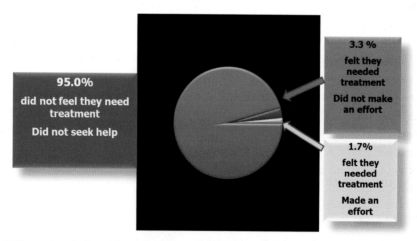

Figure 3 The vast majority of the population with a DSM-IV diagnosis of abuse or addiction, 95%, do not think they have a problem and do not seek treatment; 20.5 million people (8.1%) need and don't receive treatment for illicit drug, alcohol use.

tolerance—diminished pharmacological effects of a fixed drug dose—and intense physical signs of withdrawal during the initial phase of abstinence. In DSM-IV, these traditional criteria were retained but diluted by the preponderance of other criteria emphasizing loss of behavioral control and adverse consequences. The diminishing emphasis on these specific physical signs is justified: First, they are not uniformly applicable to all addictive drugs. For example, withdrawal from cocaine and nicotine is not manifest by physical signs (e.g. vomiting, pain, diarrhea, tremors, piloerection, hand tremor, fever or convulsions). Instead, withdrawal is manifest as dysphoria, irritability, sleep disorders and other nonphysical symptoms; Second, patients treated with high doses of addictive opioids for acute pain, upon drug cessation, can display profuse sweating, nausea, and vomiting, similar to moderate withdrawal from heroin, without manifesting an addictive disorder; and third, prescription drugs with low abuse potential also can engender tolerance and withdrawal symptoms. Neuroadaptation, manifest as psychological or physical withdrawal symptoms, is but a component that may contribute to compulsive drug-seeking.

Behavioral terminology prevails until a consensus emerges on validated biological markers for the disease of addiction, if common markers can be identified. The new DSM-V criteria discard definitions that separate substance abuse and dependence, eliminates the term dependence, and substitutes "use disorder". Severity of the disorder is based on increasing signs of loss of behavioral control and adverse consequences. A use disorder can be viewed as a chronic, relapsing disease, characterized by compulsive, uncontrollable use, despite adverse consequences, with various levels of severity, gauged by loss of control and adverse consequences (Leshner, 1997). The disease model does not supplant the role of personal responsibility in propagating or attenuating the behavior. Within the construct of a disease, patients are urged to assume responsibility for compliance with treatment, similar to patients with asthma, hypertension, and diabetes.

The biological disease model offers a platform to medicalize this public health challenge, by justifying medications research, reducing stigma, focusing on problem-solving, and increasing treatment availability, including within the criminal justice system. The disease model has penetrated national and international policy decisions. The previous Executive Director of the United Nations Office on Drugs and Crime, Antonio Costa, stated that drug use should be treated as an illness in need of medical help, and appealed for universal access to drug treatment. Yet, integrating a scientific evidence with public policy is not an easy task. It is a historical reality that public policies are forged out of a mosaic of evidence and opinions, propounded by scientific, medical, political, financial and other special interests, moral, ethical, and legal pundits.

2.1.3. Drug Use in the United States

An informed perspective on drugs owes much to factual information gleaned from a variety of reliable sources, some in primary literature, some from foundation-sponsored research, and others from federal data bases, a national repository of vast statistical surveys: Arrestee Drug Abuse Monitoring Program (ADAM), Center for Disease Control and Prevention (CDC), the National Center on Addiction and Substance Abuse at Columbia (CASA), the Drug Abuse Warning Network (DAWN), National Survey of Substance Abuse Treatment Services (DASIS), Department of Labor (DOL), Health Behavior in School-aged Children (HBSC), Monitoring the Future (MTF), National Co-morbidity Survey (NCS), National Epidemiological Survey on Alcohol and Related Conditions (NESARC), National Survey on Drug Use and Health (NSDUH), National Vital Statistics System (NVSS), Treatment Episode Data Set (TEDS), Youth Risk Behavior Surveillance System (YRBSS).

Changing norms and widespread availability has fueled a rapid expansion in drug use in the US and globally. In 1960, four million Americans had ever tried drugs. By 2009, over 100 million people had ever used marijuana alone, with 37.3–43.8% of 12th graders and 51.1% of young adults reported a lifetime use of marijuana (NSDUH, 2011; MTF, 2011). Drug use, abuse, and addiction are among the most prevalent, consequential, costliest, and deadliest of neuropsychiatric disorders. National and international drug policies reflect this public health burden. There are approximately 578,000 drug-related deaths each year, 23% of the 2.5 million deaths in the United States annually. Of these, 443,000 are attributable to tobacco smoking (annual average from 2000 to 2004), 80,000 are attributable to alcohol (annual average 2001–2005), 37,000 are attributable to unintentional drug poisoning (2008), and 18,500 suicides are related to the abuse of addictive substances (half the ~37,000 suicides in 2009). Substance use and addiction permeate every sector of society, and can burden individuals, healthcare systems, the workplace, educational environments, social welfare costs, criminal justice, highway and public safety. Current use of drugs remains high among high school students, with alcohol, marijuana, and cigarettes dominating the use (Figure 4). Trends for specific drugs are inconsistent. Since 2008, while usage of cigarettes and alcohol dropped 7.1%, and 9.3%, respectively, usage of all illicit drugs increased 16.4%, and marijuana use increased 21.6% (Figure 5(A), (B)). The steep rise in marijuana use among youth, including the highest level of daily use seen in 30 years, in marijuana potency and emergency department mentions of marijuana and escalating deaths from opioid overdoses since 1999 are a cause for concern (Figure 6). It is estimated that over 22 million people (nearly 9% of the population over 12 years, Figure 7) harbor a medical diagnosis (DSM-IV) of alcohol or illicit drug abuse/addiction (NSDUH, 2011). More than double this number are engaged in risky, problematic alcohol and other drug use, with increased risk for addiction.

2.1.3.1. The Undiagnosed

As described in Section 2.1.2 (Figure 3), the vast majority of people harboring a substance use disorder (SUD), over 20 million people according to DSM-IV criteria, do not feel they need treatment, do not seek treatment, and do not receive it. Others know they need help but do not obtain it for various reasons (e.g. not ready, insurance and payment

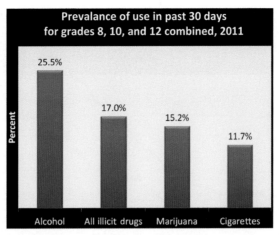

Figure 4 Prevalence of drug use in the past 30 days for grades 8, 10, 12 combined. *Monitoring the Future (2011).*

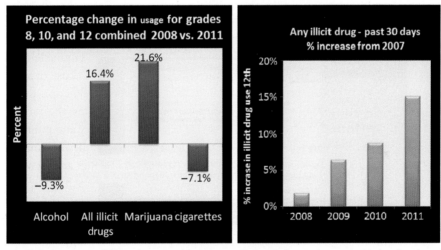

Figure 5 Trends in use of illicit drugs and marijuana (past 30 day use) among high school students. Percent change in drug use from 2008 to 2011 (left). Change in illicit drug use among 12th graders from 2008 to 2011 (right). *Monitoring the Future (2011).*

issues, NSDUH). A different statistic highlights the markedly high incidence of risky, problematic substance use. In healthcare settings, 459,599 people presenting for other medical reasons were offered SBIRT services (screening, brief intervention, referral to treatment, see chapter by Gentilello). Of these, 22.7%(!) self-reported that they engaged in a spectrum of substance use, from risky, problematic, to abuse/addiction that triggered an intervention (Figure 8). SBIRT services reduced risky alcohol use and use of other substances significantly (Figure 9, Madras et al., 2009). Diagnosis and intervention for

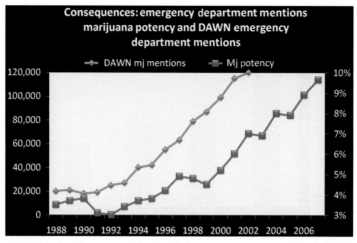

Figure 6 Marijuana potency has increased in tandem with emergency room mentions of marijuana. *Drug Abuse Warning Network (DAWN), Marijuana Potency Monitoring Project, Univ. of Mississippi (Feb 2006).*

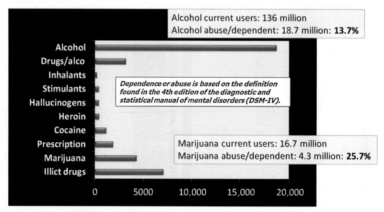

Figure 7 Number of people with abuse of and dependence on various alcohol, other drugs in the United States. *NSDUH (2010).*

the at-risk population and treatment of the disease of addiction, is as essential for this problem, as for any other medical disease requiring prevention and treatment services.

2.1.3.2. At-Risk Populations: The Adolescent

Specific populations have unique vulnerabilities. Adolescent drug use is a unique public health challenge. Aggressive prevention and intervention policies targeted to this population is warranted for these reasons: (1) At least 60% of new initiates fall below the age of 18, with a higher percentage for tobacco and alcohol use (NSDUH); (2) age

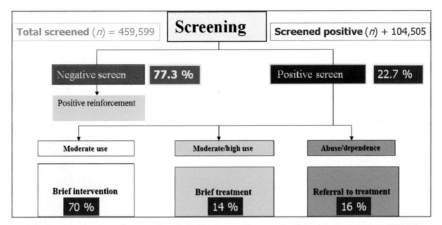

Figure 8 A federal program of screening, brief intervention and referral to treatment (SBIRT) screened 459,599 patients in healthcare settings in six states. Of these, 104,505 or 22.7% of the population presenting in healthcare facilities for other reasons, screened positive for a range of substance use disorders (risky, problematic, abuse, dependence). Of the positive screens, 70% were recommended for a brief intervention, 145 for a brief treatment and 16% were referred to specialty treatment. For an explanation of these services, please see chapter in this book by Gentilello. *Madras et al. (2009).*

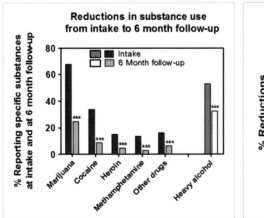 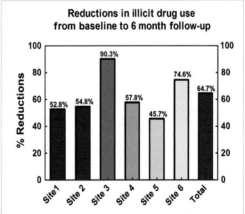

Figure 9 SBIRT services resulted in significant decreases in substance use on the basis of specific drugs (left) and in six different state-based programs (right). *Madras et al. (2009).*

of onset of drug use is declining; (3) daily use of marijuana among adolescents is at its highest level in 30 years (MTF); (4) early initiation confers a higher risk for developing addiction; (5) psychiatric symptoms are higher in adolescent users; (6) drug use is associated with risk-seeking behavior, delinquency, and criminal behavior; (7) adolescent drug use is associated with a discernibly higher likelihood of injury or death; (8) adolescent use is associated with compromised school performance, absenteeism, higher school drop-out rates, gang membership, and later involvement in criminal behaviors (YRBSS). Studies have shown a common order of drug initiation, with alcohol and tobacco leading initiation into marijuana use, followed by use of other illicit drugs (Degenhardt et al., 2011; Degenhardt and Hall, 2012; Mayet et al., 2012). A general risk for substance abuse disorders for any substance in young adulthood is predicted by involvement with alcohol, tobacco, and marijuana during adolescence. Adolescent smoking increases the odds of developing tobacco dependence, and developing abuse or dependence on alcohol and marijuana in young adulthood. Early users do not limit their use to a single drug, and a sampling of substances increases the risk for developing problems of abuse and dependence across a number of substances (Palmer et al., 2009). Another robust finding in adolescent drug use, over the past 30 years, is that almost all adolescents who have tried cocaine and heroin, first used alcohol, tobacco, and marijuana; the more regularly adolescents use marijuana, and the earlier the age at which they begin, the more likely they are to use other drugs (Kandel et al., 1992, 2006). Together, these findings call for multisubstance prevention programs (Chen et al., 2009; Madras et al., 2009; Madras, 2010).

Risk analyses demonstrate considerably higher rates of addiction with early onset of use of marijuana, cocaine, other psychostimulants (e.g. amphetamines), hallucinogens, opioids, inhalants, alcohol use, smoking, and prescription drugs. Not only are the rates higher, but progression to addiction is higher, if involvement with drugs occurs prior to 18 years of age (Anthony et al., 1994; Anthony and Petronis, 1995; Grant and Dawson, 1998; Wagner and Anthony, 2002; O'Brien and Anthony, 2005; Chen et al., 2005; Storr et al., 2005; McCabe et al., 2007; Swift et al., 2008; Palmer et al., 2009; Chen et al., 2009). Early onset of use prior to age 18, of marijuana, cocaine, other psychostimulants (e.g. amphetamines), hallucinogens, opioids, inhalants, alcohol use, smoking, prescription drugs (stimulants, opioid analgesics, sedatives, tranquilizers, anxiolytics), is associated with higher prevalence of addiction in adults.

Environment, psychiatric disorders, and other factors may elevate risk for addiction in the adolescent brain. However, drugs may play a key role in heightening susceptibility for addiction during adolescence. During brain development, a single cell gives rise to 100 billion neurons and an estimated one trillion "supporting" glial cells, in tandem with vascular growth (e.g. Colín-Castelán et al., 2011). Dopamine circuitry, implicated in additive processes, is one of many systems that are guided by proteins that shape brain circuits during development. At least four classes of proteins, slits, robos, semaphorins and ephrins, among others, guide axons toward their targets or remove redundant or unneeded connections (Vanderhaeghen and Cheng, 2010; Dugan et al., 2011; Flores, 2011), They regulate connectivity of billions of neurons within local circuitry or long distance neural networks that control sensory perception, motor activity, cognition, behavior, volitional control, and an array of other functions. This remarkably accurate orchestrated series of events (Mason, 2009) is completed by the mid-twenties (Gogtay et al., 2004; Giedd, 2008; Raznahan et al., 2011). During adolescence, brain development proceeds with new growth, pruning, and reorganization of circuits that regulate impulse control, executive function, and judgment (Chambers et al., 2003; Yurgelun-Todd, 2007). Adolescence is also a period during which dopamine neural circuitry matures, the very system affected by most drugs (Tomasi et al., 2010; Manitt et al., 2011). Drugs can modulate the expression of proteins responsible for neurodevelopment (Bahi and Dreyer, 2005; Jassen et al., 2006) conceivably altering the normal trajectory of adolescent brain and dopamine development, and raising addiction vulnerability. The risk of the adolescent brain could arise if drugs introduced during this phase of neurodevelopment interfere with the trajectory of a normal brain development.

2.1.3.3. At-Risk Populations: The 18–25 Year Old

The 18–25 year old cohort steadfastly remains the highest users and harbors the highest numbers among populations with a DSM-IV diagnosis of abuse/dependence (NSDUH, 2011). On college campuses, marijuana reportedly is associated with concentration problems, driving while high, missing class and placing oneself at risk for physical injury. A significant proportion of cannabis-using college students meet diagnostic criteria for a marijuana-use disorder (Caldeira et al., 2008; Buckner et al., 2010). As drug use is detrimental in the work place, many positions require drug testing as a condition of preemployment or random testing during employment. This requirement affects employability of substance-using populations in this critical age cohort.

2.1.3.4. At-Risk Populations: Older Adults

Drug use among 50–60 years old has nearly doubled in the past 5 years (NSDUH, 2011), demonstrating the persistence of the drug culture of the 1970s. Among the consequences are exacerbation of costly medical conditions such as heart disease and diabetes, which are more common in older adults, higher susceptibility to the damaging effects of drug/medications/alcohol interactions, because they are more likely to take multiple medications and metabolize them more slowly (SAMSHA-OAS, 2009; SAMHSA-older adults; Friedmann et al., 1999). Drinkers are at higher risk for infection because alcohol compromises the immune system, at higher risk for falls and other accidents, involving loss of balance and impaired judgment.

2.1.3.5. Psychiatric Comorbidity

In mental health settings, 20–50% of patients had a lifetime co-occurring substance-use disorder, while conversely, in a substance use treatment center, 50–75% had a co-occurring mental health problem, with the majority not severe. Overall, substance use disorders are present in more than 9% of our population, and more than 9% also have a diagnosable mood disorder. More than five million adults have a serious psychiatric illness combined with a substance use disorder (Swendsen et al., 2010).

2.1.4. Global Drug Use

The United Nations reports annually on global drug issues. Globally, approximately 210 million people, or 4.8% of the population aged 15–64, consumed illicit substances at least once in the previous year, with the overall drug use remaining stable (UNODC, 2011). Although data on marijuana are limited, marijuana remains the most widely consumed illicit substance globally (Degenhardt and Hall, 2011). In 2009, between 2.8 and 4.5% of the world population aged 15–64 (125–203 million people) had used marijuana at least once in the past year. While marijuana production is widespread, notably in the Americas and Africa, marijuana resin production (hashish) continues to be concentrated in just two countries: Morocco, supplying the West European and North African markets, and Afghanistan supplying the markets in Southwest Asia. Over the past decade, cocaine consumption in Europe doubled, but over the last few years, it has remained stable. Demand soared for substances not under international control, such as piperazine and cathinone.

2.2. CONSEQUENCES OF USE

Abuse of alcohol, tobacco, and other drugs is the most deadly and costly behavioral health problem in the United States, accounting for greater than $500 billion annually, which includes expenditures in healthcare, lost productivity, and criminal justice (NIAAA; NCJRS; CDC-alcohol; CDC-tobacco; CDC-deaths; Justice; Uhl and Grow, 2004). It adversely affects every sector of society, with a heightened burden to individuals, to children, healthcare systems, the workplace, educational environments, social welfare costs, criminal justice system, and public safety (CASA-children; SAMHSA-workplace; YRBSS, others).

2.2.1. Substance Use and Health

Medical conditions and healthcare costs are higher for people that harbor a medical diagnosis of abuse/addiction and their family members (Mertens et al., 2003, 2005, 2007; Ray et al., 2007, 2009; Prekker at al., 2009; Ramchaud et al., 2009; London et al., 2009; Buttner, 2011). Medical and psychiatric conditions occur more frequently in people diagnosed with a substance use disorder, increasing suffering and healthcare costs (Figure 10). Substance abuse can be associated with or be a causal agent for: (1) injuries, accidents, trauma, violence, drug crises, and overdose, leading to increased use of emergency departments, associated healthcare costs, lost work time, and added criminal justice costs; (2) exacerbation of medical conditions (diabetes, hypertension, sleep disorders, depression); (3) induction of medical diseases (e.g. stroke, hypertension, cancer, addiction); (4) increased risk of infections, infectious diseases (e.g. HIV-AIDS, Hepatitis C) which can impact employment, family finances and stress; (5) affect medication

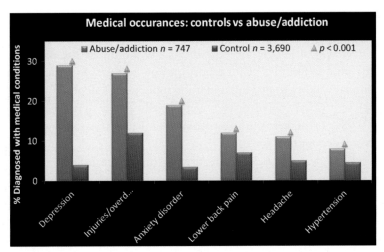

Figure 10 Medical occurrences are higher in patients with a medical diagnosis of abuse/addiction, according to DSM-IV definitions. *Data adapted from Mertens et al., Arch. Int. Med., 2002.*

efficacy; (6) nonmedical use of prescription medications can result in addiction or over-dose; (7) affect the developing fetus with low birth weight, premature deliveries, still-births, and developmental disorders; (8) be associated with a higher incidence of medical conditions in family members of drug users. Federal statistics on these consequences are provided by SAMHSA.gov website.

2.2.2. Substance Use and Crime

A majority of arrestees, probationers, and parolees test positive for illicit drugs. In the 2010 ADAM II, a majority of male arrestees tested positive for an illicit drug. State and local prisons are crowded with prisoners who test positive for drug use and engage in drug distribution. Individuals may be referred to substance abuse treatment through the criminal justice system either as part of a diversionary program before formal adjudi-cation or as part of a formal sentencing program. With steady growth of the source of referrals, adequate resources are needed to meet the needs of criminal justice referrals. Treatment completion and transfer to another level of care are predictors of longer term positive-treatment outcomes. Compared with all other referral admissions, criminal jus-tice system referral admissions were slightly more likely to complete treatment in 2007 (49% vs 46%), and less likely to drop out of treatment (22% vs 27%), indicating that coerced treatment can have positive outcomes.

2.2.3. Substance Use and Highway Safety

The National Highway Traffic Safety Administration (NHTSA) just issued its first report on the incidence of drugged driving in July 2009 (Compton and Berning, 2009). The roadside survey indicated that the percentage of individuals driving with illegal levels of blood alcohol has steadily declined over the past several decades but that a disturbingly high percentage of people are now driving while under the influence of drugs. This led to the production of a major report on how to assess for drugged driving that would have similar validity to driving under the influence of alcohol (NHTSA-drugged driving, 2011) In 1973, 7.3% of drivers were legally drunk with blood alcohol content level of 0.08% or higher; the latest study found that this rate had fallen to 2.2%, the low rates still accounting for more than 13,000 deaths each year on highways. Of nighttime, weekend drivers, 16.3%, were driving under the influence of psychoactive prescription and illegal drugs, as detected in saliva or blood. Heading the illegal drug list were marijuana (8.6%) and cocaine (3.9%), with 3.9% tested positive for prescription or over-the-counter medications. NHTSA is currently conducting research on the relationship between drug levels in motorists and traffic accidents, using the research protocol designed previously to establish hazardous levels of blood alcohol.

2.2.4. Substance Use and the Workplace

Over 75% of illegal drug users hold either full-time or part-time jobs and more than 60% of adults know someone who has worked under the influence of alcohol or drugs. Alcohol and drug abuse create significant safety and health hazards and are associated with adverse outcomes in the workplace. Intoxicants can lead to decreased productivity, fewer work hours, increased absenteeism, poor employee morale, high job turnover, and higher unemployment. Heavy alcohol use is associated with negative attitudes at work, performance problems, and poor work quality. Healthcare costs for employees with alcohol problems are twice those for other employees, with alcohol and drug abusers being 3.5 times more likely to be involved in a workplace accident. Substance abusers also add costs in healthcare claims, especially short-term disability claims. Substance abuse costs American businesses approximately $81 billion annually in lost productivity, absenteeism, poor job performance, and accidents and 500 million workdays are lost annually from employee substance abuse (SAMHSA-work). The elderly in the workplace have an added set of problems (SAMHSA-older adults). To diminish this challenge, Federal employees and certain industries (e.g. transportation, nuclear energy) mandate random drug-testing, with attendant problems.

2.3. BIOLOGY

2.3.1. Biology of Drugs: General Principles

Drawing conclusions on the association of neurobiological and pathological findings in human brain of those with addictive behaviors warrants caution. Research in human subjects is restricted to brain imaging techniques, analyses of peripheral body fluids or postmortem tissues, which are then correlated with reported history of drug use.

Conclusions consistently raise caveats of "causality or association?" Unfortunately, longitudinal studies that trace brain changes prior to initiation of drug use through onset of drug use and into adulthood are rare. Self-reports of drug consumption are confounded by accurate recall of polysubstance use, doses, exposure time, route of administration or lifestyle of the user (nutritional status, infections, sleep patterns, other diseases). Contaminants and impure street drugs of unknown doses also confound self-reports. In animal studies, the drug, dose, and dosing regimen can be carefully controlled, but myriad factors unique to humans that affect use, e.g. genetics (Miller et al., 2004), peer pressure, physical or psychological abuse, psychiatric comorbidity, depression, stress can challenge conclusions from preclinical research.

2.3.2. Drug Effects

Individuals' response to drugs is determined by their current and previous environmental experiences (NIDA), their unique biology (genetics, personality, psychiatric state) and the interplay of a specific drug with the biology and psychology of the individual (Kalivas and Brady, 2012). The genetics of individuals influence several dimensions of drug response, even success in treatment and are manifest in first to fifth-degree relatives (see chapter by Ho et al., 2010; Tyrfingsson et al., 2010; Johnson et al., 2011). Susceptibility genes overlap for different drugs, even though the starting point of addiction is activation of different receptors unique to each drug class. These findings imply that addictive processes may converge on common molecular events and neural networks in the brain. The genetic approach portends genetically based personalized prevention and treatment approaches in the future.

Addictive drugs can elicit powerful, unique, pleasant, even euphoric responses, governed by the chemistry and formulation of each substance (e.g. salt form, additives, free base), dose, dosing regimen, route of administration (intravenous, inhalation-smoking, insufflation, subcutaneous, oral), rate of transport, and reverse transport, metabolism by the intestinal tract and liver, penetration of the blood–brain barrier, access to brain targets, and user response. At equimolar doses, the intravenous route or inhalation results in faster brain entry and a more robust "high" than the oral route, as evidenced with cocaine, nicotine, methamphetamine, and others (Samaha and Robinson, 2005; Spencer et al., 2006). It is a daunting task to predict the relative addiction potential of drugs for individuals, because factors that drive use and addiction—history, environment, genetics, and epigenetic changes, psychiatric status, metabolic function, drug response—are unique to each individual.

Some drugs can be toxic to the brain, with cocaine, methamphetamine, MDMA (ecstasy), inhalants and alcohol, shown to destroy neurons (alcohol, inhalants) or their axons (amphetamines), disrupt normal blood supply (cocaine), or alter gross brain morphology (Buttner, 2011; see this book). Even in the absence of neurotoxic responses, drugs promote adaptation in cell structure, metabolism, brain signals, and circuitry, sufficiently robust to surpass, suppress, surmount, or supplant natural rewards. A drug-focused existence can attenuated by personal drive, medications assistance, behavioral or other therapies, but a "subterranean" memory can be reawakened by drug cravings during vulnerable periods months or years after withdrawal symptoms have ceased, triggered by drug cues/paraphernalia, drug-associated environments, or stress. Relapse to use discourages and disappoints, but effective relapse prevention strategies that target drug cues and craving with cognitive behavioral therapy, medication-assisted therapy, can surmount the biological triggers of relapse.

2.3.3. Drug-Induced Brain Changes

Addiction begins with unique, initial responses elicited by a small compound with 46 to less than 1000 molecular weight. Initial processes, e.g. activity at receptors or augmentation of neurotransmitters via transporter blockade (Figure 11), trigger a cascade of biological events and rewarding sensations that drive escalation of dose and frequency, transitioning to loss of control and diminished response to normal biological stimuli, (e.g. food, sex, social interaction, responsible actions in school or at work). If motivated to abstain, unpleasant withdrawal symptoms may trigger craving and relapse (Figure 12). The biological substrates associated with progression to addiction are a vast and largely uncharted route from molecule-to-mind. Nonetheless, drugs can change expression of signaling systems linked to receptor modulation that affects transcription factors, gene expression and epigenetic modifications (persistent changes in DNA that alter its capacity to be expressed), cell biology, morphology, synaptic strength, neural circuitry, and behavior (Saka et al., 2004; Russo et al., 2010; Robison and Nestler, 2011; Maze and

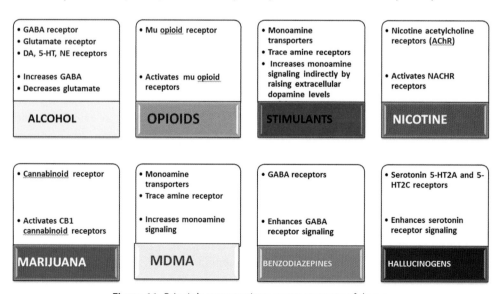

Figure 11 Principle receptor/transporter targets of drugs.

Nestler, 2011). By mechanisms not fully understood, the adaptive responses ultimately affect memory, cognition, judgment, executive function, the salience of natural rewards, and reset volitional and motivational control behavior (Koob, 2009; Goldstein et al., 2009; Koob and Volkow, 2010; Volkow et al., 2010). Our current understanding is sufficiently robust to incorporate these concepts into education, prevention programs, and provide leads for medications development to relieve withdrawal symptoms, or attenuate the rewarding effects of drugs and prevent relapse.

At least 15% of the human genome is devoted to encoding proteins involved in communication, as cell–cell communication is a key to the survival of multicellular organisms. Addictive drugs lodge in the brain's communication system, targeting primarily neurons. Neurons communicate by releasing quantal amounts of neurotransmitters into a synapse, which diffuse to adjacent neurons and activate corresponding receptors. The chemical signal triggers a conformational change in the receptor, which is coupled to interacting proteins in the intracellular milieu and propagates the signal to neurons in other brain regions. The transmitter/receptor signaling partners can initiate movement (dopamine), suppress pain (opioids), engender tranquility or fear (serotonin), imprint or erase memories (dopamine, glutamate, acetylcholine, anandamide), produce arousal, pleasurable or unpleasant sensations (dopamine, endorphins, norepinephrine, serotonin, dynorphin), induce paranoia, regulate heart rate and blood pressue (catecholamines), respiration (opioids), and a myriad of other functions. Receptors readily shed the neurotransmitter because systems have evolved to enable communication to function in ms time frames. As the transmitter binds to the receptor, it undergoes a conformational

Figure 12 A cycle of addiction.

change and coupling to a different protein that facilitates discarding of the transmitter. The transmitter is released and and sequestered by a transporter into local neurons.

Drugs bear a strong resemblance to endogenous transmitters (Figure 13). Cocaine, amphetamine and ecstasy and the endogenous neurotransmitters dopamine, serotonin, and norepinephrine have common core features. Heroin and morphine share structural elements with brain opioids (endorphins, enkephalins). The core structure of LSD is found in the neurotransmitter serotonin. Δ9-tetrahydrocannabinol (THC) produced by the marijuana plant overlaps structurally with brain cannabinoids (anandamide, 2-arachidonylglycerol). The "imposters" bind to receptors and transporters but are incapable of replicating, with fidelity, natural communication. Brain transmitters are produced by specific brain neurons, are stored in vesicles, and released in carefully regulated quantities and then sequestered by transporters into neurons following receptor activation. The blood stream delivers drugs to all brain regions at unregulated concentrations. Depending on structure, they may activate multiple receptors simultaneously, generate signals of unusually long duration, and generate abnormal signal transduction cascades, to affect normal tonic or phasic signals that maintain homeostasis.

Termination of signals by transporters is an essential process for normal communication and maintaining homeostasis. But transporters cannot accommodate the unusual structures, lipophilicity, and charges of most psychoactive drugs (e.g. LSD, morphine, cocaine, THC), resulting in persistent signals, at abnormally high strengths, for abnormally prolonged periods of time, and in unusual brain regions. The hallucinogenic effects

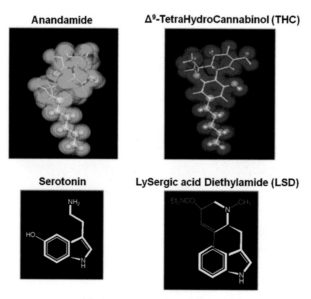

Figure 13 Drugs bear strong resemblance to neurotransmitters. Because they are not identical, they are not processed biologically as are endogenous neurotransmitters.

of LSD are mediated by serotonin receptors. Marijuana magnifies sensory perception, distorts time, produces relaxation, reduces coordination and interferes with working memory by acting at cannabinoid receptors. Cocaine's indirect activity at dopamine receptors produces stimulation and euphoria. Heroin induces tranquility and euphoria via opioid receptors. Nonetheless, the unique and diverse effects of addictive drugs converge in dopamine circuits that are associated with selective stimulus–reward learning (Flagel et al., 2011).

Dopamine is not a hedonic signal, but a signal for learning, gauging the salience of an experience and contributing to motivated behavior. Drugs of abuse raise extracelluilar dopamine levels by various mechanisms, including blockade of the dopamine transporter (e.g. Di Chiara and Imperato, 1986 Ritz et al., 1987; Madras et al., 1989; Bergman et al., 1989). Dopamine can be released to alert the body of novel, meaningful rewarding or aversive stimuli and of a pending stimulus associated with a hedonic reward. Dopamine is released in the nucleus accumbens by drugs at levels and in a time frame largely exceeding release and clearance rates of natural rewards (Willuhn et al., 2010; Sesack and Grace, 2010; Morales et al., 2012). The unregulated dopamine signal is propagated to other brain regions and recruits other signaling partners, primarily glutamate (Kalivas, 2009). Glutamate, implicated in mediating short- and long-term memories, likely encodes and stores the drug experience and carries the message to brain circuits that consolidate positive memories and regulate conditioned responding (hippocampus, amygdala), learn cues associated with the experience (hippocampus, amygdala), learn that repetition of the behaviors will reacquire these rewards (dorsal striatum, nucleus accumbens), assign a priority and value for response to these awards (orbital prefrontal cortex), and imprint cognitive control over rewarding behaviors (prefrontal cortex, striatum, thalamus), (Kalivas, 2009). Natural rewards and drugs share similar circuits of memory and alerting function. Natural rewards arouse survival behaviors. The reward circuit adapts to the dopamine surge by reducing dopamine synthesis and downregulating dopamine receptors, changes associated with anhedonic states and depression. Adjustments to abnormal signals profoundly affect every dimension of brain biology and function (Kalivas and O'Brien, 2008).

2.3.4. Postmortem Brain Changes in Human Drug Users

At the cellular level, drugs can precipitate axonal damage, neuronal loss, microglial activation, a depletion of astrocytes, vasculopathy (Buttner, 2011). Opioid abusers have depleted neurons in hippocampus, other brain regions, and ischemic lesions. Cocaine is the most frequent drug of abuse associated with fatal or nonfatal vascular ischemia or hemorrhage in various brain regions. These are attributable to vasospasm (Kaufman et al., 1998), cardiac arrhythmia with secondary cerebral ischemia, or enhanced platelet aggregation. Cocaine has robust effects on signaling systems that engender positive sensations: reductions in enkephalin mRNA/mu opioid receptor, dopamine/D1 and D2/3 dopamine receptors, conceivably accounting for dysphoria. Cocaine augments dysphoria by increasing dysphoric-producing dynorphin mRNA/kappa opioid receptor binding, and by disrupting glutamate signaling in the amygdala (Hurd et al., 1993; Okvist et al., 2011). Amphetamine and methamphetamine target both monoamine transporters and the trace amine receptor (see chapter by Kish, Bunzow et al., 2001; Miller et al., 2005; Verrico et al., 2007). Similarly, MDMA, a derivative of amphetamine, is neurotoxic to human serotonin systems, as detected by a loss of markers for serotonin axons and serotonin in postmortem brain of frequent MDMA users and with brain imaging techniques (see chapter by Ricaurte).

2.3.5. Imaging Brain Changes in Human Drug Users

Imaging techniques have revealed widespread changes in brains of drug users, ranging from selective declines in receptors and transporters, brain atrophy, decreased grey and white matter volume, reorganization of neural circuitry, ischemic lesions and others. Neuroimaging techniques applied to brains of human drug users include PET (Positron Emission Tomography), SPECT (Single Photon Emission Computed Tomography), which can view receptors, transporters, neurotransmitters, other relevant proteins, blood flow, oxygen, and glucose utilization (e.g. see Volkow et al., 2003, 2011). MRI (magnetic resonance imaging), fMRI (functional MRI), and PMRS (proton magnetic resonance spectroscopy) have revealed changes in gross brain anatomy, cortical thickness, brain activity, levels of specific chemicals in the brain, altered cerebral blood flow, and cognitive deficits in chronic drug users (Cosgrove, 2010; Lundqvist, 2010; Barrós-Loscertales et al., 2011; Prescot et al., 2011; Lopez-Larson, 2011; Jacobus et al., 2012). Intriguingly, it may be possible to predict initiation of marijuana use in adolescents, on the basis of orbitofrontal volume (Cheetham et al., 2012).

Diffusion tensor imaging MRI (DTI MRI) monitors microstructural changes and CT (computed tomography) scans can gauge changes in brain size (Licata and Renshaw, 2010). In cocaine addicts, functional brain imaging has identified increased metabolic activity in the anterior cingulate and orbitofrontal cortex. Areas of the prefrontal cortex become overactive and are thought to reflect greater motivation to seek a drug, not a natural reward, and a decreased ability to overcome this urge (review, London, 2009). The insula has been designated the "hidden island of addiction." Uniquely, it is activated in human subjects during urges to use cocaine, heroin, alcohol, or cigarettes (Naqvi et al., 2007; Naqvi and Bechara, 2009). Intriguingly, addicted smokers with lesioned or destroyed insula (but not populations with other damaged brain regions) readily quit smoking without relapse, and lose the urge to smoke. This brain region is viewed as a key location for integrating stimuli arising from the body, arousing conscious awareness of these feelings, attributing a value to them and integrating a response to them. Accordingly, signals from the insula may override the prefrontal cortex, subverting its function as a reasoning center for impulse control. Conversely, the impaired insula may permit the prefrontal cortex to reassume impulse control, executive function and judgment. Conceivably, medications that interrupt insula function may dampen the conscious urge to seek and consume drugs during the abstinence phase.

2.4. PUBLIC POLICY

2.4.1. Biology Informs Policy

Drug policy has been shaped, in part, by the perspective of biological sciences. There are no agreed upon biological markers for addiction in humans and, notwithstanding remarkable gains over the past few decades on the science of drug effects and addiction, our knowledge remains rudimentary. This book is a repository of the biological effects of drugs in the human nervous system. Among the host of research discoveries that are relevant to public policy, a few can be summarized:

1. The sensations of drug reward are not restricted to a subpopulation of vulnerable people. The mammalian, avian, and even the invertebrate brain (fruit fly) sense reward when exposed to abusable drugs (Bergman et al., 1989; Winsauer and Thomson, 1991; Willuhn et al., 2010). Brain neurochemistry and neural circuitry are uniquely sensitive to specific drugs, and proffer a rationale for imposing vigilance and restraint on their availability.

2. As a function of dose, frequency, route of administration and exposure time, drugs can produce biochemical, cellular, circuitry, physiological, behavioral, and psychological effects that become manifest as addictive behaviors. These findings justify a disease model of addiction and a public health response. A disease model should also create an impetus for incorporating this field into medical education and equipping medical professionals with knowledge of substance use biology, diagnosis, intervention, and treatment.

3. Adolescents, people with psychiatric disorders, with a history of use, poor parenting, dysfunctional families, and other comorbid factors are at high risk for continuing use and addiction. Prevention and intervention programs should address these populations with effective prevention and intervention services.

4. Drugs of abuse produce psychological and/or physiological withdrawal signs that reflect adaptation by the brain and body. Medication's assistance to suppress adaptive mechanisms that precipitate withdrawal, that attenuate drug-induced hedonic signals or drug craving, should be integrated into treatment, and research expanded to develop other effective medications that incorporate newly discovered drug mechanisms.

5. Biological research can inform legal opinions and policies. Sentencing laws governing powder vs "crack" cocaine have been controversial, with possession of 5 g of crack cocaine triggering the same mandatory minimum sentence as possession of 500 g of powder cocaine. Apparently "crack" cocaine was developed to deliver a rapid bolus of free base of cocaine if the crystallized form is smoked (The ionic bond of

the bicarbonate salt of cocaine is weaker and dissociates readily upon heating; the hydrochloride salt form of cocaine powder is a stronger bond and the heat energy required to disrupt this bond and yield cocaine-free base upon smoking is more likely to destroy the alkaloid). Some evidence suggests that crack cocaine has higher addictive potential than cocaine powder, but not by a factor of 100-fold. Users of smoked crack cocaine are more likely than other cocaine users to display clinical features of cocaine dependence, to develop more medical illness, and to have higher incidence of HIV-AIDS (Reboussin and Anthony, 2006; Schönnesson et al, 2008).

6. State ballot and legislative initiatives that circumvent FDA decisions are an unsafe and scientifically unsound mechanism for a drug approval progress and a retrogressive step in our drug safety monitoring system.

2.4.2. Policy and Systems Change

Current public policy balances support for demand reduction (prevention, intervention, treatment) and supply reduction. A balanced policy is justified. One needs only to examine the parallel rise in nonmedical use of prescription opioids, addiction, and overdose deaths with the rapid increase in opioid availability through prescriptions, to conclude that unfettered access to other drugs conceivably would have similar consequences. Public policy is shaped by scientific research, but in many cases the research is not translated into public education or to practice. For example, there is a compelling need for effective prevention programs that incorporate modern scientific discoveries, as reported in this book. There is an equally compelling need to incorporate business practices to bringing effective, evidence-based prevention and treatment models to scale (Madras, 2010). Prevention and treatment research has generated several 100 evidence-based effective programs, but scaling the most effective ones in naturalistic environments has not occurred.

Strategic plans are needed for penetration of these programs at a national, state and community level, coupled with measures of outcomes and effectiveness. Mainstreaming screening, brief interventions and referral to treatment (SBIRT) services (chapter by Gentilello) into our healthcare systems, to address early stages of substance use disorders and to identify and assist those with a sever disorder is likely to have a major impact on use and use patterns. SBIRT is a potential gateway program for assisting an estimated 23% of our population that are screened in healthcare settings and trigger the need for an intervention, brief intervention, brief treatment, or specialty treatment (Madras et al., 2009, 2010). SBIRT-type questionnaires distributed to populations for self-assessment, in the workplace, in middle and high school systems, on college campuses, and in the media, would enable individuals to gauge where along a spectrum, individual risky behavior resides, and assist if a score indicates the need for changed behavior. SBIRT research is compelling but much remains to be explored. Conceivably, but not proven, if SBIRT were mainstreamed into healthcare, could it address these additional public health challenges?

1. Reduce medical and psychiatric conditions that occur at higher frequency in DSM-IV abuse/dependent patients (injuries, trauma, depression, sleep disorders, HIV-AIDS, other infections, cancer, cardiovascular disease, and others)?

2. Reduce the progression to addiction, thereby reducing incidence of addiction and associated medical conditions in family members of patients with SUDs who suffer higher rates of medical problems?

3. Reduce drug use in subpopulations, especially in adolescents, a high-risk group for addiction?

4. Reduce the risks of drug exposure in fetuses? Exposure to heavy alcohol or drugs in utero can cause harmful developmental, behavioral, and physical effects.

5. Prevent adverse drug interactions? It is routine for physicians to inquire about all patients' medications (drugs), to prevent drug interactions and compromised effectiveness of prescribed medications, yet physicians do not routinely inquire about all nonmedical substance use.

6. Reduce prescription drug misuse and abuse? Prescription drug abuse is much more common in alcohol and illicit drug abusers.

7. Reduce overdose deaths due to prescription drug misuse? Currently these are higher than at any period in recent history and far exceed deaths due to heroin or cocaine.

8. Reduce extensive use of emergency departments and trauma centers, injuries, trauma, and violence related to intoxication?

9. Reduce healthcare costs associated with SUD, as SBI saves an estimated $4 for each $1 expended?

10. Reduce the need for the justice system to continue to be the primary source of referrals to treatment? SUDs are largely undiagnosed by medical professionals. It is time for the healthcare system to diminish the justice system.

There is a compelling need for strategic planning to provide assistance for highly vulnerable populations and to reform treatment. As with evidence-based prevention programs, effective treatment principles have not been adopted widely or brought to scale. Among needed improvements are documentation of services and finances of federally supported treatment centers, more appropriately defined treatment outcomes, improved infrastructure of treatment centers, seamless entry into treatment (e.g. reasonable access to the facility and minimal waiting time), integration of treatment with psychiatric and other healthcare services, provision of medications-assisted treatment, HIV-AIDS screening and treatment, and adoption of other practices common in healthcare delivery systems. There is also a need to integrate recovery support-social services (e.g. job training, housing assistance, child care, transportation assistance) into treatment services, to document treatment effectiveness, (admissions, drop-outs and terminations, completion, long-term outcomes), enable long-term treatment follow-up contact and care, and long-term documentation of treatment outcomes.

Expansion of drug courts and drug treatment in prisons is needed as fewer than 20% of the incarcerated in need of treatment receive treatment and fewer than 10% of people of people eligible for drug court treatment receive treatment. To generate funding for appropriate treatment, data are needed on what sectors of society pay for the costs of substance use/addiction, the cost-effectiveness of SBIRT and treatment, and what public or private sectors benefit from the cost-offsets of effective programs, to clarify how and who should help finance a massive expansion of treatment projected by SBIRT and other programs. These and other recommendations for policies and programs were recently articulated (Madras, 2010).

3. CONCLUSIONS

Psychoactive drugs, used for nonmedical purposes, can produce a range of pleasant feelings, from mild stimulation to euphoria, to altered perception and elevated mood. These effects reinforce drug-seeking and consumptive behaviors. Yet these positive sensations, intoxication or euphoria, can result in a progressive brain disease with adverse consequences to individuals, to others and from "in utero to old age". Research has been instrumental in clarifying the individual, genetic, environmental risk and protective factors associated with drug use and use disorders. It has revealed the immediate targets of drugs in the brain, dramatic adaptive changes to signaling systems, neuronal morphology, neural circuitry, genes and epigenetics, behavior after repeated exposure.

At a macrolevel, imaging research has shown a significant shift in the balance of neural circuitry that governs inhibitory control, emotional processing/regulation toward regions that promote compulsive, risk-seeking behaviors. Yet major voids in our understanding of biological effects of drug persist. If filled, these gaps could further shape public health policies for prevention, intervention, and treatment: why are adolescents more vulnerable to addiction? Is it largely because of the interaction of a drug during brain development or because of predisposing factors in the adolescent? Does the adolescent brain respond and adapt differently to drugs than the adult brain? Are drug-induced adaptive responses in adolescent brain more or less likely to be reversed in adulthood? Why are youth with psychiatric disorders (e.g. attention deficit hyperactivity disorder, oppositional defiant disorder) more likely to initiate drug use? Are specific brain changes at any age irreversible? Is irreversibility age-dependent? Do biological changes recover in tandem with cognitive and behavioral improvements during recovery? Longitudinal imaging research combined with psychosocial questionnaires from periadolescent to young adulthood would provide needed insight into these questions.

Why does a psychiatric disorder confer a high risk for a substance use disorder? Improved understanding of the reasons people with neuropsychiatric diseases are attracted to smoking, alcohol, and other drugs and changes that may uniquely affect the comorbid brain will clarify this question and conceivably offer improved intervention strategies. Given that there is a significant genetic component to drug use disorders, are there genetic markers with high predictive value for individuals? Markers may alert family members to the increased risk of initiating drug use. Can markers in biologically accessible fluids or tissues be identified for a substance use disorder that reflects disease and recovery? Such markers are needed to verify objectively the efficacy of treatment, and for a number of other purposes.

Ongoing research will provide further insight into substance use disorders, brain function and ease the coalescence of substance use disorders with mainstream medicine. The advent of CPT®, Medicare and Medicaid billing codes for SBIRT services, the extensive use of prescription opioids for nonmedical purposes and availability of office-based medications such as buprenorphine are promoting medicalization of what, in the past diverged into two separate systems of care.

REFERENCES

ADAM (Arrestee Drug Abuse Monitoring Program). http://www.ojp.usdoj.gov/nij/topics/drugs/adam. htm.

ADAM II. http://www.whitehouse.gov/sites/default/files/ondcp/policy-and-research/adam2010.pdf.

American Psychiatric Association, 2000. Diagnostic and Statistical Manual of Mental Disorders, fourth ed. . text rev. Washington, DC.

Anthony, J.C., Warner, L.A., Kessler, R.C., 1994. Comparative epidemiology of dependence on tobacco, alcohol, controlled substances and inhalants: basic findings from the national comorbidity survey. Exp. Clin. Psychopharm. 2, 244–268.

Anthony, J.C., Petronis, K.R., 1995. Early-onset drug use and risk of later drug problems. Drug Alcohol Depend. 40, 9–15.

Bahi, A., Dreyer, J.L., 2005. Cocaine-induced expression changes of axon guidance molecules in the adult rat brain. Mol. Cell. Neurosci. 28, 275–291.

Barrós-Loscertales, A., Garavan, H., Bustamante, J.C., Ventura-Campos, N., Llopis, J.J., Belloch, V., Parcet, M.A., Avila, C., 2011. Reduced striatal volume in cocaine-dependent patients. Neuroimage 56, 1021–1026.

Bergman, J., Madras, B.K., Johnson, S.E., Spealman, R.D., 1989. Effects of cocaine and related drugs in nonhuman primates. III. Self-administration by squirrel monkeys. J. Pharmacol. Exp. Ther. 251, 150–155.

Brownstein, M.J., 1993. A brief history of opiates, opioid peptides, and opioid receptors. Proc. Natl. Acad. Sci. U.S.A. 90, 5391–5393.

Buttner, A., 2011. Review: the neuropathology of drug abuse. Neuropathol. Appl. Neurobiol. 37, 118–134.

Buckner, J.D., Ecker, A.H., Cohen, A.S., 2010. Mental health problems and interest in marijuana treatment among marijuana-using college students. Addict. Behav. 35, 826–833.

Bunzow, J.R., Sonders, M.S., Arttamangkul, S., Harrison, L.M., Zhang, G., Quigley, D.I., Darland, T., Suchland, K.L., Pasumamula, S., Kennedy, J.L., Olson, S.B., Magenis, R.E., Amara, S.G., Grandy, D.K., 2001. Amphetamine, 3,4-methylenedioxymethamphetamine, lysergic acid diethylamide, and metabolites of the catecholamine neurotransmitters are agonists of a rat trace amine receptor. Mol. Pharmacol. 60, 1181–1188.

Caldeira, K.M., Arria, A.M., O'Grady, K.E., Vincent, K.B., Wish, E.D., 2008. The occurrence of cannabis use disorders and other cannabis-related problems among first-year college students. Addict. Behav. 33, 397–411.

CASA (The National Center on Addiction and Substance Abuse at Columbia University). http://www. casacolumbia.org/.

CASA-children. http://www.casacolumbia.org/templates/Home.aspx?articleid=287&zoneid=32. The National Center on Addiction and Substance Abuse at Columbia University Report: No safe haven; children of substance-abusing parents, January 1999.

CDC (Centers for Disease Control and Prevention). http://www.cdc.gov.

CDC-Alcohol. http://www.cdc.gov/Features/AlcoholConsumption/. See also: Harwood, H., 2000. Updating Estimates of the Economic Costs of Alcohol Abuse in the United States: Estimates, Update Methods, and Data. Report prepared by The Lewin Group for the National Institute on Alcohol Abuse and Alcoholism. Based on estimates, analyses, and data reported in Harwood, H., Fountain, D., Livermore, G., 1992. The Economic Costs of Alcohol and Drug Abuse in the United States. Report Prepared for the National Institute on Drug Abuse and the National Institute on Alcohol Abuse and Alcoholism, National Institutes of Health, Department of Health and Human Services. NIH Publication No. 98–4327. National Institutes of Health, Rockville, MD, 1998.

CDC-alcohol deaths. http://apps.nccd.cdc.gov/DACH_ARDI/Default/Report.aspx?T=AAM&P=de9de51e-d51b-4690-a9a4-358859b692bc&R=804296a0-ac47-41d3-a939-9df26a176186&M=E2769A53-0BFC-453F-9FD7-63C5AA6CE5D7&F=&D.

CDC-deaths. http://www.cdc.gov/nchs/data/nvsr/nvsr60/nvsr60_04.pdf.

CDC-tobacco. http://www.cdc.gov/tobacco/data_statistics/fact_sheets/fast_facts/#toll.

Chambers, R.A., Taylor, J.R., Potenza, M.N., 2003. Developmental neurocircuitry of motivation in adolescence: a critical period of addiction vulnerability. Am. J. Psychiatry 160 (6), 1041–1052.

Cheetham, A., Allen, N.B., Whittle, S., Simmons, J.G., Yücel, M., Lubman, D.I., 2012. Orbitofrontal volumes in early adolescence predict initiation of cannabis use: a 4-year longitudinal and prospective study. Biol. Psychiatry 71, 684–692.

Chen, C.Y., O'Brien, M.S., Anthony, J.C., 2005. Who becomes cannabis dependent soon after onset of use? Epidemiological evidence from the United States: 2000–2001. Drug Alcohol Depend. 79, 11–22.

Chen, C.Y., Storr, C.L., Anthony, J.C., 2009. Early-onset drug use and risk for drug dependence problems. Addict. Behav. 34, 319–322.

Cohen, S., 1975. Cocaine. JAMA 231, 74–75.

Colín-Castelán, D., Phillips-Farfán, B.V., Gutiérrez-Ospina, G., Fuentes-Farias, A.L., Báez-Saldaña, A., Padilla-Cortés, P., Meléndez-Herrera, E., 2011. EphB4 is developmentally and differentially regulated in blood vessels throughout the forebrain neurogenic niche in the mouse brain: implications for vascular remodeling. Brain Res. 1383, 90–98.

Compton, R., Berning, A., July 2009. Results of the 2007 National Roadside Survey of Alcohol and Drug Use by Drivers. Traffic Safety Facts Research Note. National Highway Traffic Safety Administration. http://www.nhtsa.gov/staticfiles/DOT/NHTSA/Traffic%20Injury%20Control/Articles/Associated%20Files/811175.pdf.

Cosgrove, K.P., 2010. Imaging receptor changes in human drug abusers. Curr. Top. Behav. Neurosci. 3, 199–217.

Dannaway, F.R., 2010. Strange fires, weird smokes and psychoactive combustibles: entheogens and incense in ancient traditions. J. Psychoactive Drugs 42, 485–497.

Dannaway, F.R., Piper, A., Webster, P., 2006. Bread of heaven or wines of light: entheogenic legacies and esoteric cosmologies. J. Psychoactive Drugs 38, 493–503.

DASIS (Drug and Alcohol Services Information System). http://www.oas.samhsa.gov/dasis.htm#nssats2.

DAWN (Drug Abuse Warning Network). https://dawninfo.samhsa.gov/default.asp.

DEA (Drug Enforcement Agency). http://www.justice.gov/dea/history.htm.

Degenhardt, L., Bucello, C., Calabria, B., Nelson, P., Roberts, A., Hall, W., Lynskey, M., Wiessing, L., GBD illicit drug use writing group, Mora, M.E., Clark, N., Thomas, J., Briegleb, C., McLaren, J., 2011. What data are available on the extent of illicit drug use and dependence globally? Results of four systematic reviews. Drug Alcohol Depend. 117, 85–101.

Degenhardt, L., Hall, W., 2012. Extent of illicit drug use and dependence, and their contribution to the global burden of disease. Lancet 379, 55–70.

Di Chiara, G., Imperato, A., 1986. Preferential stimulation of dopamine release in the nucleus accumbens by opiates, alcohol, and barbiturates: studies with transcerebral dialysis in freely moving rats. Ann. N.Y. Acad. Sci. 473, 367–381.

DOL (Department of Labor). http://www.dol.gov/asp/programs/drugs/workingpartners/stats/wi.asp.

Dugan, J.P., Stratton, A., Riley, H.P., Farmer, W.T., Mastick, G.S., 2011. Midbrain dopaminergic axons are guided longitudinally through the diencephalon by slit/robo signals. Mol. Cell. Neurosci. 46, 347–356.

DuPont, R.L., Madras, B.K., Johansson, P., 2011. Section 12: policy issues, chapter 77, drug policy: a biological science perspective. In: Lowinson, J.H., Ruiz, P. (Eds.), Substance Abuse: A Comprehensive Textbook. fifth ed. Lippincott Williams & Wilkins, 530 Walnut Street Philadelphia, PA 19106.

Flagel, S.B., Clark, J.J., Robinson, T.E., Mayo, L., Czuj, A., Willuhn, I., Akers, C.A., Clinton, S.M., Phillips, P.E., Akil, H., 2011. A selective role for dopamine in stimulus-reward learning. Nature 469, 53–57.

Flores, C., 2011. Role of netrin-1 in the organization and function of the mesocorticolimbic dopamine system. J. Psychiatry Neurosci. 36, 296–310.

Friedmann, P.D., Jin, L., Karrison, T., Nerney, M., Hayley, D.C., Mulliken, R., Walter, J., Miller, A., Chin, M.H., 1999. The effect of alcohol abuse on the health status of older adults seen in the emergency department. Am. J. Drug Alcohol. Abuse 25, 529–542.

Gay, G.R., Inaba, D.S., Sheppard, C.W., Newmeyer, J.A., 1975. Cocaine: history, epidemiology, human pharmacology, and treatment. A perspective on a new debut for an old girl. Clin. Toxicol. 8, 149–178.

Giedd, J.N., 2008. The teen brain: insights from neuroimaging. J. Adolesc. Health 42, 335–343.

Gogtay, N., Giedd, J.N., Lusk, L., Hayashi, K.M., Greenstein, D., Vaituzis, A.C., Nugent 3rd, T.F., Herman, D.H., Clasen, L.S., Toga, A.W., Rapoport, J.L., Thompson, P.M., 2004. Dynamic mapping of human cortical development during childhood through early adulthood. Proc. Natl. Acad. Sci. U.S.A. 101, 8174–8179.

Goldstein, R.Z., Craig, A.D., Bechara, A., Garavan, H., Childress, A.R., Paulus, M.P., Volkow, N.D., 2009. The neurocircuitry of impaired insight in drug addiction. Trends. Cogn. Sci. 13, 372–380.

Grant, B.F., Dawson, D.A., 1998. Age of onset of drug use and its association with DSM-IV drug abuse and dependence: results from the National Longitudinal Alcohol Epidemiologic Survey. J. Subst. Abuse 10, 163–173.

HBSC (Health Behavior in School-aged Children). http://www.hbsc.org/overview.html.

Ho, M.K., Goldman, D., Heinz, A., Kaprio, J., Kreek, M.J., Li, M.D., Munafò, M.R., Tyndale, R.F., 2010. Breaking barriers in the genomics and pharmacogenetics of drug addiction. Clin. Pharmacol. Ther. 88, 779–791.

Hurd, Y.L., Herkenham, M., 1993. Molecular alterations in the neostriatum of human cocaine addicts. Synapse 13, 357–369.

Jacobus, J., Goldenberg, D., Wierenga, C.E., Tolentino, N.J., Liu, T.T., Tapert, S.F., March 7, 2012. Altered cerebral blood flow and neurocognitive correlates in adolescent cannabis users. Psychopharmacology (Berl) (Epub ahead of print).

Jassen, A.K., Yang, H., Miller, G.M., Calder, E., Madras, B.K., 2006. Receptor regulation of gene expression of axon guidance molecules: implications for adaptation. Mol. Pharmacol. 70, 71–77.

Johnson, C., Drgon, T., Walther, D., Uhl, G.R., 2011. Genomic regions identified by overlapping clusters of nominally-positive SNPs from genome-wide studies of alcohol and illegal substance dependence. PLoS One 6, e19210.

Justice. http://www.justice.gov/ndic/pubs44/44731/44731p.pdf.

Kalivas, P.W., 2009. The glutamate homeostasis hypothesis of addiction. Nat. Rev. Neurosci. 10, 561–572.

Kalivas, P.W., O'Brien, C., 2008. Drug addiction as a pathology of staged neuroplasticity. Neuropsychopharmacology 33, 166–180.

Kalivas, P.W., Brady, K., 2012. Getting to the core of addiction: hatching the addiction egg. Nat. Med. 18, 502–503.

Kandel, D.B., Yamaguchi, K., Klein, L.C., 2006. Testing the gateway hypothesis. Addiction 101, 470–472.

Kandel, D.B., Yamaguchi, K., Chen, K., 1992. Stages of progression in drug involvement from adolescence to adulthood: further evidence for the gateway theory. J. Stud. Alcohol. 53, 447–457.

Kaufman, M.J., Levin, J.M., Ross, M.H., Lange, N., Rose, S.L., Kukes, T.J., Mendelson, J.H., Lukas, S.E., Cohen, B.M., Renshaw, P.F., 1998. Cocaine-induced cerebral vasoconstriction detected in humans with magnetic resonance angiography. JAMA 279, 376–380.

Koob, G.F., 2009. Neurobiological substrates for the dark side of compulsivity in addiction. Neuropharmacology 56 (Suppl. 1), 18–31.

Koob, G.F., Volkow, N.D., 2010. Neurocircuitry of addiction. Neuropsychopharmacology 35, 217–238.

Leshner, A.I., 1997. Addiction is a brain disease, and it matters. Science 278, 45–47.

Licata, S.C., Renshaw, P.F., 2010. Neurochemistry of drug action: insights from proton magnetic resonance spectroscopic imaging and their relevance to addiction. Ann. N.Y. Acad. Sci. 1187, 148–171.

London, J.A., Utter, G.H., Battistella, F., Wisner, D., 2009. Methamphetamine use is associated with increased hospital resource consumption among minimally injured trauma patients. J. Trauma 66, 485–490.

London, E.D., 2009. Studying addiction in the age of neuroimaging. Marian W. Fischman Lecture given at the 2008 meeting of CPDD. Drug Alcohol Depend. 100, 182–185.

Lopez-Larson, M.P., Bogorodzki, P., Rogowska, J., McGlade, E., King, J.B., Terry, J., Yurgelun-Todd, D., 2011. Altered prefrontal and insular cortical thickness in adolescent marijuana users. Behav. Brain Res. 220, 164–172.

Lundqvist, T., 2010. Imaging cognitive deficits in drug abuse. Curr. Top. Behav. Neurosci. 3, 247–275.

MTF (Monitoring the Future). http://monitoringthefuture.org/pubs/monographs/mtf-overview2011 .pdf.

Madras, B.K., Fahey, M.A., Bergman, J., Canfield, D.R., Spealman, R.D., 1989. Effects of cocaine and related drugs in nonhuman primates. I. [3H]cocaine binding sites in caudate-putamen. J. Pharmacol. Exp. Ther. 251, 131–141.

Madras, B.K., Compton, W.M., Avula, D., Stegbauer, T., Stein, J.B., Clark, H.W., 2009. Screening, brief interventions, referral to treatment (SBIRT) for illicit drug and alcohol use at multiple healthcare sites: comparison at intake and 6 months later. Drug Alcohol Depend. 99, 280–295.

Madras, B.K., 2010. Office of National Drug Control Policy: a scientist in drug policy in Washington, DC. Ann. N.Y. Acad. Sci. 1187, 370–402.

Manitt, C., Mimee, A., Eng, C., Pokinko, M., Stroh, T., Cooper, H.M., Kolb, B., Flores, C., 2011. The netrin receptor DCC is required in the pubertal organization of mesocortical dopamine circuitry. J. Neurosci. 23, 8381–8394.

Mason, C., 2009. The development of developmental neuroscience. J. Neurosci. 29, 12735–12747.

Mayet, A., Legleye, S., Falissard, B., Chau, N., 2012. Cannabis use stages as predictors of subsequent initiation with other illicit drugs among French adolescents: use of a multi-state model. Addict. Behav. 37, 160–166.

Maze, I., Nestler, E.J., 2011. The epigenetic landscape of addiction. Ann. N.Y. Acad. Sci. 1216, 99–113.

McCabe, S.E., West, B.T., Morales, M., Cranford, J.A., Boyd, C.J., 2007. Does early onset of non-medical use of prescription drugs predict subsequent prescription drug abuse and dependence? Results from a national study. Addiction 102, 1920–1930.

Mertens, J.R., Lu, Y.W., Parthasarathy, S., Moore, C., Weisner, C.M., 2003. Medical and psychiatric conditions of alcohol and drug treatment patients in an HMO: comparison with matched controls. Arch. Intern. Med. 163, 2511–2517.

Mertens, J.R., Weisner, C., Ray, G.T., Fireman, B., Walsh, K., 2005. Hazardous drinkers and drug users in HMO primary care: prevalence, medical conditions, and costs. Alcohol Clin. Exp. Res. 29, 989–998.

Mertens, J.R., Flisher, A.J., Fleming, M.F., Weisner, C.M., 2007. Medical conditions of adolescents in alcohol and drug treatment: comparison with matched controls. J. Adolesc. Health 40, 173–179.

Miller, G.M., Bendor, J., Tiefenbacher, S., Yang, H., Novak, M.A., Madras, B.K., 2004. A mu-opioid receptor single nucleotide polymorphism in rhesus monkey: association with stress response and aggression. Mol. Psychiatry 9, 99–108.

Miller, G.M., Verrico, C.D., Jassen, A., Konar, M., Yang, H., Panas, H., Bahn, M., Johnson, R., Madras, B.K., 2005. Primate trace amine receptor 1 modulation by the dopamine transporter. J. Pharmacol. Exp. Ther. 313, 983–994.

Morales, M., Pickel, V.M., 2012. Insights to drug addiction derived from ultrastructural views of the meso-corticolimbic system. Ann. N.Y. Acad. Sci. 1248, 71–88.

MTF (monitoring the future). http://www.drugabuse.gov/related-topics/trends-statistics/monitoring-future/trends-in-prevalence-various-drugs; Monitoring the Future: National Results on Adolescent Drug Use, 2011. Available from; http://www.monitoringthefuture.org/pubs/monographs/mtf-over-view2011.pdf.

Murray, B.L., Murphy, C.M., Beuhler, M.C., 2012. Death following recreational use of designer drug "bath salts" containing 3,4-Methylenedioxypyrovalerone (MDPV). J. Med. Toxicol. 8, 69–75.

Musto, D.F., 1968. A study in cocaine. Sherlock Holmes and Sigmund Freud. JAMA 204, 27–32.

Musto, D.F., 1995. Perception and regulation of drug use: the rise and fall of the tide. Ann. Intern. Med. 123, 468–469.

Musto, D.F., Korsmeyer, P., Maulucci, T.W., 2002. One Hundred Years of Heroin. Greenwood Publishing Group, Incorporated, Santa Barbara, CA.

Musto, D.F., 2002. Drugs in America. New York University Press, New York.

Naqvi, N.H., Rudrauf, D., Damasio, H., Bechara, A., 2007. Damage to the insula disrupts addiction to cigarette smoking. Science 315, 531–534.

Naqvi, N.H., Bechara, A., 2009. The hidden island of addiction: the insula. Trends Neurosci. 32, 56–67.

NCS (National Comorbidity survey). http://www.hcp.med.harvard.edu/ncs/.

NCJRS.https://www.ncjrs.gov/ondcppubs/publications/pdf/economic_costs.pdf.

NESARC (National Epidemiological Survey on Alcohol and Related Conditions). http://aspe.hhs.gov/hsp/06/Catalog-AI-AN-NA/NESARC.htm.

NHTSA-drugged driving, 2011. http://www.nhtsa.gov/staticfiles/nti/pdf/811438.pdf.

NIAAA. http://pubs.niaaa.nih.gov/publications/economic-2000/.

NIDA. http://www.drugabuse.gov/publications/preventing-drug-abuse-among-children-adolescents/chapter-1-risk-factors-protective-factors.

NSDUH (National Survey on Drug Use and Health). http://www.samhsa.gov/data/NSDUH/2k10NSDUH/2k10Results.htm.

NSDUH-addiction. http://www.samhsa.gov/data/NSDUH/2k10NSDUH/2k10Results.htm#7.3.

NVSS (National Vital statistics System). http://www.cdc.gov/nchs/nvss.htm.

O'Brien, M.S., Anthony, J.C., 2005. Risk of becoming cocaine dependent: epidemiological estimates for the United States, 2000–2001. Neuropsychopharmacology 30, 1006–1018.

Okvist, A., Fagergren, P., Whittard, J., Garcia-Osta, A., Drakenberg, K., Horvath, M.C., Schmidt, C.J., Keller, E., Bannon, M.J., Hurd, Y.L., 2011. Dysregulated postsynaptic density and endocytic zone in the amygdala of human heroin and cocaine abusers. Biol. Psychiatry 69, 245–252.

Palmer, R.H., Young, S.E., Hopfer, C.J., Corley, R.P., Stallings, M.C., Crowley, T.J., Hewitt, J.K., 2009. Developmental epidemiology of drug use and abuse in adolescence and young adulthood: evidence of generalized risk. Drug Alcohol Depend. 102, 78–87.

Paulozzi, L.J., Weisler, R.H., Patkar, A.A., 2011. A national epidemic of unintentional prescription opioid overdose deaths: how physicians can help control it. J. Clin. Psychiatry 72, 589–592.

Prekker, M.E., Miner, J.R., Rockswold, E.G., Biros, M.H., 2009. The prevalence of injury of any type in an urban emergency department population. J. Trauma 66, 1688–1695.

Prescot, A.P., Locatelli, A.E., Renshaw, P.F., Yurgelun-Todd, D.A., 2011. Neurochemical alterations in adolescent chronic marijuana smokers: a proton MRS study. Neuroimage 57, 69–75.

Prosser, J.M., Nelson, L.S., 2012. The toxicology of bath salts: a review of synthetic cathinones. J. Med. Toxicol. 8, 33–42.

Ramchand, R., Marshall, G.N., Schell, T.L., Jaycox, L.H., Hambarsoomians, K., Shetty, V., Hinika, G.S., Cryer, H.G., Meade, P., Belzberg, H., 2009. Alcohol abuse and illegal drug use among Los Angeles County trauma patients: prevalence and evaluation of single item screener. J. Trauma 66, 1461–1467.

Ray, G.T., Mertens, J.R., Weisner, C., 2007. The excess medical cost and health problems of family members of persons diagnosed with alcohol or drug problems. Med. Care 45, 116–122.

Ray, G.T., Mertens, J.R., Weisner, C., 2009. Family members of people with alcohol or drug dependence: health problems and medical cost compared to family members of people with diabetes and asthma. Addiction 104, 203–214.

Raznahan, A., Lerch, J.P., Lee, N., Greenstein, D., Wallace, G.L., Stockman, M., Clasen, L., Shaw, P.W., Giedd, J.N., 2011. Patterns of coordinated anatomical change in human cortical development: a longitudinal neuroimaging study of maturational coupling. Neuron 72, 873–884.

Reboussin, B.A., Anthony, J.C., 2006. Is there epidemiological evidence to support the idea that a cocaine dependence syndrome emerges soon after onset of cocaine use? Neuropsychopharmacology 31, 2055–2064.

Robison, A.J., Nestler, E.J., 2011. Transcriptional and epigenetic mechanisms of addiction. Nat. Rev. Neurosci. 12, 623–637.

Ritz, M.C., Lamb, R.J., Goldberg, S.R., Kuhar, M.J., 1987. Cocaine receptors on dopamine transporters are related to self-administration of cocaine. Science 237, 1219–1223.

Russo, S.J., Dietz, D.M., Dumitriu, D., Morrison, J.H., Malenka, R.C., Nestler, E.J., 2010. The addicted synapse: mechanisms of synaptic and structural plasticity in nucleus accumbens. Trends Neurosci. 33, 267–276.

Saka, E., Goodrich, C., Harlan, P., Madras, B.K., Graybiel, A.M., 2004. Repetitive behaviors in monkeys are linked to specific striatal activation patterns. J. Neurosci. 24, 7557–7565.

Samaha, A.N., Robinson, T.E., 2005. Why does the rapid delivery of drugs to the brain promote addiction? Trends Pharmacol. Sci. 26, 82–87.

SAMHSA OAS. http://www.oas.samhsa.gov/2k9/168/168OlderAdults.htm.

SAMSHA-older adults.http://www.samhsa.gov/OlderAdultsTAC/SA_MH_%20AmongolderAdultsfinal 102105.pdf.

SAMHSA-work. http://radar.boisestate.edu/pdfs/TheCostofSubstanceAbuse.pdf.

Sesack, S.R., Grace, A.A., 2010. Cortico-Basal Ganglia reward network: microcircuitry. Neuropsychopharmacology 35, 27–47.

Schönnesson, L.N., Atkinson, J., Williams, W.L., Bowen, A., Ross, M.W., Timpson, S.C., 2008. A cluster analysis of drug use and sexual HIV risks and their correlates in a sample of African-American crack cocaine smokers with HIV infection. Drug Alcohol Depend. 97, 44–53.

SOS 2011 Annual Report, 2011. S.O.S., Inc, St Petersburg, FL.

Spencer, T.J., Biederman, J., Ciccone, P.E., Madras, B.K., Dougherty, D.D., Bonab, A.A., Livni, E., Parasrampuria, D.A., Fischman, A.J., 2006. PET study examining pharmacokinetics, detection and likeability, and dopamine transporter receptor occupancy of short- and long-acting oral methylphenidate. Am. J. Psychiatry 163, 387–395.

Storr, C.L., Westergaard, R., Anthony, J.C., 2005. Early onset inhalant use and risk for opiate initiation by young adulthood. Drug Alcohol Depend. 78, 253–261.

Swendsen, J., Conway, K.P., Degenhardt, L., Glantz, M., Jin, R., Merikangas, K.R., Sampson, N., Kessler, R.C., 2010. Mental disorders as risk factors for substance use, abuse and dependence: results from the 10-year follow-up of the National Comorbidity Survey. Addiction 105, 1117–1128.

Swift, W., Coffey, C., Carlin, J.B., Degenhardt, L., Patton, G.C., 2008. Adolescent cannabis users at 24 years: trajectories to regular weekly use and dependence in young adulthood. Addiction 103, 1361–1370.

TEDS (Treatment Episode Data Set). http://www.oas.samhsa.gov/2k2/TEDS/TEDS.cfm.

TEDS-prescription Pain-relievers: Substance Abuse Treatment Admissions Involving Abuse of Pain Relievers: 1998 and 2008. http://oas.samhsa.gov/2k10/230/230PainRelvr2k10.htm.

Tomasi, D., Volkow, N.D., Wang, R., Carrillo, J.H., Maloney, T., Alia-Klein, N., Woicik, P.A., Telang, F., Goldstein, R.Z., 2010. Disrupted functional connectivity with dopaminergic midbrain in cocaine abusers. PLoS One 5, e10815.

Tyrfingsson, T., Thorgeirsson, T.E., Geller, F., Runarsdóttir, V., Hansdóttir, I., Bjornsdottir, G., Wiste, A.K., Jonsdottir, G.A., Stefansson, H., Gulcher, J.R., Oskarsson, H., Gudbjartsson, D., Stefansson, K., 2010. Addictions and their familiality in Iceland. Ann. N.Y. Acad. Sci. 1187, 208–217.

Uhl, G.R., Grow, R.W., 2004. The burden of complex genetics in brain disorders. Arch. Gen. Psychiatry 61, 223–229.

UNODC (United Nations Office on Drugs and Crime). http://idpc.net/publications/2011/06/unodc-world-drug-report-2011.

Vanderhaeghen, P., Cheng, H.J., 2010. Guidance molecules in axon pruning and cell death. Cold Spring Harb Perspect. Biol. 2, a001859.

Verrico, C.D., Miller, G.M., Madras, B.K., 2007. MDMA (ecstasy) and human dopamine, norepinephrine, and serotonin transporters: implications for MDMA-induced neurotoxicity and treatment. Psychopharmacology 189, 489–503.

Volkow, N.D., Fowler, J.S., Wang, G.-J., 2003. The addicted human brain: insights from imaging studies. J. Clin. Invest. 111, 1444–1451.

Volkow, N.D., Wang, G.J., Fowler, J.S., Tomasi, D., Telang, F., Baler, R., 2010. Addiction: decreased reward sensitivity and increased expectation sensitivity conspire to overwhelm the brain's control circuit. Bioessays 32, 748–755.

Volkow, N.D., Tomasi, D., Wang, G.-J., Fowler, J.S., Telang, F., Goldstein, R.Z., Alia-Klein, N., Wong, C., 2011. Reduced metabolism in brain "control networks" following cocaine-cues exposure in female cocaine abusers. PLoS One 6 (2), e16573.

Wagner, F.A., Anthony, J.C., 2002. From first drug use to drug dependence; developmental periods of risk for dependence upon marijuana, cocaine, and alcohol. Neuropsychopharmacology 26, 479–488.

White, W.L., 1998. Slaying the Dragon: the History of Addiction Treatment and Recovery in America. Chestnut Health Systems/Lighthouse Institute, Bloomington, IL.

Winsauer, P.J., Thompson, D.M., 1991. Cocaine self-administration in pigeons. Pharmacol. Biochem. Behav. 40, 41–52.

Willuhn, I., Wanat, M.J., Clark, J.J., Phillips, P.E., 2010. Dopamine signaling in the nucleus accumbens of animals self-administering drugs of abuse. Curr. Top. Behav. Neurosci. 3, 29–71.

YRBSS (Youth Risk Behavior Surveillance System). http://www.cdc.gov/HealthyYouth/yrbs/index.htm.

Yurgelun-Todd, D., 2007. Emotional and cognitive changes during adolescence. Curr. Opin. Neurobiol. 17, 251–257.

Zias, J., Stark, H., Sellgman, J., Levy, R., Werker, E., Breuer, A., Mechoulam, R., 1993. Early medical use of cannabis. Nature 363, 215.

CHAPTER TWO

Genetics of Substance Use, Abuse, Cessation, and Addiction: Novel Data Implicate Copy Number Variants

George R. Uhl, Donna Walther and Tomas Drgon
Molecular Neurobiology Branch, NIH-IRP (NIDA), Baltimore, MD, USA

1. INTRODUCTION

The sizable genetic influences on individual differences in vulnerability to developing dependence on an addictive substance (*see below*) are likely to overlap substantially with overall genetic influences on abilities to quit, as well as with genetic influences on a number of related phenotypes and comorbid conditions. At a molecular level, this means that DNA sequence variants that are passed from generation to generation can, in permissive environments, work to alter these vulnerabilities. We can thus start from an overview of the types of DNA variants that are likely to contribute to these vulnerabilities. We include novel data for possible roles for copy number variants (CNVs). We then move upscale to discussion of shared genetic influences on comorbidities and on the possible implications for thinking about, treating, and preventing addictions.

2.1. WORKING HYPOTHESIS I: GENOMIC VARIANTS OF SEVERAL CLASSES AND DIFFERING FREQUENCIES CONTRIBUTE TO VULNERABILITY TO ADDICTION AND ABILITY TO QUIT

Thinking about the genetic architectures of vulnerability to addiction and ability to quit can start from the microscopic level: What sorts of variants in the genome are likely to confer these risks? What are the ways in which these variants are distributed in individuals, families, and populations? (See Table 1)

Neither linkage studies, which assess the way in which DNA markers and disease vulnerability move together in families, nor most association studies, which assess the ways in which single nucleotide polymorphism (SNP) DNA markers and disease are associated in nominally unrelated individuals, provide direct assessments of the actual DNA variants that contribute to linkage or association findings. Follow-up studies from these findings, and initial data from studies of CNV probes, are beginning to paint a rich picture of the ways in which specific variants may influence vulnerability to dependence. Data from exome or genome-wide resequencing will soon provide a deluge of variants

The Effects of Drug Abuse on the Human Nervous System
http://dx.doi.org/10.1016/B978-0-12-418679-8.00002-2

Table 1 Working Hypotheses for Genetic Architecture for Human Individual Differences in the Vulnerability to Dependence and Ability to Quit Smoking

1. Genomic variants of several classes and differing frequencies contribute to vulnerability to addiction and ability to quit.
2. Genomic variants that contribute to vulnerability to addiction and ability to quit provide largely additive influences.
3. Most genomic variants that contribute to dependence or ability to quit exert effect of small size, though there are larger influences in specific populations and for addiction-associated phenotypes.
4. There are robust overall genetic influences on vulnerability to dependence, many shared across vulnerabilities to different substances. There are robust overall genetic influences on ability to quit, some of which overlap with those that determine the degree of dependence.

that have the potential for association with addiction in at least some individuals. Separating epiphenomena from truly associated variants from this work will be likely to take large amounts of effort over many years, however.

A common Mendelian disorder, cystic fibrosis, provides an analogy for the types of DNA variants that are likely to influence addiction vulnerability and ability to quit. Almost 2000 separate DNA variants that migrate in families along with cystic fibrosis are described: http://projects.tcag.ca/variation/. These include variants that influence the sequence of the cystic fibrosis transmembrane regulator (CFTR) protein in which variation causes cystic fibrosis by changing amino acids or terminating the protein prematurely. These disease-associated variants also provide a variety of different influences in intronic, 5′flanking and even 3′ flanking regions of the CFTR gene. Further, the entire constellation of CFTR variants and the disease are more abundant in individuals with European genetic backgrounds. Specific variants are much more prominent in individuals from specific regions of Europe. This degree of complexity for even a relatively common disorder with classical Mendelian inheritance patterns presages substantial molecular genetic complexity for addiction.

Some of the most detailed information about individual addiction-associated variants comes from studies of candidate genes. Association with levels of expression and with missense variants that alter receptor function has been identified for genes in the chromosome 15 nicotinic acetylcholine receptor gene cluster that are associated with measures of smoking quantity/frequency, for example (Wang et al., 2009). Some data indicate subtle differences in the binding properties associated with OPRM1 missense variants that alter its N-terminus and have been associated with differential efficacies of opiate antagonist therapies for alcohol in at least some studies (Bond et al., 1998; Oslin et al., 2003). We have identified splicing variants in NRXN3 that are associated with addiction vulnerability (Hishimoto et al., 2007); these findings have been recently supported by initial studies of mice engineered to differentially express these splicing variants (Boucard et al., 2011). It appears safe to postulate that all types of DNA variants will eventually be found to contribute to addiction vulnerability and abilities to quit.

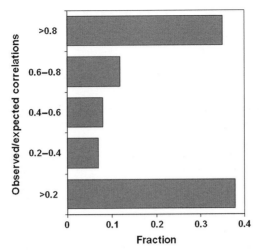

Figure 1 Validation of pooled CNV genotyping. The distribution of correlation coefficients of quantile-normalized CNV probe hybridization intensities in validation pools indicate large proportion of probes with >0.8 correlations between observed and expected hybridization intensities in validation pools, but also a large proportion of underperforming probes with correlations <0.2.

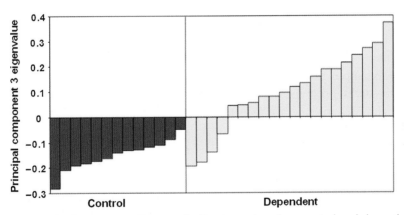

Figure 2 CNV principal component 3 scores for European American control and dependent pools distinguish dependent from control ($p = 0.0000037$; t test for PC3 eigenvalues).

We specifically sought evidence for overrepresentation of CNVs in Affymetrix 6.0 genotyping studies of DNAs pooled from groups of 20 dependent and 20 control individuals recruited in Baltimore, MD (Drgon et al., 2009a). Validating studies for use of the CNV probes in pooled samples provided evidence for poorer overall performance of some of the CNV probes than of the SNP probes assessed on the same arrays (Figure 1). Nevertheless, principal components analyses of data from all CNV probes provided interesting components that are strongly associated with addiction phenotypes (Figures 2 and 3).

This data provide some of the only evidence of which we are aware that CNVs are enriched, as an overall class, in dependence. CNVs that were identified in both

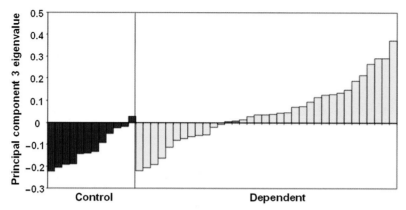

Figure 3 CNV principal component 3 scores for African American control and dependent pools distinguish dependent from control ($p = 0.00079$; t test).

independent African- and European-American samples that compared individuals with (1) substantial use of and, almost always, dependence on at least one illegal substance to (2) controls with no significant lifetime use of any addictive substance are listed in Table 2.

Studies of the frequency of variants associated with dependence can also be considered in light of recent evidence for the extent to which the same genomic regions are identified in independent addiction vulnerability samples of the same vs different racial/ethnic backgrounds (Drgon et al., 2011; Johnson et al., 2011). We and others have identified association signals in samples from both African- and European-American genetic backgrounds. A single quit-success genotype scores can differ between quitters and nonquitters from both European- and African-American genetic backgrounds (Uhl et al., 2010a). Nevertheless, when we have recently compared addiction associations in independent samples of European or African genetic backgrounds, there are much stronger overlaps between different studies of European samples than there are in comparison of European vs African samples (Drgon et al., 2011; Johnson et al., 2011). Taken together, these findings provide support for the idea that the majority of the variants that contribute to addiction vulnerability, and perhaps to ability to quit, are present at significantly different frequencies in individuals with different genetic backgrounds. Another way of thinking about this is that many of the addiction-associated variants may have entered into and/or been differentially propagated in these populations after they were largely genetically separated from each other.

2.2. WORKING HYPOTHESIS II: GENOMIC VARIANTS THAT CONTRIBUTE TO VULNERABILITY TO ADDICTION AND ABILITY TO QUIT PROVIDE LARGELY ADDITIVE INFLUENCES

There are many ways in which the genetic variants that contribute to dependence and ability to quit could combine, interact, and interact with the strong environmental

Table 2 Genomic Regions that Contain Clustered PC3-Positive CNV Probes in Both African- and European American Samples

Ch	Bp Start	Size (bp)	AfAm Clust p<0.05 probes	EuAm Clust p<0.05 probes	p	Gene	Description	Toronto CNV
2	81,650,309	31	6	6	<0.00001			0020,2397
3	138,509,946	6164	5	7	<0.00001			38469,7366
3	180,409,224	9809	7	5	0.00014	PIK3CA	α catalyst sub PI3 kinase	
4	9,837,869	5483	7	7	<0.00001			0571,1664,23216,31165,3479,36326, 36327,38803,7459,8443,8996
4	104,963,004	20,042	4	8	0.00025			1400,1678,32629,38511,39388, 1680,31185,32632,3507,36179
4	115,391,016	16,730	5	4	0.00042			36180,36181,36182,37429,38773
7	13,943,468	7607	4	5	0.00033	ETV1	ets variant gene 1	0292,1742,23125,23827,32948, 36615
7	125,826,938	7167	7	6	<0.00001			36616,36617,38794
11	106,836,636	265	4	4	0.00044	CWF19L2	CWF19–like 2	2957
13	103,205,380	17,318	4	4	0.00085			3031,8758
14	63,766,115	7604	5	4	0.00045	ESR2	Estrogen receptor 2	

Columns list the chromosome, starting basepair and size of the region, numbers of clustered, nominally positive SNPs in African–American and in European–American samples, p values for the probability that the observed convergence occurs by chance, gene/description (where applicable) and numbers for corresponding previously reported CNVs whose coordinates overlap (http://projects.tcag.ca/variation/).

influences that are also present to contribute to an addiction vulnerability. In some cases, there are likely to be gene–gene interactions (g × g) and gene–environment interactions (g × e). Groups of addiction-associated genes do fall into specific biochemically defined pathways, and they are often expressed in overlapping neuroanatomically defined pathways. g × g interactions could thus occur. There are clear epidemiological comorbidities between dependence and other disorders that may share overlapping environmental features with dependence. A permissive environment is important: twin studies of smoking in Scandinavian women clearly document the emergence of heritability as the environment became more permissive for expression of smoking vulnerability genetics in women during the course of the twentieth century (Kendler et al., 2000b).

It is nevertheless important to note that several sorts of datasets provide little support for large influences of g × g or g × e interactions in addiction vulnerability or abilities to quit smoking. Conversely, additive models receive cautious support from initial data from our abilities to predict success in quitting smoking using an additive score.

Family datasets support a linear, step-wise falloff of risk of dependence with degree of relatedness. Recently reported data from Iceland, in which ascertainment of addiction and its treatment provide an approximate population sample, identify an almost mathematically precise linear falloff of addiction vulnerability with degree of relatedness (Tyrfingsson et al., 2010). Identification of roughly similar extents of heritability for addiction and abilities to quit smoking in data from a number of environments supports modest overall contributions of g × e interactions (Uhl et al., 1995). None of these datasets invalidate additive models for genetic influences on dependence, at least as a first approximation.

We have also developed an initial "v1.0" additive SNP model for the genetic bases of ability to quit smoking (Peng and Yin, 2009; Uhl et al., 2010a). The scores provided by this additive model have been able to provide gratifying, highly significant ability to predict abilities to quit when applied prospectively and blindly to subsequent samples (e.g. samples not used to generate the scores). These data are again consistent with the fact that additive models provide a good first approximation of the genetic architectures for dependence and ability to quit.

2.3. WORKING HYPOTHESIS III: MOST GENOMIC VARIANTS THAT CONTRIBUTE TO DEPENDENCE OR ABILITY TO QUIT EXERT EFFECTS OF SMALL SIZE, THOUGH THERE ARE LARGER INFLUENCES IN SPECIFIC POPULATIONS AND FOR ADDICTION-ASSOCIATED PHENOTYPES

2.3.1 Large Influence in a Specific Population

The largest characterized genetic influence on addiction is provided by differential responses to alcohol in members of specific Asian populations. Flushing occurs when vasoactive toxic ethanol metabolites accumulate in individuals who cannot metabolize alcohol efficiently. Skin flushing is accompanied by other aversive symptoms that dramatically reduce the vulnerability to dependence on alcohol in affected individuals. As recently reviewed by Peng et al., homozygosity of variant allele ADH1B★2 per se can reduce the risk of alcohol dependence by eightfold in East Asians, whereas ALDH2★2 homozygosity appears to be almost completely protective (Peng and Yin, 2009). This large effect on vulnerability to alcohol dependence is based on missense variants that change amino acids at important sites in the products of these genes that encode major alcohol detoxifying enzymes. Indeed, these influences provided a basis for development of the moderately specific inhibitor of these enzymes, disulfiram, for treatment of alcoholism.

2.3.2 Oligogenic Influences on Addiction-Associated Traits

nAChR and nicotine metabolism and smoking quantity/frequency: Beirut, Saccone, and their colleagues used pooled and candidate gene strategies, coupled with analyses of a number of smoking-related traits, to identify effects of several distinct variants in a cluster of nicotinic acetylcholine receptor genes on chromosome 15 on smoking quantity/frequency (Bierut et al., 2007; Saccone et al., 2007). Much work has gone into larger and larger analyses and metanalyses that have provided strong statistical support for oligogenic influences of variants in this gene cluster on levels of the nicotine metabolite cotinine, as well as more modest influences on numbers of cigarettes smoked/day (Liu et al., 2010; Liu et al., 2009; Thorgeirsson et al., 2008; Thorgeirsson et al., 2010; Tobacco and GeneticsConsortium, 2010). Large effects on DSM (diagnostic and statistical manual) nicotine dependence have been absent, however, from most of this work.

Effects of haplotypes in the cytochrome oxidases that metabolize nicotine have also been reported in candidate gene studies, though these have not been as reliably identified in genome-wide association datasets (Benowitz et al., 2006; Sellers and Tyndale, 2000). It is thus not clear if these CYP variants provide modest or oligogenic effects on nicotine dependence or abilities to quit smoking.

TRPA1 and smoking mentholated cigarettes: We have recently identified association between haplotypes in one of the menthol receptor genes, TRPA1, and preference for mentholated cigarettes among smokers (Uhl et al., 2011). In smokers of >15 cigarettes/day from both research volunteer and clinical trials participants, there was a robust "oligogenic" influence of a TRPA1 haplotype on menthol preference, providing about a 1.3-fold odds ratio of menthol preference in European-American samples. As this is replicated in other samples, we will have increasing confidence in this genetic effect on this interesting smoking-associated phenotype.

2.3.3 Negative Evidence for Other Common "Oligogenic" Influences on DSM Dependence

If there were large common influences of variants elsewhere in the genome, linkage studies should identify them consistently. Linkage assesses the ways in which genomic markers and disease are coinherited among members of families in which disease is present, usually in multiple members. Linkage identifies loci at which variants in single genes cause diseases that display Mendelian inheritance patterns, and can identify oligogenic effects in many disorders, such as those exerted by haplotypes at the human leukocyte antigen (HLA) locus in disorders with prominent inflammatory components (Wellcome Trust Consortium, 2007).

When we and others have assembled linkage data for vulnerability to dependence on addictive substances, largely in samples of European heritage, there has been no evidence for consistent linkage (Uhl et al., 2008a). Many papers have reported numerous nominally significant linkage results. However, for phenotypes related to smoking, these nominally significant results are distributed through the genome. Overall, they do not display clustering at specific genomic loci more often than anticipated by chance. Such results do not provide support for large, common effects at any single genomic locus for addiction vulnerability. These negative observations are important, since they appear to place an upper bound on the magnitude of influence that we might expect from variation at any locus.

The support for the idea that we have not identified consistent linkage evidence for oligogenic influences on substance dependence is buttressed by data from genome-wide association studies. These studies compare allele frequencies in "unrelated" individuals with disease to those manifest in controls. When we began to use arrays to compare allele frequencies between addicted and control samples in about 2000 using the 1500 SNP arrays available at the time, we obtained few striking positive results, but instead observed modest sized association signals that were present in each of pairs of trios of independent case vs control samples (Uhl et al., 2001). A possible initial explanation for the modest p values observed in these studies was that the modest density of SNPs available on these arrays sampled only modest parts of the genome. However, we and others have now used higher and higher SNP densities to study almost a dozen independent samples. The results provide good support for the idea that there are no common, large influences on vulnerability to substance dependence in members of most populations (Bierut et al., 2010; Drgon et al., 2011; Drgon et al., 2009b; Drgon et al., 2010; Johnson et al., 2006; Johnson et al., 2008; Johnson et al., 2011; Lind et al., 2010; Liu et al., 2006; Liu et al., 2005; Treutlein et al., 2009; Uhl et al., 2008c).

Smoking cessation has also been approached with genome-wide association. We have largely studied ability to quit in participants in clinical trials of smoking cessation (Drgon et al., 2009a; Drgon et al., 2009b; Rose et al., 2010; Uhl et al., 2010a; Uhl et al., 2009a; Uhl et al., 2010b; Uhl et al., 2009b; Uhl et al., 2007; Uhl, et al. 2008d). Other studies have compared current to former smokers (Caporaso et al., 2009; Liu et al., 2010; Liu et al., 2009; Thorgeirsson et al., 2008; Thorgeirsson et al., 2010; Tobacco and GeneticsConsortium, 2010). Genome-wide studies with 500,000–1M SNPs have identified modest association signals, with no evidence for large effects on cessation based on variation at any single locus. There is thus relatively strong evidence against common, large influences of any single variant on success in quitting smoking, though virtually all of this evidence is derived from genome-wide association datasets.

2.4. WORKING HYPOTHESIS IV: THERE ARE ROBUST OVERALL GENETIC INFLUENCES ON VULNERABILITY TO DEPENDENCE, MANY SHARED ACROSS VULNERABILITIES TO DIFFERENT SUBSTANCES. THERE ARE ROBUST OVERALL GENETIC INFLUENCES ON ABILITY TO QUIT, SOME OF WHICH OVERLAP WITH THOSE THAT DETERMINE DEGREE OF DEPENDENCE

The support for sizable heritable components for vulnerability to addiction comes from a wealth of data from classical genetic studies. There is information from: (1) family study data (in which risk to relatives of addicted individuals is compared to risks in members of the general population) (Merikangas et al., 1998; Uhl et al., 1995), (2) adoption studies (in which adoptees' similarities to biological relatives vs adoptive family members are compared) (Uhl et al., 1995) and (3) twin study data (in which concordance in genetically identical monozygotic vs genetically half-identical dizygotic twins are compared) (Agrawal et al., 2004; Gynther et al., 1995; Karkowski et al., 2000; Kendler and Prescott, 1998; Kendler et al., 2000b; Tsuang et al., 1998).

Twin data, in particular, allow us to assign relative fractions of the vulnerability to addiction that come from genetic and from environmental influences. There is robust evidence that about half of addiction vulnerability comes from heritable influences, and that about half comes from environmental influences (Broms et al., 2006; Uhl et al., 2009b). Interestingly, the majority of the environmental influences are likely to come from sources that are outside of the (e.g. home) environment that is shared by sibs. Disorders such as addiction vulnerability that result from many influences are termed "complex traits". Most common brain disorders are, in fact, complex traits (Uhl and Grow, 2004).

Addictive substances fall into classes that are generally based on their different primary receptor targets in the brain and elsewhere. Addictive substances primarily recognize G protein coupled receptors (e.g. morphine and mu opiate receptors), ligand-gated ion channels (e.g. nicotine and nicotinic acetylcholine receptors) and neurotransmitter transporters (e.g. cocaine and dopamine transporter). Despite these differences in primary targets of addictive substances, virtually all addictive substances share the abilities to augment release of dopamine from terminals of cells in the ventral midbrain (Di Chiara and Imperato, 1988).

Twin data from several datasets support the idea that much of the genetic vulnerability to dependence on addictive substances is shared across classes of drugs (Karkowski et al., 2000; Kendler et al., 2000a; Tsuang et al., 1998). Though there is also likely to be substance-specific genetic influence, these data accord with common clinical observations that many addicted individuals abuse addictive substances from several different classes.

There is also relatively strong evidence from twin studies for strong genetic, as well as environmental, contributions to the ability to quit smoking (Broms et al., 2006; Lessov et al., 2004). This data come from comparisons of current and former smokers among pairs of fraternal and identical twins. Again, environmental influences are largely those not shared by sibs. These substantial data for heritability for ability to quit smoking are not accompanied by sizable datasets for abilities to quit other substances. Nevertheless, there is no reason to expect that genetic influences on abilities to quit smoking might not overlap with genetic influences on individuals' abilities to achieve and sustain abstinence from other addictive substances to which they have become dependent. Further, in recent work using a quit success molecular genetic score, we have identified differences in patterns of development of use of a variety of addictive substances, not just cigarettes, providing initial evidence for the likely ability of genetic influences on individual differences in smoking cessation to generalize to influences on other addictive substances as well (Uhl et al., 2012).

There are also modest to moderate overlaps between heritabilities for vulnerability to develop DSM dependence on addictive substances, quantity/frequency of substance use, nicotine dependence as assessed by the Fagerstrom test for nicotine dependence (FTND) and abilities to quit smoking. Modest to moderate amounts of evidence, largely from twin datasets, support these ideas. Some of the highest reported heritabilities for substance use (>0.6) derive from assessments of the "use more than 5 times" data from the Vietnam era twin study (Tsuang et al., 1998), in ways that overlap with genetics of DSM dependence in these samples. There is also significant data for heritable influences on FTND dependence. Twin data, derived from comparisons between monozygous and dizygous current vs former smokers, support substantial overlaps between smoking quantity/frequency assessments and abilities to quit (Broms et al., 2007). These data fit with clinical observations that degree of nicotine dependence is a strong predictor of ability to quit smoking (Berlin et al., 2003). The FTND scale item that assesses time from awakening to first morning cigarette provides an especially prominent contribution to this prediction of quit success.

3. CONCLUSIONS

3.1 Overall Genetic Architecture

The data presented above can be summarized to form a current picture of genetic influences on addiction and ability to quit smoking:

1. Each of these phenotypes is influenced by genetic and environmental influences.
2. The genetic influences for addiction and ability to quit overlap, as do the genetic influences for dependence on substances of different classes.
3. Most of the genetic influences, in permissive environments, come from modest-sized, polygenic contributions from variants in a number of genes at a number of chromosomal locations.
4. There are likely to be numerous addiction-associated variants in each gene that harbors addiction vulnerability or quit-success variants.

5. Single nucleotide, repeat sequence and CNVs are likely to contribute to vulnerability to addiction. Resequencing efforts that are currently underway will be likely to add many individually rare variants to the list of these variations.

6. Additive models provide a good first approximation for the way in which the effects of these variations can be assembled to provide overall risk assessments for vulnerability to dependence of ability to quit.

7. Studies need to take genetic background into account in assessment of the effects of large-sized variants (flushing syndrome, TRPA1 menthol preference, nAChR chromosome 15 genetics), as well as variants of more modest sizes.

Each of these features provides complexity that may be daunting at first glance. However, this complexity provides a good fit with the ways that addiction moves through families and the population. Indeed, this sort of complexity is now likely to underpin genetic aspects of many, if not most, brain disorders (Uhl and Grow, 2004).

3.2 Implications of Molecular Genetic Observations in Addiction and Ability to Quit

Current molecular genetic data available thus provide a number of reassuring features that fit with aspects of classical genetic data. Variations in basic pharmacokinetic and pharmodynamic properties of addictive substances provided by variation in alcohol and nicotine metabolism can provide influences on dependence. Variations in primary receptors for addictive substances and their comcomitants, such as those in nicotinic acetylcholine receptor genes and TRPA1 channel "menthol receptor" genes, do play substantial roles. There is increasing data, however, that implicate genes that regulate cell to cell connections, "cell adhesion molecules" in modest, polygenic influences on substance dependence (Uhl et al., 2008b). Though such effects can be modest, we and others have identified robust influences of variants in one of these genes, cadherin 13 (CDH13) on detailed dose–response relationships in mouse model and human laboratory administration challenges, respectively. Taken together, the results of modest association signals in many studies also provide strong support for roles for monoaminergic and glutamatergic systems, especially in addiction. These data, overall, thus provide a substrate to improve understanding of substance dependence and the ability to quit smoking.

Data from initial smoking cessation studies have allowed us to formulate a complex genetic predictive score, to model ways in which this would be able to enhance the power of clinical trials (Uhl et al., 2009a), and to use it prospectively in subsequent trials (Uhl et al., 2010a). Each of these lines of evidence supports the likely clinical utility of such complex genetic scores. In clinical trials, in which per subject costs are so large, there is a strong economic rationale to include such scores at this time. As genotyping costs decline, there is also a better and better rationale for use of such a score in clinical practice.

No one needs to become addicted; better targeted prevention efforts that could identify the individuals at highest risk could have disproportionate benefits to at-risk individuals and society. Our recent data are consistent with the idea that individuals with the most rapidly accelerating course of substance use during adolescence might be those who have some of the most difficulty in quitting smoking once their addictions are established. Genetic markers would allow selection of the individuals most likely to benefit from focused prevention efforts in this regard. Thus, despite the difficult genetic architecture of addiction and ability to quit, the results of addiction genetics are likely to inform understanding, treatment, and prevention efforts for this difficult disorder.

REFERENCES

Agrawal, A., Neale, M.C., Prescott, C.A., Kendler, K.S., 2004. A twin study of early cannabis use and subsequent use and abuse/dependence of other illicit drugs. Psychol. Med. 34, 1227–1237.

Benowitz, N.L., Swan, G.E., Jacob P., 3rd, Lessov-Schlaggar, C.N., Tyndale, R.F., 2006. CYP2A6 genotype and the metabolism and disposition kinetics of nicotine. Clin. Pharmacol. Ther. 80, 457–467.

Berlin, I., Singleton, E.G., Pedarriosse, A.M., Lancrenon, S., Rames, A., Aubin, H.J., Niaura, R., 2003. The modified reasons for smoking scale: factorial structure, gender effects and relationship with nicotine dependence and smoking cessation in French smokers. Addiction 98, 1575–1583.

Bierut, L.J., Madden, P.A., Breslau, N., Johnson, E.O., Hatsukami, D., Pomerleau, O.F., Swan, G.E., Rutter, J., Bertelsen, S., Fox, L., et al., 2007. Novel genes identified in a high-density genome wide association study for nicotine dependence. Hum. Mol. Genet. 16, 24–35.

Bierut, L.J., Agrawal, A., Bucholz, K.K., Doheny, K.F., Laurie, C., Pugh, E., Fisher, S., Fox, L., Howells, W., Bertelsen, S., et al., 2010. A genome-wide association study of alcohol dependence. Proc. Natl. Acad. Sci. U.S.A. 107, 5082–5087.

Bond, C., LaForge, K.S., Tian, M., Melia, D., Zhang, S., Borg, L., Gong, J., Schluger, J., Strong, J.A., Leal, S.M., et al., 1998. Single-nucleotide polymorphism in the human mu opioid receptor gene alters beta-endorphin binding and activity: possible implications for opiate addiction. Proc. Natl. Acad. Sci. U.S.A. 95, 9608–9613.

Boucard, Sudhof, et al., 2011. Society for Neuroscience. NIDA. (Premeeting).

Broms, U., Silventoinen, K., Madden, P.A., Heath, A.C., Kaprio, J., 2006. Genetic architecture of smoking behavior: a study of Finnish adult twins. Twin Res. Hum. Genet. 9, 64–72.

Broms, U., Madden, P.A., Heath, A.C., Pergadia, M.L., Shiffman, S., Kaprio, J., 2007. The nicotine dependence syndrome scale in Finnish smokers. Drug Alcohol Depend. 89, 42–51.

Caporaso, N., Gu, F., Chatterjee, N., Sheng-Chih, J., Yu, K., Yeager, M., Chen, C., Jacobs, K., Wheeler, W., Landi, M.T., et al., 2009. Genome-wide and candidate gene association study of cigarette smoking behaviors. PloS One 4, e4653.

Di Chiara, G., Imperato, A., 1988. Drugs abused by humans preferentially increase synaptic dopamine concentrations in the mesolimbic system of freely moving rats. Proc. Natl. Acad. Sci. U.S.A. 85, 5274–5278.

Drgon, T., Johnson, C., Walther, D., Albino, A.P., Rose, J.E., Uhl, G.R., 2009a. Genome-wide association for smoking cessation success: participants in a trial with adjunctive denicotinized cigarettes. Mol. Med. 15, 268–274.

Drgon, T., Montoya, I., Johnson, C., Liu, Q.R., Walther, D., Hamer, D., Uhl, G.R., 2009b. Genome-wide association for nicotine dependence and smoking cessation success in NIH research volunteers. Mol. Med. 15, 21–27.

Drgon, T., Zhang, P.W., Johnson, C., Walther, D., Hess, J., Nino, M., Uhl, G.R., 2010. Genome wide association for addiction: replicated results and comparisons of two analytic approaches. PLoS One 5, e8832.

Drgon, T., Johnson, C.A., Nino, M., Drgonova, J., Walther, D.M., Uhl, G.R., 2011. "Replicated" genome wide association for dependence on illegal substances: genomic regions identified by overlapping clusters of nominally positive SNPs. Am. J. Med. Genet., Part B 156, 125–138.

Gynther, L.M., Carey, G., Gottesman II, Vogler, G.P., 1995. A twin study of non-alcohol substance abuse. Psychiatry Res. 56, 213–220.

Hishimoto, A., Liu, Q.R., Drgon, T., Pletnikova, O., Walther, D., Zhu, X.G., Troncoso, J.C., Uhl, G.R., 2007. Neurexin 3 polymorphisms are associated with alcohol dependence and altered expression of specific isoforms. Hum. Mol. Genet. 16, 2880–2891.

Johnson, C., Drgon, T., Liu, Q.R., Walther, D., Edenberg, H., Rice, J., Foroud, T., Uhl, G.R., 2006. Pooled association genome scanning for alcohol dependence using 104,268 SNPs: validation and use to identify alcoholism vulnerability loci in unrelated individuals from the collaborative study on the genetics of alcoholism. Am. J. Med. Genet., Part B 141, 844–853.

Johnson, C., Drgon, T., Liu, Q.R., Zhang, P.W., Walther, D., Li, C.Y., Anthony, J.C., Ding, Y., Eaton, W.W., Uhl, G.R., 2008. Genome wide association for substance dependence: convergent results from epidemiologic and research volunteer samples. BMC Med. Genet. 9, 113.

Johnson, C., Drgon, T., Walther, D., Uhl, G.R., 2011. Genomic regions identified by overlapping clusters of nominally-positive SNPs from genome-wide studies of alcohol and illegal substance dependence. PloS One 6, e19210.

Karkowski, L.M., Prescott, C.A., Kendler, K.S., 2000. Multivariate assessment of factors influencing illicit substance use in twins from female-female pairs. Am. J. Med. Genet. 96, 665–670.

Kendler, K.S., Prescott, C.A., 1998. Cocaine use, abuse and dependence in a population-based sample of female twins. Br. J. Psychiatry 173, 345–350.

Kendler, K.S., Karkowski, L.M., Neale, M.C., Prescott, C.A., 2000a. Illicit psychoactive substance use, heavy use, abuse, and dependence in a US population-based sample of male twins. Arch. Gen. Psychiatry 57, 261–269.

Kendler, K.S., Thornton, L.M., Pedersen, N.L., 2000b. Tobacco consumption in Swedish twins reared apart and reared together. Arch. Gen. Psychiatry 57, 886–892.

Lessov, C.N., Martin, N.G., Statham, D.J., Todorov, A.A., Slutske, W.S., Bucholz, K.K., Heath, A.C., Madden, P.A., 2004. Defining nicotine dependence for genetic research: evidence from Australian twins. Psychol. Med. 34, 865–879.

Lind, P.A., Macgregor, S., Vink, J.M., Pergadia, M.L., Hansell, N.K., de Moor, M.H., Smit, A.B., Hottenga, J.J., Richter, M.M., Heath, A.C., et al., 2010. A genomewide association study of nicotine and alcohol dependence in Australian and Dutch populations. Twin Res. Hum. Genet. 13, 10–29.

Liu, Q.R., Drgon, T., Walther, D., Johnson, C., Poleskaya, O., Hess, J., Uhl, G.R., 2005. Pooled association genome scanning: validation and use to identify addiction vulnerability loci in two samples. Proc. Natl. Acad. Sci. U.S.A. 102, 11864–11869.

Liu, Q.R., Drgon, T., Johnson, C., Walther, D., Hess, J., Uhl, G.R., 2006. Addiction molecular genetics: 639,401 SNP whole genome association identifies many "cell adhesion" genes. Am. J. Med. Genet., Part B 141, 918–925.

Liu, Y.Z., Pei, Y.F., Guo, Y.F., Wang, L., Liu, X.G., Yan, H., Xiong, D.H., Zhang, Y.P., Levy, S., Li, J., et al., 2009. Genome-wide association analyses suggested a novel mechanism for smoking behavior regulated by IL15. Mol. Psychiatry 14, 668–680.

Liu, J.Z., Tozzi, F., Waterworth, D.M., Pillai, S.G., Muglia, P., Middleton, L., Berrettini, W., Knouff, C.W., Yuan, X., Waeber, G., et al., 2010. Meta-analysis and imputation refines the association of 15q25 with smoking quantity. Nat. Genet. 42, 436–440.

Merikangas, K.R., Stolar, M., Stevens, D.E., Goulet, J., Preisig, M.A., Fenton, B., Zhang, H., O'Malley, S.S., Rounsaville, B.J., 1998. Familial transmission of substance use disorders. Arch. Gen. Psychiatry 55, 973–979.

Oslin, D.W., Berrettini, W., Kranzler, H.R., Pettinati, H., Gelernter, J., Volpicelli, J.R., O'Brien, C.P., 2003. A functional polymorphism of the mu-opioid receptor gene is associated with naltrexone response in alcohol-dependent patients. Neuropsychopharmacology 28, 1546–1552.

Peng, G.S., Yin, S.J., 2009. Effect of the allelic variants of aldehyde dehydrogenase ALDH2*2 and alcohol dehydrogenase ADH1B*2 on blood acetaldehyde concentrations. Hum. Genomics 3, 121–127.

Rose, J.E., Behm, F.M., Drgon, T., Johnson, C., Uhl, G.R., 2010. Personalized smoking cessation: interactions between nicotine dose, dependence and quit-success genotype score. Mol. Med. 16, 247–253.

Saccone, S.F., Hinrichs, A.L., Saccone, N.L., Chase, G.A., Konvicka, K., Madden, P.A., Breslau, N., Johnson, E.O., Hatsukami, D., Pomerleau, O., et al., 2007. Cholinergic nicotinic receptor genes implicated in a nicotine dependence association study targeting 348 candidate genes with 3713 SNPs. Hum. Mol. Genet. 16, 36–49.

Sellers, E.M., Tyndale, R.F., 2000. Mimicking gene defects to treat drug dependence. Ann. N.Y. Acad. Sci. 909, 233–246.

Thorgeirsson, T.E., Geller, F., Sulem, P., Rafnar, T., Wiste, A., Magnusson, K.P., Manolescu, A., Thorleifsson, G., Stefansson, H., Ingason, A., et al., 2008. A variant associated with nicotine dependence, lung cancer and peripheral arterial disease. Nature 452, 638–642.

Thorgeirsson, T.E., Gudbjartsson, D.F., Surakka, I., Vink, J.M., Amin, N., Geller, F., Sulem, P., Rafnar, T., Esko, T., Walter, S., et al., 2010. Sequence variants at CHRNB3-CHRNA6 and CYP2A6 affect smoking behavior. Nat. Genet. 42, 448–453.

TobaccoAndGeneticsConsortium, 2010. Genome-wide meta-analyses identify multiple loci associated with smoking behavior. Nat. Genet. 42, 441–447.

Treutlein, J., Cichon, S., Ridinger, M., Wodarz, N., Soyka, M., Zill, P., Maier, W., Moessner, R., Gaebel, W., Dahmen, N., et al., 2009. Genome-wide association study of alcohol dependence. Arch. Gen. Psychiatry 66, 773–784.

Tsuang, M.T., Lyons, M.J., Meyer, J.M., Doyle, T., Eisen, S.A., Goldberg, J., True, W., Lin, N., Toomey, R., Eaves, L., 1998. Co-occurrence of abuse of different drugs in men: the role of drug-specific and shared vulnerabilities. Arch. Gen. Psychiatry 55, 967–972.

Tyrfingsson, T., Thorgeirsson, T.E., Geller, F., Runarsdottir, V., Hansdottir, I., Bjornsdottir, G., Wiste, A.K., Jonsdottir, G.A., Stefansson, H., Gulcher, J.R., et al., 2010. Addictions and their familiality in Iceland. Ann. N.Y. Acad. Sci. 1187, 208–217.

Uhl, G.R., Grow, R.W., 2004. The burden of complex genetics in brain disorders. Arch. Gen. Psychiatry 61, 223–229.

Uhl, G.R., Elmer, G.I., Labuda, M.C., Pickens, R.W., 1995. Genetic influences in drug abuse. In: Gloom, F.E., Kupfer, D.J. (Eds.), Psychopharmacology: The Fourth Generation of Progress. Raven Press, New York, pp. 1793–2783.

Uhl, G.R., Liu, Q.R., Walther, D., Hess, J., Naiman, D., 2001. Polysubstance abuse-vulnerability genes: genome scans for association, using 1,004 subjects and 1,494 single-nucleotide polymorphisms. Am. J. Hum. Genet. 69, 1290–1300.

Uhl, G.R., Liu, Q.R., Drgon, T., Johnson, C., Walther, D., Rose, J.E., 2007. Molecular genetics of nicotine dependence and abstinence: whole genome association using 520,000 SNPs. BMC Genetics 8, 10.

Uhl, G.R., Drgon, T., Johnson, C., Fatusin, O.O., Liu, Q.R., Contoreggi, C., Li, C.Y., Buck, K., Crabbe, J., 2008a. "Higher order" addiction molecular genetics: convergent data from genome-wide association in humans and mice. Biochem. Pharmacol. 75, 98–111.

Uhl, G.R., Drgon, T., Johnson, C., Li, C.Y., Contoreggi, C., Hess, J., Naiman, D., Liu, Q.R., 2008b. Molecular genetics of addiction and related heritable phenotypes: genome-wide association approaches identify "connectivity constellation" and drug target genes with pleiotropic effects. Ann. N.Y. Acad. Sci. 1141, 318–381.

Uhl, G.R., Drgon, T., Liu, Q.R., Johnson, C., Walther, D., Komiyama, T., Harano, M., Sekine, Y., Inada, T., Ozaki, N., et al., 2008c. Genome-wide association for methamphetamine dependence: convergent results from 2 samples. Arch. Gen. Psychiatry 65, 345–355.

Uhl, G.R., Liu, Q.R., Drgon, T., Johnson, C., Walther, D., Rose, J.E., David, S.P., Niaura, R., Lerman, C., 2008d. Molecular genetics of successful smoking cessation: convergent genome-wide association study results. Arch. Gen. Psychiatry 65, 683–693.

Uhl, G.R., Drgon, T., Johnson, C., Rose, J.E., 2009a. Nicotine abstinence genotyping: assessing the impact on smoking cessation clinical trials. Pharmacogenomics J. 9, 111–115.

Uhl, G.R., Drgon, T., Li, C.Y., Johnson, C., Liu, Q.R., 2009b. Smoking and smoking cessation in disadvantaged women: assessing genetic contributions. Drug Alcohol Depend. 104 (Suppl. 1), S58–S63.

Uhl, G.R., Drgon, T., Johnson, C., Ramoni, M.F., Behm, F.M., Rose, J.E., 2010a. Genome-wide association for smoking cessation success in a trial of precessation nicotine replacement. Mol. Med. 16, 513–526.

Uhl, G.R., Drgon, T., Johnson, C., Walther, D., David, S.P., Aveyard, P., Murphy, M., Johnstone, E.C., Munafo, M.R., 2010b. Genome-wide association for smoking cessation success: participants in the patch in practice trial of nicotine replacement. Pharmacogenomics 11, 357–367.

Uhl, G.R., Walther, D., Behm, F.M., Rose, J.E., 2011. Menthol preference among smokers: association with TRPA1 variants. Nicotine Tob. Res. 13 (12), 1311–1315.

Uhl, G.R., Walther, D., Musci, R., Fisher, C., Storr, C.L., Behm, F.M., Eaton, W.W., Ialongo, N., Rose, J.E., 2012. Mol Psychiatry. http://dx.doi.org/ 10.1038/mp.2012.155. [Epub ahead of print] PMID:23128154.

Wang, J.C., Cruchaga, C., Saccone, N.L., Bertelsen, S., Liu, P., Budde, J.P., Duan, W., Fox, L., Grucza, R.A., Kern, J., et al., 2009. Risk for nicotine dependence and lung cancer is conferred by mRNA expression levels and amino acid change in CHRNA5. Hum. Mol. Genet. 18, 3125–3135.

WellcomeTrustConsortium, 2007. Genome-wide association study of 14,000 cases of seven common diseases and 3,000 shared controls. Nature 447, 661–678.

CHAPTER THREE

Epidemiology of Drug Abuse: Building Blocks for Etiologic Research

Naimah Weinberg, Marsha Lopez and Wilson M. Compton

Division of Epidemiology, Services and Prevention Research, National Institute on Drug Abuse, Bethesda, MD, USA

1. INTRODUCTION

Addictions and related misuse of intoxicants are costly, distressing, and challenging problems on personal, clinical, and public health levels. Advances in the neuroscience of these conditions depend, in large part, on foundational understanding of the epidemiology of the phenomena. For instance, the recognition that addictions have a genetic basis derives from family, twin, and adoption epidemiology studies. The onset of addictions in adolescence with roots much earlier in childhood experiences also derives from observational studies of the conditions in broad populations. In the past 10 years, significant new epidemiologic approaches have been developed and applied to study the patterns, origins, and consequences of developmental behavior problems that point the way for fruitful basic science research.

This review covers both substance use (SU) and substance use disorders (SUD), drawing almost exclusively on studies using population-based rather than clinical samples. Such study samples (sometimes called "epidemiologic" or "community-based") offer greater generalizability and freedom from the biases and confounders that are inherent in the use of clinic-based samples (Armstrong and Costello, 2002). Most of the citations are drawn from longitudinal studies. Innovative approaches further enhance the utility of epidemiologic studies for understanding causality (Rutter, 2009).

An important caveat is that SU is not a precise term and is used differently by various researchers (Armstrong and Costello, 2002). When a cited study focuses particularly on substance initiation, that is, the age at which a substance is first used, this is indicated. For SUD, the cited studies generally employ standardized instruments to assess abuse and dependence based on criteria in the Diagnostic and Statistical Manual Version IIIR (American Psychiatric Association, 1987) or IV (American Psychiatric Association, 1994). Further, SU and SUD encompass a variety of licit and illicit substances. In this paper, most findings apply to multiple substances and when a finding relates only to a specific drug, this is indicated.

The Effects of Drug Abuse on the Human Nervous System
http://dx.doi.org/10.1016/B978-0-12-418679-8.00003-4

References for this review were drawn from literature searches conducted using PubMed and other databases. However, this review presents new findings and innovative approaches in the epidemiology and etiology of SU/SUD, rather than a comprehensive review of all relevant research. As an increasingly strong body of research on SU and SUD is emerging using population-based, high-risk, and clinical approaches, only a fraction can be cited herein, and further reading can be found in the references.

2.1. EPIDEMIOLOGY

Tracking the shifting landscape of SU and SUD, a key goal of epidemiologic research, provides clues about etiology and informs prevention, treatment, and policy planning. On a frequent recurrent timeframe, three national surveys assess and report the prevalence of drug-using behavior and attitudes toward drugs in the general US population: the Monitoring the Future (MTF) study of U.S. 8th, 10th and 12th graders and linked longitudinal follow-up study (Johnston et al., 2013), The National Survey of Drug Use and Health (NSDUH) study of the U.S. populations ages 12 and older (Substance Abuse and Mental Health Services Administration, 2011), and the Youth Risk Behavior Survey (YRBS) of adolescent students (Centers for Disease Control and Prevention, 2010). While each survey tends to report slightly different estimates in any particular year, the overall trends in SU are consistent across the surveys over time. After peaking in the late 1970s, overall drug-use rates declined during the 1980s and then rose again during the first half of the 1990s. Since the mid/late 1990s, overall drug-use rates declined slowly or remained generally stable. Of course, within this overall pattern, many specific drug epidemics were observed. For example, rates of MDMA (ecstasy) use increased markedly in the late 1990s and then declined rapidly in the early 2000s.

In addition to the surveys that document trends in overall drug-use rates across the USA, periodic surveys of national samples also inform our understanding of drug using adult populations, especially the National Epidemiologic Survey of Alcohol and Related Conditions (NESARC) and the National Comorbidity Surveys (NCS). These are also highlighted as they emphasize the relationship of SU and SUD to other mental illnesses (Compton et al., 2007; Swendson et al., 2008).

Initiation: The age of initiation for specific illicit drugs varies by substance, although SU tends to start during adolescence. The NSDUH finds that among the nearly three million people ages 12 and older who reported using an illicit drug for the first time in the past year, 56.7% were under the age of 18 and the average age at initiation, among persons ages 12 to 49, was just under 19 years old (Substance Abuse and Mental Health Services Administration, 2011). According to the long-standing MTF study (Johnston et al., 2013), alcohol, tobacco, and inhalants demonstrate earlier patterns of initiation in comparison to illicit drugs such as cocaine powder and hallucinogens, which show a pattern of initiation in later adolescence. Additionally, among the MTF respondents, age of initiation is generally earlier for the drugs perceived to be less risky. For most substances, age of onset has been reported as early as the 4th grade (about 10 years old), albeit at very low rates, with more rapidly increasing rates starting in the teenage years. Estimates of reported age of illicit drug initiation should be considered in the context of the limitations of this type of information. For example, the MTF study surveys 8th, 10th, and 12th graders who are enrolled and attending school; thus, dropouts and absentees who are more likely to have SU and early onset drug use are more likely to be captured in the 8th grade sample but not in the 12th grade sample (Johnston et al., 2013). Also, retrospective reports of age of onset are subject to recall biases. Despite limitations, however, these data provide useful information on trends in use over time.

Patterns of SU: After recent peaks in the mid- to late-1990s, a slow but steady decline in the overall reported SU among adolescents was seen until 2008. Since that time, overall rates of SU in adolescents have remained stubbornly steady. In 2012, nearly half of high school seniors reported having used an illicit drug at some point in their lives, and 40% had used in the past year (Johnston et al., 2013). Although marijuana is the most prevalent illicit drug reported, it is important to note that the while cigarette and alcohol use are legal among adults and thus not included in the "any illicit drug" category, their use is illegal among adolescents. Roughly, one out of every eight 8th graders, one third of 10th graders, and nearly half of high school seniors reported having been drunk in the year prior to survey. While the prevalence of cigarette use has dropped quite dramatically over the years, reductions in illicit drug use have been less substantial, such that estimates of reported illicit drug use now match or exceed those for cigarette use among adolescents (Figure 1, Johnston, et al., 2013). Estimates for specific illicit drug use reported by high school seniors in 2011 are presented in Table 1. Of note, the MTF study examined rates of so-called synthetic marijuana (sold under such trade names as "K-2" or "Spice") for the first time in 2011 and found that 11.4% of 12th graders reported using these substances during the past year. In 2012 past year use held at 11.3% of 12th graders. This important result provides evidence that newly created substances can quickly make their way into widespread consumption.

The patterns of drug use shift across different ages, with inhalants more commonly abused by children and early adolescents than by older adolescents and adults (Johnston et al., 2013). Most other substances are used more frequently by older adolescents and young adults with rates peaking in the twenties (Substance Abuse and Mental Health Services Administration, 2011). These patterns are meaningful as drug users who report inhalant use appear to have worse outcomes, including early initiation of other drugs and a higher prevalence of drug-use disorders, compared to users who did not report inhalant use (Storr et al., 2005; Wu et al., 2008). As seen in Table 1, after alcohol and tobacco, the most commonly used drugs reported by youth are marijuana, amphetamines, nonmedical use of prescription drugs, and hallucinogens; for those who go on to use other drugs such as cocaine, that use emerges more commonly in later years.

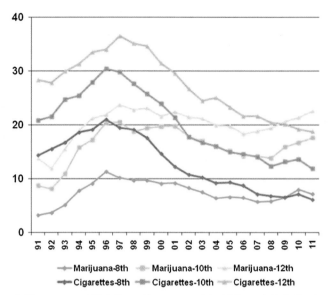

Figure 1 Past Month Cigarette and Illicit Drug Use among 8th, 10th, and 12th Graders in the US, Monitoring the Future Study (Johnston, et al., 2013).

Table 1 Prevalence of Past Year Drug Use[1] among 12th Graders, 2012 Monitoring the Future Study (Johnston, et al., 2013)

Drug	%
Alcohol	63.5
Marijuana/Hashish	36.4
Synthetic Marijuana	11.3
Amphetamines[1]	7.9
Vincodin[1]	7.5
Adderall[1]	7.6
Salvia	4.4
Tranquilizers[1]	5.3
Cough Medicine[1]	5.6
MDMA (Ecstasy)	3.8
Hallucinogens	4.8
OxyContin[1]	4.3
Sedatives[1]	4.5
Inhalants	2.9
Cocaine (any form)	2.7
LSD	2.4
Ritalin[1]	2.6
Ketamine	1.5
GHB	1.4
Methamphetamine	1.4

[1]Categories not mutually exclusive.

Polysubstance use: Research also suggests that polysubstance use—using multiple substances within a defined period of time—is a typical pattern of SU, especially among youth (Mitchell and Plunkett, 2000; Whitesell et al., 2006; Dierker et al., 2007; Connell et al., 2009; Connell et al., 2010). Among adolescents in secondary school, between 13% and 39% of adolescents may be described as polysubstance users (though the actual estimates vary by sample, reporting period, and definition of polysubstance use). For example, the 2005 YRBS found four types of users based upon past-month patterns of use among adolescents in grades 9–12: alcohol experimenters (38%), nonusers/abstainers (27%), occasional polysubstance users (23%), and frequent polysubstance users (13%) (Connell et al., 2009). Analyses of the 2001–02 National Study of Adolescent Health data identified five classes of students in grades 7–12 based upon use during the past month/year: low use (55%), alcohol only (15%), alcohol-marijuana (8%), cigarettes (8%), and three-substance use (14%) (Dierker et al., 2007). In addition, adolescent polysubstance use has been associated with significant deleterious outcomes, including substance dependence (Whitesell et al., 2006), smoking in adulthood (Dierker et al., 2007), and risky sexual behavior (Connell et al., 2009). A recent study using MTF data from senior high school students between 2002 and 2006 found that nearly 7 in 10 nonmedical users of prescription opioids reported co-ingestion of prescription opioids and other drugs (marijuana, alcohol, cocaine, tranquilizers, and amphetamines) in the past year (McCabe et al., 2012). Using data from a genetically informative longitudinal sample, Vrieze and colleagues (Vrieze et al., 2012) examined changes in correlations among symptoms of nicotine, alcohol, and marijuana abuse and dependence from ages 11–29; at younger ages use of these substances showed higher correlations, suggesting indiscriminate use, which declined with age as individuals showed a great preference for one substance over the others.

Differential patterns of SU: Patterns of drug use vary significantly by gender and ethnic group as well as by drug. For many specific drug types, males tend to report more use than females particularly as they get older (Johnston et al., 2013). Previous literature suggests that some of this gender difference may stem from males having greater opportunities to initiate drug use; once opportunity to use was controlled for, males and females had equal likelihood of using drugs (Van Etten et al., 1999). One notable exception is inhalants, which is reported by more females in the younger grades than by males (Swendson et al., 2008). White and Hispanic youth tend to report more overall drug use than African American youth; however, these differences vary by drug type and grade (Johnston et al., 2013). Additional details regarding racial and ethnic differences can be found in the respective reports of the major national surveys cited above.

Medication abuse: The profile of drug use has shifted over the past 10 years, with the recent emergence of medication abuse. As noted in Table 1, in 2012, many of the most prevalent drugs used by 12th graders are nonprescribed prescription or over-the-counter medications (Johnston et al., 2013) such as pain medications, stimulants, and sedatives.

Results from the 2009 YRBS indicate that one out of five U.S. high school students reported having taken a medication without a doctor's prescription, with rates of lifetime medication misuse rising from 15% among 9th graders to 26% among 12th graders (Centers for Disease Control and Prevention, 2010). These data also show that medication misuse is highest among white students, followed by Hispanic students, and lowest among African Americans, with no gender differences found; however, analyses of other data suggest that these trends may vary by drug (Sung et al., 2004). A review of both community-based and clinical studies of adolescents and college-aged subjects found higher risk for diverting and misusing stimulant medications by those who were white, had lower grade point averages, used immediate release rather than delayed-release formulas, and had symptoms of attention deficit hyperactivity disorder (ADHD), although stimulant misuse was found among those with and without ADHD (Wilens et al., 2008).

The problem of prescription opioid misuse has been particularly acute, with markedly increasing rates of nonfatal and fatal overdoses (Paulozzi et al., 2011). Of note, this report suggests an association between increasing rates of overall availability of prescription opioids with increasing rates of overdose mortality as well as geographic associations such that states with higher rates of prescriptions for opioids tended to have higher rates of overdose mortality. In addition, the increases in mortality related to prescription opioids have paralleled sharp increases in rates of treatment admission for opioid addiction over the past decade (Centers for Disease Control and Prevention, 2011).

A key concern is the source of prescription medications being diverted for nonmedical use. Data from all subjects in the combined 2009–10 NSDUH show that "friends or relatives" were the primary source for pain relievers (the most commonly abused prescription medication class), while physicians served as the direct source about 20% of the time (Figure 2). In turn, these friends and relatives reportedly obtained the medications from one doctor 80% of the time (Substance Abuse and Mental Health Services Administration, 2011). An analysis of adolescent subjects in the combined 2005-6 NSDUH data (Schepis and Krishnan-Sarin, 2008) found that those purchasing prescription medications illegally were more likely to be abusing other substances, while those who obtained prescription opioids from a physician were more likely to have had a major depressive episode in the past year.

Associated risks and consequences: Alcohol and drug use are associated with a variety of risky behaviors and related problems, even when a use does not progress to SUD. Drug and alcohol use serve as both distal and proximal risk factors for suicidal behavior and attempts (Esposito-Smythers and Spirito, 2004), possibly mediated by impulsive personality traits (Dougherty et al., 2004). Alcohol use in particular may have a causal effect on violence and vandalism (Felson et al., 2008), and is associated with other delinquent acts, such as shoplifting (Felson et al., 2008) and gambling (Duhig et al., 2007). Furthermore, SU is associated with increased risky sexual activity (Anderson and Mueller, 2008) and serious driving injuries or fatalities (National Highway Traffic Safety

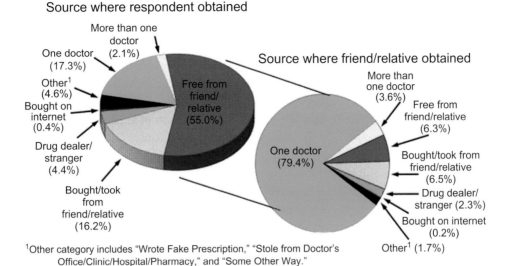

Source where respondent obtained

More than one
doctor
One doctor (2.1%)
(17.3%)
Other[1]
(4.6%)
Bought on
internet
(0.4%)
Drug dealer/
stranger
(4.4%)
Bought/took
from
friend/relative
(16.2%)
Free from
friend/
relative
(55.0%)

Source where friend/relative obtained

More than
one doctor
(3.6%)
Free from
friend/relative
(6.3%)
One doctor
(79.4%)
Bought/took from
friend/relative
(6.5%)
Drug dealer/
stranger (2.3%)
Bought on internet
(0.2%)
Other[1] (1.7%)

[1]Other category includes "Wrote Fake Prescription," "Stole from Doctor's
Office/Clinic/Hospital/Pharmacy," and "Some Other Way."

Figure 2 People abusing analgesics mostly do not obtain them by prescription: Most recent source of abused analgesic for household residents in the USA ages 12 and older who report abusing analgesics *(SAMHSA, 2010–2011 National Survey on Drug Use and Health).*

Administration, 2010). One longitudinal study found increased risks for adult substance dependence, herpes infection, early pregnancy, and crime associated with early SU, even in the absence of conduct disorder (Ogders et al., 2008).

2.1.1. Substance Use Disorders

Adolescence is not only a period of experimental drug usage, but it is also a period during which young people have high potential to develop drug-use disorders and other longer term negative outcomes. Studies of adult samples using DSM-IV criteria for SUDs have found similar patterns of onset across all substances during adolescence and early adulthood, peaking at about 20 years of age, with onsets after 25 quite rare (Compton et al., 2007). Thus, adolescence is a prime period of risk for the development of SUDs.

Recent findings from the NSDUH documented consistent rates of DSM-IV drug-use disorders (i.e. DSM-IV abuse or dependence on illicit drugs) over the past decade, varying little from 2002 (3.1%) to 2010 (2.8%) among those ages 12 or older (Substance Abuse and Mental Health Services Administration, 2011). While the prevalence of abuse or dependence on illicit drugs in 2010 was higher among all males (3.6%) than females (2.0%) in the survey, the rate was higher for females (4.9%) than males (4.6%) among subjects between the ages of 12–17. Rates of SUD were also found to vary by ethnicity: Asian Americans had lower rates of abuse or dependence on illicit drugs (1.3%) than African Americans (4.0%), Hispanic or Latino (3.5%), Whites (2.5%), and American Indian or Alaska Native (6.3%) populations ages 12 and older.

2.1.2. Psychiatric Comorbidity

Psychiatric disorders commonly co-occur with SU and SUD and are discussed further below in the section on etiology. The risk of developing SU/SUD in the presence of comorbid psychiatric disorder varies by study, but has consistently been found to be about three times greater in comparison to those without a psychiatric disorder (Armstrong and Costello, 2002). Among adults, strong associations of SU/SUD and psychiatric illnesses are the rule rather than the exception (Swendson et al., 2008). Associations are strong and specific, especially for antisocial personality disorder (and other personality disorders) and bipolar disorder (BPD).

As noted above, SU is frequently comorbid with psychiatric illnesses and may increase the risk for psychiatric disorders in vulnerable individuals. For both depression and anxiety disorders, there is evidence for bidirectional relationships with alcohol disorders, where internalizing disorders may precede or follow adolescent alcohol use (Nurnberger et al., 2002; Marmorstein, 2009; Kushner et al., 2000). While substances, particularly alcohol, may have depressive effects or may elicit anxiety symptoms on withdrawal, this comorbidity may also be the result of common liability factors leading to both SU and internalizing disorders (Nurnberger et al., 2002; Kuo et al., 2006). Cigarette smoking in adolescence may have a similar bidirectional relationship with depression, *contributing* to the onset of depression in some individuals (Windle and Windle, 2001) although possibly *resulting* from depression in others, as discussed further below. Cannabis use has been linked in numerous studies to an increased risk for psychotic disorders, and use in adolescence may pose a particularly elevated risk (Henquet et al., 2008), although accounting for only a small fraction of the onset of psychosis; a variety of mechanisms have been suggested, including gene–environment interaction (GxE) (Henquet et al., 2008). In fact, a recent population-based Dutch study has found a bidirectional relationship between vulnerability to psychosis and cannabis use in adolescents (Griffith-Lendering et al., 2013). The gravity and costs of psychiatric outcomes underscore the clinical and public health importance of understanding and preventing SU and SUD.

2.2. ETIOLOGY

SU and SUD result from a complex interplay of individual-level, environmental, and developmental factors. In discussing etiology, the term "risk factor" indicates a variable that has been found to be associated with an increased risk of an adverse condition; however, even if a risk factor precedes an outcome temporally, it cannot be assumed to be causal. For this discussion, the term "causal" is used if there is evidence that the risk factor contributes directly to the development of SU or SUD, and "risk indicator" is used for factors that may serve as proxies for a causal risk factor but which have no demonstrated role themselves in the causal chain.

2.2.1. Gene–Environment Interplay

While a full discussion of the role of genetics and epigenetics in SU and SUD is beyond the scope of this review, a few key concepts of gene–environment interplay are essential for the ensuing discussion of etiology; further reading can be found in the references.

Gene–environment correlation (rGE) is a complex concept that needs to be considered in studies of psychopathology, in order to help discriminate true causal risk factors from indicators of risk. In rGE, genetic factors contribute to the likelihood that an individual will be exposed to particular types of environments. Three types of rGE have been described (Dick et al., 2009a; Rutter et al., 2006). In passive rGE, parental genes set the stage for the type of environments to which children are exposed; for example, parents with alcohol dependence are more likely to provide a chaotic home life. In active rGE, a child may seek out certain environments based on genetic propensities (Kendler et al., 2007). In evocative rGE, a child's characteristics (such as genetically based temperament features) may evoke certain responses from others, such as harsh treatment of children with difficult temperaments. Thus, factors that appear to be environmental in origin may actually arise from genetic predispositions in the parents or children.

In GxE, the effects of environmental exposures vary by an individual's genotype, or conversely, genetic effects are moderated by environment (Dick et al., 2009a); this occurs in normal and pathological development. In a large study of adoptees in Sweden, both genetic and environmental factors were implicated in the onset of SU/SUD, with important synergies documented such that those with the highest genetic risk showed a stronger association of environmental risk factors with SU/SUD outcomes compared to those with lower genetic risk (Kendler et al., 2012). Environmental effects can also enhance resilience in vulnerable youth. Recent findings on the effects of parental monitoring interacting with genotype (Brody et al., 2009a) exemplify GxE and are discussed below in the section on environmental risk factors.

Heritability refers to the proportion of variance of a trait in a population that can be attributed to genetic influences. For SUD, heritability estimates in adults range from 45% to 79% (Dick and Agrawal, 2008). However, it is important to remember that heritability is a population statistic, and does not indicate the degree of genetic contribution in any individual case (Rutter et al., 2006); in fact, as a complex disorder, SUD most likely involves a large number of genes (Dick and Agrawal, 2008) many of which are likely to be of small effect (Rutter et al., 2006). Furthermore, heritability estimates will vary with the degree of exposure to relevant environmental conditions in a study population (Rutter et al., 2006); thus, heritability and environment are confounded and there is never a fixed level of heritability for a particular disorder.

Numerous studies have found that heritability of SUD rises with age and with progression of SUD (Dick et al., 2009a; Hopfer et al., 2003), suggesting that genetic factors play an increasing role in the etiology of SUD as individuals pass through later adolescence into adulthood and as their SU advances to heavier use and addiction. Heritability studies have also shown that much of the genetic risk for SUD is nonspecific in regard to choice of substance (Hopfer et al., 2003; Iacono et al., 2008), although there are also genetic vulnerabilities to dependence on licit vs illicit substances or specific substances (Dick and Agrawal, 2008; Iacono et al., 2008), perhaps most strongly for nicotine (Hopfer et al., 2003). In addition to genetic vulnerability to SUD, there is evidence for genetic predisposition to a constellation of externalizing behaviors across development, including conduct disorder (CD), ADHD, and SUD (Dick and Agrawal, 2008; Iacono et al., 2008); Figure 3 presents a simplified model of these findings.

Twin studies provide much of the data concerning the genetic liability for SU and SUD because they can explicate the causal nature of environmental factors separate from inherited in the etiology of psychopathologic disorders (Rutter, 2009; Hopfer et al., 2003). For instance, comparisons of monozygotic and dizygotic twins can yield estimates of the relative contributions of genetic, shared, and nonshared environmental factors at various developmental stages. Comparisons of exposed and unexposed twins (co-twin control method) can offer insights into the causal role of environmental factors, as can studies of the offspring of monozygotic twins, who are genetic half-siblings raised in different environments.

While a discussion of epigenetic mechanisms in addiction is beyond the scope of this review, it represents another example of gene–environment interplay with a potentially key role in drug-abuse trajectories, including in the effects of early initiation of drugs as discussed below (Volkow, 2011; for a review, see Satterlee, 2013).

2.2.2. Psychiatric Risk Factors

As noted above, psychiatric disorders are frequently comorbid with SUD, less so with SU. Examining each major cluster of psychiatric disorders illuminates the varying nature of the relationship between these disparate conditions.

Conduct disorder (CD): Early CD has consistently been found to constitute a risk factor for later SU and SUD in both males and females (e.g. Sung et al., 2004; Fergusson et al., 2007; Disney et al., 1999), regardless of the presence of other comorbid disorders. However, it is not clear whether CD is a causal risk factor for SUD or an indicator of risk. On one hand, some of the relationship appears to have a strong genetic base, where both CD and SUD constitute manifestations of an inherited predisposition toward behavioral disinhibition (Iacono et al., 2008). On the causal side, CD may predispose to SU through association with deviant peers; however, social factors would not account for the progression to SUD, and must be evaluated in light of studies showing heritable influences on a peer choice (Kendler et al., 2007). Several prevention studies are underway that may ultimately help clarify both the role of CD and opportunities for intervention; for example, a manualized behavior therapy program for disruptive behaviors in middle childhood demonstrated reduced SU in adolescence (Zonnevylle-Bender et al., 2007), although it remains to be seen whether later SUDs were affected. Another

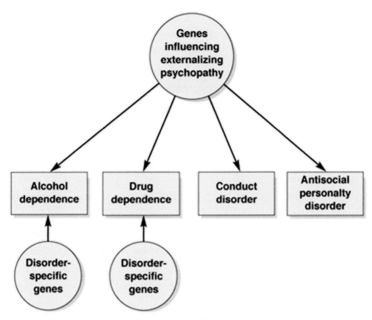

Figure 3 Common Genetic Factors for the Externalizing Disorders.

study found that an intervention to improve parenting skills had a particularly strong impact on reducing high-risk behaviors in adolescents at greatest genetic risk (Brody et al., 2009b), thus supporting the role of gene–environment interplay in both risk and resilience in high-risk populations (Iacono et al., 2008). Despite the strong risk for SUD that CD poses, it should be noted that the majority of those with SUD do not have a history of CD (Sung et al., 2004).

Oppositional defiant disorder (ODD): Most prior research on ODD and SUD has treated ODD as a comorbid condition with other externalizing disorders (e.g. King et al., 2004; August et al., 2006), and it is not clear what role if any ODD plays in the risk for SUD. One longitudinal study found that ODD was associated with neither SUD risk nor with progression from SU to SUD (Sung et al., 2004). A retrospective study found that childhood and adolescent ODD predicted SUD in adulthood, even in the absence of a history of CD and even when the ODD remitted (Nock et al., 2007). Given these and other findings on the psychiatric risks associated with ODD (Copeland et al., 2009), it seems that further prospective study is warranted, particularly to clarify whether ODD is an indicator of risk and whether interventions could possibly reduce the chances of adverse outcomes.

Attention deficit hyperactivity disorder (ADHD): Given the high prevalence of child-hood ADHD and of the use of stimulant medications in the U.S., the possible role of childhood ADHD and of ADHD treatments in the etiology and course of SUD are of particular public health importance and clinical significance. In contrast to many clinical studies, numerous population-based prospective studies have *not* found ADHD to constitute a risk factor for SUD, once comorbid CD and ODD are taken into account (Fergusson et al., 2007; Disney et al., 1999; August et al., 2006; Pardini et al., 2007; Bussing et al., 2010). While this may offer some reassurance regarding children with uncomplicated ADHD, significant subgroups of children with ADHD may be at greater risk, and these may be the populations that more often present clinically. Clearly, those with comorbid CD face an increased risk (e.g. Fergusson et al., 2007), and the ADHD–ODD combination may pose an elevated risk as well (August et al., 2006). Findings are contradictory as to whether ADHD augments the risk from CD (Fergusson et al., 2007; Flory et al., 2003) or whether the presence of multiple externalizing disorders may indicate problems with behavioral disinhibition, often of a familial nature, which may be manifested by these childhood disorders and SUD (Iacono et al., 2003; Tarter et al., 2003). Additionally, certain features of ADHD may relate to specific outcomes: hyperactivity/impulsivity appears to pose particular risk for SUD (Elkins et al., 2007; Tarter et al., 2007), while inattention may pose a greater risk specifically for tobacco use (Burke et al., 2007) or nicotine dependence (Elkins et al., 2007); these findings underscore the potential utility of studying dimensional approaches to risk. Those whose ADHD persists beyond adolescence may also face higher risk (Kessler et al., 2006). Several longitudinal studies of clinical populations have examined SUD outcomes of childhood ADHD and associated features; for a detailed review, see (Looby, 2008).

Contradiction and controversy surround the question of the potential role of prescription stimulant medication in the onset and course of SU and SUD. Numerous methodological challenges make it very difficult to distinguish either a causal or protective role from correlations. These include referral biases; differential attrition (wherein those most likely to use substances are more likely to be lost to follow up); systematic differences between subjects and controls (such as age or socioeconomic status); changes in drug-abuse trends over time; differences between study samples on relevant factors such as comorbidity; and the nonrandomized nature of most treatment studies (self-selection factors) (Armstrong and Costello, 2002).

Concerns about stimulant medications arose from the sensitization hypothesis, wherein exposure to stimulant medication may induce neurochemical changes that lead to drug craving, based on alterations in the dopamine system (Looby, 2008; Volkow and Swanson, 2008). One community-based longitudinal study (Winters et al., 2011), a recent meta-analysis (Humphreys et al., 2013), and a number of clinical follow-up studies (see reviews by Looby, 2008; Volkow and Swanson, 2008) have not found elevated rates of SUD in those who were prescribed stimulant medications in childhood, which should provide reassurance that stimulant medications do not appear to increase the risk for a later SUD. Alternatively, some authors have found reduced rates of SUD among those receiving stimulant medication, and have posited a protective effect of stimulant medications in ADHD based on reductions in impulsivity and in social and school failure, as well as stimulant effects on reward circuits in the brain (reviewed in Faraone and Wilens, 2007). Unfortunately, a number of doubts have been raised about these findings derived from nonrandomized studies that are subject to the methodological challenges mentioned above, particularly systematic differences between subjects and controls (Volkow and Swanson, 2008). Moreover, by adulthood, prior treatment with stimulants was no longer associated with decreased risk for alcohol, drug, or nicotine use disorders (Faraone and Wilens, 2007; Biederman et al., 2008), suggesting that, at best, the treatment delayed an SUD. Age at initiation of stimulant treatment has been hypothesized to affect later risk for SUD (Mannuzza et al., 2008), but findings thus far are contradictory (Biederman et al., 2008; Mannuzza et al., 2008). Based on this review, prevention of SUD cannot currently be supported as a justification for the prescription of stimulant medication (Swanson and Volkow, 2009).

Further data, ideally from randomized study designs, and creative approaches to interpreting available data, such as trajectory analyses, are needed to disentangle this question; for example, subjects in the Multimodal Treatment of ADHD (MTA) study (Molina et al., 2007) will be followed through early adulthood with SUD assessments, and a wealth of covariates, including randomized ADHD treatments at ages 7–9, will be studied, although this study, too, is subject to methodological concerns (Hazell, 2009). Thus far, the MTA cohort, followed up naturalistically for eight years (mean age at follow up of 17), shows neither a protective effect nor an increase in SU or SUDs in adolescence related to medication history (Molina et al., 2013).

Bipolar disorder: Both epidemiologic (Glantz et al., 2009) and clinical (Strakowski et al., 2000) studies have found high rates of comorbidity between SUD and BPD in adults, but population-based data on BPD as a precursor to SUD are few, probably due to low rates of BPD in the general population (Lewinsohn et al., 2004). Evidence from clinical samples supports SUD as either an antecedent (Strakowski et al., 2000) or consequence (Wilens et al., 2004) of BPD, and the association may in some cases be familial (for a review of possible mechanisms, see Strakowski et al., 2000). A bidirectional pattern was found in a population-based study using NESARC data, wherein nicotine dependence and BPD each predicted the onset of the other, and the trajectories differed significantly based on which disorder developed first (Martínez-Ortega et al., 2013). These data suggest that individuals with adolescent BPD merit monitoring and education regarding risk for SUD.

Depression: Findings on depression as a risk factor for SU and SUD have been quite contradictory. For example, one population-based study (Costello et al., 1999) reported that depression was associated with higher rates and earlier onset of SU and SUD, while another (Rohde et al., 2001) found that adolescent alcohol use disorder predicted depression in early adulthood rather than the reverse. Analyses of the National Study of Adolescent Health data showed a bidirectional relationship (Marmorstein, 2009), where depressive symptoms and alcohol problems predicted each other. Similar findings were reported using data from NESARC, where alcohol use disorders and cannabis use disorders, individually or comorbidly, showed a bidirectional relationship with major depressive disorder (Pacek et al., 2013). Numerous factors may account for these inconsistencies: using categorical vs dimensional approaches to define the outcomes, age at assessment, gender differences, and comorbidity (for a detailed discussion see Marmorstein et al., 2010a); moreover, much of the literature studies only alcohol or nicotine, and it is not clear whether these relationships differ by substance. In some cases, this bidirectional relationship may result from genetic factors that predispose individuals to both depression and SU (Nurnberger et al., 2002; Kuo et al., 2006), but this merits further exploration. One factor that has not been addressed in these studies is caffeine use in early adolescence, and its possible associations with sleep and mood disturbances (Whalen et al., 2008). The clinical implications of all these relationships are significant and more research in this area is clearly needed.

Anxiety disorders: As with depression, findings regarding anxiety disorders and SUD have been inconsistent. Several studies have found that anxiety disorders are not associated with later SUD (e.g. Pardini et al., 2007; Goodwin et al., 2004) after controlling for potential confounders, while another recent study found that generalized anxiety predicted early use and greater progression to problematic marijuana use in boys (Marmorstein et al., 2010b). One review found evidence for bidirectional relationships between anxiety and alcohol disorders, wherein the presence of either increased the risk for the other (Kushner et al., 2000). Analysis of data from the NCS replication study found that social phobia, panic disorder, and agoraphobia tended to occur prior to SUDs, while generalized anxiety disorder tended to occur after onset of at least one SUD (Marmorstein, 2012). Thus, it may be that particular types of anxiety disorders pose risk for, or provide protection from, SU and SUD, and that the risk varies by substance and by developmental stage. Further studies are needed that consider these factors.

Personality disorders: One longitudinal population-based study found that borderline, histrionic, and narcissistic personality disorders in early adolescence predicted later SUD, after controlling for relevant risk factors (Cohen et al., 2007). Studies drawing on follow-up data in the NESARC found consistent associations in adults of several personality disorders with progression to problematic SU (Compton et al., 2013) and specifically of antisocial, borderline, and schizotypal personality disorders with persistent SUD (Hasin et al., 2011).

Learning disabilities: A population-based study found that learning disabilities (LD) were associated with later SUD as well as other psychiatric disorders at age 19 (Beitchman et al., 2001); however, the risk may be nonspecific (Beitchman et al., 2001) and potentially confounded by biological and psychosocial risk factors (Weinberg, 2001). Thus, while children with LD may be at increased risk for SUD, the LD may be an indicator of general risk rather than a specific causal factor.

Multiple comorbid disorders: Few studies have examined the impact of comorbid internalizing and externalizing disorders on SUD risk, but there are suggestions that this constellation represents a particularly high risk; for a review of this potentially important area, see (Pardini et al., 2007; Tully and Iacono, in press).

Self-medication: The use of alcohol and drugs, particularly by those with depression or anxiety, is often characterized as "self-medication", but little scientific evidence exists to support this theory. While some individuals with anxiety disorder may find temporary relief in SU (Bolton et al., 2006), it is important to approach such explanations with caution: most individuals with internalizing disorders do not develop SUD; many substances, particularly alcohol, are CNS depressants that can exacerbate depression or cause anxiety symptoms on withdrawal; and retrospective reports of "self-medication" may actually represent efforts to rationalize addiction (Vaillant, 1983). Work on integrating the neurobiology of addiction with affective and motivational states is potentially illuminating and needed.

Alternative approaches to understanding risk: New and exciting approaches have been developed to help delineate the developmental psychopathologic roots of SU and SUD. In contrast to categorical approaches, where a subject is classified as having or not having a disorder, dimensional approaches may analyze individual symptoms or other quantifiable variables to capture the dimensions of risk (e.g. Iacono et al., 2008; Elkins et al., 2007). Trajectory analyses test causal hypotheses by classifying individuals by their risk for a particular outcome and analyzing whether an environmental circumstance alters the trajectory (Rutter, 2009), and this type of analysis has been applied to drug-abuse research. Propensity score matching aims to correct for selection biases through scores that reflect the probability of being exposed to a hypothesized causal agent (Rutter, 2009). Endophenotypes are hereditary characteristics associated with risk for a condition that are not symptoms of the condition themselves; one example is reductions in the amplitude of P300 brain waves associated with behavioral disinhibition (Iacono et al., 2008). Integrating imaging and other neurobiological approaches with epidemiologic research may also yield important insights (Compton et al., 2005). It is hoped that these approaches will be applied more widely to enhance understanding of the mechanisms of risk and opportunities for intervention.

2.2.3. Early Use

The role of early SU in later onset of SUD is very important but not yet clear. It is well-established that early initiation of SU is significantly associated with future SUD (Chen et al., 2009), even after controlling for comorbid psychiatric disorders (Sung et al., 2004). Less clear is whether early use plays a causal role in later SUD, such that delay in use might protect against SUD, or whether early use is a risk indicator. A number of studies suggest that early use is a nonspecific risk indicator associated with multiple forms of adult psychopathology (McGue and Iacono, 2005) and that common genetic factors account for both early initiation and later SUD (e.g. Sartor et al., 2009). Early cannabis use in particular has been associated with progression to other illicit drug use (Lessem et al., 2006; Lynskey et al., 2003), however, a recent retrospective twin study suggests that cannabis use itself, regardless of age of initiation, is associated with increased risk for SUD (Grant et al., 2010). A natural experiment design compared prevalence of alcohol and SUDs in adults who were exposed to different minimum legal drinking age laws in the 1970s and 1980s (Norberg et al., 2009); those for whom it had been legal to purchase alcohol before age 21 were more likely to report symptoms of an alcohol or drug-use disorder in adulthood, even decades later, and the association seemed less related to age of initiation of drinking than to patterns of drinking in late adolescence. Another recent study using retrospective twin data found evidence for a GxE between first use of alcohol and later alcohol dependence (Agrawal et al., 2009); this finding merits replication with other samples and drugs as a fruitful way to understand this risk and the potential impact of preventive interventions. An interaction between early smoking and genetic vulnerability was also found in a large meta-analysis (Hartz et al., 2012), where early onset smokers with a known risk allele for smoking were significantly more likely to become heavy smokers in adulthood than carriers of the risk allele who were late-onset smokers. It is possible that other neurobiological and epigenetic changes take place with early use, and further work in this important area is needed. Although it is not clear whether delaying early use will prevent later SUD, there are ample other reasons to delay or prevent the use of substances by adolescents, and to treat early use as a risk indicator.

A particular pattern of early use has generated the "gateway hypothesis" of drug abuse in which the first exposures to alcohol, tobacco, and marijuana are causally linked later use of other drugs (e.g. cocaine, amphetamines or opioids, Kandel, 2003). Overall, the trajectory in which cocaine, amphetamine, and other drugs are nearly universally preceded by use of alcohol, tobacco and marijuana is well-established but the reasons for this trajectory are less certain. Common vulnerability to the multiple substances has been suggested (Morral et al., 2002). In contrast, a more recent animal study (in mice) generally supports a causal gateway pathway in which prior exposure to nicotine induces epigenetic changes that enhance responses to cocaine (Levine et al., 2011). This work included a demonstration that these basic research findings are consistent with human population data, and if substantiated, the suggestion is that reducing nicotine exposure by youth may reduce cocaine (and possibly other) drug addictions (Volkow, 2011).

2.2.4. Environmental Factors

A large and long-established literature has documented numerous environmental factors that correlate with SU and SUD (Galea et al., 2004). Applying newer strategies to distinguish causal risk factors and moderators of risk from simple correlations illuminates the mechanisms by which the interplay of environmental factors with individual-level factors can increase or ameliorate risk over the course of development. These methods include studies of gene–environment interplay, natural experiments, and prevention studies that can inform etiology.

In discussing the impact of environmental factors, it is important to distinguish those that may impact SU (initiation or low-level use) from factors that influence the progression to SUD. While environmental factors contribute significantly to onset of SU, genetic factors play an increasingly strong role in vulnerable individuals who initiate SU, as use progresses to abuse and dependence and as individuals progress through adolescence to early adulthood (Dick et al., 2009a; Hopfer et al., 2003). However, those who are genetically vulnerable may also be both the most vulnerable to environmental adversity (Hicks et al., 2009; Kendler et al., 2012) and the most likely to derive benefit from protective features in the environment (Brody et al., 2009b; Dick et al., 2009b).

Family circumstances: While family socioeconomic status alone does not clearly relate to drug initiation (Galea et al., 2004), family adversity is associated with a greater prevalence of drug use and drug-use disorders later in life (Galea et al., 2004). Two recent studies have started to shed light on the potential role of family circumstances in behavioral problems that can lead to an SU. A natural experiment found that adults whose families were moved out of poverty due to tribal casino profits (thus, unrelated to family characteristics) during adolescence experienced reduced levels of CD and ODD in adolescence, and lower rates of alcohol and cannabis abuse and dependence in adulthood, providing support for a social causation role in these high-risk disorders (Costello et al., 2010). A twin study found a GxE between stressful life events (such as parental divorce or family financial or legal problems) and offspring externalizing disorders (Hicks et al., 2009), offering further support for a causal role for family adversity in high-risk behaviors; further follow up to assess for effects on SUD is anticipated.

Early stress and child abuse: Although a discussion of the psychobiological effects of early stress on development and risk for SU/SUD is beyond the scope of this review (Sinha, 2008), a few epidemiological findings will be highlighted. Individuals presenting for treatment often report a history of early trauma; however, clarifying the role played by abuse and neglect in the etiology of SU/SUD is very challenging, given that these early stressors tend to occur in families that confer genetic risk for SU/SUD, along with other adverse environmental conditions, including prenatal substance exposure (De Bellis, 2005). Some groups have applied innovative twin designs to control for the psychosocial effects of child abuse (Sartor et al., 2008). One study using an offspring of twins approach reported that child sexual abuse, but not child physical abuse, is associated with an increased risk for cannabis abuse and dependence in adolescence and young adulthood, beyond that posed by genetic and other environmental risk factors (Duncan et al., 2008); this is consistent with findings for other substances. Further integration of the roles of genetic, environmental, and developmental factors in response to stress is anticipated.

Parenting: Parenting qualities, and in particular parental monitoring, have long been associated with SU (Eaton et al., 2009). A recent meta-analysis describes the methodologic challenges in studying this important area and confirms a strong link between parental monitoring and reduced marijuana use in adolescents (Lac and Crano, 2009). However, it has been difficult to assess whether poor parental monitoring is a causal risk factor or a demonstration of rGE. On one hand, a recent study found that "parental monitoring" was more a function of the child's personality relating to a tendency to disclose information to parents, rather than an active behavior by the parents (Eaton et al., 2009), supporting a reconceptualization of parental monitoring. On the other hand, another recent study has shown that high levels of involved-supportive parenting were associated with reduced SU in youth who carried a genetic allele associated with increased risk (Brody et al., 2009a); a test of the data ruled out rGE as an explanation. A similar moderating effect of parental monitoring on behavioral outcomes in those with a different high-risk genotype was found in another community sample (Dick et al., 2009b). Moreover, an intervention to improve parenting reduced risk behaviors in adolescents at increased genetic risk (Brody et al., 2009b), providing further support for a protective role for positive parenting, particularly for those at highest risk. The role of parental monitoring as either a marker of risk or an opportunity for intervention is an important and potentially fruitful subject for future research.

Sibling influences: Although previously little studied, recent findings using genetically informative approaches have reported significant environmental influence by siblings on the development of externalizing disorders including drug use (Hicks et al., 2013), and that siblings closer in age are more likely to resemble one another in drug-use risk compared with those further apart in age, with older siblings more likely to transmit risk to those younger than the reverse (Kendler et al., 2012)

Religiosity: Various measures of religiosity among teens are strongly associated with reduced nicotine and alcohol use (Kendler and Myers, 2009; Timberlake et al., 2006), but the mechanisms remain unclear. Genetically informed approaches suggest that there are both genetic and environmental components to religious practice (Kendler and Myers, 2009), and that religiosity may moderate these factors (Timberlake et al., 2006), but more study is warranted.

Prenatal exposure: A full discussion of the risks associated with prenatal exposure to alcohol, nicotine, and illicit drugs is beyond the scope of this review. Clearly these environmental factors are experienced in high-risk families, and sorting through the relative roles of genes, exposure, and their interplay is complex and challenging (Glantz and Chambers, 2006). A number of studies are underway to follow prenatally exposed samples into adolescence and to apply sophisticated methodology to understand the contributions of these exposures to adverse outcomes (see National Institute on Drug Abuse, 2010).

Peers: Peer deviance has long been regarded as an environmental risk factor for SU (Gillespie et al., 2009), but recent work has started to question this role in the causal chain.

Based on retrospective data and twin modeling, peer group deviance appears to result from genetic as well as environmental factors, and in fact may be more a result of cannabis use than a cause (Gillespie et al., 2009). Furthermore, similar analyses have found that genetics play an increasing role in the choice of peer groups as adolescent's age and move away from parental influence (Kendler et al., 2007). Thus, what appears to be peer influence may be in part the result of active rGE, wherein an individual seeks out deviant peers based on genetic predisposition as well as environmental influences. More work in this key area is anticipated.

Schools: While schools are often the site of preventive interventions for SU/SUD, there is little research on the role of school factors in the etiologic chain. One cross-sectional study found that aggregated school attitudes disapproving of drug use (school norms) were related to reduced probably of SU, more strongly for 8th and 10th graders than 12th grade students (Kumar et al., 2002), even among those who personally did not disapprove of SU. In one intriguing study, student norms regarding smoking were found to moderate the heritability of daily smoking in high school, although genetic factors remained strong contributors (Boardman et al., 2008). Further investigations of school interventions are underway and needed to clarify the potential role of school norms.

Neighborhood factors and drug availability: While neighborhood disadvantage has been reported to be related to SU (Fite et al., 2009), it is not clear whether neighborhood factors play a direct causal role or serve as risk indicators. A recent study emphasizing the interrelationships among influences on adolescent behavior found that neighborhood influences are mediated through parenting and peer factors (Chuang et al., 2005). More-over, drug availability, long considered the quintessential environmental factor, may (like peer influences) be due in part to active rGE, where high-risk youth seek out environ-ments where they are more likely to be exposed to drugs (Gillespie et al., 2007).

2.2.5. Development

SU and SUDs are clearly developmental phenomena, with onset typically in adolescence, and changing expression associated with developmental trajectories, as individual-level factors interact with environmental. This review has incorporated a developmental per-spective, but does not include a full discussion of the roles of pubertal changes, adoles-cent brain development, and shifts in roles and relationships as they impact SU and SUD (Windle et al., 2008; Brown et al., 2008).

3. CONCLUSIONS

Epidemiologic research has contributed vital findings on patterns, comorbidity, and etiology of SU/SUD, which can help guide future neuroscientific investigations. Ongoing surveillance and analyses of existing data will continue to provide such needed information. Increasingly, sophisticated methods for parsing out the causal roles of iden-tified risk factors, and incorporation of genetic, neuroimaging, and psychophysiological approaches into prospective and developmentally sensitive epidemiological studies, will further enhance the value and relevance of these studies. This review highlights four key sets of findings with implications for neuroscience:

1. Comorbidity: Psychiatric disorders commonly co-occur with drug-use disorders; this is the rule, rather than the exception. In some cases, both types of disorders may be manifestations of a common underlying genetic diathesis, particularly for the externalizing spectrum as discussed above. In others, SU may precipitate the onset of psychiatric disorder to which the individual was already vulnerable. For many disorders, there may be a bidirectional relationship, where the presence of either a psychiatric disorder or a drug-use disorder increases the risk for the other. These relationships are very important to investigate to understand risk and treatment, and comorbidity should be incorporated into studies of both animal models and human subjects.

2. Polysubstance use: While much of the substance abuse research literature focuses on single drugs of abuse, the reality as evidenced from epidemiologic studies is that individuals tend to use multiple drugs, either in sequence as shown in the gateway and progression studies, or within the same time period (i.e. within the same day, week or month). Both animal models and human studies need to incorporate polysubstance use into their approaches to forward true understanding of the phenomenon of SUD, and methodologies need to be developed to overcome the significant conceptual and feasibility challenges to doing so.

3. The role of environment: As discussed above, environmental factors play key roles in the exposure to and initiation and progress of SU as well as vulnerability to SUDs, through such mechanisms as stress responses, GxE, and epigenetics. The findings that those at genetic risk may be most vulnerable to environmental adversity highlight the significance of this interplay. Including environmental factors in etiologic and animal models will help refine understanding of vulnerability to SUDs.

4. The role of development: The findings cited above on the impact of timing of drug initiation (early use), gateway drugs and progression, and epigenetics, all highlight the importance of taking developmental stage and developmental sensitivities into account in human laboratory studies and animal models of drug abuse.

ACKNOWLEDGMENTS

The views expressed in this paper are those of the authors and do not necessarily represent the views of the National Institute on Drug Abuse, the National Institutes of Health, the United States Department of Health and Human Services of the United States Federal Government.

REFERENCES

Agrawal, A., Sartor, C., Lynskey, M.T., et al., 2009. Evidence for an interaction between age at first drink and genetic influences on DSM-IV alcohol dependence symptoms. Alcohol Clin. Exp. Res. 33, 2047–2056.
American Psychiatric Association (APA), 1987. Diagnostic and Statistical Manual of Mental Disorders, third ed., Text Revision, American Psychological Association, Washington, DC.

American Psychiatric Association (APA), 1994. Diagnostic and Statistical Manual of Mental Disorders, fourth ed. American Psychological Association, Washington, DC.

Anderson, J.E., Mueller, T.E., 2008. Trends in sexual risk behavior and unprotected sex among high school students, 1991–2005: the role of substance use. J. Sch. Health 78, 575–580.

Armstrong, T.D., Costello, E.J., 2002. Community studies on adolescent substance use, abuse, or dependence and psychiatric comorbidity. J. Consult. Clin. Psychol. 70, 1224–1239.

August, G., Winters, K.C., Realmuto, G.M., Fahnhorst, T., Botzet, A., Lee, S., 2006. Prospective study of adolescent drug use among community samples of ADHD and non-ADHD participants. J. Am. Acad. Child Adolesc. Psychiatry 45, 824–832.

Beitchman, J.H., Wilson, B., Douglas, L., Young, A., Adlaf, E., 2001. Substance use disorders in young adults with and without LD: predictive and concurrent relationships. J. Learn. Disabil. 34, 317–332.

Biederman, J., Monteaux, M.C., Spencer, T., Wilens, T.E., Macpherson, H.A., Faraone, S.V., 2008. Stimulant therapy and risk for subsequent substance use disorders in male adults with ADHD: a naturalistic controlled 10-year follow-up study. Am. J. Psychiatry 165, 597–603.

Boardman, J.D., Saint Onge, J.M., Haberstick, B.C., Timberlake, D.S., Hewitt, J.K., 2008. Do schools moderate the genetic determinants of smoking? Behav. Genet. 38, 234–246.

Bolton, J., Cox, B., Clara, I., Sareen, J., 2006. Use of alcohol and drugs to self-medicate anxiety disorders in a nationally representative sample. J. Nerv. Ment. Dis. 194, 818–825.

Brody, G.H., Beach, S.R., Philibert, R.A., et al., 2009a. Parenting moderates a genetic vulnerability factor in longitudinal increases in youths' substance use. J. Consult. Clin. Psychol. 77, 1–11.

Brody, G.H., Chen, Y.F., Beach, S.R., Philibert, R.A., Kogan, S., 2009b. Participation in a family-centered prevention program decreases genetic risk for adolescents' risky behaviors. Pediatrics 124, 911–917.

Brown, S.A., McGue, M., Maggs, J., et al., 2008. Developmental perspective on alcohol and youths 16 to 20 years of age. Pediatrics 121 (Suppl. 4), S290–S310.

Burke, J.D., Loeber, R., White, H.R., Stouthamer-Loeber, M., Pardini, D.A., 2007. Inattention as a key predictor of tobacco use in adolescence. J. Abnorm. Psychol. 116, 249–259.

Bussing, R., Mason, D.M., Bell, L., Porter, P., Garvan, C., 2010. Adolescent outcomes of childhood attention-deficit/hyperactivity disorder in a diverse community sample. J. Am. Acad. Child Adolesc. Psychiatry 49, 595–605.

Centers for Disease Control and Prevention (CDC), 2010. Youth risk behavior surveillance-United States, 2009. MMWR Surveill. Summ. 2010 (SS-5), 59.

Centers for Disease Control and Prevention (CDC), 2011. Prescription Pain Killer Overdoses in the US. CDC Vital Signs. http://www.cdc.gov/VitalSigns/PainkillerOverdoses/. (accessed 18.04.12.).

Chen, C.Y., Storr, C.L., Anthony, J.C., 2009. Early-onset drug use and risk for drug dependence problems. Addict. Behav. 34, 319–322.

Chuang, Y.G., Ennett, S.T., Bauman, K.E., Foshee, V.A., 2005. Neighborhood influences on adolescent cigarette and alcohol use: mediating effects through parent and peer behaviors. J. Health Soc. Behav. 46, 187–204.

Cohen, P., Chen, H., Crawford, T.N., Brook, J.S., Gordon, K., 2007. Personality disorders in early adolescence and the development of later substance use disorders in the general population. Drug Alcohol Depend. 88 (Suppl. 1), S71–S84.

Compton, W.M., Thomas, Y.F., Conway, K.P., Colliver, J.D., 2005. Developments in the epidemiology of drug use and drug use disorders. Am. J. Psychiatry 162, 1494–1502.

Compton, W.M., Thomas, Y.F., Stinson, F.S., Grant, B.F., 2007. Prevalence, correlates, disability, and comorbidity of DSM-IV drug abuse and dependence in the United States: results from the National Epidemiologic Survey on Alcohol and Related Conditions. Arch. Gen. Psychiatry 64, 566–576.

Compton, W.M., Dawson, D.A., Conway, K.P., Brodsky, M., Grant, B.F., 2013. Transitions in illicit drug use status over 3 years: a prospective analysis of a general population sample. Am. J. Psychiatry 170, 660–670.

Connell, C., Gilreath, T., Aklin, W., Brex, R., 2010. Social-ecological influences on patterns of substance use among non-metropolitan high school students. Am. J. Community Psychol. 45, 36–48.

Connell, C.M., Gilreath, T.D., Hansen, N.B., 2009. A multiprocess latent class analysis of the co-occurrence of substance use and sexual risk behavior among adolescents. J. Stud. Alcohol Drugs 70, 943–951.

Copeland, W.E., Shanahan, L., Costello, E.J., Angold, A., 2009. Childhood and adolescent psychiatric disorders as predictors of young adult disorders. Arch. Gen. Psychiatry 66, 764–772.

Costello, E.J., Erkanli, A., Copeland, W.E., Angold, A., 2010. Association of family income supplements in adolescence with development of psychiatric and substance use disorders in adulthood among an American Indian population. JAMA 303, 1954–1960.

Costello, E.J., Erkanli, A., Federman, E., Angold, A., 1999. Development of psychiatric comorbidity with substance abuse in adolescents: effects of timing and sex. J. Clin. Child Psychol. 28, 298–311.

De Bellis, M.D., 2005. The psychobiology of neglect. Child Maltreat. 10, 150–172.

Dick, D., Prescott, C.A., McGue, M., 2009a. The genetics of substance use and substance use disorders. In: Kim, Y.-K. (Ed.), Handbook of Behavioral Genetics, vol. 1. Springer, New York, NY.

Dick, D.M., Agrawal, A., 2008. The genetics of alcohol and other drug dependence. Alcohol Res. Health 31, 111–118.

Dick, D.M., Latendresse, S., Lansford, J.E., et al., 2009b. Role of GABRA2 in trajectories of externalizing behavior across development and evidence of moderation by parental monitoring. Arch. Gen. Psychiatry 66, 649–657.

Dierker, L.C., Vesel, F., Sledjeski, E.M., Costello, D., Perrine, N., 2007. Testing the dual pathway hypothesis to substance use in adolescence and young adulthood. Drug Alcohol Depend. 87, 83–93.

Disney, E.R., Elkins, I.J., McGue, M., Iacono, W.G., 1999. Effects of ADHD, conduct disorder, and gender on substance use and abuse in adolescence. Am. J. Psychiatry 156, 1515–1521.

Dougherty, D.M., Mathias, C.W., Marsh, D.M., Moeller, F.G., Swann, A.C., 2004. Suicidal behaviors and drug abuse: impulsivity and its assessment. Drug Alcohol Depend. 76, S93–S105.

Duhig, A.M., Maciejewski, P.K., Desai, R.A., Krishnan-Sarin, S., Potenza, M.N., 2007. Characteristics of adolescent past-year gamblers and non-gamblers in relation to alcohol drinking. Addict. Behav. 32, 80–89.

Duncan, A.E., Sartor, C.E., Scherrer, J.F., et al., 2008. The association between cannabis abuse and dependence and childhood physical and sexual abuse: evidence from an offspring of twins design. Addiction 103, 990–997.

Eaton, N.R., Krueger, R.F., Johnson, W., McGue, M., Iacono, W.G., 2009. Parental monitoring, personality, and delinquency: further support for a reconceptualization of monitoring. J. Res. Pers. 43, 49–59.

Elkins, I.J., McGue, M., Iacono, W.G., 2007. Prospective effects of attention-deficit/hyperactivity disorder, conduct disorder, and sex on adolescent substance use and abuse. Arch. Gen. Psychiatry 64, 1145–1152.

Esposito-Smythers, C., Spirito, A., 2004. Adolescent substance use and suicidal behavior: a review with implications for treatment research. Alcohol Clin. Exp. Res. 28, 77S–88S.

Faraone, S.V., Wilens, T.E., 2007. Effect of stimulant medications for attention-deficit/hyperactivity disorder on later substance use and the potential for stimulant misuse, abuse, and diversion. J. Clin. Psychiatry 68 (Suppl. 11), 15–22.

Felson, R., Savolainen, J., Aaltonen, M., Moustgaard, H., 2008. Is the association between alcohol use and delinquency causal or spurious? Criminology 46, 785–808.

Fergusson, D.M., Horwood, L.J., Ridder, E.M., 2007. Conduct and attentional problems in childhood and adolescence and later substance use, abuse, and dependence: results of a 25-year longitudinal study. Drug Alcohol Depend. 88, S14–S26.

Fite, P.J., Wynn, P., Lochman, J.E., Wells, K.C., 2009. The influence of neighborhood disadvantage and perceived disapproval on early substance use initiation. Addict. Behav. 34, 769–771.

Flory, K., Milich, R., Lynam, D.R., Leukefeld, C., Clayton, R., 2003. Relation between childhood disruptive behavior disorders and substance use and dependence symptoms in young adulthood: individuals with symptoms of attention-deficit/hyperactivity disorder and conduct disorder are uniquely at risk. Psychol. Addict. Behav. 17, 151–158.

Galea, S., Nandi, A., Vlahov, D., 2004. The social epidemiology of substance use. Epidemiol. Rev. 26, 36–52.

Gillespie, N., Kendler, K.S., Prescott, C., et al., 2007. Longitudinal modeling of genetic and environmental influences on self-reported availability of psychoactive substances: alcohol, cigarettes, marijuana, cocaine and stimulants. Psychol. Med. 37, 947–959.

Gillespie, N., Neale, M.C., Jacobson, K.C., Kendler, K.S., 2009. Modeling the genetic and environmental association between peer group deviance and cannabis use in male twins. Addiction 104, 420–429.

Glantz, M.D., Chambers, J.C., 2006. Prenatal drug exposure on subsequent vulnerability to drug abuse. Dev. Psychopathol. 18, 893–922.

Glantz, M.D., Anthony, J.C., Berglund, P.A., et al., 2009. Mental disorders as risk factors for later substance dependence: estimates of optimal prevention and treatment benefits. Psychol. Med. 39, 1365–1377.

Goodwin, R.D., Fergusson, D.M., Horwood, L.J., 2004. Association between anxiety disorders and substance use disorders among young persons: results of a 21-year longitudinal study. J. Psychiatr. Res. 38, 295–304.

Grant, J.D., Lynskey, M.T., Scherrer, J.F., Agrawal, A., Heath, A.C., Bucholz, K.K., 2010. A cotwin-control analysis of drug use and abuse/dependence risk associated with early-onset cannabis use. Addict. Behav. 35, 35–41.

Griffith-Lendering, M.F.H., Wigman, J.T.W., Prince van Leeuwen, A., Huijbregts, S.C.J., Huizink, A.C., Ormel, J., Verhulst, F.C., et al., 2013. Cannabis use and vulnerability for psychosis in early adolescence–a TRAILS study. Addiction 108 (4), 733–740.

Hartz, S.M., Short, S.E., Saccone, N.L., Culverhouse, R., Chen, L., Schwantes-An, T.-H., Coon, H., et al., 2012. Increased genetic vulnerability to smoking at CHRNA5 in early-onset smokers. Arch. Gen. Psychiatry 69 (8), 854–860.

Hasin, D., Fenton, M.C., Skodol, A., Krueger, R., Keyes, K., Geier, T., Greenstein, E., Blanco, C., Grant, B., 2011. Personality disorders and the 3-year course of alcohol, drug, and nicotine use disorders. Arch. Gen. Psychiatry 68 (11), 1158–1167.

Hazell, P.L., 2009. 8-year follow-up of the MTA sample. J. Am. Acad. Child Adolesc. Psychiatry 48, 461–462.

Henquet, C., Di Forti, M., Morrison, P., Kuepper, R., Murray, R.M., 2008. Gene-environment interplay between cannabis and psychosis. Schizophr. Bull. 34, 1111–1121.

Hicks, B.M., South, S.C., Dirago, A.C., Iacono, W.G., McGue, M., 2009. Environmental adversity and increasing genetic risk for externalizing disorders. Arch. Gen. Psychiatry 66, 640–648.

Hicks, B.M., Foster, K.T., Iacono, W.G., McGue, M., 2013. Genetic and environmental influences on the familial transmission of externalizing disorders in adoptive and twin offspring. JAMA Psychiatry 70 (10), 1076–1083. http://dx.doi.org/10.1111/jamapsychiatry.2013.258.

Hopfer, C.J., Crowley, T.J., Hewitt, J.K., 2003. Review of twin and adoption studies of adolescent substance use. J. Am. Acad. Child Adolesc. Psychiatry 42, 710–719.

Humphreys, K.L., Eng, T., Lee, S.S., 2013. Stimulant medication and substance use outcomes: a meta-analysis. JAMA Psychiatry.

Iacono, W.G., Malone, S.M., McGue, M., 2008. Behavioral disinhibition and the development of early-onset addiction: common and specific influences. Annu. Rev. Clin. Psychol. 4, 325–348.

Iacono, W.G., Malone, S.M., McGue, M., 2003. Substance use disorders, externalizing psychopathology, and P300 event-related potential amplitude. Int. J. Psychophysiol. 48, 147–178.

Johnston, L.D., O'Malley, P.M., Bachman, J.G., Schulenberg, J.E., 2013. Monitoring the Future National Survey Results on Drug Use, 1975–2012. Secondary School Studentsvol. I. Institute for Social Research, The University of Michigan, Ann Arbor.

Kandel, D.B., 2003. Does marijuana use cause the use of other drugs? JAMA 289 (4), 482–483.

Kendler, K.S., Myers, J., 2009. A developmental twin study of church attendance and alcohol and nicotine consumption: a model for analyzing the changing impact of genes and environment. Am. J. Psychiatry 166, 1150–1155.

Kendler, K.S., Jacobson, K.C., Gardner, C.O., Gillespie, N., Aggen, S.A., Prescott, C.A., 2007. Creating a social world: a developmental twin study of peer-group deviance. Arch. Gen. Psychiatry 64, 958–965.

Kendler, K.S., Sundquist, K., Ohlsson, H., Palmér, K., Maes, H., Winkleby, M.A., Sundquist, J., 2012. Genetic and familial environmental influences on the risk for drug abuse: a national Swedish adoption study. Arch. Gen. Psychiatry online March 5, 2012.

Kessler, R.C., Adler, L., Barkley, R., et al., 2006. The prevalence and correlates of adult ADHD in the United States: results from the National Comorbidity Survey Replication. Am. J. Psychiatry 163, 716–723.

King, S.M., Iacono, W.G., McGue, M., 2004. Childhood externalizing and internalizing psychopathology in the prediction of early substance use. Addiction 99, 1548–1559.

Kumar, R., O'Malley, P.M., Johnston, L.D., Schulenberg, J.E., Bachman, J.G., 2002. Effects of school-level norms on student substance use. Prev. Sci. 3, 105–124.

Kuo, P.H., Gardner, C.O., Kendler, K.S., Prescott, C.A., 2006. The temporal relationship of the onsets of alcohol dependence and major depression: using a genetically informative study design. Psychol. Med. 36, 1153–1162.

Kushner, P.H., Abrams, K., Borchardt, C., 2000. The relationship between anxiety disorders and alcohol use disorders: a review of major perspectives and findings. Clin. Psychol. Rev. 20, 149–171.

Lac, A., Crano, W.D., 2009. Monitoring matters: meta-analytic review reveals the reliable linkage of parental monitoring with adolescent marijuana use. Perspect. Psychol. Sci. 4, 578–586.

Lessem, J.M., Hopfer, C.J., Haberstick, B.C., et al., 2006. Relationship between adolescent marijuana use and young adult illicit drug use. Behav. Genet. 36, 498–506.

Levine, A., Huang, Y.Y., Drisaldi, B., Griffin Jr., E.A., Pollak, D.D., Xu S, Yin D., Schaffran, C., Kandel, D.B., Kandel, E.R., 2011. Molecular mechanism for a gateway drug: epigenetic changes Initiated by nicotine prime gene expression by cocaine. Sci. Transl. Med. 107 (3), 107–109.

Lewinsohn, P.M., Klein, D.N., Seeley, J.R., 2004. Bipolar disorder during adolescence and young adulthood in a community sample. Bipolar Disord. 2, 281–293.

Looby, A., 2008. Childhood attention deficit hyperactivity disorder and the development of substance use disorders: valid concern or exaggeration? Addict. Behav. 33, 451–463.

Lynskey, M.T., Andrew, C., Heath, A.C., Bucholz, K.K., Slutske, W.S., Madden, P.A.F., Nelson, E.C., Statham, D.J., Martin, N.G., 2003. Escalation of drug use in early-onset cannabis users vs co-twin controls. JAMA 289 (4), 427–433.

Mannuzza, S., Klein, R.G., Truong, N.L., et al., 2008. Age of methylphenidate treatment initiation in children with ADHD and later substance abuse: prospective follow-up into adulthood. Am. J. Psychiatry 165, 604–609.

Marmorstein, N.R., 2012. Anxiety disorders and substance use disorders: different associations by anxiety disorder. J. Anxiety Disord. 26 (1), 88–94.

Marmorstein, N.R., 2009. Longitudinal associations between alcohol problems and depressive symptoms: early adolescence through early adulthood. Alcohol Clin. Exp. Res. 33, 49–59.

Marmorstein, N.R., Iacono, W.G., Malone, S.M., 2010a. Longitudinal associations between depression and substance dependence from adolescence through early adulthood. Drug Alcohol Depend. 107, 154–160.

Marmorstein, N.R., White, H.R., Loeber, R., Stouthamer-Loeber, M., 2010b. Anxiety as a predictor of age at first use of substances and progression to substance use problems among boys. J. Abnorm. Child Psychol. 38, 211–224.

Martínez-Ortega, J.M., Goldstein, B.I., Gutiérrez-Rojas, L., Sala, R., Wang, S., Blanco, C., 2013. Temporal sequencing of nicotine dependence and bipolar disorder in the National Epidemiologic Survey on Alcohol and Related Conditions (NESARC). J. Psychiatr. Res. 47 (7), 858–864.

McCabe, S.E., West, B.T., Teter, C.J., Boyd, C.J., 2012. Co-ingestion of prescription opioids and other drugs among high school seniors: results from a national study. Drug Alcohol Depend. 126 (1–2), 65–70.

McGue, M., Iacono, W.G., 2005. The association of early adolescent problem behavior with adult psychopathology. Am. J. Psychiatry 162, 1118–1124.

Mitchell, C.M., Plunkett, M., 2000. The latent Structure of substance Use among American Indian adolescents: an example using categorical variables. Am. J. Community Psychol. 28, 105–125.

Molina, B.S., Flory, K., Hinshaw, S.P., et al., 2007. Delinquent behavior and emerging substance use in the MTA at 36 months: prevalence, course, and treatment effects. J. Am. Acad. Child Adolesc. Psychiatry 46, 1028–1040.

Molina, B.S.G., Hinshaw, S.P., Eugene Arnold, L., Swanson, J.M., Pelham, W.E., Hechtman, L., Hoza, B., et al., 2013. Adolescent substance use in the multimodal treatment study of attention-deficit/hyperactivity disorder (ADHD) (MTA) as a function of childhood ADHD, random assignment to childhood treatments, and subsequent medication. J. Am. Acad. Child Adolesc. Psychiatry 52 (3), 250–263.

Morral, A.R., McCaffrey, D.F., Paddock, S.M., 2002. Reassessing the marijuana gateway effect. Addiction 97, 1493–1503.

National Highway Traffic Safety Administration (NHTSA), Traffic safety facts 2005/2006. National Center for Statistics and Analysis. http://www-nrd.nhtsa.dot.gov/pubs/810803.pdf (accessed 05.01.10.).

National Institute on Drug Abuse (NIDA). Prenatal exposure to drugs of abuse. http://www.drugabuse.gov/tib/prenatal.html (accessed 13.02.10.).

Nock, M.K., Kazdin, A.E., Hiripi, E., Kessler, R.C., 2007. Lifetime prevalence, correlates, and persistence of oppositional defiant disorder: results from the National Comorbidity Survey Replication. J. Child Psychol. Psychiatry 48, 703–713.

Norberg, K.E., Bierut, L.J., Grucza, R.A., 2009. Long-term effects of minimum drinking age laws on past-year alcohol and drug use disorders. Alcohol Clin. Exp. Res. 33, 2180–2190.

Nurnberger, J.I., Foroud, T., Flury, L., Meyer, E.T., Wiegand, R., 2002. Is there a genetic relationship between alcoholism and depression? Alcohol Res. Health 26, 233–240.

Ogders, C.L., Caspi, A., Nagin, D.S., et al., 2008. Is it important to prevent early exposure to drugs and alcohol among adolescents? Psychol. Sci. 19, 1037–1044.

Pacek, L.R., Martins, S.S., Crum, R.M., 2013. The bidirectional relationships between alcohol, cannabis, co-occurring alcohol and cannabis use disorders with major depressive disorder: results from a national sample. J. Affect. Disord. 148 (2–3), 188–195.

Pardini, D.A., White, H.R., Stouthamer-Loeber, M., 2007. Early adolescent psychopathology as a predictor of alcohol disorders by young adulthood. Drug Alcohol Depend. 88 (Suppl. 1), S38–S49.

Paulozzi, L.J., Jones, C.M., Mack, C.A., Rudd, R.A., 2011. Vital signs: overdoses of prescription opioid pain relievers – United States, 1999–2008. MMWR Morb. Mortal. Wkly. Rep. 60 (43), 1487–1492.

Rohde, P., Clarke, G.N., Lewinsohn, P.M., Seeley, J.R., Kaufman,., N.K., 2001. Impact of comorbidity on a cognitive-behavioral group treatment for adolescent depression. J. Am. Acad. Child Adolesc. Psychiatry 40, 795–802.

Rutter, M., 2009. Epidemiological methods to tackle causal questions. Int. J. Epidemiol. 38, 3–6.

Rutter, M., Moffitt, T.E., Caspi, A., 2006. Gene-environment interplay and psychopathology: multiple varieties but real effects. J. Child Psychol. Psychiatry 47, 226–261.

Sartor, C.E., Agrawal, A., McCutcheon, V.V., Duncan, A.E., Lynskey, M.T., 2008. Disentangling the complex association between childhood sexual abuse and alcohol-related problems: a review of methodological issues and approaches. J. Stud. Alcohol Drugs 69, 718–727.

Sartor, C.E., Lynskey, M.T., Bucholz, K.K., Madden, P., Martin, N., Heath, A.C., 2009. Timing of first alcohol use and alcohol dependence: evidence of common genetic influences. Addiction 104, 1512–1518.

Satterlee, J.S., 2013. Epigenomic and non-coding RNA regulation in addictive processes. In: Jirtle, R.L., Tyson, F.L. (Eds.), Environmental Epigenomics in Health and Disease: Epigenetics and Complex Diseases Origins. Springer, Heidelberg.

Schepis, T.S., Krishnan-Sarin, S., 2008. Characterizing adolescent prescription misusers: a population-based study. J. Am. Acad. Child Adolesc. Psychiatry, 745–754.

Sinha, R., 2008. Chronic stress, drug use, and vulnerability to addiction. Ann. N.Y. Acad. Sci. 1141, 105–130.

Storr, C.L., Westergaard, R., Anthony, J.C., 2005. Early onset inhalant use and risk for opiate initiation by young adulthood. Drug Alcohol Depend. 78, 253–261.

Strakowski, S.M., DelBello, M.P., Fleck, D.E., Arndt, S., 2000. The impact of substance abuse on the course of bipolar disorder. Biol. Psychiatry 48, 477–485.

Substance Abuse and Mental Health Services Administration (SAMHSA), 2011. Results from the 2010 National Survey on Drug Use and Health: National Findings. Rockville, MD. http://www.samhsa.gov/data/NSDUH/2k10Results/Web/HTML/2k10Results.htm. (accessed 20.03.12.).

Sung, M., Erkanli, A., Angold, A., Costello, E.J., 2004. Effects of age at first substance use and psychiatric comorbidity on the development of substance use disorders. Drug Alcohol Depend. 75, 287–299.

Swanson, J.M., Volkow, N.D., 2009. Psychopharmacology: concepts and opinions about the use of stimulant medications. J. Child Psychol. Psychiatry 50, 180–193.

Swendson, J., Anthony, J.C., Conway, K.P., et al., 2008. Improving targets for the prevention of drug use disorders: sociodemographic predictors of transitions across drug use stages in the National Comorbidity Survey Replication. Prev. Med. 47, 629–634.

Tarter, R.E., Kirisci, L., Mezzich, A., et al., 2003. Neurobehavioral disinhibition in childhood predicts early age at onset of substance use disorder. Am. J. Psychiatry 160, 1078–1085.

Tarter, R.E., Kirisci, L., Feske, U., Vanyukov, M., 2007. Modeling the pathways linking childhood hyperactivity and substance use disorder in young adulthood. Psychol. Addict. Behav. 21, 266–271.

Timberlake, D.S., Rhee, S.H., Haberstick, B.V., et al., 2006. The moderating effects of religiosity on the genetic and environmental determinants of smoking initiation. Nicotine Tob. Res. 8, 123–133.

Tully, E.C., Iacono, W.G., in press. An integrative common liabilities model for the comorbidity of substance use disorders with externalizing and internalizing disorders. In: Sher, K. (Ed.), Oxford Handbook of Substance Use Disorders. Oxford University Press, New York, NY.

Vaillant, G.E., 1983. The Natural History of Alcoholism. Harvard University Press, Cambridge, MA.

Van Etten, M.L., Neumark, Y.D., Anthony, J.C., 1999. Male-female differences in the earliest stages of drug involvement. Addiction 94, 1413–1419.

Volkow, N.D., Swanson, J.M., 2008. Does childhood treatment of ADHD with stimulant medication affect substance abuse in childhood? Am. J. Psychiatry 165, 553–555.

Volkow, N.D., 2011. Epigenetics of nicotine: another nail in the coughing. Sci. Transl. Med. 3 (107), 107.

Vrieze, S.I., Hicks, B.M., Iacono, W.G., McGue, M., 2012. Decline in genetic influence on the co-occurrence of alcohol, marijuana, and nicotine dependence symptoms from age 14 to 29. Am. J. Psychiatry 169 (10), 1073–1081.

Weinberg, N.Z., 2001. Risk factors for adolescent substance abuse. J. Learn. Disabil. 34, 343–351.

Whalen, D.J., Silk, J.S., Semel, M., et al., 2008. Caffeine consumption, sleep, and affect in the natural environments of depressed youth and healthy controls. J. Pediatr. Psychol. 33, 358–367.

Whitesell, N.R., Beals, J., Mitchell, C.M., Novins, D.K., Spicer, P., Manson, S.M., 2006. Latent class analysis of substance use: comparison of two American Indian reservation populations and a national sample. J. Stud. Alcohol 67, 32–43.

Wilens, T.E., Adler, L.A., Adams, J., et al., 2008. Misuse and diversion of stimulants prescribed for ADHD: a systematic review of the literature. J. Am. Acad. Child Adolesc. Psychiatry 47, 21–31.

Wilens, T.E., Biederman, J., Kwon, A., et al., 2004. Risk of substance use disorders in adolescents with bipolar disorder. J. Am. Acad. Child Adolesc. Psychiatry 43, 1380–1386.

Windle, M., Spear, L.P., Fuligni, A.J., et al., 2008. Transitions into underage and problem drinking: developmental processes and mechanisms between 10 and 15 years of age. Pediatrics 121 (Suppl. 4), S273–S289.

Windle, M., Windle, R.C., 2001. Depressive symptoms and cigarette smoking among middle adolescents: prospective associations and intrapersonal and interpersonal influences. J. Consult. Clin. Psychol. 69, 215–226.

Winters, K.C., Lee, S., Botzet, A., Fahnhorst, T., Realmuto, G.M., August, G., 2011. A prospective examination of the association of stimulant medication history and drug use outcomes among community samples of ADHD youth. J. Child Adolesc. Subst. Abuse 20, 314–329.

Wu, L.-T., Howard, M.O., Pilowsky, D.J., 2008. Substance use disorders among inhalant users: results from the National Epidemiologic Survey on Alcohol and Related Conditions. Addict. Behav. 33, 968–973.

Zonnevylle-Bender, M.J., Matthys, W., van de Wiel, N.M., Lochman, J.E., 2007. Preventive effects of treatment of disruptive behavior disorder in middle childhood on substance use and delinquent behavior. J. Am. Acad. Child Adolesc. Psychiatry 46, 33–39.

CHAPTER FOUR

Detection of Populations At-Risk or Addicted: Screening, Brief Intervention, and Referral to Treatment (SBIRT) in Clinical Settings

Larry Gentilello
Department of Surgery, University of Texas, Dallas, TX, USA

1. INTRODUCTION

Substance misuse is one of the most neglected medical problems in the United States. A study that screened hundreds of thousands of patients in clinics, hospital wards, and emergency departments in six states found that 23% of patients screened positive for an alcohol problem, a drug use problem, or both (Summary Health Statistics for US Adults, 2008; Madras et al., 2009). Despite its prevalence most physicians, nurses and other medical professionals receive little training in detecting substance misuse, and most lack confidence and skills in managing it. In a nationwide health-care survey, the majority of patients with a substance use problem indicated that they are not asked about substance use during appointments with their primary care physician (Hingson et al., 2011). Fifty percent of patients who have a severe substance use problem and who voluntarily seek and obtain treatment indicate that their primary care physician is not aware that they have a problem (Saitz et al., 1997).

When a substance use problem is detected, problems with the delivery of effective care are worse than those affecting any other common, serious medical condition. A RAND Corporation study analyzed the treatment of 25 common medical conditions in 12 U.S. metropolitan areas. Evidence-based treatments were provided to 70–80% of patients with cataracts, breast cancer, prenatal care, lower back pain, and coronary care. The rate of delivery of evidence-based treatment for substance use disorders ranked last, with only 10% of patients receiving care that was in compliance with acceptable standards (McGlynn et al., 2003).

2.1. USE OF ALCOHOL

2.1.1. Alcohol and Drug Misuse and Mortality

The amount of emphasis placed on alcohol and drug misuse in health-care settings and in medical school curricula is not proportionate to the number of deaths associated with

The Effects of Drug Abuse on the Human Nervous System
http://dx.doi.org/10.1016/B978-0-12-418679-8.00004-6

their use. By causing or exacerbating over 50 different acute and chronic medical conditions, alcohol misuse results in 34,833 deaths due to chronic diseases, and 75,766 deaths due to acute diseases such as injuries. The total (110,599 U.S. deaths per year) is more than the number of deaths caused by diabetes, influenza and pneumonia, influenza, all forms of kidney diseases, septicemia, and Alzheimer's Disease (Kochanek et al., 2011).

Injuries are the single leading cause of these deaths. Mortality, however, is not always the best measure of disease burden because it primarily occurs in elderly patients near the end of their life expectancy. The years of potential life lost (YPLL's), which subtracts the age that a person dies from their life expectancy, is often a better measure (Tables 1 and 2). Injuries account for nearly half of all alcohol-related deaths, and since they primarily occur in young people, the number of YPLL's due to alcohol-related injuries is almost three times the number caused by all other acute and chronic alcohol-related medical conditions combined.

Heart disease is considered the leading cause of death in the United States, followed by cancer, stroke, pulmonary diseases, and injuries. A study conducted by the Centers for Disease Control (CDC) determined the actual underlying causes of these deaths (Mokdad et al., 2004). They found that tobacco use was the leading cause of death in the United States (435,000 per year; 18.1% of total U.S. deaths), followed by obesity (400,000; 6.6%). Excessive alcohol consumption ranked third (85,000 deaths; 3.5%), and illicit drug use ranked ninth (17,000 deaths).

The number of deaths caused by alcohol and drug misuse combined exceeded the number of deaths caused by infections, motor vehicle crashes, firearms, and all sexually transmitted diseases combined. The CDC study used year 2000 data, and likely underestimated the number of deaths currently caused by substance misuse because mortality from heart disease, cancer, stroke, and injuries have declined in the past decade, while the incidence of drug-related deaths has more than doubled. In 2006 a total of 38,396 people died of drug overdose in the United States, more than twice the number that had been reported by the CDC six years earlier (Heron et al., 2009). The number of deaths due to drug overdoses is four to five times higher than it was at the height of the black tar heroin epidemic in the 1970s, and more than twice as high as it was at the height of the crack cocaine epidemic in the 1990s (Figure 1).

Motor vehicle crashes have been the leading cause of accidental death (as opposed to suicides or assaults) for the last half-century, followed by falls, then firearm injuries. Deaths due to drug overdoses have increased to the point where they are now the second leading cause of accidental death in the U.S., and in 15 states have surpassed car crashes to become the number one cause. Illegal, illicit, or "street drugs" are no longer the primary cause of fatal drug overdoses. The increase is almost entirely related to a six-fold increase in the amount of opiates and benzodiazepines prescribed by physicians and dentists over the past 10 years (Paulozzi, 2008, Figure 2).

Along with the increase in overdose deaths, drugged driving has become an increasingly severe problem, and is as large, if not larger, than the drunk-driving problem. The

Table 1 Number of Deaths and Years of Potential Life Lost (YPLL's) Attributable to the Harmful Effects of Alcohol Use, by Cause and Sex—United States, 2001

Cause	Deaths			YPLLs		
	Male	Female	Total	Male	Female	Total
Chronic conditions						
Acute pancreatitis	370	364	734	7138	6054	13,192
Alcohol abuse	1804	517	2321	50,375	16,433	66,808
Alcohol cardiomyopathy	443	56	499	10,195	1552	11,747
Alcohol dependence syndrome	2770	750	3250	71,782	22,017	93,799
Alcohol polyneuropathy	3	0	3	86	0	86
Alcohol-induced chronic pancreatitis	224	71	295	6209	2135	8344
Alcoholic gastritis	6	2	8	130	46	1116
Alcoholic liver disease	8927	3274	12,201	221,368	94,952	316,321
Alcoholic myopathy	2	0	2	49	0	49
Alcoholic psychosis	564	178	742	12,609	3996	16,605
Breast cancer	N/A[1]	352	352	N/A	6786	6786
Cholelithiases	0	0	0	0	0	0
Chronic hepatitis	3	3	6	55	63	119
Chronic pancreatitis	126	106	232	2608	1952	4560
Degeneration of nervous system attributable to alcohol	93	21	114	1668	486	2154
Epilepsy	96	81	177	2912	2235	5147
Esophageal cancer	394	53	447	6213	788	7000
Esophageal varices	50	21	71	1063	342	1405
Fetal alcohol syndrome	3	2	5	210	137	347
Fetus and newborn affected by maternal use of alcohol	0	1	1	0	80	80
Gastroesophageal hemorrhage	19	9	28	301	139	440
Hypertension	632	552	1184	9458	6,460	15,918
Ischemic heart disease	635	273	908	8012	2,898	10,909
Laryngeal cancer	203	30	233	3146	519	3665
Liver cancer	518	172	690	8640	2633	11,273
Liver cirrhosis, unspecified	3917	2802	6719	80,616	54,528	135,144
Low birth weight, prematurity, and intrauterine growth retardation	96	50	146	7139	3961	11,100
Oropharyngeal cancer	303	57	360	5280	889	6169

Continued

Table 1 Number of Deaths and Years of Potential Life Lost (YPLLs) Attributable to the Harmful Effects of Alcohol Use, by Cause and Sex—United States, 2001—cont'd

Cause	Deaths			YPLLs		
	Male	Female	Total	Male	Female	Total
Portal hypertension	23	14	37	451	298	750
Prostate cancer	233	N/A	233	2224	N/A	2224
Psoriasis	0	0	0	1	1	2
Spontaneous abortion	N/A	0	0	N/A	0	0
Stroke, hemorrhagic	1399	290	1690	22,476	4592	27,068
Stroke, ischemic	520	191	711	5331	1853	7184
Superventricular cardiac dysrythmia	73	92	165	639	796	1435
Total†	24,448	10,385	34,833	548,386	239,619	788,005
Acute conditions						
Air-space transport	122	37	159	3917	1404	5321
Alcohol poisoning	253	78	331	8798	2,952	11,750
Aspiration	97	99	196	1865	1692	3557
Child maltreatment	100	71	171	7086	5,386	12,472
Drowning	671	141	812	25,461	4,633	30,093
Excessive blood alcohol concentration	1	1	2	35	25	61
Fall injuries	2560	2206	4,766	41,627	24,288	65,914
Fire injuries	702	465	1,167	18,991	11,729	30,720
Firearm injuries	113	18	131	4434	695	5129
Homicide	5963	1692	7,655	262,379	71,543	333,922
Hypothermia	164	83	247	3692	1343	5035
Motor-vehicle—nontraffic injuries	171	33	204	5712	1072	6784
Motor-vehicle—nontraffic injuries	10,674	3000	13,674	442,943	135,558	520,501
Occupational and machine injuries	121	6	127	3467	151	3619
Injuries from other road vehicle crashes	178	53	231	6139	1709	7849
Poisoning (not alcohol)	2782	1182	3964	103,917	45,127	149,043
Suicide	5617	1352	6969	186,568	49,297	235,865
Suicide by and exposure to alcohol	21	5	26	777	231	1008
Water transport	90	10	100	3220	454	3674
Total	30,399	10,534	40,933	1,131,028	360,289	1,491,317
Total	54,847	20,918	75,766	1,679,414	599,908	2,279,322

*Because of rounding, numbers might not sum to totals.
†Not applicable.
Source: Centers for Disease Prevention and Control, 2004. Alcohol-Attributable Deaths and Years of Potential Life Lost—United States 2001. MMWR. Morb. Mortal. Wkly. Rep. (S3), 866–870.

Table 2 Actual Causes of Death in the U.S in 1990 and 2000

Actual Cause	No, (%) in 1990*	No, (%) in 2000
Tobacco	400,000, (19)	435,000, (18,1)
Poor died and physical inactivity	300,000, (14)	400,000, (16,6)
Alcohol consumption	100,000, (5)	85,000, (3,5)
Microbial agents	90,000, (4)	75,000, (3,1)
Toxic agents	60,000, (3)	55,000, (2,3)
Motor vehicle	25,000, (1)	43,000, (1,8)
Firearms	35,000, (2)	29,000, (1,2)
Sexual behavior	30,000, (1)	20,000, (0,8)
Illicit drug use	20,000, (<1)	17,000, (0,7)
Total	1,060,000, (50)	1,159,000, (48,2)

*The percentages are for all deaths. The data are from Actual Causes of Death. McGinnis JM, Foege WH. JAMA 270;2207-2212, 1993.
The parentheses are for all deaths.
Source: Mokdad et al., 2004

National Highway Traffic Safety Administration conducts periodic national surveys to estimate the prevalence of driving while under the influence. These roadside surveys use random vehicle stoppage at 300 sites across the United States (Compton and Berning, 2009). The percentage of weekend nighttime drivers testing positive for alcohol dropped from 36.1% in 1973, to 12.4% in 2007. Drivers with a blood alcohol concentration (BAC) over the legal limit of 0.08 g/dl dropped from 7.5%, to 2.2%. However, 16.3% of drivers tested positive for drugs; 11.3% were positive for illegal drugs, and 1.1% tested positive for a combination of illegal drugs and prescription narcotics. With increasing drugged driving, the number of traffic fatalities in which the driver tested positive for drugs, has also increased (Traffic Safety Facts, 2010).

2.1.2. Safe Alcohol and Drug Use Limits

In clinical settings, detection of substance misuse requires familiarity with the thresholds that are associated with risk or harm. With alcohol, it must take into account the usual amount consumed during each drinking occasion, and the overall amount consumed over a period of time. Excessive consumption on single occasions, referred to as binge drinking, results in acute intoxication, and generally occurs in males after intake of four standard size drinks, or in females after three standard drinks. Each standard drink contains the same amount of alcohol, despite differing volumes (12 oz. of beer, 5 oz. of wine, 1.5 oz. of 80 proof spirit). In most patients, this amount of consumption results in a BAC greater than the legal limit for driving in the U.S. (>0.08 mg/dl). Binge drinking is the leading cause of alcohol-related consequences such as car crashes, falls, violence, suicide, and other acute medical problems. Half of all alcohol-related deaths are the result of injuries caused by the acute effects of alcohol.

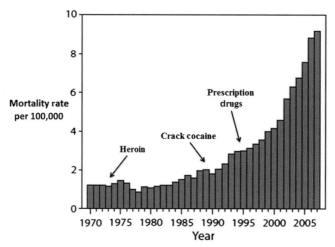

Figure 1 Drug overdose mortality rate during the height of the heroin epidemic in the 1970's, the crack cocaine epidemic in the 1980's, and the rapid increase in drug-related deaths during the period when prescription narcotic use accelerated during the 1990's

In addition to acute effects, alcohol is toxic to virtually every organ system, and continued heavy use over time may result in chronic conditions such as liver and gastrointestinal disease, upper airway and digestive track malignancies, dementia, cardiomyopathy, and other problems. Excessive use that carries a risk of chronic disease is present when intake exceeds 14 drinks per week in a male, or seven drinks per week in a female NIH (NIH Publication, 2007).

These recommended daily and weekly limits should be reduced in certain clinical situations such as in pregnant women, where no amount of alcohol use is considered safe. In patients age 65 and older, it is recommended that daily and weekly intakes should not exceed three and seven drinks, respectively, for both genders. Patients taking medications that interact with alcohol, or with medical problems that are exacerbated by alcohol, should also reduce intake to lower levels, or avoid drinking.

Definitions of unsafe drug use are more complicated because alcohol has the same acute and chronic effects despite the type of beverage consumed, whereas there are many different types of drugs, each with unique risks of causing acute poisoning or overdose, or long-term harm. Also, different types of drugs are often combined, along with varying quantities of alcohol. Therefore, screening that reveals use of "street" drugs or misuse of prescription drugs deserves further assessment, along with a brief intervention and possibly a referral to treatment.

2.2. SUBSTANCE ABUSE

2.2.1. Substance Use Pyramid

The theoretical basis for understanding and addressing substance use problems has undergone sweeping changes in recent years. In the past, in what has been referred to as the

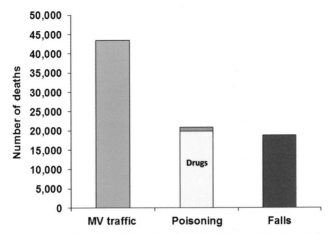

Figure 2 Three leading causes of accidental death in the United States, with the number of deaths due to drug overdose exceeded only by the number of deaths caused by motor vehicle and pedestrian injuries.

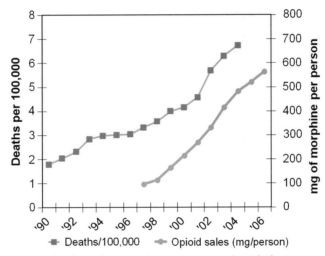

Figure 3 Increase in mortality from drug overdosage corresponds with the increase in the number of narcotic pain relievers prescribed by physicians.

"two-bucket model", patients who used alcohol were placed into one of two categories. They were considered normal ("social") drinkers, even if they often drank to the point of intoxication, or consumed alcohol in quantities that over time are sufficient to cause serious mental and physical harm. The other bucket contained patients that had passed through an irreversible threshold, due to what was thought to be a lack of will, and became "alcoholics". A similar classification system was often applied to individuals who used drugs; some were considered occasional users who used drugs as a form of "recreation", while others were thought to be unwilling to control themselves and became addicts (Figure 3).

The categorical two-bucket approach has been replaced by the recognition that substance use disorders exist across a broad spectrum of problem severity, often involve alcohol and multiple other intoxicating substances, and are often intermixed with mental and physical health problems. No distinct boundary between acceptable use and dependence or addiction is apparent in many patients. The trajectory of substance use problems is usually bidirectional with periods of exacerbation and remission, based on biological, social, environmental, and other factors, and are best understood using continuous, rather than categorical constructs. The "two-bucket model" has effectively been replaced by a "severity pyramid" (Figure 4).

Based on current data, approximately 30% of the population is composed of nondrinkers. Another 40–45% use alcohol in a way that does not pose any health risk. The remaining 25–30% consume it in quantities that are above safe limits and are at-risk for an acute injury or illness, or over time, the development of chronic alcohol-related medical conditions. Drinking above safe limits is often referred to as unhealthy alcohol use, a nonstigmatizing, nonjudgmental term that reflects a health-care provider's appropriate concern about their patient's well-being.

Patients can usually be placed in one of several categories within the severity pyramid. Some are considered at-risk because they consume an unhealthy amount of alcohol, but have never experienced any harmful consequences. Others are engaged in harmful drinking because their use has already resulted in an injury, damaged their health, or caused legal, employment, relationship, financial, or other problems. Roughly one out of six (3–5% of all patients) whose alcohol use is unhealthy have an addiction, or alcohol dependence. Thus, patients with dependence or addiction represent the smallest group of patients with unhealthy use who warrant an intervention within health-care settings (Madras et al., 2009).

Similarly, it is estimated that two-thirds of drug related harm occurs in patients who use drugs, but are not addicted. Drug use problems also exist along a severity spectrum, and often co-exist with alcohol misuse. Appropriately designed screening and

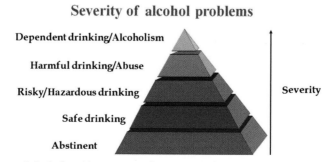

Figure 4 Spectrum of alcohol problem severity, from none to from none to severe, with severe problems being less common than mild or moderate problems.

intervention programs must take into account that patients with unhealthy use often have problems with more than one substance, and nearly half have a concurrent anxiety, depressive, or other mental health disorder that may require attention. It is unlikely that patients will be able to successfully change their alcohol or drug use pattern if they struggle with these problems, and that are not detected and addressed.

2.2.2. The Preventive Paradox

Although dependent patients are at highest risk for injury or illness, the majority of alcohol-related problems occur in the nondependent population because they constitute the largest portion of the problem of drinking population. A large number of patients at moderate risk usually create more cases than a small population of patients at high-risk patients. For example, the risk of a fatal myocardial infarction is five times higher among patients with a very high cholesterol level (≥ 300 mg/dl), compared to those with only a borderline elevation (200–225 mg/dl). However, most myocardial infarctions occur in patients with borderline elevations because such patients are far more common.

The majority of deaths, YPLL's and hospital admissions attributable to alcohol misuse do not occur in those who consume the greatest amounts of alcohol, and the majority of drug use problems do not occur in patients who are addicted (Poikolainen et al., 2007). For these reasons, from a public health perspective, preventive interventions focus on the whole population at-risk, not just on detecting "alcoholism" or addiction, or those patients who are at highest risk.

2.2.3. New Terminology

In the early 1980s, the Diagnostic and Statistics Manual III (DSM III) replaced the word alcoholism with two separate terms: alcohol abuse and alcohol dependence. Alcohol abuse defined patients who experience repeated adverse consequences from their drinking, but do not display signs of dependence. These patients can often quit on their own if they are sufficiently motivated to do so.

Patients with dependence have repeated consequences, but also demonstrate tolerance, withdrawal symptoms, and loss of control. They are often unable to stop drinking without treatment, even after experiencing catastrophic consequences. It should be noted that according to the current DSM-IV classification system, alcohol abuse and alcohol dependence are mutually exclusive terms.

It is anticipated that this terminology will be updated in the DSM-V, scheduled for publication in 2013. The term "abuse" of alcohol or drugs will be eliminated in part because it is stigmatizing, is clinically detrimental because patients resist being so labeled, and whether or not a patient experiences consequences depends not only on their substance use but on whether they own a car, how driving laws are enforced in their state, whether they are employed, are socially engaged, and other factors.

In the DSM-V update, the term dependence will be replaced by the word addiction, which refers to the physiologic changes that take place with chronic use of intoxicating substances, and the mental and physical effects that occur when intake of the substance is abruptly discontinued.

Additionally, changes to the DSM-V definitions will take into account the multiple risky and harmful patterns of alcohol and drug use that exist besides abuse and dependence. New diagnostic terms will refer to addictions and related disorders, with subgroups characterized by substances used such as alcohol, cannabis, or cocaine. There will also be a set of modifiers that describe the severity and duration of symptoms.

2.3. PREVENTION AND INTERVENTION

2.3.1. Preventive Interventions

A commonly cited reason for not providing clinical preventive services is that there are too many competing time demands placed upon clinicians (Gentilello, 2005). Time, staffing, and financial constraints are significant concerns in all medical settings. In order to facilitate prioritization, the National Commission on Prevention Priorities ranked the practices currently recommended by the U.S. Preventive Services Task Force based upon the magnitude of the problem targeted by the intervention, the effectiveness of the intervention, and the cost of delivering it (Maciosek et al., 2006). Three preventive interventions were tied for the highest rating: aspirin chemoprophylaxis in patients at high risk for myocardial infarction, childhood immunizations, and tobacco-use screening and brief intervention (Table 3).

Alcohol screening and brief interventions received the next highest rating, and were tied with screening for hypertension, screening for colorectal cancer, and providing influenza and pneumococcal immunizations to elderly at-risk patients. Alcohol screening and intervention ranked higher than screening for cervical cancer, breast cancer, hypercholesterolemia, obesity, osteoporosis, chlamydia, depression, and calcium and folic acid chemoprophylaxis, and higher than many other services that are often provided that have less evidence of impact, and cost-effectiveness.

2.3.2. Screening, Brief Intervention, and Referral to Treatment

Routine screening, the provision of a brief intervention, and a referral to treatment for those with more severe problems (SBIRT), has been recommended as a comprehensive, evidence-based, public health approach to addressing alcohol and drug misuse (Maciosek et al., 2006; SAMSHA News; Office of National Drug Control Policy; CDC Guide; Neumann et al., 2006; SAMHSA-HRSA; Babor et al., 2007; Vaca et al., 2011). SBIRT has been evaluated in over 100 clinical trials, dozens of which were randomized, most of which have demonstrated positive effects. Meta-analysis of

Table 3 Ranking of Current U.S. Preventive Services Task Force Recommendations

Aspirin chemoprophylaxis	Discuss the benefits/harms of daily aspirin use for the prevention of cardiovascular events with men ≥40, women ≥50, and others at increased risk.	5	5	10
Childhood immunization series	Immunize children: Diphtheria, tetanus, pertussis, measles, mumps, rubella, varicella, pneumococcal conjugate, influenzae.	5	5	10
Tobacco-use screening and brief intervention	Screen adults for tobacco use, provide brief counseling, and offer pharmacotherapy.	5	5	10
Colorectal cancer screening	Screen adults aged ≥50 years routinely with FOBT, sigmoidoscopy, or colonoscopy.	4	4	8
Hypertension screening	Measure blood pressure routinely in all adults and treat with antihypertensive medication to prevent incidence of cardiovascular disease.	5	3	8
Influenza immunization	Immunize adults aged ≥50 against influenza annually.	4	4	8
Pneumococcal immunization	Immunize adults aged ≥65 against pneumococcal disease with one does for most in this population.	3[1]	5	8
Problem drinking screening and brief counseling	Screen adults routinely to identify those whose alcohol use places them at increased risk and provide brief counseling with follow-up.	4	4[1]	8
Vision screening—adults	Screen adults aged ≥65 routinely for diminished visual acuity with Snellen visual acuity chart.	3	5	8
Cervical cancer screening	Screen women who have been sexually active and have a cervix within 3 years of onset of sexual activity or age 21 routinely with cervical cytology (Pap smears).	4	3	7
Cholesterol screening	Screen routinely for lipid disorders among men aged ≥35 and women aged ≥45 and treat with lipid-lowering drugs to prevent the incidence of cardiovascular disease.	5[1]	2[1]	7
Breast cancer screening	Screen women aged ≥50 routinely with mammography alone or with clinical breast examination, and discuss screening with women aged 40 to 49 to choose an age to initiate screening.	4	2	6
Chlamydia screening	Screen sexually active women aged <25 routinely.	2	4	6
Calcium chemoprophylaxis	Counsel adolescent and adult women to use calcium supplements to prevent fractures.	3[1]	3[1]	6

Continued

Table 3 Ranking of Current U.S. Preventive Services Task Force Recommendations—cont'd

Vision screening-children	Screen children aged <5 years routinely to detect amblyopia, strabismus, and defects in visual acuity.	2	4[1]	6
Folic acid chemoprophylaxis	Counsel women of childbearing age routinely on use of folic acid supplements to prevent birth defects.	2	3	5
Obesity screening	Screen all adults patients routinely for obesity and offer obese patients high-intensity counseling about died, exercise, or both together with behavioral interventions for at least 1 year.	3	2	5
Depression screening.	Screen adults for depression in clinical practices that have system in place to assure accurate diagnosis, treatment, and follow-up.	3	1	4
Hearing screening	Screen for hearing impairment in adults aged ≥65 and make referrals to specialists.	2	2	4
Injury prevention counseling	Assess safety practices of parents of children aged <5 years and provide counseling on child safety seats, window/stair guards, pool fence, poison control, hot water temperature, and bicycle helmets.	1	3[1]	4
Osteoporosis screening	Screen women aged ≥65 and women aged ≥60 at increased risk routinely for osteoporosis and discuss benefits and harms of treatments options.	2	2	4
Cholesterol screening-high risk	Screen men aged 20–35 and women aged 20–45 routinely for lipid disorders if they have other risk factors for coronary heart disease, and treat with lipid-lowering drugs to prevent incidence of cardiovascular disease.	1	1[1]	2
Diabetes screening	Screen for diabetes in adults with high cholesterol or hypertension, and treat with a goal of lowering levels below conventional target values.	1	1	2
Diet counseling	Offer intensive behavioral dietary counseling to adult patients with hyperlipidemia and other known risk factors for cardiovascular and diet-related chronic disease.	1	1	2
Tetanus–diphtheria booster	Immunize adults every 10 years.	1	1	2

CE = Cost effectiveness, CPB = Clinically preventable burden.
[1]Services in boldface are those with scores of 6+ for which data indicate that delivery to the U.S. population eligible for the services is likely ≥50%.
Source: Maciosek et al., 2006

randomized trials has shown that patients who receive a brief intervention are twice as likely to reduce their heavy drinking as "usual medical care" patients who serve as controls (SAMHSA News, 2009; Wilk et al., 1997).

Given the relationship between alcohol and injury, trauma centers have established alcohol screening and intervention programs in hundreds of hospitals across the United States, and must provide these services as a condition for receipt of their Level 1 Trauma Center designation (COT Quick Guide, Centers for Disease Control). Studies have shown that brief interventions reduce alcohol intake, injury recidivism, trauma center readmission, emergency department visits, drunk-driving arrests, and reduce health-care costs (COT Quick Guide, CDC Guide, Gentilello et al., 1999; Gentilello et al., 2005; Schermer et al, 2006).

A study of injured adolescents showed that at six-month follow-up intervention group patients had a significant reduction in drinking and driving, moving violations, alcohol-related problems, and less than half as many alcohol-related injuries as control group patients (Monti et al., 1999).

Another study, also a randomized, prospective trial, was performed on 762 patients admitted to a Level 1 Trauma Center (Gentilello et al., 1999). Patients who screened positive were randomized to a single 15–30 min brief intervention, or to a no-intervention control group. At a one-year follow-up, the intervention group had decreased their alcohol intake by 22 drinks per week, compared to an only a two-drink reduction in the conventional care group. A statewide registry was used to detect readmission to any hospital in the state for treatment of an injury. At a three-year follow-up, there was a 47% reduction in hospital readmission in the intervention group, compared to controls. There was also a 48% reduction in return visits to the emergency department for treatment of another injury (Figures 5 and 6).

Through their effect on reducing trauma recurrence, brief interventions have been shown to result in cost-savings to trauma centers of over $320 per intervention performed, and $3.81 for every dollar invested in the program (Gentilello et al., 2005).

Although brief interventions were designed for nondependent heavy drinkers, they have been shown to have some effectiveness even when provided to patients with dependence. A randomized controlled trial conducted in a trauma center showed that at three-year follow-up one DUI had been prevented for every nine interventions performed, even though the majority of patients were dependent and had a history of prior Driving Under the Influence (DUI) arrests (Schermer et al, 2006; Guth et al., 2008; Cobain et al., 2011).

The World Health Organization (WHO) and the United States government supported the growth of SBIRT services in a broad range of health-care facilities and the expansion of screening for drugs other than alcohol. The largest study of SBIRT to date was conducted across six states and one tribal nation (Madras et al., 2009). The study included trauma centers, primary care clinics, emergency departments, general medical

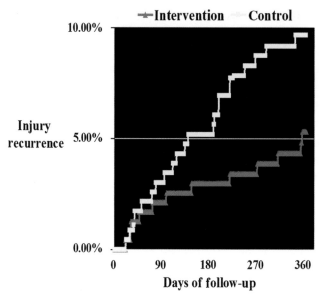

Figure 5 Rate of injury recurrence in requiring an emergency department visit in patients randomized to receive a brief intervention after trauma center admission, or to a control (no intervention) group (Gentilello, 1999).

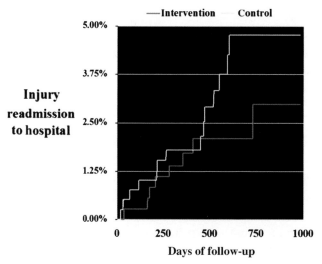

Figure 6 Rate of injury recurrence requiring admission to a hospital in patients randomized to receive a brief intervention after trauma center admission, or to a control (no intervention) group (Gentilello, 1999).

and surgery wards, college health clinics, and enrolled a diverse population. Screening was administered to all adult patients who presented to the study sites for medical care unless they were too severely ill or injured. There were 459,599 patients screened using standardized questionnaires, making it the largest study of its kind.

Nearly one out of four patients who presented for medical care (22.7%) screened positive for unhealthy alcohol or drug use, or both. Screening results indicated that 15.9% of patients were nondependent, and in need of a brief intervention only. An additional 3.2% were assessed as needing two to three additional sessions that could be administered during return visits. Only 3.7% of patients, or one out of six patients who screened positive, were considered dependent, and in need of formal treatment. Thus, difficulties with accessing treatment do not affect the majority of patients in health-care settings who need help for their substance use problems. At six-month follow-up drug misuse was reduced by 67.7%, and heavy alcohol use was reduced by 38.6%. There was a 64.3% reduction in arrests, and 45.8% of patients who were homeless at study entry had found a place to live. In systematic analyses of specific sites within the United States government-sponsored SBIRT programs, SBIRT promoted significant reductions in drugs (marijuana, cocaine, heroin, methamphetamine) at follow-up periods of 6 months, with some sites reporting cost-effectiveness data (InSight Project Research Group, 2009; Estee et al., 2010; Gryczynski et al., 2011).

A large multinational randomized controlled trial, sponsored by the WHO, evaluated the effectiveness of a brief intervention for illicit drugs (cannabis, cocaine, amphetamine-type stimulants, and opioids) linked to the WHO Alcohol, Smoking, and Substance Involvement Screening Test or ASSIST (Humeniuk et al., 2011). The ASSIST screens for problem or risky use of 10 psychoactive substances, producing a score for each substance that falls into either a "low-", "moderate-" or "high-" risk category. In this prospective trial, participants were assigned either to a 3-month control group or received brief motivational counseling lasting an average of 13.8 min for the drug receiving the highest ASSIST score. Primary health-care settings in four countries, Australia, Brazil, India, and the United States, recruited subjects (731) within the moderate-risk range. Omnibus analyses indicated that those receiving the brief intervention had significantly reduced scores for all measures, compared with control participants.

Country-specific analyses showed that, with the exception of the site in the United States, participants provided brief interventions had significantly lower ASSIST total illicit substance involvement scores at follow-up compared with the control participants. In India and Brazil, subjects demonstrated a very strong brief intervention effect for cannabis scores ($p < 0.005$ for both sites), sites in Australia ($p < 0.005$) and Brazil ($p < 0.05$) for Stimulant Scores, and the Indian site for Opioid Scores ($p < 0.01$). The study concluded that ASSIST-linked brief intervention aimed at reducing illicit substance use and related risks is effective at least in the short-term, with the effect generalizing across countries.

Severity of problematic use, setting, and type of provider has also been investigated, particularly for alcohol-related problems. Studies comparing brief intervention effects on nondependent problem drinkers and dependent drinkers conducted within in-patient medical units showed both cohorts benefited equally (Guth et al.,

2008; Cobain et al., 2011), but not in an earlier analysis (Emmen et al., 2004). Brief interventions also were effective when performed by hospital nurses who did or did not have special training in managing substance use or mental health disorders. Brief interventions in dependent drinkers performed by dedicated alcohol special-ist nurses showed that at six months follow-up 49% of patients no longer exhibited dependence, and 40% reported abstinence (Cobain et al., 2011). Nurse delivered interventions in emergency departments also were effective (Désy et al., 2010). Brief interventions increase the likelihood that dependent patients will follow through on a referral to treatment, and they improve Alcohol Use Disorders Identification Test (AUDIT) and Severity of Alcohol Dependence Questionnaire scores in patients admitted to the hospital with severe medical problems (Guth et al., 2008; Cobain et al., 2011).

Patients with unhealthy use may respond to a skillfully delivered brief interven-tion whether it is conducted in a trauma center, primary care clinic, or general hospital setting. Health-care workers who do not specialize in addiction or mental health disorders can be taught to provide a brief intervention, and can have a list of treatment resources available in the community in order to provide a referral to patients who need it. Brief interventions can be provided by anyone who is able to speak to substance misusing patients in an empathic, nonjudgmental, nonconfronta-tional manner.

2.3.3. Screening

The immediate goal of screening is to determine if a patient has an unhealthy alcohol or drug use pattern. Additionally, an effort should be made to determine if an addic-tion is present so that in addition to a brief intervention, a referral to treatment can be made.

Many clinicians are mistrustful of patient self-report. However, when asked in a respectful, concerned and confidential manner that does not put patients on the defen-sive, patients with a substance use problem usually do not underreport their drinking when responding to a screening questionnaire (Donovan et al., 2004).

The CAGE is a four-item questionnaire whose name is an acronym based on its four questions (Ewing, 1984). Have you ever tried to *C*ut down on your drinking? Have people *A*ngered you by criticizing your drinking? Have you ever felt *G*uilty about your drinking? Have you ever had a drink in the morning (*E*ye-opener) to steady your nerves or to reduce a hangover? It is considered positive if there are two or more "yes" answers.

The Michigan Alcohol Screening Test (MAST) is a 25-item questionnaire that focuses on the patient's perceptions about their drinking, and the consequences they have experienced (Selzer, 1971). There are shorter forms available such as the Short MAST (SMAST, 13 questions) and the Brief-MAST (BMAST, 10 questions).

The CAGE and MAST questionnaires have been used for decades, and were primarily designed to detect dependence. They are not very sensitive to less severe drinking problems, and thus have less utility given the current understanding of the need to address alcohol and drug problems not just when they have progressed all the way to addiction. The CAGE and MAST may also not be appropriate in certain patient groups. For example, the first question of the CAGE questionnaire asks the patient if they ever tried to cut down on their drinking. Teenagers and college students who drink with a binge pattern are unlikely to respond with a "yes" answer to this question.

The AUDIT is a 10-question, self-report screening tool developed by the WHO, and has been validated in multiple settings, and in numerous languages and cultures (Saunders et al., 1993). It was designed to be sensitive to a broad spectrum of harmful drinking levels. It assesses drinking quantity and frequency (3 questions), consequences caused by alcohol (4 questions), and symptoms of dependence (3 questions). It takes less than two to 4 min to administer, and seconds to score the results. A score of 8 or more is considered a positive screen. Brief intervention (Figure 7).

As the safe recommended level of alcohol intake is lower for females than for males, using the same cut-off to the three quantity and frequency questions for both genders decreases the sensitivity of the AUDIT in women. In order to equalize sensitivity and specificity a cut-off score of five points has been recommended when using the AUDIT with women (Neumann et al., 2004). Patients who score over 20 points are likely to have dependence or addiction and require further assessment or a referral to treatment. The AUDIT can often be abbreviated because many patients drink infrequently or not at all. If the response to the first question that asks how often the patient has a drink containing alcohol is "never", no further questioning is necessary. The AUDIT is quickly becoming the standard in most clinical settings, and is widely available in downloadable form (AUDIT, 2001).

Preliminary prescreening can also be performed by asking just one question, "When was the last time you had five or more drinks in a day (four drinks for women)?" If the patient states that it has happened within the last six months the prescreen is considered positive, and the remaining nine questions of the AUDIT should be asked. The single question has acceptable sensitivities and specificities of 85% and 70% in males, and 82% and 77% in females (Canagasaby and Vinson, 2005).

The CAGE questionnaire has been modified to enable screening for drug use by changing the word "alcohol" to "drugs" in each of the four questions. The Drug Abuse Screening Test (DAST) is an alternative 28-item questionnaire that asks about illicit drugs, over the counter drugs, and prescription drug misuse, and is modeled after the MAST. The DAST has been reduced to 20 questions and even to 10 questions without much loss of accuracy (Skinner, 1982).

Although more time consuming, the ASSIST is an instrument that was developed for the WHO and is designed to detect and score problems with three harmful substances;

Questions	0 Points	1 Point	2 Points	3 Points	4 Points	Score
1. How often do you have a drink containing alcohol?	Never	Monthly or less	2 to 4 times a month	2 to 3 times a week	4 or more times a week	
2. How many drinks containing alcohol do you have on a typical day when you are drinking?	1 or 2	3 or 4	5 or 6	7 to 9	10 or more	
3. How often do you have five or more drinks on one occasion?	Never	Less than monthly	Monthly	Weekly	Daily or almost daily	
4. How often during the last year have you found that you were not able to stop drinking once you had started?	Never	Less than monthly	Monthly	Weekly	Daily or almost daily	
5. How often during the last year have you failed to do what was normally expected of you because of drinking?	Never	Less than monthly	Monthly	Weekly	Daily or almost daily	
6. How often during the last year have you needed a first drink in the morning to get yourself going after a heavy drinking session?	Never	Less than monthly	Monthly	Weekly	Daily or almost daily	
7. How often during the last year have you had a feeling of guilt or remorse after drinking?	Never	Less than monthly	Monthly	Weekly	Daily or almost daily	
8. How often during the last year have you been unable to remember what happened the night before because of your drinking?	Never	Less than monthly	Monthly	Weekly	Daily or almost daily	
9. Have you or someone else been injured because of your drinking?	No		Yes, but not in the last year		Yes, during the last year	
10. Has a relative, friend, doctor, or other healthcare worker been concerned about your drinking or suggested you cut down?	No		Yes, but not in the last year		Yes, during the last year	
Male or Female	Under 21 or 21+				Total Score=	

My Score = _____

Figure 7 Alcohol Use Identification Test (AUDIT).

Risk Level	Feedback	Score
I	Your score indicates that at this time you seem to be making relatively low risk drinking choices. You may benefit from programs like AlcoholEdu, Training for Intervention Procedures (TIPS), and Because I Care.	0-7
II	Your score indicates that at this time you seem to be drinking in excess of low risk guidelines. You may want to consider meeting with a substance abuse counselor to discuss ways to lower your risk and decrease negative consequences. It is recommended that you participate in programs such as AlcoholEdu, Training for Intervention Procedures (TIPS), and Because I Care.	8-15
III	Your score indicates that at this time you are making high risk drinking choices that are causing some negative consequences. Higher scores indicate higher risk drinking and greater likelihood of negative consequences. It is strongly recommended that you meet with a substance abuse for an assessment and to develop a plan to lower risk and minimize negative consequences. It is also recommended that you participate in programs such as AlcoholEdu, Training for Intervention Procedures (TIPS), and Because I Care.	16-40

NOTE: If you suspect that you have a drinking or drug problem, you should seek help from a professional regardless of how you score on this screening test.

Figure 7 *Continued*

alcohol, drugs, and tobacco (Humeniuk et al., 2008). It was recently modified and made accessible on the internet (NIDAMED).

2.3.4. Guidelines for Brief Interventions

Change occurs when people perceive a large enough discrepancy between the way things are, and the way they would like them to be. The clinician's demeanor during the intervention strongly influences the patient's willingness to discuss and to contemplate this discrepancy. A polite and nonjudgmental style facilitates openness, while an intervention style that the patient perceives as attacking or confrontational provokes resistance or denial.

The belief that patients with alcohol problems have poor motivation and rigid defense mechanisms (denial) is not supported by research. Denial is often a product of the way the clinician chooses to interact with the patient. Patients who are considered to be in denial often argue against change because they have an emotional need to defend themselves against a confrontational clinician whom they perceive as insulting or accusatory. Confrontational tactics are particularly counterproductive during a brief intervention because the time needed to overcome the resulting "denial" leaves little time for the brief intervention itself.

2.3.5. There are a Variety of Approaches to Brief Interventions, and Their Key Component Parts Have Been Summarized Using a Variety of Acronyms, Some of Which Are Below

2.3.5.1. "Frames"

The elements of a brief intervention have been summarized by the acronym, *FRAMES*. They consist of providing nonpersonal, concerned *F*eedback about the risks associated with the patient's substance misuse, placing an emphasis on their personal *R*esponsibility to change rather than threatening or demanding that they do so, providing clear *A*dvice that from a medical perspective change is needed to avoid harm, and presenting a *M*enu of alternative change goals and strategies.

Many patients considering changing their substance use pattern are discouraged. They lack confidence in their ability to quit because they have failed in their attempts before, or they have quit, sometimes for several years, but ultimately relapsed. Brief interventions should be conducted in an *E*mpathetic manner that is intended to increase the patient's sense of *S*elf-efficacy, or optimism, and confidence that change is possible by discussing successful prior periods of abstinence, and reviewing the patient's strengths that made it possible. In addition to empathy, closing on good terms is important because in the event of relapse, it is important for the patient to trust that in the future, when clinicians approach them to discuss their substance misuse, it will not be confrontational and uncomfortable.

2.3.5.2. "Sum"

Another acronym that has been used to summarize the key components of a brief intervention is *SUM*. This refers to providing feedback about *S*creening results, *U*nderstanding the patient's views about their drinking rather than demanding that they accept the clinician's interpretation, and providing a *M*enu of possible change options.

A key source of feedback can be the results obtained using a screening questionnaire. It is important for clinicians to convey to patients that they are not concerned with labeling them (e.g. having alcoholism). It is important to indicate, "I am concerned about whether or not your use of alcohol (and/or drugs) is hurting you." The goal of an intervention is to establish empathy, not to make the patient feel upset. Empathy is primary tool that clinicians have to avoid denial or resistance, and to increase the likelihood that the patient will rely on the advice and support of the interventionist.

Understanding the patient's views can be performed by asking them about the "pros and cons" of their drinking. The patient's response may be, "It helps me to relax," "It makes it easier to talk to others," or "It helps me to have a good time". After acknowledging the legitimacy of these feelings, the clinician may summarize them and ask the patient about the "cons" by asking, "What are some of the things you don't like about drinking?" For example, if the patient indicates that it causes arguments with family members the clinician might ask, "How does your alcohol use affect your ability to have a stable family life?" If the clinician and patient have established appropriate rapport the patient will usually acknowledge the role their substance use has in causing their medical and other problems, and will often spontaneously begin to discuss the benefits of quitting or cutting down ("change talk").

Most patients are stuck between the reasons why they should quit, and the reasons why they would like to maintain the status quo. The clinician should heighten the patient's awareness of this situation. This can be done by making statements such as, "On the one hand, drinking helps you to relax and to enjoy hanging out with your friends. On the other hand it is causing fights and arguments at home, is causing you to have financial problems, has been causing you to miss work, and is causing your stomach problems. So, where does that leave you?" In many cases with respect to the pro's and con's, these discussions will affect the patient's decisional balance, will result in change talk, and in many cases, lead the patient to contemplate change, or to make a commitment to reduce their drinking or to stop altogether.

Asking for, then summarizing the pro's and con's, is a key brief intervention technique. It is designed to cause the patient to articulate the reasons why they might want to change, and how their life might improve as a result.

Menus of potential change strategies are also offered when the *SUMS* technique is used. When offered a menu of choices the patient is more likely to find an approach that is acceptable to their own unique situation. For example, a patient may refuse a referral to treatment, but may be willing to accept an incremental change such as setting specific limits on their alcohol intake, and agree to seek further help if they are unable to stay within agreed upon limits.

The "menu of options" is presented in ascending order of commitment, from no change whatsoever, merely thinking about it and noticing more about their drinking pattern in the future, to cutting down or quitting. If the patient selects one of these options plans for meeting their goal can be discussed.

Also, according to change theory, change will not occur if the patient believes that change is not possible, or if they believe that even if they do change, their life will not improve. A corollary is that change occurs only when two conditions are met. The patient must believe that change is possible, and that if they do change, their life will be better off. If either one or both of these two conditions are missing, the interventionist should try to establish the missing piece.

2.3.5.3. *Ask, Advise, Assist, Arrange*

The National Institute of Alcohol Abuse and Alcoholism, and the National Institute on Drug Abuse recommend the Ask, Advise, Assist, Arrange method. Clinicians should *A*sk about alcohol and drug intake using a single question, "How many times in the past year have you had five or more drinks in a day (four or more for a woman)?" They should also ask, "In the past six months have you used any street drugs, or used any prescription medications in order to get high, or in a way that wasn't prescribed by your doctor?" If the single question preliminary screen is positive the full AUDIT, DAST, or similar questionnaire is administered and scored. If the questionnaire result is positive, the clinician should provide *A*dvice such as, "In my medical opinion you are drinking more than is safe," or "your use of drugs is not safe and is putting you at-risk for serious health consequences."

If the patient is ready to change the clinician should *A*ssist them in setting goals, and help to *A*rrange mechanisms to support these efforts by providing follow-up visits, a treatment referral, and a list of available resources for self or treatment-assisted change.

3. CONCLUSIONS

Brief intervention studies with positive outcomes have been performed using physicians, nurses, psychologists, social workers, health educators, counselors, peers, chaplains, and other types of staff who are not specialists in mental health, or in addiction medicine. They can be performed by anyone who is willing to learn the techniques, and who is capable of showing respect and concern for patients who drink too much or who misuse drugs.

Although physicians may not be the primary providers of SBIRT in their own clinical setting, their role is critical in providing administrative support and institutional advocacy for these services. The brief intervention procedures discussed in this chapter have been tested and applied in a variety of health-care settings, and a large number of research studies have established their effectiveness. There is a wealth of downloadable information on line including scripts, manuals, and videos (Alcohol Alert, NIAAA, NIDA). There are also frequent seminars where the basic methods can be learned in one or two sessions.

REFERENCES

Alcohol Alert: Brief Interventions, July 2005. National Institute on Alcohol Abuse and Alcoholism Number 66. Accessible at http://pubs.niaaa.nih.gov/publications/AA66/AA66.htm.

AUDIT, 2001. The Alcohol Use Disorders Identification Test. Version 2. World Health Organization, Geneva, Switzerland. http://whqlibdoc.who.int/hq/2001/who_msd_msb_01.6a.pdf.

Babor, T.F., McRee, B.G., Kassebaum, P.A., Grimaldi, P.L., Ahmed, K., Bray, J., 2007. Screening, brief intervention, and referral to treatment (SBIRT): toward a public health approach to the management of substance abuse. Subst. Abus. 28, 7–30.

Canagasaby, A., Vinson, D.C., 2005. Screening for hazardous or harmful drinking using one or two quantity-frequency questions. Alcohol 40, 208–213.

CDC Guide: Screening and Brief Interventions (SBI) for Unhealthy Alcohol Use. A Step-by-step Implementation Guide for Trauma Centers. http://www.cdc.gov/InjuryResponse/alcohol-screening/pdf/SBI-Implementation-Guide-a.pdf/ (accessed 01.09.12.).

Cobain, K., Owens, L., Kolamunnage-Dona, R., Fitzgerald, R., Gilmore, I., Pirmohamed, M., 2011. Brief interventions in dependent drinkers: a comparative prospective analysis in two hospitals. Alcohol Alcohol. 46, 434–440.

Compton, R., Berning, A., 2009. Results of the 2007 National Roadside Survey of Alcohol and Drug Use by Drivers. http://www.nhtsa.gov/DOT/NHTSA/Traffic%20Injury%20Control/Articles/Associated%20Files/811175.pdf.

COT Quick Guide. Alcohol screening and brief intervention (SBI) for Trauma Patients. American College of Surgeons, Committee on Trauma. http://www.facs.org/trauma/publications/sbirtguide.pdf.

Centers for Disease Prevention and Control, 2004. Alcohol-Attributable Deaths and Years of Potential Life Lost—United States 2001. MMWR Morb. Mortal. Wkly. Rep. (S3), 866–870.

Désy, P.M., Howard, P.K., Perhats, C., Li, S., 2010. Alcohol screening, brief intervention, and referral to treatment conducted by emergency nurses: an impact evaluation. J. Emerg. Nurs. 36, 538–545.

Donovan, D.M., Dunn, C.W., Rivara, F.P., Jurkovich, G.J., Ries, R.R., Gentilello, L.M., 2004. Comparison of trauma center patient self-reports and proxy reports on the alcohol use identification test (AUDIT). J. Trauma 56, 873–882.

Emmen, M.J., Schippers, G.M., Bleijenberg, G., Wollersheim, H., 2004. Effectiveness of opportunistic brief interventions for problem drinking in a general hospital setting: systematic review. BMJ 328, 318–323.

Estee, S., Wickizer, T., He, L., Shah, M.F., Mancuso, D., 2010. Evaluation of the Washington state screening, brief intervention, and referral to treatment project: cost outcomes for Medicaid patients screened in hospital emergency departments. Med. Care 48, 18–24.

Ewing, J.A., 1984. Detecting alcoholism: the CAGE questionnaire. JAMA 252, 1905–1907.

Gentilello, L.M., Rivara, F.P., Donovan, D.M., Jurkovich, G.J., Daranciang, E., Dunn, C.W., Villaveces, A., Copass, M., Ries, R.R., 1999. Alcohol interventions in a trauma center as a means of reducing the risk of injury recurrence. Ann. Surg. 230, 473–483.

Gentilello, L.M., 2005. Confronting the obstacles to screening and interventions for alcohol problems in trauma centers. J. Trauma 59, S137–S143.

Gentilello, L.M., Ebel, B.E., Wickizer, T.M., Salkever, D.S., Rivara, F.P., 2005. Alcohol interventions for trauma patients treated in emergency departments and hospitals: a cost benefit analysis. Ann. Surg. 241, 541–550.

Gryczynski, J., Mitchell, S.G., Peterson, T.R., Gonzales, A., Moseley, A., Schwartz, R.P., 2011. The relationship between services delivered and substance use outcomes in New Mexico's screening, brief intervention, referral and treatment (SBIRT) initiative. Drug Alcohol Depend. 118, 152–157.

Guth, S., Lindberg, S.A., Badger, G.J., Thomas, C.S., Rose, G.L., Helzer, J.E., 2008. Brief intervention in alcohol-dependent versus nondependent individuals. J. Stud. Alcohol Drugs 69, 243–250.

Heron, M.P., Hoyert, D.L., Murphy, S.L., Xu, J.Q., Kochanek, K.D., Tejada-Vera, B., 2009. Deaths: final data for 2006. Natl. Vital Stat. Rep. 57, 1–134.

Hingson, R.W., Heeren, T., Edwards, E.M., Saitz, R., September 21, 2011. Young adults at risk for excess alcohol consumption are often not asked or counseled about drinking alcohol. J. Gen. Intern. Med. (Epub ahead of print).

Humeniuk, R.E., Ali, R.A., Babor, T.F., Farrell, M., Formigoni, M.L., Jittiwutikarn, J., de Lacerda, R.B., Ling, W., Marsden, J., Monteiro, M., Nhiwatiwa, S., Pal, H., Poznyak, V., Simon, S., 2008. Validation of the alcohol smoking and substance involvement screening test (ASSIST). Addiction 103, 1039–1047.

Humeniuk, R., Ali, R., Babor, T., Souza-Formigoni, M.L., de Lacerda, R.B., Ling, W., McRee, B., Newcombe, D., Pal, H., Poznyak, V., Simon, S., Vendetti, J., November 30, 2011. A randomized controlled trial of a brief intervention for illicit drugs linked to the alcohol, smoking, and substance involvement screening test (ASSIST) in clients recruited from primary health care settings in four countries. Addiction. http://dx.doi.org/10.1111/j.1360-0443.2011.03740.x (Epub ahead of print).

InSight Project Research Group, 2009. SBIRT outcomes in Houston: final report on InSight, a hospital district-based program for patients at risk for alcohol or drug use problems. Alcohol Clin. Exp. Res. 33, 1374–1381.

Kochanek, K.D., Xu, J., Murphy, S.L., Miniño, A.M., Kung, H.C., 2011. Division of vital statistics. Deaths: preliminary data for 2009. Natl. Vital Stat. Rep. 59, 1–51.

Maciosek, M.V., Coffield, A.B., Edwards, N.M., Flottemesch, T.J., Goodman, M.J., Solberg, L.I., 2006. Priorities among effective clinical preventive services results of a systematic review and analysis. Am. J. Prev. Med. 31, 52–61.

Madras, B.K., Compton, W.M., Avula, D., Stegbauer, T., Stein, J.B., Clark, H.W., 2009. Screening, brief interventions, referral to treatment (SBIRT) for illicit drug and alcohol use at multiple healthcare sites: comparison at intake and 6 months later. Drug Alcohol Depend. 99, 280–295.

McGlynn, E.A., Asch, S.A., Adams, J., Keesey, J., Hicks, J., DeCristofaro, A., Kerr, E.A., 2003. The quality of health care delivered to adults in the United States. N. Engl. J. Med. 348, 2635–2645.

Mokdad, A.H., Marks, J.S., Stroup, D.S., Gerberding, J.L., 2004. Actual causes of death in the United States, 2000. JAMA 291, 1238–1245.

Monti, P.M., Colby, S.M., Barnett, N.P., Spirito, A., Rohsenow, D.J., Myers, M., Woolard, R., Lewander, W., 1999. Brief intervention for harm reduction with alcohol-positive older adolescents in a hospital emergency department. J. Consult. Clin. Psychol. 67, 1989–1994.

Neumann, T., Neuner, B., Gentilello, L.M., Weiss-Gerlach, E., Mentz, H., Rettig, J.S., Schröder, T., Wauer, H., Müller, C., Schütz, M., Mann, K., Siebert, G., Dettling, M., Müller, J.M., Kox, W.J., Spies, C.D., 2004. Gender differences in the performance of a computerized version of the alcohol use disorders identification test in subcritically injured patients who are admitted to the emergency department. Alcohol Clin. Exp. Res. 28, 1693–1701.

Neumann, T., Neuner, B., Weiss-Gerlach, E., Tønnesen, H., Gentilello, L.M., Wernecke, K.D., Schmidt, K., Schröder, T., Wauer, H., Heinz, A., Mann, K., Müller, J.M., Haas, N., Kox, W.J., Spies, C.D., 2006. The effect of computerized tailored brief advice on at-risk drinking in sub-critically injured trauma patients. J. Trauma 61, 805–814.

NIAAA: Helping Patients Who Drink Too Much. National Institute on Alcohol Abuse and Alcoholism, Bethesda, Maryland. Accessible at: http://pubs.niaaa.nih.gov/publications/Practitioner/CliniciansGuide2005/clinicians_guide.htm.

NIDA, 2008. Screening for Drug Use in General Medical Settings. National Institute on Drug Abuse, Bethesda, Maryland. http://www.nida.nih.gov/nidamed/resguide/resourceguide.pdf.

NIDAMED. Resources for Medical and Health Professionals. National Institute on Drug Abuse, Bethesda, Maryland. Available at: http://www.drugabuse.gov/nidamed.

NIH Publication, 2007. Helping Patients Who Drink Too Much: A Clinician's Guide. National Institute on Alcohol Abuse and Alcoholism. NIH Publication 07–3769, Rockville, MD Reprinted May, 2007.

Office of National Drug Control Policy. Seek Early Intervention Opportunities in Health Care. http://www.whitehouse.gov/ondcp/chapter-seek-early-intervention-opportunities-in-health-care/ (accessed 01.09.12.).

Paulozzi, L.J., March 12, 2008. Trends in Unintentional Drug Overdose Deaths. Testimony by before Senate Judiciary Subcommittee on Crime and Drugs. Accessible at: www.hhs.gov/asl/testify/2008/03/t20080312b.html.

Poikolainen, K., Paljarvi, T., Makela, P., 2007. Alcohol and the preventive paradox: serious harms and drinking patterns. Addiction 102, 571–578.

Saitz, R., Mulvey, K.P., Plough, A., Samet, J.H., 1997. Physician unawareness of serious substance abuse. Am. J. Drug Alcohol Abuse 23, 343–354.

SAMHSA-HRSA: Center for Integrated Health Solutions. SBIRT: Screening, Brief Intervention, and Referral to Treatment. http://www.integration.samhsa.gov/clinical-practice/sbirt/ (accessed 01.09.12.).

SAMHSA News, November/December 2009. Screening, brief intervention, and referral to treatment (SBIRT): new populations, effectiveness data. SAMHSA News 17 (6). (accessed 01.09.12.) http://store.samhsa.gov/product/SAMHSA-News-Screening-Brief-Intervention-and-Referral-to-Treatment-SBIRT-New-Populations-Effectiveness-Data/SAM09-176/.

Saunders, J.B., Aasland, O.F., Babor, T.F., de la Fuente, J.R., Grant, M., 1993. Development of the alcohol use disorders identification test (AUDIT): WHO collaborative project on early detection of persons with harmful alcohol consumption-II. Addiction 88, 791–804.

Schermer, C.R., Moyers, T.B., Miller, W.R., Bloomfield, L.A., 2006. Trauma center brief interventions for alcohol disorders decrease subsequent driving under the influence arrests. J. Trauma 60, 29–34.

Selzer, M.L., 1971. The Michigan alcoholism screening test (MAST): the quest for a new diagnostic instrument. Am. J. Psychiatry 127, 1653–1658.

Skinner, H.A., 1982. The drug abuse screening test. Addict. Behav. 7, 363–371.

Summary Health Statistics for U.S. Adults: National Health Interview Survey, 2008, table 37. Centers for Disease Control and Prevention. http://www.cdc.gov/nchs/fastats/docvisit.htm/.

Traffic Safety Facts, 2010. From the National Highway and Traffic Safety Administration, U.S. Department of Transportation, DOT HS 811 415 Nov 2010. http://www-nrd.nhtsa.dot.gov/Pubs/811415.pdf (accessed 26.01.12.).

Vaca, F.E., Winn, D., Anderson, C.L., Kim, D., Arcila, M., 2011. Six-month follow-up of computerized alcohol screening, brief intervention, and referral to treatment in the emergency department. Subst. Abus. 32, 144–152.

Wilk, A.I., Jensen, N.M., Havighurst, T.C., 1997. Meta-analysis of randomized control trials addressing brief interventions in heavy alcohol drinkers. J. Gen. Intern. Med. 12, 274–283.

CHAPTER FIVE

Cocaine: Mechanism and Effects in the Human Brain

Pierre Trifilieff[1] and Diana Martinez[2]
[1]School of Public Health, Columbia University, New York, NY, USA
[2]Department of Biological Sciences, Rutgers University, Newark, NJ, USA

1. INTRODUCTION

Cocaine is a tropane alkaloid derived from the coca plant (*Erythroxylum coca* and *Erythroxylum novogranatense*), which is indigenous to western South America. The coca bush has been cultivated for millennia, and 2000-year-old mummies from Nazca, Peru, have been found with small bags used for coca leaves in their possession (Stolberg, 2011). When the Spanish conquistador Francisco Pizarro encountered the Inca empire in the 1530s, coca leaves were a crucial component of the social structure and economy, and were used both for payment and religious tribute (Stolberg, 2011). Coca leaves were transported back to Europe, where there was intermittent interest in their psychoactive properties. However, it wasn't until the nineteenth century, with the discovery of methods to extract and concentrate the cocaine alkaloid, that cocaine use became more widespread in Europe (Stolberg, 2011).

Among the most successful in this venture was Angelo Mariani, who became quite wealthy with the creation of Vin Mariani, the combination of Bordeaux wine and cocaine extract. Pope Leo XIII, the nineteenth century pope known for his diplomacy and being the first pope to appear on film, was also devoted to Vin Mariani (and awarded Mariani a gold medal). Other advocates included Ulysses S Grant, Queen Victoria, and Alexandre Dumas. The use of cocaine powder also became prominent, and Sigmund Freund was among the more well known of Victorian-era cocaine users. In 1884, Freud published the paper "Uber Coca", in which he described the effects of cocaine, having experimented on himself extensively. These included euphoria, increased attention, and the ability to work long hours without fatigue (Freud, 1884). Based on his observations in others, Freud believed that the treatment of morphine dependence, depression, anxiety, and asthma were possible using cocaine, the only impediment being its cost. Freud missed the use of cocaine as an anesthetic, particularly for eye surgery, which was a major advancement since there was, at the time, no other suitable anesthesia for cataract removal. Instead, this discovery was made by Karl Koller, a colleague who worked at the same hospital as Freud, much to Freud's displeasure.

In the nineteenth century, cocaine was included in a number of pharmaceuticals and consumer products, such as Coca-Cola and cigarettes. But by the 1890s, there were

The Effects of Drug Abuse on the Human Nervous System
http://dx.doi.org/10.1016/B978-0-12-418679-8.00005-8

numerous medical reports on its toxic effects and on cocaine dependence (Gootenberg, 2006). In the United States, the Harrison Narcotics Tax Act of 1914 was passed to regulate the production and distribution of opium and cocaine, and was followed by a series of laws that made cocaine illegal. Today, the countries of the former Inca empire, Colombia, Peru and Bolivia, remain the major source of the world's cocaine, and current consumption is estimated to be over 800 metric tons (UNODC, 2009). The United States is the major consumer, although increasingly cocaine is being transported throughout the Europe Union (UNODC, 2009). For decades, the U.S. has engaged in an interdiction campaign to impede the flow of cocaine, and cocaine trafficking continues to contribute to instability in Latin America (RAND, 2011). The Colombian armed conflict was fought over cocaine distribution, and Pablo Escobar, leader of the largest cocaine trafficking organization, was known for ordering countless murders, including the bombing of an airline flight with 110 passengers on board. He was also listed as the 7th richest man in the world by Forbes magazine in 1989. Cocaine also plays a prominent role in the drug war in Mexico, which has a current estimated death toll of over 47,000 (RAND, 2011). In the most recent Mexican presidential election, the candidates endorsed a major shift in policy by reducing the role of the military in pursuing traffickers and impeding the flow of drugs, in an attempt to reduce the violence. Recently, there also appears to be a shift in the United States policy. The Office of National Drug Control Policy has stated that the current administration views drug addiction as a brain disorder that should be managed with treatment, a view supported by decades of neuroscience research.

2.1. COCAINE USE IN HUMANS

2.1.1. Overview

Cocaine hydrochloride, which is the powdered form of cocaine, is produced by converting coca paste obtained from the leaves, into the water-soluble salt form that is either snorted or injected into a vein. Cocaine base, or "crack" cocaine, is the base form of salt, which is usually made using sodium bicarbonate and can be smoked. Cocaine can also be taken orally, which is not common in developed countries, but remains widespread in the Andes. Injection, snorting, and the smoking of cocaine increase the rapidity with which cocaine enters the brain over the oral form, contributing to the euphoria and its reinforcing effects. Once ingested, cocaine is metabolized in two major inactive metabolites, benzoylecgonine (detected by most drug testing) and ecgonine methyl ester.

In the United States, approximately 2–3% of the population reports recent cocaine use, but the majority of cocaine users do not develop dependence. Current estimate of the lifetime cumulative probability of developing cocaine dependence among cocaine users is 20.9%. compared to 67.5% of nicotine users, 22.7% of alcohol users, and 8.9% of cannabis users (Lopez-Quintero et al., 2011). Among the risk factors for developing cocaine dependence are being young at first use, male, with injected or smoked cocaine use (Lopez-Quintero et al., 2011; O'Brien and Anthony, 2005). Only a small percentage of cocaine abusers are in treatment at a given time, and at present, no pharmacologic treatment is effective for cocaine dependence. Behavioral treatments for cocaine dependence are available, but most are of limited effect (Knapp et al., 2007). Numerous previous clinical trials have investigated many medications for the treatment of cocaine dependence, yet none have shown true success (Montoya and Vocci, 2008). One reason for this is that many of these medications that have been tested were originally developed for other disorders, and there is a need to develop pharmacologic treatments that target the alterations in neurochemistry occur in the human brain specifically following exposure to cocaine.

2.1.2. Cocaine, Dopamine, and the Brain

Like most other drugs of abuse, cocaine acts on the brain by changing the signaling of the endogenous neurotransmitters. The major neurotransmitter systems of the brain, gamma-aminobutyric acid (GABA) and glutamate, are responsible for signaling perception, cognition, and motor output that control behavior. But the GABA and glutamate systems are modulated by other neurotransmitters, including the opioid, serotonin, and dopamine systems. Among these, the dopamine system is the most closely associated with modulating motivated, reward-driven behavior and it is also the most directly associated with addiction.

The dopamine system of the brain originates in the midbrain, and the cell bodies of the dopamine neurons are located in the substantia nigra and ventral tegmental area. The dopamine projections consist of four pathways: (1) mesolimbic, which projects from the ventral tegmental area to the nucleus accumbens of the striatum, plays a crucial role reward-driven behavior and addiction; (2) mesostriatal, consisting of projections from the substantia nigra to the dorsal striatum (caudate and putamen), which largely plays a role in modulating cognitive processes and motor output; (3) mesocortical, which projects from the ventral tegmental area to the prefrontal cortex, directs behavior toward abstract goals (Cools, 2008); and (4) tubero-infundibular, which regulates the hypothalamus-pituitary system. While the function of these projection systems can be segregated based on the brain regions they project to, it's important to recognize that there is overlap across these pathways. In addition, although the limbic pathway is crucial for reward-driven behavior, both preclinical and human studies show that the dopamine systems involved in cognitive processes also show involvement in addiction.

2.1.3. Cocaine at the Synapse

Cocaine has a very direct effect on dopamine signaling. At the dopamine nerve terminal of the striatum, cocaine binds the dopamine transporter (DAT) and blocks the re-uptake of dopamine by the presynaptic dopaminergic neurons. Since re-uptake via the DAT is the major mechanism for terminating the dopamine signal in striatum, cocaine administration results in a significant increase in extracellular dopamine, several orders of magnitude over baseline levels, based on microdialysis studies (Bradberry, 2000; Bradberry and

Roth, 1989; Di Chiara and Imperato, 1988a; Sorg et al., 1997). The excess in extracellular dopamine is dependent on the firing of the neuron, since dopamine is released from the presynaptic dopamine neuron as it fires, but then is not reabsorbed by the DAT.

Cocaine actually blocks the uptake of the monoamine neurotransmitters, not just dopamine, and also increases the concentrations of serotonin and norepinephrine (Carroll et al., 1992; Lewin et al., 1992). In fact, cocaine has a similar affinity for each of these monoamine transporters (for review see Jayanthi and Ramamoorthy, 2005). However, it is the effect on extracellular dopamine that has been shown to produce most of the behavioral effects that are associated with addiction. For example, it is the increase in dopamine, rather than cocaine's effect on serotonin or norepinephrine, that is most directly associated with the reinforcing effects of cocaine. A crucial study of cocaine analogues in monkeys demonstrated that the potency of DAT inhibition (rather than inhibition of the norepinephrine or serotonin transporters) correlated with their ability to support cocaine self-administration (Ritz et al., 1987). Similarly, inhibition of the norepinephrine transporter (NET) and serotonin reuptake transporter (SERT) has little effect on the behavioral effects of cocaine in nonhuman primates, while drugs that block the DAT also inhibit cocaine's effects (Spealman et al., 1989; Tella, 1995). In addition, cocaine self-administration is dependent on the mesolimbic dopamine pathway being intact, which is not the case for noradrenergic or serotonergic neurons (Kuhar et al., 1991).

2.1.4. The Role of Dopamine

The recognition that dopamine plays a crucial role in addiction has been established for decades, but the actual role of dopamine in mediating the reinforcing effects of drugs of abuse has been an evolving concept. Dopamine does not appear to simply signal "reward" in the setting of a drug or food reward, although dopamine neurons fire in response to, or in the expectation of, a reward. Instead, dopamine appears to mediate the reinforcing effects of natural rewards (and drugs of abuse), in that the dopamine signaling makes the behavior required to obtain the reward more likely to be repeated. Dopamine neurons fire in response to a reward, which results in higher levels of dopamine being released in the nucleus accumbens, and increases the likelihood that the organism will repeat the behavior in an attempt to receive more of the reward.

The exact mechanism by which dopamine does this has been the subject of a long debate. Salamone et al. (Salamone, 2009) showed that dopamine transmission in the nucleus accumbens influences effort-related behavior and that dopamine depletion results in an animal being less likely to exert effort to receive a food reward. Thus, dopamine can be understood as mediating the behavioral economics of motivated behavior and regulating the reinforcement value of a reward (i.e. the extent to which a given reward is worth the effort required). Dopamine can also be viewed as mediating the "incentive salience" of a reward, meaning the extent to which the reward is wanted by the animal (Berridge and Robinson, 1998).

In a series of electrophysiological studies in nonhuman primates, Schultz et al. (Schultz, 2010) have shown that dopamine serves as a reward prediction error, and codes for reward as it differs from prediction, thus plays a crucial role in reward-based learning. While the debate over the mechanism by which dopamine mediates the reinforcing effects of a reward, whether it is by signaling the willingness to work for a reward, the degree to which a reward is wanted, or reward-related learning, the fact that drugs of abuse, such as cocaine, are associated with dysregulation of dopamine transmission that affects decision making is a crucial starting point for understanding the neurobiology of this disorder.

2.2. IMAGING COCAINE ABUSE IN THE HUMAN BRAIN

2.2.1. Positron Emission Tomography Brain Imaging Methods

Imaging the neurochemistry associated with cocaine abuse in the human brain can be performed with Positron Emission Tomography (PET) or Single Photon Emission Tomography (SPECT). These imaging modalities use radionuclide-labeled molecules, usually an agonist or antagonist specific for a receptor, and detect the signal emitted by the radiotracer. The radiotracers most frequently used in substance abuse research label the dopamine type 2/3 (D2/3) receptors of the striatum, and include the D2/3 antagonists $[^{18}F]N$-methylspiroperidol (labeled with a positron emitting fluorine) and $[^{11}C]$raclopride (labeled with a positron-emitting carbon). In addition to labeling the D2/3 receptor, other radiotracers are available to label other molecular targets, such as dopamine D1 receptors, the monoamine transporters, serotonin receptors, GABA, and opioid receptors.

In the brain, the radiotracer binds to the brain receptor (referred to as specific binding) in addition to other nonreceptor proteins in the brain (nonspecific binding). For example, when imaging the D2/3 receptor, the radiotracer accumulates in the striatum and the measured signal consists of specific binding to the D2/3 receptor and nonspecific binding to other proteins (Figure 1). Nonspecific binding can be corrected by measuring the binding of the radiotracer to a brain region known to have negligible receptor levels (often the cerebellum when imaging the dopamine receptor). In most human studies of addiction the outcome measure is referred to as the binding potential relative to nonspecific binding (BPND), which is the ratio of specific binding to nonspecific binding (Innis et al., 2007).

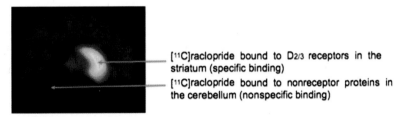

$[^{11}C]$raclopride bound to D2/3 receptors in the striatum (specific binding)

$[^{11}C]$raclopride bound to nonreceptor proteins in the cerebellum (nonspecific binding)

Figure 1 Sagittal View of a $[^{11}C]$raclopride PET Scan of the D2/3 Receptors Human Brain. The striatum shows the highest uptake and represents specific binding of the radiotracer to the receptor. The cerebellum is largely without D2/3 receptors, and is used to estimate nonspecific binding.

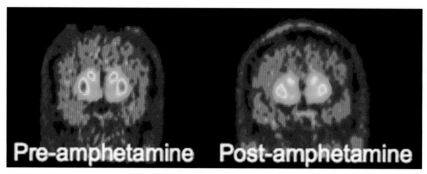

Figure 2 PET Scans with [^{11}C]raclopride in a Healthy Control, Obtained before (left) and after (right) the Administration of Amphetamine (0.3 mg/kg iv). The scan obtained after amphetamine shows a reduction in radiotracer binding, which correlates with increased endogenous dopamine, providing an indirect measure of presynaptic dopamine release.

In addition to measuring the D2/3 receptor in the baseline (resting) condition, the PET radiotracer [^{11}C]raclopride can also be used to measure changes in extracellular dopamine in the striatum. This occurs because [^{11}C]raclopride binding is sensitive to changes in the level of endogenous dopamine. The administration of a drug that increases dopamine, such as cocaine, amphetamine, or methylphenidate, results in a decreased [^{11}C]raclopride binding (Figure 2). In other words, the administration of a stimulant increases endogenous dopamine, which binds to the D2/3 receptors, resulting in fewer receptors available to bind to [^{11}C]raclopride, and a reduction in the measured PET signal. Thus, a comparison of baseline BPND (i.e. baseline, or prestimulant administration) and BPND following the stimulant provides an indirect measure of the increase in striatal dopamine. This comparison is the percent change in BPND, or ΔBPND, which is defined as (BPNDbaseline − BPNDchallenge)/BPNDbaseline.

In nonhuman primates, studies using imaging combined with microdialysis have shown that there is a linear correlation between the percent change in BPND (ΔBPND) and the percent change in extracellular dopamine due to stimulant administration (Breier et al., 1997; Laruelle et al., 1997). However, there is a significant loss of sensitivity: each 1% decrease in [^{11}C]raclopride binding corresponds to a 40–50% increase in extracellular dopamine measured with microdialysis (Breier et al., 1997; Laruelle et al., 1997). Thus, in healthy controls the administration of amphetamine (0.3 mg/kg iv) produces a decrease in [^{11}C]raclopride binding of the order of 10–20% (Drevets et al., 2001; Martinez et al., 2003; Munro et al., 2006) despite the fact that animal studies show that similar doses of amphetamine produces a several hundred-fold increase in extrasynaptic dopamine (Breier et al., 1997; Laruelle et al., 1997). The exact mechanism behind the decrease in [^{11}C]raclopride binding in response to a stimulant challenge is not known, and neither is the mechanism behind the loss in sensitivity. While competition between extracellular dopamine and the radiotracer for the receptor has been used as the model to explain the decrease in [^{11}C]raclopride binding, more recent data show that other phenomena, such as internalization or oligomerization of the receptors, have also been implicated (Laruelle, 2000; Logan et al., 2001; Skinbjerg et al., 2010).

Using these techniques, PET studies in cocaine abusers have been performed to characterize the changes in dopamine signaling that occur in cocaine abuse/dependence. Overall, these imaging studies show that cocaine abuse, following years of exposure, is associated with a loss in the responsiveness of the dopamine system compared to controls. In some ways, this is counterintuitive, since stimulant administration is more reinforcing in stimulant abusers compared to nonabusing controls, and reinforced behavior is usually associated with dopamine transmission in the striatum. However, as discussed below, blunted presynaptic dopamine release in cocaine abusers has significant behavioral consequences that contribute to addiction and relapse.

2.2.2. Imaging Dopamine in Cocaine Abuse

In 1983, the first PET scan of D2/3 receptors was performed in human subjects (Wagner et al., 1983) and in 1990, the first PET imaging study of dopamine D2/3 receptor binding in cocaine abusers was performed (Volkow et al., 1990). The results showed that cocaine abuse was associated with a significant reduction in D2/3 receptor binding compared to matched controls. Subsequent studies, performed with the newer radiotracer [^{11}C]raclopride, have shown similar results, with decreases of 10–15% in cocaine abusers compared to healthy controls (Malison et al., 1999; Martinez et al., 2004; Volkow et al., 1993, 1997). Thus, the decrease in D2/3 receptor binding seen with PET is one of the most reproducible findings in human addiction research. In addition, it appears that the decrease in D2/3 receptor binding is long lasting. A PET imaging study in rhesus monkeys showed that cocaine exposure resulted in a decrease in D2/3 receptor binding that persisted for up to a year (Nader et al., 2006). In humans, only one study has been performed scanning cocaine abusers after prolonged abstinence, and this study showed that D2/3 receptor binding remains reduced after three months of inpatient rehabilitation (Volkow et al., 1993).

Given that [^{11}C]raclopride binding is sensitive to endogenous dopamine, there is a possibility that [^{11}C]raclopride BPND is lower in cocaine abusers compared to controls due to increases in endogenous dopamine in the cocaine addicts. In other words, if cocaine abusers have higher levels of endogenous dopamine in the resting condition when the scan is obtained, compared to controls, this would be measured as lower D2 receptor binding, when there may be no actual difference in actual D2 receptor levels. However, a previous study showed that cocaine dependence is actually associated with lower levels of endogenous dopamine in the baseline condition compared to control subjects, measured as the change in [^{11}C]raclopride binding following depletion of endogenous dopamine (Martinez et al., 2009a). Thus, this study confirms that cocaine abuse is associated with decrease in striatal D2/3 receptor binding that cannot be attributed to higher levels of endogenous dopamine.

Although cocaine acts at the DAT, PET imaging studies in cocaine abuse have not shown a consistent effect on DAT binding. A previous SPECT reported a 20% increase in cocaine abusers who had been abstinent for 96 h (Malison et al., 1998). However, subsequent PET studies imaging the DAT showed no difference between healthy controls and cocaine abusers who had been abstinent either more than 5 days (Volkow et al., 1996; Wang et al., 1997). Together, these studies suggest that the DAT may be elevated only in very early abstinence.

In addition to the D2/3 receptor, PET imaging can be used to image the D1 receptor. Only one PET imaging study has measured D1 receptor binding in cocaine abuse (Martinez et al., 2009b), and this study showed no difference in D1 receptor BPND in cocaine abusers compared to controls. This finding is consistent with a postmortem study showing that striatal D1 receptor mRNA was unchanged in cocaine abusers compared to matched controls (Meador-Woodruff et al., 1993).

2.2.3. Imaging Presynaptic Dopamine Release in Cocaine Abuse

PET studies have also been performed to measure presynaptic dopamine release, using a stimulant challenge, in cocaine abuse. Four studies have studied this, and all four show that cocaine abuse is associated with less [^{11}C]raclopride displacement following a stimulant challenge compared to healthy controls, suggesting that cocaine abuse is associated with a reduction in presynaptic dopamine release (Malison et al., 1999; Martinez et al., 2007, 2011; Volkow et al., 1997), as shown in Figure 3. In addition, PET imaging studies using other radiotracers have also shown that cocaine abuse is associated with a reduction in presynaptic dopamine. Using the radiotracer 6-[^{18}F]-fluoro-L-DOPA [^{18}F] DOPA, which provides a measure of presynaptic nerve terminal function, a previous study showed that cocaine abusers had lower [^{18}F]DOPA uptake compared to healthy controls (Wu et al., 1997). More recently, Narendran et al. (Narendran et al., 2012) performed an imaging study of the vesicular monoamine transporter 2 (VMAT2), and showed that cocaine abuse is associated with decrease in VMAT2 binding compared to matched controls, which is consistent with previous postmortem studies (Little et al., 1999, 2003; Wilson et al., 1996). Since striatal VMAT2 binding provides a measure of presynaptic dopaminergic vesicles, these data also show that cocaine abuse is associated with reduced stores of presynaptic dopamine. Taken together, these imaging studies show that cocaine abuse is associated with a reduction in presynaptic dopamine, measured as dopamine synthesis and storage capacity and as the response to a stimulant challenge.

2.2.4. Sensitization and the Human Brain

Previous preclinical studies have explored the effect of chronic cocaine exposure on dopamine signaling in the striatum, and have shown opposite findings. A number of rodent studies show dopaminergic sensitization, which is the process by which repeated cocaine exposure produces an excess exaggerated dopamine response compared to the initial

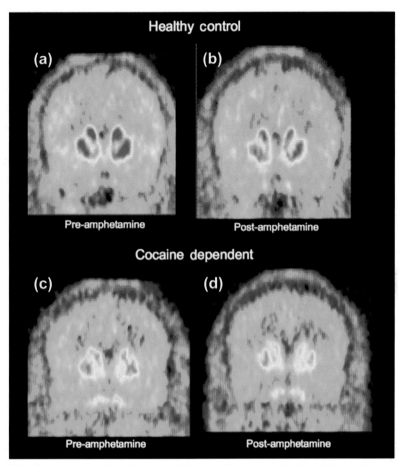

Figure 3 [¹¹C]raclopride Scans in a Control and Cocaine-Dependent Subject. Prior to the amphetamine administration, there is higher D2/3 receptor binding in the healthy control compared to the cocaine-dependent subject. With respect to amphetamine, there is a decrease in [¹¹C]raclopride binding in the control, and little change in the cocaine-dependent subject, the percent decrease in [¹¹C]raclopride binding in the control subjects is—9.5% compared to—3.0% in cocaine abusers *(From Martinez et al., 2007)*. This is consistent with higher dopamine release in controls and blunted dopamine in cocaine dependence.

response (Kalivas and Stewart, 1991; Robinson et al., 1988; Robinson, 1993; Vezina, 2004). To elicit sensitization of dopamine, an animal is exposed repeatedly to cocaine, which elicits dopamine in the nucleus accumbens (Di Chiara and Imperato, 1988a). When this same animal then undergoes a period of abstinence and subsequent reexposure to the drug, this results in dopamine release of a greater magnitude than that seen with initial drug exposure.

This exaggerated dopaminergic release is referred to as dopaminergic sensitization response (Kalivas and Stewart, 1991; Robinson et al., 1988; Robinson, 1993; Vezina, 2004). In addition to sensitization of the dopamine system, behavioral sensitization has been shown, in which repeated cocaine exposure results in exaggerated locomotor behaviors (Kalivas and Stewart, 1991). The opposite finding has also been reported in rodent studies, where repeated cocaine exposure results in blunted dopamine release in the striatum (for review see Melis et al., 2005). In this setting, a hypodopaminergic state is induced, where repeated cocaine exposure results in both marked decrease in the ability of cocaine to inhibit dopamine uptake and reduced basal levels of extracellular dopamine (Jones et al., 1996; Maisonneuve et al., 1995; Mateo et al., 2005; Zhang et al., 2001).

The human imaging studies in cocaine abusers are in agreement with studies showing a hypodopaminergic state, given that dopamine release, measured as radiotracer displacement, is blunted compared to matched control subjects (Malison et al., 1999; Martinez et al., 2004; Volkow et al., 1993, 1997). Studies in nonhuman primates have also not shown that chronic cocaine exposure is associated with a sensitized dopamine response (Bradberry, 2000; Castner et al., 2000). The results of these PET studies are consistent with behavioral studies in human cocaine abusers, which have also failed to show evidence for sensitization, measured with vital signs, motor activity, subjective effects, hormone levels, pupil diameter, and EEG, following a cocaine challenge (Gorelick and Rothman, 1997; Rothman et al., 1994; Strakowski et al., 1996).

However, a recent imaging study did demonstrate the phenomena of sensitization in healthy controls without a history of stimulant abuse or dependence (Boileau et al., 2006). In this study, 10 healthy men underwent PET scans with [^{11}C]raclopride before and after the intermittent, but repeated, administration of oral amphetamine (0.3 mg/kg, on five occasions with 14 days of abstinence). The first dose of amphetamine (the presensitized condition) resulted in an 18% decrease in [^{11}C]raclopride BPND, whereas the last dose of amphetamine (sensitized condition) produced a 28% decrease in the ventral, but not dorsal, striatum (Boileau et al., 2006, 2008). These results are consistent with the rodent studies, which also show that sensitization is limited to the nucleus accumbens. But, in terms of the human brain, these studies suggest that sensitization may occur early on following stimulant exposure, but that once addiction has developed, the opposite phenomena is seen, where the dopamine pathways are hyporesponsive.

2.3. IMAGING COCAINE: BEHAVIORAL CORRELATES

2.3.1. Overview

As described above, imaging studies in cocaine abuse consistently show a reduction in both dopamine D2/3 receptor binding and presynaptic dopamine release in response to a stimulant challenge, compared to healthy controls. However, it is crucial to also understand the behavior that is associated with these alterations in dopamine transmission.

2.3.2. 2D2/3 Receptor BPND and Behavior

A question that arises is whether the reduction in D2/3 receptor BPND seen in cocaine abuse results from cocaine exposure or whether this serves as a risk factor for dependence.

Preclinical studies have investigated this question by comparing D2/3 binding in animals that differ in characteristics that are often associated with the development of addiction. In rhesus monkeys, lower striatal D2/3 BPND is associated with higher levels of stress and greater levels of cocaine self-administration, compared to animals with higher D2 receptor binding (Morgan et al., 2002). Similarly, studies in rodents show that lower D2/3 receptor binding in the striatum is associated with impulsive behavior and predictive of a greater susceptibility to the reinforcing effects of cocaine (Dalley et al., 2007; Michaelides et al., 2012). Importantly, D2/3 receptor binding was measured prior to any drug exposure, yet the animals had a higher propensity to self-administer cocaine, suggesting that low D2/3 receptor BPND could serve as a risk factor for cocaine abuse.

There is a less direct evidence of this phenomenon in humans. In healthy control, nonaddicted individuals, low striatal D2/3 receptor BPND is predictive of a pleasurable response to intravenous methylphenidate, while high D2/3 BPND was associated with an unpleasant experience (Volkow et al., 1999b; Volkow et al., 2002). To the extent that a pleasurable experience with a drug is indicative of a risk for addiction, these results suggest that high D2/3 receptor BPND may be protective. Another imaging study showed that healthy controls with a family history of cocaine abuse had higher D2 receptor BPND compared to those without this risk factor, again suggesting that this may serve as a marker for resilience (Volkow et al., 2009).

Finally, although there are data suggesting that low D2 BPND may serve as a risk factor for the development of addiction, other studies show that cocaine exposure itself is associated with decreased D2/3 receptor BPND. An imaging study in monkeys, scanned before and after cocaine self-administration, showed that chronic cocaine exposure itself produces a decrease in D2/3 receptor binding (Nader et al., 2006). This is in agreement with previous postmortem studies in nonhuman primates showing that D2/3 receptor density is unaffected by a short-term cocaine administration but reduced after prolonged exposure (Moore et al., 1998; Nader et al., 2002). In addition, the study of Nader et al. (Nader et al., 2006) showed that a long-term abstinence from cocaine is associated with a recovery of D2/3 receptors to control levels, suggesting that low D2/3 receptor BPND is a consequence of cocaine exposure (Maggos et al., 1998; Nader et al., 2006). In summary, there is evidence to support both the theory that low D2/3 receptor BPND results from extended cocaine exposure, and that low D2/3 receptor binding could serve as a marker for a predisposition to addiction.

2.3.3. Cocaine and the Prefrontal Cortex

Imaging studies have also been performed in cocaine abuse to characterize the alterations in the prefrontal cortex that occur in cocaine abuse. The majority of these studies have used PET imaging with [18F]fluorodeoxy-glucose, which measures regional brain metabolic activity, and functional magnetic resonance imaging, which measures changes in blood flow via blood oxygen level-dependent signals. Overall, the PET studies show that cocaine abuse is associated with decreases in glucose metabolism in the prefrontal cortex, and that these decreases correlate with the decreases in striatal D2/3 receptor BPND. Volkow et al. (Volkow et al., 1993) showed that decreases in D2 binding were associated with decreased metabolism in orbitofrontal cortex and cingulate gyrus, leading the authors to propose that dysregulation of dopamine is involved in channeling of drive and affect, and leads to compulsive drug-taking behavior.

In a subsequent study, this group showed that D2 receptor binding correlated with changes in glucose metabolism induced by methylphenidate administration: subjects with higher D2 BPND increased metabolism, and those with the lower D2 BPND decreased metabolism in the cingulate and orbitofrontal cortex (Volkow et al., 1999a). Although these studies did not measure behavior per se, these brain regions of the frontal cortex have been implicated in salience attribution (orbitofrontal cortex) and inhibitory control (cingulate gyrus) (Volkow et al., 2004), which are involved in drug-seeking behavior.

Functional MR imaging studies in cocaine abuse have shown alterations in prefrontal cortical blood flow in response to several psychological tasks (for a detailed review, see Goldstein and Volkow, 2011). Tasks that probe inhibitory control, such as the go/no-go task and the stop-signal reaction-time task show that cocaine abuse is associated with more task errors and an inhibitory behavioral deficit compared to control that was attributed to hypoactivation of the anterior cingulate (Hester and Garavan, 2004; Kaufman et al., 2003). Inhibitory control can also be assessed using the Stroop task, where more errors and slower performance are associated with dysfunction of the prefrontal cortex, particularly the anterior cingulate and dorsolateral frontal cortex (Leung et al., 2000; Pardo et al., 1990).

Functional MR imaging studies of cocaine abusers using this task have also shown lower activation of these regions of the prefrontal cortex (Bolla et al., 2004) and a recent study showed that activation of the ventromedial prefrontal cortex with this task predicted treatment outcome (Brewer et al., 2008). Lastly, Goldstein et al. (Goldstein et al., 2009) showed hypoactivation of the anterior cingulate in response to a rewarded drug cue-reactivity task in cocaine abusers compared to healthy controls, suggesting that this brain region is hypoactive in response to reward-related tasks as well. Thus, taken together, these imaging studies in cocaine abuse show that there are alterations in information processing in the prefrontal cortex that regulate cognitive function and emotion regulation, which play a key role in the development and maintenance of addiction.

2.3.4. Presynaptic Dopamine and Cocaine-Seeking Behavior

Imaging studies in humans can also be used to correlate behavior with dopamine transmission in the striatum. Using a laboratory model of cocaine self-administration in human volunteers, a previous PET imaging study showed that blunted dopamine release in the ventral striatum correlated with the choice to self-administer cocaine (Martinez et al., 2007). In this study, cocaine abusers were scanned with [^{11}C]raclopride and an amphetamine challenge, in order to obtain both D2/3 BPND and delta BPND. The imaging results showed that both of these parameters of dopamine transmission were blunted in cocaine abusers compared to controls, across the subdivisions of the striatum. Following the PET scans, the cocaine abusers underwent self-administration sessions, where they chose between a dose of smoked cocaine vs money, a competing reinforcer. The results showed that the cocaine-abusing subjects with the greatest deficit in dopamine release were more likely to choose the cocaine over money, while those with higher dopamine release chose money. Thus, although dopamine release was blunted in both the ventral and dorsal striatum in the cocaine abusers, the correlation with cocaine self-administration was seen in the ventral striatum only.

More recently, a similar study was performed imaging these same parameters of dopamine transmission in cocaine abusers who were enrolled in a motivation-based treatment that uses positive reinforcement incorporated into behavioral therapy (Martinez et al., 2011). In this study, cocaine abusers were scanned prior to enrollment in the treatment program, using contingency management combined with the community reinforcement approach (Higgins et al., 1994, 2003). This treatment uses monetary vouchers to induce abstinence from cocaine, which is similar to the choice presented in the cocaine self-administration study (Martinez et al., 2007). The results of this study showed that both D2/3 receptor BPND and ΔBPND were lower in the ventral cocaine abusers who failed to respond to treatment compared to those who responded to treatment (Figure 4, and Table 3). Similar results have also been reported in methamphetamine abusers, where both D2/3 receptor BPND and ΔBPND were higher in subjects who successfully completed detoxification compared to those who relapsed (Wang et al., 2011).

Thus, these imaging studies show that low dopamine release in the ventral striatum is associated with the choice to consume cocaine over alternative reinforcers, both in the laboratory and the treatment clinic. In the laboratory study, subjects were given the choice between a dose of cocaine and money, and the choices were weighted toward the money (the amount of money was slightly lower than the street values of the dose of cocaine). In the treatment study, cocaine-dependent subjects could earn money for pursuing their goal of abstaining from cocaine. Thus, in both cohorts, the more adaptive response is to choose money over cocaine. But, in both studies there were a number of subjects who still chose cocaine. The failure of the cocaine-dependent subjects with low dopamine release to alter their behavior can be viewed as a blunted brain reward system that is unable to respond to alternative sources of reward.

2.4. DOPAMINE TRANSMISSION IN STRIATAL SUBDIVISIONS

2.4.1. Overview

The first studies measuring dopamine release in cocaine abusers measured the striatum as a whole, due to the resolution of PET scanners available at the time (Malison et al., 1999; Volkow et al., 1997). Subsequent studies performed on higher resolution scanners are able to separate the signal from the ventral and dorsal striatum (Martinez et al., 2003).

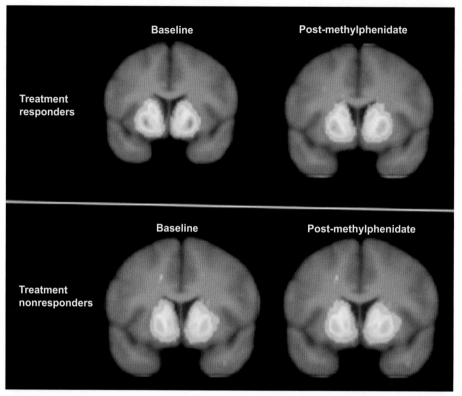

Figure 4 [^{11}C]raclopride D2/3 Receptor Binding (BPND) in the Treatment Responders (top) and Non-responders (bottom). The scans shown are before (left) and after (right) methylphenidate administration (60 mg PO). Treatment responders had higher values D2/3 receptor binding and higher dopamine release compared to nonresponders. *(From Martinez et al., 2011).*

The striatum is a heterogeneous structure that has several anatomical and functional subregions. Corticostriatal-thalamocortical circuits provide a framework for correlating the anatomic subregions of the striatum with their cognitive and/or behavioral functions. These circuits involve projections from the cortex to the striatum, then to globus pallidus/substantia nigra/subthalamic nucleus, to the thalamus, and back to the cortex. The circuits have been functionally classified into limbic, associative, and sensorimotor networks that are largely parallel but have some overlap, which may allow for information in one circuit to be modulated by another (Alexander et al., 1986; Ferry et al., 2000; Hoover and Strick, 1993; Joel and Weiner, 2000; Parent and Hazrati, 1995). Thus, the limbic, associative, and sensorimotor corticostriatal-thalamocortical provide a framework that directs information from the cerebral cortex to subcortical structures, and then back again to relevant regions of the cortex (Leckman et al., 2010).

Human PET imaging studies using a high resolution scanner have developed methods to measure D2/3 receptors and presynaptic dopamine release (with [^{11}C]raclopride

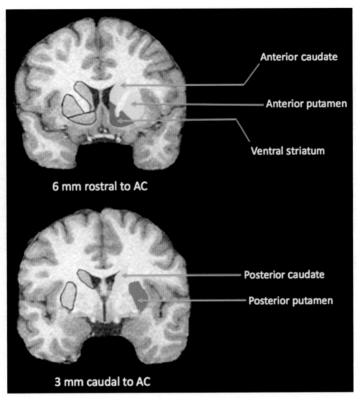

Figure 5 Subdivisions of the Striatum, Based on the Limbic, Associative, and Sensorimotor Circuits. The subdivisions are defined on a subject's MRI scan using anatomic criteria. The MRI slices shown here are rostral and caudal to AC. The limbic striatum is shown in red, the associative is shown in yellow, and the sensorimotor striatal subdivision is depicted in blue.

and a stimulant challenge) in the subregions of the striatum contained within each of the corticostriatal-thalamocortical circuits. Using these circuits as a guide to subdivide the striatum, the following subregions were developed:

1. The ventral, limbic striatal subregion represents the affective circuit. This region receives dopaminergic input from the ventral tegmental area and it is the projection region for the mesolimbic dopamine pathway. It also receives glutamatergic input from the amygdala/hippocampus. In humans and nonhuman primates, the ventral striatum includes the nucleus accumbens, ventral caudate and ventral putamen (shown in red in Figure 5).

2. The associative striatum is more closely involved in cognition, and contains the subregions of the striatum that receive input from the association regions of the dorsolateral and medial prefrontal cortex. The dopamine input comes from the substantia nigra and the mesostriatal pathway. The associative striatum includes the anterior putamen (the putamen rostral to the anterior commissure (AC)) and the entire caudate (anterior and posterior caudate, yellow in Figure 5).

3. The sensorimotor subregion of the striatum receives mesostriatal dopamine input from the substantia nigra and glutamate input from the premotor cortex and supplementary motor area. This subdivision consists of the putamen caudal to AC (shown in blue in Figure 5).

PET imaging studies in healthy controls show that dopamine signaling in response to a stimulant challenge is not uniform within the striatum but varies across subdivisions. The administration of a stimulant to measure presynaptic dopamine shows that [^{11}C] raclopride displacement is higher in the limbic and sensorimotor striatum compared to the associative striatum (Martinez et al., 2003), as shown in Table 1. Thus, in response to the same challenge (stimulant administration) the striatal subregions contained within the limbic and sensorimotor circuits show higher displacement of [^{11}C]raclopride compared to the associative striatum (Clatworthy et al., 2009; Martinez et al., 2003; Riccardi et al., 2006; Slifstein et al., 2010). In theory, this would seem to result from greater dopamine release in ventral and sensorimotor subregions.

In accordance, previous microdialysis studies in rodents and nonhuman primates have shown that the administration of a stimulant results in higher dopamine release in the ventral striatum compared to the dorsal striatum (Bradberry, 2000; Di Chiara and Imperato, 1988a; Sharp et al., 1987). This differential dopamine response has not been studied for the sensorimotor (postcommissural) putamen. However, neuroanatomical studies of the sensorimotor putamen show that this brain region shares histochemical features with the ventral striatum, such as similar tyrosine hydroxylase immunoreactivity (Fudge and Haber, 2002). In addition, the ventral portion of the postcommissural putamen has similar glutamatergic input (from the amygdala and limbic regions) as the ventral striatum (Fudge et al., 2004; Fudge and Haber, 2002). Thus, these data show that the postcommissural, sensorimotor striatum actually contains a limbic component, which may explain why dopamine signaling is similar between the ventral striatum and the sensorimotor putamen.

Table 1 Comparison of Amphetamine-Induced Dopamine Release in the Functional Subdivisions of the Striatum. Notably, the Limbic (Ventral) Striatum and Sensorimotor Striatum (Posterior Putamen) Displace More [^{11}C]raclopride Following Amphetamine (0.3 mg/kg) Compared to Regions of the Associative Striatum. These Findings Show that There Is a Differential Control of Dopamine Release across the Functional Subdivisions, and Suggests that the Limbic and Sensorimotor Striatum Share Commonalities in This Respect

Subdivision	Brain Region	Difference (%)
Limbic	Ventral striatum	-15.3 ± 11.8
Associative		-8.1 ± 7.2
	Anterior caudate	-6.1 ± 7.6
	Anterior putamen	-10.2 ± 7.9
	Posterior caudate	-7.6 ± 11.0
Sensorimotor	Posterior putamen	-16.1 ± 9.6
Whole striatum		-10.3 ± 7.2

Source: From Martinez et al., 2003

In addition to the differential dopamine response, studies in control subjects have also shown that dopamine signaling in the striatal subregions correlates with different behavioral measures. Stimulant-induced [^{11}C]raclopride displacement in the ventral striatum correlates with measures of "high" following the administration of intravenous amphetamine or methylphenidate (Clatworthy et al., 2009; Drevets et al., 1999; Martinez et al., 2003; Volkow et al., 1999c). Clatworthy et al. (Clatworthy et al., 2009) showed that performance on a cognitive task that probes the prefrontal cortex correlated with dopamine release in the associative striatum.

Similarly, Strafella et al. (Strafella et al., 2003) reported a decrease in [^{11}C]raclopride binding in the left dorsal caudate following repetitive transcranial magnetic stimulation of the left dorsolateral prefrontal cortex, showing that changing neuronal activity in one part of the associative corticostriatal-thalamocortical circuit affects dopamine signaling in the associative striatum. These observations point out that the changes in dopaminergic transmission within the different subdivisions of the striatum can support various cognitive functions, an idea largely supported by animal studies (see below). Therefore, with the improvement of imaging resolution, one challenge of brain imaging in addiction has been to study dopaminergic transmission in various striatal subdivisions.

2.4.2. Signaling in the Ventral and Dorsal Striatum and Cocaine

As described above, imaging studies using [^{11}C]raclopride and a stimulant challenge to image dopamine release have shown that cocaine abuse is associated with blunted dopamine release across the ventral, associative, and sensorimotor striatum (Martinez et al., 2007, 2011) (also see Table 2). Similarly, a PET imaging study measuring presynaptic dopamine vesicles (VMAT2) also showed that BPND was also reduced across each of the striatal subregions in cocaine abuse (Narendran et al., 2012). Thus, blunted dopamine transmission appears to involve each of the functional subdivisions of the striatum. However, two studies in cocaine abusers suggest that alteration of dopaminergic transmission in addiction might implicate specific striatal subdivisions. Wong et al. (Wong et al., 2006) showed that [^{11}C]raclopride displacement occurred in the dorsal putamen of participants who reported cue-elicited cocaine craving compared to those who did not. Similarly, Volkow et al. (Volkow et al., 2006) reported that a cocaine-associated cue displaced [^{11}C]raclopride binding in the dorsal caudate and putamen, but not the ventral striatum, compared to a neutral cue. Taken together, these studies suggest that cue-induced cocaine craving is associated with dopamine transmission in the dorsal striatum, which has been implicated in habit learning and in action initiation (Volkow et al., 2006).

These imaging studies in addiction show that these parameters of dopamine transmission can be correlated with behavior, and that suggest that neurochemistry can predict clinically relevant outcomes. Much of our understanding of the brain circuits associated with these processes comes from preclinical animal studies. These studies support the idea that addiction can be viewed as the result of an alteration of the reward system by

Table 2 Comparison of [¹¹C]raclopride Binding (BPND) and Amphetamine (0.3 mg/kg) Induced [¹¹C]raclopride Displacement (ΔBPND) in Cocaine Abusers vs Healthy Control Subject. Cocaine Abuse Is Associated with a Decrease in Dopamine D2/3 Receptors and Presynaptic Dopamine across All the Subdivisions

Subdivision	Brain Region	HC BPND	CD BPND	P*
Limbic	Ventral striatum	2.06 ± 0.30	1.79 ± 0.26	0.002
Associative	Anterior caudate	2.38 ± 0.28	2.12 ± 0.20	0.03
	Anterior putamen	2.95 ± 0.33	2.55 ± 0.24	0.001
	Posterior caudate	1.73 ± 0.31	1.58 ± 0.25	0.13
Sensorimotor	Posterior putamen	2.96 ± 0.36	2.57 ± 0.29	0.002
Whole striatum		2.57 ± 0.28	2.26 ± 0.20	0.001

Subdivision	Brain Region	HC ΔBPND	CD ΔBPND	P*
Limbic	Ventral striatum	−12.4 ± 9.0%	−1.2 ± 7.3%	<0.001
Associative	Anterior caudate	−4.6 ± 6.2%	−2.8 ± 7.8%	0.39
	Anterior putamen	−8.7 ± 7.0%	−1.0 ± 6.5%	<0.001
	Posterior caudate	−6.9 ± 7.8%	−6.3 ± 10.7%	0.82
Sensorimotor	Posterior putamen	−14.1 ± 7.8%	−4.3 ± 7.5%	<0.001
Whole striatum		−9.5 ± 5.9%	−3.0 ± 6.5%	0.001

Source: From Martinez et al., 2007

Table 3 Comparison of Baseline [¹¹C]raclopride Binding (BPND) and Methylphenidate (60 mg PO)-Induced ΔBPND in Cocaine Abusers Who Respond to a Behavioral Treatment Compared to Those Who Do Not. Response Is Associated with Higher Values for BPND and ΔBPND (in the Ventral Striatum) Compared to Cocaine-Dependent Subjects Who Relapse

Subdivision	Brain Region	Responders BPND	Nonresponders BPND	P*
Limbic	Ventral striatum	1.94 ± 0.27	1.75 ± 0.17	0.05
Associative	Anterior caudate	2.11 ± 0.31	1.96 ± 0.17	0.12
	Anterior putamen	2.51 ± 0.34	2.26 ± 0.20	0.03
	Posterior caudate	1.40 ± 0.26	1.32 ± 0.19	0.40
Sensorimotor	Posterior putamen	2.59 ± 0.39	2.39 ± 0.13	0.09
Whole striatum		2.23 ± 0.29	2.05 ± 0.13	0.05

Subdivision	Brain Region	HC ΔBPND	CD ΔBPND	P*
Limbic	Ventral striatum	−12.1 ± 6.8%	−1.3 ± 6.7%	<0.001
Associative	Anterior caudate	−8.5 ± 10.7%	−2.6 ± 9.9%	0.318
	Anterior putamen	−8.7 ± 7.0%	−1.0 ± 6.5%	0.04
	Posterior caudate	−7.4 ± 11.2%	−1.9 ± 6.0%	0.13
Sensorimotor	Posterior putamen	−11.0 ± 7.9%	−6.7 ± 5.8%	0.15
Whole striatum		−8.1 ± 6.5%	−6.6 ± 6.5%	0.5

which an initial recreational and "goal-directed" drug use, during which the drug has prominent hedonic effects, can lead to a "compulsive", habitual process with loss of control over drug consumption (Everitt et al., 2008). Similarly to reward learning, addiction relies largely on instrumental processes and Pavlovian conditioning.

In the context of addiction, instrumental learning refers to the "action" of drug seeking and taking, whereas the pavlovian components correspond to the associations between the drug and environmental cues that can facilitate drug consumption, craving, or relapse. The brain circuits that are involved in these processes have been fairly characterized in animal studies and have led to the theory that addiction involves dysfunctional prefrontal cortical executive control over abnormally strong Pavlovian and instrumental learning mechanisms, which depend upon the basal ganglia and their limbic cortical afferents (Belin et al., 2009; Everitt et al., 2001; Everitt and Robbins, 2005).

While the more ventral part of the basal ganglia has been related to emotional learning and is involved in Pavlovian conditioning and its interaction with instrumental learning, the dorsal striatum and its subdivisions have been implicated in more cognitive and motor functions and play a central role in goal-directed and habit behaviors in instrumental responding (Belin et al., 2009; Haber, 2003). Thus, the ventral part of the striatum is largely involved in acquisition of instrumental behavior whereas the dorsal striatum is mainly implicated in action-outcome, habit formation, and pavlovian associations. With respect to drug addiction, the acquisition of cue-controlled cocaine seeking depends on signaling within the NAc, whereas the cue-controlled cocaine seeking habit involves intrastriatal connectivity and correlates with neuronal firing in the dorsolateral striatum (Belin and Everitt, 2008; Haber, 2003; Ito et al., 2002; Takahashi et al., 2007; Vanderschuren et al., 2005). Similarly, cocaine self-administration (in monkeys) induces a progressive change in DAT binding sites from an increase in ventral parts of the striatum during initial phases of cocaine-seeking behavior, to an increase in the entire striatum after prolonged self-administration (Letchworth et al., 2001; Porrino et al., 2004). Therefore, it is believed that whereas the early stages of drug use depend on signaling in the ventral striatum, the transition to compulsive or "habitual" cocaine consumption involves more dorsal subdivisions of the striatum.

2.4.3. Direct and Indirect Pathways

Within the stratum, the medium spiny neurons (MSNs) can be categorized into subgroups according to their projection sites and protein's expression (Gerfen, 2000). The direct or striatonigral pathway is composed of MSNs that project monosynaptically to the medial globus pallidus and substantia nigra pars reticulata. MSNs from the direct pathway express the dopaminergic D1 receptor (D1 receptor), M4 muscarinic acetylcholine receptor, substance P, and dynorphin. The "indirect" striatopallidal pathway is made up of MSNs that reach the substantia nigra reticulata through synaptic relays in the globus pallidus and subthalamic nucleus. These MSNs express the dopaminergic D2 receptor, adenosine receptors A2A, and enkephalin. It should be noted that, whereas the segregation of these two MSN populations has been established in the dorsal striatum, recent studies have shown that a subpopulation of MSNs coexpress the D1 receptor and D2 receptor the nucleus accumbens (George and O'Dowd, 2007; Valjent et al., 2009).

2.4.3.1. Molecular and Cellular Mechanisms Downstream of the Dopaminergic Receptors

Dopamine, through D1 receptor and D2 receptor, activates or inhibits cyclic AMP-dependent signaling as reviewed below. Therefore, dopamine has different effects on D1 receptor and D2 receptor expressing MSNs. Thus, cocaine-induced increases on extracellular dopamine would be expected to trigger long-term cellular alterations via different mechanisms in the two MSNs populations. Dopamine release does not act alone to trigger long-term cellular alteration in response to cocaine and a growing body of evidence support the idea that it is the coordinated signaling of dopaminergic and glutamatergic systems that plays an essential role in shaping synaptic configurations and in altering the activity of neural ensembles (Kelley, 2004). The interaction between dopaminergic and glutamatergic signaling on MSNs can occur at the level of the receptors via oligomerization processes (Lee et al., 2011; Pascoli et al., 2011; Pei et al., 2004) or downstream through cross-talk and cooperation of signaling pathways.

The most characterized intracellular pathways activated by cocaine in the NAc are the cAMP-PKA (McClung and Nestler, 2003) and the ERK/DARPP-32 (Girault et al., 2007) pathways and their inhibition suppress sensitization to cocaine. Interestingly, the ERK/DARPP-32 pathway acts as a coincidence detector of activation of glutamate and DA inputs on D1 receptor-expressing neurons (Girault et al., 2007). Downstream of these pathways, cocaine administration triggers a complex transcriptional program in the striatum by recruiting transcription factors such as cAMP response element-binding protein (CREB) and Elk-1 (Brami-Cherrier et al., 2002; Lavaur et al., 2007; Valjent et al., 2000) and epigenetic mechanisms through histones and DNA modifications (Day and Sweatt, 2011).

Histone modifications are capable of being both gene- and site-specific within a given chromatin molecule, meaning that they are in an ideal position to selectively modulate gene expression (Day and Sweatt, 2011). These modifications can persist for several days, which could account for the cocaine-induced increase in some genes levels that are maintained for long periods of time after withdrawal (Grimm et al., 2003). Importantly, altering histones modifications modulate the rewarding properties of cocaine (Day and Sweatt, 2011). Another epigenetic mechanism induced by cocaine consists in the direct methylation of DNA and site-specific inhibition of such mechanism in the nucleus accumbens alters preference for cocaine (Anier et al., 2010; LaPlant et al., 2010).

The aforementioned mechanisms consist in rapid neuronal alteration that constitute a molecular switch to long-term cellular changes in part by triggering the transcription of several genes implicated in a multitude of neuronal functions. One of the best-characterized gene product upregulated by cocaine in the striatum is ΔFosB, which is a transcription factor encoded by the fosB gene. Because of its long half-life, ΔFosB accumulates in the NAc after repeated drug exposure and persists for at least several weeks after withdrawal (Nestler, 2008). It has been proposed that ΔFosB constitutes a "molecular switch" that helps initiate and then maintain an addicted state (McClung et al., 2004; Nestler, 2008; Nestler et al., 2001). ΔFosB modulates neuronal plasticity by controlling the expression of several gene products such as GluR2, an α-amino-3-hydroxy-5-methyl-4-isoxazolepropionic acid (AMPA) glutamate receptor subunit (Kelz et al., 1999), the opioid peptide dynorphin (Zachariou et al., 2006), cyclin-dependent kinase-5 and its cofactor P35 (Bibb et al., 2001; McClung and Nestler, 2003), and NFκB (Ang et al., 2001; Peakman et al., 2003). The use of microarray as well as proteomic approach have allowed the identification of a number of genes and proteins relevant for synaptic function that are altered by cocaine exposure such as proteins involved in axonal guidance and neurotransmission, receptors—including the dopaminergic receptors—and signal transduction proteins (for review: Lull et al., 2008).

These molecular changes induced by cocaine have a direct impact on neuronal function by altering synaptic and wiring plasticity. Indeed, cocaine produces long-lasting reorganization of synaptic connectivity in the NAc (Robinson and Kolb, 2004) as well as changes at excitatory synapses in the NAc (for reviews: Luscher and Malenka, 2011; Surmeier et al., 2007) supporting the existence of a "drug-induced synaptic plasticity" (Luscher and Malenka, 2011). In a very recent study, Pascoli et al. (Pascoli et al., 2012) showed that cocaine exposure occludes the induction of synaptic plasticity specifically in D1 receptor—but not D2 receptor—expressing neurons in the NAc and that artificially reversing this cocaine-induced long term potentiation (LTP) prevented sensitization to cocaine. This elegant study further supports the idea of specific cocaine-induced plasticity and directly demonstrates for the first time that cocaine-induced plasticity is responsible for the concomitant behavioral alterations.

In addition, to better understand the molecular and cellular changes induced by cocaine, animal studies also offer the opportunity to identify new targets for the development of potential treatments for addiction. One example is the dynorphin/kappa pathway. As mentioned above, dynorphin is upregulated in response to cocaine (Daunais and McGinty, 1995, 1996; Daunais et al., 1993; Daunais et al., 1995; Fagergren et al., 2003; Jenab et al., 2003; Schlussman et al., 2003, 2005; Sivam, 1989; Smiley et al., 1990; Spangler et al., 1993, 1996; Zhou et al., 2002). The administration of a kappa agonist was shown to potentiate cocaine-induced place preference (McLaughlin et al., 2006) and reinstate cocaine-conditioned place preference in mice (Redila and Chavkin, 2008). Importantly, these studies have also shown that signaling at the Kappa-opioid receptor (KOR) plays a significant role in cocaine-seeking behavior following a physical stress. Importantly, these studies demonstrate that KOR activation is selective for stress-induced place preference, and that cocaine-induced drug-seeking behavior is not affected.

One role of the kappa/dynorphin system in the striatum is to regulate dopamine signaling in the direct pathway by inhibiting D1 receptor-dependent signaling. In addition to dampening the downstream effects of D1 receptor activation, kappa receptors are also located presynaptically on the dopamine neurons where they inhibit presynaptic dopamine release and reduce extracellular dopamine levels (Di Chiara and Imperato, 1988b; Donzanti et al., 1992; Heijna et al., 1990, 1992; Spanagel et al., 1992; Thompson et al., 2000; Zhang et al., 2004). Kappa receptor activation inhibits electrically evoked [^{3}H] dopamine release in the nucleus accumbens, only when administered in the striatum and not in the midbrain (Chefer et al., 2005; Heijna et al., 1992; Margolis et al., 2006; Spanagel et al., 1992; Yokoo et al., 1992). Alternatively, the deletion of KOR is associated with an enhancement of basal dopamine release (Chefer et al., 2005). Taken together, these studies demonstrate that striatal KOR activation significantly reduces dopamine transmission in the striatum.

These findings in preclinical studies raise the question about a potential role for dynorphin signaling at the kappa receptor in the human striatum in cocaine abuse. As described above, imaging studies in humans show that both basal levels of extracellular dopamine (Martinez et al., 2009a) and presynaptic dopamine (Malison et al., 1999; Martinez et al., 2007, 2011; Narendran et al., 2012; Volkow et al., 1997; Wu et al., 1997) are reduced in cocaine abusers compared to controls. Thus, one potential mechanism by which this could occur would be through excess activation of the kappa/dynorphin system.

In humans, three postmortem studies have shown an increase in kappa receptor binding in cocaine abusers compared to matched controls. In the first study, Hurd et al. (Hurd and Herkenham, 1993) showed a twofold increase in kappa receptor binding in the caudate, and then Mash et al. (Mash and Staley, 1999) also showed a twofold increase in the anterior and ventral sectors of the caudate and putamen, and nucleus accumbens compared to controls. Lastly, Staley et al. (Staley et al., 1997) showed a similar increase in kappa receptors in the caudate, putamen, and nucleus accumbens in cocaine abuse. In addition to the kappa receptor, postmortem studies have also shown increases in dynorphin levels. Hurd et al. (Hurd and Herkenham, 1993) demonstrated an increase in pre-prodynorphin mRNA in the putamen and recently, Frankel et al. (Frankel et al., 2008) showed increase in dynorphin in the caudate (92%). To date, no imaging studies have yet been published measuring the kappa receptor in human cocaine abusers.

3. CONCLUSIONS

Imaging studies in humans as well as animal models have helped understanding the neurobiological effects of cocaine and identify substrates of cocaine addiction. Cocaine induces a massive increase in extracellular dopamine in the mesoaccumbens and nigrostriatal pathways, an effect that is likely to enhance the rewarding properties of cocaine.

Animal studies suggest that this effect favors the abnormal consolidation of distinct learning processes that depend on interconnected striatal subdivisions, and facilitates the transition from a recreational use to compulsive consumption. Imaging data in humans clearly demonstrate that addiction to cocaine is associated with long-lasting modifications of the dopaminergic transmission and more particularly with a decrease in dopamine release and D2R availability in the different subdivisions of the striatum. Surprisingly, despite the extensive characterization of the molecular and cellular mechanisms responsible for the induction of cocaine-induced neuronal alterations and addictive-like behaviors in animal models, little is known about their precise role for long-lasting behavioral adaptations induced by cocaine. More importantly, there is a lack of evidence for a possible reversibility of these changes. Considering its central role in reward processing, which is altered in cocaine addicts, it is crucial to better understand the mechanisms involved in the dysregulation of dopaminergic transmission in cocaine abusers. Of particular interest are the mechanisms responsible for the decrease in D2 receptor expression in the striatum and the downstream impact of such alteration on dopaminergic transmission.

REFERENCES

Alexander, G.E., Delong, M.R., Stick, P.L., 1986. Parallel organization of functionally segregated circuits linking basal ganglia and cortex. Annu. Rev. Neurosci. 9, 357–381.

Ang, E., Chen, J., Zagouras, P., Magna, H., Holland, J., Schaeffer, E., Nestler, E.J., 2001. Induction of nuclear factor-kappaB in nucleus accumbens by chronic cocaine administration. J. Neurochem. 79, 221–224.

Anier, K., Malinovskaja, K., Aonurm-Helm, A., Zharkovsky, A., Kalda, A., 2010. DNA methylation regulates cocaine-induced behavioral sensitization in mice. Neuropsychopharmacology 35, 2450–2461.

Belin, D., Everitt, B.J., 2008. Cocaine seeking habits depend upon dopamine-dependent serial connectivity linking the ventral with the dorsal striatum. Neuron 57, 432–441.

Belin, D., Jonkman, S., Dickinson, A., Robbins, T.W., Everitt, B.J., 2009. Parallel and interactive learning processes within the basal ganglia: relevance for the understanding of addiction. Behav. Brain Res. 199, 89–102.

Berridge, K.C., Robinson, T.E., 1998. What is the role of dopamine in reward: hedonic impact, reward learning, or incentive salience? Brain Res. Brain Res. Rev. 28, 309–369.

Bibb, J.A., Chen, J., Taylor, J.R., Svenningsson, P., Nishi, A., Snyder, G.L., Yan, Z., Sagawa, Z.K., Ouimet, C.C., Nairn, A.C., Nestler, E.J., Greengard, P., 2001. Effects of chronic exposure to cocaine are regulated by the neuronal protein Cdk5. Nature 410, 376–380.

Boileau, I., Dagher, A., Leyton, M., Gunn, R.N., Baker, G.B., Diksic, M., Benkelfat, C., 2006. Modeling sensitization to stimulants in humans: an [^{11}C]raclopride/positron emission tomography study in healthy men. Arch. Gen. Psychiatry 63, 1386–1395.

Boileau, I., Warsh, J.J., Guttman, M., Saint-Cyr, J.A., McCluskey, T., Rusjan, P., Houle, S., Wilson, A.A., Meyer, J.H., Kish, S.J., 2008. Elevated serotonin transporter binding in depressed patients with Parkinson's disease: a preliminary PET study with [^{11}C]DASB. Mov. Disord. 23, 1776–1780.

Bolla, K., Ernst, M., Kiehl, K., Mouratidis, M., Eldreth, D., Contoreggi, C., Matochik, J., Kurian, V., Cadet, J., Kimes, A., Funderburk, F., London, E., 2004. Prefrontal cortical dysfunction in abstinent cocaine abusers. J. Neuropsychiatry Clin. Neurosci. 16, 456–464.

Bradberry, C.W., 2000. Acute and chronic dopamine dynamics in a nonhuman primate model of recreational cocaine use. J. Neurosci. 20, 7109–7115.

Bradberry, C.W., Roth, R.H., 1989. Cocaine increases extracellular dopamine in rat nucleus accumbens and ventral tegmental area as shown by in vivo microdialysis. Neurosci. Lett. 103, 97–102.

Brami-Cherrier, K., Valjent, E., Garcia, M., Pages, C., Hipskind, R.A., Caboche, J., 2002. Dopamine induces a PI3-kinase-independent activation of Akt in striatal neurons: a new route to cAMP response element-binding protein phosphorylation. J. Neurosci. 22, 8911–8921.

Breier, A., Su, T.P., Saunders, R., Carson, R.E., Kolachana, B.S., deBartolomeis, A., Weinberger, D.R., Weisenfeld, N., Malhotra, A.K., Eckelman, W.C., Pickar, D., 1997. Schizophrenia is associated with elevated amphetamine-induced synaptic dopamine concentrations: evidence from a novel positron emission tomography method. Proc. Natl. Acad. Sci. U.S.A. 94, 2569–2574.

Brewer, J.A., Worhunsky, P.D., Carroll, K.M., Rounsaville, B.J., Potenza, M.N., 2008. Pretreatment brain activation during stroop task is associated with outcomes in cocaine-dependent patients. Biol. Psychiatry 64, 998–1004.

Carroll, F.I., Lewin, A.H., Boja, J.W., Kuhar, M.J., 1992. Cocaine receptor: biochemical characterization and structure-activity relationships of cocaine analogues at the dopamine transporter. J. Med. Chem. 35, 969–981.

Castner, S.A., al-Tikriti, M.S., Baldwin, R.M., Seibyl, J.P., Innis, R.B., Goldman-Rakic, P.S., 2000. Behavioral changes and [^{123}I]IBZM equilibrium SPECT measurement of amphetamine-induced dopamine release in rhesus monkeys exposed to subchronic amphetamine. Neuropsychopharmacology 22, 4–13.

Chefer, V.I., Czyzyk, T., Bolan, E.A., Moron, J., Pintar, J.E., Shippenberg, T.S., 2005. Endogenous kappa-opioid receptor systems regulate mesoaccumbal dopamine dynamics and vulnerability to cocaine. J. Neurosci. 25, 5029–5037.

Clatworthy, P.L., Lewis, S.J., Brichard, L., Hong, Y.T., Izquierdo, D., Clark, L., Cools, R., Aigbirhio, F.I., Baron, J.C., Fryer, T.D., Robbins, T.W., 2009. Dopamine release in dissociable striatal subregions predicts the different effects of oral methylphenidate on reversal learning and spatial working memory. J. Neurosci. 29, 4690–4696.

Cools, R., 2008. Role of dopamine in the motivational and cognitive control of behavior. Neuroscientist 14, 381–395.

Dalley, J.W., Fryer, T.D., Brichard, L., Robinson, E.S., Theobald, D.E., Laane, K., Pena, Y., Murphy, E.R., Shah, Y., Probst, K., Abakumova, I., Aigbirhio, F.I., Richards, H.K., Hong, Y., Baron, J.C., Everitt, B.J., Robbins, T.W., 2007. Nucleus accumbens D2/3 receptors predict trait impulsivity and cocaine reinforcement. Science 315, 1267–1270.

Daunais, J.B., McGinty, J.F., 1995. Cocaine binges differentially alter striatal preprodynorphin and zif/268 mRNAs. Brain Res. Mol. Brain Res. 29, 201–210.

Daunais, J.B., McGinty, J.F., 1996. The effects of D1 or D2 dopamine receptor blockade on zif/268 and preprodynorphin gene expression in rat forebrain following a short-term cocaine binge. Brain Res. Mol. Brain Res. 35, 237–248.

Daunais, J.B., Roberts, D.C., McGinty, J.F., 1993. Cocaine self-administration increases preprodynorphin, but not c-fos, mRNA in rat striatum. Neuroreport 4, 543–546.

Daunais, J.B., Roberts, D.C., McGinty, J.F., 1995. Short-term cocaine self administration alters striatal gene expression. Brain Res. Bull. 37, 523–527.

Day, J.J., Sweatt, J.D., 2011. Epigenetic mechanisms in cognition. Neuron 70, 813–829.

Di Chiara, G., Imperato, A., 1988a. Drugs abused by humans preferentially increase synaptic dopamine concentrations in the mesolimbic system of freely moving rats. Proc. Natl. Acad. Sci. U.S.A. 85, 5274–5278.

Di Chiara, G., Imperato, A., 1988b. Opposite effects of mu and kappa opiate agonists on dopamine release in the nucleus accumbens and in the dorsal caudate of freely moving rats. J. Pharmacol. Exp. Ther. 244, 1067–1080.

Donzanti, B.A., Althaus, J.S., Payson, M.M., Von Voigtlander, P.F., 1992. Kappa agonist-induced reduction in dopamine release: site of action and tolerance. Res. Commun. Chem. Pathol. Pharmacol. 78, 193–210.

Drevets, W.C., Gautier, C., Price, J.C., Kupfer, D.J., Kinahan, P.E., Grace, A.A., Price, J.L., Mathis, C.A., 2001. Amphetamine-induced dopamine release in human ventral striatum correlates with euphoria. Biol. Psychiatry 49, 81–96.

Drevets, W.C., Price, J.C., Kupfer, D.J., Kinahan, P.E., Lopresti, B., Holt, D., Mathis, C., 1999. PET measures of amphetamine-induced dopamine release in ventral versus dorsal striatum. Neuropsychopharmacology 21, 694–709.

Everitt, B.J., Belin, D., Economidou, D., Pelloux, Y., Dalley, J.W., Robbins, T.W., 2008. Review. Neural mechanisms underlying the vulnerability to develop compulsive drug-seeking habits and addiction. Philos. Trans. R. Soc. Lond., B, Biol. Sci. 363, 3125–3135.

Everitt, B.J., Dickinson, A., Robbins, T.W., 2001. The neuropsychological basis of addictive behaviour. Brain Res. Brain Res. Rev. 36, 129–138.

Everitt, B.J., Robbins, T.W., 2005. Neural systems of reinforcement for drug addiction: from actions to habits to compulsion. Nat. Neurosci. 8, 1481–1489.

Fagergren, P., Smith, H.R., Daunais, J.B., Nader, M.A., Porrino, L.J., Hurd, Y.L., 2003. Temporal upregulation of prodynorphin mRNA in the primate striatum after cocaine self-administration. Eur. J. Neurosci. 17, 2212–2218.

Ferry, A.T., Ongur, D., An, X., Price, J.L., 2000. Prefrontal cortical projections to the striatum in macaque monkeys: evidence for an organization related to prefrontal networks. J. Comp. Neurol. 425, 447–470.

Frankel, P.S., Alburges, M.E., Bush, L., Hanson, G.R., Kish, S.J., 2008. Striatal and ventral pallidum dynorphin concentrations are markedly increased in human chronic cocaine users. Neuropharmacology 55, 41–46.

Freud, S., 1884. Uber coca. Centralblatt für die ges. Therapie 2, 289–314.

Fudge, J.L., Breitbart, M.A., McClain, C., 2004. Amygdaloid inputs define a caudal component of the ventral striatum in primates. J. Comp. Neurol. 476, 330–347.

Fudge, J.L., Haber, S.N., 2002. Defining the caudal ventral striatum in primates: cellular and histochemical features. J. Neurosci. 22, 10078–10082.

George, S.R., O'Dowd, B.F., 2007. A novel dopamine receptor signaling unit in brain: heterooligomers of D1 and D2 dopamine receptors. ScientificWorldJournal 7, 58–63.

Gerfen, C.R., 2000. Molecular effects of dopamine on striatal-projection pathways. Trends Neurosci. 23, S64–S70.

Girault, J.A., Valjent, E., Caboche, J., Herve, D., 2007. ERK2: a logical AND gate critical for drug-induced plasticity? Curr. Opin. Pharmacol. 7, 77–85.

Goldstein, R.Z., Alia-Klein, N., Tomasi, D., Carrillo, J.H., Maloney, T., Woicik, P.A., Wang, R., Telang, F., Volkow, N.D., 2009. Anterior cingulate cortex hypoactivations to an emotionally salient task in cocaine addiction. Proc. Natl. Acad. Sci. U.S.A. 106, 9453–9458.

Goldstein, R.Z., Volkow, N.D., 2011. Dysfunction of the prefrontal cortex in addiction: neuroimaging findings and clinical implications. Nat. Rev. Neurosci. 12, 652–669.

Gootenberg, P., 2006. Cocaine: Global Histories. Routledge, New York, NY.

Gorelick, D.A., Rothman, R.B., 1997. Stimulant sensitization in humans. Biol. Psychiatry 42, 230–231.

Grimm, J.W., Lu, L., Hayashi, T., Hope, B.T., Su, T.P., Shaham, Y., 2003. Time-dependent increases in brain-derived neurotrophic factor protein levels within the mesolimbic dopamine system after withdrawal from cocaine: implications for incubation of cocaine craving. J. Neurosci. 23, 742–747.

Haber, S.N., 2003. The primate basal ganglia: parallel and integrative networks. J. Chem. Neuroanat. 26, 317–330.

Heijna, M.H., Bakker, J.M., Hogenboom, F., Mulder, A.H., Schoffelmeer, A.N., 1992. Opioid receptors and inhibition of dopamine-sensitive adenylate cyclase in slices of rat brain regions receiving a dense dopaminergic input. Eur. J. Pharmacol. 229, 197–202.

Heijna, M.H., Padt, M., Hogenboom, F., Portoghese, P.S., Mulder, A.H., Schoffelmeer, A.N., 1990. Opioid receptor-mediated inhibition of dopamine and acetylcholine release from slices of rat nucleus accumbens, olfactory tubercle and frontal cortex. Eur. J. Pharmacol. 181, 267–278.

Hester, R., Garavan, H., 2004. Executive dysfunction in cocaine addiction: evidence for discordant frontal, cingulate, and cerebellar activity. J. Neurosci. 24, 11017–11022.

Higgins, S.T., Budney, A.J., Bickel, W.K., Foerg, F.E., Donham, R., Badger, G.J., 1994. Incentives improve outcome in outpatient behavioral treatment of cocaine dependence. Arch. Gen. Psychiatry 51, 568–576.

Higgins, S.T., Sigmon, S.C., Wong, C.J., Heil, S.H., Badger, G.J., Donham, R., Dantona, R.L., Anthony, S., 2003. Community reinforcement therapy for cocaine-dependent outpatients. Arch. Gen. Psychiatry 60, 1043–1052.

Hoover, J.E., Strick, P.L., 1993. Multiple output channels in the basal ganglia. Science 259, 819–821.

Hurd, Y.L., Herkenham, M., 1993. Molecular alterations in the neostriatum of human cocaine addicts. Synapse 13, 357–369.

Innis, R.B., Cunningham, V.J., Delforge, J., Fujita, M., Gjedde, A., Gunn, R.N., Holden, J., Houle, S., Huang, S.C., Ichise, M., Iida, H., Ito, H., Kimura, Y., Koeppe, R.A., Knudsen, G.M., Knuuti, J., Lammertsma, A.A., Laruelle, M., Logan, J., Maguire, R.P., Mintun, M.A., Morris, E.D., Parsey, R., Price, J.C., Slifstein, M., Sossi, V., Suhara, T., Votaw, J.R., Wong, D.F., Carson, R.E., 2007. Consensus nomenclature for in vivo imaging of reversibly binding radioligands. J. Cereb. Blood Flow Metab. 27, 1533–1539.

Ito, R., Dalley, J.W., Robbins, T.W., Everitt, B.J., 2002. Dopamine release in the dorsal striatum during cocaine-seeking behavior under the control of a drug-associated cue. J. Neurosci. 22, 6247–6253.

Jayanthi, L.D., Ramamoorthy, S., 2005. Regulation of monoamine transporters: influence of psychostimulants and therapeutic antidepressants. AAPS J. 7, E728–E738.

Jenab, S., Festa, E.D., Russo, S.J., Wu, H.B., Inturrisi, C.E., Quinones-Jenab, V., 2003. MK-801 attenuates cocaine induction of c-fos and preprodynorphin mRNA levels in Fischer rats. Brain Res. Mol. Brain Res. 117, 237–239.

Joel, D., Weiner, I., 2000. The connections of the dopaminergic system with the striatum in rats and primates: an analysis with respect to the functional and compartmental organization of the striatum. Neuroscience 96, 451–474.

Jones, S.R., Lee, T.H., Wightman, R.M., Ellinwood, E.H., 1996. Effects of intermittent and continuous cocaine administration on dopamine release and uptake regulation in the striatum: in vitro voltammetric assessment. Psychopharmacology 126, 331–338.

Kalivas, P.W., Stewart, J., 1991. Dopamine transmission in the initiation and expression of drug- and stress-induced sensitization of motor activity. Brain Res. Brain Res. Rev. 16, 223–244.

Kaufman, J.N., Ross, T.J., Stein, E.A., Garavan, H., 2003. Cingulate hypoactivity in cocaine users during a GO-NOGO task as revealed by event-related functional magnetic resonance imaging. J. Neurosci. 23, 7839–7843.

Kelley, A.E., 2004. Memory and addiction: shared neural circuitry and molecular mechanisms. Neuron 44, 161–179.

Kelz, M.B., Chen, J., Carlezon Jr., W.A., Whisler, K., Gilden, L., Beckmann, A.M., Steffen, C., Zhang, Y.J., Marotti, L., Self, D.W., Tkatch, T., Baranauskas, G., Surmeier, D.J., Neve, R.L., Duman, R.S., Picciotto, M.R., Nestler, E.J., 1999. Expression of the transcription factor deltaFosB in the brain controls sensitivity to cocaine. Nature 401, 272–276.

Knapp, W.P., Soares, B.G., Farrel, M., Lima, M.S., 2007. Psychosocial interventions for cocaine and psychostimulant amphetamines related disorders. Cochrane Database Syst. Rev., CD003023.

Kuhar, M.J., Ritz, M.C., Boja, J.W., 1991. The dopamine hypothesis of the reinforcing properties of cocaine. Trends Neurosci. 14, 299–302.

LaPlant, Q., Vialou, V., Covington 3rd, H.E., Dumitriu, D., Feng, J., Warren, B.L., Maze, I., Dietz, D.M., Watts, E.L., Iniguez, S.D., Koo, J.W., Mouzon, E., Renthal, W., Hollis, F., Wang, H., Noonan, M.A., Ren, Y., Eisch, A.J., Bolanos, C.A., Kabbaj, M., Xiao, G., Neve, R.L., Hurd, Y.L., Oosting, R.S., Fan, G., Morrison, J.H., Nestler, E.J., 2010. Dnmt3a regulates emotional behavior and spine plasticity in the nucleus accumbens. Nat. Neurosci. 13, 1137–1143.

Laruelle, M., 2000. Imaging synaptic neurotransmission with in vivo binding competition techniques: a critical review. J. Cereb. Blood Flow Metab. 20, 423–451.

Laruelle, M., Iyer, R.N., Al-Tikriti, M.S., Zea-Ponce, Y., Malison, R., Zoghbi, S.S., Baldwin, R.M., Kung, H.F., Charney, D.S., Hoffer, P.B., Innis, R.B., Bradberry, C.W., 1997. Microdialysis and SPECT measurements of amphetamine-induced dopamine release in nonhuman primates. Synapse 25, 1–14.

Lavaur, J., Bernard, F., Trifilieff, P., Pascoli, V., Kappes, V., Pages, C., Vanhoutte, P., Caboche, J., 2007. A TAT-DEF-Elk-1 peptide regulates the cytonuclear trafficking of Elk-1 and controls cytoskeleton dynamics. J. Neurosci. 27, 14448–14458.

Leckman, J.F., Bloch, M.H., Smith, M.E., Larabi, D., Hampson, M., 2010. Neurobiological substrates of Tourette's disorder. J. Child Adolesc. Psychopharmacol. 20, 237–247.

Lee, D.K., Ahn, S.M., Shim, Y.B., Koh, W.C., Shim, I., Choe, E.S., 2011. Interactions of dopamine D1 and N-methyl-d-aspartate receptors are required for acute cocaine-evoked nitric oxide efflux in the dorsal striatum. Exp. Neurobiol. 20, 116–122.

Letchworth, S.R., Nader, M.A., Smith, H.R., Friedman, D.P., Porrino, L.J., 2001. Progression of changes in dopamine transporter binding site density as a result of cocaine self-administration in rhesus monkeys. J. Neurosci. 21, 2799–2807.

Leung, H.C., Skudlarski, P., Gatenby, J.C., Peterson, B.S., Gore, J.C., 2000. An event-related functional MRI study of the stroop color word interference task. Cereb. Cortex 10, 552–560.

Lewin, A.H., Gao, Y.G., Abraham, P., Boja, J.W., Kuhar, M.J., Carroll, F.I., 1992. 2β-Substituted analogues of cocaine. Synthesis and inhibition of binding to the cocaine receptor. J. Med. Chem. 35, 135–140.

Little, K.Y., Krolewski, D.M., Zhang, L., Cassin, B.J., 2003. Loss of striatal vesicular monoamine transporter protein (VMAT2) in human cocaine users. Am. J. Psychiatry 160, 47–55.

Little, K.Y., Zhang, L., Desmond, T., Frey, K.A., Dalack, G.W., Cassin, B.J., 1999. Striatal dopaminergic abnormalities in human cocaine users. Am. J. Psychiatry 156, 238–245.

Logan, J., Fowler, J.S., Dewey, S.L., Volkow, N.D., Gatley, S.J., 2001. A consideration of the dopamine D2 receptor monomer-dimer equilibrium and the anomalous binding properties of the dopamine D2 receptor ligand, N-methyl spiperone. J. Neural. Transm. 108, 279–286.

Lopez-Quintero, C., Perez de los Cobos, J., Hasin, D.S., Okuda, M., Wang, S., Grant, B.F., Blanco, C., 2011. Probability and predictors of transition from first use to dependence on nicotine, alcohol, cannabis, and cocaine: results of the National Epidemiologic Survey on Alcohol and Related Conditions (NESARC). Drug Alcohol Depend. 115, 120–130.

Lull, M.E., Freeman, W.M., Vrana, K.E., Mash, D.C., 2008. Correlating human and animal studies of cocaine abuse and gene expression. Ann. N.Y. Acad. Sci. 1141, 58–75.

Luscher, C., Malenka, R.C., 2011. Drug-evoked synaptic plasticity in addiction: from molecular changes to circuit remodeling. Neuron 69, 650–663.

Maggos, C.E., Tsukada, H., Kakiuchi, T., Nishiyama, S., Myers, J.E., Kreuter, J., Schlussman, S.D., Unterwald, E.M., Ho, A., Kreek, M.J., 1998. Sustained withdrawal allows normalization of in vivo $[^{11}C]N$-methyl-spiperone dopamine D2 receptor binding after chronic binge cocaine: a positron emission tomography study in rats. Neuropsychopharmacology 19, 146–153.

Maisonneuve, I.M., Ho, A., Kreek, M.J., 1995. Chronic administration of a cocaine "binge" alters basal extracellular levels in male rats: an in vivo microdialysis study. J. Pharmacol. Exp. Ther. 272, 652–657.

Malison, R.T., Best, S.E., van Dyck, C.H., McCance, E.F., Wallace, E.A., Laruelle, M., Baldwin, R.M., Seibyl, J.P., Price, L.H., Kosten, T.R., Innis, R.B., 1998. Elevated striatal dopamine transporters during acute cocaine abstinence as measured by $[^{123}I]$ beta-CIT SPECT. Am. J. Psychiatry 155, 832–834.

Malison, R.T., Mechanic, K.Y., Klummp, H., Baldwin, R.M., Kosten, T.R., Seibyl, J.P., Innis, R.B., 1999. Reduced amphetamine-stimulated dopamine release in cocaine addicts as measured by $[^{123}I]$IBZM SPECT. J. Nucl. Med. 40, P. 110.

Margolis, E.B., Lock, H., Chefer, V.I., Shippenberg, T.S., Hjelmstad, G.O., Fields, H.L., 2006. Kappa opioids selectively control dopaminergic neurons projecting to the prefrontal cortex. Proc. Natl. Acad. Sci. U.S.A. 103, 2938–2942.

Martinez, D., Broft, A., Foltin, R.W., Slifstein, M., Hwang, D.R., Huang, Y., Perez, A., Frankle, W.G., Cooper, T., Kleber, H.D., Fischman, M.W., Laruelle, M., 2004. Cocaine dependence and d2 receptor availability in the functional subdivisions of the striatum: relationship with cocaine-seeking behavior. Neuropsychopharmacology 29, 1190–1202.

Martinez, D., Carpenter, K.M., Liu, F., Slifstein, M., Broft, A., Friedman, A.C., Kumar, D., Van Heertum, R., Kleber, H.D., Nunes, E., 2011. Imaging dopamine transmission in cocaine dependence: link between neurochemistry and response to treatment. Am. J. Psychiatry 168, 634–641.

Martinez, D., Greene, K., Broft, A., Kumar, D., Liu, F., Narendran, R., Slifstein, M., Van Heertum, R., Kleber, H.D., 2009a. Lower level of endogenous dopamine in patients with cocaine dependence: findings from PET imaging of D(2)/D(3) receptors following acute dopamine depletion. Am. J. Psychiatry 166, 1170–1177.

Martinez, D., Narendran, R., Foltin, R.W., Slifstein, M., Hwang, D.R., Broft, A., Huang, Y., Cooper, T.B., Fischman, M.W., Kleber, H.D., Laruelle, M., 2007. Amphetamine-induced dopamine release: markedly blunted in cocaine dependence and predictive of the choice to self-administer cocaine. Am. J. Psychiatry 164, 622–629.

Martinez, D., Slifstein, M., Broft, A., Mawlawi, O., Hwang, D.R., Huang, Y., Cooper, T., Kegeles, L., Zarahn, E., Abi-Dargham, A., Haber, S.N., Laruelle, M., 2003. Imaging human mesolimbic dopamine transmission with positron emission tomography. Part II: amphetamine-induced dopamine release in the functional subdivisions of the striatum. J. Cereb. Blood Flow Metab. 23, 285–300.

Martinez, D., Slifstein, M., Narendran, R., Foltin, R.W., Broft, A., Hwang, D.R., Perez, A., Abi-Dargham, A., Fischman, M.W., Kleber, H.D., Laruelle, M., 2009b. Dopamine D1 receptors in cocaine dependence measured with PET and the choice to self-administer cocaine. Neuropsychopharmacology.

Mash, D.C., Staley, J.K., 1999. D3 dopamine and kappa opioid receptor alterations in human brain of cocaine-overdose victims. Ann. N.Y. Acad. Sci. 877, 507–522.

Mateo, Y., Lack, C.M., Morgan, D., Roberts, D.C., Jones, S.R., 2005. Reduced dopamine terminal function and insensitivity to cocaine following cocaine binge self-administration and deprivation. Neuropsychopharmacology 30, 1455–1463.

McClung, C.A., Nestler, E.J., 2003. Regulation of gene expression and cocaine reward by CREB and DeltaFosB. Nat. Neurosci. 6, 1208–1215.

McClung, C.A., Ulery, P.G., Perrotti, L.I., Zachariou, V., Berton, O., Nestler, E.J., 2004. DeltaFosB: a molecular switch for long-term adaptation in the brain. Brain Res. Mol. Brain Res. 132, 146–154.

McLaughlin, J.P., Land, B.B., Li, S., Pintar, J.E., Chavkin, C., 2006. Prior activation of kappa opioid receptors by U50,488 mimics repeated forced swim stress to potentiate cocaine place preference conditioning. Neuropsychopharmacology 31, 787–794.

Meador-Woodruff, J.H., Little, K.Y., Damask, S.P., Mansour, A., Watson, S.J., 1993. Effects of cocaine on dopamine receptor gene expression: a study in the postmortem human brain. Biol. Psychiatry 34, 348–355.

Melis, M., Spiga, S., Diana, M., 2005. The dopamine hypothesis of drug addiction: hypodopaminergic state. Int. Rev. Neurobiol. 63, 101–154.

Michaelides, M., Thanos, P.K., Kim, R., Cho, J., Ananth, M., Wang, G.J., Volkow, N.D., 2012. PET imaging predicts future body weight and cocaine preference. Neuroimage 59, 1508–1513.

Montoya, I.D., Vocci, F., 2008. Novel medications to treat addictive disorders. Curr. Psychiatry Rep. 10, 392–398.

Moore, R.J., Vinsant, S.L., Nader, M.A., Porrino, L.J., Friedman, D.P., 1998. Effect of cocaine self-administration on dopamine D2 receptors in rhesus monkeys. Synapse 30, 88–96.

Morgan, D., Grant, K.A., Gage, H.D., Mach, R.H., Kaplan, J.R., Prioleau, O., Nader, S.H., Buchheimer, N., Ehrenkaufer, R.L., Nader, M.A., 2002. Social dominance in monkeys: dopamine D2 receptors and cocaine self-administration. Nat. Neurosci. 5, 169–174.

Munro, C.A., McCaul, M.E., Wong, D.F., Oswald, L.M., Zhou, Y., Brasic, J., Kuwabara, H., Kumar, A., Alexander, M., Ye, W., Wand, G.S., 2006. Sex differences in striatal dopamine release in healthy adults. Biol. Psychiatry 59, 966–974.

Nader, M.A., Daunais, J.B., Moore, T., Nader, S.H., Moore, R.J., Smith, H.R., Friedman, D.P., Porrino, L.J., 2002. Effects of cocaine self-administration on striatal dopamine systems in rhesus monkeys. Initial and chronic exposure. Neuropsychopharmacology 27, 35–46.

Nader, M.A., Morgan, D., Gage, H.D., Nader, S.H., Calhoun, T.L., Buchheimer, N., Ehrenkaufer, R., Mach, R.H., 2006. PET imaging of dopamine D2 receptors during chronic cocaine self-administration in monkeys. Nat. Neurosci. 9, 1050–1056.

Narendran, R., Lopresti, B.J., Martinez, D., Mason, N.S., Himes, M., May, M.A., Daley, D.C., Price, J.C., Mathis, C.A., Frankle, W.G., 2012. In vivo evidence for low striatal vesicular monoamine transporter 2 (VMAT2) availability in cocaine abusers. Am. J. Psychiatry 169, 55–63.

Nestler, E.J., 2008. Review. Transcriptional mechanisms of addiction: role of DeltaFosB. Philos. Trans. R. Soc. Lond., B, Biol. Sci. 363, 3245–3255.

Nestler, E.J., Barrot, M., Self, D.W., 2001. DeltaFosB: a sustained molecular switch for addiction. Proc. Natl. Acad. Sci. U.S.A. 98, 11042–11046.

O'Brien, M.S., Anthony, J.C., 2005. Risk of becoming cocaine dependent: epidemiological estimates for the United States, 2000–2001. Neuropsychopharmacology 30, 1006–1018.

Pardo, J.V., Pardo, P.J., Janer, K.W., Raichle, M.E., 1990. The anterior cingulate cortex mediates processing selection in the Stroop attentional conflict paradigm. Proc. Natl. Acad. Sci. U.S.A. 87, 256–259.

Parent, A., Hazrati, L.N., 1995. Functional anatomy of the basal ganglia. I. The cortico-basal ganglia-thalamo-cortical loop. Brain Res. Brain Res. Rev. 20, 91–127.

Pascoli, V., Besnard, A., Herve, D., Pages, C., Heck, N., Girault, J.A., Caboche, J., Vanhoutte, P., 2011. Cyclic adenosine monophosphate-independent tyrosine phosphorylation of NR2B mediates cocaine-induced extracellular signal-regulated kinase activation. Biol. Psychiatry 69, 218–227.

Pascoli, V., Turiault, M., Luscher, C., 2012. Reversal of cocaine-evoked synaptic potentiation resets drug-induced adaptive behaviour. Nature 481, 71–75.

Peakman, M.C., Colby, C., Perrotti, L.I., Tekumalla, P., Carle, T., Ulery, P., Chao, J., Duman, C., Steffen, C., Monteggia, L., Allen, M.R., Stock, J.L., Duman, R.S., McNeish, J.D., Barrot, M., Self, D.W., Nestler, E.J., Schaeffer, E., 2003. Inducible, brain region-specific expression of a dominant negative mutant of c-Jun in transgenic mice decreases sensitivity to cocaine. Brain Res. 970, 73–86.

Pei, L., Lee, F.J., Moszczynska, A., Vukusic, B., Liu, F., 2004. Regulation of dopamine D1 receptor function by physical interaction with the NMDA receptors. J. Neurosci. 24, 1149–1158.

Porrino, L.J., Lyons, D., Smith, H.R., Daunais, J.B., Nader, M.A., 2004. Cocaine self-administration produces a progressive involvement of limbic, association, and sensorimotor striatal domains. J. Neurosci. 24, 3554–3562.

RAND, 2011. The Latin American Drug Trade: Scope, Dimensions, Impact, and Response. RAND Corporation.

Redila, V.A., Chavkin, C., 2008. Stress-induced reinstatement of cocaine seeking is mediated by the kappa opioid system. Psychopharmacology 200, 59–70.

Riccardi, P., Li, R., Ansari, M.S., Zald, D., Park, S., Dawant, B., Anderson, S., Doop, M., Woodward, N., Schoenberg, E., Schmidt, D., Baldwin, R., Kessler, R., 2006. Amphetamine-induced displacement of [^{18}F] fallypride in striatum and extrastriatal regions in humans. Neuropsychopharmacology 31, 1016–1026.

Ritz, M.C., Lamb, R.J., Goldberg, S.R., Kuhar, M.J., 1987. Cocaine receptors on dopamine transporters are related to self-administration of cocaine. Science 237, 1219–1223.

Robinson, P.H., Jurson, P.A., Bennet, J.A., Bentgen, K.M., 1988. Persistent sensitization of dopamine neurotransmission in ventral striatum (nucleus accumbens) produced by prior experience with (+)-amphetamine: a microdialysis study in freely moving rats. Brain Res. 462, 211–222.

Robinson, T.E., 1993. Persistent sensitizing effects of drugs on brain dopamine systems and behavior: implication for addiction and relapse. In: Korenman, S.G., Brarchas, J.D. (Eds.), Biological Basis of Substance Abuse. Raven press, New York.

Robinson, T.E., Kolb, B., 2004. Structural plasticity associated with exposure to drugs of abuse. Neuropharmacology 47 (Suppl. 1), 33–46.

Rothman, R.B., Gorelick, D.A., Baumann, M.H., Guo, X.Y., Herning, R.I., Pickworth, W.B., Gendron, T.M., Koeppl, B., Thomson 3rd, L.E., Henningfield, J.E., 1994. Lack of evidence for context-dependent cocaine-induced sensitization in humans: preliminary studies. Pharmacol. Biochem. Behav. 49, 583–588.

Salamone, J.D., 2009. Dopamine, effort, and decision making: theoretical comment on Bardgett et al. (2009). Behav. Neurosci. 123, 463–467.

Schlussman, S.D., Zhang, Y., Yuferov, V., LaForge, K.S., Ho, A., Kreek, M.J., 2003. Acute 'binge' cocaine administration elevates dynorphin mRNA in the caudate putamen of C57BL/6J but not 129/J mice. Brain Res. 974, 249–253.

Schlussman, S.D., Zhou, Y., Bailey, A., Ho, A., Kreek, M.J., 2005. Steady-dose and escalating-dose "binge" administration of cocaine alter expression of behavioral stereotypy and striatal preprodynorphin mRNA levels in rats. Brain Res. Bull. 67, 169–175.

Schultz, W., 2010. Dopamine signals for reward value and risk: basic and recent data. Behav. Brain Funct. 6, 24.

Sharp, T., Zetterstrom, T., Ljungberg, T., Ungerstedt, U., 1987. A direct comparison of amphetamine-induced behaviours and regional brain dopamine release in the rat using intracerebral dialysis. Brain Res. 401, 322–330.

Sivam, S.P., 1989. Cocaine selectively increases striatonigral dynorphin levels by a dopaminergic mechanism. J. Pharmacol. Exp. Ther. 250, 818–824.

Skinbjerg, M., Liow, J.S., Seneca, N., Hong, J., Lu, S., Thorsell, A., Heilig, M., Pike, V.W., Halldin, C., Sibley, D.R., Innis, R.B., 2010. D2 dopamine receptor internalization prolongs the decrease of radioligand binding after amphetamine: a PET study in a receptor internalization-deficient mouse model. Neuroimage 50, 1402–1407.

Slifstein, M., Kegeles, L.S., Xu, X., Thompson, J.L., Urban, N., Castrillon, J., Hackett, E., Bae, S.A., Laruelle, M., Abi-Dargham, A., 2010. Striatal and extrastriatal dopamine release measured with PET and [(18)F] fallypride. Synapse 64, 350–362.

Smiley, P.L., Johnson, M., Bush, L., Gibb, J.W., Hanson, G.R., 1990. Effects of cocaine on extrapyramidal and limbic dynorphin systems. J. Pharmacol. Exp. Ther. 253, 938–943.

Sorg, B.A., Davidson, D.L., Kalivas, P.W., Prasad, B.M., 1997. Repeated daily cocaine alters subsequent cocaine-induced increase of extracellular dopamine in the medial prefrontal cortex. J. Pharmacol. Exp. Ther. 281, 54–61.

Spanagel, R., Herz, A., Shippenberg, T., 1992. Opposing tonically active endogenous opioid systems modulate the mesolimbic dopaminergic pathway. Proc. Natl. Acad. Sci. U.S.A. 89, 2046–2050.

Spangler, R., Ho, A., Zhou, Y., Maggos, C.E., Yuferov, V., Kreek, M.J., 1996. Regulation of kappa opioid receptor mRNA in the rat brain by "binge" pattern cocaine administration and correlation with preprodynorphin mRNA. Brain Res. Mol. Brain Res. 38, 71–76.

Spangler, R., Unterwald, E., Kreek, M., 1993. Binge cocaine administration induces a sustained increase of prodynorphin mRNA in rat caudate-putamen. Brain Res. Mol. Brain Res. 19, 323–327.

Spealman, R.D., Madras, B.K., Bergman, J., 1989. Effects of cocaine and related drugs in nonhuman primates. II. Stimulant effects on schedule-controlled behavior. J. Pharmacol. Exp. Ther. 251, 142–149.

Staley, J.K., Rothman, R.B., Rice, K.C., Partilla, J., Mash, D.C., 1997. Kappa2 opioid receptors in limbic areas of the human brain are upregulated by cocaine in fatal overdose victims. J. Neurosci. 17, 8225–8233.

Stolberg, V.B., 2011. The use of coca: prehistory, history, and ethnography. J. Ethn. Subst. Abuse 10, 126–146.

Strafella, A.P., Paus, T., Fraraccio, M., Dagher, A., 2003. Striatal dopamine release induced by repetitive transcranial magnetic stimulation of the human motor cortex. Brain 126, 2609–2615.

Strakowski, S.M., Sax, K.W., Setters, M.J., Keck Jr., P.E., 1996. Enhanced response to repeated d-amphetamine challenge: evidence for behavioral sensitization in humans. Biol. Psychiatry 40, 872–880.

Surmeier, D.J., Ding, J., Day, M., Wang, Z., Shen, W., 2007. D1 and D2 dopamine-receptor modulation of striatal glutamatergic signaling in striatal medium spiny neurons. Trends Neurosci. 30, 228–235.

Takahashi, Y., Roesch, M.R., Stalnaker, T.A., Schoenbaum, G., 2007. Cocaine exposure shifts the balance of associative encoding from ventral to dorsolateral striatum. Front. Integr. Neurosci. 1, 11.

Tella, S.R., 1995. Effects of monoamine reuptake inhibitors on cocaine self-administration in rats. Pharmacol. Biochem. Behav. 51, 687–692.

Thompson, A., Zapata, A., Justice, J., Vaughan, R., Sharpe, L., Shippenberg, T., 2000. Kappa-opioid receptor activation modifies dopamine uptake in the nucleus accumbens and opposes the effects of cocaine. J. Neurosci. 20, 9333–9340.

UNODC, 2009. World Drug Report 2009. UNODC, Vienna.

Valjent, E., Bertran-Gonzalez, J., Herve, D., Fisone, G., Girault, J.A., 2009. Looking BAC at striatal signaling: cell-specific analysis in new transgenic mice. Trends Neurosci. 32, 538–547.

Valjent, E., Corvol, J.C., Pages, C., Besson, M.J., Maldonado, R., Caboche, J., 2000. Involvement of the extracellular signal-regulated kinase cascade for cocaine-rewarding properties. J. Neurosci. 20, 8701–8709.

Vanderschuren, L.J., Di Ciano, P., Everitt, B.J., 2005. Involvement of the dorsal striatum in cue-controlled cocaine seeking. J. Neurosci. 25, 8665–8670.

Vezina, P., 2004. Sensitization of midbrain dopamine neuron reactivity and the self-administration of psychomotor stimulant drugs. Neurosci. Biobehav. Rev. 27, 827–839.

Volkow, N., Wang, J., Fowler, J., Logan, J., Hitezmann, R., Ding, Y., Pappas, N., C., S., K., P., 1996. Decreases in dopamine receptors but not in dopamine transporters in alcoholics. Alcohol. Clin. Exp. Res. 20, 1594–1598.

Volkow, N.D., Fowler, J.S., Wang, G.J., 2004. The addicted human brain viewed in the light of imaging studies: brain circuits and treatment strategies. Neuropharmacology 47 (Suppl. 1), 3–13.

Volkow, N.D., Fowler, J.S., Wang, G.J., Baler, R., Telang, F., 2009. Imaging dopamine's role in drug abuse and addiction. Neuropharmacology 56 (Suppl. 1), 3–8.

Volkow, N.D., Fowler, J.S., Wang, G.J., Hitzemann, R., Logan, J., Schlyer, D.J., Dewey, S.L., Wolf, A.P., 1993. Decreased dopamine D2 receptor availability is associated with reduced frontal metabolism in cocaine abusers. Synapse 14, 169–177.

Volkow, N.D., Fowler, J.S., Wolf, A.P., Schlyer, D., Shiue, C.Y., Alpert, R., Dewey, S.L., Logan, J., Bendriem, B., Christman, D., et al., 1990. Effects of chronic cocaine abuse on postsynaptic dopamine receptors. Am. J. Psychiatry 147, 719–724.

Volkow, N.D., Wang, G.J., Fowler, J.S., Hitzemann, R., Angrist, B., Gatley, S.J., Logan, J., Ding, Y.S., Pappas, N., 1999a. Association of methylphenidate-induced craving with changes in right striato-orbitofrontal metabolism in cocaine abusers: implications in addiction. Am. J. Psychiatry 156, 19–26.

Volkow, N.D., Wang, G.J., Fowler, J.S., Logan, J., Gatley, S.J., Gifford, A., Hitzemann, R., Ding, Y.S., Pappas, N., 1999b. Prediction of reinforcing responses to psychostimulants in humans by brain dopamine D2 receptor levels. Am. J. Psychiatry 156, 1440–1443.

Volkow, N.D., Wang, G.J., Fowler, J.S., Logan, J., Gatley, S.J., Hitzemann, R., Chen, A.D., Dewey, S.L., Pappas, N., 1997. Decreased striatal dopaminergic responsiveness in detoxified cocaine-dependent subjects. Nature 386, 830–833.

Volkow, N.D., Wang, G.J., Fowler, J.S., Logan, J., Gatley, S.J., Wong, C., Hitzemann, R., Pappas, N.R., 1999c. Reinforcing effects of psychostimulants in humans are associated with increases in brain dopamine and occupancy of D(2) receptors. J. Pharmacol. Exp. Ther. 291, 409–415.

Volkow, N.D., Wang, G.J., Fowler, J.S., Thanos, P., Logan, J., Gatley, S.J., Gifford, A., Ding, Y.S., Wong, C., Pappas, N., 2002. Brain DA D2 receptors predict reinforcing effects of stimulants in humans: replication study. Synapse 46, 79–82.

Volkow, N.D., Wang, G.J., Telang, F., Fowler, J.S., Logan, J., Childress, A.R., Jayne, M., Ma, Y., Wong, C., 2006. Cocaine cues and dopamine in dorsal striatum: mechanism of craving in cocaine addiction. J. Neurosci. 26, 6583–6588.

Wagner Jr., H., Burns, H.D., Dannals, R.F., Wong, D.F., Langstrom, B., Duelfer, T., Frost, J.J., Ravert, H.T., Links, J.M., Rosenbloom, S.B., Lukas, S.E., Kramer, A.V., Kuhar, M.J., 1983. Imaging dopamine receptors in the human brain by positron tomography. Science 221, 1264–1266.

Wang, G.J., Geliebter, A., Volkow, N.D., Telang, F.W., Logan, J., Jayne, M.C., Galanti, K., Selig, P.A., Han, H., Zhu, W., Wong, C.T., Fowler, J.S., 2011. Enhanced striatal dopamine release during food stimulation in binge eating disorder. Obesity (Silver Spring) 19, 1601–1608.

Wang, G.J., Volkow, N.D., Fowler, J.S., Fischman, M., Foltin, R., Abumrad, N.N., Logan, J., Pappas, N.R., 1997. Cocaine abusers do not show loss of dopamine transporters with age. Life Sci. 61, 1059–1065.

Wilson, J.M., Levey, A.I., Bergeron, C., Kalasinsky, K., Ang, L., Peretti, F., Adams, V.I., Smialek, J., Anderson, W.R., Shannak, K., Deck, J., Niznik, H.B., Kish, S.J., 1996. Striatal dopamine, dopamine transporter, and vesicular monoamine transporter in chronic cocaine users. Ann. Neurol. 40, 428–439.

Wong, D.F., Kuwabara, H., Schretlen, D.J., Bonson, K.R., Zhou, Y., Nandi, A., Brasic, J.R., Kimes, A.S., Maris, M.A., Kumar, A., Contoreggi, C., Links, J., Ernst, M., Rousset, O., Zukin, S., Grace, A.A., Lee, J.S., Rohde, C., Jasinski, D.R., Gjedde, A., London, E.D., 2006. Increased occupancy of dopamine receptors in human striatum during cue-elicited cocaine craving. Neuropsychopharmacology 31, 2716–2727.

Wu, J.C., Bell, K., Najafi, A., Widmark, C., Keator, D., Tang, C., Klein, E., Bunney, B.G., Fallon, J., Bunney, W.E., 1997. Decreasing striatal 6-FDOPA uptake with increasing duration of cocaine withdrawal. Neuropsychopharmacology 17, 402–409.

Yokoo, H., Yamada, S., Yoshida, M., Tanaka, M., Nishi, S., 1992. Attenuation of the inhibitory effect of dynorphin on dopamine release in the rat nucleus accumbens by repeated treatment with methamphetamine. Eur. J. Pharmacol. 222, 43–47.

Zachariou, V., Bolanos, C.A., Selley, D.E., Theobald, D., Cassidy, M.P., Kelz, M.B., Shaw-Lutchman, T., Berton, O., Sim-Selley, L.J., Dileone, R.J., Kumar, A., Nestler, E.J., 2006. An essential role for DeltaFosB in the nucleus accumbens in morphine action. Nat. Neurosci. 9, 205–211.

Zhang, Y., Butelman, E.R., Schlussman, S.D., Ho, A., Kreek, M.J., 2004. Effect of the kappa opioid agonist R-84760 on cocaine-induced increases in striatal dopamine levels and cocaine-induced place preference in C57BL/6J mice. Psychopharmacology 173, 146–152.

Zhang, Y., Loonam, T.M., Noailles, P.A., Angulo, J.A., 2001. Comparison of cocaine- and methamphetamine-evoked dopamine and glutamate overflow in somatodendritic and terminal field regions of the rat brain during acute, chronic, and early withdrawal conditions. Ann. N.Y. Acad. Sci. 937, 93–120.

Zhou, Y., Spangler, R., Schlussman, S.D., Yuferov, V.P., Sora, I., Ho, A., Uhl, G.R., Kreek, M.J., 2002. Effects of acute "binge" cocaine on preprodynorphin, preproenkephalin, proopiomelanocortin, and corticotropin-releasing hormone receptor mRNA levels in the striatum and hypothalamic-pituitary-adrenal axis of mu-opioid receptor knockout mice. Synapse 45, 220–229.

CHAPTER SIX

Stress, Anxiety, and Cocaine Abuse

Caryne P. Craige, Nicole M. Enman and Ellen M. Unterwald
Department of Pharmacology and Center for Substance Abuse Research, Temple University School of Medicine, Philadelphia, PA, USA

1. INTRODUCTION

1.1. Cocaine Addiction

A major obstacle in the treatment of cocaine addiction is high susceptibility to relapse. Anxiety and enhanced sensitivity to stress are two major contributors to the maintenance of the repetitive cycles of cocaine abuse. Early stages of cocaine addiction are dominated by bingeing and intoxication as a result of acute drug intake, and are largely driven by positive reinforcement mechanisms or reward. Later stages of cocaine dependence manifest in a negative affective state during drug withdrawal and the development of intense craving for the drug that increases during prolonged abstinence. The withdrawal syndrome resulting from chronic cocaine abuse in humans is characterized by the emergence of dysphoria, irritability, and anxiety, which may be potentiated in response to stress. Individuals may return to drug use in order to remove these aversive emotional or physical states during withdrawal or following the experience of stress. Hence, individuals may return to drug use in order to remove these aversive emotional or physical states during withdrawal or following the experience of stress, resulting in negative reinforcement of drug-seeking behavior (Koob and Volkow, 2010). The negative affective state that manifests during withdrawal is mediated by multiple neurotransmitter circuits in the central nervous system; the contribution of these neurotransmitters and neuromodulators to withdrawal-induced anxiety and stress-induced relapse will be discussed in this chapter. Investigation of the mechanisms regulating anxiety and atypical responsivity to stress during cocaine abstinence is necessary for the development of pharmacological interventions that help prevent relapse to cocaine use.

1.2. Cocaine Withdrawal-Induced Anxiety

Anxiety is a prominent symptom of the negative affective state experienced during acute cessation from cocaine use (Coffey et al., 2000; Cottler et al., 1993; Margolin et al., 1996; Satel et al., 1991; Weddington, 1992; Weddington et al., 1990). Human cocaine abusers report heightened anxiety during the so-called "crash" period following chronic cocaine use or cocaine "binges" (Gawin and Kleber, 1986; Gawin, 1991; Resnick and Resnick,

The Effects of Drug Abuse on the Human Nervous System
http://dx.doi.org/10.1016/B978-0-12-418679-8.00006-X

135

1984). The state of anxiety during cocaine withdrawal has been shown to improve over time; however, anxiety during early withdrawal or in response to stress during protracted withdrawal is hypothesized to drive individuals to resume drug use to remove the aversive emotional and physiological state. Indeed, during cessation from drug use, cocaine-dependent individuals exhibit heightened cocaine craving and anxiety-like symptoms like elevated heart rate and salivary cortisol levels in response to stress or drug cues (Sinha, 2008). Thus, anxiety during cocaine withdrawal is a motivating stimulus for continuous drug use, which may impede the ability of cocaine users to abstain from drug seeking and taking and may strengthen negative reinforcement mechanisms in a self-medication manner. Thus, it is important to study the underlying systems mediating anxiety during withdrawal from cocaine.

Rodent models have proven useful in the study of cocaine withdrawal, as heightened anxiety and arousal during acute cocaine withdrawal have been demonstrated by a number of methods. One example is the elevated plus maze. The rodent's natural aversion to open elevated spaces and preference for enclosed dark spaces versus its desire to explore novel areas underlies the theory behind testing anxiety behavior on the elevated plus maze (Barnett, 1975; Pellow et al., 1985). The paradigm has been validated both pharmacologically and physiologically. Pharmacological studies demonstrate that prototypical anxiolytic agents (e.g. benzodiazepines) increase the time spent on and entries onto the open arms, whereas anxiogenic agents (e.g. pentylenetetrazole and yohimbine) decrease open arm time and entries (Pellow et al., 1985). Likewise, confinement to the open arms increases freezing, defecation, and plasma corticosterone levels (Pellow et al., 1985). The elevated plus maze has been widely used in studies investigating anxiety during cocaine withdrawal. For example, early withdrawal (24–48 h) from chronic cocaine results in increased anxiety-like behaviors on the elevated plus maze, reflected as decreases in open arm entries and time (Basso et al., 1999; Perrine et al., 2008; Sarnyai et al., 1995). The defensive burying paradigm is another model commonly used for measuring anxiety-like traits. In this test, the response to a noxious stimulus, such as a shock probe, is assessed. As such, the defensive burying test is often considered a measure of stress-induced anxiety. The duration of time an animal spends burying the electrified probe provide a measure of anxiety-like behavior, such that an increase in the duration of burying indicates an elevation in anxiety (Pinel and Treit, 1978). Studies show that anxiolytic agents reduce the burying time in the defensive burying test, proving it to be a useful procedure to study anxiety-like behaviors (Rohmer et al., 1990; Treit et al., 1981). In the context of cocaine, repeated noncontingent injections of cocaine consistently enhance anxiety-like behaviors during acute withdrawal in the elevated plus maze and defensive burying paradigms (Basso et al., 1999; DeVries and Pert, 1998; Harris and Aston-Jones, 1993; Perrine et al., 2008; Rudoy and Van Bockstaele, 2007). Anxiety is also heightened during withdrawal from self-administered cocaine as assessed by the defensive burying test (Aujla et al., 2007).

Rats also demonstrate heightened startle reflexes and ultrasonic distress calls upon exposure to startling tactile stimuli during withdrawal from repeated contingent or noncontingent administration of cocaine, suggesting hypervigilance and heightened anxiety-like states (Gordon and Rosen, 1999; Mutschler and Miczek, 1998a,b).

Studies of anxiety during extended-withdrawal periods suggest that rodents may be more likely to demonstrate anxiety if reactivated by a stressful stimulus or exposure to drug-associated cues. For example, heightened anxiety-like behavior was not apparent after two to three weeks of withdrawal from escalated cocaine self-administration (Mantsch et al., 2008); however, enhanced stress-induced anxiety was demonstrated after 42 days of abstinence in similar cocaine-escalated rats (Aujla et al., 2007). In addition, rodents exhibit heightened anxiety-like behavior following 10 days of cocaine withdrawal in the elevated plus maze and light–dark transition tests only upon exposure to cocaine-associated contextual cues (Erb et al., 2006). Together, these results are consistent with human studies demonstrating that a state of heightened anxiety occurs during early cocaine withdrawal, which may decay over time or be reactivated in response to drug-related environmental cues or in response to stress. These clinical and preclinical studies provide a basis in which the contributions of specific neurotransmitter systems in different periods of cocaine withdrawal-induced anxiety can be studied.

1.3. Stress-Induced Relapse

Although the response to stress is essential for life, when it is chronic, repeated, or excessive, stress is associated with a variety of serious disorders. Importantly, stress has been shown to lead to an increased propensity of an individual to engage in substance abuse. Human studies show that life stress is not only a risk factor in the development of addiction, but also a trigger for relapse to drug use (Brown et al., 1990; Dewart et al., 2006; McFall et al., 1992; Ouimette et al., 2007), suggesting that treatment aimed at reducing the effects of stress may be therapeutically useful for drug addiction.

The direct involvement of stress in addiction has been established with both human and animal studies (Briand and Blendy, 2010). Substance abuse is more prevalent in persons with anxiety disorders and depression—conditions that are often preceded by stressful life events (Ford et al., 2004; Herrero et al., 2008; Shaffer and Eber, 2002). For example, childhood trauma such as sexual abuse has been shown to increase vulnerability to drug addiction (DeWit et al., 1999; Teusch, 2001; Triffleman et al., 1995; Walker et al., 1998). Also, rates of substance abuse and stress-induced relapse are significantly higher in combat veterans with posttraumatic stress disorder, which is precipitated by traumatic stress, than veterans without the disorder (Donovan et al., 2001; McFall et al., 1992; Ouimette et al., 2007; Penk et al., 1988; Zaslav, 1994). Likewise, preclinical studies have shown that repeated stress exposure (i.e. maternal deprivation, footshock, restraint, swim stress, and social stress) facilitates the acquisition of drug self-administration and enhances the rewarding properties of drugs of abuse including cocaine (Der-Avakian et al., 2007; Haney et al., 1995; Kabbaj et al., 2001; Piazza and Le Moal, 1998; Tidey and Miczek, 1997).

Stress-induced reinstatement of drug-seeking behavior is a rodent model used to mimic relapse to drug addiction in the context of stress. Typically in this paradigm, animals are trained to self-administer a drug for a prolonged period of time, and then are forced to go through a period of abstinence in which the drug-seeking behavior is extinguished by withholding the drug in the presence of a previously reinforced response. After this period of extinction training, the animals are exposed to an acute stressor and tested in the drug-seeking paradigm in which they were initially trained (i.e. self-administration). Acute stress elicits reinstatement to cocaine-seeking behavior in this paradigm (Ahmed and Koob, 1997; Erb et al., 1996). A second related procedure measures reinstatement to cocaine-conditioned reward using the conditioned place preference paradigm (Kreibich and Blendy, 2004; Lu et al., 2002; Sanchez et al., 2003). Taken together, human and rodent studies support the concept that exposure to stress can enhance vulnerability to the development of addiction and can perpetuate the cycle of drug addiction triggering relapse to drug-seeking behaviors.

The mechanisms by which stress increases the risk of relapse involves a complex network of neurotransmitters and neuropeptides that contribute to the subjective effects of cocaine withdrawal and associated maladaptive stress responses including intense anxiety, craving, and dysphoria. Specifically, the role of stress as a trigger for relapse during withdrawal appears to involve mainly the mesocorticolimbic reward pathway and its interaction with brain stress circuits, particularly the corticotropin releasing factor (CRF), norepinephrine, and serotonin systems. These systems undergo neuroadaptations with repeated cocaine use, which may contribute to augmented sensitivity to stress, and thereby enhance the predisposition to relapse. The role of these neurotransmitter systems in the context of stress-induced relapse to cocaine seeking will be discussed in this chapter.

2.1. NEUROTRANSMITTER SYSTEMS IN COCAINE WITHDRAWAL-INDUCED ANXIETY AND STRESS-INDUCED RELAPSE

2.1.1. Dopamine

Substantial evidence exists linking midbrain dopamine function to the generally negative affective state that occurs during cocaine withdrawal. Mesolimbic dopamine neurons undergo persistent and fluctuating changes during withdrawal. Cocaine withdrawal in animals causes a transient increase followed by a long-lasting decrease in mesolimbic dopamine efflux, as seen in Figure 1, and the number of dopamine cells firing spontaneously (Koeltzow and White, 2003; Peris et al., 1990; Rossetti et al., 1992). Preclinical studies show reductions in basal extracellular dopamine levels in the nucleus accumbens during withdrawal from chronic self-administered cocaine (Parsons et al., 1991).

In conjunction with reduced dopamine levels, brain stimulation reward thresholds are elevated during cocaine withdrawal, indicating an anhedonic state (Markou and Koob, 1992). Elevated reward thresholds during cocaine withdrawal can be attenuated by pre-treatment with the dopamine agonist, bromocriptine, providing further support for the notion that cocaine withdrawal-induced anhedonia is due to deficits in dopamine (Markou and Koob, 1992). Other studies have found reductions in the activity of midbrain ventral tegmental area neurons during withdrawal from repeated cocaine (Ackerman and White, 1992; Henry et al., 1989). The decreased neuronal activity in the ventral tegmental area likely contributes to the reduction in synaptic dopamine in the nucleus accumbens reported in cocaine-withdrawn rats, which may be related to the accompanying anhedonia.

In addition to dopamine levels, other components of the dopamine system are altered during cocaine withdrawal. The levels and function of dopamine transporters, which terminate dopamine transmission via reuptake mechanisms, are altered during cocaine withdrawal. For example, dopamine transporter levels in the nucleus accumbens are elevated three days after the last cocaine administration, and then significantly reduced 10 days after the last exposure to cocaine (Pilotte et al., 1994; Sharpe et al., 1991), a decrease which is potentially dependent on an increase in dopamine transporter protein turnover rate (Meiergerd et al., 1994). Likewise, there is a corresponding decrease in dopamine transporter mRNA 10 days after withdrawal of cocaine in similarly treated animals, suggesting that altered gene expression may underlie the changes in transporter levels (Cerruti et al., 1994). An alternative pattern of dopamine transporter expression is evident in the caudate putamen at up to 3 weeks of cocaine withdrawal. Surface

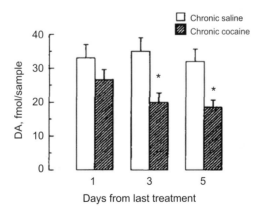

Figure 1 Time Course Data of Extracellular Dopamine Concentrations after Withdrawal of Chronic Cocaine (15 mg/kg twice a day for 18 days, Intraperitoneally) in Rats. Values represent means + SEM (*n* = 5–7/group). Cocaine-withdrawn rats exhibited reduced striatal extracellular dopamine levels as compared to saline controls at 1 day, 3 days, and 5 days following the last cocaine injection (*$p < 0.01$, student's *t*-test). *(Figure adapted from Rossetti et al., 1992; with permission from the publisher).*

expression of the dopamine transporter and levels of dopamine transporter serine phosphorylation are increased, leading to an overall increase in dopamine transporter activity and dopamine reuptake in the caudate putamen (Samuvel et al., 2008).

Changes in dopamine receptors during cocaine withdrawal have been reported; however, the results of these studies are conflicting. Some reports suggest that dopamine receptors are altered in the nucleus accumbens and frontal cortex during cocaine withdrawal. Such studies show an increase in cell surface D1 receptors and a decrease in D2 receptors in the nucleus accumbens after 24 h of withdrawal (Conrad et al., 2010). Conversely, in the premotor cortex, D2 receptors are upregulated at 30 min, 24 h, and 5 days of cocaine withdrawal, whereas premotor cortex D1 receptors are upregulated at 30 min of cocaine withdrawal, and then return to normal levels at 24 h and 5 days (Macêdo et al., 2001). Other studies have found that following 7–14 days of withdrawal from repeated cocaine administration, there is a transient decrease in D1 receptor density in the frontal cortex, caudate putamen, and nucleus accumbens (Kleven et al., 1990; Neisewander et al., 1994) and increase in D1 receptor sensitivity (Henry and White, 1991). Additionally, a microdialysis study shows that increases in medial prefrontal cortex D2 autoreceptor activity on dopamine inhibition during a 1-day cocaine withdrawal are evident, and demonstrate persistent increased function following 30 days of cocaine withdrawal compared to saline controls, which is inconsistent with other studies that show decreased D2 receptor levels in the nucleus accumbens on day 45 of cocaine withdrawal (Conrad et al., 2010; Liu and Steketee, 2011). Another study shows that altered activity of midbrain dopamine neurons following 7 days of withdrawal from intermittent cocaine is normalized by D2 receptor stimulation (Lee et al., 1999). These alterations likely contribute to dysfunction of the dopamine system during cocaine withdrawal. Evidence shows a reduction in inhibitory D2-mediated regulation of local glutamate release and cortical excitability during short-term cocaine withdrawal, a reduction that affects downstream targets of the prefrontal cortex, such as the nucleus accumbens and ventral tegmental area following short- and long-term cocaine withdrawals (Liu and Steketee, 2011; Nogueira et al., 2006).

However, other reports suggest conflicting patterns of dopamine receptor alterations during cocaine withdrawal. For example, studies show that increases in D1 and D2 receptor mRNA exhibited 2 h following the last cocaine infusion are normalized by day 7 of cocaine withdrawal in the limbic region. The receptor binding portion of this study found that no changes in D2 receptor number were found in the limbic region or frontal cortex; however, frontal cortex and limbic region D1 receptor density decreased 2 h following the last cocaine infusion, an alteration that persisted in only the limbic region following 7 days of withdrawal (Laurier et al., 1994). An additional study using positron emission tomography shows that striatal D1 receptors are decreased immediately following chronic binge cocaine administration, and these levels return to normal during 10–21 days of withdrawal. The same study found that striatal D2 receptors are decreased at 10 days of withdrawal from chronic binge cocaine administration, a reduction which returns to baseline levels at 21 days of withdrawal (Maggos et al., 1998). These studies demonstrate that D1 and D2 receptors may undergo differing alterations in various brain regions as a result of cocaine administration and subsequent withdrawal. This change in pattern of dopamine receptor expression may contribute to the negative symptoms occurring at the time points discussed during cocaine withdrawal.

Human brain imaging studies provide support for preclinical data, with reduced dopamine transmission and D2 receptors seen in the frontal cortex and striatal regions in cocaine abusers during both acute and protracted withdrawal (up to 3–4 months) (Volkow et al., 1993, 1997). These studies show an association between the low levels of D2 receptors with a reduction in brain glucose metabolism in frontal lobe regions in these individuals. The change in metabolic function could indicate cocaine-induced alterations of neurotransmitter function, and could also account for the blunted response of cocaine abusers to dopamine receptor agonists. Additional imaging studies have shown that cerebral blood flow in the prefrontal cortex is decreased during cocaine withdrawal (Volkow et al., 1988). The frontal cortex is hypothesized to inhibit anxiety circuits and dopamine-mediated behaviors such as decision making (de Visser et al., 2011; Jackson and Moghaddam, 2001; Maier et al., 2012; Ohata and Shibasaki, 2011; Sullivan and Gratton, 1998; Wang and Pickel, 2002). Therefore, dysfunction of the frontal cortex may lead to an inability to inhibit both anxiety during withdrawal and drug-seeking behavior in response to craving, thereby contributing to compulsive cocaine use.

Chronic or repeated stress leads to alterations in the regulation of brain reward circuitry mediated by the mesolimbic dopamine pathway, resulting in heightened sensitivity to the reinforcing properties of drugs (Piazza and Le Moal, 1998). Likewise, long-term changes in brain reward pathways that are a result of chronic drug abuse also play a key role in craving, drug-seeking behavior, and relapse (Kalivas et al., 1998), increasing the motivation to use drugs compulsively. Therefore, interactions of stress neurocircuitry and dopamine dysregulation occur in brain areas largely involved in behaviors associated with drive, motivation, and decision making. These effects of stress and chronic drug use act to increase an individual's vulnerability to drug use or susceptibility to relapse.

Studies have identified a role for the prefrontal cortex dopamine circuitry in regulating stress-induced reinstatement of cocaine seeking. Footshock-induced reinstatement of cocaine seeking is blocked by injection of flupenthixol, a D1/D2 receptor antagonist, into the dorsal prefrontal cortex (McFarland et al., 2004) and by injection of D1-like (SCH 23390), but not D2-like receptor antagonists (raclopride) into the medial prefrontal cortex or orbitofrontal cortex (Capriles et al., 2003). Likewise, injection of a D1-like receptor antagonist into the medial prefrontal cortex blocks the effects of immobilization stress on the reinstatement of extinguished cocaine-conditioned place preference (Sanchez et al., 2003). Some studies identify a role for glutamate and dopamine interactions in stress-induced reinstatement to cocaine seeking. It is established that stress increases glutamate release in the ventral tegmental area, which in turn enhances dopamine activity, identifying a potential mechanism by which the D1/D2 receptor antagonist acts in blockade of stress-induced reinstatement by attenuating heightened dopamine levels (McFarland et al., 2004; Wang et al., 2005). In addition, inactivation of the prefrontal cortex blocks stress-induced glutamate increases while concurrently blocking footshock-induced reinstatement of cocaine-seeking behavior (McFarland et al., 2004). Therefore, the evident rise in glutamate following exposure to stress could mediate the rise in dopamine on which stress-induced reinstatement depends.

Consideration of the interactions of the dopamine system with other stress-activated systems like the glucocorticoid, CRF, and noradrenergic systems is important for understanding the regulation of stress-induced reinstatement. It has been shown that stress exposure and increased levels of glucocorticoids enhance dopamine release in the nucleus accumbens and ventral striatum (Sinha, 2008). However, chronic glucocorticoids inhibit dopamine synthesis and turnover (Pacak et al., 2002) suggesting that alterations in the hypothalamic pituitary adrenal axis (HPA) axis and glucocorticoids can affect dopamine neurotransmission, and potentially contribute to the development of a negative affective state. In addition, it has been shown that a D1/5 receptor antagonist, but not D2/3 receptor antagonist, blocks reinstatement of cocaine self-administration induced by either the stress peptide CRF or pharmacological stressor yohimbine, an alpha 2-adrenergic receptor antagonist, which enhances norepinephrine release, suggesting a role for D1/5 receptors in mediating stress-induced reinstatement (Brown et al., 2012).

2.2. SEROTONIN

The role of serotonin in the etiology and treatment of anxiety is well established. Selective serotonin reuptake inhibitors are approved for use in a variety of anxiety disorders. This class of drugs has proven to be effective in the alleviation of anxiety symptoms through their resultant increase in extracellular serotonin levels (Davidson, 2009; Gartside et al., 1995). In fact, in vivo microdialysis experiments show reductions in serotonin levels during cocaine withdrawal in the nucleus accumbens (Parsons et al., 1995). Reduced serotonin neurotransmission has been associated with many psychiatric disorders and symptoms including depression (Meltzer and Maes, 1995), compulsive behavior (Dolberg et al., 1996), and panic disorder (Charney and Heninger, 1986), all of which have been linked to addictive behaviors (Koob et al., 1998; Markou et al., 1998; O'Brien et al., 1998). Thus, reductions in the serotonin system potentially contribute to the heightened anxiety during cocaine withdrawal, and targeting components of the serotonin pathways during abstinence may prove therapeutically advantageous in the prevention of relapse to addiction.

The withdrawal-related deficiency in serotonin neurotransmission is consistent with the development of neural adaptations during the course of chronic cocaine exposure. Preclinical studies demonstrating changes in serotonin receptor function and transport show that with repeated cocaine administration, inhibitory 5-HT$_{1A}$ receptors increase in sensitivity (Cunningham et al., 1992; King et al., 1993; Levy et al., 1992), particularly in the dorsal raphe (Cunningham et al., 1992), whereas excitatory 5-HT$_{2A}$ receptors show decreased sensitivity (Levy et al., 1992). These studies indicate that following chronic cocaine administration, serotonin receptors undergo specific alterations in sensitivity levels that render each serotonin receptor to act in a dysfunctional state. In addition, increases in the density of prefrontal cortex and frontal cortex serotonin uptake sites have been associated with repeated cocaine exposure (Cunningham et al., 1992) and may further decrease the levels of serotonin during withdrawal. Other studies have demonstrated deficits in presynaptic central serotonin function lasting as long as 10 days after withdrawal from cocaine (Baumann et al., 1995; Darmani et al., 1997). Upon the cessation of repeated cocaine, dysfunction of serotonin receptors, uptake mechanisms, and presynaptic serotonin function may persist and contribute to reduced serotonin neurotransmission and increased anxiety.

Dysfunction at the level of serotonin receptors and neurotransmission may contribute to the aforementioned dysregulation of the dopamine system. The 5-HT$_{2A}$ receptors have excitatory influences on cocaine-induced dopamine release, and the 5-HT$_{1A}$ receptors and 5-HT$_{2C}$ receptors act in an oppositional inhibitory manner on cocaine-induced dopamine regulation (Bubar and Cunningham, 2006). The integration of activity of each of these receptors contributes to the overall control of dopamine regulation by the serotonin system in the context of cocaine addiction and withdrawal. Data from our laboratory show that antagonism of 5-HT$_{2C}$ receptors with SB 242084 in cocaine-withdrawn mice blocks the expression of anxiety-like behavior on the elevated plus maze, as seen in Figure 2 (Craige et al., 2012). As 5-HT$_{2C}$ receptors are expressed specifically on GABA neurons, this mechanism of 5-HT$_{2C}$ receptor control is thought to arise via a negative feedback mechanism involving the effects of GABA on the serotonin system. In fact, electrophysiology recordings of cells from mice undergoing cocaine withdrawal demonstrate increased GABA frequency, and thus, heightened GABA receptor activity in dorsal raphe serotonin neurons as compared to cells from saline-injected controls and this effect is diminished upon bath application of the 5-HT$_{2C}$ receptor antagonist SB 242084 (Craige et al., 2012). In this schema, withdrawal from chronic cocaine leads to increased anxiety and concomitant increased inhibitory GABA activity, which are both attenuated upon 5-HT$_{2C}$ receptor blockade. This mechanism is proposed to contribute to the downregulation of the serotonin and dopamine systems associated with anxiety during withdrawal.

The major source of serotonin in the brain originates in the dorsal raphe nucleus. Projections from the dorsal raphe innervate areas of the forebrain, contributing to the regulation of stress systems in these brain regions (Lowry, 2002). In addition, the dorsal raphe has been identified as a brain region that regulates anxiety-like behaviors associated with cocaine exposure. Interestingly, inactivation of the dorsal raphe, and thus,

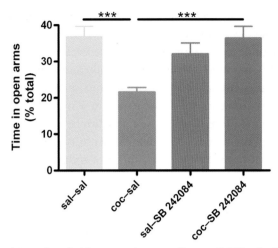

Figure 2 Mice exposed to a chronic binge cocaine paradigm and 24 h of withdrawal with a saline injection 1 h prior to elevated plus maze testing (coc–sal) spent significantly less time in the open arms than the saline-administered controls (sal–sal), indicating heightened anxiety (Two way ANOVA, ***$p < 0.001$). Administration of 1 mg/kg SB 242084 1 h prior to elevated plus maze testing resulted in a blockade of the expression of anxiety-like behavior in cocaine-withdrawn mice (Two way ANOVA, ***$p < 0.001$). *(Figure adapted from Craige et al., 2012; Society for Neuroscience abstract).*

blockade of serotonin signaling, blocks the expression of anxiety in cocaine-withdrawn rats (Ettenberg et al., 2011). This study is contradictory to other reports showing that reductions in serotonin signaling increase anxiety during cocaine withdrawal, thus the exact mechanism by which serotonin regulates anxiety during cocaine withdrawal is debated and may be brain region-specific. In addition, the effect of dorsal raphe inactivation on other neurotransmitter circuits like the GABA system should be considered when interpreting these results.

It is likely that the serotonin system is important in anxiety during cocaine withdrawal, potentially through its interactions with the dopamine system, which is dysregulated during cocaine withdrawal. Indeed, the dorsal raphe nucleus serotonin system plays a complex role in addiction through its ability to regulate activity of dopamine neurons and dopamine release in specific target areas such as the nucleus accumbens and ventral tegmental area (De Deurwaerdère and Spampinato, 1999; Gervais and Rouillard, 2000). In further support of the importance of serotonin-dopamine interactions, it has been established that dysregulation of the serotonin system is associated with decreased mesolimbic function during cocaine withdrawal (Parsons et al., 1995). This study suggests that the decline in serotonin levels, which precedes the decrease in dopamine neurotransmission, may correspond to the onset of postcocaine reward deficits associated with anxiety during withdrawal. This interpretation is supported by the finding that brain stimulation reward thresholds are sensitive to serotonergic manipulations (Gibson et al., 1970). In addition, there is evidence that selective serotonin reuptake inhibitors used to treat anxiety disorders exert their effects through not only increasing serotonin levels, but increasing dopamine levels as well (Clark et al., 1996).

Biochemical alterations in serotonin regulation have been identified in the context of stress and addiction. For example, stressed rats have increased serotonin transporter densities in the nucleus accumbens and ventral tegmental area, suggesting possible compensatory effects of elevated serotonin in these regions (Kohut et al., 2012). Another study investigated the p38α MAPK signaling cascade downstream of serotonin receptor stimulation in stress-induced reinstatement to cocaine seeking. A role for p38α MAPK in serotonin cells was identified, such that animals lacking p38α MAPK in serotonin-producing cells of the dorsal raphe do not show reinstatement of cocaine place preference following social defeat stress. In addition, stress-induced activation of p38α MAPK translocates the serotonin transporter to the plasma membrane and increases the rate of transmitter reuptake at serotonergic nerve terminals (Bruchas et al., 2011). It is known that stress-induced activation of the dynorphin/kappa opioid (KOP) receptor system by stimulation of p38α MAPK increases serotonin transport into the striatum, which may mediate some of the negative influences of stress exposure leading to cocaine relapse (Schindler et al., 2012).

In addition to the recognized effects of serotonin on dopamine, it appears that serotonin and CRF and/or norepinephrine interactions are also critical in the stress response in drug addiction. According to Kirby and colleagues, a hyposerotonergic state results from CRF1 receptor-mediated inhibition of the serotonin system via enhanced GABA release onto serotonin neurons (Kirby et al., 2008). This decrease in serotonin levels via CRF influence is hypothesized to promote impulsive behavior leading to the initiation of drug abuse, or may be a contributing factor in relapse to drug abuse (Virrkunen and Linnoila, 1990). In addition, a preclinical study used the pharmacological noradrenergic stressor, yohimbine, to elicit stress-induced reinstatement of cocaine seeking. In this study, the 5-HT_{2C} receptor agonist Ro 60-0175 dose-dependently attenuates yohimbine-induced reinstatement, identifying a role for 5-HT_{2C} receptor activity in mediating the effects of the noradrenergic stress system on cocaine reinstatement (Fletcher et al., 2007).

2.3. CORTICOTROPIN RELEASING FACTOR

Studies demonstrate a major role for CRF in anxiety and the autonomic and behavioral responses to stress (Bale and Vale, 2004), as well as the behavioral effects of cocaine (Sarnyai et al., 1995). As such, these studies have warranted investigation of this stress peptide in the robust anxiogenic effects of early cocaine withdrawal and stress-induced cocaine reinstatement. CRF is expressed in the hypothalamus and extrahypothalamic brain regions modulating stress, emotion, and cognition (Owens and Nemeroff, 1991), where cocaine- and stress-induced alterations in the CRF system can be observed. It has been shown in vivo that acute cocaine decreases tissue levels of CRF in the hypothalamus, basal forebrain, frontal cortex, and hippocampus, but increases CRF in the amygdala (Gardi et al., 1997; Sarnyai et al., 1993). In contrast, extracellular levels of CRF are decreased in the central amygdala during the first 12 h of unrestricted cocaine-access in self-administering rats (Richter and Weiss, 1999). Repeated cocaine injections initially induce short-lasting increases in CRF gene expression in the hypothalamus, amygdala, and frontal cortex (Maj et al., 2003; Zhou et al., 1996), whereas decreased amygdalar and forebrain CRF binding sites are observed shortly following repeated cocaine administration (Ambrosio et al., 1997; Goeders et al., 1990). A microdialysis study shows that repeated cocaine also results in augmented levels of extracellular CRF in the central amygdala upon cocaine challenge (Richter et al., 1995), while self-administering rats exhibit differential regulation of tissue CRF in the amygdala and dorsal raphe depending on cocaine availability (Zorrilla et al., 2012).

Cessation of cocaine administration, specifically 24–48 h of withdrawal, induces a reliable and robust increase in anxiety-like behavior, which is associated with altered CRF function within the amygdala. For example, extracellular levels of CRF are robustly increased after 12 h of withdrawal in cocaine self-administering rats, as seen in Figure 3 (Richter and Weiss, 1999). Similarly, gene expressions of amygdalar CRF and CRF-binding protein are also increased after 24 h of cocaine withdrawal (Erb et al., 2004; Maj et al., 2003). In contrast, decreases in tissue levels of CRF were observed in the amygdala after 24 (Zorrilla et al., 2001) and 48 h of withdrawal, concurrently with increased anxiety-like behavior (Sarnyai et al., 1995). A more direct role for CRF in the anxiogenic effects of cocaine withdrawal has been established with CRF antiserum and CRF receptor antagonists. Administration of CRF antiserum prior to daily cocaine injections blocks the development of anxiety during cocaine withdrawal (Sarnyai et al., 1995), while CRF antagonists administered acutely within 48 h of cocaine withdrawal also attenuate anxiety-like behavior (Basso et al., 1999; DeVries and Pert, 1998).

The duration of anxiety and the role of CRF beyond the first 48 h of cocaine withdrawal is not clearly established (Fontana and Commissaris, 1989; Gordon and Rosen, 1999; Mantsch et al., 2008; Mutschler and Miczek, 1998a). Studies generally report that alterations in CRF gene expression and receptor binding sites return to basal levels within 10 days of cocaine abstinence (Ambrosio et al., 1997; Zhou et al., 1996). However, elevated levels of CRF in the amygdala and frontal cortex have been shown to persist as long as 6 weeks of cocaine withdrawal (Zorrilla et al., 2001). Some studies suggest that a cocaine history may alter central nervous system responsivity to CRF after longer withdrawal periods, as centrally administered CRF enhanced c-fos expression in the amygdala following 10 days of cocaine withdrawal, but did not have the same effect in cocaine naïve animals (Erb et al., 2005). Enhanced CRF-induced long-term potentiation in the central amygdala and corticostriatal circuit is also present in animals following two weeks of cocaine withdrawal (Fu et al., 2007; Guan et al., 2010; Krishnan et al.,

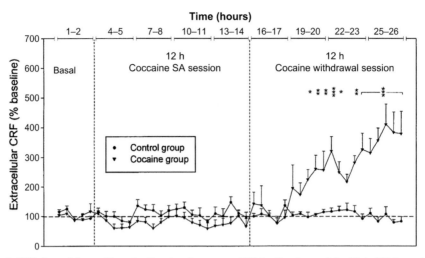

Figure 3 Withdrawal from Chronic Cocaine Increases CRF in the Amygdala. Male Wistar rats were trained to self-administer cocaine (0.25 mg/infusion) on a limited access schedule of reinforcement during daily 3-h sessions. Following 14–16 days of limited access to cocaine, rats received 12-h of continuous access to cocaine (Cocaine group) or were connected to the infusion system without access to cocaine infusions (Control group). Dialysates were collected from the central nucleus of the amygdala at baseline, during the cocaine self-administration session, and during subsequent withdrawal. Extracellular concentrations of CRF were significantly increased following termination of continuous cocaine access (*$p < 0.05$, **$p < 0.01$, ***$p < 0.001$). *(Figure adapted from Richter and Weiss, 1999; with permission from the publisher).*

2010; Pollandt et al., 2006). Another study suggests that the role of CRF in anxiety during extended-withdrawal may only be apparent upon presentation of cocaine-associated cues to induce anxiety-like behavior (Erb et al., 2006).

It is well recognized that chronic cocaine enhances sensitivity to stress and that stress has the ability to provoke relapse during extended periods of withdrawal. Pre-clinical studies have identified a role for CRF in mediating stress-induced reinstatement to cocaine-seeking behavior following protracted withdrawal. For example, centrally administered CRF induces reinstatement to cocaine-seeking behavior, and stress-induced reinstatement is attenuated by CRF antagonists, as seen in Figure 4 (Erb et al., 1998; Erb and Stewart, 1999; Lu et al., 2001; Shaham et al., 1998). Interestingly, CRF and stressors both reinstate rats with a history of unlimited access to cocaine self-administration more effectively than those with limited access to cocaine (Mantsch et al., 2008). Similarly, CRF receptor antagonists decrease self-administration more effectively in rats that have escalated their cocaine intake upon unlimited access to the drug (Specio et al., 2008). The role of CRF in stress-induced reinstatement is mediated in part by the extended amygdala, namely the bed nucleus of the stria terminalis (BNST), as well as the mesolimbic dopamine pathway (Wang et al., 2006; Wise and Morales, 2010). CRF administered locally to the ventrolateral BNST or ventral

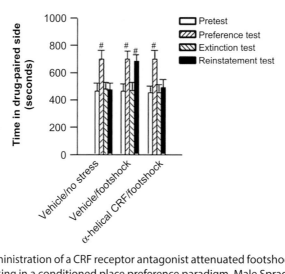

Figure 4 Central administration of a CRF receptor antagonist attenuated footshock-induced reinstatement to cocaine seeking in a conditioned place preference paradigm. Male Sprague-Dawley rats were tested for initial preference for either side of a conditioning chamber on day 0 (Pretest). Rats were conditioned on alternating days with cocaine (10 mg/kg, i.p.) or saline on days 2–7, followed by a Preference test for the drug-paired side of the conditioning chamber on day 7. Rats underwent a drug-free period of 28 days followed by an Extinction test on day 36, exposure to intermittent footshocks or no stress on day 37, and a test for Reinstatement to cocaine seeking on day 38. Rats conditioned with cocaine showed an increase in time spent in the cocaine-paired side of the conditioning chamber in the Preference test compared to the Pretest, which was abolished in the Extinction test. Rats exposed to footshocks demonstrated significant reinstatement to cocaine-seeking behavior as indicated by an increase in time spent on the cocaine-paired side compared to the Pretest. Footshock-induced reinstatement to cocaine-seeking behavior was blocked by i.c.v administration of α-helical CRF (10 μg), a nonselective CRF receptor antagonist, as the time spent in the cocaine-paired side of the chamber was similar to the Pretest ($\#p < 0.05$). *(Figure adapted from Lu et al., 2001; with permission from the publisher).*

tegmental area induces reinstatement to cocaine-seeking behavior, and stress-induced reinstatement is blocked by CRF receptor antagonists microinjected into either of these regions (Blacktop et al., 2011; Erb and Stewart, 1999; Sahuque et al., 2006; Walker et al., 2009; Wang et al., 2006, 2007; Wise and Morales, 2010). It has been demonstrated that a CRF projection from the central amygdala to the BNST contributes to stress-induced reinstatement (Cassell et al., 1986; Cummings et al., 1983; Sakanaka et al., 1986). It is also hypothesized that the BNST may mediate reinstatement behavior via projections to the mesolimbic dopamine pathway (Georges and Aston-Jones, 2001, 2002; Rodaros et al., 2007; Shaham et al., 2000). The ventral tegmental area receives both excitatory and inhibitory CRF inputs from the limbic forebrain and expresses CRF receptors on dopaminergic neurons. Importantly, stress causes the release of CRF in the ventral tegmental area (Wang et al., 2005; Wise and Morales, 2010), where CRF is known to increase the firing rate of dopamine neurons (Wanat et al., 2008). Therefore, these effects in the ventral tegmental area are hypothesized to modulate behavioral stress responses and enhanced cocaine-seeking behavior (Haass-Koffler and Bartlett, 2012; Korotkova et al., 2006; Rodaros et al., 2007; Van Pett et al., 2000; Wanat et al., 2008; Wang et al., 2007; Wise and Morales, 2010).

Interactions between CRF and other neurotransmitter circuits including norepinephrine (Brown et al., 2009; Shaham et al., 2000) and serotonin (Valentino et al., 2010; Zorrilla et al., 2012) have also been hypothesized as putative mechanisms by which CRF modulates anxiety during acute cocaine withdrawal or stress-induced reinstatement to cocaine-seeking behavior.

2.4. NOREPINEPHRINE

Heightened anxiety during acute withdrawal from cocaine has been attributed in part to dysregulation of the noradrenergic system. Clinical support for this hypothesis is provided by human studies in which yohimbine elicits panic attacks during acute cocaine withdrawal (McDougle et al., 1994). In addition to this finding, the β-adrenergic antagonist propranolol and partial α1 receptor agonist buspirone are effective in ameliorating the symptoms of acute cocaine withdrawal (Giannini et al., 1993; Kampman et al., 2001). Altered noradrenergic function has also been observed in rodents following repeated cocaine exposure (Baumann et al., 2004), and multiple noradrenergic receptor subtypes have been implicated in the etiology of anxiety manifested during acute cocaine withdrawal. For example, activation of the α2 adrenergic receptor, which inhibits norepinephrine cell firing and release (Abercrombie et al., 1988; Aghajanian and VanderMaelen, 1982; Mongeau et al., 1997) attenuates anxiety-like behavior following 48 h of cocaine withdrawal (Buffalari et al., 2012). After one week of cocaine withdrawal, neurons in the locus coeruleus are sensitized to cocaine and show enhanced sensitivity to the α2 adrenergic autoreceptor agonist clonidine (Harris and Williams, 1992). The nonselective β-adrenergic receptor antagonist propranolol and selective β1 receptor antagonists atenolol and betaxolol diminish anxiety-like behavior following 48 h of withdrawal from cocaine, as seen in Figure 5 (Harris and Aston-Jones, 1993; Rudoy and Van Bockstaele, 2007). The effectiveness of β-adrenergic receptor antagonists during acute withdrawal may be mediated by the central amygdala, which receives noradrenergic input from the locus coeruleus (Clayton and Williams, 2000; Pitkanen, 2000), as β1 receptors and downstream signaling through protein kinase A are upregulated in this region following chronic cocaine administration and withdrawal (Rudoy et al., 2009). Administration of betaxolol during early withdrawal from cocaine blocks this increase in amygdalar β1 receptor protein, as well as enhanced downstream adrenergic signaling (Rudoy et al., 2009; Rudoy and Van Bockstaele, 2007). Evidence also suggests an interaction between noradrenergic and CRF systems at the level of the amygdala, as β1 receptors are localized on CRF-containing neurons and betaxolol attenuates cocaine withdrawal-induced increases in CRF gene expression in this region (Rudoy et al., 2009).

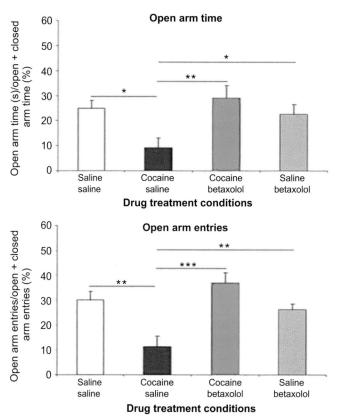

Figure 5 Betaxolol, a β1 Adrenergic Receptor Antagonist, Reversed Anxiety-Like Behaviors during Cocaine Withdrawal. Male Sprague-Dawley rats received daily cocaine (20 mg/kg, i.p.) or saline injections for 14 days. Betaxolol (5 mg/kg, i.p.) or saline was administered twice (24- and 44-h) after the last cocaine injection, followed by anxiety testing on the elevated plus maze 2 h later. Rats that received chronic cocaine exhibited significantly less time spent and entries onto the open arms of the plus maze compared to saline-injected controls, thereby demonstrating an increase in anxiety-like behavior during early cocaine withdrawal. Pretreatment with betaxolol significantly reversed the anxiety induced by cocaine withdrawal, indicated by a significant increase in time spent and entries onto the open arms compared to cocaine-injected rats pretreated with saline before testing (**$p < 0.01$, ***$p < 0.001$). *(Figure adapted from Rudoy and Van Bockstaele, 2007; with permission from the publisher).*

A role for the noradrenergic system has also been indicated during extended-withdrawal, a time when both humans and rodents exhibit heightened sensitivity to stress and susceptibility to relapse to cocaine-seeking behavior. Pharmacological inhibition of α2 adrenergic autoreceptors by yohimbine increases norepinephrine cell firing and release (Cooper et al., 1991), and reinstates cocaine-seeking behavior in rodents and nonhuman primates (Brown et al., 2012; Feltenstein and See, 2006; Lee et al., 2004). In agreement, activation of central α2 adrenergic autoreceptors by clonidine relieves stress-induced craving in human cocaine addicts (Jobes et al., 2011), while multiple α2 receptor agonists attenuate stress-induced norepinephrine release and prevent reinstatement to cocaine-seeking behavior in rodents (Erb et al., 2000). Also, central administration of norepinephrine induces reinstatement to cocaine seeking and this can be blocked by CRF antagonists, whereas the α2 receptor agonist clonidine fails to interfere with CRF-induced reinstatement (Brown et al., 2009). As in acute cocaine withdrawal, evidence also suggests that CRF mediates the downstream effects of norepinephrine during protracted withdrawal. While altered β1 adrenergic receptor levels are no longer apparent after 12 days of cocaine withdrawal (Rudoy and Van Bockstaele, 2007), β-adrenergic neurotransmission still appears to mediate the effects of stress on reinstatement behavior. These effects have been shown to occur via the central amygdala and BNST, as administration of a β1/2 receptor antagonist cocktail into either region attenuates stress-induced reinstatement (Leri et al., 2002).

2.5. ENDOGENOUS OPIOIDS

The endogenous opioid system has been shown to influence basal levels of anxiety, as well as to play an etiological role in cocaine-induced anxiety.

Endogenous opioids can inhibit the HPA axis and stress responses (Kreek et al., 2002). Interestingly, different opioid peptides and opioid receptors play opposing role in regulating anxiety-like responses. In the case of the delta opioid receptor (DOP receptor), evidence suggests that DOP receptors mediate anxiolytic responses and may supply a positive endogenous tone over emotional states, as DOP receptor activation by endogenous enkephalins maintains the homeostasis of emotional behaviors (Nieto et al., 2005). Mice with genetic deletion of DOP receptors show higher levels of anxiety-like behaviors compared with wild-type controls (Filliol et al., 2000). Likewise, pharmacological antagonism of DOP receptors can be anxiogenic in rodents (Marin et al., 2003; Perrine et al., 2006; Saitoh et al., 2005) and corticosterone levels are increased by DOP receptor antagonists (Saitoh et al., 2005). Mice lacking enkephalins have elevated levels of anxiety (Bilkei-Gorzo et al., 2004) and are hypersensitive to stressful stimuli (Kung et al., 2010). Further support of the importance of the DOP system in anxiety comes from pharmacological studies of DOP receptor agonists. The selective DOP receptor agonist SNC80 dose-dependently reduces anxiety-like behaviors in rodents (Perrine et al., 2006; Saitoh et al., 2004), as does administration of the DOP receptor-selective enkephalin derivative, D-Penicillamine(2,5)-enkephalin (DPDPE), when microinjected directly into the central nucleus of the amygdala (Randall–Thompson et al., 2010). Therefore, activation of DOP receptors by endogenous opioid peptides appears to be critical in the homeostatic control of stress responses and anxiety.

Data from our laboratory demonstrate a dysregulation of DOP receptor function after chronic cocaine exposure, and this dysregulation may contribute to heightened anxiety that accompanies cocaine dependence and withdrawal. Two weeks of repeated cocaine administration to male rats attenuates DOP receptor function in the caudate putamen and nucleus accumbens as shown by the ability of DOP receptor agonists to regulate adenylyl cyclase activity (Perrine et al., 2008; Unterwald et al., 1993). Twenty-four hours after withdrawal from chronic administration of cocaine, DOP receptor function is downregulated in the striatum and amygdala and heightened anxiety-like responses are seen. Acute administration of a selective DOP receptor agonist reverses the heightened anxiety seen during cocaine withdrawal, as seen in Figure 6 (Perrine et al., 2008). Similar findings were found in female rats (Ambrose-Lanci et al., 2010). These data provide evidence that early withdrawal from cocaine can increase anxiety-like behaviors, which may be mediated in part by reduced DOP receptor function.

In opposition to the DOP receptor system, KOP receptors and dynorphin contribute to anhedonia and dyphoria, and potentially, anxiety. Prodynorphin deficient mice show reduced levels of anxiety-like behaviors compared with wild-type controls (Kastenberger et al., 2012; Wittmann et al., 2008). Chronic exposure to cocaine in both humans and rodents is associated with increases in dynorphin and preprodynorphin mRNA levels in the nucleus accumbens and caudate putamen (Daunais et al., 1995; Frankel et al., 2008; Hurd and Herkenham, 1993; Spangler et al., 1993; Steiner and Gerfen, 1993). Further, genetic ablation of KOP receptors from dopamine neurons produces an anxiolytic effect, and increases the sensitizing effects of cocaine (Van't Veer et al., 2013). The KOP/dynorphin system, together with its interactions with CRF, has been shown to be a significant regulator of the neurochemical and behavioral responses to cocaine (Bruchas et al., 2010, 2011). The heightened KOP tone during repeated cocaine exposure likely contributes to anxiety, anhedonia, and depression seen during cocaine withdrawal.

Substantial data indicate that the KOP system contributes to stress-induced cocaine reinstatement. For example, administration of KOP receptor agonists alone can reinstate cocaine seeking (Redila and Chavkin, 2008; Valdez et al., 2007). KOP receptor antagonists can reduce reinstatement of cocaine seeking following stressful stimuli (Beardsley et al., 2005, 2010; Land et al., 2009). Specifically, direct blockade of ventral tegmental area KOP receptors results in prevention of stress-induced reinstatement to cocaine seeking (Graziane et al., 2013). Although KOP receptor antagonists can block stress-induced cocaine reinstatement, they are not effective in blocking cocaine-primed reinstatement (Carey et al., 2007; Redila and Chavkin, 2008). Stress-induced reinstatement to cocaine place preference is absent in KOP receptor or prodynorphin knockout mice (Redila and Chavkin, 2008) further demonstrating the critical role of the KOP system in stress-responsivity in cocaine-exposed mice.

2.6. OTHER NEUROPEPTIDES

2.6.1. Overview

Several neuropeptides have been implicated in the pathophysiology of anxiety and stress disorders, however their role in anxiety or stress related to cocaine abuse has been less well-studied. Despite this, there are some important indications that several neuropeptides may contribute to anxiety associated with cocaine dependence and/or withdrawal and stress-induced relapse to drug taking.

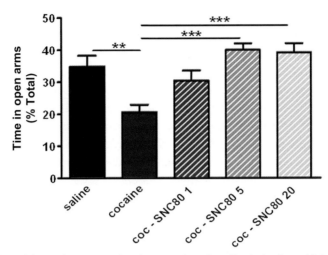

Figure 6 Withdrawal from chronic cocaine increased anxiety-like behavior, which was reversed by acute administration of a DOP receptor agonist. Male Sprague-Dawley rats received cocaine or saline for 14 days in a binge-pattern. SNC80, a selective DOP receptor agonist was administered 1 h prior to testing on the elevated plus maze after 24 h of cocaine withdrawal. Time spent on the open arms was significantly lower in rats undergoing cocaine withdrawal as compared with chronic saline-injected controls. SNC80 (1, 5, and 20 mg/kg) reduced the anxiety produced by cocaine withdrawal as shown by greater time spent in the open arms. *(Figure adapted from Perrine et al., 2008).*

2.6.2. Cholecystokinin

Cholecystokinin (CCK) peptides can elicit panic attacks in humans (Bradwejn et al., 1990; de Montigny, 1989) and anxiety-like responses in rodents (Harro, 2006; Li et al., 2013). Further, abnormalities in the endogenous CCK system have been documented in persons with panic disorders (Akiyoshi et al., 1996; Lydiard et al., 1992), indicating a potential role of CCK in the pathophysiology of anxiety disorders. CCK-B receptor antagonists have anxiolytic properties in rodent models (Li et al., 2013), although their clinical efficacy is less well-established (Bowers et al., 2012; Sramek et al., 1994–1995). As related to the potential role of the CCK system in cocaine abuse, it has been demonstrated that Wistar rats that voluntarily drink high amounts of cocaine have heightened levels of CCK in the nucleus accumbens and further, a CCK-B antagonist is effective in reducing cocaine drinking in these cocaine-preferring rats (Crespi, 1998). The author suggests that there is a CCK-B receptor mechanism in the regulation of individual sensitivity towards cocaine and craving for cocaine. Evidence also supports the role of CCK in anxiety during cocaine withdrawal. Costall and colleagues demonstrated that withdrawal from 14 days of cocaine administration in mice produces an anxiogenic response that is reversed by acute administration of CCK-B antagonists. The CCK-B antagonists were also effective in alleviating anxiety produced by withdrawal from repeated nicotine, ethanol, and diazepam (Costall et al., 1991). Exposure to cocaine or stress readily reinstates cocaine-seeking behaviors after a period of abstinence and extinction. Using the model of conditioned place preference to measure cocaine reinstatement behaviors, one group found that a CCK-B antagonist could block stress-induced reinstatement to cocaine-conditioned place preference, whereas a CCK-A antagonist blocked cocaine-primed reinstatement (Lu et al., 2002). In addition, the CCK-A antagonist was effective in blocking cocaine's actions when delivered directly into the nucleus accumbens, whereas the CCK-B antagonist attenuated stress-induced reinstatement when injected into either the accumbens or amygdala (Lu et al., 2002). These data collectively suggest that the endogenous CCK system is important not only in anxiety, but specifically in cocaine-induced anxiety. The CCK-B receptor may contribute to stress responses including stress-induced drug seeking and hence may be a potential target of therapeutics to treat cocaine dependence.

2.6.3. Neuropeptide Y

Endogenous neuropeptide Y (NPY) is recognized for its role in reducing anxiety and counteracting the behavioral effects of stress. Exogenously administered NPY and other agonists of the Y1 receptor are consistently anxiolytic in several rodent models such as the elevated plus maze and fear-potentiated startle (Britton et al., 1997; Broqua et al., 1995; Heilig et al., 1989, 1992). In agreement, genetic deletion of NPY results in an anxiogenic phenotype (Sørensen and Woldbye, 2012). Downregulation or blockade of NPY receptors, particularly the Y1 receptor, increases anxiety-like behaviors and can produce conditioned place aversion (Kask and Harro, 2000; Kask et al., 1996, 1999, 2002; Wahlestedt et al., 1993). Research on the potential clinical utility of NPY agonists as anxiolytics has been hindered by the lack of selective nonpeptide agonists for NPY receptors. Despite this, the potential role of NPY in cocaine abuse should not be overlooked. Mice lacking NPY have enhanced sensitivity to cocaine in three behavioral tests—cocaine self-administration, locomotion, and conditioned place preference (Sørensen and Woldbye, 2012). However, central administration of NPY is shown to modestly increase cocaine-induced hyperlocomotion and self-administration (Maric et al., 2009). Mice with genetic deletion of the Y5 receptor have reduced responses to cocaine (Sørensen and Woldbye, 2012). Of potent relevance to cocaine-induced anxiety, Wahlstedt and colleagues demonstrated that repeated administration of cocaine to rats results in reduced levels of NPY and NPY mRNA in the nucleus accumbens and prefrontal cortex (Wahlestedt et al., 1991). As clinical studies have noted that mood and anxiety disorders are associated with low levels of NPY (Holmes et al., 2003) as are suicide attempts (Widdowson et al., 1992), it is possible that cocaine-induced reductions in NPY contribute to heightened anxiety in cocaine-dependent persons. Further studies in this area are needed to firmly establish the contribution of the NPY system in cocaine-induced anxiety.

2.6.4. Additional Neuropeptides

Several other neuropeptides have been suggested to play a role in anxiety produced by withdrawal from repeated cocaine or in stress-induced reinstatement to cocaine-seeking behaviors. Although there is currently insufficient evidence to draw firm conclusions about these peptides, these peptides are worth considering further.

Orexin-A and –B (also referred to as hypocretin-1 and -2) are synthesized in a limited number of neurons in the lateral hypothalamus. These neurons project widely throughout the brain including the basal forebrain structures described above, which are involved in stress, reward, and motivation. The role of orexins in stress-induced cocaine reinstatement was investigated by Boutrel and colleagues (Boutrel et al., 2005). These authors found that orexin-A reinstates cocaine-seeking behaviors. Orexin-induced reinstatement can be blocked by antagonists of noradrenergic and CRF systems, suggesting that orexin-A-induced reinstatement is due to engagement of stress circuits. In addition, an orexin-A antagonist can block footshock-induced reinstatement of cocaine-seeking behaviors. Taken together, these results suggest that orexins have a role in facilitating cocaine-seeking through activation of stress pathways.

Another neuropeptide that has been implicated in the negative symptoms that accompany cocaine withdrawal is arginine vasopressin (AVP). Data show that stressors can increase the secretion of AVP together with CRF from terminals of the paraventricular nucleus of the hypothalamus (Knepel et al., 1985) and that AVP has a role in anxiogenic responses (Wigger et al., 2003). AVP potentiates CRF-induced adrenocorticotropic hormone (ACTH) sections thus enhancing the physiological mediators of the stress response (Antoni, 1993; Gillies et al., 1982; Rivier and Vale, 1983). V1a receptor antagonists have been shown to be effective anxiolytics in preclinical models (Bleickardt et al., 2009). Kreek and colleagues (Zhou et al., 2005, 2011) have investigated AVP in the setting of repeated cocaine exposure. Their results demonstrate that AVP mRNA levels in the paraventricular nucleus of the hypothalamus are persistently increased during acute to protracted (1–14 days) withdrawal from escalating-dose cocaine. Elevations of plasma ACTH and corticosterone levels were also found, indicating persistent elevations of basal HPA activity. Intriguingly, a V1b receptor antagonist was effective in blocking stress-induced heroin reinstatement (Zhou et al., 2007) providing further support for the contribution of AVP to stress responses and addictive behaviors.

3. CONCLUSIONS

Heightened anxiety and sensitivity to stress following chronic cocaine use are considered major risk factors for relapse to drug-seeking behavior. Thus, it is important to investigate the neural mechanisms responsible for the regulation of anxiety during cocaine withdrawal and stress-induced relapse in order to identify potential targets for pharmacological intervention. Although it is known that dopamine, serotonin, CRF, norepinephrine, and opioid systems all play distinct roles in regulating the negative consequences of cocaine withdrawal and the effects of stress in eliciting relapse to drug use, the integration of all of these systems is important to consider. During cocaine withdrawal, decreased levels of dopamine and serotonin are thought to contribute to anxiety-like symptoms, as biochemical and behavioral paradigms demonstrate dysfunction at the level of serotonin and dopamine receptors that mediate anxiety during cocaine withdrawal. Serotonin and dopamine systems also mediate the effects identified in stress-induced reinstatement studies. With regard to the CRF system during withdrawal, CRF receptor antagonists attenuate anxiety during withdrawal and prevent stress-induced reinstatement, suggesting that CRF signaling is amplified following abstinence from chronic cocaine exposure. Likewise, the norepinephrine system demonstrates a role in regulating anxiety during withdrawal and stress-induced reinstatement in varying ways depending on receptor subtype. For example, β-receptor antagonists, and α-receptor agonists produce similar attenuation of anxiety during cocaine withdrawal and attenuation of stress-induced reinstatement. In addition, endogenous opioids and neuropeptides have been shown to undergo alterations in relation to anxiety and stress neurocircuitry. Agents targeting these systems have proven to be anxiogenic or anxiolytic; however, their role in the context of anxiety during cocaine withdrawal or stress-induced reinstatement requires further study. Likewise, preclinical studies on other neuropeptides such as AVP, NPY, CCK, and orexin indicate their potential involvement in anxiety, stress, and/or cocaine withdrawal, suggesting that they may be worthwhile targets for pharmacological intervention. Further investigation of these systems in the context of chronic cocaine use, withdrawal, and relapse are warranted.

Several studies highlight the significance of the interactions between systems in regulating anxiety during withdrawal. For example, manipulation of specific serotonin receptor subtypes alters activity in mesolimbic dopamine areas known to be important in mediating the effects of cocaine. Likewise, both the serotonin and dopamine systems form networks with CRF, norepinephrine, and opioid systems in modulating anxiety during withdrawal and stress-induced reinstatement. The interactions of the stress systems and the pathways involved in drug addiction are complex; however, identifying potential targets for pharmacological intervention within these systems may contribute to the prevention of drug abuse and relapse to addiction.

REFERENCES

Abercrombie, E.D., Keller, R.W., Zigmond, M.J., 1988. Characterization of hippocampal norepinephrine release as measured by microdialysis perfusion: pharmacological and behavioral studies. Neuroscience 27, 897–904.

Ackerman, J.M., White, F.J., 1992. Decreased activity of rat A10 dopamine neurons following withdrawal from repeated cocaine. Eur. J. Pharmacol. 218, 171–173.

Aghajanian, G.K., VanderMaelen, C.P., 1982. Alpha 2-adrenoceptor-mediated hyperpolarization of locus coeruleus neurons: intracellular studies in vivo. Science 215, 1394–1396.

Ahmed, S.H., Koob, G.F., 1997. Cocaine-but not food-seeking behavior is reinstated by stress after extinction. Psychopharmacology 132, 289–295.

Akiyoshi, J., Moriyama, T., Isogawa, K., Miyamoto, M., Sasaki, I., Kuga, K., Yamamoto, H., Yamada, K., Fujii, I., 1996. CCK-4-induced calcium mobilization in T cells is enhanced in panic disorder. J. Neurochem. 66, 1610–1616.

Ambrose-Lanci, L.M., Sterling, R.C., Van Bockstaele, E.J., 2010. Cocaine withdrawal-induced anxiety in females: impact of circulating estrogen and potential use of delta-opioid receptor agonists for treatment. J. Neurosci. Res. 88, 816–824.

Ambrosio, E., Sharpe, L.G., Pilotte, N.S., 1997. Regional binding to corticotropin releasing factor receptors in brain of rats exposed to chronic cocaine and cocaine withdrawal. Synapse 25, 272–276.

Antoni, F.A., 1993. Vasopressinergic control of pituitary adrenocorticotropin secretion comes of age. Front. Neuroendocrinol. 14, 76–122.

Aujla, H., Martin-Fardon, R., Weiss, F., 2007. Rats with extended access to cocaine exhibit increased stress reactivity and sensitivity to the anxiolytic-like effects of the mGluR 2/3 agonist LY379268 during abstinence. Neuropsychopharmacology 33, 1818–1826.

Bale, T.L., Vale, W.W., 2004. CRF and CRF receptors: role in stress responsivity and other behaviors. Annu. Rev. Pharmacol. Toxicol. 44, 525–557.

Barnett, S.A., 1975. The Rat – a Study in Behavior. Univ. Chicago Press.

Basso, A.M., Spina, M., Rivier, J., Vale, W., Koob, G.F., 1999. Corticotropin-releasing factor antagonist attenuates the "anxiogenic-like" effect in the defensive burying paradigm but not in the elevated plus-maze following chronic cocaine in rats. Psychopharmacology (Berl) 145, 21–30.

Baumann, M.H., Becketts, K.M., Rothman, R.B., 1995. Evidence for alterations in presynaptic serotonergic function during withdrawal from chronic cocaine in rats. Eur. J. Pharmacol. 282, 87–93.

Baumann, M.H., Milchanowski, A.B., Rothman, R.B., 2004. Evidence for alterations in alpha2-adrenergic receptor sensitivity in rats exposed to repeated cocaine administration. Neuroscience 125, 683–690.

Beardsley, P., Howard, J., Shelton, K., Carroll, F.I., 2005. Differential effects of the novel kappa opioid receptor antagonist, JDTic, on reinstatement of cocaine-seeking induced by footshock stressors vs cocaine primes and its antidepressant-like effects in rats. Psychopharmacology 183, 118–126.

Beardsley, P.M., Pollard, G.T., Howard, J.L., Carroll, F.I., 2010. Effectiveness of analogs of the kappa opioid receptor antagonist (3R)-7-hydroxy-N-((1S)-1-{[(3R,4R)-4-(3-hydroxyphenyl)-3,4-dimethyl-1-piperidinyl]methyl}-2-methylpropyl)-1,2,3,4-tetrahydro-3-isoquinolinecarboxamide (JDTic) to reduce U50,488-induced diuresis and stress-induced cocaine reinstatement in rats. Psychopharmacology 210, 189–198.

Bilkei-Gorzo, A., Racz, I., Michel, K., Zimmer, A., Klingmüller, D., Zimmer, A., 2004. Behavioral phenotype of pre-proenkephalin-deficient mice on diverse congenic backgrounds. Psychopharmacology 176, 343–352.

Blacktop, J.M., Seubert, C., Baker, D.A., Ferda, N., Lee, G., Graf, E.N., Mantsch, J.R., 2011. CRF delivered into the ventral tegmental area following long-access self-administration is mediated by CRF receptor type 1 but not CRF receptor type 2. J. Neurosci. 31, 11396–11403.

Bleickardt, C.J., Mullins, D.E., MacSweeney, C.P., Werner, B.J., Pond, A.J., Guzzi, M.F., Martin, F.D.C., Varty, G.B., Hodgson, R.A., 2009. Characterization of the V1a antagonist, JNJ-17308616, in rodent models of anxiety-like behavior. Psychopharmacology 202, 711–718.

Boutrel, B., Kenny, P.J., Specio, S.E., Martin-Fardon, R. m., Markou, A., Koob, G.F., de Lecea, L., 2005. Role for hypocretin in mediating stress-induced reinstatement of cocaine-seeking behavior. Proc. Natl. Acad. Sci. U.S.A. 102, 19168–19173.

Bowers, M.E., Choi, D.C., Ressler, K.J., 2012. Neuropeptide regulation of fear and anxiety: implications of cholecystokinin, endogenous opioids, and neuropeptide Y. Physiol. Behav. 107, 699–710.

Bradwejn, J., Koszycki, D., Meterissian, G., 1990. Cholecystokinin-tetrapeptide induces panic attacks in patients with panic disorder. Can. J. Psychiatry 35, 83–85.

Briand, L.A., Blendy, J.A., 2010. Molecular and genetic substrates linking stress and addiction. Brain Res. 1314, 219–234.

Britton, K.T., Southerland, S., Uden, E.V., Kirby, D., Rivier, J., Koob, G., 1997. Anxiolytic activity of NPY receptor agonists in the conflict test. Psychopharmacology 132, 6–13.

Broqua, P., Wettstein, J.G., Rocher, M.N., Gauthier-Martin, B., Junien, J.L., 1995. Behavioral effects of neuropeptide Y receptor agonists in the elevated plus-maze and fear-potentiated startle procedures. Behav. Pharmacol. 6, 215–222.

Brown, S.A., Vik, P.W., McQuaid, J.R., Patterson, T.L., Irwin, M.R., Grant, I., 1990. Severity of psychosocial stress and outcome of alcoholism treatment. J. Abnorm. Psychol. 99, 344–348.

Brown, Z.J., Kupferschmidt, D.A., Erb, S., 2012. Reinstatement of cocaine seeking in rats by the pharmacological stressors, corticotropin-releasing factor and yohimbine: role for D1/5 dopamine receptors. Psychopharmacology 224, 431–440.

Brown, Z.J., Tribe, E., D'Souza, N.A., Erb, S., 2009. Interaction between noradrenaline and corticotrophin-releasing factor in the reinstatement of cocaine seeking in the rat. Psychopharmacology (Berl) 203, 121–130.

Bruchas, M.R., Land, B.B., Chavkin, C., 2010. The dynorphin/kappa opioid system as a modulator of stress-induced and pro-addictive behaviors. Brain Res. 1314, 44–55.

Bruchas, Michael R., Schindler, Abigail G., Shankar, H., Messinger, Daniel I., Miyatake, M., Land, Benjamin B., Lemos, Julia C., Hagan, C.E., Neumaier, John F., Quintana, A., Palmiter, Richard D., Chavkin, C., 2011. Selective p38α MAPK deletion in serotonergic neurons produces stress resilience in models of depression and addiction. Neuron 71, 498–511.

Bubar, M., Cunningham, K., 2006. Serotonin 5-HT$_{2A}$ and 5-HT$_{2C}$ receptors as potential targets for modulation of psychostimulant use and dependence. Curr. Top. Med. Chem. 6, 1971–1985.

Buffalari, D.M., Baldwin, C.K., See, R.E., 2012. Treatment of cocaine withdrawal anxiety with guanfacine: relationships to cocaine intake and reinstatement of cocaine seeking in rats. Psychopharmacology (Berl) 223, 179–190.

Capriles, N., Rodaros, D., Sorge, R., Stewart, J., 2003. A role for the prefrontal cortex in stress- and cocaine-induced reinstatement of cocaine seeking in rats. Psychopharmacology 168, 66–74.

Carey, A.N., Borozny, K., Aldrich, J.V., McLaughlin, J.P., 2007. Reinstatement of cocaine place-conditioning prevented by the peptide kappa-opioid receptor antagonist arodyn. Eur. J. Pharmacol. 569, 84–89.

Cassell, M.D., Gray, T.S., Kiss, J.Z., 1986. Neuronal architecture in the rat central nucleus of the amygdala: a cytological, hodological, and immunocytochemical study. J. Comp. Neurol. 246, 478–499.

Cerruti, C., Pilotte, N.S., Uhl, G., Kuhar, M.J., 1994. Reduction in dopamine transporter mRNA after cessation of repeated cocaine administration. Brain Res. Mol. Brain Res. 22, 132–138.

Charney, D.S., Heninger, G.R., 1986. Serotonin function in panic disorders: the effect of intravenous tryptophan in healthy subjects and patients with panic disorder before and during alprazolam treatment. Arch. Gen. Psychiatry 43, 1059–1065.

Clark, R., Ashby, C., Dewey, S., Ramachandran, P., Strecker, R., 1996. Effect of acute and chronic fluoxetine on extracellular dopamine levels in the caudate-putamen and nucleus accumbens of rat. Synapse 23, 123–131.

Clayton, E.C., Williams, C.L., 2000. Adrenergic activation of the nucleus tractus solitarius potentiates amygdala norepinephrine release and enhances retention performance in emotionally arousing and spatial memory tasks. Behav. Brain Res. 112, 151–158.

Coffey, S.F., Dansky, B.S., Carrigan, M.H., Brady, K.T., 2000. Acute and protracted cocaine abstinence in an outpatient population: a prospective study of mood, sleep and withdrawal symptoms. Drug Alcohol Depend. 59, 277–286.

Conrad, K.L., Ford, K., Marinelli, M., Wolf, M.E., 2010. Dopamine receptor expression and distribution dynamically change in the rat nucleus accumbens after withdrawal from cocaine self-administration. Neuroscience 169, 182–194.

Cooper, J.R., Bloom, F.E., Roth, R.H., 1991. In: The Biochemical Basis of Neuropharmacology. eighth ed. , pp. 218–219.

Costall, B., Domeney, A.M., Hughes, J., Kelly, M.E., Naylor, R.J., Woodruff, G.N., 1991. Anxiolytic effects of CCK-B antagonists. Neuropeptides 19 (Suppl).

Cottler, L.B., Shillington, A.M., Compton III, W.M., Mager, D., Spitznagel, E.L., 1993. Subjective reports of withdrawal among cocaine users: recommendations for DSM-1V. Drug Alcohol Depend. 33, 97–104.

Craige, C., Lewandowski, S., Kirby, L., Unterwald, E.M., 2012. Regulation of Serotonin Circuits through 5-HT$_{2C}$ Receptor and Gaba-Mediated Networks during Cocaine Withdrawal-Induced Anxiety. Program No. 456.22/R17 2012 Neuroscience Meeting Planner. Society for Neuroscience, New Orleans, Louisiana Online.

Crespi, F., 1998. The role of cholecystokinin (CCK), CCK-A or CCK-B receptor antagonists in the spontaneous preference for drugs of abuse (alcohol or cocaine) in naive rats. Methods Find. Exp. Clin. Pharmacol. 20, 679–697.

Cummings, S., Elde, R., Ells, J., Lindall, A., 1983. Corticotropin-releasing factor immunoreactivity is widely distributed within the central nervous system of the rat: an immunohistochemical study. J. Neurosci. 3, 1355–1368.

Cunningham, K., Paris, J., Goeders, N.E., 1992. Chronic cocaine enhances serotonin autoregulation and serotonin uptake binding. Synapse 11, 112–123.

Darmani, N.A., Shaddy, J., Elder, E.L., 1997. Prolonged deficits in presynaptic serotonin function following withdrawal from chronic cocaine exposure as revealed by 5-HTP-induced head-twitch response in mice. J. Neural Transm. 104, 1229–1247.

Daunais, J.B., Roberts, D.C.S., McGinty, J.F., 1995. Short-term cocaine self administration alters striatal gene expression. Brain Res. Bull. 37, 523–527.

Davidson, J., 2009. First-line pharmacotherapy approaches for generalized anxiety disorder. J. Clin. Psychiatry 70, 25–31.

De Deurwaerdère, P., Spampinato, U., 1999. Role of serotonin$_{2A}$ and serotonin$_{2B/2C}$ receptor subtypes in the control of accumbal and striatal dopamine release elicited in vivo by dorsal raphe nucleus electrical stimulation. J. Neurochem. 73, 1033–1042.

de Montigny, C., 1989. Cholecystokinin tetrapeptide induces panic-like attacks in healthy volunteers: preliminary findings. Arch. Gen. Psychiatry 46, 511–517.

de Visser, L., Baars, A., van 't Klooster, J., van den Bos, R., 2011. Transient inactivation of the medial prefrontal cortex affects both anxiety and decision-making in male wistar rats. Front. Neurosci. 5.

Der-Avakian, A., Bland, S., Rozeske, R., Tamblyn, J., Hutchinson, M., Watkins, L., Maier, S., 2007. The effects of a single exposure to uncontrollable stress on the subsequent conditioned place preference responses to oxycodone, cocaine, and ethanol in rats. Psychopharmacology 191, 909–917.

DeVries, A.C., Pert, A., 1998. Conditioned increases in anxiogenic-like behavior following exposure to contextual stimuli associated with cocaine are mediated by corticotropin-releasing factor. Psychopharmacology (Berl) 137, 333–340.

Dewart, T., Frank, B., Schmeidler, J., 2006. The impact of 9/11 on patients in New York city's substance abuse treatment programs. Am. J. Drug Alcohol Abuse 32, 665–672.

DeWit, D., MacDonald, K., Offord, D., 1999. Childhood stress and symptoms of drug dependence in adolescence and early adulthood: social phobia as a mediator. Am. J. Orthopsychiatry 69, 61–72.

Dolberg, O.T., Iancu, I., Sasson, Y., Zohar, J., 1996. The pathogenesis and treatment of obsessive-compulsive disorder. Clin. Neuropharmacol. 19, 129–147.

Donovan, B., Padin-Rivera, E., Kowaliw, S., 2001. "Transcend": initial outcomes from a posttraumatic stress disorder/substance abuse treatment program. J. Trauma. Stress 14, 757–772.

Erb, S., Funk, D., Borkowski, S., Watson, S.J., Akil, H., 2004. Effects of chronic cocaine exposure on corticotropin-releasing hormone binding protein in the central nucleus of the amygdala and bed nucleus of the stria terminalis. Neuroscience 123, 1003–1009.

Erb, S., Funk, D., Lê, A.D., 2005. Cocaine pre-exposure enhances CRF-induced expression of c-fos mRNA in the central nucleus of the amygdala: an effect that parallels the effects of cocaine pre-exposure on CRF-induced locomotor activity. Neurosci. Lett. 383, 209–214.

Erb, S., Hitchcott, P.K., Rajabi, H., Mueller, D., Shaham, Y., Stewart, J., 2000. Alpha-2 adrenergic receptor agonists block stress-induced reinstatement of cocaine seeking. Neuropsychopharmacology 23, 138–150.

Erb, S., Kayyali, H., Romero, K., 2006. A study of the lasting effects of cocaine pre-exposure on anxiety-like behaviors under baseline conditions and in response to central injections of corticotropin-releasing factor. Pharmacol. Biochem. Behav. 85, 206–213.

Erb, S., Shaham, Y., Stewart, J., 1996. Stress reinstates cocaine-seeking behavior after prolonged extinction and a drug-free period. Psychopharmacology 128, 408–412.

Erb, S., Shaham, Y., Stewart, J., 1998. The role of corticotropin-releasing factor and corticosterone in stress- and cocaine-induced relapse to cocaine seeking in rats. J. Neurosci. 18, 5529–5536.

Erb, S., Stewart, J., 1999. A role for the bed nucleus of the stria terminalis, but not the amygdala, in the effects of corticotropin-releasing factor on stress-induced reinstatement of cocaine seeking. J. Neurosci. 19, RC35.

Ettenberg, A., Ofer, O.A., Mueller, C.L., Waldroup, S., Cohen, A., Ben-Shahar, O., 2011. Inactivation of the dorsal raphe nucleus reduces the anxiogenic response of rats running an alley for intravenous cocaine. Pharmacol. Biochem. Behav. 97, 632–639.

Feltenstein, M.W., See, R.E., 2006. Potentiation of cue-induced reinstatement of cocaine-seeking in rats by the anxiogenic drug yohimbine. Behav. Brain Res. 174, 1–8.

Filliol, D., Ghozland, S., Chluba, J., Martin, M., Matthes, H.W., Simonin, F., Befort, K., Gavériaux-Ruff, C., Dierich, A., LeMeur, M., Valverde, O., Maldonado, R., Kieffer, B.L., 2000. Mice deficient for delta- and mu-opioid receptors exhibit opposing alterations of emotional responses. Nat. Genet. 25, 195–200.

Fletcher, P.J., Rizos, Z., Sinyard, J., Tampakeras, M., Higgins, G.A., 2007. The 5-HT$_{2C}$ receptor agonist Ro60-0175 reduces cocaine self-administration and reinstatement induced by the stressor yohimbine, and contextual cues. Neuropsychopharmacology 33, 1402–1412.

Fontana, D.J., Commissaris, R.L., 1989. Effects of cocaine on conflict behavior in the rat. Life Sci. 45, 819–827.

Ford, J.D., Trestman, R.L., Steinberg, K., Tennen, H., Allen, S., 2004. Prospective association of anxiety, depressive, and addictive disorders with high utilization of primary, specialty and emergency medical care. Soc. Sci. Med. 58, 2145–2148.

Frankel, P.S., Alburges, M.E., Bush, L., Hanson, G.R., Kish, S.J., 2008. Striatal and ventral pallidum dynorphin concentrations are markedly increased in human chronic cocaine users. Neuropharmacology 55, 41–46.

Fu, Y., Pollandt, S., Liu, J., Krishnan, B., Genzer, K., Orozco-Cabal, L., Gallagher, J.P., Shinnick-Gallagher, P., 2007. Long-term potentiation (LTP) in the central amygdala (CeA) is enhanced after prolonged withdrawal from chronic cocaine and requires CRF1 receptors. J. Neurophysiol. 97, 937–941.

Gardi, J., Bíró, E., Sarnyai, Z., Vecsernyés, M., Julesz, J., Telegdy, G., 1997. Time-dependent alterations in corticotropin-releasing factor-like immunoreactivity in different brain regions after acute cocaine administration to rats. Neuropeptides 31, 15–18.

Gartside, S., Umbers, V., Hajós, M., Sharp, T., 1995. Interaction between a selective 5-HT$_{1A}$ receptor antagonist and an SSRI in vivo: effects on 5-HT cell firing and extracellular 5-HT. Br. J. Pharmacol. 115, 1064–1070.

Gawin, F., Kleber, H., 1986. Abstinence symptomatology and psychiatric diagnosis in cocaine abusers: clinical observations. Arch. Gen. Psychiatry 43, 107–113.

Gawin, F.H., 1991. Cocaine addiction: psychology and neurophysiology. Science 251, 1580–1586.

Georges, F., Aston-Jones, G., 2001. Potent regulation of midbrain dopamine neurons by the bed nucleus of the stria terminalis. J. Neurosci. 21, RC160.

Georges, F., Aston-Jones, G., 2002. Activation of ventral tegmental area cells by the bed nucleus of the stria terminalis: a novel excitatory amino acid input to midbrain dopamine neurons. J. Neurosci. 22, 5173–5187.

Gervais, J., Rouillard, C., 2000. Dorsal raphe stimulation differentially modulates dopaminergic neurons in the ventral tegmental area and substantia nigra. Synapse 35, 281–291.

Giannini, A.J., Loiselle, R.H., Graham, B.H., Folts, D.J., 1993. Behavioral response to buspirone in cocaine and phencyclidine withdrawal. J. Subst. Abuse Treat. 10, 523–527.

Gibson, S., McGeer, E.G., McGeer, P.L., 1970. Effect of selective inhibitors of tyrosine and tryptophan hydroxylases on self-stimulation in the rat. Exp. Neurol. 27, 283–290.

Gillies, G.E., Linton, E.A., Lowry, P.J., 1982. Corticotropin releasing activity of the new CRF is potentiated several times by vasopressin. Nature 299, 355–357.

Goeders, N.E., Bienvenu, O.J., De Souza, E.B., 1990. Chronic cocaine administration alters corticotropin-releasing factor receptors in the rat brain. Brain Res. 531, 322–328.

Gordon, M.K., Rosen, J.B., 1999. Lasting effect of repeated cocaine administration on acoustic and fear-potentiated startle in rats. Psychopharmacology (Berl) 144, 1–7.

Graziane, N.M., Polter, A.M., Briand, L.A., Pierce, R.C., Kauer, J.A., 2013. Kappa opioid receptors regulate stress-induced cocaine seeking and synaptic plasticity. Neuron 77, 942–954.

Guan, X., Wang, L., Chen, C.L., Guan, Y., Li, S., 2010. Roles of two subtypes of corticotrophin-releasing factor receptor in the corticostriatal long-term potentiation under cocaine withdrawal condition. J. Neurochem. 115, 795–803.

Haass-Koffler, C.L., Bartlett, S.E., 2012. Stress and addiction: contribution of the corticotropin releasing factor (CRF) system in neuroplasticity. Front. Mol. Neurosci. 5, 91.

Haney, M., Maccari, S., Le Moal, M., Simon, H., Vincenzo Piazza, P., 1995. Social stress increases the acquisition of cocaine self-administration in male and female rats. Brain Res. 698, 46–52.

Harris, G.C., Aston-Jones, G., 1993. Beta-adrenergic antagonists attenuate withdrawal anxiety in cocaine- and morphine-dependent rats. Psychopharmacology (Berl) 113, 131–136.

Harris, G.C., Williams, J.T., 1992. Sensitization of locus ceruleus neurons during withdrawal from chronic stimulants and antidepressants. J. Pharmacol. Exp. Ther. 261, 476–483.

Harro, J., 2006. CCK and NPY as anti-anxiety treatment targets: promises, pitfalls, and strategies. Amino Acids 31, 215–230.

Heilig, M., McLeod, S., Koob, G.K., Britton, K.T., 1992. Anxiolytic-like effect of neuropeptide Y (NPY), but not other peptides in an operant conflict test. Regul. Pept. 41, 61–69.

Heilig, M., Söderpalm, B., Engel, J.A., Widerlöv, E., 1989. Centrally administered neuropeptide Y (NPY) produces anxiolytic-like effects in animal anxiety models. Psychopharmacology (Berl) 98, 524–529.

Henry, D.J., Greene, M.A., White, F.J., 1989. Electrophysiological effects of cocaine in the mesoaccumbens dopamine system: repeated administration. J. Pharmacol. Exp. Ther. 251, 833–839.

Henry, D.J., White, F.J., 1991. Repeated cocaine administration causes persistent enhancement of D1 dopamine receptor sensitivity within the rat nucleus accumbens. J. Pharmacol. Exp. Ther. 258, 882–890.

Herrero, M.J., Domingo-Salvany, A., Torrens, M., Brugal, M.T., ITINERE Investigators, 2008. Psychiatric comorbidity in young cocaine users: induced versus independent disorders. Addiction 103, 284–293.

Holmes, A., Heilig, M., Rupniak, N.M.J., Steckler, T., Griebel, G., 2003. Neuropeptide systems as novel therapeutic targets for depression and anxiety disorders. Trends Pharmacol. Sci. 24, 580–588.

Hurd, Y., Herkenham, M., 1993. Molecular alterations in the neostriatum of human cocaine addicts. Synapse 13, 357–369.

Jackson, M.E., Moghaddam, B., 2001. Amygdala regulation of nucleus accumbens dopamine output is governed by the prefrontal cortex. J. Neurosci. 21, 676–681.

Jobes, M.L., Ghitza, U.E., Epstein, D.H., Phillips, K.A., Heishman, S.J., Preston, K.L., 2011. Clonidine blocks stress-induced craving in cocaine users. Psychopharmacology (Berl) 218, 83–88.

Kabbaj, M., Norton, C., Kollack-Walker, S., Watson, S., Robinson, T., Akil, H., 2001. Social defeat alters the acquisition of cocaine self-administration in rats: role of individual differences in cocaine-taking behavior. Psychopharmacology 158, 382–387.

Kalivas, P.W., Chris Pierce, R., Cornish, J., Sorg, B.A., 1998. A role for sensitization in craving and relapse in cocaine addiction. J. Psychopharmacol. 12, 49–53.

Kampman, K.M., Volpicelli, J.R., Mulvaney, F., Alterman, A.I., Cornish, J., Gariti, P., Cnaan, A., Poole, S., Muller, E., Acosta, T., Luce, D., O'Brien, C., 2001. Effectiveness of propranolol for cocaine dependence treatment may depend on cocaine withdrawal symptom severity. Drug Alcohol Depend. 63, 69–78.

Kask, A., Harro, J., 2000. Inhibition of amphetamine- and apomorphine-induced behavioural effects by neuropeptide Y Y(1) receptor antagonist BIBO 3304. Neuropharmacology 39, 1292–1302.

Kask, A., Harro, J., von Hörsten, S., Redrobe, J.P., Dumont, Y., Quirion, R., 2002. The neurocircuitry and receptor subtypes mediating anxiolytic-like effects of neuropeptide Y. Neurosci. Biobehav. Rev. 26, 259–283.

Kask, A., Kivastik, T., Rägo, L., Harro, J., 1999. Neuropeptide Y Y1 receptor antagonist BIBP3226 produces conditioned place aversion in rats. Prog. Neuropsychopharmacol. Biol. Psychiatry 23, 705–711.

Kask, A., Rägo, L., Harro, J., 1996. Anxiogenic-like effect of the neuropeptide Y Y1 receptor antagonist BIBP3226: antagonism with diazepam. Eur. J. Pharmacol. 317, R3–R4.

Kastenberger, I., Lutsch, C., Herzog, H., Schwarzer, C., 2012. Influence of sex and genetic background on anxiety-related and stress-induced behaviour of prodynorphin-deficient mice. PLoS ONE 7, e34251.

King, G.R., Joyner, C.M., Ellinwood Jr., E.H., 1993. Withdrawal from continuous or intermittent cocaine: behavioral responsivity to 5-HT$_1$ receptor agonists. Pharmacol. Biochem. Behav. 45, 577–587.

Kirby, L.G., Freeman-Daniels, E., Lemos, J.C., Nunan, J.D., Lamy, C., Akanwa, A., Beck, S.G., 2008. Corticotropin-releasing factor increases GABA synaptic activity and induces inward current in 5-hydroxytryptamine dorsal raphe neurons. J. Neurosci. 28, 12927–12937.

Kleven, M.S., Perry, B.D., Woolverton, W.L., Seiden, L.S., 1990. Effects of repeated injections of cocaine on D1 and D2 dopamine receptors in rat brain. Brain Res. 532, 265–270.

Knepel, W., Vlaskovska, M., Meyer, D.K., 1985. Release of prostaglandin E2 and β-endorphin-like immunoreactivity from rat adenohypophysis in vitro: variations after adrenalectomy or lesions of the paraventricular nuclei. Brain Res. 326, 87–94.

Koeltzow, T.E., White, F.J., 2003. Behavioral depression during cocaine withdrawal is associated with decreased spontaneous activity of ventral tegmental area dopamine neuron. Behav. Neurosci. 117, 860–865.

Kohut, S.J., Decicco-Skinner, K.L., Johari, S., Hurwitz, Z.E., Baumann, M.H., Riley, A.L., 2012. Differential modulation of cocaine's discriminative cue by repeated and variable stress exposure: relation to monoamine transporter levels. Neuropharmacology 63, 330–337.

Koob, G.F., Rocio, M., Carrera, A., Gold, L.H., Heyser, C.J., Maldonado-Irizarry, C., Markou, A., Parsons, L.H., Roberts, A.J., Schulteis, G., Stinus, L., Walker, J.R., Weissenborn, R., Weiss, F., 1998. Substance dependence as a compulsive behavior. J. Psychopharmacol. 12, 39–48.

Koob, G.F., Volkow, N.D., 2010. Neurocircuitry of addiction. Neuropsychopharmacology 35, 217–238.

Korotkova, T.M., Brown, R.E., Sergeeva, O.A., Ponomarenko, A.A., Haas, H.L., 2006. Effects of arousal- and feeding-related neuropeptides on dopaminergic and GABAergic neurons in the ventral tegmental area of the rat. Eur. J. Neurosci. 23, 2677–2685.

Kreek, M.J., LaForge, K.S., Butelman, E., 2002. Pharmacotherapy of addictions. Nat. Rev. Drug Discov. 1, 710–726.

Kreibich, A.S., Blendy, J.A., 2004. cAMP response element-binding protein is required for stress but not cocaine-induced reinstatement. J. Neurosci. 24, 6686–6692.

Krishnan, B., Centeno, M., Pollandt, S., Fu, Y., Genzer, K., Liu, J., Gallagher, J.P., Shinnick-Gallagher, P., 2010. Dopamine receptor mechanisms mediate corticotropin-releasing factor-induced long-term potentiation in the rat amygdala following cocaine withdrawal. Eur. J. Neurosci. 31, 1027–1042.

Kung, J.-C., Chen, T.-C., Shyu, B.-C., Hsiao, S., Huang, A.C.W., 2010. Anxiety- and depressive-like responses and c-fos activity in preproenkephalin knockout mice: oversensitivity hypothesis of enkephalin deficit-induced posttraumatic stress disorder. J. Biomed. Sci. 17, 29.

Land, B.B., Bruchas, M.R., Schattauer, S., Giardino, W.J., Aita, M., Messinger, D., Hnasko, T.S., Palmiter, R.D., Chavkin, C., 2009. Activation of the kappa opioid receptor in the dorsal raphe nucleus mediates the aversive effects of stress and reinstates drug seeking. Proc. Natl. Acad. Sci. U.S.A. 106, 19168–19173.

Laurier, L.G., Corrigall, W.A., George, S.R., 1994. Dopamine receptor density, sensitivity and mRNA levels are altered following self-administration of cocaine in the rat. Brain Res. 634, 31–40.

Lee, B., Tiefenbacher, S., Platt, D.M., Spealman, R.D., 2004. Pharmacological blockade of alpha2-adrenoceptors induces reinstatement of cocaine-seeking behavior in squirrel monkeys. Neuropsychopharmacology 29, 686–693.

Lee, T.H., Gao, W.-Y., Davidson, C., Ellinwood, E.H., 1999. Altered activity of midbrain dopamine neurons following 7-day withdrawal from chronic cocaine abuse is normalized by D2 receptor stimulation during the early withdrawal phase. Neuropsychopharmacology 21, 127–136.

Leri, F., Flores, J., Rodaros, D., Stewart, J., 2002. Blockade of stress-induced but not cocaine-induced reinstatement by infusion of noradrenergic antagonists into the bed nucleus of the stria terminalis or the central nucleus of the amygdala. J. Neurosci. 22, 5713–5718.

Levy, A.D., Li, Q., Alvarez Sanz, M.C., Rittenhouse, P.A., Brownfield, M.S., Van de Kar, L.D., 1992. Repeated cocaine modifies the neuroendocrine responses to the 5-HT$_{1C}$/5-HT$_2$ receptor agonist DOI. Eur. J. Pharmacol. 221, 121–127.

Li, H., Ohta, H., Izumi, H., Matsuda, Y., Seki, M., Toda, T., Akiyama, M., Matsushima, Y., Goto, Y.-i., Kaga, M., Inagaki, M., 2013. Behavioral and cortical EEG evaluations confirm the roles of both CCKA and CCKB receptors in mouse CCK-induced anxiety. Behav. Brain Res. 237, 325–332.

Liu, K., Steketee, J.D., 2011. Repeated exposure to cocaine alters medial prefrontal cortex dopamine D$_2$-like receptor modulation of glutamate and dopamine neurotransmission within the mesocorticolimbic system. J. Neurochem. 119, 332–341.

Lowry, C.A., 2002. Functional subsets of serotonergic neurones: implications for control of the hypothalamic-pituitary-adrenal axis. J. Neuroendocrinol. 14, 911–923.

Lu, L., Liu, D., Ceng, X., 2001. Corticotropin-releasing factor receptor type 1 mediates stress-induced relapse to cocaine-conditioned place preference in rats. Eur. J. Pharmacol. 415, 203–208.

Lu, L., Zhang, B., Liu, Z., Zhang, Z., 2002. Reactivation of cocaine conditioned place preference induced by stress is reversed by cholecystokinin-B receptors antagonist in rats. Brain Res. 954, 132–140.

Lydiard, R.B., Ballenger, J.C., Laraia, M.T., Fossey, M.D., Beinfeld, M.C., 1992. CSF cholecystokinin concentrations in patients with panic disorder and in normal comparison subjects. Am. J. Psychiatry 149, 691–693.

Macêdo, D.S., Sousa, F.C.F., Vasconcelos, S.M.M., Lima, V.T.M., Viana, G.S.B., 2001. Different times of withdrawal from cocaine administration cause changes in muscarinic and dopaminergic receptors in rat premotor cortex. Neurosci. Lett. 312, 129–132.

Maggos, C.E., Tsukada, H., Kakiuchi, T., Nishiyama, S., Myers, J.E., Kreuter, J., Schlussman, S.D., Unterwald, E.M., Ho, A., Kreek, M.J., 1998. Sustained withdrawal allows normalization of in vivo [11C]N-methyl-spiperone dopamine D_2 receptor binding after chronic binge cocaine: a positron emission tomography study in rats. Neuropsychopharmacology 19, 146–153.

Maier, S., Szalkowski, A., Kamphausen, S., Perlov, E., Feige, B., Blechert, J., Philipsen, A., van Elst, L.T., Kalisch, R., Tüscher, O., 2012. Clarifying the role of the rostral dmPFC/dACC in fear/anxiety: learning, appraisal or expression? PLoS ONE 7, e50120.

Maj, M., Turchan, J., Smiałowska, M., Przewłocka, B., 2003. Morphine and cocaine influence on CRF biosynthesis in the rat central nucleus of amygdala. Neuropeptides 37, 105–110.

Mantsch, J.R., Baker, D.A., Francis, D.M., Katz, E.S., Hoks, M.A., Serge, J.P., 2008. Stressor- and corticotropin releasing factor-induced reinstatement and active stress-related behavioral responses are augmented following long-access cocaine self-administration by rats. Psychopharmacology (Berl) 195, 591–603.

Margolin, A., Avants, S., Kosten, T., 1996. Abstinence symptomatology associated with cessation of chronic cocaine abuse among methadone-maintained patients. Am. J. Drug Alcohol Abuse 22, 377–388.

Maric, T., Cantor, A., Cuccioletta, H., Tobin, S., Shalev, U., 2009. Neuropeptide Y augments cocaine self-administration and cocaine-induced hyperlocomotion in rats. Peptides 30, 721–726.

Marin, S., Marco, E., Biscaia, M., Fernández, B., Rubio, M., Guaza, C., Schmidhammer, H., Viveros, M.P., 2003. Involvement of the κ-opioid receptor in the anxiogenic-like effect of CP 55,940 in male rats. Pharmacol. Biochem. Behav. 74, 649–656.

Markou, A., Koob, G.F., 1992. Bromocriptine reverses the elevation in intracranial self-stimulation thresholds observed in a rat model of cocaine withdrawal. Neuropsychopharmacology 7, 213–224.

Markou, A., Kosten, T.R., Koob, G.F., 1998. Neurobiological similarities in depression and drug dependence: a self-medication hypothesis. Neuropsychopharmacology 18, 135–174.

McDougle, C.J., Black, J.E., Malison, R.T., Zimmermann, R.C., Kosten, T.R., Heninger, G.R., Price, L.H., 1994. Noradrenergic dysregulation during discontinuation of cocaine use in addicts. Arch. Gen. Psychiatry 51, 713–719.

McFall, M., Mackay, P., Donovan, D., 1992. Combat-related posttraumatic stress disorder and severity of substance abuse in vietnam veterans. J. Stud. Alcohol Drugs 53, 357–363.

McFarland, K., Davidge, S.B., Lapish, C.C., Kalivas, P.W., 2004. Limbic and motor circuitry underlying footshock-induced reinstatement of cocaine-seeking behavior. J. Neurosci. 24, 1551–1560.

Meiergerd, S.M., McElvain, J.S., Schenk, J.O., 1994. Effects of cocaine and repeated cocaine followed by withdrawal: alterations of dopaminergic transporter turnover with no changes in kinetics of substrate recognition. Biochem. Pharmacol. 47, 1627–1634.

Meltzer, H.Y., Maes, M., 1995. Effects of ipsapirone on plasma cortisol and body temperature in major depression. Biol. Psychiatry 38, 450–457.

Mongeau, R., Blier, P., de Montigny, C., 1997. The serotonergic and noradrenergic systems of the hippocampus: their interactions and the effects of antidepressant treatments. Brain Res. Brain Res. Rev. 23, 145–195.

Mutschler, N.H., Miczek, K.A., 1998a. Withdrawal from a self-administered or non-contingent cocaine binge: differences in ultrasonic distress vocalizations in rats. Psychopharmacology (Berl) 136, 402–408.

Mutschler, N.H., Miczek, K.A., 1998b. Withdrawal from i.v. cocaine "binges" in rats: ultrasonic distress calls and startle. Psychopharmacology (Berl) 135, 161–168.

Neisewander, J.L., Lucki, I., McGonigle, P., 1994. Time-dependent changes in sensitivity to apomorphine and monoamine receptors following withdrawal from continuous cocaine administration in rats. Synapse 16, 1–10.

Nieto, M.M., Guen, S.L.E., Kieffer, B.L., Roques, B.P., Noble, F., 2005. Physiological control of emotion-related behaviors by endogenous enkephalins involves essentially the delta opioid receptors. Neuroscience 135, 305–313.

Nogueira, L., Kalivas, P.W., Lavin, A., 2006. Long-term neuroadaptations produced by withdrawal from repeated cocaine treatment: role of dopaminergic receptors in modulating cortical excitability. J. Neurosci. 26, 12308–12313.

O'Brien, C.P., Childress, A.R., Ehrman, R., Robbins, S.J., 1998. Conditioning factors in drug abuse: can they explain compulsion? J. Psychopharmacol. 12, 15–22.

Ohata, H., Shibasaki, T., 2011. Microinjection of different doses of corticotropin-releasing factor into the medial prefrontal cortex produces effects opposing anxiety-related behavior in rats. J. Nippon Med. Sch. 78, 286–292.

Ouimette, P., Coolhart, D., Funderburk, J.S., Wade, M., Brown, P.J., 2007. Precipitants of first substance use in recently abstinent substance use disorder patients with PTSD. Addict. Behav. 32, 1719–1727.

Owens, M.J., Nemeroff, C.B., 1991. Physiology and pharmacology of corticotropin-releasing factor. Pharmacol. Rev. 43, 425–473.

Pacak, K., Tjurmina, O., Palkovits, M., Goldstein, D.S., Koch, C.A., Hoff, T., Chrousos, G.P., 2002. Chronic hypercortisolemia inhibits dopamine synthesis and turnover in the nucleus accumbens: an in vivo microdialysis study. Neuroendocrinology 76, 148–157.

Parsons, L.H., Koob, G.F., Weiss, F., 1995. Serotonin dysfunction in the nucleus accumbens of rats during withdrawal after unlimited access to intravenous cocaine. J. Pharmacol. Exp. Ther. 274, 1182–1191.

Parsons, L.H., Smith, A.D., Justice, J.B.J., 1991. Basal extracellular dopamine is decreased in the rat nucleus accumbens during abstinence from chronic cocaine. Synapse 9, 60–65.

Pellow, S., Chopin, P., File, S.E., Briley, M., 1985. Validation of open: closed arm entries in an elevated plus-maze as a measure of anxiety in the rat. J. Neurosci. Methods 14, 149–167.

Penk, W., Peck, R., Robinowitz, R., Bell, W., Little, D., 1988. Coping and defending styles among vietnam combat veterans seeking treatment for posttraumatic stress disorder and substance use disorder. Recent Dev. Alcohol. 6, 69–88.

Peris, J., Boyson, S.J., Cass, W.A., Curella, P., Dwoskin, L.P., Larson, G., Lin, L.H., Yasuda, R.P., Zahniser, N.R., 1990. Persistence of neurochemical changes in dopamine systems after repeated cocaine administration. J. Pharmacol. Exp. Ther. 253, 38–44.

Perrine, S.A., Hoshaw, B.A., Unterwald, E.M., 2006. Delta opioid receptor ligands modulate anxiety-like behaviors in the rat. Br. J. Pharmacol. 147, 864–872.

Perrine, S.A., Sheikh, I.S., Nwaneshiudu, C.A., Schroeder, J.A., Unterwald, E.M., 2008. Withdrawal from chronic administration of cocaine decreases delta opioid receptor signaling and increases anxiety- and depression-like behaviors in the rat. Neuropharmacology 54, 355–364.

Piazza, P.V., Le Moal, M., 1998. The role of stress in drug self-administration. Trends Pharmacol. Sci. 19, 67–74.

Pilotte, N.S., Sharpe, L.G., Kuhar, M.J., 1994. Withdrawal of repeated intravenous infusions of cocaine persistently reduces binding to dopamine transporters in the nucleus accumbens of Lewis rats. J. Pharmacol. Exp. Ther. 269, 963–969.

Pinel, J.P., Treit, D., 1978. Burying as a defensive response in rats. J. Comp. Physiol. Psychol. 92, 708–712.

Pitkanen, A., 2000. Connectivity of the rat amygdaloid complex. In: Aggleton, J.P. (Ed.), The Amygdala: A Functional Analysis. Oxford University Press, New York, pp. 31–115.

Pollandt, S., Liu, J., Orozco-Cabal, L., Grigoriadis, D.E., Vale, W.W., Gallagher, J.P., Shinnick-Gallagher, P., 2006. Cocaine withdrawal enhances long-term potentiation induced by corticotropin-releasing factor at central amygdala glutamatergic synapses via CRF, NMDA receptors and PKA. Eur. J. Neurosci. 24, 1733–1743.

Randall-Thompson, J.F., Pescatore, K.A., Unterwald, E.M., 2010. A role for delta opioid receptors in the central nucleus of the amygdala in anxiety-like behaviors. Psychopharmacology 212, 585–595.

Redila, V., Chavkin, C., 2008. Stress-induced reinstatement of cocaine seeking is mediated by the kappa opioid system. Psychopharmacology 200, 59–70.

Resnick, R.B., Resnick, E.B., 1984. Cocaine abuse and its treatment. Psychiatr. Clin. North Am. 7, 713–728.

Richter, R.M., Pich, E.M., Koob, G.F., Weiss, F., 1995. Sensitization of cocaine-stimulated increase in extracellular levels of corticotropin-releasing factor from the rat amygdala after repeated administration as determined by intracranial microdialysis. Neurosci. Lett. 187, 169–172.

Richter, R.M., Weiss, F., 1999. In vivo CRF release in rat amygdala is increased during cocaine withdrawal in self-administering rats. Synapse 32, 254–261.

Rivier, C., Vale, W., 1983. Interaction of corticotropin-releasing factor and arginine vasopressin on adrenocorticotropin secretion in vivo. Endocrinology 113, 939–942.

Rodaros, D., Caruana, D.A., Amir, S., Stewart, J., 2007. Corticotropin-releasing factor projections from limbic forebrain and paraventricular nucleus of the hypothalamus to the region of the ventral tegmental area. Neuroscience 150, 8–13.

Rohmer, J.-G., Scala, G.D., Sandner, G., 1990. Behavioral analysis of the effects of benzodiazepine receptor ligands in the conditioned burying paradigm. Behav. Brain Res. 38, 45–54.

Rossetti, Z.L., Melis, F., Carboni, S., Gessa, G.L., 1992. Dramatic depletion of mesolimbic extracellular dopamine after withdrawal from morphine, alcohol or cocaine: a common neurochemical substrate for drug dependence. Ann. N.Y. Acad. Sci. 654, 513–516.

Rudoy, C.A., Reyes, A.R., Van Bockstaele, E.J., 2009. Evidence for beta1-adrenergic receptor involvement in amygdalar corticotropin-releasing factor gene expression: implications for cocaine withdrawal. Neuropsychopharmacology 34, 1135–1148.

Rudoy, C.A., Van Bockstaele, E.J., 2007. Betaxolol, a selective beta(1)-adrenergic receptor antagonist, diminishes anxiety-like behavior during early withdrawal from chronic cocaine administration in rats. Prog. Neuropsychopharmacol. Biol. Psychiatry 31, 1119–1129.

Sahuque, L.L., Kullberg, E.F., McGeehan, A.J., Kinder, J.R., Hicks, M.P., Blanton, M.G., Janak, P.H., Olive, M.F., 2006. Anxiogenic and aversive effects of corticotropin-releasing factor (CRF) in the bed nucleus of the stria terminalis in the rat: role of CRF receptor subtypes. Psychopharmacology (Berl) 186, 122–132.

Saitoh, A., Kimura, Y., Suzuki, T., Kawai, K., Nagase, H., Kamei, J., 2004. Potential anxiolytic and antidepressant-like activities of SNC80, a selective δ-opioid agonist, in behavioral models in rodents. J. Pharmacol. Sci. 95, 374–380.

Saitoh, A., Yoshikawa, Y., Onodera, K., Kamei, J., 2005. Role of delta-opioid receptor subtypes in anxiety-related behaviors in the elevated plus-maze in rats. Psychopharmacology 182, 327–334.

Sakanaka, M., Shibasaki, T., Lederis, K., 1986. Distribution and efferent projections of corticotropin-releasing factor-like immunoreactivity in the rat amygdaloid complex. Brain Res. 382, 213–238.

Samuvel, D.J., Jayanthi, L.D., Manohar, S., Kaliyaperumal, K., See, R.E., Ramamoorthy, S., 2008. Dysregulation of dopamine transporter trafficking and function after abstinence from cocaine self-administration in rats: evidence for differential regulation in caudate putamen and nucleus accumbens. J. Pharmacol. Exp. Ther. 325, 293–301.

Sanchez, C.J., Bailie, T.M., Wu, W.R., Li, N., Sorg, B.A., 2003. Manipulation of dopamine d1-like receptor activation in the rat medial prefrontal cortex alters stress- and cocaine-induced reinstatement of conditioned place preference behavior. Neuroscience 119, 497–505.

Sarnyai, Z., Bíró, E., Gardi, J., Vecsernyés, M., Julesz, J., Telegdy, G., 1993. Alterations of corticotropin-releasing factor-like immunoreactivity in different brain regions after acute cocaine administration in rats. Brain Res. 616, 315–319.

Sarnyai, Z., Bíró, E., Gardi, J., Vecsernyés, M., Julesz, J., Telegdy, G., 1995. Brain corticotropin-releasing factor mediates 'anxiety-like' behavior induced by cocaine withdrawal in rats. Brain Res. 675, 89–97.

Satel, S., Southwick, S., Gawin, F.H., 1991. Clinical features of cocaine-induced paranoia. Am. J. Psychiatry 148, 495–498.

Schindler, A.G., Messinger, D.I., Smith, J.S., Shankar, H., Gustin, R.M., Schattauer, S.S., Lemos, J.C., Chavkin, N.W., Hagan, C.E., Neumaier, J.F., Chavkin, C., 2012. Stress produces aversion and potentiates cocaine reward by releasing endogenous dynorphins in the ventral striatum to locally stimulate serotonin reuptake. J. Neurosci. 32, 17582–17596.

Shaffer, H.J., Eber, G.B., 2002. Temporal progression of cocaine dependence symptoms in the us national comorbidity survey. Addiction 97, 543–554.

Shaham, Y., Erb, S., Leung, S., Buczek, Y., Stewart, J., 1998. CP-154,526, a selective, non-peptide antagonist of the corticotropin-releasing factor1 receptor attenuates stress-induced relapse to drug seeking in cocaine- and heroin-trained rats. Psychopharmacology (Berl) 137, 184–190.

Shaham, Y., Erb, S., Stewart, J., 2000. Stress-induced relapse to heroin and cocaine seeking in rats: a review. Brain Res. Brain Res. Rev. 33, 13–33.

Sharpe, L.G., Pilotte, N.S., Mitchell, W.M., De Souza, E.B., 1991. Withdrawal of repeated cocaine decreases autoradiographic [3H]mazindol-labelling of dopamine transporter in rat nucleus accumbens. Eur. J. Pharmacol. 203, 141–144.

Sinha, R., 2008. Chronic stress, drug use, and vulnerability to addiction. Ann. N.Y. Acad. Sci. 1141, 105–130.

Sørensen, G., Woldbye, D.P.D., 2012. Mice lacking neuropeptide Y show increased sensitivity to cocaine. Synapse 66, 840–843.

Spangler, R., Unterwald, E.M., Kreek, M.J., 1993. 'Binge' cocaine administration induces a sustained increase of prodynorphin mRNA in rat caudate-putamen. Brain Res. Mol. Brain Res. 19, 323–327.

Specio, S.E., Wee, S., O'Dell, L.E., Boutrel, B., Zorrilla, E.P., Koob, G.F., 2008. CRF(1) receptor antagonists attenuate escalated cocaine self-administration in rats. Psychopharmacology (Berl) 196, 473–482.

Sramek, J.J., Kramer, M.S., Reines, S.A., Cutler, N.R., 1994–1995. Pilot study of a CCKB antagonist in patients with panic disorder: preliminary findings. Anxiety 1, 141–143.

Steiner, H., Gerfen, C.R., 1993. Cocaine-induced c-fos messenger RNA is inversely related to dynorphin expression in striatum. J. Neurosci. 13, 5066–5081.

Sullivan, R.M., Gratton, A., 1998. Relationships between stress-induced increases in medial prefrontal cortical dopamine and plasma corticosterone levels in rats: role of cerebral laterality. Neuroscience 83, 81–91.

Teusch, R., 2001. Substance abuse as a symptom of childhood sexual abuse. Psychiatr. Serv. 52.

Tidey, J.W., Miczek, K.A., 1997. Acquisition of cocaine self-administration after social stress: role of accumbens dopamine. Psychopharmacology 130, 203–212.

Treit, D., Pinel, J.P.J., Fibiger, H.C., 1981. Conditioned defensive burying: a new paradigm for the study of anxiolytic agents. Pharmacol. Biochem. Behav. 15, 619–626.

Triffleman, E., Marmar, C., Delucchi, K., Ronfeldt, H., 1995. Childhood trauma and posttraumatic stress disorder in substance abuse inpatients. J. Nerv. Ment. Dis. 183, 172–176.

Unterwald, E.M., Cox, B.M., Kreek, M.J., Cote, T.E., Izenwasser, S., 1993. Chronic repeated cocaine administration alters basal and opioid-regulated adenylyl cyclase activity. Synapse 15, 33–38.

Valdez, G.R., Platt, D.M., Rowlett, J.K., Rüedi-Bettschen, D., Spealman, R.D., 2007. κ agonist-induced reinstatement of cocaine seeking in squirrel monkeys: a role for opioid and stress-related mechanisms. J. Pharmacol. Exp. Ther. 323, 525–533.

Valentino, R.J., Lucki, I., Van Bockstaele, E., 2010. Corticotropin-releasing factor in the dorsal raphe nucleus: linking stress coping and addiction. Brain Res. 1314, 29–37.

Van't Veer, A., Bechtholt, A.J., Onvani, S., Potter, D., Wang, Y., Liu-Chen, L.Y., Shütz, G., Chartoff, E.H., Rudolph, U., Cohen, B.M., Carlezon, W.A.J., 2013. Ablation of kappa-opioid receptors from brain dopamine neurons has anxiolytic-like effects and enhances cocaine-induced plasticity. Neuropsychopharmacology Epub ahead of print.

Van Pett, K., Viau, V., Bittencourt, J.C., Chan, R.K., Li, H.Y., Arias, C., Prins, G.S., Perrin, M., Vale, W., Sawchenko, P.E., 2000. Distribution of mRNAs encoding CRF receptors in brain and pituitary of rat and mouse. J. Comp. Neurol. 428, 191–212.

Virrkunen, M., Linnoila, M., 1990. Serotonin in early onset, male alcoholics with violent behaviour. Ann. Med. 22, 327–331.

Volkow, N.D., Fowler, J.S., Wang, G.J., Hitzemann, R., Logan, J., Schlyer, D.J., Dewey, S.L., Wolf, A.P., 1993. Decreased dopamine D2 receptor availability is associated with reduced frontal metabolism in cocaine abusers. Synapse 14, 169–177.

Volkow, N.D., Mullani, N., Gould, K.L., Adler, S., Krajewski, K., 1988. Cerebral blood flow in chronic cocaine users: a study with positron emission tomography. Br. J. Psychiatry 152, 641–648.

Volkow, N.D., Wang, G.J., Fowler, J.S., Logan, J., Gatley, S.J., Hitzemann, R., Chen, A.D., Dewey, S.L., Pappas, N., 1997. Decreased striatal dopaminergic responsiveness in detoxified cocaine-dependent subjects. Nature 386, 830–833.

Wahlestedt, C., Karoum, F., Jaskiw, G., Wyatt, R.J., Larhammar, D., Ekman, R., Reis, D.J., 1991. Cocaine-induced reduction of brain neuropeptide Y synthesis dependent on medial prefrontal cortex. Proc. Natl. Acad. Sci. U.S.A. 88, 2078–2082.

Wahlestedt, C., Pich, E.M., Koob, G.F., Yee, F., Heilig, M., 1993. Modulation of anxiety and neuropeptide Y-Y1 receptors by antisense oligodeoxynucleotides. Science 259, 528–531.

Walker, D.L., Miles, L.A., Davis, M., 2009. Selective participation of the bed nucleus of the stria terminalis and CRF in sustained anxiety-like versus phasic fear-like responses. Prog. Neuropsychopharmacol. Biol. Psychiatry 33, 1291–1308.

Walker, G., Scott, P., Koppersmith, G., 1998. The impact of child sexual abuse on addiction severity: an analysis of trauma processing. J. Psychosoc. Nurs. Ment. Health Serv. 36, 10–18.

Wanat, M.J., Hopf, F.W., Stuber, G.D., Phillips, P.E., Bonci, A., 2008. Corticotropin-releasing factor increases mouse ventral tegmental area dopamine neuron firing through a protein kinase C-dependent enhancement of Ih. J. Physiol. 586, 2157–2170.

Wang, B., Shaham, Y., Zitzman, D., Azari, S., Wise, R.A., You, Z.B., 2005. Cocaine experience establishes control of midbrain glutamate and dopamine by corticotropin-releasing factor: a role in stress-induced relapse to drug seeking. J. Neurosci. 25, 5389–5396.

Wang, B., You, Z.-B., Rice, K., Wise, R., 2007. Stress-induced relapse to cocaine seeking: roles for the CRF2 receptor and CRF-binding protein in the ventral tegmental area of the rat. Psychopharmacology 193, 283–294.

Wang, H., Pickel, V.M., 2002. Dopamine D2 receptors are present in prefrontal cortical afferents and their targets in patches of the rat caudate-putamen nucleus. J. Comp. Neurol. 442, 392–404.

Wang, J., Fang, Q., Liu, Z., Lu, L., 2006. Region-specific effects of brain corticotropin-releasing factor receptor type 1 blockade on footshock-stress- or drug-priming-induced reinstatement of morphine conditioned place preference in rats. Psychopharmacology (Berl) 185, 19–28.

Weddington, W.W., 1992. Cocaine abstinence: "withdrawal" or residua of chronic intoxication? Am. J. Psychiatry 149, 1761–1762.

Weddington, W.W., Brown, B.S., Haertzen, C.A., Cone, E.J., Dax, E.M., Herning, R.I., Michaelson, B.S., 1990. Changes in mood, craving, and sleep during short-term abstinence reported by male cocaine addicts. A controlled, residential study. Arch. Gen. Psychiatry 47, 861–868.

Widdowson, P.S., Ordway, G.A., Halaris, A.E., 1992. Reduced neuropeptide Y concentrations in suicide brain. J. Neurochem. 59, 73–80.

Wigger, A., Sanchez, M.M., Mathys, K.C., Ebner, K., Frank, E., Liu, D., Kresse, A., Neumann, I.D., Holsboer, F., Plotsky, P.M., Landgraf, R., 2003. Alterations in central neuropeptide expression, release, and receptor binding in rats bred for high anxiety: critical role of vasopressin. Neuropsychopharmacology 29, 1–14.

Wise, R.A., Morales, M., 2010. A ventral tegmental CRF-glutamate-dopamine interaction in addiction. Brain Res. 1314, 38–43.

Wittmann, W., Schunk, E., Rosskothen, I., Gaburro, S., Singewald, N., Herzog, H., Schwarzer, C., 2008. Prodynorphin-derived peptides are critical modulators of anxiety and regulate neurochemistry and corticosterone. Neuropsychopharmacology 34, 775–785.

Zaslav, M., 1994. Psychology of comorbid posttraumatic stress disorder and substance abuse: lessons from combat veterans. J. Psychoactive Drugs 26, 393–400.

Zhou, L.L., Ming, L., Ma, C.G., Cheng, Y., Jiang, Q., 2005. Antidepressant-like effects of BCEF0083 in the chronic unpredictable stress models in mice. Chin. Med. J. 118, 903–908.

Zhou, Y., Leri, F., Cummins, E., Hoeschele, M., Kreek, M.J., 2007. Involvement of arginine vasopressin and V1b receptor in heroin withdrawal and heroin seeking precipitated by stress and by heroin. Neuropsychopharmacology 33, 226–236.

Zhou, Y., Litvin, Y., Piras, A.P., Pfaff, D.W., Kreek, M.J., 2011. Persistent increase in hypothalamic arginine vasopressin gene expression during protracted withdrawal from chronic escalating-dose cocaine in rodents. Neuropsychopharmacology 36, 2062–2075.

Zhou, Y., Spangler, R., LaForge, K.S., Maggos, C.E., Ho, A., Kreek, M.J., 1996. Corticotropin-releasing factor and type 1 corticotropin-releasing factor receptor messenger RNAs in rat brain and pituitary during "binge"-pattern cocaine administration and chronic withdrawal. J. Pharmacol. Exp. Ther. 279, 351–358.

Zorrilla, E.P., Valdez, G.R., Weiss, F., 2001. Changes in levels of regional CRF-like-immunoreactivity and plasma corticosterone during protracted drug withdrawal in dependent rats. Psychopharmacology (Berl) 158, 374–381.

Zorrilla, E.P., Wee, S., Zhao, Y., Specio, S., Boutrel, B., Koob, G.F., Weiss, F., 2012. Extended access cocaine self-administration differentially activates dorsal raphe and amygdala corticotropin-releasing factor systems in rats. Addict. Biol. 17, 300–308.

CHAPTER SEVEN

The Neuropathology of Drug Abuse

Andreas Büttner
Institute of Forensic Medicine, University of Rostock, Rostock, Germany

1. INTRODUCTION

Drug abuse represents a substantial problem worldwide and is associated with significant morbidity and mortality. The predominant substances taken alone or in combination include cannabis, opiates, cocaine, amphetamines, methamphetamine and "designer drugs" (Karch, 2008; Quinn et al., 1997). In addition, alcohol and other central nervous system (CNS) depressants are frequently consumed by drug abusers (polysubstance abuse) (Bird, 2010; Coffin et al., 2003; Karch, 2008; Molina and Hargrove, 2011; Preti et al., 2002; Quinn et al., 1997). Besides cardiovascular complications, psychiatric and neurological disturbances are the most common manifestations of drug toxicity (Brust, 2004; Cardoso and Jankovic, 1993; Goforth et al., 2010; Gruber et al., 2007; Mackesy-Amiti et al., 2012; Neiman et al., 2000). Although most of our knowledge of the effects of drugs has been derived from animal models, a variety of CNS alterations is observed in human drug abusers (Büttner and Weis, 2004; Büttner, 2011; Karch, 2008).

However, since polysubstance abuse is observed in the majority of cases, it is difficult to relate the human CNS findings to a specific substance. Another difficulty is to differentiate between substance-specific and secondary effects by adulterants or the lifestyle of the drug abuser (Büttner and Weis, 2004; Büttner, 2011; Karch, 2008). In injecting drug abusers there is also a high risk for CNS infections leading to additional brain pathologies (Anthony et al., 2008; Backmund et al., 2005; Büttner and Weis, 2004; Büttner, 2011; Karch, 2008; Wang et al., 2011).

In the past, detailed postmortem studies of human drug abusers were scarce and the reports predominantly focused on the consequences of hypoxia-ischemia or cerebrovascular events. However, recent studies could demonstrate fundamental drug-induced effects on the cellular elements of the human CNS. In addition, various alterations of neurotransmitters, receptors and second messengers have been detected (Büttner and Weis, 2004; Büttner, 2011; Karch, 2008).

The present chapter gives an updated overview on the neuropathology of the most prevalent abused illicit drugs based on former articles (Büttner and Weis, 2004; Büttner, 2011) followed by a synthesis of the recent postmortem findings.

The Effects of Drug Abuse on the Human Nervous System
http://dx.doi.org/10.1016/B978-0-12-418679-8.00007-1

2.1. NEUROBIOLOGICAL BASICS OF DRUG ABUSE

2.1.1. Overview

The reinforcing properties and the addictive potential of most drugs of abuse are predominantly mediated by the brain "reward-system" with the mesolimbic dopaminergic system, the orbitofrontal cortex, and the extended amygdala as its major anatomical basis (Feltenstein and See, 2008; Goldstein and Volkow, 2002; Guo et al., 2009; Hyman and Malenka, 2001; Koob and Volkow, 2010; Koob and Le Moal, 2006; Leshner and Koob, 1999; Lucantonio et al., 2012; Nestler, 2001; Rapaka and Sadée, 2008; Sell et al., 1999; Shalev et al., 2002; Stewart, 2000; Urban and Martinez, 2012; Volkow and Fowler, 2000; Weiss and Koob, 2000; Wise, 2009). Moreover, the locus coeruleus seems to be an important key nucleus for the mediation of drug abuse (Dyuizen and Lamash, 2009; Gold, 1993; Lane-Ladd et al., 1997; McClung et al., 2005; Van Bockstaele et al., 2010). Alterations of various intracellular messenger pathways, transcription factors, and immediate early genes in these circuits seem to be responsible for the development of chronic drug abuse and addiction (Bowers et al., 2004; Harlan and Garcia, 1998; Hyman and Malenka, 2001; Kelz and Nestler, 2000; Kirby et al., 2011; Nestler, 2001; Rapaka and Sadée, 2008; Traynor, 2010; Uzbay and Oglesby, 2001; Van Bockstaele et al., 2010).

Moreover, environmental and genetic factors are influencing variables (Kendler et al., 2012). However, possible genetic risk factors for drug addiction are still under investigation (Hemby, 2010; Kendler et al., 2012; Kuhar et al., 2001; Kumar et al., 2011; Levran et al., 2008; Lull et al., 2010; Nestler and Landsman, 2001; Pietrzykowski, 2010; Robison and Nestler, 2011; Stallings et al., 2003; Torres and Horowitz, 1999; Zill et al., 2011).

2.1.2. Neuroimaging Studies in Drug Abuse

Neuroimaging studies of human drug abusers have demonstrated brain atrophy, focal demyelinations and hyperintense lesions, focal perfusions deficits, metabolic disturbances, neurochemical abnormalities, and structural abnormalities in the white matter. Details of these alterations are described elsewhere (Büttner and Weis, 2004; Geibprasert et al., 2010; Kaufman, 2001; Netrakom et al., 1999; Tamrazi and Almast, 2012). Although most of these changes have been attributed to ischemic-hypoxic conditions, the underlying cause or possible neuropathological correlates of these findings are still unclear.

2.2. CNS ALTERATIONS OF THE MAJOR DRUGS OF ABUSE

2.2.1. Cannabis

Cannabis is the most frequently drug in use today (Iversen, 2003; Karch, 2008). The primary psychoactive component of cannabis is Δ9-tetrahydrocannabinol (THC). THC is a lipophilic substance and is distributed heterogeneously within the brain, with predominance in neocortical, limbic, sensory, and motor areas (Ashton, 2001). Since cannabis increases the activity of dopaminergic neurons in the mesolimbic dopaminergic system, the drug has reinforcing and abuse potential (Ameri, 1999; Diana et al., 1998; French et al., 1997; Hoffman and Lupica, 2001; Tanda et al., 1997) and may lead to psychological dependency (Abood and Martin, 1992; Ambrosio et al., 1999; Ashton, 2001; Johns, 2001; Nahas, 2001; Smith, 2002). THC and other cannabinoids exert their effects by the interaction with specific cannabinoid (CB) receptors that are members of the G protein-coupled receptor family (Ameri, 1999; Breivogel and Sim-Selley, 2009; Console-Bram et al., 2012; Howlett et al., 2002; Köfalvi, 2008; Pertwee, 1997). Two CB receptors, CB1 and CB2, have been well characterized (Breivogel and Sim-Selley, 2009; Console-Bram et al., 2012; Fride, 2002; Glass et al., 1997; Howlett et al., 2002; Köfalvi, 2008): CB1 receptors are present predominantly in the central and peripheral nervous system, whereby CB2 receptors are present predominantly on immune cells (Ameri, 1999; Childers and Breivogel, 1998; Fride, 2002; Köfalvi, 2008). Both receptors are involved in various signal transduction pathways (Ameri, 1999; Childers and Breivogel, 1998; Console-Bram et al., 2012; Howlett et al., 2002; Köfalvi, 2008; Fride, 2002; Wilson and Nicoll, 2002). After the identification of specific receptors, the discovery of endogenous cannabinoid agonists ("endocannabinoids") followed subsequently. These lipid mediators are involved in a variety of physiological functions and pathological conditions (Bari et al., 2010; Console-Bram et al., 2012; Di Marzo, 2011; Di Marzo et al., 1998; Fattore et al., 2007; Fride, 2002; Katona and Freund, 2012; Köfalvi, 2008; López-Moreno et al., 2008; Maccarrone, 2010; Maldonado et al., 2006; Orgado et al., 2009; Parolaro et al., 2010; Pazos et al., 2005; Pava and Woodward, 2012; Rodriguez de Fonseca et al., 2005; Sidhpura and Parsons, 2011; Solinas et al., 2008; Wilson and Nicoll, 2002).

The CB1 receptors are distributed heterogeneously within the brain with the highest density in the substantia nigra, basal ganglia, hippocampus and cerebellum (Devane et al., 1988; Glass et al., 1997; Herkenham, 1992; Herkenham et al., 1990; Mailleux et al., 1992). In the neocortex the highest density of CB1 receptors is found in the frontal cortex, dentate gyrus, mesolimbic dopaminergic system, and temporal lobe (Glass et al., 1997; Herkenham, 1992; Herkenham et al., 1990; Mailleux et al., 1992). This distribution correlates well with the effects of cannabis on memory, perception, and motor control. The very low density of CB1 receptors in the brainstem explains the low acute toxicity and the lack of lethality of cannabis (Abood and Martin, 1992; Ameri, 1999; Herkenham et al., 1990). However, recent findings revealed a THC-induced generation of free radicals with toxic effect on cultured hippocampal, cortical and neonatal neurons indicating a neurotoxic potential of cannabis (Scallet, 1991; Campbell, 2001; Chan et al., 1998; Guzmán et al., 2001; Hampson and Deadwyler, 1999). These findings might be responsible for the cognitive deficits seen in chronic cannabis abusers (Hampson and Deadwyler, 1999).

2.2.2. CNS Complications

Cardiovascular and CNS complications are the most frequently seen consequences of acute cannabis toxicity (Johns, 2001). The latter may manifest as panic attacks, anxiety, depression, or psychosis (Hollister, 1986; Johns, 2001; Maykut, 1985). In addition, THC may affect cognition and impair verbal and memory skills (Ashton, 2001; Bolla et al., 2002; Pope et al., 2001; Schwartz, 2002). These impairments seem to be reversible and neuropathological changes are not visible (Iversen, 2003). In the context of polysubstance abuse cannabis increases the depressive effect of alcohol, sedatives, and opiates (Nahas, 2001; Reid and Bornheim, 2001). The interaction of cannabis with psychostimulants can be either additive or antagonistic (Nahas, 2001; Reid and Bornheim, 2001).

Although cannabis is widespread abused, there are only very few reported cases of CNS complications. Cerebrovascular events, e.g. cerebral infarction or transitory ischemic attacks, have been related to a cannabis-induced vasospasm or a cannabis-induced hypotension (Barnes et al., 1991; Mouzak et al., 2000; Thanvi and Treadwell, 2009; Zachariah, 1991). However, whether these events are truly associated with cannabis abuse or purely coincidental remains to be established.

2.3. OPIOIDS AND DERIVATIVES

2.3.1. Overview

Opioids, particularly heroin, are the leading substances causing death in drug abusers. The main opioid derivatives include morphine, hydrocodone, oxycodone, hydromorphine, codeine, and other narcotics such as fentanyl, meperidene, methadone, and opium (Bird, 2010; Karch, 2008; Quinn et al., 1997). Heroin is manufactured from opium and usually taken intravenously (Karch, 2008). Intranasal and subcutaneous administration is also common. Heroin alkaloid may be inhaled by heating on metallic foil ("chasing the dragon") (Karch, 2008; Krinsky and Reichard, 2012). Heroin crosses the blood–brain barrier (BBB) faster than morphine leading to an extreme euphoria lasting for several minutes (Brust, 1995; Karch, 2008).

In opioid overdose, coma, respiratory depression, and miosis represent the classical clinical triad (Brust, 1995). A nonfatal overdose has been reported by up to 70% of intravenous heroin abusers during their lives (Coffin et al., 2003). Risk factors for opioid-induced deaths include overdose, concomitant use of other CNS depressants, and loss of tolerance after a period of abstinence (Bird, 2010; Coffin et al., 2003; Preti et al., 2002; Darke et al., 2010; Darke, 2003; Darke and Zador, 1996; Gerostamoulos et al., 2001; Minett et al., 2010; Polettini et al., 1999; Püschel et al., 1993; Quaglio et al., 2001; Sporer, 1999; Warner-Smith et al., 2001).

2.3.2. Neuropathological Findings

Rapid death after lethal heroin ingestion usually does not lead to any morphological evidence of cell injury. In cases of delayed death, ischemic-hypoxic neuronal damage will be apparent. In up to 90% of all deaths due to opioids, brain edema and vascular congestion are seen at autopsy (Adelman and Aronson, 1969; Büttner and Weis, 2006; Gosztonyi et al., 1993; Metter, 1978; Oehmichen et al., 1996; Pearson et al., 1972a; Richter et al., 1973). After a survival period of 5 h or longer, ischemic-hypoxic neuronal damage with cytoplasmic eosinophilia, loss of Nissl substance, and nuclear shrinkage are seen on histological examination (Oehmichen et al., 1996; Pearson et al., 1972a).

Bilateral ischemic-hypoxic lesions of the globus pallidus, as seen in Figure 1, have been observed in 5–10% of heroin abusers (Andersen and Skullerud, 1999; Daras et al., 2001; Ginsberg et al., 1976; Pearson et al., 1975; Richter et al., 1973; Riße and Weiler, 1984; Yee et al., 1994; Zuckerman et al., 1996).

Cerebral infarction in heroin abusers has rarely been described (Quaglio et al., 2001; Adle-Biassette et al., 1996; Bartolomei et al., 1992; Brust and Richter, 1976; Caplan et al., 1982; Herskowitz and Gross, 1973; Jensen et al., 1990; Kelly et al., 1992; Niehaus and Meyer, 1998; Sloan et al., 1991; Vila and Chamorro, 1997). As the main pathogenetic mechanism a global cerebral hypoxia (Adle-Biassette et al., 1996; Brust and Richter, 1976; Jensen et al., 1990; Kelly et al., 1992; Niehaus and Meyer, 1998; Vila and Chamorro, 1997), a focal decrease of the perfusion pressure (Niehaus and Meyer, 1998), a cerebral arteritis, necrotizing angiitis (Kelly et al., 1992; Halpern and Citron, 1971; King et al., 1978; Woods and Strewler, 1972), or vasculitis (Brust, 1997; Brust and Richter, 1976; Niehaus and Meyer, 1998; Rumbaugh et al., 1971), and an embolism from adulterants (Adle-Biassette et al., 1996; Bartolomei et al., 1992; Kelly et al., 1992; Sloan et al., 1991; Vila and Chamorro, 1997) has been proposed. Other concepts implied a hypersensitivity reaction of the blood vessels to heroin in persons who were re-exposed to heroin after a period of abstinence (Caplan et al., 1982; Kelly et al., 1992; Rumbaugh et al., 1971; Woods and Strewler, 1972), or a positional vascular compression (Karch, 2008).

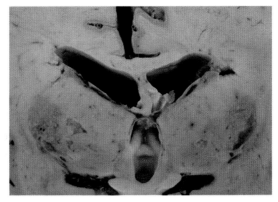

Figure 1 Gross Section of Bilateral Ischemic-Hypoxic Lesions of the Basal Ganglia with Concomitant Laminar Necrosis after 3 Months of Persistent Vegetative State.

Hypoxic-ischemic leukoencephalopathy results from hypoxia secondary to respiratory depression (Ginsberg et al., 1976; Zuckerman et al., 1996; O'Brien and Todd, 2009; Protass, 1971; Shprecher and Mehta, 2010). It is characterized by different degrees of myelin damage (Ginsberg et al., 1976; Pearson and Richter, 1979; Protass, 1971). In addition, loss of neurons in the hippocampal formation, Purkinje cell layer and/or olivary nucleus is frequently seen (Oehmichen et al., 1996). In surviving persons suffering from persistent vegetative state, laminar and subcortical necroses, with extensive nerve cell loss and reactive astrogliosis are observed (O'Brien and Todd, 2009; Protass, 1971; Shprecher and Mehta, 2010).

CNS infections predominantly result from high-risk injection techniques and from the immunosuppression caused by chronic opiate abuse (Adelman and Aronson, 1969; Büttner and Weis, 2004; Karch, 2008; Pearson and Richter, 1979; Roy et al., 2006; Wang et al., 2011). Besides HIV-1 or hepatitis virus, bacterial abscess, meningitis, or ventriculitis (Büttner and Weis, 2004; Karch, 2008; Richter et al., 1973) as well as fungal infections (Büttner and Weis, 2004; Karch, 2008) have been reported. Endocarditis might lead to septic foci in the brain (Büttner and Weis, 2004; Karch, 2008; Louria et al., 1967; Richter et al., 1973; Pearson and Richter, 1979) or to intracranial mycotic aneurysms (Adelman and Aronson, 1969; Brust, 1995; Pearson and Richter, 1979; Richter et al., 1973).

In single instances, transverse myelitis/myelopathy has been reported in heroin abusers (Bernasconi et al., 1996; Brust, 2004; Büttner and Weis, 2004; Ell et al., 1981; Goodhart et al., 1982; Karch, 2008; McCreary et al., 2000; Nyffeler et al., 2003; Pearson et al., 1972b; Richter and Rosenberg, 1968; Riva et al., 2007; Sahni et al., 2008).

Spongiform leukoencephalopathy has been described almost exclusively after inhalation of preheated heroin (Büttner and Weis, 2004; Karch, 2008; Kriegstein et al., 1997; Krinsky and Reichard, 2012; Nuytten et al., 1998; Rizzuto et al., 1997; Wolters et al., 1982). Lipophilic toxic contaminants in conjunction with ischemia-hypoxia are assumed to be the cause, but a definite toxin has not yet been identified (Nuytten et al., 1998; Krinsky and Reichard, 2012; Wolters et al., 1982). The affected brain shows a diffuse white matter spongiform and vacuolar degeneration with loss of oligodendrocytes, axonal reduction, astrogliosis, and subcortical U fiber sparing (Krinsky and Reichard, 2012; Rizzuto et al., 1997; Wolters et al., 1982). The gray matter, brainstem, spinal cord and peripheral nerves are usually unremarkable (Kriegstein et al., 1997; Krinsky and Reichard, 2012; Nuytten et al., 1998; Rizzuto et al., 1997; Wolters et al., 1982). Spongiform leukoencephalopathy distinguishes from delayed leukoencephalopathy following severe hypoxia by the presence of spongiosis with astrogliosis and the absence of typical ischemic-hypoxic lesions (Rizzuto et al., 1997).

2.3.3. Alterations of Neurotransmitters, Receptors, and Second Messengers

The effects of opioids are mediated by specific CNS receptors (Akil et al., 1998; Gold, 1993; Miotto et al., 1996; Mayer and Höllt, 2006; Nestler, 2001; Van Bockstaele et al.,

2010; Waldhoer et al., 2004). Whether chronic opioid abuse leads to a reduced density of these receptors is not yet fully resolved. Some authors reported a similar density of μ- and δ-opioid in the brain of heroin abusers as compared to controls (Gabilondo et al., 1994; García-Sevilla et al., 1997a; Meana et al., 2000; Nestler, 1997; Schmidt et al., 2000, 2001). Others found an increased density of μ-opioid receptor-immunoreactive neurons (Schmidt et al., 2003).

In the development of opioid addiction, the second messenger-signaling system, involving guanosine triphosphate binding (G-) proteins, seems to play a crucial role (Escriba et al., 1994; Hashimoto et al., 1996; Law et al., 2000; Maher et al., 2005; Nestler, 2001; Shichinohe et al., 1998, 2001; Yao et al., 2005). After acute administration to laboratory animals, opioids inhibit adenylyl cyclase activity that converts ATP to cAMP, resulting in decreased cellular cAMP levels (Nestler, 2001). Chronically, opioid exposure induces an upregulation in this adenylyl cyclase-cAMP system mediated via the transcription factor CREB (cAMP response element-binding protein, which is interpreted to be a compensatory mechanism in order to maintain homeostasis (Nestler, 2001; Shichinohe et al., 1998, 2001; Lane-Ladd et al., 1997)). Postmortem studies could demonstrate that chronic heroin abuse causes an increase in certain G-proteins in different regions of the brain (Escriba et al., 1994; Hashimoto et al., 1996). Based on these studies, opioid addiction seems to be associated with abnormalities in second messenger and signal transduction systems involving G-proteins (Escriba et al., 1994; Meana et al., 2000; Hashimoto et al., 1996).

Further findings in the brain of heroin abusers include a decreased level of Ca2+-dependent protein kinase C (PKC)-α (García-Sevilla et al., 1997a), an increased level of a membrane-associated G-protein-coupled receptor kinase (Ozaita et al., 1998), a downregulation of the adenylyl-cyclase subtype I (Shichinohe et al., 1998), a decrease in the density of alpha 2-adrenoreceptors (Gabilondo et al., 1994), a decrease in the immunoreactivity of PKC-αβ (Busquets et al., 1995), decreased levels of immunoreactive neurofilament proteins (García-Sevilla et al., 1997b), and alterations of striatal dopaminergic and serotonergic markers (Kish et al., 2001).

Furthermore, a postmortem study could demonstrate that chronic opioid abuse induces a downregulation of I2-imidazoline receptors in astrocytes and, thereby, presumably downregulates the functions associated with these receptors (Sastre et al., 1996).

2.3.4. Substances for Maintenance Therapy

Codeine, dihydrocodeine, methadone, fentanyl, buprenorphine, and other opioids, sometimes clinically prescribed for maintenance therapy, are often detected in deaths associated for heroin addiction. Monointoxication with one of these substances is the exception and additional CNS depressants, mainly alcohol and benzodiazepines, can be detected frequently (Auriacombe et al., 2001; Barrett et al., 1996; Bell et al., 2009; Bryant et al., 2004; Cooper et al., 1999; Corkery et al., 2004; Darke et al., 2011; Drummer et al.,

1992; Gaulier et al., 2000; Gerostamoulos et al., 1996; Graß et al., 2003; Harding-Pink, 1993; Heinemann et al., 2000; Hickman et al., 2003; Hull et al., 2007; Jumbelic, 2010; Karch and Stephens, 2000; Kintz, 2001; Kreek, 1997; Krinsky et al., 2011; Milroy and Forrest, 2000; Pelissier-Alicot et al., 2010; Pirnay et al., 2004; Seldén et al., 2012; Seymour et al., 2003; Stoops et al., 2010; Worm et al., 1993; Wong et al., 2010; Zamparutti et al., 2011). The neuropathological findings in these cases are similar to those seen in heroin-associated deaths.

2.4. COCAINE

2.4.1. Overview

Cocaine is a potent CNS stimulant and crosses the BBB very rapidly due to its high lipophilic properties (Büttner and Weis, 2004; Karch, 2008; Oyesiku et al., 1993; Prakash and Das, 1993; Quinn et al., 1997; Spiehler and Reed, 1985; White and Lambe, 2003). In the presence of alcohol, cocaine is metabolized to cocaethylene, which also crosses the BBB rapidly. With a longer half-life time, cocaethylene accumulates to a 4-times higher concentration and has a similar pharmacologic profile to cocaine (Andrews, 1997; Hearn et al., 1991; Horowitz and Torres, 1999; Quinn et al., 1997). Within the brain, receptors with varying affinities are present for cocaine and its major metabolites (Biegon et al., 1992; Calligaro and Eldefrawi, 1987; Kalasinsky et al., 2000; Volkow et al., 1995).

The CNS effects of cocaine are mainly mediated through interactions and alterations of the neurotransmitters dopamine, norepinephrine, serotonin, acetylcholine, and gamma aminobutyric acid (Karch, 2008; Quinn et al., 1997; Prakash and Das, 1993; Strang et al., 1993; White and Lambe, 2003). After binding to specific receptors at presynaptic sites, cocaine prevents neurotransmitter reuptake (Büttner and Weis, 2004; Karch, 2008; Quinn et al., 1997; White and Lambe, 2003). The predominant synaptic effect of cocaine is the release of dopamine from the synaptic vesicles and the blocking of dopamine reuptake resulting in an enhanced dopaminergic neurotransmission (Büttner and Weis, 2004; Karch, 2008; Quinn et al., 1997; White and Lambe, 2003).

The abuse potential of cocaine is based on its reinforcing properties and the rapid development of tolerance to the euphoric effects (Bowers et al., 2004; Hemby, 2010; Kalivas and McFarland, 2003; Lucantonio et al., 2012; Nestler, 2001; Quinn et al., 1997; Shalev et al., 2002; Stewart, 2000; Strang et al., 1993; Weiss and Koob, 2000).

2.4.2. Cerebrovascular Complications

Cocaine is the most common drug of abuse associated with cerebrovascular events (Kaku and Lowenstein, 1990; Kelly et al., 1992; Levine et al., 1991). Intracerebral and sub-arachnoidal hemorrhages as well as hemorrhagic or ischemic stroke have been reported frequently (Aggarwal et al., 1996; Bhattacharya et al., 2011; Brown et al., 1992; Brust, 1993; Cregler and Mark, 1986; Daras et al., 1994; Davis and Swalwell, 1996; Fessler et al.,

1997; Herning et al., 1999; Jacobs et al., 1989; Klonoff et al., 1989; Konzen et al., 1995; Levine and Welch, 1988; Lundberg et al., 1977; Mangiardi et al., 1988; Martin-Schild et al., 2010; Merkel et al., 1995; Mittleman and Wetli, 1987; Mody et al., 1988; Nanda et al., 2000, 2006; Nolte et al., 1996; Oyesiku et al., 1993; Peterson et al., 1991; Petty et al., 1990; Qureshi et al., 1997, 2001; Sen et al., 1999; Sloan et al., 1991; Tardiff et al., 1989; Toossi et al., 2010; Treadwell and Robinson, 2007; Vannemreddy et al., 2008; Wojak and Flamm, 1987). Cocaine-associated cerebrovascular complications occur primarily in young adults with a peak in the early thirties (Brown et al., 1992; Klonoff et al., 1989; Karch, 2008; Levine et al., 1991; McEvoy et al., 2000; Petitti et al., 1998).

Cocaine-associated ischemic infarctions are attributed to cerebral vasospasm as a result of the vasoconstrictive effects of cocaine and its metabolites (Albuquerque and Kurth, 1993; Andrews, 1997; Covert et al., 1994; Cregler and Mark, 1986; Daras et al., 1994; Fessler et al., 1997; Gottschalk and Kosten, 2002; He et al., 1994; Karch, 2008; Kaufman et al., 1998; Kelly et al., 1992; Konzen et al., 1995; Levine et al., 1991; Madden et al., 1995; Schreiber et al., 1994; Spiehler and Reed, 1985; Strickland et al., 1993). Other mechanisms include cocaine-induced cardiac arrhythmia with secondary embolic or hemodynamic cerebral ischemia basis (Brust, 1993; Cregler and Mark, 1986; Konzen et al., 1995; Levine et al., 1991; Petty et al., 1990), or the effects of cocaine on hemostasis with increased platelet aggregation (Jennings et al., 1993; Konzen et al., 1995; Kugelmass et al., 1993; Rinder et al., 1994; Treadwell and Robinson, 2007; Yao et al., 2011).

Cocaine-associated intracerebral and subarachnoidal hemorrhages as seen in Figures 2 and 3 are associated with underlying arteriovenous malformations or aneurysms in about half of the affected persons (Cregler and Mark, 1986; Davis and Swalwell, 1996; Fessler et al., 1997; Klonoff et al., 1989; Levine and Welch, 1988; Levine et al., 1991; Mangiardi et al., 1988; McEvoy et al., 2000; Mittleman and Wetli, 1987; Nolte et al., 1996; Oyesiku et al., 1993; Peterson et al., 1991; Sloan et al., 1991; Tardiff et al., 1989; Vannemreddy et al., 2008). A sudden elevation of blood pressure is believed to be the cause (Brown et al., 1992; Daras et al., 1994; Davis and Swalwell, 1996; Fessler et al., 1997; Jacobs et al., 1989; Kelly et al., 1992; Klonoff et al., 1989; Levine et al., 1991; Mangiardi et al., 1988; Nolte et al., 1996; Oyesiku et al., 1993; Tardiff et al., 1989). Compared to the nondrug-using population, cocaine abuse has been shown to predispose to aneurysm rupture at an earlier age and in much smaller aneurysms (McEvoy et al., 2000; Nanda et al., 2000; Oyesiku et al., 1993; Vannemreddy et al., 2008).

As another pathogenetic mechanism for the occurrence of cerebrovascular events, a cocaine-induced vasculitis has been proposed but could only be demonstrated in rare instances (Brown et al., 1992; Brust, 1997; Diez-Tejedor et al., 1998; Fredericks et al., 1991; Kaye and Fainstat, 1987; Klonoff et al., 1989; Krendel et al., 1990; Levine and Welch, 1988; Martin et al., 1995; Merkel et al., 1995; Morrow and McQuillen, 1993; Peterson et al., 1991).

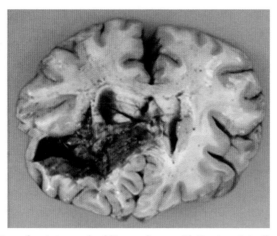

Figure 2 Gross Section of an Intracerebral Hemorrhage with Rupture into the Ventricular System.

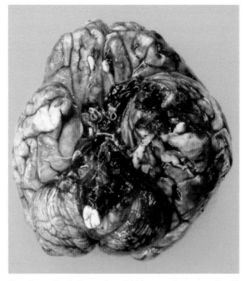

Figure 3 Gross Section of a Subarachnoid Hemorrhage at the Base of the Brain.

2.4.3. Alterations of Neurotransmitters, Receptors, and Gene Expression

In cocaine-related deaths marked reductions in the levels of dopamine, enkephalin mRNA, μ opioid receptor binding, and dopamine uptake site binding as well as elevation in levels of dynorphin mRNA and κ opioid receptor and receptor binding have been described in various brain regions (Hurd and Herkenham, 1993; Little et al., 1993, 1996, 1998, 2003; Staley et al., 1997; Wilson et al., 1996a). The decrease in the levels of dopamine was not paralleled by an increase of a dopamine D1 and D2 gene expression (Meador-Woodruff et al., 1993). There was also an increase of cocaine binding sites on the dopamine transporter with a decrease of dopamine D1 and D3 receptor density in specific regions of the brain reward system (Little et al., 1993, 1996, 1998; Mash et al., 2002; Segal et al., 1997; Staley et al., 1994). A cocaine-induced damage to the dopaminergic system has also been suggested by a marked reduction in vesicular monoamine transporter-2 immunoreactivity (Little et al., 2003), and of the transcription factor NURR1 (Bannon et al., 2002). An increase of the serotonin transporter has been detected in the striatum, substantia nigra, and the limbic system (Mash et al., 2000). Moreover, a disturbed phospholipid metabolism could be demonstrated in cocaine abusers (Ross et al., 1996, 2002).

2.5. AMPHETAMINES, METHAMPHETAMINE, AND DESIGNER DRUGS

2.5.1. Overview

Over the past years, the abuse of amphetamines, methamphetamine, and designer drugs has significantly increased worldwide (Albertson et al., 1999; Carroll et al., 2012; Freese et al., 2002; Karch, 2008; Koesters et al., 2002; Rome, 2001; Smith et al., 2002). Amphetamines and methamphetamine are psychostimulants and include a broad spectrum of substances (Büttner and Weis, 2004; Karch, 2008; Kish, 2008; Quinn et al., 1997). Their potent sympathomimetic effects include an elevation of pulse rate and blood pressure, an increased level of alertness, and decreased fatigue (Cruickshank and Dyer, 2009; Quinn et al., 1997; Sulzer et al., 2005). Adverse effects include seizures, agitation and psychosis, often accompanied by aggressive behavior and suicide (Baskin-Sommers and Sommers, 2006; Derlet et al., 1989; Hart and Wallace, 1975; Logan et al., 1998; Martin et al., 2009; Zhu et al., 2000). Deaths after amphetamines and methamphetamine abuse have also been reported in several autopsy series (Darke et al., 2008; Gould et al., 2009; Karch et al., 1999; Kaye et al., 2008; Logan et al., 1998; Lora-Tamayo et al., 1997; Raikos et al., 2002; Shaw, 1999; Zhu et al., 2000).

2.5.2. Cerebrovascular Complications

After cocaine, amphetamines and methamphetamine are the second-most-common causes of ischemic or hemorrhagic stroke occurring in persons younger than 45 years (Heye and Hankey, 1996; Kaku and Lowenstein, 1990; Karch, 2008; Perez et al., 1999; Rothrock et al., 1988; Yen et al., 1994). In addition, subarachnoidal and intracerebral hemorrhages have been described (Caplan et al., 1982; Davis and Swalwell, 1996; Delaney and Estes, 1980; D'Souza and Shraberg, 1981; Harrington et al., 1983; Imanse and Vanneste, 1990; Karch et al., 1999; Kelly et al., 1992; Klys et al., 2005; Lessing and Hyman, 1989; Lukes, 1983; McEvoy et al., 2000; Moriya and Hashimoto, 2002; Petitti et al., 1998; Sloan et al., 1991; Selmi et al., 1995; Shibata et al., 1991; Zhu et al., 2000).

A sudden elevation in blood pressure (Heye and Hankey, 1996; Kelly et al., 1992; Logan et al., 1998) or the vasoconstrictive effects of both substances (Perez et al., 1999) are the principal cause. In single cases, a cerebral vasculitis has also been reported (Bostwick, 1981; Brust, 1997; Kelly et al., 1992; Margolis and Newton, 1971; Matick et al., 1983; Zhu et al., 2000). The neuropathological findings in these cases are similar to those seen in cocaine-associated cerebrovascular events as seen in Figures 2 and 3.

2.5.3. Neurotoxicity

The neurotoxic consequences of amphetamines and methamphetamine on the dopaminergic system have been described in various animal species and in humans by numerous neuroimaging and postmortem studies (Bennett et al., 1997; Broening et al., 1997; Brown et al., 2000; Cadet and Krasnova, 2009; Davidson et al., 2001; Ernst et al., 2000; Friedman et al., 1998; Frost and Cadet, 2000; Hanson et al., 1998; Iacovelli et al., 2006; Kuperman et al., 1997; McCann et al., 1998; Melega et al., 2000; Metzger et al., 2000; O'Dell et al., 1991; Ricaurte and McCann, 1992; Robinson and Becker, 1986; Seiden and Sabol, 1996; Sekine et al., 2001; Tong et al., 2003; Trulson et al., 1985; Villemagne et al., 1998; Volkow et al., 2001a,b; Wagner et al., 1980a,b; Wilson et al., 1996b). Similar alterations have been detected in the serotonergic system (Axt and Molliver, 1991; Cadet and Krasnova, 2009; Frost and Cadet, 2000; Fukui et al., 1989; Haughey et al., 2000). However, whether these neurochemical alterations are irreversible and reflect neuroadaptation or neurotoxicity is still unclear (Harvey et al., 2000).

2.6. AMPHETAMINE AND METHAMPHETAMINE DERIVATIVES
2.6.1. Overview

Common amphetamine and methamphetamine derivatives include DOM (4-methyl-2,5-dimethoxyamphetamine), DOB (4-bromo-2,5-dimethoxyamphetamine), MDA (3,4-methylenedioxyamphetamine), MDE (3,4-methylenedioxyethylamphetamine), MDMA (3,4-methylenedioxymethamphetamine), mephedrone, 4-MTA (4-methyl-thioamphetamine), and PMA (4-para-methoxyamphetamine) (Christophersen, 2000; Felgate et al., 1998; Freese et al., 2002; James and Dinan, 1998; Karch, 2008; Koesters et al., 2002; Winstock et al., 2002, 2011). The street name "ecstasy" subsumes different hallucinogenic amphetamine derivatives with MDMA and MDE being the main components (Cole et al., 2002; Morefield et al., 2011; Wolff et al., 1995).

2.6.2. MDMA

MDMA is the most commonly abused substance of the above mentioned derivatives. MDMA acts predominantly on the serotonergic system with sympathomimetic properties and modulates psychomotor and neuroendocrine functions (Battaglia et al., 1988; Christophersen, 2000; De la Torre et al., 2004; Downing, 1986; Freese et al.,

2002; Green et al., 1995; Kalant, 2001; Liester et al., 1992; Liechti and Vollenweider, 2001; McCann et al., 2000; Rochester and Kirchner, 1999; White et al., 1996). The unique effect of MDMA is the feeling of intimacy and closeness, designated as "entactogenic" (Nichols, 1986).

Fatalities after "ecstasy" consumption have been reported, which were frequently associated with coagulopathy and hyperthermia (Arimany et al., 1998; Byard et al., 1998; Chadwick et al., 1991; Dowling et al., 1987; Fineschi et al., 1999; Forrest et al., 1994; Gill et al., 2002; Henry et al., 1992; Libiseller et al., 2005; Milroy, 2011; Milroy et al., 1996; Schifano, 2004; Turillazzi et al., 2010; Walubo and Seger, 1999).

Moreover, ischemic and hemorrhagic stroke (Hanyu et al., 1995; Harries and De Silva, 1992; Hughes et al., 1993; Manchanda and Connolly, 1993; Muntan and Tuckler, 2006; Schlaeppi et al., 1999) as well as subarachnoidal hemorrhage (Gledhill et al., 1993), and leukoencephalopathy (Bertram et al., 1999) have been described in single cases after "ecstasy" abuse.

In one postmortem study, necrosis of the globus pallidus, diffuse astrogliosis, and spongiform changes of the white matter have been noted (Squier et al., 1995). Further findings in deaths after "ecstasy" consumption consisted of the complications of hyperthermia with disseminated intravascular coagulopathy and included cerebral edema, focal hemorrhages, and nerve cell loss (Milroy et al., 1996).

2.6.3. Neurotoxicity

After acute and long-term MDMA exposure, neurotoxic effects have been demonstrated in animals and nonhuman primates (Battaglia et al., 1988; Hatzidimitriou et al., 1999; Huether et al., 1997; Insel et al., 1989; McKenna and Peroutka, 1990; Ricaurte et al., 2000a; Ricaurte et al., 2000b; Schmidt, 1987; Scallet et al., 1988). The serotonergic system seems to be predominantly affected (Hatzidimitriou et al., 1999; Ricaurte et al., 2000b; Scallet et al., 1988). However, the underlying pathogenetic mechanisms are not fully resolved (Cadet, 1998; Curran, 2000; Lyles and Cadet, 2003; Seiden and Sabol, 1996; Sprague et al., 1998; Turner and Parrott, 2000). In humans, there is also evidence that MDMA might be neurotoxic (Bolla et al., 1998; Buchert et al., 2003; Chang et al., 1999; Gerra et al., 2000; Green and Goodwin, 1996; Hegadoren et al., 1999; McCann et al., 2000; Obrocki et al., 2002; Parrott, 2001, 2002; Reneman et al., 2001; Ricaurte et al., 2000a; Soar et al., 2001), since impaired cognitive performance and an increased incidence of neuropsychiatric disorders have been reported in MDMA abusers (Reneman et al., 2001; Soar et al., 2001). Nevertheless, it is still not fully resolved how to extrapolate animal and nonhuman primate data to the human condition and whether persistent neurotoxicity occurs in humans (Cadet, 1998; Curran, 2000; Gouzoulis-Mayfrank and Daumann, 2006; Kish, 2002; Lyles and Cadet, 2003; McCann et al., 2001; McGuire, 2000; Turner and Parrott, 2000).

2.7. NEUROPATHOLOGICAL INVESTIGATIONS OF (POLY-) DRUG ABUSERS

In the past, there were only few reports of pathological findings in the brains of human drug abusers. The authors described predominantly brain edema, vascular congestion, ischemic nerve cell damage, and neuronal loss (Adelman and Aronson, 1969; Gosztonyi et al., 1993; Makrigeorgi-Butera et al., 1996; Metter, 1978; Oehmichen et al., 1996; Pearson et al., 1972a; Richter et al., 1973). These changes have been ascribed as secondary findings due to drug-induced respiratory depression (Makrigeorgi-Butera et al., 1996; Oehmichen et al., 1996). However, in recent studies, widespread morphological alterations within the cellular network of the brain have been detected in (poly) substance abusers with heroin as the leading cause of death (Büttner and Weis, 2006). The findings consisted of a marked neuronal loss (Büttner and Weis, 2006), as shown in Figure 4. There is also a reduced density of GFAP-positive astrocytes (Büttner and Weis, 2006), and an axonal damage (Büttner and Weis, 2006; Büttner et al., 2006) that was accompanied by a microglia activation (Anthony et al., 2005; Büttner and Weis, 2006; Büttner et al., 2006; Gosztonyi et al., 1993; Makrigeorgi-Butera et al., 1996; Tomlinson et al., 1999). Cerebrovascular changes consisted of reactive endothelial cell proliferation, endothelial swelling and endothelial cell hyperplasia, degenerative hyaline vessel wall thickening, and a decrease of the collagen type IV content of the basal lamina (Büttner

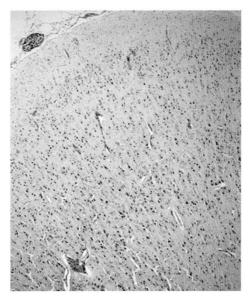

Figure 4 Microscopic Section of the Parietal Lobe Showing Nerve Cell Loss (Luxol-Fast-Blue Stain, Original Magnification x40).

and Weis, 2006; Büttner et al., 2005). In addition, there is a disruption of the tight junctions at the BBB (Bell et al., 2006).

Besides agonal ischemic-hypoxic conditions, the neuronal loss seems to be induced directly by drugs of abuse and, indirectly, by a drug-induced damage of astrocytes, axons, and the cerebral microvessels (Büttner and Weis, 2006).

The widespread axonal damage as demonstrated with β-APP immunohistochemistry and the activation of microglia predominantly in the white matter indicates a chronic progressive process (Anthony et al., 2005; Büttner and Weis, 2006; Büttner et al., 2006; Gosztonyi et al., 1993; Makrigeorgi-Butera et al., 1996; Tomlinson et al., 1999) and is suggestive for a toxic-metabolic drug effect. These findings might represent the morphological correlate of the demyelinations and hyperintense areas observed on neuroimaging.

The numerical reduction of GFAP-positive astrocytes (Büttner and Weis, 2006) might be caused by the interference of drugs with the GFAP-gene transcription, or the generation of free radicals (Anderson et al., 2003; Miguel-Hidalgo, 2009; Sastre et al., 1996; Stadlin et al., 1988).

The observation of concentric small-vessel wall thickening, perivascular space dilatation, perivascular pigment deposition, vessel wall mineralization, and occasional perivascular inflammatory cell infiltrates (Bell et al., 2002, 2006; Büttner and Weis, 2006) in the brains of drug abusers were similar to those observed in AIDS-patients (Connor et al., 2000). Besides, the above mentioned cerebrovascular changes represent a noninflammatory vasculopathy that is indicative for a disturbed BBB and might be another morphological substrate of the alterations seen on neuroimaging.

2.8. NEURODEGENERATION AND DRUGS OF ABUSE

Recent neuroimaging and postmortem studies indicate that drug abusers might develop neurodegeneration or Parkinsonism as they age (Callaghan et al., 2012; Davidson et al., 2001; Guilarte, 2001; Iacovelli et al., 2006; Kish, 2002, 2003; Kuniyoshi and Jankovic, 2003; Little et al., 2009; Mash et al., 2003; Tang et al., 2003; Thrash et al., 2009). In younger opiate abusers, an increase in hyperphosphorylated tau-positive neurofibrillary pretangles, fully formed tangles, and neuritic threads was found (Anthony et al., 2010; Ramage et al., 2005). In these cases, ubiquitin-positive neuronal inclusions were also present. Within the substantia nigra of drug abusers, the numerical density of the pigmented neurons was decreased (Büttner and Weis, 2006) and ubiquitinated cytoplasmic neuronal inclusions have been detected (Bell et al., 2002). The accumulation of α-synuclein protein in long-term cocaine abusers is suggestive for an increased risk for the development of Parkinsonism (Mash et al., 2003). In "ecstasy" and methamphetamine abusers, there is evidence for axonal damage and subsequent reduction (Davidson et al., 2001; Guilarte, 2001; Iacovelli et al., 2006; Thrash et al., 2009).

These studies strongly suggest a drug-related neurodegeneration. However, whether or to what extent neurodegeneration occurs in human drug abusers remains to be established.

3. CONCLUSIONS

There is a broad spectrum of neuropathological alterations in the brains of drug abusers involving nearly all types of CNS cells. However, a clear demarcation between the primary neuropathological effects of a specific drug and the nonspecific changes occurring secondarily is rarely possible. Nevertheless, recent studies have demonstrated morphological correlates of the neuroimaging findings seen in the brains of drug abusers and provided evidence for a drug-related neurodegeneration. Based on these findings future studies may further elucidate the profound disturbances of drugs of abuse on the complex network of CNS cell-interactions.

REFERENCES

Abood, M.A., Martin, B.R., 1992. Neurobiology of marijuana abuse. Trends Pharmacol. Sci. 13, 301–306.

Adelman, L.S., Aronson, S.M., 1969. The neuropathologic complications of narcotic drug addiction. Bull. N. Y. Acad. Med. 45, 225–234.

Adle-Biassette, H., Marc, B., Benhaiem-Sigaux, N., Durigon, M., Gray, F., 1996. Infarctus cérébraux chez un toxicomane inhalant l'héroïne. Arch. Anat. Cytol. Pathol. 44, 12–17.

Aggarwal, S.K., Williams, V., Levine, S.R., Cassin, B.J., Garcia, J.H., 1996. Cocaine-associated intracranial hemorrhage: absence of vasculitis in 14 cases. Neurology 46, 1741–1743.

Akil, H., Owens, C., Gutstein, H., Taylor, L., Curran, E., Watson, S., 1998. Endogenous opioids: overview and current issues. Drug Alcohol Depend. 51, 127–140.

Albertson, T.E., Derlet, R.W., van Hoozen, B.E., 1999. Methamphetamine and the expanding complications of amphetamines. West. J. Med. 170, 214–219.

Albuquerque, M.L., Kurth, C.D., 1993. Cocaine constricts immature cerebral arterioles by a local anesthetic mechanism. Eur. J. Pharmacol. 249, 215–220.

Ambrosio, E., Martin, S., García-Lecumberri, C., Osta, A., Crespo, J.A., 1999. The neurobiology of cannabinoid dependence: sex differences and potential interactions between cannabinoid and opioid systems. Life Sci. 65, 687–694.

Ameri, A., 1999. The effects of cannabinoids on the brain. Prog. Neurobiol. 58, 315–348.

Andersen, S.N., Skullerud, K., 1999. Hypoxic/ischaemic brain damage, especially pallidal lesions, in heroin addicts. Forensic Sci. Int. 102, 51–59.

Anderson, C.E., Tomlinson, G.S., Pauly, B., Brannan, F.W., Chiswick, A., Brack-Werner, R., Simmonds, P., Bell, J.E., 2003. Relationship of Nef-positive and GFAP-reactive astrocytes to drug use in early and late HIV infection. Neuropathol. Appl. Neurobiol. 29, 378–388.

Andrews, P., 1997. Cocaethylene toxicity. J. Addict. Dis. 16, 75–84.

Anthony, I.C., Norrby, K.E., Dingwall, T., Carnie, F.W., Millar, T., Arango, J.C., Robertson, R., Bell, J.E., 2010. Predisposition to accelerated Alzheimer-related changes in the brains of human immunodeficiency virus negative opiate abusers. Brain 133, 3685–3698.

Anthony, I.C., Arango, J.C., Stephens, B., Simmonds, P., Bell, J.E., 2008. The effects of illicit drugs on the HIV infected brain. Front. Biosci. 13, 1294–1307.

Anthony, I.C., Ramage, S.N., Carnie, F.W., Simmonds, P., Bell, J.E., 2005. Does drug abuse alter microglial phenotype and cell turnover in the context of advancing HIV infection? Neuropathol. Appl. Neurobiol. 31, 325–338.

Arimany, J., Medallo, J., Pujol, A., Vingut, A., Borondo, J.C., Valverde, J.L., 1998. Intentional overdose and death with 3,4-methylenedioxyamphetamine (MDEA; "Eve"). Am. J. Forensic Med. Pathol. 19, 148–151.

Ashton, C.H., 2001. Pharmacology and effects of cannabis: a brief review. Br. J. Psychiatry 178, 101–106.

Auriacombe, M., Franques, P., Tignol, J., 2001. Deaths attributable to methadone vs buprenorphine in France. JAMA 285, 45.

Axt, K.J., Molliver, M.E., 1991. Immunocytochemical evidence for methamphetamine-induced serotonergic axon loss in the rat brain. Synapse 9, 302–313.

Backmund, M., Reimer, J., Meyer, K., Gerlach, J.T., Zachoval, R., 2005. Hepatitis C virus infection and injection drug users: prevention, risk factors, and treatment. Clin. Infect. Dis. 40 (Suppl. 5), S330–S335.

Bannon, M.J., Pruetz, B., Manning-Bog, A.B., Whitty, C.J., Michelhaugh, S.K., Sacchetti, P., Granneman, J.G., Mash, D.C., Schmidt, C.J., 2002. Decreased expression of the transcription factor NURR1 in dopamine neurons of cocaine abusers. Proc. Natl. Acad. Sci. USA 99, 6382–6385.

Bari, M., Rapino, C., Mozetic, P., Maccarrone, M., 2010. The endocannabinoid system in gp120-mediated insults and HIV-associated dementia. Exp. Neurol. 224, 74–84.

Barnes, D., Palace, J., O'Brien, M.D., 1991. Stroke following marijuana smoking. Stroke 22, 1381.

Barrett, D.H., Luk, A.J., Parrish, R.G., Jones, T.S., 1996. An investigation of medical examiner cases in which methadone was detected, Harris County, Texas, 1987–1992. J. Forensic Sci. 41, 442–448.

Bartolomei, F., Nicoli, F., Swiader, L., Gastaut, J.L., 1992. Accident vasculaire cérébral ischémique après prise nasale d'héroïne. Une nouvelle observation. Presse Med. 21, 983–986.

Baskin-Sommers, A., Sommers, I., 2006. Methamphetamine use and violence among young adults. J. Crim. Just. 34, 661–674.

Battaglia, G., Brooks, B.P., Kulsakdinun, C., De Souza, E.B., 1988. Pharmacologic profile of MDMA (3,4-methylenedioxymethamphetamine) at various brain recognition sites. Eur. J. Pharmacol. 149, 159–163.

Bell, J.R., Butler, B., Lawrance, A., Batey, R., Salmelainen, P., 2009. Comparing overdose mortality associated with methadone and buprenorphine treatment. Drug Alcohol. Depend. 104, 73–77.

Bell, J.E., Arango, J.C., Anthony, I.C., 2006. Neurobiology of multiple insults: HIV-1-associated brain disorders in those who use illicit drugs. J. Neuroimmune Pharmacol. 1, 182–191.

Bell, J.E., Arango, J.C., Robertson, R., Brettle, R.P., Leen, C., Simmonds, P., 2002. HIV and drug misuse in the Edinburgh cohort. J. Acquir. Immune Defic. Syndr. 31 (Suppl. 2), S35–S42.

Bennett, B.A., Hollingsworth, C.K., Martin, R.S., Harp, J.J., 1997. Methamphetamine-induced alterations in dopamine transporter function. Brain Res. 782, 219–227.

Bernasconi, A., Kuntzer, T., Ladbon, N., Janzer, R.C., Yersin, B., Regli, F., 1996. Complications neurologiques périphériques et médullaires de la toxicomanie intraveineuse à l'héroïne. Rev. Neurol. 152, 688–694.

Bertram, M., Egelhoff, T., Schwarz, S., Schwab, S., 1999. Toxic leukoencephalopathy following "ecstasy" ingestion. J. Neurol. 246, 617–618.

Bhattacharya, P., Taraman, S., Shankar, L., Chaturvedi, S., Madhavan, R., 2011. Clinical profiles, complications, and disability in cocaine-related ischemic stroke. J. Stroke Cerebrovasc. Dis. 20, 443–449.

Biegon, A., Dillon, K., Volkow, N.D., Hitzemann, R.J., Fowler, J.S., Wolf, A.P., 1992. Quantitative autoradiography of cocaine binding sites in human brain postmortem. Synapse 10, 126–130.

Bird, S.M., 2010. Over 1200 drugs-related deaths and 190,000 opiate-user-years of follow-up: relative risks by sex and age group. Addict. Res. Theory 18, 194–207.

Bolla, K.I., Brown, K., Eldreth, D., Tate, K., Cadet, J.L., 2002. Dose-related neurocognitive effects of marijuana use. Neurology 59, 1337–1343.

Bolla, K.I., McCann, U.D., Ricaurte, G.A., 1998. Memory impairment in abstinent MDMA ("ecstasy") users. Neurology 51, 1532–1537.

Bostwick, D.G., 1981. Amphetamine induced cerebral vasculitis. Hum. Pathol. 12, 1031–1033.

Bowers, M.S., McFarland, K., Lake, R.W., Peterson, Y.K., Lapish, C.C., Gregory, M.L., Lanier, S.M., Kalivas, P.W., 2004. Activator of G protein signaling 3: a gatekeeper of cocaine sensitization and drug seeking. Neuron 42, 269–281.

Breivogel, C.S., Sim-Selley, L.J., 2009. Basic neuroanatomy and neuropharmacology of cannabinoids. Int. Rev. Psychiatry 21, 113–121.

Broening, H.W., Pu, C., Vorhees, C.V., 1997. Methamphetamine selectively damages dopaminergic innervation to the nucleus accumbens core while sparing the shell. Synapse 27, 153–160.

Brown, J.M., Hanson, G.R., Fleckenstein, A.E., 2000. Methamphetamine rapidly decreases vesicular dopamine uptake. J. Neurochem. 74, 2221–2223.

Brown, E., Prager, J., Lee, H.-Y., Ramsey, R.G., 1992. CNS complications of cocaine abuse: prevalence, pathophysiology, and neuroradiology. AJR Am. J. Roentgenol. 159, 137–147.

Brust, J.C.M., 1993. Clinical, radiological, and pathological aspects of cerebrovascular disease associated with drug abuse. Stroke. 24, 129–133.

Brust, J.C.M., 2004. Neurological Aspects of Substance Abuse, second ed. Elsevier Butterworth-Heinemann Ltd, Philadelphia.

Brust, J.C.M., 1997. Vasculitis owing to substance abuse. Neurol. Clin. 15, 945–957.

Brust, J.C.M., 1995. Opiate addiction and toxicity. In: Handbook of Clinical Neurology. In: Vinkens, P.J. (Ed.), Intoxications of the Nervous System, Part II, vol. 65. Elsevier, Amsterdam, pp. 349–361.

Brust, J.C.M., Richter, R.W., 1976. Stroke associated with addiction to heroin. J. Neurol. Neurosurg. Psychiatry 39, 194–199.

Bryant, W.K., Galea, S., Tracy, M., Markham Piper, T., Tardiff, K., Vlahov, D., 2004. Overdose deaths attributed to methadone and heroin in New York City, 1990–1998. Addiction 99, 846–854 168.

Buchert, R., Thomasius, R., Nebeling, B., Petersen, K., Obrocki, J., Jenicke, L., Wilke, F., Wartberg, L., Zapletalova, P., Clausen, M., 2003. Long-term effects of "ecstasy" use on serotonin transporters of the human brain investigated by PET. J. Nucl. Med. 44, 375–384.

Büttner, A., 2011. The neuropathology of drug abuse. Neuropathol. Appl. Neurobiol. 37, 118–134.

Büttner, A., Rohrmoser, K., Mall, G., Penning, R., Weis, S., 2006. Widespread axonal damage in the brain of drug abusers as evidenced by accumulation of β-amyloid precursor protein (β-APP): an immunohisto-chemical investigation. Addiction 101, 1339–1346.

Büttner, A., Weis, S., 2006. Neuropathological alterations in drug abusers: the involvement of neurons, glial, and vascular systems. Forensic Sci. Med. Pathol. 2, 115–126.

Büttner, A., Kroehling, C., Mall, G., Penning, R., Weis, S., 2005. Alterations of the vascular basal lamina in the cerebral cortex in drug abuse: a combined morphometric and immunohistochemical investigation. Drug Alcohol. Depend. 69, 63–79.

Büttner, A., Weis, S., 2004. Central nervous system alterations in drug abuse. In: Tsokos, M. (Ed.), Forensic Pathology Reviews, vol. 1. Humana Press, Totowa, NJ, pp. 79–136.

Busquets, X., Escriba, P.V., Sastre, M., García-Sevilla, J.A., 1995. Loss of protein kinase C-αβ in brain of heroin addicts and morphine-dependent rats. J. Neurochem. 64, 247–252.

Byard, R.W., Gilbert, J., James, R., Lokan, R.J., 1998. Amphetamine derivative fatalities in South Australia – is "ecstasy" the culprit? Am. J. Forensic Med. Pathol. 19, 261–265.

Cadet, J.L., Krasnova, I.N., 2009. Molecular bases of methamphetamine-induced neurodegeneration. Int. Rev. Neurobiol. 88, 101–119.

Cadet, J.L., 1998. Neurotoxicity of drugs of abuse. In: Koliatsos, V.E., Ratan, R. (Eds.), Cell Death and Diseases of the Nervous System. Humana Press, Totowa, NJ, pp. 521–526.

Callaghan, R.C., Cunningham, J.K., Sykes, J., Kish, S.J., 2012. Increased risk of Parkinson's disease in individuals hospitalized with conditions related to the use of methamphetamine or other amphetamine-type drugs. Drug Alcohol. Depend. 120, 35–40.

Calligaro, D.O., Eldefrawi, M.E., 1987. Central and peripheral cocaine receptors. J. Pharmacol. Exp. Ther. 243, 61–68.

Campbell, V.A., 2001. Tetrahydrocannabinol-induced apoptosis of cultured cortical neurones is associated with cytochrome c release and caspase-3 activation. Neuropharmacology 40, 702–709.

Caplan, L.R., Hier, D.B., Banks, G., 1982. Current concepts of cerebrovascular disease – stroke: stroke and drug abuse. Stroke 13, 869–872.

Cardoso, F.E.C., Jankovic, J., 1993. Cocaine-related movement disorders. Mov. Disord. 8, 175–178.

Carroll, F.I., Lewin, A.H., Mascarella, S.W., Seltzman, H.H., Reddy, P.A., 2012. Designer drugs: a medicinal chemistry perspective. Ann. N.Y. Acad. Sci. 1248, 18–38.

Chadwick, I.S., Linsley, A., Freemont, A.J., Doran, B., 1991. Ecstasy, 3,4-methylenedioxymethamphetamine (MDMA), a fatality associated with coagulopathy and hyperthermia. J. R. Soc. Med. 84, 371.

Chan, G.C.K., Hinds, T.R., Impey, S., Storm, D.R., 1998. Hippocampal neurotoxicity of Δ9-tetrahydrocannabinol. J. Neurosci. 18, 5322–5332.

Chang, L., Ernst, T., Grob, C.S., Poland, R.E., 1999. Cerebral 1H MRS alterations in recreational 3,4-methylenedioxymethamphetamine (MDMA, "ecstasy") users. J. Magn. Reson. Imaging 10, 521–526.

Childers, S.R., Breivogel, C.S., 1998. Cannabis and endogenous cannabinoid systems. Drug Alcohol Depend. 51, 173–187.

Christophersen, A.S., 2000. Amphetamine designer drugs an overview and epidemiology. Toxicol. Lett. 112, 127–131.

Coffin, P.O., Galea, S., Ahern, J., Leon, A.C., Vlahov, D., Tardiff, K., 2003. Opiates, cocaine and alcohol combinations in accidental drug overdose deaths in New York City, 1990–1998. Addiction 98, 739–747.

Cole, J.C., Bailey, M., Sumnall, H.R., Wagstaff, G.F., King, L.A., 2002. The content of ecstasy tablets: implications for the study of their long-term effects. Addiction 97, 1531–1536.

Connor, M.D., Lammie, G.A., Bell, J.E., Warlow, C.P., Simmonds, P., Brettle, R.P., 2000. Cerebral infarction in adult AIDS patients: observations from the Edinburgh HIV autopsy cohort. Stroke 31, 2117–2126.

Console-Bram, L., Marcu, J., Abood, M.E., 2012. Cannabinoid receptors: nomenclature and pharmacological principles. Prog. Neuropsychopharmacol. Biol. Psychiatry 38, 4–15.

Cooper, G.A.A., Seymour, A., Cassidy, M.T., Oliver, J.S., 1999. A study of methadone fatalities in the Strathclyde Region, 1991–1996. Med. Sci. Law 39, 233–241.

Corkery, J.M., Schifano, F., Ghodse, A.H., Oyefeso, A., 2004. The effects of methadone and its role in fatalities. Hum. Psychopharmacol. Clin. Exp. 19, 565–576.

Covert, R.F., Schreiber, M.D., Tebbett, I.R., Torgerson, L.J., 1994. Hemodynamic and cerebral blood flow effects of cocaine, cocaethylene and benzoylecgonine in conscious and anesthetized fetal lambs. J. Pharmacol. Exp. Ther. 270, 118–126.

Cregler, L.L., Mark, H., 1986. Medical complications of cocaine abuse. N. Engl. J. Med. 315, 1495–1500.

Cruickshank, C.C., Dyer, K.R., 2009. A review of the clinical pharmacology of methamphetamine. Addiction 104, 1085–1099.

Curran, H.V., 2000. Is MDMA ("ecstasy") neurotoxic in humans? An overview of evidence and of methodological problems in research. Neuropsychobiology 42, 34–41.

Daras, M., Tuchman, A.J., Koppel, B.S., Samkoff, L.M., Weitzner, I., Marc, J., 1994. Neurovascular complications of cocaine. Acta Neurol. Scand. 90, 124–129.

Daras, M.D., Orrgeo, J.J., Akfirat, G.L., Samkoff, L.M., Koppel, B.S., 2001. Bilateral symmetrical basal ganglia infarction after intravenous use of cocaine and heroin. J. Clin. Imag. 25, 12–14.

Darke, S., Duflou, J., Torok, M., 2011. Toxicology and characteristics of fatal oxycodone toxicity cases in New South Wales, Australia 1999–2008. J. Forensic Sci. 56, 690–693.

Darke, S., Duflou, J., Torok, M., 2010. Comparative toxicology of intentional and accidental heroin overdose. J. Forensic Sci. 55, 1015–1018.

Darke, S., Kaye, S., McKetin, R., Duflou, J., 2008. Major physical and psychological harms of methamphetamine use. Drug Alcohol Rev. 27, 253–262.

Darke, S., 2003. Polydrug use and overdose: overthrowing old myths. Addiction 98, 711.

Darke, S., Zador, D., 1996. Fatal heroin "overdose": a review. Addiction 91, 1765–1772.

Davidson, C., Gow, A.J., Lee, T.H., Ellinwood, E.H., 2001. Methamphetamine neurotoxicity: necrotic and apoptotic mechanisms and relevance to human abuse and treatment. Brain Res. Rev. 36, 1–22.

Davis, G.D., Swalwell, C.I., 1996. The incidence of acute cocaine or methamphetamine intoxication in deaths due to ruptured cerebral (berry) aneurysms. J. Forensic Sci. 41, 626–628.

De la Torre, R., Farre, M., Roset, P.N., Pizarro, N., Abanades, S., Segura, M., Segura, J., Cami, J., 2004. Human pharmacology of MDMA: pharmacokinetics, metabolism and disposition. Ther. Drug Monit. 26, 137–144.

Delaney, P., Estes, M., 1980. Intracranial hemorrhage with amphetamine abuse. Neurology 30, 1125–1128.

Derlet, R.W., Rice, P., Horowitz, B.Z., Lord, R.V., 1989. Amphetamine toxicity: experience with 127 cases. J. Emerg. Med. 7, 157–161.

Devane, W.A., Dysarz III, F.A., Johnson, M.R., Melvin, L.S., Howlett, A.C., 1988. Determination and characterization of a cannabinoid receptor in rat brain. Mol. Pharmacol. 34, 605–613.

Diana, M., Melis, M., Muntoni, A.L., Gessa, G.L., 1998. Mesolimbic dopaminergic decline after cannabinoid withdrawal. Proc. Natl. Acad. Sci. USA 95, 10269–10273.

Di Marzo, V., Melck, D., Bisogno, T., De Petrocellis, L., 1998. Endocannabinoids: endogenous cannabinoid receptor ligands with neuromodulatory action. Trends Neurosci. 21, 521–528.

Di Marzo, V., 2011. Endocannabinoid signaling in the brain: biosynthetic mechanisms in the limelight. Nat. Neurosci. 14, 9–15.

Diez-Tejedor, E., Frank, A., Gutierrez, M., Barreiro, P., 1998. Encephalopathy and biopsy-proven cerebrovascular inflammatory changes in a cocaine abuser. Eur. J. Neurol. 5, 103–107.

Dowling, G.P., McDonough, E.T.I., Bost, R.O., 1987. "Eve" and "Ecstasy": a report of five deaths associated with the use of MDEA and MDMA. JAMA 257, 1615–1617.

Downing, J., 1986. The psychological and physiological effects of MDMA on normal volunteers. J. Psychoactive Drugs 18, 335–340.

Drummer, O.H., Opeskin, K., Syrjanen, M., Cordner, S.M., 1992. Methadone toxicity causing death in ten subjects starting on a methadone maintenance program. Am. J. Forensic Med. Pathol. 13, 346–350.

Dyuizen, I., Lamash, N.E., 2009. Histo- and immunocytochemical detection of inducible NOS and TNF-α in the locus coeruleus of human opiate addicts. J. Chem. Neuroanat. 37, 65–70.

D'Souza, T., Shraberg, D., 1981. Intracranial hemorrhage associated with amphetamine use. Neurology 31, 922–923.

Ell, J.J., Uttley, D., Silver, J.R., 1981. Acute myelopathy in association with heroin addiction. J. Neurol. Neurosurg. Psychiatry 44, 448–450.

Ernst, T., Chang, L., Leonido-Yee, M., Speck, O., 2000. Evidence for long-term neurotoxicity associated with methamphetamine abuse. A 1H MRS study. Neurology 54, 1344–1349.

Escriba, P.V., Sastre, M., García-Sevilla, J.A., 1994. Increased density of guanine nucleotide-binding proteins in the postmortem brains of heroin addicts. Arch. Gen. Psychiatry 51, 494–501.

Fattore, L., Spano, M.S., Deiana, S., Melis, V., Cossu, G., Fadda, P., Fratta, W., 2007. An endocannabinoid mechanism in relapse to drug seeking: a review of animal studies and clinical perspectives. Brain Res. Rev. 53, 1–16.

Felgate, H.E., Felgate, P.D., James, R.A., Sims, D.N., Vozzo, D.C., 1998. Recent paramethoxymethamphetamine deaths. J. Anal. Toxicol. 22, 169–172.

Feltenstein, M.W., See, R.E., 2008. The neurocircuitry of addiction: an overview. Br. J. Pharmacol. 154, 261–274.

Fessler, R.D., Esshaki, C.M., Stankewitz, R.C., Johnson, R.R., Diaz, F.G., 1997. The neurovascular complications of cocaine. Surg. Neurol. 47, 339–345.

Fineschi, V., Centini, F., Mazzeo, E., Turillazzi, E., 1999. Adam (MDMA) and Eve (MDEA) misuse: an immunohistochemical study on three fatal cases. Forensic Sci. Int. 104, 65–74.

Forrest, A.R.W., Galloway, J.H., Marsh, I.D., Strachan, G.A., Clark, J.C., 1994. A fatal overdose with 3,4-methylenedioxyamphetamine derivatives. Forensic Sci. Int. 64, 57–59.

Fredericks, R.K., Lefkowitz, D.S., Challa, V.R., Troost, B.T., 1991. Cerebral vasculitis associated with cocaine abuse. Stroke 22, 1437–1439.

Freese, T.E., Miotto, K., Reback, C.J., 2002. The effects and consequences of selected club drugs. J. Subst. Abuse Treat. 23, 151–156.

Friedman, S.D., Castañeda, E., Hodge, G.K., 1998. Long-term monoamine depletion, differential recovery, and subtle behavioral impairment following methamphetamine-induced neurotoxicity. Pharmacol. Biochem. Behav. 61, 35–44.

Fride, E., 2002. Endocannabinoids in the central nervous system – an overview. Prostaglandins Leukot. Essent. Fatty Acids 66, 221–233.

Frost, D.O., Cadet, J.L., 2000. Effects of methamphetamine-induced neurotoxicity on the development of neural circuits: a hypothesis. Brain Res. Rev. 34, 103–118.

French, E.D., Dillon, K., Wu, X., 1997. Cannabinoids excite dopamine neurons in the ventral tegmentum and substantia nigra. Neuroreport 8, 649–652.

Fukui, K., Nakajima, T., Kariyama, H., Kashiba, A., Kato, N., Tohyama, I., Kimura, H., 1989. Selective reduction of serotonin immunoreactivity in some forebrain regions of rats induced by acute methamphetamine treatment; quantitative morphometric analysis by serotonin immunocytochemistry. Brain Res. 482, 198–203.

Gabilondo, A.M., Meana, J.J., Barturen, F., Sastre, M., García-Sevilla, J.A., 1994. μ-Opioid receptor and α2-adrenoreceptor agonist binding sites in the postmortem brain of heroin addicts. Psychopharmacology 115, 135–140.

García-Sevilla, J.A., Ventayol, P., Busquets, X., La-Harpe, R., Walzer, C., Guimón, J., 1997a. Regulation of immunolabelled mu-opioid receptors and protein kinase C-alpha and zeta isoforms in the frontal cortex of human opiate addicts. Neurosci. Lett. 226, 29–32.

García-Sevilla, J.A., Ventayol, P., Busquets, X., La-Harpe, R., Walzer, C., Guimón, J., 1997b. Marked decrease of immunolabelled 68 kDa neurofilament (NF-L) proteins in brains of opiate addicts. Neuroreport 8, 1561–1570.

Gaulier, J.M., Marquet, P., Lacassie, E., Dupuy, J.L., Lachatre, G., 2000. Fatal intoxication following self-administration of a massive dose of buprenorphine. J. Forensic Sci. 45, 226–228.

Geibprasert, S., Gallucci, M., Krings, T., 2010. Addictive illegal drugs: structural neuroimaging. AJNR Am. J. Neuroradiol. 31, 803–808.

Gerostamoulos, J., Staikos, V., Drummer, O.H., 2001. Heroin-related deaths in Victoria: a review of cases for 1997 and 1998. Drug Alcohol Depend. 61, 123–127.

Gerostamoulos, J., Burke, M.P., Drummer, O.H., 1996. Involvement of codeine in drug-related deaths. Am. J. Forensic Med. Pathol. 17, 327–335.

Gerra, G., Zaimovic, A., Ferri, M., Zambelli, U., Timpano, M., Neri, E., Marzocchi, F., Delsignore, R., Brambilla, F., 2000. Long-lasting effects of (±)3,4-methylenedioxymethamphetamine (ecstasy) on serotonin system functions in humans. Biol. Psychiatry 47, 127–136.

Gill, J.R., Hayes, J.A., deSouza, I.S., Marker, E., Stajic, M., 2002. Ecstasy (MDMA) deaths in New York City: a case series and review of the literature. J. Forensic Sci. 47, 121–126.

Ginsberg, M.D., Hedley-Whyte, E.T., Richardson Jr., E.P., 1976. Hypoxic-ischemic leukoencephalopathy in man. Arch. Neurol. 33, 5–14.

Glass, M., Dragunow, M., Faull, R.L.M., 1997. Cannabinoid receptors in the human brain: a detailed anatomical and quantitative autoradiographic study in the fetal neonatal and adult human brain. Neuroscience 77, 299–318.

Gledhill, J.A., Moore, D.F., Bell, D., Henry, J.A., 1993. Subarachnoid haemorrhage associated with MDMA abuse. J. Neurol. Neurosurg. Psychiatry 56, 1036–1037.

Goforth, H., Murtaugh, R., Fernandez, F., 2010. Neurologic aspects of drug abuse. Neurol. Clin. 28, 199–215.

Gold, M.S., 1993. Opiate addiction and the locus coeruleus. The clinical utility of clonidine, naltrexone, methadone, and buprenorphine. Psychiatr. Clin. North Am. 16, 61–73.

Goldstein, R.Z., Volkow, N.D., 2002. Drug addiction and its underlying neurobiological basis: neuroimaging evidence for the involvement of the frontal cortex. Am. J. Psychiatry 159, 1642–1652.

Goodhart, L.C., Loizou, L.A., Anderson, M., 1982. Heroin myelopathy. J. Neurol. Neurosurg. Psychiatry 45, 562–563.

Gosztonyi, G., Schmidt, V., Nickel, R., Rothschild, M.A., Camacho, S., Siegel, G., Zill, E., Pauli, G., Schneider, V., 1993. Neuropathologic analysis of postmortal brain samples of HIV-seropositive and -seronegative i.v. drug addicts. Forensic Sci. Int. 62, 101–105.

Gottschalk, P.C., Kosten, T.R., 2002. Cerebral perfusion defects in combined cocaine and alcohol dependence. Drug Alcohol. Depend. 68, 95–104.

Gould, M.S., Walsh, B.T., Munfakh, J.L., Kleinman, M., Duan, N., Olfson, M., Greenhill, L., Cooper, T., 2009. Sudden death and use of stimulant medications in youths. Am. J. Psychiatry 166, 992–1001.

Gouzoulis-Mayfrank, E., Daumann, J., 2006. Neurotoxicity of methylenedioxyamphetamines (MDMA; ecstasy) in humans: how strong is the evidence for persistent brain damage? Addiction 101, 348–361.

Graß, H., Behnsen, S., Kimont, H.-G., Staak, M., Käferstein, H., 2003. Methadone and its role in drug-related fatalities in Cologne 1989–2000. Forensic Sci. Int. 132, 195–200.

Green, A.R., Goodwin, G.M., 1996. Ecstasy and neurodegeneration. Br. Med. J. 312, 1493–1494.

Green, A.R., Cross, A.J., Goodwin, G.M., 1995. Review of the pharmacology and clinical pharmacology of 3,4-methylenedioxymethamphetamine (MDMA or "ecstasy"). Psychopharmacology 119, 247–260.

Gruber, S.A., Silveri, M.M., Yurgelun-Todd, D.A., 2007. Neuropsychological consequences of opiate use. Neuropsychol. Rev. 17, 299–315.

Guilarte, T.R., 2001. Is methamphetamine abuse a risk factor in parkinsonism? Neurotoxicology 22, 725–731.

Guo, Y., Wang, H.-L., Xiang, X.-H., Zhao, Y., 2009. The role of glutamate and its receptors in mesocortico-limbic dopaminergic regions in opioid addiction. Neurosci. Biobehav. Rev. 33, 864–873.

Guzmán, M., Sánchez, C., Galve-Roperh, I., 2001. Control of the cell survival/death decision by cannabinoids. J. Mol. Med. 78, 613–625.

Halpern, M., Citron, B.P., 1971. Necrotizing angiitis associated with drug abuse. AJR Am. J. Roentgenol. 111, 663–671.

Hampson, R.E., Deadwyler, S.A., 1999. Cannabinoids, hippocampal function and memory. Life Sci. 65, 715–723.

Hanson, G.R., Gibb, J.W., Metzger, R.R., Kokoshka, J.M., Fleckenstein, A.E., 1998. Methamphetamine-induced rapid and reversible reduction in the activities of tryptophan hydroxylase and dopamine transporters: oxidative consequences? Ann. N.Y. Acad. Sci. 844, 103–107.

Hanyu, S., Ikeguchi, K., Imai, H., Imai, N., Yoshida, M., 1995. Cerebral infarction associated with 3,4-methylenedioxymethamphetamine ("ecstasy") abuse. Eur. Neurol. 35, 173.

Harding-Pink, D., 1993. Methadone: one person's maintenance dose is anothers poison. Lancet 341, 665–666.

Harlan, R.E., Garcia, M.M., 1998. Drugs of abuse and immediate-early genes in the forebrain. Mol. Neurobiol. 16, 221–267.

Harries, D.P., De Silva, R., 1992. "Ecstasy" and intracerebral haemorrhage. Scott. Med. J. 37, 150–152.

Harrington, H., Heller, H.A., Dawson, D., Caplan, L., Rumbaugh, C., 1983. Intracerebral hemorrhage and oral amphetamine. Arch. Neurol. 40, 503–507.

Hart, J.B., Wallace, J., 1975. The adverse effects of amphetamines. Clin. Toxicol. 8, 179–190.

Harvey, D.C., Lacan, G., Tanious, S.P., Melega, W.P., 2000. Recovery from methamphetamine induced long-term nigrostriatal dopaminergic deficits without substantia nigra cell loss. Brain Res. 871, 259–270.

Hashimoto, E., Frölich, L., Ozawa, H., Saito, T., Shichinohe, S., Takahata, N., Riederer, P., 1996. Alteration of guanosine triphosphate binding proteins in postmortem brains of heroin addicts. Alcohol. Clin. Exp. Res. 20, 301A–304A.

Hatzidimitriou, G., McCann, U.D., Ricaurte, G.A., 1999. Altered serotonin innervation patterns in the forebrain of monkeys treated with (±)3,4-methylenedioxymethamphetamine seven years previously: factors influencing abnormal recovery. J. Neurosci. 19, 5096–5107.

Haughey, H.M., Fleckenstein, A.E., Metzger, R.R., Hanson, G.R., 2000. The effects of methamphetamine on serotonin transporter activity: role of dopamine and hyperthermia. J. Neurochem. 75, 1608–1617.

He, G.Q., Zhang, A., Altura, B.T., Altura, B.M., 1994. Cocaine-induced cerebrovasospasm and its possible mechanism of action. J. Pharmacol. Exp. Ther. 268, 1532–1539.

Hearn, W.L., Flynn, D.D., Hime, G.W., Rose, S., Cofino, J.C., Mantero-Atienza, E., Wetli, C.V., Mash, D.C., 1991. Cocaethylene: a unique cocaine metabolite displays high affinity for the dopamine transporter. J. Neurochem. 56, 698–701.

Hegadoren, K.M., Baker, G.B., Bourin, M., 1999. 3,4-Methylenedioxy analogues of amphetamine: defining the risks to humans. Neurosci. Biobehav. Rev. 23, 539–553.

Heinemann, A., Iwersen-Bergmann, S., Stein, S., Schmoldt, A., Püschel, K., 2000. Methadone-related fatalities in Hamburg 1990–1999: implications for quality standards in maintenance treatment? Forensic Sci. Int. 113, 449–455.

Hemby, S.E., 2010. Cocainomics: new insights into the molecular basis of cocaine addiction. J. Neuroimmune Pharmacol. 5, 70–82.

Henry, J.A., Jeffreys, K.J., Dawling, S., 1992. Toxicity and deaths from 3,4-methylenedioxymethamphetamine ("ecstasy"). Lancet 340, 384–387.

Herkenham, M., 1992. Cannabinoid receptor localization in brain: relationship to motor and reward systems. Ann. N.Y. Acad. Sci. 654, 19–32.

Herkenham, M., Lynn, A.B., Little, M.D., Johnson, M.R., Melvin, L.S., de Costa, B.R., Rice, K.C., 1990. Cannabinoid receptor localization in brain. Proc. Natl. Acad. Sci. USA 87, 1932–1936.

Herning, R.I., King, D.E., Better, W.E., Cadet, J.L., 1999. Neurovascular deficits in cocaine abusers. Neuropsychopharmacology 21, 110–118.

Herskowitz, A., Gross, E., 1973. Cerebral infarction associated with heroin sniffing. South. Med. J. 66, 778–784.

Heye, N., Hankey, G.J., 1996. Amphetamine-associated stroke. Cerebrovasc. Dis. 6, 149–155.

Hickman, M., Madden, P., Henry, J., Baker, A., Wallace, C., Wakefield, J., Stimson, G., Elliott, P., 2003. Trends in drug overdose deaths in England and Wales 1993–98: methadone does not kill more people than heroin. Addiction 98, 419–425.

Hoffman, A.F., Lupica, C.R., 2001. Direct actions of cannabinoids on synaptic transmission in the nucleus accumbens: a comparison with opioids. J. Neurophysiol. 85, 72–83.

Hollister, L.E., 1986. Health aspects of cannabis. Pharmacol. Rev. 38, 1–20.

Horowitz, J.M., Torres, G., 1999. Cocaethylene: effects on brain systems and behavior. Addict. Biol. 4, 127–140.

Howlett, A.C., Barth, F., Bonner, T.I., Cabral, G., Casellas, P., Devane, W.A., Felder, C.C., Herkenham, M., Mackie, K., Martin, B.R., Mechoulam, R., Pertwee, R.G., 2002. Classification of cannabinoid receptors. Pharmacol. Rev. 54, 161–202.

Huether, G., Zhou, D., Rüther, E., 1997. Causes and consequences of the loss of serotonergic presynapses elicited by the consumption of 3,4-methylenedioxymethamphetamine (MDMA, "ecstasy") and its congeners. J. Neural Transm. 104, 771–794.

Hughes, J.C., McCabe, M., Evans, R.J., 1993. Intracranial haemorrhage associated with ingestion of "ecstasy". Arch. Emerg. Med. 10, 372–374.

Hull, M.J., Juhascik, M., Mazur, F., Flomenbaum, M.A., Behonick, G.S., 2007. Fatalities associated with fentanyl and co-administered cocaine or opiates. J. Forensic Sci. 52, 1383–1388.

Hurd, Y.L., Herkenham, M., 1993. Molecular alterations in the neostriatum of human cocaine addicts. Synapse 13, 357–369.

Hyman, S.E., Malenka, R.C., 2001. Addiction and the brain: the neurobiology of compulsion and its persistence. Nat. Rev. Neurosci. 2, 695–703.

Iacovelli, L., Fulceri, F., De Blasi, A., Nicoletti, F., Ruggieri, S., Fornai, F., 2006. The neurotoxicity of amphetamines: bridging drugs of abuse and neurodegenerative disorders. Exp. Neurol. 201, 24–31.

Imanse, J., Vanneste, J., 1990. Intraventricular hemorrhage following amphetamine abuse. Neurology 40, 1318–1319.

Insel, T.R., Battaglia, G., Johannessen, J.N., Marra, S., De Souza, E.B., 1989. 3,4-Methylenedioxymethamphetamine ("ecstasy") selectively destroys brain serotonin terminals in rhesus monkeys. J. Pharmacol. Exp. Ther. 249, 713–720.

Iversen, L., 2003. Cannabis and the brain. Brain 126, 1252–1270.

Jensen, R., Olsen, T.S., Winther, B.B., 1990. Severe non-occlusive ischemic stroke in young heroin addicts. Acta Neurol. Scand. 81, 354–357.

Jacobs, I.G., Roszler, M.H., Kelly, J.K., Klein, M.A., Kling, G.A., 1989. Cocaine abuse: neurovascular complications. Radiology 170, 223–227.

James, R.A., Dinan, A., 1998. Hyperpyrexia associated with fatal paramethoxyamphetamine (PMA) abuse. Med. Sci. Law 38, 83–85.

Jennings, L.K., White, M.M., Sauer, C.M., Mauer, A.M., Robertson, J.T., 1993. Cocaine-induced platelets defects. Stroke 24, 1352–1359.

Johns, A., 2001. Psychiatric effects of cannabis. Br. J. Psychiatry 178, 116–122.

Jumbelic, M.I., 2010. Deaths with transdermal fentanyl patches. Am. J. Forensic Med. Pathol. 31, 18–21.

Kaku, D.A., Lowenstein, D.H., 1990. Emergence of recreational drug abuse as a major risk factor for stroke in young adults. Ann. Intern. Med. 113, 821–827.

Kalant, H., 2001. The pharmacology and toxicology of "ecstasy" (MDMA) and related drugs. CMAJ 165, 917–928.

Kalasinsky, K.S., Bosy, T.Z., Schmunk, G.A., Ang, L., Adams, V., Gore, S.B., Smialek, J., Furukawa, Y., Guttman, M., Kish, S.J., 2000. Regional distribution of cocaine in postmortem brain of chronic human cocaine users. J. Forensic Sci. 45, 1041–1048.

Kalivas, P.W., McFarland, K., 2003. Brain circuitry and the reinstatement of cocaine-seeking behavior. Psychopharmacology 168, 55–56.

Karch, S.B., 2008. Karch's Pathology of Drug Abuse, fourth ed. CRC Press, Boca Raton, FL.

Karch, S.B., Stephens, B.G., 2000. Toxicology and pathology of deaths related to methadone: retrospective review. West. J. Med. 172, 11–14.

Karch, S.B., Stephens, B.G., Ho, C.H., 1999. Methamphetamine-related deaths in San Francisco: demographic, pathologic, and toxicologic profiles. J. Forensic Sci. 44, 359–368.

Katona, I., Freund, T.F., 2012. Multiple functions of endocannabinoid signaling in the brain. Annu. Rev. Neurosci. 35, 529–558.

Kaufman, M.J., 2001. Brain Imaging in Substance Abuse: Research, Clinical, and Forensic Applications, first ed. Humana Press Inc., Totowa, NJ.

Kaufman, M.J., Levin, J.M., Ross, M.H., Lange, N., Rose, S.L., Kukes, T.J., Mendelson, J.H., Lukas, S.E., Cohen, B.M., Renshaw, P.F., 1998. Cocaine-induced cerebral vasoconstriction detected in humans with magnetic resonance angiography. JAMA 279, 376–380.

Kaye, S., Darke, S., Duflou, J., McKetin, R., 2008. Methamphetamine-related fatalities in Australia: demographics, circumstances, toxicology and major organ pathology. Addiction 103, 1353–1360.

Kaye, B.R., Fainstat, M., 1987. Cerebral vasculitis associated with cocaine abuse. JAMA 258, 2104–2106.

Kelly, M.A., Gorelick, P.B., Mirza, D., 1992. The role of drugs in the etiology of stroke. Clin. Neuropharmacol. 15, 249–275.

Kelz, M.B., Nestler, E.J., 2000. ΔFosB: a molecular switch underlying long-term neural plasticity. Curr. Opin. Neurol. 13, 715–720.

Kendler, K.S., Chen, X., Dick, D., Maes, H., Gillespie, N., Neale, M.C., Riley, B., 2012. Recent advances in the genetic epidemiology and molecular genetics of substance use disorders. Nat. Neurosci. 15, 181–189.

King, J., Richards, M., Tress, B., 1978. Cerebral arteritis associated with heroin abuse. Med. J. Aust. 2, 444–445.

Kintz, P., 2001. Deaths involving buprenorphine: a compendium of French cases. Forensic Sci. Int. 121, 65–69.

Kirby, L.G., Zeeb, F.D., Winstanley, C.A., 2011. Contributions of serotonin in addiction vulnerability. Neuropharmacology 61, 421–432.

Kish, S.J., 2008. Pharmacologic mechanisms of crystal meth. CMAJ 178, 1679–1682.

Kish, S.J., 2002. How strong is the evidence that brain serotonin neurons are damaged in human users of ecstasy? Pharmacol. Biochem. Behav. 71, 845–855.

Kish, S.J., 2003. What is the evidence that ecstasy (MDMA) can cause Parkinson's disease? Mov. Disord. 18, 1219–1223.

Kish, S.J., Kalasinsky, K.S., Derkach, P., Schmunk, G.A., Guttman, M., Ang, L., Adams, V., Furukawa, Y., Haycock, J.W., 2001. Striatal dopaminergic and serotonergic markers in human heroin users. Neuropsychopharmacology 24, 561–567.

Klonoff, D.C., Andrews, B.T., Obana, W.G., 1989. Stroke associated with cocaine use. Arch. Neurol. 46, 989–993.

Klys, M., Konopka, T., Rojek, S., 2005. Intracerebral hemorrhage associated with amphetamine. J. Anal. Toxicol. 29, 577–581.

Koesters, S.C., Rogers, P.D., Rajasingham, C.R., 2002. MDMA ("ecstasy") and other "club drugs": the new epidemic. Pediatr. Clin. North Am. 49, 415–433.

Konzen, J.P., Levine, S.R., Garcia, J.H., 1995. Vasospasm and thrombus formation as possible mechanism of stroke related to alkaloidal cocaine. Stroke 26, 1114–1118.

Koob, G.F., Volkow, N.D., 2010. Neurocircuitry of addiction. Neuropsychopharmacology 35, 217–238.

Koob, G.F., Le Moal, M., 2006. Neurobiology of Addiction. Academic Press, Amsterdam.

Köfalvi, A., 2008. Cannabinoids and the Brain. Springer Science + Business Media, LLC, Berlin.

Kreek, M.J., 1997. Clinical update of opioid agonist and partial agonist medications for the maintenance treatment of opioid addiction. Semin. Neurosci. 9, 140–157.

Krendel, D.A., Ditter, S.M., Frankel, M.R., Ross, W.K., 1990. Biopsy-proven cerebral vasculitis associated with cocaine abuse. Neurology 40, 1092–1094.

Kriegstein, A.R., Armitage, B.A., Kim, P.Y., 1997. Heroin inhalation and progressive spongiform leukoencephalopathy. N. Engl. J. Med. 336, 589–590.

Krinsky, C.S., Reichard, R.R., 2012. Chasing the Dragon: a review of toxic leukoencephalopathy. Acad. Forensic Pathol. 2, 67–73.

Krinsky, C.S., Lathrop, S.L., Crossey, M., Baker, G., Zumwalt, R., 2011. A toxicology-based review of fentanyl-related deaths in New Mexico (1986–2007). Am. J. Forensic Med. Pathol. 32, 347–351.

Kuhar, M.J., Joyce, A., Dominguez, G., 2001. Genes in drug abuse. Drug Alcohol Depend. 62, 157–162.

Kugelmass, A.D., Oda, A., Monahan, K., Cabral, C., Ware, J.A., 1993. Activation of human platelets by cocaine. Circulation 88, 876–883.

Kumar, D., Deb, I., Chakraborty, J., Mukhopadhyay, S., Das, S., 2011. A polymorphism of the CREB binding protein (CREBBP) gene is a risk factor for addiction. Brain Res. 1406, 59–64.

Kuniyoshi, S.M., Jankovic, J., 2003. MDMA and Parkinsonism. N. Engl. J. Med. 349, 96–97.

Kuperman, D.I., Freyaldenhoven, T.E., Schmued, L.C., Ali, S.F., 1997. Methamphetamine-induced hyperthermia in mice: examination of dopamine depletion and heat-shock protein induction. Brain Res. 771, 221–227.

Lane-Ladd, S.B., Pineda, J., Boundy, V.A., Pfeuffer, T., Krupinski, J., Aghajanian, G.K., Nestler, E.J., 1997. CREB (cAMP response element-binding protein) in the locus coeruleus: biochemical, physiological, and behavioral evidence for a role in opiate dependence. J. Neurosci. 17, 7890–7901.

Law, P.-Y., Wong, Y.H., Loh, H.H., 2000. Molecular mechanisms and regulation of opioid receptor signaling. Annu. Rev. Pharmacol. Toxicol. 40, 389–430.

Leshner, A.I., Koob, G.F., 1999. Drugs of abuse and the brain. Proc. Assoc. Am. Phys. 111, 99–108.

Lessing, M.P.A., Hyman, N.M., 1989. Intracranial hemorrhage caused by amphetamine abuse. J. R. Soc. Med. 82, 766–767.

Levine, S.R., Brust, J.C.M., Futrell, N., Brass, L.M., Blake, D., Fayad, P., Schultz, L.R., Millikan, C.H., Ho, K.-L., Welch, K.M.A., 1991. A comparative study of the cerebrovascular complications of cocaine: alkaloidal versus hydrochloride a review. Neurology 41, 1173–1177.

Levine, S.R., Welch, K.M.A., 1988. Cocaine and stroke. Stroke 19, 779–783.

Levran, O., Londono, D., O'Hara, K., Nielsen, D.A., Peles, E., Rotrosen, J., Casadonte, P., Linzy, S., Randesi, M., Ott, J., Adelson, M., Kreek, M.J., 2008. Genetic susceptibility to heroin addiction: a candidate gene association study. Genes Brain Behav. 7, 720–729.

Libiseller, K., Pavlic, M., Grubwieser, P., Rabl, W., 2005. Ecstasy – deadly risk even outside rave parties. Forensic Sci. Int. 153, 227–230.

Liechti, M.E., Vollenweider, F.X., 2001. Which neuroreceptors mediate the subjective effects of MDMA in humans? A summary of mechanistic studies. Hum. Psychopharmacol. Clin. Exp. 16, 589–598.

Liester, M.B., Grob, C.S., Bravo, G.L., Walsh, R.N., 1992. Phenomenology and sequelae of 3,4-methylenedioxymethamphetamine use. J. Nerv. Ment. Dis. 180, 345–352.

Little, K.Y., Ramssen, E., Welchko, R., Volberg, V., Roland, C.J., Cassin, B., 2009. Decreased brain dopamine cell numbers in human cocaine users. Psychiatry Res. 168, 173–180.

Little, K.Y., Krolewski, D.M., Zhang, L., Cassin, B.J., 2003. Loss of striatal vesicular monoamine transporter protein (VMAT2) in human cocaine users. Am. J. Psychiatry 160, 47–55.

Little, K.Y., McLaughlin, D.P., Zhang, L., McFinton, P.R., Dalack, G.W., Cook Jr., E.H., Cassin, B.J., Watson, S.J., 1998. Brain dopamine transporter messenger RNA and binding sites in cocaine users: a postmortem study. Arch. Gen. Psychiatry 55, 793–799.

Little, K.Y., Patel, U.N., Clark, T.B., Butts, J.D., 1996. Alterations of brain dopamine and serotonin levels in cocaine users: a preliminary report. Am. J. Psychiatry 153, 1216–1218.

Little, K.Y., Kirkman, J.A., Carroll, F.I., Clark, T.B., Duncan, G.E., 1993. Cocaine use increases [3H]WIN 35428 binding sites in human striatum. Brain Res. 628, 17–25.

Logan, B.K., Fligner, C.L., Haddix, T., 1998. Cause and manner of death in fatalities involving methamphetamine. J. Forensic Sci. 43, 28–34.

López-Moreno, J.A., González-Cuevas, G., Moreno, G., Navarro, M., 2008. The pharmacology of the endocannabinoid system: functional and structural interactions with other neurotransmitter systems and their repercussions in behavioral addiction. Addict. Biol. 13, 160–187.

Lora-Tamayo, C., Tena, T., Rodríguez, A., 1997. Amphetamine derivative related deaths. Forensic Sci. Int. 85, 149–157.

Louria, D.B., Hensle, T., Rose, J., 1967. The major medical complications of heroin addiction. Ann. Intern. Med. 67, 1–22.

Lucantonio, F., Stalnaker, T.A., Shaham, Y., Niv, Y., Schoenbaum, G., 2012. The impact of orbitofrontal dysfunction on cocaine addiction. Nat. Neurosci. 15, 358–366.

Lukes, S.A., 1983. Intracerebral hemorrhage from an arteriovenous malformation after amphetamine injection. Arch. Neurol. 40, 60–61.

Lull, M.E., Freeman, W.M., VanGuilder, H.D., Vrana, K.E., 2010. The use of neuroproteomics in drug abuse research. Drug Alcohol. Depend. 107, 11–22.

Lundberg, G.D., Garriott, J.C., Reynolds, P.C., Cravey, R.H., Shaw, R.F., 1977. Cocaine-related death. J. Forensic Sci. 22, 402–408.

Lyles, J., Cadet, J.L., 2003. Methylenedioxymethamphetamine (MDMA, Ecstasy) neurotoxicity: cellular and molecular mechanisms. Brain Res. Rev. 42, 155–168.

Maccarrone, M., 2010. Endocannabinoid signaling in healthy and diseased brain. Exp. Neurol. 224, 1–2.

Manchanda, S., Connolly, M.J., 1993. Cerebral infarction in association with ecstasy abuse. Postgrad. Med. J. 69, 874–875.

Mackesy-Amiti, M.E., Donenberg, G.R., Ouellet, L.J., 2012. Prevalence of psychiatric disorders among young injection drug users. Drug Alcohol. Depend. 124, 70–78.

Madden, J.A., Konkol, R.J., Keller, P.A., Alvarez, T.A., 1995. Cocaine and benzoylecgonine constrict cerebral arteries by different mechanisms. Life Sci. 56, 679–686.

Maher, C.E., Martin, T.J., Childers, S.R., 2005. Mechanisms of mu opioid receptor/G-protein desensitization in brain by chronic heroin administration. Life Sci. 77, 1140–1154.

Mailleux, P., Parmentier, M., Vanderhaeghen, J.J., 1992. Distribution of cannabinoid receptor messenger RNA in the human brain: an in situ hybridization histochemistry with oligonucleotides. Neurosci. Lett. 143, 200–204.

Makrigeorgi-Butera, M., Hagel, C., Laas, R., Püschel, K., Stavrou, D., 1996. Comparative brain pathology of HIV-seronegative and HIV-infected drug addicts. Clin. Neuropathol. 15, 324–329.

Maldonado, R., Valverde, O., Berrendero, F., 2006. Involvement of the endocannabinoid system in drug addiction. Trends Neurosci. 29, 225–232.

Mangiardi, J.R., Daras, M., Geller, M.E., Weitzner, I., Tuchman, A.J., 1988. Cocaine-related intracranial hemorrhage. Report of nine cases and review. Acta Neurol. Scand. 77, 177–180.

Margolis, M.T., Newton, T.H., 1971. Methamphetamine ("speed") arteritis. Neuroradiology 2, 179–182.

Martin, I., Palepu, A., Wood, E., Li, K., Montaner, J., Kerr, T., 2009. Violence among street-involved youth: the role of methamphetamine. Eur. Addict. Res. 15, 32–38.

Martin, K., Rogers, T., Kavanaugh, A., 1995. Central nervous system angiopathy associated with cocaine abuse. J. Rheumatol. 22, 780–782.

Martin-Schild, S., Albright, K.C., Hallevi, H., Barreto, A.D., Philip, M., Misra, V., Grotta, J.C., Savitz, S.I., 2010. Intracerebral hemorrhage in cocaine users. Stroke 41, 680–684.

Mash, D.C., Ouyang, Q., Pablo, J., Basile, M., Izenwasser, S., Lieberman, A., Perrin, R.J., 2003. Cocaine abusers have an overexpression of α-synuclein in dopamine neurons. J. Neurosci. 23, 2564–2571.

Mash, D.C., Pablo, J., Ouyang, Q., Hearn, W.L., Itzenwasser, S., 2002. Dopamine transport function is elevated in cocaine users. J. Neurochem. 81, 292–300.

Mash, D.C., Staley, J.K., Itzenwasser, S., Basile, M., Ruttenber, A.J., 2000. Serotonin transporters upregulate with chronic cocaine use. J. Chem. Neuroanat. 20, 271–280.

Matick, H., Anderson, D., Brumlik, J., 1983. Cerebral vasculitis associated with oral amphetamine overdose. Arch. Neurol. 40, 253–254.

Mayer, P., Höllt, V., 2006. Pharmacogenetics of opioid receptors and addiction. Pharmacogenet. Genomics 16, 1–7.

Maykut, M.O., 1985. Health consequences of acute and chronic marihuana use. Prog. Neuropsychopharmacol. Biol. Psychiatry 9, 209–238.

McCann, U.D., Ricaurte, G.A., Molliver, M.E., 2001. "Ecstasy" and serotonin neurotoxicity. New findings raise more questions. Arch. Gen. Psychiatry 58, 907–908.

McCann, U.D., Eligulashvili, V., Ricaurte, G.A., 2000. (±)3,4-Methylenedioxymethamphetamine ("ecstasy")-induced serotonin neurotoxicity: clinical studies. Neuropsychobiology 42, 11–16.

McCann, U.D., Wong, D.F., Yokoi, F., Villemagne, V., Dannals, R.F., Ricaurte, G.A., 1998. Reduced striatal dopamine transporter density in abstinent methamphetamine and methcathinone users: evidence from positron emission tomography studies with [11C]WIN-35,428. J. Neurosci. 18, 8417–8422.

McClung, C.A., Nestler, E.J., Zachariou, V., 2005. Regulation of gene expression by chronic morphine and morphine withdrawal in the locus ceruleus and ventral tegmental area. J. Neurosci. 25, 6005–6015.

McCreary, M., Emerman, C., Hanna, J., Simon, J., 2000. Acute myelopathy following intranasal insufflation of heroin: a case report. Neurology 55, 316–317.

McEvoy, A.W., Kitchen, N.D., Thomas, D.G.T., 2000. Intracerebral haemorrhage in young adults: the emerging importance of drug misuse. Br. Med. J. 320, 1322–1324.

McGuire, P., 2000. Long term psychiatric and cognitive effects of MDMA use. Toxicol. Lett. 112–113, 153–156.

McKenna, D.J., Peroutka, S.J., 1990. Neurochemistry and neurotoxicity of 3,4-methylenedioxymethamphetamine (MDMA, "Ecstasy"). J. Neurochem. 54, 14–22.

Meador-Woodruff, J.H., Little, K.Y., Damask, S.P., Mansour, A., Watson, S.J., 1993. Effects of cocaine on dopamine receptor gene expression: a study in the postmortem human brain. Biol. Psychiatry 34, 348–355.

Meana, J.J., González-Maeso, J., García-Sevilla, J.A., Guimón, J., 2000. μ-Opioid receptor and α2-adrenoreceptor agonist stimulation of [35S]GTPγS binding to G-proteins in postmortem brains of opioid addicts. Mol. Psychiatry 5, 308–315.

Melega, W.P., Laćan, G., DeSalles, A.A.F., Phelps, M.E., 2000. Long term methamphetamine-induced decreases of [11C]WIN 35,428 binding in striatum are reduced by GDNF: PET studies in the vervet monkey. Synapse 35, 243–249.

Merkel, P.A., Koroshetz, W.J., Irizarry, M.C., Cudkowicz, M.E., 1995. Cocaine-associated cerebral vasculitis. Semin. Arthritis Rheum. 25, 172–183.

Metter, D., 1978. Pathologisch-anatomische Befunde bei Heroinvergiftung. Beitr. Gerichtl. Med. 36, 433–437.

Metzger, R.R., Haughey, H.M., Wilkins, D.G., Gibb, J.W., Hanson, G.R., Fleckenstein, A.E., 2000. Methamphetamine-induced rapid decrease in dopamine transporter function: role of dopamine and hyperthermia. J. Pharmacol. Exp. Ther. 295, 1077–1085.

Miguel-Hidalgo, J.J., 2009. The role of glial cells in drug abuse. Curr. Drug Abuse Rev. 2, 76–82.

Milroy, C.M., 2011. "Ecstasy" associated deaths: what is a fatal concentration? Analysis of a case series. Forensic Sci. Med. Pathol. 7, 248–252.

Milroy, C.M., Forrest, A.R.W., 2000. Methadone deaths: a toxicological analysis. J. Clin. Pathol. 53, 277–281.

Milroy, C.M., Clark, J.C., Forrest, A.R.W., 1996. Pathology of deaths associated with "ecstasy" and "eve" misuse. J. Clin. Pathol. 49, 149–153.

Minett, W.J., Moore, T.L., Juhascik, M.P., Nields, H.M., Hull, M.J., 2010. Concentrations of opiates and psychotropic agents in polydrug overdoses: a surprising correlation between morphine and antidepressants. J. Forensic Sci. 55, 1319–1325.

Miotto, K., Kaufman, D., Anton, B., Keith Jr., D.E., Evans, C.J., 1996. Human opioid receptors: chromosomal mapping and mRNA localization. NIDA Res. Monogr. 161, 72–82.

Mittleman, R.E., Wetli, C.V., 1987. Cocaine and sudden "natural" death. J. Forensic Sci. 32, 11–19.

Mody, C.K., Miller, B.L., McIntyre, H.B., Cobb, S.K., Goldberg, M.A., 1988. Neurologic complications of cocaine abuse. Neurology 38, 1189–1193.

Molina, D.K., Hargrove, V.M., 2011. Fatal cocaine interactions: a review of cocaine-related deaths in Bexar County, Texas. Am. J. Forensic Med. Pathol. 32, 71–77.

Morefield, K.M., Keane, M., Felgate, P., White, J.M., Irvine, R.J., 2011. Pill content, dose and resulting plasma concentrations of 3,4-methylendioxymethamphetamine (MDMA) in recreational "ecstasy" users. Addiction 106, 1293–1300.

Moriya, F., Hashimoto, Y., 2002. A case of fatal hemorrhage in the cerebral ventricles following intravenous use of methamphetamine. Forensic Sci. Int. 129, 104–109.

Morrow, P.L., McQuillen, J.B., 1993. Cerebral vasculitis associated with cocaine abuse. J. Forensic Sci. 38, 732–738.

Mouzak, A., Agathos, P., Kerezoudi, E., Mantas, A., Vourdeli-Yiannakoura, E., 2000. Transient ischemic attack in heavy cannabis smokers – how 'safe' is it? Eur. Neurol. 44, 42–44.

Muntan, C.D., Tuckler, V., 2006. Cerebrovascular accident following MDMA ingestion. J. Med. Toxicol. 2, 16–18.

Nahas, G.G., 2001. The pharmacokinetics of THC in fat and brain: resulting functional responses to marihuana smoking. Hum. Psychopharmacol. 16, 247–255.

Nanda, A., Vannemreddy, P., Willis, B., Kelley, R., 2006. Stroke in the young: relationship of active cocaine use with stroke mechanism and outcome. Acta Neurochir. Suppl. 96, 91–96.

Nanda, A., Vannemreddy, P.S.S.V., Polin, R.S., Willis, B.K., 2000. Intracranial aneurysms and cocaine abuse: analysis of prognostic indicators. Neurosurgery 46, 1063–1069.

Neiman, J., Haapaniemi, H.M., Hilbom, M., 2000. Neurological complications of drug abuse: pathophysiological mechanisms. Eur. J. Neurol. 7, 595–606.

Nestler, E.J., 2001. Molecular basis of long-term plasticity underlying addiction. Nat. Rev. Neurosci. 2, 119–128.

Nestler, E.J., Landsman, D., 2001. Learning about addiction from the genome. Nature 409, 834–835.

Nestler, E.J., 1997. Molecular mechanisms underlying opiate addiction: implications for medications development. Semin. Neurosci. 9, 84–93.

Netrakom, P., Krasuki, J.S., Miller, N.S., O'Tuama, L.A., 1999. Structural and functional neuroimaging findings in substance-related disorders. Psychiatr. Clin. North Am. 22, 313–329.

Nichols, D.E., 1986. Differences between the mechanism of action of MDMA, MBDB, and the classic hallucinogens. Identification of a new therapeutic class: entactogens. J. Psychoactive Drugs 18, 305–313.

Niehaus, L., Meyer, B.-U., 1998. Bilateral borderzone brain infarction in association with heroin abuse. J. Neurol. Sci. 160, 180–182.

Nuytten, D., Wyffels, E., Michiels, K., Ferrante, M., Verbraeken, H., Daelemans, R., Baeck, E., Cras, P., 1998. Drug-induced spongiform leucoencephalopathy, a case report with review of the literature. Acta Neurol. Belg. 98, 32–35.

Nolte, K.B., Brass, L.M., Fletterick, C.F., 1996. Intracranial hemorrhage associated with cocaine abuse: a prospective autopsy study. Neurology 46, 1291–1296.

Nyffeler, T., Stabba, A., Sturzenegger, M., 2003. Progressive myelopathy with selective involvement of the lateral and posterior columns after inhalation of heroin vapour. J. Neurol. 250, 496–498.

O'Dell, S.J., Weihmuller, F.B., Marshall, J.F., 1991. Multiple methamphetamine injections induce marked increases in extracellular striatal dopamine which correlate with subsequent neurotoxicity. Brain Res. 564, 256–260.

O'Brien, P., Todd, J., 2009. Hypoxic brain injury following heroin overdose. Brain Imp. 10, 169–179.

Obrocki, J., Schmoldt, A., Buchert, R., Andresen, B., Petersen, K., Thomasius, R., 2002. Specific neurotoxicity of chronic use of ecstasy. Toxicol. Lett. 127, 285–297.

Oehmichen, M., Meißner, C., Reiter, A., Birkholz, M., 1996. Neuropathology in non-human immunodeficiency virus-infected drug addicts: hypoxic brain damage after chronic intravenous drug abuse. Acta Neuropathol. 91, 642–646.

Orgado, J.M., Fernández-Ruiz, J., Romero, J., 2009. The endocannabinoid system in neuropathological states. Int. Rev. Psychiatry 21, 172–180.

Ozaita, A., Escriba, P.V., Ventayol, P., Murga, C., Mayor Jr., F., García-Sevilla, J.A., 1998. Regulation of G protein-coupled receptor kinase 2 in brains of opiate-treated rats and human opiate addicts. J. Neurochem. 70, 1249–1257.

Oyesiku, N.M., Colohan, A.R.T., Barrow, D.L., Reisner, A., 1993. Cocaine-induced aneurysmal rupture: an emergent negative factor in the natural history of intracranial aneurysms? Neurosurgery 32, 518–526.

Parolaro, D., Realini, N., Vigano, D., Guidali, C., Rubino, T., 2010. The endocannabinoid system and psychiatric disorders. Exp. Neurol. 224, 3–14.

Parrott, A.C., 2002. Recreational ecstasy/MDMA, the serotonin syndrome, and serotonergic neurotoxicity. Pharmacol. Biochem. Behav. 71, 837–844.

Parrott, A.C., 2001. Human psychopharmacology of ecstasy (MDMA): a review of 15 years of empirical research. Hum. Psychopharmacol. Clin. Exp. 16, 557–577.

Pava, M.J., Woodward, J.J., 2012. A review of the interactions between alcohol and the endocannabinoid system: Implications for alcohol dependence and future directions for research. Alcohol 46, 185–204.

Pazos, M.R., Núñez, E., Benito, C., Tolón, R.M., Romero, J., 2005. Functional neuroanatomy of the endocannabinoid system. Pharmacol. Biochem. Behav. 81, 239–247.

Pearson, J., Baden, M.B., Richter, R.W., 1975. Neuronal depletion in the globus pallidus of heroin addicts. Drug Alcohol. Depend. 1, 349–356.

Pearson, J., Richter, R.W., 1979. Addiction to opiates: neurologic aspects. In: Handbook of Clinical Neurology, Vinken, P.J., Bruyn, G.W., (Eds.), Intoxications of the Nervous System, Part II, vol. 65. North-Holland Publishing Company, Amsterdam, 365–400.

Pearson, J., Challenor, Y.B., Baden, M., Richter, R.W., 1972a. The neuropathology of heroin addiction. J. Neuropathol. Exp. Neurol. 31, 165–166.

Pearson, J., Richter, R.W., Baden, M.M., Challenor, Y.B., Bruun, B., 1972b. Transverse myelopathy as an illustration of the neurologic and neuropathologic features of heroin addiction. Hum. Pathol. 3, 107–113.

Pelissier-Alicot, A.-L., Sastre, C., Baillif-Couniou, V., Gaulier, J.-M., Kintz, P., Kuhlmann, E., Perich, P., Bartoli, C., Piercecchi-Marti, M.-D., Leonetti, G., 2010. Buprenorphine-related deaths: unusual forensic situations. Int. J. Legal Med. 124, 644–651.

Perez Jr., J.A., Arsura, E.L., Strategos, S., 1999. Methamphetamine-related stroke: four cases. J. Emerg. Med. 17, 469–471.

Pertwee, R.G., 1997. Pharmacology of cannabinoid CB1 and CB2 receptors. Pharmacol. Ther. 74, 129–180.

Peterson, P.L., Roszler, M., Jacobs, I., Wilner, H.I., 1991. Neurovascular complications of cocaine abuse. J. Neuropsychiatry Clin. Neurosci. 3, 143–149.

Petitti, D.B., Sidney, S., Quesenberry, C., Bernstein, A., 1998. Stroke and cocaine or amphetamine use. Epidemiology 9, 596–600.

Petty, G.W., Brust, J.C.M., Tatemichi, T.K., Barr, M.L., 1990. Embolic stroke after smoking "crack" cocaine. Stroke 21, 1632–1635.

Pietrzykowski, A.Z., 2010. The role of microRNAs in drug addiction – a big lesson from tiny molecules. Int. Rev. Neurobiol. 91, 1–24.

Pirnay, S., Borron, S.W., Giudicelli, C.P., Tourneau, J., Baud, F.J., Ricordel, I., 2004. A critical review of the causes of death among post-mortem toxicological investigations: analysis of 34 buprenorphine-associated and 35 methadone-associated deaths. Addiction 99, 978–988.

Polettini, A., Groppi, A., Montagna, M., 1999. The role of alcohol abuse in the etiology of heroin-related deaths. Evidence for pharmacokinetic interactions between heroin and alcohol. J. Anal. Toxicol. 23, 570–576.

Pope Jr., H.G., Gruber, A.J., Hudson, J.I., Huestis, M.A., Yurgelun-Todd, D., 2001. Neuropsychological performance in long-term cannabis users. Arch. Gen. Psychiatry 58, 909–915.

Prakash, A., Das, G., 1993. Cocaine and the nervous system. Int. J. Clin. Pharmacol. Ther. Toxicol. 31, 575–581.

Preti, A., Miotto, P., De Coppi, M., 2002. Deaths by unintentional illicit drug overdose in Italy, 1984–2000. Drug Alcohol. Depend. 66, 275–282.

Protass, L.M., 1971. Delayed postanoxic encephalopathy after heroin use. Ann. Intern. Med. 74, 738–739.

Püschel, K., Teschke, F., Castrup, U., 1993. Etiology of accidental/unexpected overdose in drug-induced deaths. Forensic Sci. Int. 62, 129–134.

Quaglio, G., Talamini, G., Lechi, A., Venturini, L., Lugoboni, F., Mezzelani, P., 2001. Study of 2708 heroin-related deaths in north-eastern Italy 1985–98 to establish the main causes of death. Addiction 96, 1127–1137.

Qureshi, A.I., Suri, M.F.K., Guterman, L.R., Hopkins, L.N., 2001. Cocaine use and the likelihood of non-fatal myocardial infarction and stroke. Data from the Third National Health and Nutrition Examination Survey. Circulation 103, 502–506.

Qureshi, A.I., Akbar, M.S., Czander, E., Safdar, K., Janssen, R.S., Frankel, M.R., 1997. Crack cocaine use and stroke in young patients. Neurology 48, 341–345.

Quinn, D.I., Wodak, A., Day, R.O., 1997. Pharmacokinetic and pharmacodynamic principles of illicit drug use and treatment of illicit drug users. Clin. Pharmacokinet. 33, 344–400.

Raikos, N., Tsoukali, H., Psaroulis, D., Vassiliadis, N., Tsoungas, M., Njau, S.N., 2002. Amphetamine derivative deaths in northern Greece. Forensic Sci. Int. 128, 31–34.

Ramage, S.N., Anthony, I.C., Carnie, F.W., Busuttil, A., Robertson, R., Bell, J.E., 2005. Hyperphosphorylated tau and amyloid precursor protein deposition is increased in the brains of young drug abusers. Neuropathol. Appl. Neurobiol. 31, 439–448.

Rapaka, R.S., Sadée, W., 2008. Drug Addiction: From Basic Research to Therapy. Springer, Berlin.

Reid, M.J., Bornheim, L.M., 2001. Cannabinoid-induced alterations in brain disposition of drugs of abuse. Biochem. Pharmacol. 61, 1357–1367.

Reneman, L., Booij, J., Majoie, C.B.L., van den Brink, W., den Heeten, G.J., 2001. Investigating the potential neurotoxicity of ecstasy (MDMA): an imaging approach. Hum. Psychopharmacol. Clin. Exp. 16, 579–588.

Ricaurte, G.A., McCann, U.D., Szabo, Z., Scheffel, U., 2000a. Toxicodynamics and long-term toxicity of the recreational drug, 3,4-methylenedioxymethamphetamine (MDMA, "ecstasy"). Toxicol. Lett. 112, 143–146.

Ricaurte, G.A., Yuan, J., McCann, U.D., 2000b. (±)3,4-Methylenedioxymethamphetamine ("ecstasy")-induced serotonin neurotoxicity: studies in animals. Neuropsychobiology 42, 5–10.

Ricaurte, G.A., McCann, U.D., 1992. Neurotoxic amphetamine analogues: effects in monkeys and implications for humans. Ann. N.Y. Acad. Sci. 654, 371–382.

Ricaurte, G.A., Seiden, L.S., Schuster, C.R., 1984. Further evidence that amphetamines produce long-lasting dopamine neurochemical deficits by destroying dopamine nerve fibers. Brain Res. 303, 359–364.

Richter, R.W., Pearson, J., Bruun, B., 1973. Neurological complications of addiction to heroin. Bull. N.Y. Acad. Med. 49, 3–21.

Richter, R.W., Rosenberg, R.N., 1968. Transverse myelitis associated with heroin addiction. JAMA 206, 1255–1257.

Rinder, H.M., Ault, K.A., Jatlow, P.I., Kosten, T.R., Smith, B.R., 1994. Platelet alpha-granule release in cocaine users. Circulation 90, 1162–1167.

Riße, M., Weiler, G., 1984. Heroinsucht als seltene Ursache einer symmetrischen Pallidumnekrose. Z. Rechtsmed. 93, 227–235.

Riva, N., Morana, P., Cerri, F., Gerevini, S., Amadio, S., Formaglio, F., Comi, G., Comola, M., Del Carro, U., 2007. Acute myelopathy selectively involving lumbar anterior horns following intranasal insufflation of ecstasy and heroin. J. Neurol. Neurosurg. Psychiatry 78, 908–909.

Rizzuto, N., Morbin, M., Ferrari, S., Cavallaro, T., Sparaco, M., Boso, G., Gaetti, L., 1997. Delayed spongiform leukoencephalopathy after heroin abuse. Acta Neuropathol. 94, 87–90.

Robinson, T.E., Becker, J.B., 1986. Enduring changes in brain and behavior produced by chronic amphetamine administration: a review and evaluation of animal models of amphetamine psychosis. Brain Res. Rev. 11, 157–198.

Robison, A.J., Nestler, E.J., 2011. Transcriptional and epigenetic mechanisms of addiction. Nat. Rev. Neurosci. 12, 623–637.

Rochester, J.A., Kirchner, J.T., 1999. Ecstasy (3,4-methylenedioxymethamphetamine): history, neurochemistry, and toxicology. J. Am. Board Fam. Pract. 12, 137–142.

Rodriguez de Fonseca, F., Del Arco, I., Bermudez-Silva, F.J., Bilbao, A., Cippitelli, A., Navarro, M., 2005. The endocannabinoid system: physiology and pharmacology. Alcohol Alcohol. 40, 2–14.

Rome, E.S., 2001. It's a rave new world: rave culture and illicit drug use in the young. Cleve. Clin. J. Med. 68, 541–550.

Ross, B.M., Moszczynska, A., Kalasinsky, K., Kish, S.J., 1996. Phospholipase A2 activity is selectively decreased in the striatum of chronic cocaine users. J. Neurochem. 67, 2620–2623.

Ross, B.M., Moszczynska, A., Peretti, F.J., Adams, V., Schmunk, G.A., Kalasinsky, K., Ang, L., Mamalis, N., Turenne, S.D., Kish, S.J., 2002. Decreased activity of brain phospholipid metabolic enzymes in human users of cocaine and methamphetamine. Drug Alcohol Depend. 67, 73–79.

Rothrock, J.F., Rubenstein, R., Lyden, P.D., 1988. Ischemic stroke associated with methamphetamine inhalation. Neurology 38, 589–592.

Roy, S., Wang, J., Kelschenbach, J., Koodie, L., Martin, J., 2006. Modulation of immune function by morphine: implications for susceptibility to infection. J. Neuroimmune Pharmacol. 1, 77–89.

Rumbaugh, C.L., Bergeron, T., Fang, H.C.H., McCormick, R., 1971. Cerebral angiographic changes in the drug abuse patient. Radiology 101, 335–344.

Sastre, M., Ventayol, P., García-Sevilla, J.A., 1996. Decreased density of I2-imidazoline receptors in the postmortem brains of heroin addicts. Neuroreport 7, 509–512.

Sahni, V., Garg, D., Garg, S., Agarwal, S.K., Singh, N.P., 2008. Unusual complications of heroin abuse: transverse myelitis, rhabdomyolysis, compartment syndrome, and ARF. Clin. Toxicol. 46, 153–155.

Scallet, A.C., 1991. Neurotoxicology of cannabis and THC: a review of chronic exposure studies in animals. Pharmacol. Biochem. Behav. 40, 671–676.

Scallet, A.C., Lipe, G.W., Ali, S.F., Holson, R.R., Frith, C.H., Slikker Jr., W., 1988. Neuropathological evaluation by combined immunohistochemistry and degeneration-specific methods: application to methylenedioxymethamphetamine. Neurotoxicology 9, 529–538.

Segal, D.M., Moraes, C.T., Mash, D.C., 1997. Up-regulation of D3 dopamine receptor mRNA in the nucleus accumbens of human cocaine fatalities. Mol. Brain Res. 45, 335–339.

Sell, L.A., Morris, J., Bearn, J., Frackowiak, R.S.J., Friston, K.J., Dolan, R.J., 1999. Activation of reward circuitry in human opiate addicts. Eur. J. Neurosci. 11, 1042–1048.

Schifano, F., 2004. A bitter pill. Overview of ecstasy (MDMA, MDA) related fatalities. Psychopharmacology 173, 242–248.

Schlaeppi, M., Prica, A., de Torrentém, A., 1999. Hémorragie cérébrale et "ecstasy". Praxis 88, 568–572.

Schmidt, P., Schmolke, C., Musshoff, F., Menzen, M., Prohaska, C., Madea, B., 2003. Area-specific increased density of μ-opioid receptor immunoreactive neurons in the cerebral cortex of drug-related fatalities. Forensic Sci. Int. 133, 204–211.

Schmidt, P., Schmolke, C., Mußhoff, F., Prohaska, C., Menzen, M., Madea, B., 2001. Numerical density of μ opioid receptor expressing neurons in the frontal cortex of drug related fatalities. Forensic Sci. Int. 115, 219–229.

Schmidt, P., Schmolke, C., Mußhoff, F., Menzen, M., Prohaska, C., Madea, B., 2000. Numerical density of δ-opioid receptor expressing neurons in the frontal cortex of drug-related fatalities. Forensic Sci. Int. 113, 423–433.

Schmidt, C.J., 1987. Neurotoxicity of the psychedelic amphetamine, methylenedioxymethamphetamine. J. Pharmacol. Exp. Ther. 240, 1–7.

Schreiber, M.D., Madden, J.A., Covert, R.F., Torgerson, L.J., 1994. Effects of cocaine, benzoylecgonine, and cocaine metabolites in cannulated pressurized fetal sheep cerebral arteries. J. Appl. Physiol. 77, 834–839.

Schwartz, R.H., 2002. Marijuana: a decade and a half later, still a crude drug with underappreciated toxicology. Pediatrics 109, 284–289.

Seiden, L.S., Sabol, K.E., 1996. Methamphetamine and methylenedioxymethamphetamine neurotoxicity: possible mechanisms of cell destruction. NIDA Res. Monogr. 163, 251–276.

Sekine, Y., Iyo, M., Ouchi, Y., Matsunaga, T., Tsukada, H., Okada, H., Yoshikawa, E., Futatsubashi, M., Takei, N., Mori, N., 2001. Methamphetamine-related psychiatric symptoms and reduced brain dopamine transporters studied with PET. Am. J. Psychiatry 158, 1206–1214.

Seldén, T., Ahlner, J., Druid, H., Kronstrand, R., 2012. Toxicological and pathological findings in a series of buprenorphine related deaths. Possible risk factors for fatal outcome. Forensic Sci. Int. 220, 284–290.

Selmi, F., Davies, K.G., Sharma, R.R., Neal, J.W., 1995. Intracerebral haemorrhage due to amphetamine abuse: report of two cases with underlying arteriovenous malformations. Br. J. Neurosurg. 9, 93–96.

Sen, S., Silliman, S.L., Braitman, L.E., 1999. Vascular risk factors in cocaine users with stroke. J. Stroke Cerebrovasc. Dis. 8, 254–258.

Seymour, A., Black, M., Jay, J., Cooper, G., Weir, C., Oliver, J., 2003. The role of methadone in drug-related deaths in the west of Scotland. Addiction 98, 995–1002.

Shalev, U., Grimm, J.W., Shaham, Y., 2002. Neurobiology of relapse to heroin and cocaine seeking: a review. Pharmacol. Rev. 54, 1–42.

Shaw, K.P., 1999. Human methamphetamine-related fatalities in Taiwan during 1991–1996. J. Forensic Sci. 44, 27–31.

Shibata, S., Mori, K., Sekine, I., Suyama, H., 1991. Subarachnoid and intracerebral hemorrhage associated with necrotizing angiitis due to methamphetamine abuse. An autopsy case. Neurol. Med. Chir. (Tokyo) 31, 49–52.

Shichinohe, S., Ozawa, H., Hashimoto, E., Tatschner, T., Riederer, P., Saito, T., 2001. Changes in the cAMP-related signal transduction mechanism in postmortem human brains of heroin addicts. J. Neural Transm. 108, 335–347.

Shichinohe, S., Ozawa, H., Saito, T., Hashimoto, E., Lang, C., Riederer, P., Takahata, N., 1998. Differential alteration of adenyl cyclase subtypes I, II, and V/VI in postmortem human brains of heroin addicts. Alcohol. Clin. Exp. Res. 22, 84S–87S.

Shprecher, D., Mehta, L., 2010. The syndrome of delayed post-hypoxic leukoencephalopathy. NeuroRehabilitation 26, 65–72.

Sidhpura, N., Parsons, L.H., 2011. Endocannabinoid-mediated synaptic plasticity and addiction-related behavior. Neuropharmacology 61, 1070–1087.

Sloan, M.A., Kittner, S.J., Rigamonti, D., Price, T.R., 1991. Occurrence of stroke associated with use/abuse of drugs. Neurology 41, 1358–1364.

Smith, N.T., 2002. A review of the published literature into cannabis withdrawal symptoms in human users. Addiction 97, 621–632.

Smith, K.M., Larive, L.L., Romanelli, F., 2002. Club drugs: methylenedioxymethamphetamine, flunitrazepam, ketamine hydrochloride, and γ-hydroxybutyrate. Am. J. Health Syst. Pharm. 59, 1067–1076.

Soar, K., Turner, J.J.D., Parrott, A.C., 2001. Psychiatric disorders in ecstasy (MDMA) users: a literature review focusing on personal predisposition and drug history. Hum. Psychopharmacol. Clin. Exp. 16, 641–645.

Solinas, M., Goldberg, S.R., Piomelli, D., 2008. The endocannabinoid system in brain reward processes. Br. J. Pharmacol. 154, 369–383.

Spiehler, V.R., Reed, D., 1985. Brain concentrations of cocaine and benzoylecgonine in fatal cases. J. Forensic Sci. 30, 1003–1011.

Sporer, K.A., 1999. Acute heroin overdose. Ann. Intern. Med. 130, 584–590.

Sprague, J.E., Everman, S.L., Nichols, D.E., 1998. An integrated hypothesis for the serotonergic axonal loss induced by 3,4-methylenedioxymethamphetamine. Neurotoxicology 19, 427–442.

Squier, M.V., Jalloh, S., Hilton-Jones, D., Series, H., 1995. Death after ecstasy ingestion: neuropathological findings. J. Neurol. Neurosurg. Psychiatry 58, 756.

Stadlin, A., Lau, J.W.S., Szeto, Y.K., 1998. A selective regional response of cultured astrocytes to methamphetamine. Ann. N.Y. Acad. Sci. 844, 108–121.

Staley, J.K., Rothman, R.B., Rice, K.C., Partilla, J., Mash, D.C., 1997. κ2 opioid receptors in limbic areas of the human brain are upregulated by cocaine in fatal overdose victims. J. Neurosci. 17, 8225–8233.

Staley, J.K., Hearn, W.L., Ruttenber, A.J., Wetli, C.V., Mash, D.C., 1994. High affinity cocaine recognition sites on dopamine transporter are elevated in fatal cocaine overdose victims. J. Pharmacol. Exp. Ther. 271, 1678–1685.

Stallings, M.C., Corley, R.P., Hewitt, J.K., Krauter, K.S., Lessem, J.M., Mikulich, S.K., Rhee, S.H., Smolen, A., Young, S.E., Crowley, T.J., 2003. A genome-wide search for quantitative trait loci influencing substance dependence vulnerability in adolescence. Drug Alcohol. Depend. 70, 295–307.

Stewart, J., 2000. Pathways to relapse: the neurobiology of drug- and stress-induced relapse to drug-taking. J. Psychiatr. Neurosci. 25, 125–136.

Stoops, W.W., Hatton, K.W., Lofwall, M.R., Nuzzo, P.A., Walsh, S.L., 2010. Intravenous oxycodone, hydrocodone, and morphine in recreational opioid users: abuse potential and relative potencies. Psychopharmacology 212, 193–203.

Strang, J., Johns, A., Caan, W., 1993. Cocaine in the UK – 1991. Br. J. Psychiatry 162, 1–13.

Strickland, T.L., Mena, I., Villanueva-Meyer, J., Miller, B.L., Cummings, J., Mehringer, C.M., Satz, P., Myers, H., 1993. Cerebral perfusion and neuropsychological consequences of chronic cocaine use. J. Neuropsychiatry Clin. Neurosci. 5, 419–427.

Sulzer, D., Sonders, M.S., Poulsen, N.W., Galli, A., 2005. Mechanisms of neurotransmitter release by amphetamines: a review. Prog. Neurobiol. 75, 406–433.

Tamrazi, B., Almast, J., 2012. Your brain on drugs: imaging of drug-related changes in the central nervous system. Radiographics 32, 701–719.

Tanda, G., Pontieri, F.E., Di Chiara, G., 1997. Cannabinoid and heroin activation of mesolimbic dopamine transmission by a common μ1 opioid receptor mechanism. Science 276, 2048–2050.

Tang, W.-X., Fasulo, W.H., Mash, D.C., Hemby, S.E., 2003. Molecular profiling of midbrain dopamine regions in cocaine overdose victims. J. Neurochem. 85, 911–924.

Tardiff, K., Gross, E., Wu, J., Stajic, M., Millman, R., 1989. Analysis of cocaine-positive fatalities. J. Forensic Sci. 34, 53–63.

Thanvi, B.R., Treadwell, S.D., 2009. Cannabis and stroke: is there a link? Postgrad. Med. J. 85, 80–83.

Thrash, B., Thiruchelvan, K., Ahuja, M., Suppiramaniam, V., Dhanasekaran, M., 2009. Methamphetamine-induced neurotoxicity: the road to Parkinson's disease. Pharmacol. Rep. 61, 966–977.

Tomlinson, G.S., Simmonds, P., Busuttil, A., Chiswick, A., Bell, J.E., 1999. Upregulation of microglia in drug users with and without pre-symptomatic HIV infection. Neuropathol. Appl. Neurobiol. 25, 369–379.

Tong, J., Ross, B.M., Schmunk, G.A., Peretti, F.J., Kalasinsky, K.S., Furukawa, Y., Ang, L.-C., Aiken, S.S., Wickham, D.J., Kish, S.J., 2003. Decreased striatal dopamine D1 receptor-stimulated adenylyl cyclase activity in human methamphetamine users. Am. J. Psychiatry 160, 896–903.

Toossi, S., Hess, C.P., Hills, N.K., Josephson, S.A., 2010. Neurovascular complications of cocaine use at a tertiary stroke center. J. Stroke Cerebrovasc. Dis. 19, 273–278.

Torres, G., Horowitz, J.M., 1999. Drugs of abuse and brain gene expression. Psychosom. Med. 61, 630–650.

Traynor, J., 2010. Regulator of G protein–signaling proteins and addictive drugs. Ann. N.Y. Acad. Sci. 1187, 341–352.

Treadwell, S.D., Robinson, T.G., 2007. Cocaine use and stroke. Postgrad. Med. J. 83, 389–394.

Trulson, M.E., Cannon, M.S., Faegg, T.S., Raese, J.D., 1985. Effects of chronic methamphetamine on the nigral-striatal dopamine system in rat brain: tyrosine hydroxylase immunochemistry and quantitative light microscopic studies. Brain Res. Bull. 15, 569–577.

Turillazzi, E., Riezzo, I., Neri, M., Bello, S., Fineschi, V., 2010. MDMA toxicity and pathological consequences: a review about experimental data and autopsy findings. Curr. Pharm. Biotechnol. 11, 500–509.

Turner, J.J.D., Parrott, A.C., 2000. "Is MDMA a human neurotoxin?" Diverse views from the discussions.
Neuropsychobiology 42, 42–48.

Urban, N.B., Martinez, D., 2012. Neurobiology of addiction: insight from neurochemical imaging. Psychiatr.
Clin. North Am. 35, 521–541.

Uzbay, I.T., Oglesby, M.W., 2001. Nitric oxide and substance dependence. Neurosci. Biobehav. Rev. 25,
43–52.

Van Bockstaele, E.J., Reyes, B.A.S., Valentino, R.J., 2010. The locus coeruleus: a key nucleus where stress and
opioids intersect to mediate vulnerability to opiate abuse. Brain Res. 1314, 162–174.

Vannemreddy, P., Caldito, G., Willis, B., Nanda, A., 2008. Influence of cocaine on ruptured intracranial aneurysms: a case control study of poor prognostic indicators. J. Neurosurg. 108, 470–476.

Vila, N., Chamorro, A., 1997. Ballistic movements due to ischemic infarcts after intravenous heroin overdose:
report of two cases. Clin. Neurol. Neurosurg. 99, 259–262.

Villemagne, V.L., Yuan, J., Wong, D.F., Dannals, R.F., Hatzidimitriou, G., Mathews, W.B., Ravert, H.T., Musachio, J., McCann, U.D., Ricaurte, G.A., 1998. Brain dopamine neurotoxicity in baboons treated with
doses of methamphetamine comparable to those recreationally abused by humans: evidence from [11C]
WIN-35,428 positron emission tomography studies and direct in vitro determinations. J. Neurosci. 18,
419–427.

Volkow, N.D., Chang, L., Wang, G.-J., Fowler, J.S., Ding, Y.-S., Sedler, M., Logan, J., Franceschi, D., Gatter, J.,
Hitzemann, R., Gifford, A., Wong, C., Pappas, N., 2001a. Low level of brain dopamine D2 receptors in
methamphetamine abusers: association with metabolism in the orbitofrontal cortex. Am. J. Psychiatry
158, 2015–2021.

Volkow, N.D., Chang, L., Wang, G.-J., Fowler, J.S., Leonido-Yee, M., Franceschi, D., Sedler, M.J., Gatley, S.J.,
Hitzemann, R., Ding, Y.-S., 2001b. Association of dopamine transporter reduction with psychomotor
impairment in methamphetamine abusers. Am. J. Psychiatry 158, 377–382.

Volkow, N.D., Fowler, J.S., 2000. Addiction, a disease of compulsion and drive: involvement of the orbitofrontal cortex. Cereb. Cortex 10, 318–325.

Volkow, N.D., Fowler, J.S., Logan, J., Gatley, S.J., Dewey, S.L., MacGregor, R.R., Schlyer, D.J., Pappas, N.,
Wang, G.-J., Wolf, A.P., 1995. Carbon-11-cocaine binding compared at subpharmacological and pharmacological doses: a PET study. J. Nucl. Med. 36, 1289–1297.

Wagner, G., Ricaurte, G., Johanson, C., Schuster, C., Seiden, L., 1980a. Amphetamine induces depletion of
dopamine and loss of dopamine uptake sites in caudate. Neurology 30, 547–550.

Wagner, G.C., Ricaurte, G.A., Seiden, L.S., Schuster, C.R., Miller, R.J., Westley, J., 1980b. Long-lasting
depletions of striatal dopamine and loss of dopamine uptake sites following repeated administration of
methamphetamine. Brain Res. 181, 151–160.

Waldhoer, M., Bartlett, S.E., Whistler, J.L., 2004. Opioid receptors. Annu. Rev. Biochem. 73, 953–990.

Walubo, A., Seger, D., 1999. Fatal multi-organ failure after suicidal overdose with MDMA, "ecstasy": case
report and review of the literature. Hum. Exp. Toxicol. 18, 119–125.

Wang, X., Zhang, T., Ho, W.-Z., 2011. Opioids and HIV/HCV infection. J. Neuroimmune Pharmacol. 6,
477–489.

Warner-Smith, M., Darke, S., Lynskey, M., Hall, W., 2001. Heroin overdose: causes and consequences.
Addiction 96, 1113–1125.

Weiss, F., Koob, G.F., 2000. Drug addiction: functional neurotoxicity of the brain reward systems. Neurotox.
Res. 3, 145–156.

White, S.M., Lambe, C.J.T., 2003. The pathophysiology of cocaine abuse. J. Clin. Forensic Med. 10, 27–39.

White, S.R., Obradovic, T., Imel, K.M., Wheaton, M.J., 1996. The effects of methylenedioxymethamphetamine (MDMA, "Ecstasy") on monoaminergic neurotransmission in the central nervous system. Prog.
Neurobiol. 49, 455–479.

Wilson, R.I., Nicoll, R.A., 2002. Endocannabinoid signaling in the brain. Science 296, 678–682.

Wilson, J.M., Levey, A.I., Bergeron, C., Kalasinsky, K., Ang, L., Peretti, F., Adams, V.I., Smialek, J., Anderson,
W.R., Shannak, K., Deck, J., Niznik, H.B., Kish, S.J., 1996a. Striatal dopamine, dopamine transporter,
and vesicular monoamine transporter in chronic cocaine users. Ann. Neurol. 40, 428–439.

Wilson, J.M., Kalasinsky, K.S., Kalasinsky, K.S., Levey, A.I., Bergeron, C., Reiber, G., Anthony, R.M., Schmunk, G.A., Shannak, K., Haycock, J.W., Kish, S.J., 1996b. Striatal dopamine nerve terminal markers in human, chronic methamphetamine users. Nat. Med. 2, 699–703.

Winstock, A., Mitcheson, L., Ramsey, J., Davies, S., Puchnarewicz, M., Marsden, J., 2011. Mephedrone: use, subjective effects and health risks. Addiction 106, 1991–1996.

Winstock, A.R., Wolff, K., Ramsey, J., 2002. 4-MTA: a new synthetic drug on the dance scene. Drug Alcohol Depend. 67, 111–115.

Wise, R.A., 2009. Roles for nigrostriatal – not just mesocorticolimbic – dopamine in reward and addiction. Trends Neurosci. 32, 517–524.

Wojak, J.C., Flamm, E.S., 1987. Intracranial hemorrhage and cocaine use. Stroke 18, 712–715.

Wolff, K., Hay, A.W.M., Sherlock, K., Conner, M., 1995. Contents of "ecstasy". Lancet 346, 1100–1101.

Wolters, E.C., Stam, F.C., Lousberg, R.J., van Wijngaarden, G.K., Rengelink, H., Schipper, M.E.I., Verbeeten, B., 1982. Leucoencephalopathy after inhalating "heroin" pyrolysate. Lancet 320, 1233–1237.

Wong, S.C., Mundy, L., Drake, R., Curtis, J.A., Wingert, W.E., 2010. The prevalence of fentanyl in drug-related deaths in Philadelphia 2004–2006. J. Med. Toxicol. 6, 9–11.

Woods, B.T., Strewler, G.J., 1972. Hemiparesis occuring six hours after intravenous heroin injection. Neurology 22, 863–866.

Worm, K., Steentoft, A., Kringsholm, B., 1993. Methadone and drug addicts. Int. J. Legal Med. 106, 119–123.

Yao, H., Duan, M., Buch, S., 2011. Cocaine-mediated induction of platelet-derived growth factor: implication for increased vascular permeability. Blood 117, 2538–2547.

Yao, L., McFarland, K., Fan, P., Jiang, Z., Inoue, Y., Diamond, I., 2005. Activator of G protein signaling 3 regulates opiate activation of protein kinase A signaling and relapse of heroin-seeking behavior. Proc. Natl. Acad. Sci. U.S.A 102, 8746–8751.

Yee, T., Gronner, A., Knight, R.T., 1994. CT findings of hypoxic basal ganglia damage. South. Med. J. 87, 624–626.

Yen, D.J., Wang, S.J., Ju, T.H., Chen, C.C., Liao, K.K., Fuh, J.L., Hu, H.H., 1994. Stroke associated with methamphetamine inhalation. Eur. Neurol. 34, 16–22.

Zachariah, S.B., 1991. Stroke after heavy marijuana smoking. Stroke 22, 406–409.

Zamparutti, G., Schifano, F., Corkery, J.M., Oyefeso, A., Ghodse, A.H., 2011. Deaths of opiate/opioid misusers involving dihydrocodeine, UK, 1997–2007. Br. J. Clin. Pharmacol. 72, 330–337.

Zhu, B.L., Oritani, S., Shimotouge, K., Ishida, K., Quan, L., Fujita, M.Q., Ogawa, M., Maeda, H., 2000. Methamphetamine-related fatalities in forensic autopsy during 5 years in southern half of Osaka city and surrounding areas. Forensic Sci. Int. 113, 443–447.

Zill, P., Vielsmeier, V., Büttner, A., Eisenmenger, W., Siedler, F., Scheffer, B., Möller, H.J., Bondy, B., 2011. Postmortem proteomic analysis in human amygdala of drug addicts: possible impact of tubulin on drug abusing behavior. Eur. Arch. Psychiatry Clin. Neurosci. 261, 121–131.

Zuckerman, G.B., Ruiz, D.C., Keller, I.A., Brooks, J., 1996. Neurologic complications following intranasal administration of heroin in an adolescent. Ann. Pharmacother. 30, 778–781.

CHAPTER EIGHT

The Pathology of Methamphetamine Use in the Human Brain

Stephen J. Kish
Human Brain Laboratory, Centre for Addiction and Mental Health, Departments of Psychiatry and Pharmacology, University of Toronto, Toronto ON, Canada

1. INTRODUCTION

1.1. Objective: To Evaluate Evidence of a Characteristic Brain Pathology and Drug Neurotoxicity of Recreational Methamphetamine Users

Methamphetamine (METH) and its metabolite amphetamine (Baselt, 2004) are stimulant drugs used worldwide for a variety of purposes (for an older, but still excellent book review on METH, its metabolism and effects on brain neurochemistry and behavior see Cho and Segal, 1994). The stimulants are used for their pleasurable effects and also for therapeutic purposes, for example, to maintain wakefulness (e.g. in patients with narcolepsy), help in weight reduction, and to treat patients with attention deficit hyperactivity disorder (ADHD), (see Kish, 2008).

The primary *objective* of this review is to summarize the evidence for (and against) the presence of a "characteristic" brain pathology and drug neurotoxicity in human recreational METH users, with only limited mention of the results of experimental animal study findings as these have been dealt with in detail in other review articles (see especially McCann and Ricaurte, 2004 for historical perspective and Krasnova and Cadet, 2009 for completeness of literature review).

The *scope of this review* is to focus primarily on the question whether METH exposure for psychoactive purposes might cause damage to dopamine neurons and, to a lesser extent, whether the drug might cause changes in measures that can be associated with brain damage, gliosis, and reduction or enlargement in brain size. Although there is no consensus yet on any of these questions, there is now an emerging literature comprising both postmortem brain and brain imaging investigations.

The review will be especially focused on whether the literature indicates that reliable, replicated differences are likely to exist in brain of human METH users and whether changes are "robust" (i.e. minimal overlap between control and drug-user values; finding typically replicated in independent studies), with only limited speculation on the nature of possible behavioral consequences of the changes (i.e. functional "impairment"). This is, in part, because it is not possible in correlational studies to establish a cause and effect

The Effects of Drug Abuse on the Human Nervous System
http://dx.doi.org/10.1016/B978-0-12-418679-8.00008-3

and because, in the opinion of this reviewer, most of the changes reported to exist in brain of METH users have yet to be conclusively established.

1.2. There Are Significant Underappreciated Policy Issues for Possible Harm Caused by Methamphetamine

Findings suggestive of a human brain pathology associated with METH exposure will be used in part by governments to set public policy related to use of the stimulant drug. As recreational use and production of METH is generally prohibited, such information related to harm will also be used, for example, by the trial judge at a sentencing hearing to set the penalty for a drug-related offense. For these and the above reasons, an opinion on the question of METH harm to the brain must aim to be accurate, balanced, appropriately qualified, and derived from the product of sound and reliable scientific methodology (see Daubert v Merrell Dow Pharmaceuticals, Inc, 1993; Berger, 2005).

The popular press now emphasizes harm to society caused by those scientists who are considered to exaggerate brain toxicity caused by methamphetamine (Szalavitz, 2011). On the other hand, scientific reviewers such as myself can also inadvertently "misquote", by word omission, scientists who had in fact appropriately "qualified" a conclusion related to drug toxicity: e.g., Reviewer: "…the investigators concluded that '…chronic methamphetamine abuse causes a selective pattern of cerebral deterioration…'" (Hart et al., 2012); The actual text: "MRI-based maps suggest that chronic methamphetamine abuse causes a selective pattern of cerebral deterioration…"; Thompson et al., 2004).

1.2.1. Conflicts of Interest and Public Policy Issues on Brain Pathology Research in Methamphetamine Users

There can be conflicts of interest when dealing with the question of brain pathology related to recreational use of an illegal drug that is already known to cause significant harm (e.g. addiction) to at least some users.

The reader needs to appreciate that the current author, by receiving research monies from a governmental funding body (see Acknowledgments) has an incentive (financial and career-wise) to skew his own research findings and opinions in one direction: to find evidence of harm caused by the drug. Thus, as a practical matter, if my research demonstrates *lack* of brain toxicity of an abused drug, grant funding for a study of brain pathology will likely be terminated; if harm is demonstrated, funding might continue. The public also has an opinion on harm caused by recreational drugs, which could also influence my opinion on the extent of harm caused by METH.

Notwithstanding the above recommendations, the current author personally acknowledges the real difficulty in providing a balanced, unbiased opinion in an area of science in which the author has expended considerable effort and has already developed conclusions on the subject.

1.3. Stylistic Issues Regarding Terminology

For the sake of convenience, a number of terms will be employed without specific implications:

1.3.1. Assume for the Purpose of This Review that Methamphetamine and Amphetamine are Substantially the Same

Despite years of use of both METH and metabolite amphetamine, which differs from METH by a methyl group (Figure 1), it is not clear whether there exist any substantial differences in the actions or behavioral effects of the two drugs in the human. An early clinical study found oral METH and amphetamine to be "equipotent" with respect to most physiologic and behavioral parameters (Martin et al., 1971; see also Lamb and Henningfield, 1994; Sevak et al., 2009) whereas a recent investigation comparing intranasal METH vs amphetamine found qualitatively similar behavioral effects but with METH somewhat more potent (not necessarily having higher efficacy) on some measures (Kirkpatrick et al., 2012).

It will be assumed for the purpose of this review that the actions and consequent brain pathology and neurotoxicity of METH and amphetamine (if any) in the human are most likely highly similar.

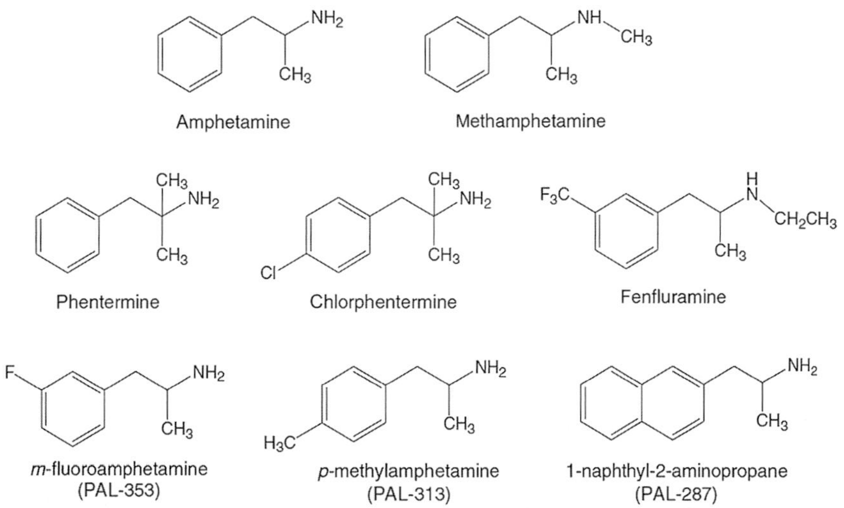

Figure 1 Chemical structure of methamphetamine, amphetamine, and related drugs. Methamphetamine and its metabolite amphetamine differ by the presence of a methyl group. Both stimulant drugs have similar or identical mechanisms of action in brain, e.g. release of monoamine neurotransmitters. Although methamphetamine is the primary amphetamine used for recreational purposes in North America, the psychoactive effects of methamphetamine and amphetamine in humans are also similar if not identical. Also included for comparison in the figure *(Reproduced with permission from Rothman et al. (2008).)* are structurally related compounds including phentermine (dopamine releaser) and fenfluramine (serotonin releaser) (see Rothman et al., 2008 for extensive discussion).

1.3.2. The Terms "Increased," and "Decreased" Should Not be Overinterpreted by the Reader

I will often use, inappropriately, the terms "increased" and "decreased" when describing differences in brain measures between control and METH user groups. This is done only for the sake of simplicity and does not imply that level of the marker actually changed as compared to a state that preexisted drug use.

1.3.3. Changes in Levels of Markers Inferred by Brain Imaging Will Not always Equal Differences in Actual Levels of the Target in Living Brain

For simplicity I may state, for example, that a brain protein (e.g. the dopamine receptor) was "measured" by brain imaging in living METH users. I emphasize, however, that measurement of neurochemical indices in a brain imaging study may not necessarily reflect actual levels in brain.

For example, brain imaging positron emission tomography (PET) studies typically employ only a single very-low tracer dose of the radiolabelled probe in which binding differences between drug user and control groups might not relate to a difference in actual level of the protein target, but rather, for example, to altered affinity of the receptor for the probe or to presence of other substances in brain that might compete with binding. As an example, ^{11}C-raclopride binding, used as a measure of striatal concentration of the dopamine D2/D3 receptor, is influenced by changes in levels of dopamine that compete with raclopride for the binding site: thus, if striatal ^{11}C-raclopride binding in a PET investigation is lower than normal in brain of a METH user, is the change explained by decreased dopamine receptor protein or by increased occupancy of dopamine at the receptor? Confirmation that a difference in binding levels (e.g. to a receptor protein) of a radiolabelled probe in a brain PET imaging study can only be accomplished in a postmortem brain investigation.

In the human METH literature, as mentioned below, such postmortem brain confirmation of a brain-imaging finding has only occurred for the single observation of low striatal dopamine transporter (and possibly for normal/near-normal striatal VMAT2 concentration) in METH users.

Similar uncertainty also applies to brain-imaging measurements employing, for example, measures of activated microgliosis. Readers should have high skepticism of the meaningfulness of such imaging investigations until confirmation that such a change actually was evident in a postmortem brain study. Thus, *proof* of increased glial number or glial activation (markers often associated with a toxic event) cannot be provided simply by demonstration of increased levels of a protein (e.g. GFAP) in a test tube or by brain imaging, showing increased binding to a reputed microgliosis marker. Confirmation of brain gliosis changes can only be accomplished by an immunohistochemical procedure.

1.3.4. There are Multiple, Reasonable Definitions of "Neurotoxicity" and "Brain Damage"; Choose One but be Clear as to What Your Definition Does or Does Not Include

The definition of the terms drug-induced "brain damage" and "neurotoxicity" continue to be debated, with much disagreement whether actual "physical" damage to a neurone is required. For example, Wikipedia teaches (I believe correctly) that the term is generally used (by the public, including neuroscientists) to mean actual physical brain damage: "The term is generally used to describe a condition or substance that has been shown to result in observable physical damage." (http://en.wikipedia.org/wiki/Neurotoxicity).

Scientists evaluating evidence that METH causes harmful changes in brain need to be explicit as to whether the harm is limited to physical brain damage.

Forensic toxicologists I have consulted are most impressed (and content) when the basis of toxicity is unequivocal structural damage to brain neurons ("show me only holes in the brain"). In sharp contrast, a longlasting (months) drug-induced downregulation of a functionally important protein within a structurally intact brain neuron is not considered by the same toxicologists to represent "neurotoxicity" even though the brain abnormality might have more severe functional consequences (i.e. harm) than a slight loss of tissue.

Similarly, what should be done with Robinson and Kolb's interesting findings (1997, 1999), (Crombag et al., 2005) of persistently altered dendritic length, number, and branching in striatum and cerebral cortex of amphetamine-treated rats (Figure 2)? If one assumes, for the sake of argument, that this (unquestionably) "structural" change is also present in brain of human METH users and is responsible, as the authors suggest, for explaining aspects of drug addiction (a harm), how could we not argue that this is indeed a neurotoxic, brain damaging, event? However, the typical response of neuroscientists, and probably the reader, to this argument is that: "This is 'neuroplasticity'—not toxicity."

The definition of "neurotoxicity" by Ricaurte, the pioneer and expert in experimental animal studies employing METH, can perhaps be criticized for being too restrictive to effects of the drug on dopamine and serotonin axon terminals (what about possible damage to nonmonoamine neurotransmitter neurons?) and would appear to require some proof that the changes could not possibly be explained by a "neuroadaptive response": "...neurotoxicity simply refers to the potential for methamphetamine and some of its analogs to produce long-lasting reductions in brain markers of DA and/or 5-HT axon terminals (e.g. a distal axotomy) that cannot be accounted for by acute pharmacological effects of drug or neuroadaptive responses to these pharmacological effects." (McCann and Ricaurte, 2004). Unfortunately, such proof of a "negative" is, on practical grounds, impossible to obtain.

My own position is that all of the above possible definitions of neurotoxicity are generally reasonable, as long as the inclusion criteria are sufficiently stated. I take a broad definition of methamphetamine-induced brain neurotoxicity ("a persistent change in the brain, for which methamphetamine use is responsible, that causes or has the potential to cause harm") that does *not* require physical loss of part or all of the structure of a neuron, and for which the harm may well be "preclinical." For example, METH would satisfy my inclusion

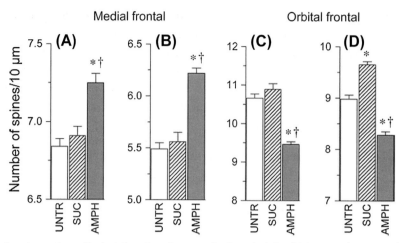

Figure 2 Amphetamine self-administration alters cerebral cortical dendritic spine density in the rat: can we call this neurotoxicity or only neuroplasticity? Rats were allowed to chronically self-administer amphetamine (AMPH) or sucrose (SUC), or receive no treatment (UNTR) and were then sacrificed 30 days after last drug session *(Reproduced with permission from Crombag et al. (2005).)*. Histological analysis showed increased spine density in apical (A, C) and basilar (B, D) dendrites of pyramidal neurones in the medial prefrontal cortex but decreased density in the orbital frontal cortex of the amphetamine-treated animals (The asterisk and daggers indicated a significant difference from UNTR and SUC conditions, respectively; $P < 0.05$). Let us assume that these changes (which certainly are structural and persistent) underlie harm (addiction) to the animals. Why then will most readers argue that this structural, persistent, and harmful change does not fulfill criteria for "neurotoxicity" and prefer "neuroplasticity" or "pathological learning"?

criteria as a brain neurotoxin if the persistent (but potentially reversible) harm caused by the drug (e.g. addiction) was not associated with any neuronal loss but with a neurochemical or physiological change within intact neurons that is harmful to the subject.

It should be highly emphasized, however, that the public will always equate methamphetamine-induced brain damage with "holes in the brain."

1.4. There are many interpretational problems in human brain METH research.

There are a variety of issues in brain studies of human METH users that make interpretation of some findings difficult or impossible.

1.4.1. Can a brain abnormality in a METH user be attributed unequivocally to use of METH—or to a preexisting difference?

Because almost all brain studies of METH users are retrospective investigations, it is not possible to conclude absolutely that an observed brain abnormality is necessarily a consequence of METH use, as the finding in brain might have preexisted use of the drug.

An example of this problem faced by medical examiners is the situation when death arises that is associated with recent and chronic use of METH and the cause of death needs to be established. Because METH can cause activation of the cardiovascular system (increased heart rate, blood pressure, possibility of cardiac arrhythmia), use of the drug could result in overactivation of the cardiovascular system causing death in a young drug user. However, it can also be argued in individual cases that death might have occurred because of a preexisting cardiac abnormality (as can be the case in young individuals who do not use drugs) that was unrelated to use of METH, or perhaps METH was a contributing factor to death related to a preexisting cardiac problem. In such cases, proof cannot be established with absolute certainty and cause of death can only be "suspected" to a reasonable medical probability.

The above considerations could in fact argue that the title to the present book "The Effects of Drug Abuse on the Human Nervous System" is somewhat misleading because of the real possibility that some or even most differences in brains may have preexisted use of the drugs.

1.4.2. There Should be Some Proof that the Methamphetamine User Actually Used the Drug in Brain Studies

Readers need to suspect brain findings from studies in which there is no proof that the subject ever used the drug.

Proof of recent use can be provided by positive urine or blood METH levels in living users and by presence of METH in tissues in postmortem cases. Chronic use over a period of time related to length of scalp hair (e.g. growth of about about one half inch per month) can now easily be obtained by measurement of METH in sequential segments of scalp hair.

Similarly, if an investigation purports to demonstrate the presence of a brain abnormality still present in long-term (e.g. years) abstinence, there should be evidence of original use of the drug as well as evidence by drug testing of abstinence over the extended period. Such information, however, is quite difficult to obtain and rarely provided.

1.4.3. There Should be Information by Drug Testing on Use or Lack of Use of Drugs Other than Methamphetamine in Brain Studies

It is our experience that most stimulant users in our present investigations knowingly or unknowingly (e.g. Kalasinsky et al., 2004) use a variety of other substances (e.g. cocaine, cannabis, alcohol, nicotine), although in our early postmortem brain investigation of METH users from California (Wilson et al., 1996a), drug testing findings suggested that these subjects might actually have been "pure" METH users. Given that exposure to other drugs (e.g. cocaine; see below) might influence the integrity of the brain, this represents an important confound that can be addressed in part by ensuring that measurement by drug hair and repeated urine testing is conducted.

1.4.4. Changes in Brain Neurochemical Markers Do Not Necessarily Equal Changes in Neuronal Number: Will We Never Have a Definitive Answer?

A longstanding question is whether METH exposure might damage or cause loss of dopamine nerve endings in human brain, as is the case in experimental animal studies.

I take the conservative position (the "Kish uncertainty principle"; Kish, 2003) that it may never be possible to have a conclusive answer on the question of dopamine nerve ending loss in human METH users, since investigators rely on neurochemical markers (e.g. dopamine transporter) that are known, or have the capacity to, change *independently* of changes in neuronal number. It is hoped that new technical developments will make my position unjustified.

Perhaps the best example of this problem facing "neurotoxicologists," which affects even postmortem brain studies in experimental animals, is provided by the investigation of Harvey and colleagues (2000) who, in a rather heroic study, examined by histological procedures the entire length of the nigrostriatal dopamine system in brain of methamphetamine-exposed vervet monkeys. Their aim in part was to establish, by a quantitative cell counting procedure, whether METH might cause loss of dopamine cell bodies in the midbrain area. In animals examined 1 month following METH exposure, extensive loss of immunohistochemical staining for three different dopamine neuronal markers (dopamine transporter, tyrosine hydroxylase, VMAT2), was observed in midbrain as well as in striatum (see Figure 3), certainly suggestive of dopaminergic changes in the cell bodies. However, Harvey was unable to quantitate the *number* of midbrain dopamine cell bodies because decreases in the dopamine markers "…precluded the unambiguous identification needed for accurate cell counts." In other words, how can one count the number of dopamine cell bodies if expression of the biochemical marker for dopamine neurons is reduced?

We are then faced with the unanswered question whether METH caused actual loss of dopamine cell bodies in the monkeys or only, as the investigators suggest, severe loss of expression of dopamine markers in intact neurons (see also Section 2.1).

1.4.5. Should We Only Care if Methamphetamine Values Fall outside of the Control Range?

There is always the generic question whether differences (especially when only slight) between levels of brain measures in a test group (e.g. METH users) vs normal controls are *likely* to be "meaningful."

In this regard, readers are encouraged to read the recent article of Hart and colleagues (2012) who review the difficult methamphetamine-cognition and related brain imaging literature. Hart argues (in part) that many or most of the findings might not have clinical significance because METH values often fall well within the range of normal levels. Hart further emphasizes that because of a lack of appreciation of this and other factors, investigators have overinterpreted their findings on possible harm caused by METH, thereby potentially causing harm to the public (see Szalavitz, 2011 for a similar but more strongly worded position and also a commentary on the Hart article (Payer et al., 2012)).

I agree with Hart that there has been some overinterpretation of brain findings in the METH literature (see below) and strongly share the feeling that brain differences are much more likely to be meaningful (functionally significant) if most drug user values fall outside the normal control range—and if values are associated with actual documented disability and negative quality of life. Moreover, my position is that assessment of the

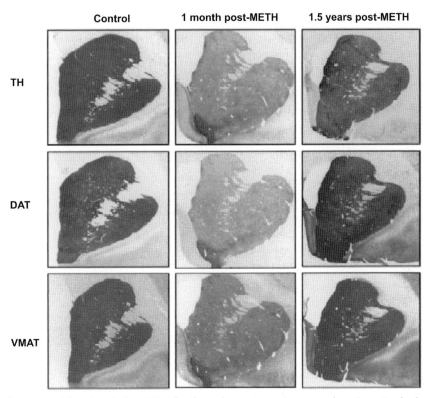

Figure 3 Immunohistochemical staining for three dopamine neurone markers (tyrosine hydroxylase, TH; dopamine transporter, DAT; and the vesicular monoamine transporter-2 (VMAT2)) in postmortem striatum of normal vervet monkeys (Left) and monkeys examined 1 month (Middle) and 1.5 Years (Right) following methamphetamine exposure: Marker vs Neuronal Loss. Note that dopamine marker immunostaining in the dopamine nerve terminal striatum area is severely reduced at 1 month following methamphetamine, but has recovered by 1.5 years *(Reproduced with permission from Harvey et al. (2000).)*. Somewhat surprisingly, immunostaining was similarly decreased in the dopamine cell body area, generally assumed to unaffected by the drug. Does loss of dopamine markers equal only loss of the markers or actual (but reversible) physical loss of dopamine neurones containing the markers? How can one count the number of dopamine cell bodies at 1 month in the near-absence of biochemical markers for dopamine neurones (see Harvey et al., 2000 for discussion)?

extent to which drug user values fall within or outside of the range of the normal controls is *as important and meaningful* (if not more so) as assessment of the magnitude of a group mean difference.

Part of the difficulty is that, with the exception of dopamine marker loss in Parkinson's disease (see below), we do not have information on the extent of loss or change of most brain markers that is required for behavioral consequence. Thus, for other estimates of harm to the brain (e.g. cerebral cortical gray matter density "reduction", increased gliosis) we do not, and might never, have such information.

Nevertheless, I do consider a validated, consistent finding of even a quantitatively "small" difference in a brain outcome measure between METH users and controls potentially "important" even if the difference has not yet shown to be functionally significant because of the possibility that the difference (at present preclinical) might compromise normal function in the presence of additional (e.g. age-related) insults. Similarly, consistent histopathological evidence of, for example, slightly above-normal activated micro-gliosis without obvious cell loss in brain does suggest possibility of an early toxic process that should be considered.

1.4.6. What is Meant by the Terms "Characteristic" or "Robust" Difference, Finding, or Pathology in the Methamphetamine Literature?

The preferred definition of a "characteristic" brain pathology must be that the pathology is an obligatory, defining feature of *all* subjects with the human brain condition. In this regard, low putamen dopamine, always below the lower limit of the control range, is a defining characteristic pathology of patients with idiopathic Parkinson's disease (Hornykiewicz and Kish, 1987). Thus, in 30 years of postmortem biochemical examination of such cases, I have not encountered a single case that does not have such a massive putamen dopamine loss in which the level falls outside of the range of the controls.

I suspect that such an absolute characteristic—no overlap with control values whatsoever and defining brain pathology—will never be found that will encompass all chronic users of METH for nonmedical purposes. Based on my review of the literature, such an absolute brain pathology might also not ever be found even for high-dose METH users who meet criteria for dependence, but I am much less certain on this point. However, the data do suggest that there are likely to be reliable *group mean* differences in levels of some brain markers (e.g. striatal dopamine transporter) between drug users and controls that probably will be replicable in future investigations and, in this limited respect, can be considered to be a characteristic group difference.

Regarding qualification of the differences between drug users and controls in a *single study* as "robust," my definition of the term must require "little or no overlap" between individual METH user and control values (How could it be otherwise?). A robust finding *across different studies* is an observation that is highly replicated in independent investigations. Thus, for example, lower cerebral cortical gray matter density might not be a robust finding in an individual study, because of extensive overlap between drug user and control values; however, the finding itself may be robust across studies, as a mean (small) group difference is typically observed in independent investigations.

My primary aim in this review is to suggest whether, in METH studies, differences (from control levels) reported in selected brain measures are likely to be real and consistent/characteristic findings and to provide information on the extent of overlap between drug user and control values. The question of functional significance is mostly left to speculation.

2.1. DOES METHAMPHETAMINE, A DOPAMINE RELEASER, CAUSE LOSS OF DOPAMINE NEURONAL MARKERS IN HUMAN BRAIN AS OBSERVED IN ANIMAL STUDIES?

Human brain imaging data suggest that METH/amphetamine can cause release of the neurotransmitter dopamine from its nerve endings in the dopamine-rich striatum (caudate, putamen, ventral striatum/nucleus accumbens) (Laruelle et al., 1995; see Figure 4; Cardenas et al., 2004).

The human brain findings are consistent with animal data showing that METH can increase striatal extracellular levels of dopamine, as well as those of the neurotransmitters noradrenaline and serotonin (see Rothman and Baumann, 2003). The mechanism is not known with certainty but probably involves an action of the stimulant on both the plasma membrane dopamine transporter and the vesicular monoamine transporter 2(VMAT2)—the latter responsible for transporting monoamine neurotransmitters into the synaptic vesicle (Pifl et al., 1995; Sulzer et al., 2005; Partilla et al., 2006).

Brain imaging data also suggest that some other drugs of abuse also acutely increase levels of dopamine at its receptors (cocaine, opiates, nicotine, cannabinoids, alcohol) and dopamine is now considered to be a critical component in some as yet unspecified aspect (incentive motivation?) of addiction (see Volkow et al., 2011). However, pharmacological studies employing dopaminergic agents have not yet had success in preventing relapse in human METH users.

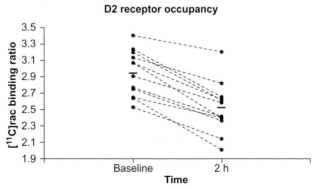

Figure 4 Evidence that amphetamine can cause release of dopamine in living human brain. [^{11}C]raclopride is a radiotracer that binds to the dopamine D2/D3 receptor and its binding in the dopamine-rich striatum is affected by changes in the concentration of dopamine. In this PET (positron emission tomography) brain imaging study, striatal raclopride binding was decreased in 12 subjects for 2 h following oral amphetamine (30 mg) administration *(Reproduced with permission from Cardenas et al. (2004).)*. The decreased binding is interpreted as evidence of increased amphetamine-induced release of dopamine competing with raclopride at its binding site.

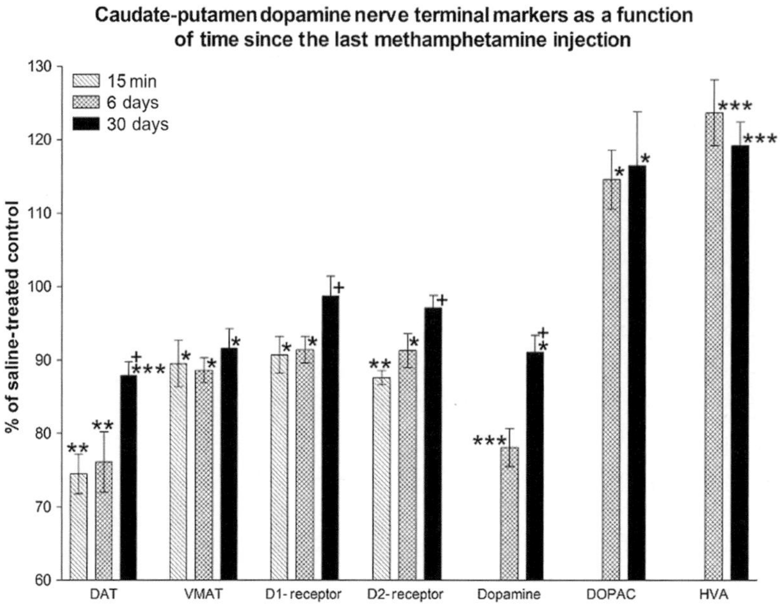

Figure 5 Time course of changes in striatal dopaminergic markers in methamphetamine-exposed rats. This animal study (n = 8–11 per group) shows that not all striatal dopamine markers change or recover to the same extent following intravenous exposure to methamphetamine *(Reproduced with permission from Segal et al. (2005).)*. Concentrations of all of the markers, except those of the dopamine metabolites DOPAC (dihydroxyphenylacetic acid) and HVA (homovanillic acid) and VMAT2 (vesicular monoamine transporter 2) were decreased at an early time point and then showed some recovery at 30 days. Levels of dopamine, VMAT2, and DAT (dopamine transporter) were still lower than normal 30 days following methamphetamine exposure. Does loss of striatal markers equal loss of dopamine nerve terminals or only loss of the markers? See below for assessment of the same dopamine markers in brain of human methamphetamine users. (*, **, ***, $P < 0.05, 0.01, 0.001$ vs saline treated animals; +$P < 0.05$ vs 15 min values).

Much interest in the question of METH toxicity to human brain is focused on the possibility that high doses of the stimulant damage dopamine axons and nerve endings but sparing cell bodies. This is suggested by results of experimental animal studies showing, in part, persistent loss of striatal neurochemical markers of dopamine neurones following high-dose METH or amphetamine exposure: dopamine, dopamine transporter, tyrosine hydroxylase (rate limiting enzyme of dopamine biosynthesis), and VMAT2 (Kogan et al., 1976; Seiden et al., 1976; for review see McCann and Ricaurte, 2004; Krasnova and Cadet, 2009). The magnitude and persistence of the changes in striatal markers are likely to be dependent upon the dosing protocol of METH and on the animal species used (McCann and Ricaurte, 2004; Segal et al., 2005).

Figure 5 provides a good example of the time-dependent changes in dopamine-related markers in striatum of rats treated chronically with intravenous METH (Segal et al., 2005). Note that 30 days following the last drug exposure striatal D1 and D2 binding, but not dopamine transporter, VMAT2 binding, or dopamine levels had returned to normal levels. Interestingly, striatal concentrations of the dopamine metabolites were not decreased at the examined time points (see Section 4.2 for discussion).

From a structural toxicity standpoint, such investigations have demonstrated in brain of methamphetamine-exposed animals the presence of "swollen and distorted" axons, Fink–Heimer silver staining (considered to reflect neurodegeneration), and changes also in microgliosis and astrogliosis that commonly accompany an acute brain damage event (Ellison et al., 1978; Lorez, 1981; Ricaurte et al., 1982, 1984; Hess et al., 1990; Ryan et al., 1990; Bowyer et al., 1994; O'Callaghan and Miller, 1994; LaVoie et al., 2004; see also below).

Nevertheless, there is still a question, amongst some investigators, whether the reduction in nigrostriatal dopamine neuronal markers reflects loss of part or all of the neurone or just decreased expression of the markers. As noted by Melega (Harvey et al., 2000) the supposition in METH animal studies is that "…degeneration of striatal terminals occurred while indeterminate lengths of their remaining axons and corresponding cell bodies remained intact." However, in their vervet monkey study (Harvey et al., 2000), levels of all examined dopamine markers in animals examined 1 month following METH were severely decreased *throughout the entirety* of the dopamine neuron (cell body, axon, striatal terminal regions; see Figure 3) but had recovered by 1.5 years. Although Melega acknowledges that their findings do not exclude possibility of some neuronal degeneration (as tests of silver staining and anterograde labeling of tracts were not conducted) the group prefers the explanation of decreased marker expression in intact neurons because of the implausibility that axons actually migrated appropriately into the striatum from the far distant cell body region.

There is also the question whether lower doses of METH might also cause changes in the brain of experimental animals. This issue has been addressed in animal studies of rodents allowed to self-administer lower doses of the stimulant drug although the findings from the different self-administration investigations are not entirely consistent (see McFadden et al., 2012; Section 7.5 for discussion).

2.1.1. Striatal Dopamine is Low in Some Methamphetamine Users: Results of a 1996 Postmortem Brain Investigation Still Awaiting Replication

To date (2012), postmortem brain neurochemical investigations of dopamine markers in METH users have been conducted in only two laboratories—why only two studies?

Given the interest from a public health point of view in establishing whether METH might damage brain dopamine neurons, it is somewhat surprising that, to my knowledge, postmortem brain neurochemical investigation of METH users have been conducted by only two groups, our team in Toronto Canada and Kitamura's group in Tokoshima Japan.

The paucity of studies is related in part to the practical difficulty in obtaining autopsied brain material (in which the interval from death to autopsy is reasonably short—less than 24 h) from medical examiners/coroners who are mandated to establish the cause of death.

Medical examiners are often not aware that a subject at autopsy was a METH user until blood toxicology becomes available, often only after a two to three-week period. In addition, there are also (reasonable) generic concerns from funding agencies that pre- and postmortem factors (e.g. postmortem interval and stability of marker, manner of death) are likely to confound significantly interpretation of any autopsied brain finding. Sharing of postmortem brain between institutions has also become much more problematic because of the now common institutional requirement of a formal "material transfer agreement" between the parties that, if taken to extreme, requires purchase of additional indemnification insurance to cover unlimited costs (e.g. attorney fees, court costs) of "any and all" harm related to negligent use by provider or recipient of the samples. Some institutions have now developed privacy concerns in sending samples (e.g. anonymized samples derived from drug users) to United States drug testing laboratories because of possible consequences related to the USA Patriot Act. These and other issues will, at a minimum, increase the time spent by the investigator in the Contracts office. Nevertheless, postmortem brain investigations of brain dopamine markers have been useful and have identified, over the past 50 years (see Kish et al., 1988) a reliable neurochemical characteristic of a deficiency of dopamine markers in striatum of patients with Parkinson's disease (Hornykiewicz and Kish, 1987). Further, it can easily be argued that a postmortem brain study is the only approach that can be used to confirm presence of, for example, activated microgliosis of reactive astrocytes in users of METH.

A plus in the Toronto and Kitamura autopsied brain investigations is the demonstration of METH in hair, which was available in most of the subjects of the Toronto study and in all of the subjects of the Japanese investigation; this suggests chronic use of the stimulant. Although drugs (e.g. cocaine, opiates) other than METH were not detected in brain, blood, or hair of most of the METH users of the Toronto study, it can never be possible to establish conclusively that such subjects chronically used only METH.

Assay procedures confirm that striatal dopamine markers are low in Parkinson's disease, the gold standard dopamine deficiency disorder.

Prior to measurement of striatal dopamine markers in postmortem brain of METH users, we confirmed that, employing the same assay procedures, levels of the markers were low in striatum of patients with idiopathic Parkinson's disease, a disorder characterized by loss of entire dopamine neurons (see Hornykiewicz and Kish, 1987; Kish et al., 1988). Protein levels (by western blotting) of two enzymes (tyrosine hydroxylase, dopa decarboxylase) and of two transporters (dopamine transporter, VMAT2), rather than activity measurements, were determined because of high variability of the activities in postmortem brain.

As expected, striatal levels of dopamine and those of all of other dopamine markers were markedly to severely decreased in the Parkinson's disease group, with the mean reduction in putamen exceeding that in caudate (Wilson et al., 1996b). Notably, all of the dopamine values in putamen of the Parkinson's disease patients fell well below the lower limit of the control range whereas in the less-affected caudate all but one of the dopamine values fell below the control lower limit and with caudate levels more scattered than those of the putamen.

Dopamine levels are markedly decreased in some METH users: A finding still awaiting independent replication.

Moszczynska et al., 2004 summarize results of our measurement of dopamine markers in our first study of 12 chronic METH users (who all tested positive at autopsy only for METH; cases from Wilson et al., 1996a) and in an additional eight METH-positive subjects, four of which tested positive for other abused drugs.

Histopathological (qualitative only) analysis of the formalin fixed half brain disclosed no obvious loss of cell bodies in the substantia nigra, consistent with most animal findings (see McCann and Ricaurte, 2004). In contrast, as shown in Figure 6, striatal dopamine levels were markedly and similarly decreased in both $n=12$ and $n=8$ METH groups, with the mean dopamine reduction in the total group being 61% in caudate and 50% in putamen—a pattern opposite to that in Parkinson's disease in which the putamen bears the brunt of the dopamine loss (Kish et al., 1988). Most striatal dopamine values (Figure 6) fell below the lower limit of the control range, especially in caudate, suggesting that the difference may be robust, with two subjects having near total loss of dopamine in the caudate (−95%, −97%) and a slightly less severe reduction in putamen (−89%, −90%).

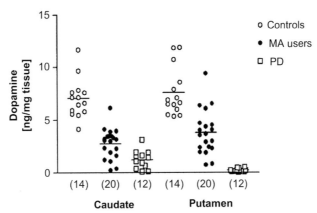

Figure 6 Caudate and putamen dopamine levels in autopsied brain of methamphetamine (METH) users vs those in patients with Parkinsons's disease (PD). Patients with Parkinson's disease, the established brain dopamine deficiency disorder, have near total loss of dopamine in the putamen portion of the striatum, with less severe loss in caudate. In contrast, methamphetamine users have, on average, a marked loss of dopamine somewhat more severe, on average, in the caudate nucleus than in putamen. Most individual methamphetamine user values in the caudate fall below the lower limit of the control range, suggesting that the difference in this subdivision of the striatum might be "robust." However, this study is still preliminary as it is awaiting replication. *Reproduced with permission from Moszczynska et al. (2004).*

The marked magnitude of striatal dopamine reduction in METH users together with minimal overlap between METH user and control values has not yet been observed with any of the other examined dopamine markers (see below).

To establish whether a marked dopamine reduction might be a feature of use of other recreational drugs, dopamine levels were also measured in autopsied striatum of users of cocaine and users of heroin and in a subject who used a high dose of MDMA (ecstasy, 3,4-methylenedioxymethamphetamine), a METH derivative considered to act preferentially on the brain serotonin (vs dopamine) neurotransmitter system (see McCann and Ricaurte, 2004). Mean striatal dopamine concentrations in cocaine users were decreased by 33% (caudate; a finding almost identical to that later reported by Little and colleagues, 2003) and 15% (putamen, trend only), (Wilson et al., 1996c) whereas levels in these brain areas of heroin users were nonsignificantly decreased by 10–11% (Kish et al., 2001) and were normal in the MDMA user (Kish et al., 2000). In contrast to the METH findings in which there was only minimal overlap between dopamine levels in the drug user and control groups (especially for caudate), we found marked overlap in control and drug user ranges for the cocaine and heroin users. These findings provide some (but not absolute) specificity to the marked caudate and putamen dopamine reduction in postmortem brain of the METH users.

Are striatal dopamine levels decreased because of excessive dopamine release and metabolism caused by METH—or by loss of dopamine neurons? All of the chronic METH users tested positive for METH in brain, allowing for the possibility that the low dopamine concentrations could be explained by either a recent or chronic effect of the drug.

In principle, low striatal dopamine could be due to an actual loss of dopaminergic innervation to striatum (but sparing the dopamine cell bodies), a metabolic abnormality involving dopamine synthesis or catabolism (e.g. decreased synthesis as part of a compensatory downregulation), altered (increased) neurotransmitter release and consequent depletion, or a by a combination of these possibilities. Parkinson's disease is an example of a condition in which low dopamine is explained in large part by actual loss of entire dopamine neurons (including the cell body); however, we suspect that part (a minority?) of the loss of the neurotransmitter in this degenerative condition is also consequent to increased turnover of dopamine as part of a compensatory process aimed at increasing neuronal activity to maintain homeostasis. This is suggested by findings of increased ratio dopamine metabolite to dopamine in autopsied brain of Parkinson's disease patients (see Hornykiewicz and Kish, 1987). In this context, the dopamine reduction in Parkinson's disease probably overestimates somewhat the actual loss of dopamine neurons (e.g. an 80% loss of dopamine might be associated with a 60% loss of dopamine neurons).

In the absence of histopathological evidence of actual dopamine nerve damage or loss in the examined METH users, it is not possible to conclude whether the dopamine reduction is explained by structural damage to dopamine neurons. My speculation is that, given that subjects had recently used METH, and that several of the other striatal dopamine markers were not decreased (dopamine metabolites, dopa decarboxylase,

VMAT2; see below), most of the dopamine reduction is explained by drug-induced massive release and subsequent breakdown of dopamine (too excessive to have been compensated for by increased dopamine biosynthesis)—with only a "modest" (if any) percentage of dopamine loss due to loss of dopaminergic innervation.

If correct, our findings suggest that some METH users take such a high dose of the drug that stored tissue levels of dopamine can actually be depleted due to excessive dopamine release and metabolism.

Might very low dopamine levels in some METH users explain reports of cognitive problems and subnormal performance on a motor task in a subgroup of drug users?

In the absence of behavioral data on the METH users of our study, we can only speculate as to the possible behavioral significance of low dopamine levels.

It is interesting that whereas dopamine loss in idiopathic Parkinson's disease is more pronounced in the putamen subdivision of the striatum (Hornykiewicz and Kish, 1987; Kish et al., 1988), the reverse was observed in the METH users in which the reduction was more marked in caudate nucleus. This difference is unlikely to be explained by intrastriatal differences in METH concentrations in view of our finding that METH levels are similar in postmortem brain of users of the drug (Kalasinsky et al., 2001; see below).

The more pronounced dopamine decrease in caudate nucleus suggests that functions subserved by the caudate might be more affected than those of the putamen. Although an oversimplification, the putamen is considered to be especially involved in aspects of motor control whereas the caudate subserves cognitive or psychomotor function (Alexander et al., 1986; Grahn et al., 2008). Thus, the defining feature of idiopathic Parkinson's disease is motor dysfunction accompanied by a severe putamenal dopamine deficit (Hornykiewicz and Kish, 1987). Some cognitive difficulties in Parkinson's disease, in contrast, might more likely be related to disturbed, but less severely affected, dopamine function in the caudate. (This issue could, of course, be better addressed by behavioral examination of, yet to be discovered, patients having "pure" caudate and pure putamen dopamine deficits.)

The question of clinically significant cognitive problems in METH users, based on the available literature, is, in my opinion, still uncertain and certainly much debated. Hart and colleagues (2012) argue that evidence for cognitive problems in METH users is greatly overinterpreted as drug user scores often fall within the control range. However, it is at least the (anecdotal) impression of some providers of treatment for METH users at large drug rehabilitation centers that cognitive difficulties in some METH users can be sufficiently severe to decrease retention in the rehabilitation program (see Moszczynska et al., 2004).

Regarding the question of motor deficits, rehabilitation centers have also advised me (anecdotally) that clinically obvious parkinsonism is not a characteristic of the METH user (see Moszczynska et al., 2004), although subtle motor problems (values within the normal range) have been reported in some polydrug METH users (e.g. on Purdue

Pegboard test; see Boileau et al., 2008). None of the dopamine values in putamen of the METH users fell within the range of those we have observed in Parkinson's disease, whereas "Parkinsonian" dopamine values were observed in caudate in some of the subjects (Moszczynska et al., 2004). The threshold for dopamine loss in the METH user (in which there is likely to be many compensatory processes) required to impact negatively on behavior is not known. However, the two (of 20) METH users having very severe loss (by 89–95%) of dopamine in caudate and putamen—if such loss were carried over into more extended abstinence—might well have had clinically significant cognitive difficulties and some (slight subclinical?) slowness of movement that could be demonstrated on formal testing. Given the suspected involvement of dopamine in motivation and hedonic status (Berridge and Kringelbach, 2008), low dopamine might also explain some of the unpleasant aspects of the METH abstinence syndrome.

Can the as-yet unreplicated finding low dopamine in METH users be believed?

The major uncertainty of our finding relates to the surprising lack of any independent replication investigation in the literature since the data were first reported in 1996 (at least two to three independent replications would be required to make a more compelling case) and also the absence of any behavioral data (e.g. presence or absence of preclinical Parkinsonism) from our study that might correlate with dopamine level differences.

2.1.2. Levels of Three Striatal Dopamine Metabolites are *Normal* in Postmortem Brain of Methamphetamine Users: Does this Mean that Substantial Loss of Dopamine Innervation is Unlikely?

In Parkinson's disease, concentrations of all three dopamine metabolites, HVA, dihidroxyphenylacetic acid (the latter showing high scatter of values in postmortem brain), and 3-methoxytyramine, are decreased in striatum, with the changes, as expected from the dopamine loss pattern, more marked in putamen than in caudate (Wilson et al., 1996b). As mentioned above, the finding that dopamine metabolite concentrations are decreased less than those of dopamine and the markedly higher ratio of dopamine metabolite/dopamine in Parkinson's disease has been interpreted as reflecting increased dopamine turnover and activity of the remaining dopamine neurones to compensate for loss of dopamine neurons (Hornykiewicz and Kish, 1987).

In postmortem brain of the METH users, however, we found that striatal concentrations of the three dopamine metabolites were not decreased (see Figure 7 and Table 1), even in the subgroup of six METH users having very low caudate dopamine (dopamine, −85% vs homovanillic acid, +13%) (Moszczynska et al., 2004). This finding could be explained by excessive dopamine release and metabolism (to the measured metabolites) caused by acute exposure to METH. However, the lack of *any* reduction of the metabolites, as occurs in Parkinson's disease, does not support the notion of a substantial (e.g. 50%) loss of striatal dopamine innervation in the examined METH users.

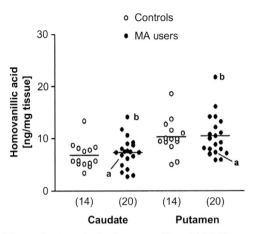

Figure 7 Striatal levels of dopamine metabolite homovanillic acid (HVA) are normal in autopsied brain of methamphetamine (METH) users. Levels of HVA are similar in caudate and putamen of 14 normal subjects and 20 methamphetamine users *(Figure reproduced with permission from Moszczynska et al. (2004).)*. Note that HVA concentrations are normal even in two subjects ("a" and "b") who had near-total loss of dopamine.

Table 1 Levels of Dopamine Metabolites are Markedly Decreased in Postmortem Striatum of Patients with Parkinson's Disease but are Normal in METH Users

	Dopamine	DOPAC	HVA	3-MT
Caudate				
Controls	7.03±0.50	0.41±0.05	6.78±0.66	1.99±0.16
METH users	2.76±0.31★ (−61%)	0.36±0.04 (−12%)	7.33±0.65 (+8%)	1.96±0.16 (−2%)
Parkinson's disease group	1.24±0.27★ (−82%)	0.16±0.03 (−61%)	3.19±0.53 (−53%)	1.06±0.22 (−47%)
Putamen				
Controls	7.61±0.61	0.52±0.07	10.3±0.09	2.51±0.19
METH users	3.78±0.47★ (−50%)	0.44±0.05 (−15%)	10.5±0.9 (+2%)	2.40±0.15 (+4%)
Parkinson's disease group	0.21±0.05★ (−97%)	0.07±0.01★ (−87%)	3.09±0.43★ (−70%)	0.35±0.08★ (−86%)

Data (ng/mg tissue; mean, SEM) are derived from 14 normal subjects, 20 METH users, and 12 patients with Parkinson's disease. Mean levels of dopamine are low in both groups whereas levels of dopamine metabolites 3,4-dihydroxyphenylacetic acid (DOPAC), homovanillic acid (HVA), and 3-methoxytyramine (3-MT) are low only in the patients with Parkinson's disease. Because of increased turnover of dopamine in a dopamine deficient state, dopamine metabolite concentrations probably underestimate any loss of dopamine neurones. However, normal dopamine metabolite concentrations suggest that there might be no substantial loss of dopamine neurones in the examined METH users (★$P<0.05$). *Source:* Reproduced with permission from Moszczynska et al. (2004).

Animal findings (Segal et al., 2005; [see Figure 5]; Krasnova et al., 2010) of an absence, 14–30 days following the last exposure to METH, of any reduction of two striatal dopamine metabolites (dihydroxyphenylacetic acid, homovanillic acid), but decreased dopamine are similar to our human data (Moszczynska et al., 2004). The lack of any decrease in striatal dopamine metabolite concentration in the animal studies suggests that any loss of dopamine innervation is unlikely to be substantial in these rat model studies as well.

2.1.3. Striatal Dopamine Biosynthetic Enzymes are Normal or Only Slightly Decreased in Postmortem Brain of Methamphetamine Users: Further Evidence against Substantial Dopamine Neuron Loss?

We previously reported a severe loss of protein levels of the dopamine marker biosynthetic enzyme tyrosine hydroxylase (−96%) and the enzyme dopa decarboxylase (−89%) in autopsied putamen of patients with Parkinson's disease (Zhong et al., 1995). Although dopa decarboxylase is not specific to dopamine neurons, being present in other monoamine neurotransmitter neurons, animal data suggest that it is preferentially localized to dopamine neurons in the dopamine-rich striatum (see Zhong et al., 1995). Other data suggest that dopa decarboxylase is probably upregulated somewhat in a state of partial dopamine neuron loss, consistent with our finding of a slightly less severe reduction in Parkinson's disease striatum of dopa decarboxylase vs that of tyrosine hydroxylase (see Kish et al., 1995).

As shown in Table 2, in postmortem brain of METH users in our Toronto study, striatal tyrosine hydroxylase protein was normal (putamen) to slightly decreased (by 22%) in caudate and nucleus accumbens whereas protein concentration of dopa decarboxylase was nonsignificantly decreased by 13% in caudate and nucleus accumbens (no reduction in putamen) (Wilson et al., 1996a).

Similarly, Kitamura et al. (2007) reported normal tyrosine hydroxylase protein (by immunohistochemistry) in caudate and putamen but decreased (by 53%) protein levels in nucleus accumbens of their METH case material (Figure 8).

The generally modest changes in the two enzyme dopaminergic markers in most of the striatal subdivisions do not support the possibility of a substantial loss of striatal dopaminergic innervation in the examined METH users.

2.1.4. Lower Striatal Dopamine Transporter Levels in Methamphetamine Users: A Finding Replicated in a Postmortem Brain Investigation and in Numerous Imaging Studies: Transporter Downregulation or Neuronal Loss?

A mean decrease (vs controls) in striatal levels (protein, radioligand binding) of the dopamine transporter is the most consistently reported neurochemical finding in brain of human METH users.

Mean striatal dopamine transporter reduction in postmortem (2 studies) and in brain imaging (6 studies) independent investigations of METH users.

In order to establish as conclusively as possible whether dopamine transporter protein levels might be different in autopsied striatum of METH users, we measured estimates of transporter concentrations in homogenates using two different approaches: a Western

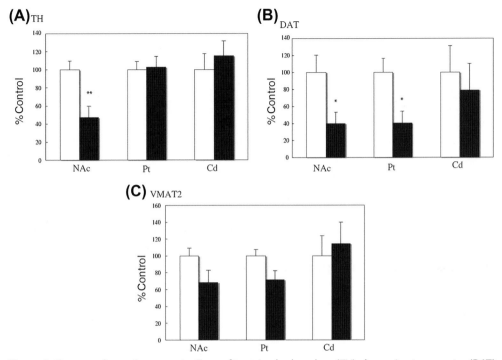

Figure 8 Decreased protein concentrations of tyrosine hydroxylase (TH), dopamine transporter (DAT) and vesicular monoamine transporter 2 (VMAT2) in postmortem striatum of human methamphetamine users: The Kitamura study. Proteins (mean, SEM) were measured by immunohistochemistry in 11 methamphetamine users (black bars) and 11 normal subjects (white bars) in nucleus accumbens (NAc), putamen (Pt) and caudate (Cd) striatal subdivisons *(Reproduced with permission from Kitamura et al. (2007).).* The pattern of dopamine marker changes is similar to that observed by Wilson et al., 1996a (see Tables 2 and 3) (*, **, *P* < 0.05, 0.01).

blot procedure, in which the molecular mass of the protein could be confirmed, and also by radioligand binding technique (at saturation) employing a tritiated compound (WIN 35,428). This compound is also used, when ^{11}C-labelled, for transporter measurements in PET brain imaging studies.

As shown in Table 3, in our postmortem brain investigation of 12 METH users (Wilson et al., 1996a) the magnitude of the dopamine transporter reduction ranged from approximately 25–50% in the different striatal subdivisions with the caudate difference not significantly different (as assessed using WIN 35,428).

In the only other postmortem brain investigation (11 METH users examined—all positive for METH in hair), Kitamura and colleagues (2007) reported almost identical results to ours, with reductions in dopamine transporter protein (by immunohistochemistry) in nucleus accumbens and putamen (−59 to −60%) with the decrease in caudate (−21%) not reaching statistical significance (see Figure 8). (Interestingly, whereas, as

Table 2 Striatal Levels of Dopamine Metabolic Enzymes in Autopsied Brain of Methamphetamine (METH) Users ($n=12$) and Control Subjects ($n=9–11$)

	Caudate	Putamen	Nucleus accumbens
dopamine			
Control	6.62 ± 0.55	7.16 ± 0.57	2.44 ± 0.27
METH	2.97 ± 0.43★★	3.49 ± 0.42★★	1.48 ± 0.17★
Noradrenaline			
Control	0.13 ± 0.05	0.13 ± 0.03	1.38 ± 0.45
METH	0.27 ± 0.06	0.12 ± 0.02	1.23 ± 0.28
Serotonin			
Control	0.35 ± 0.07	0.34 ± 0.07	0.41 ± 0.05
METH	0.26 ± 0.05	0.35 ± 0.09	0.37 ± 0.09
TH			
Control	1.08 ± 0.09	1.03 ± 0.05	1.37 ± 0.05
METH	0.84 ± 0.08	1.01 ± 0.05	1.07 ± 0.06★
DDC			
Control	1.52 ± 0.10	1.00 ± 0.06	1.29 ± 0.05
METH	1.32 ± 0.07	0.99 ± 0.06	1.12 ± 0.08

Note that whereas striatal dopamine levels (mean, SEM; ng/mg tissue) are markedly decreased in the drug users, protein concentrations of dopamine marker enzymes tyrosine hydroxylase (TH, ng tissue standard/ng protein) and dopa decarboxylase (DDC, ng tissue standard/ng protein) are normal or only modestly affected, unlike the severe reduction of the enzymes observed in Parkinson's disease (★$P<0.01$; ★★$P<0.001$)
Source: Reproduced with permission from Wilson et al. (1996a).

Table 3 Striatal Levels of Dopamine Transporter (DAT), but Not Vesicular Monoamine Transporter (VMAT2) are Decreased in Autopsied Brain of METH Users ($n=12$) and Control Subjects ($n=7–9$)

	Caudate	Putamen	Nucleus Accumbens
Dopamine transporter			
DAT protein			
Control	1.08 ± 0.10	0.66 ± 0.10	0.99 ± 0.12
METH	0.77 ± 0.11★	0.31 ± 0.05★★★	0.58 ± 0.09★★
[3H] WIN 35,428			
Control	2586 ± 309	2659 ± 160	2580 ± 299
METH	1937 ± 182	1836 ± 126★★★	1657 ± 176★★
[3H] GBR 12.935			
Control	1897 ± 324	1636 ± 147	n.e.
METH	1142 ± 192★	1082 ± 149★	n.e.
Vesicular monoamine transporter			
[3H] dihydrotetrabenazine			
Control	1955 ± 95	1850 ± 81	1404 ± 101
METH	1840 ± 150	1754 ± 59	1489 ± 129

Levels (mean, SE) of DAT, estimated by Western blotting (ng tissue standard/ng protein) and by WIN or GBR binding (B_{max}, fmol/mg protein) are decreased in drug users whereas those of VMAT2 (B_{max}, fmol/mg protein) are normal. Mean Kd values (nM) for striatal binding for the transporter measurements were not significantly different for the two groups (★,★★,★★★, $P<0.05, 0.01, 0.001$), n.e. (not estimated)
Source: Reproduced with permission from Wilson et al. (1996a).

mentioned above, the striatal dopamine decrease is most marked in the caudate [vs puta-men] of METH users, the reverse was true for the dopamine transporter, being more affected in putamen than in caudate, in the two postmortem studies.)

Results of most neuroimaging investigations of the dopamine transporter in METH users (Chou et al., 2007; McCann et al., 1998, 2008; Volkow et al., 2001a, b; Sekine et al., 2001; Johanson et al., 2006) are summarized and discussed in McCann and colleagues (2008). Reneman and colleagues (2002) also provide an interesting report in which users of MDMA (ecstasy) who also use amphetamine have lower striatal transporter binding than those who do not co-use amphetamine.

In the brain imaging investigations [PET and SPECT (single photon emission computed tomography)] employing subjects having short, up to very extended (e.g. 25 years) abstinence time, a modest mean striatal dopamine transporter binding reduction is typically observed, which ranges from approximately 10–25%. This magnitude is less than that observed in the two postmortem brain investigations but is similar to that reported in rats that have self-administered METH (Krasnova et al., 2010). Of interest is the report of a slight (13%) but statistically significant reduction of striatal dopamine transporter in a group of METH users reputedly abstinent for an average period of approximately 3 years (Johanson et al., 2006, Figure 9). This suggests that a transporter reduction might be quite persistent in some METH users.

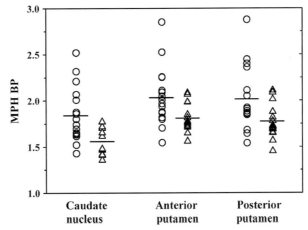

Figure 9 Decreased dopamine transporter binding in striatum of methamphetamine users in extended abstinence: A brain imaging study. The special feature of this investigation is that lower [11C] methylphenidate binding to the dopamine transporter was observed in methamphetamine users reputedly withdrawn from the drug on average for 3 years. This suggests that the change might be very long lasting *(Reproduced with permission from Johanson et al. (2006).).* However, note the almost complete overlap between drug user (triangles, *n* = 15) and control subjects (open circles, *n* = 16) and that the magnitude of the mean reduction is only slight. Thus, the authors suggest that the functional significance might be "questionable."

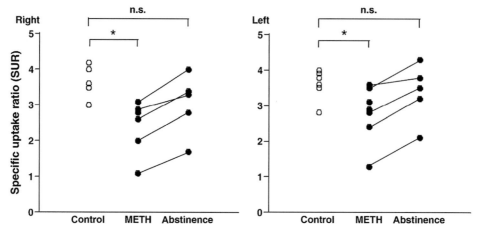

Figure 10 Striatal dopamine transporter binding partially normalizes in methamphetamine users after 2 Weeks abstinence: A brain scan study. This study shows lower striatal binding (using Tc-99m TRODAT and SPECT measurement) to the dopamine transporter in right and left striatum of seven methamphetamine (METH) users vs 7 controls and a partial normalization of values in the five drug users who had repeat scans following 2 weeks of abstinence. The preliminary data suggest that normalization of striatal dopamine transporter binding in methamphetamine users might be rapid in some users of the drug. *, P < 0.05; figure reproduced with permission from Chou et al. (2007).

The neurochemical profile of changes observed in striatum of rats allowed to self-administer METH, and sacrificed 14 days following last exposure (Krasnova et al., 2010), is similar to that observed in our postmortem brain investigation (low dopamine and dopamine transporter but normal dopamine metabolites homovanillic acid and dihydroxyphenylacetic acid). This suggests that METH exposure (vs a preexisting characteristic) offers a reasonable explanation for the low transporter levels in the human users of the drug.

Preliminary findings (consistent with animal data; see Figure 5) in two longitudinal studies (Volkow et al., 2001b; Chou et al., 2007) suggest that transporter binding might "recover" to some extent in abstinence, perhaps as early as 2 weeks following the last exposure to the drug (Figure 10).

It is difficult to establish whether the dopamine transporter findings in METH users are specific to METH as striatal transporter measurements in users of the dopaminergic stimulant cocaine have been variably reported to be normal (Staley et al., 1993; Volkow et al., 1996), elevated (Little et al., 1993; Staley et al., 1994; Malison et al., 1998) or decreased (Hurd and Herkenham, 1993; Wilson et al., 1996c).

There is an absence of drug hair data in brain imaging studies.

A deficiency of the dopamine transporter investigations, which is generic to all studies of METH users, is the inability to control for past use of other drugs. In this regard, no hair analysis data were provided indicating chronic use of METH or lack of use, or co-use, of other CNS-acting abused drugs, especially cocaine, in any of the dopamine transporter brain imaging investigations. Further, what evidence could realistically be provided that, for example, the subject who had a reputed abstinence time of 25 years (McCann et al., 2008) actually abstained from METH for the entirety of this lengthy period of time? Nevertheless, the consistency of the transporter reduction in the different postmortem brain and brain imaging studies, and the observation that the striatal dopamine transporter is decreased in animal models of METH exposure (see Krasnova and Cadet, 2009) do suggest that the finding is probably "real." This does not mean that striatal dopamine transporter levels are likely to be low in all METH users, but that the general finding has been sufficiently replicated such that future investigations of representative groups of chronic users of the drug will likely find a mean group difference as compared to matched controls (see also below).

Scientists still argue over the reason why the striatal dopamine transporter is low.

Because dopamine transporter concentrations can up- and downregulate independently of any change in density of dopaminergic neuropil (e.g. Wilson et al., 1994), low dopamine transporter concentration could represent a "neuroadaptive" change consequent to drug exposure in intact dopamine neurons (to maximize dopamine levels at the synapse) or an actual loss of dopamine nerve endings or portions thereof that contain the transporter. The latter is the position of some scientists who assume (perhaps not incorrectly) that if reduced level of a dopamine marker is very long-lasting (e.g. years), it is unlikely that this represents a "neuroadaptation" phenomenon.

Employing Ricaurte's definition of METH "neurotoxicity" (Section 1.3.4), how could one prove that a very persistent reduction of the striatal dopamine transporter in living brain of a METH user could not possibly be explained by a neuroadaptive response?

Does low striatal dopamine transporter have any functional consequence?

A stronger case for functional impairment can be made if drug users have values outside the range of the controls. Along these lines, Hart and colleagues (2012) suspect extensive overlap between control and drug user values in the McCann et al., 1998 study and reasonably question "…whether an approximately 20% difference in DAT density, as measure[d] with conventional PET-imaging techniques is within the normal range of human variability…"

In brain imaging studies in which scatterplots have been provided, overlap in transporter values between METH user and control groups was complete/almost complete in two studies (Johansen et al., 2006; McCann et al., 2008) and partial in one study (Volkow et al., 2001c) in which about 60% of the METH users had values below the lower limit of the control range in putamen, suggesting that the difference in individual studies might not always be robust. However, the Sekine et al. (2001) investigation reported almost complete separation in transporter values (in combined caudate/putamen) between the two groups, a finding more suggestive of a functionally significant change. Further, in our postmortem brain investigation, most dopamine transporter values in putamen fell below the lower limit of the control range (Wilson et al., 1996a;

Kish, unpublished observations). Perhaps the more robust difference in the Sekine and Toronto investigations could be explained by the possibility of higher doses taken by the drug users in these studies.

A functionally significant consequence of dopamine transporter loss depends in part on whether the change is limited only to the transporter (in which it might be expected that any released dopamine would remain longer at the dopamine receptor) or is representative of an actual loss of dopamine nerve endings. Assuming that low transporter levels in METH users are actually reflective of damage to dopamine neurons, brain imaging findings in different studies (Guttman et al., 1997; Lee et al., 2000; Booij et al., 2001) suggest that a loss of about 50–70% of putamen dopamine transporter binding (as estimated by PET/SPECT imaging) is required for the emergence of Parkinsonian features. Using this threshold, only a small subgroup of one to two subjects in the different imaging investigations (and in our postmortem brain investigation employing the same probe) would meet the 50% criteria in caudate/putamen. Thus, the available brain imaging studies suggest that this threshold is not typically met in METH users. In one investigation, correlations were observed between striatal dopamine transporter binding and performance on several motor and memory tasks (Volkow et al., 2001c); however, in a small subgroup of subjects examined during later abstinence, although transporter levels partially recovered, there was not parallel improvement in test performance (Volkow et al., 2001b). This raises further uncertainty as to the clinical significance of transporter changes in METH users.

2.1.5. VMAT2, the Reputed Gold Standard Dopamine Neuronal Marker, Fails to Settle the Question of Dopamine Neuron Damage in Methamphetamine Users

Because levels of the dopamine transporter and other dopamine neuronal markers can change without any change in dopamine neuronal density (e.g. Wilson et al., 1994), we and the Ann Arbor group (Frey et al., 1996, 2001) began a search for a more "stable" marker and concluded that VMAT2 might have the desired features. Although VMAT2 is not specific for dopamine neurons, as it transports monoamine neurotransmitters serotonin and noradrenaline, as well as dopamine, from the neuronal cytoplasm into storage vesicles, most VMAT2 in the mammalian dopamine-rich striatum is localized to dopamine neurons (see Wilson et al., 1996b; Frey et al., 2001).

Striatal VMAT2, decreased in Parkinson's disease and in animal models of high dose METH exposure, is considered to be a "stable" striatal dopamine neuronal marker.

As expected, we and others showed that striatal VMAT2 binding (using radiolabelled tetrabenazine analogues) is decreased in postmortem (Wilson et al., 1996b) and living striatum (see Frey et al., 2001; Lee et al., 2000) of patients with Parkinson's disease and is also decreased in striatum of experimental animals exposed to a high (but not low) dose of METH (Frey et al., 1997). However, in one investigation striatal VMAT2 binding was also reported to be decreased in nonhuman primates exposed to presumably lower doses of amphetamine "similar" to those used for the treatment of adult ADHD (Ricaurte et al., 2005).

In experimental animals exposed to cocaine (but see below for results of human studies) or to other dopamine-related drugs that do not appear to damage dopamine neurones, striatal VMAT2 binding did not change (Naudon et al., 1994; Vander Borght et al., 1995; Kilbourn et al., 1996; Wilson and Kish, 1996; Frey et al., 1997), suggesting resistance of VMAT2 to drug-induced up- and downregulations. For these reasons, VMAT2 has been considered by the scientific community to be a useful and stable striatal dopamine neuronal marker that could be used to establish whether METH might damage dopamine neurones in humans (e.g. Lee et al., 2000).

Striatal VMAT2 in postmortem brain studies: Normal in METH users but decreased in users of cocaine?

In our original 1996 postmortem brain study, we found that although concentrations of dopamine and those of the dopamine transporter were decreased in striatum of METH users, mean VMAT2 binding at saturation (B_{max}) was strikingly normal (Wilson et al., 1996a; see Table 3). However, analysis of the individual values disclosed one subject who had a marked (by 57%) reduction (outside of control range) in binding in the nucleus accumbens striatal subdivision (Kish, unpublished observations). In the second autopsied brain study Kitamura and colleagues (2007) confirmed lack of a statistically significant mean reduction in striatal VMAT2 (by immunohistochemistry) in (dopamine transporter-deficient) METH users (although a trend for reduction was observed in nucleus accumbens and putamen) (Figure 8) but did report a severe (>90%) decrease in protein immunostaining in the nucleus accumbens striatal subdivision of two cases, similar to our finding in the Toronto study.

In order to establish specificity of the lack of VMAT2 change in METH users, striatal VMAT2 was also measured in autopsied striatum of cocaine users.

Since animal studies suggest that cocaine, unlike METH, is not neurotoxic to dopamine neurons (Wilson and Kish, 1996; Ryan et al., 1988), it was expected that VMAT2 would be normal in brain of human cocaine users. Somewhat surprisingly (and still unexplained), however, are the results of two postmortem brain investigations finding slightly decreased VMAT2 levels (binding and/or protein) (Wilson et al., 1996c; Little et al., 2003) in striatum of cocaine users but with one study reporting normal concentrations of the transporter (Staley et al., 1997). In a more recent investigation, Little's group (Little et al., 2009) also reported slightly lower number of anterior midbrain dopamine cell body estimates (employing melanin as a marker) in autopsied brain of cocaine users (Figure 11). Thus, the quite paradoxical (but still preliminary) findings from the postmortem stimulant studies to date are that striatal VMAT2 is generally normal in humans exposed to METH (see also below), a stimulant considered to damage brain dopamine neurons at high doses, whereas VMAT2 might be decreased in users of cocaine, for which animal data suggest is *not* neurotoxic (Ryan et al., 1988; Thomas et al., 2004a).

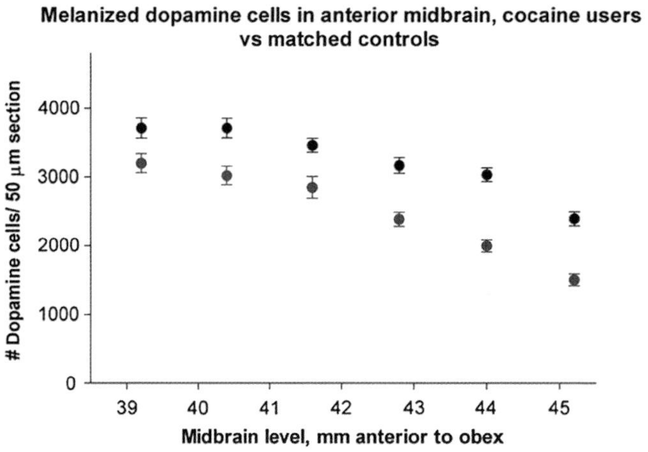

Figure 11 Density of melanized cell bodies is decreased in postmortem midbrain of human cocaine users. Although animal data suggest that cocaine is not toxic to dopamine neurones, the question in the human is not yet resolved. This preliminary, and rarely cited, study shows that the density of melanized, presumably dopaminergic, cell bodies along the anterior–posterior axis of the anterior midbrain are lower in cocaine users ($n = 10$; red circles) than in control subjects ($n = 9$; black circles) *(Reproduced with permission from Little et al. (2009).).* Cells were counted in 50 mm thick sections every 400 mm over a total of 18 levels. See Little et al. (2009) for details. The findings are consistent with the unexpected possibility that cocaine exposure might damage dopamine neurones in the human. Unfortunately, such quantitative cell counting data are not yet available for human methamphetamine users.

Striatal VMAT2 binding is slightly (by 10–11%) decreased in living brain following long METH abstinence: Schuster brain imaging study.

Given the generic uncertainties of postmortem brain investigations, it was expected that a brain PET study of VMAT2 binding in living abstinent METH users (for which much higher quality and verified information on drug use are available) would resolve the question of METH and VMAT2 in human brain. This brain imaging study, published by the Schuster group (Johanson et al., 2006), reported only a slight but statistically significant (10–11%) reduction in striatal VMAT2 binding in abstinent METH users (Figure 12).

Several key issues of the Schuster study were the absence of any forensic data proving that the subjects had used METH or other drugs that might have influenced VMAT2 and (as pointed out by the authors) the very long period of time (mean 3.4 years) between reputed last use of drug and brain scan for which no verification of long abstinence was provided. Further, the long abstinence time could have permitted marked, perhaps near total recovery of any damage (as suggested by imaging findings of Volkow et al., 2001b; Chou et al., 2007). However, the imaging data (assuming drug users actually maintained the lengthy abstinence) did suggest possibility of a longlasting but very modest reduction of a reputedly stable dopamine nerve terminal integrity marker and associated neurotoxicity. Following this logic, we in Toronto were "convinced" that if a 10% striatal VMAT2 reduction is actually present 3 years following last use of METH, a much more

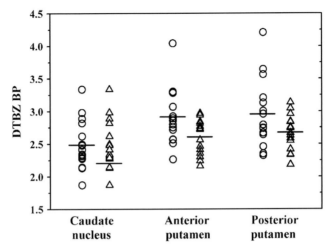

Figure 12 Decreased vesicular monoamine transporter 2 (VMAT2) binding in striatum of methamphetamine users in extended abstinence: A brain imaging study. Lower [11C]DTBZ binding to VMAT2 was observed in methamphetamine users reputedly withdrawn from the drug on average for 3 years, suggesting that the change might be very long lasting *(Reproduced with permission from Johanson et al. (2006).)*. However, note the almost complete overlap between drug user (triangles, *n* = 15) and control subjects (open circles, *n* = 16) and that, as with the authors's dopamine transporter finding (Figure 9), the magnitude of the mean reduction is only slight. If striatal VMAT2 is slightly decreased in methamphetamine users 3 years since last use of drug, what would the reader expect to find in drug users in early abstinence?

substantial decrease would have occurred in early abstinence. Our unpublished observations in an ongoing imaging study confirm the finding of a mean increase (vs. controls) in striatal VMAT2 binding in METH users during early abstinence.

Striatal VMAT2 binding is not decreased (and is slightly elevated!) in very early METH abstinence: preliminary brain imaging findings.

To address the question of abstinence time and confirmation of drug use arising from the Schuster study, we recruited METH users (who also used other drugs) who were confirmed by drug analysis to have been using METH and who were in much earlier abstinence (days to months) than in the Schuster investigation.

Contrary to our hypothesis, mean VMAT2 binding in the METH group was *increased* by 14% in the whole striatum, with the caudate nucleus subdivision showing the greatest increase (by 22%), (Boileau et al., 2008). None of the METH users (but all three patients with Parkinson's disease, the positive control) had low VMAT2 values or values below the lower limit of the control range. Examination of the relationship between reported abstinence time and transporter binding revealed that the increase was primarily limited to the METH users who were recently (1–3 days) withdrawn from the drug (Figure 13).

We believe that increased striatal VMAT2 binding in the METH users is related to the possibility that binding of the VMAT2 radiotracer (^{11}C–dihydrotetrabenazine; DTBZ) is sensitive to endogenous (intravesicular) dopamine in which both the radiotracer and

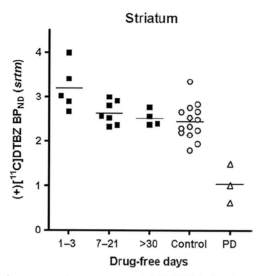

Figure 13 Striatal vesicular monoamine transporter 2 (VMAT2) binding is increased in early metham-phetamine abstinence: A preliminary brain imaging study. Based on animal data (e.g. Figure 5) and the Johanson brain PET imaging findings (Figure 12), we had expected that striatal VMAT2 binding (using [11C]DTBZ) would be markedly decreased in methamphetamine users (solid squares) as com-pared to that in control subject (open circles) during early abstinence. Striatal binding was lower in brain of patients with Parkinson's disease (PD, triangles), the positive disease control. However, mean values were either normal or slightly elevated in very early methamphetamine abstinence (1–3 days) *(Figure reproduced with permission from Boileau et al. (2008).)*. We believe that this finding is explained by decreased competition from dopamine (in the dopamine-deficient drug users) at the DTBZ binding site. If correct, this provides further evidence for decreased (vesicular) dopamine in methamphetamine users but also suggests that differences in vesicular dopamine levels might introduce a confound in interpretation of brain imaging VMAT2 binding studies.

dopamine are competing for the same site on VMAT2. In this scenario, *increased* DTBZ binding to VMAT2 could be explained by *decreased* intravesicular dopamine in METH users (as suggested by our postmortem brain finding of a severe striatal dopamine reduction in some METH users (Wilson et al., 1996a; Moszyznska et al., 2004, Figure 6)) competing for binding at the VMAT2 site (see Boileau et al., 2008; Tong et al., 2008 for discussion). The finding that VMAT2 binding elevation occurred in very early, but not later, abstinence is consistent with the possibility that low striatal dopamine is primarily a consequence of the acute action of METH in causing excessive release and depletion of tissue stores of dopamine, with some dopamine replenishment occurring in later abstinence.

The notion that VMAT2 binding is sensitive to changes in intraneuronal dopamine concentration is suggested by the (under-appreciated) brain imaging finding of increased striatal VMAT2 binding in patients having a metabolic, i.e. not structural, brain dopa-mine deficiency (dopa responsive dystonia; De La Fuente-Fernandez et al., 2003) and by our animal findings showing that striatal VMAT2 binding (in which a low tracer dose of 11C-dihydrotetrabenazine was injected intravenously) can be elevated by drugs

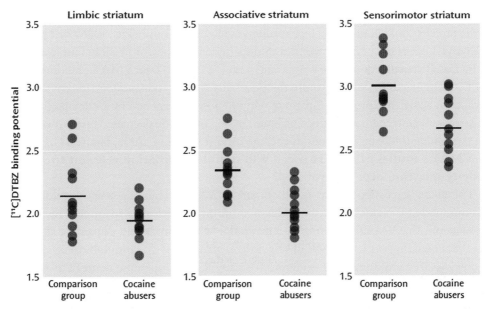

Figure 14 Striatal VMAT2 binding is slightly decreased in brain of human cocaine users: A PET brain imaging study. Because animal data suggest that cocaine, unlike methamphetamine, is not toxic to brain dopamine neurones, it would be predicted that striatal VMAT2 levels should be normal in human cocaine users. However, this brain imaging study by Narendran and colleagues (2012) finds slightly decreased striatal [^{11}C]DTBZ VMAT2 binding in "recently" abstinent cocaine users, a finding consistent with our own postmortem brain observation (Wilson et al., 1996c). The VMAT2 cocaine literature is now becoming more consistent than the methamphetamine literature. Does low VMAT2 in cocaine users mean loss of VMAT2, VMAT2-containing vesicles, or actual loss of entire VMAT2-containing dopamine neurones (see Figure 11)? *Figure reproduced by permission of the American Psychiatric Association.*

decreasing brain dopamine levels (amphetamine, biosynthetic enzyme inhibitor) and decreased by a dopamine elevating drug (levodopa), (Tong et al., 2008). [Extending the logic, we have proposed that VMAT2 binding assessed by brain imaging might provide an index of changes in vesicular dopamine levels.]

Notwithstanding contrary animal findings, striatal VMAT2 binding is slightly decreased in striatum of abstinent cocaine users: Preliminary brain imaging findings.

As mentioned above, animal data suggest that cocaine, unlike METH, is not toxic to brain dopamine neurones. To address this question in living human brain, Narendran and colleagues (2012) measured VMAT2 (using [^{11}-C]DTBZ) binding by PET imaging in recently abstinent (for at least 14 days) chronic cocaine users. In contradistinction to results of the Boileau PET study of METH users, VMAT2 binding was slightly (by approximately 10–16%) decreased in striatum of the cocaine group. A scatterplot disclosed marked overlap in control and drug user values in limbic striatum whereas about one-half of the cocaine user values fell below the lower limit of the control range in associative and sensorimotor striatum (Figure 14).

Assuming subnormal VMAT2 binding might be caused by cocaine exposure, low levels of the protein could in principle be explained by downregulation/loss of VMAT2 protein, or VMAT2-containing synaptic vesicles or dopamine neurones, or possibly (but less likely) by increased vesicular dopamine competing with [^{11}-C]DTBZ for VMAT2 binding. Here, the literature needs to consider more the possibility that the dopamine marker is reduced because of actual, but modest, dopamine neurone loss as suggested by Little's preliminary (but generally unrecognized) autopsied brain finding of slightly decreased number of melanin containing midbrain neurones in cocaine users (Little et al., 2009, Figure 11).

VMAT2 studies of METH users: Will useful information relevant to neurotoxicity question ever be provided?

The above data suggest that results of striatal VMAT2 binding *imaging* studies will be confounded to some extent by differences in brain levels of dopamine. This means, for example, that VMAT2 binding in a drug user who is markedly dopamine-deficient (e.g. because of recent use of METH or downregulation of dopamine biosynthesis due to chronic drug exposure) will underestimate any loss of dopamine neurons as inferred from VMAT2 binding.

The two *postmortem brain* studies of METH users employed procedures in which the maximal binding (B_{max}; Wilson et al., 1996a) or protein levels of VMAT2 (Kitamura et al., 2007) would *not* be influenced by changes in dopamine levels and show, generally, normal striatal levels of the rtransporter, although several of the subjects did demonstrate low values in nucleus accumbens. The autopsied brain VMAT2 findings suggest that a substantial loss of dopamine neurons is unlikely to be a typical characteristic of recreational METH exposure.

2.1.6. Slightly Decreased Dopamine Receptor (D2) Binding in Methamphetamine Users: What Does It Mean and Why is Binding Decreased?

Investigation of striatal dopamine receptor status in METH users is relevant, in part, because of animal data suggesting that METH might damage dopamine receptor-containing neurons in striatum (see Krasnova and Cadet, 2009). In addition, should striatal dopamine receptor changes underlie aspects of METH addiction, as have been suggested (Volkow et al., 2001a; Wang et al., 2011), such an abnormality can be considered "pathological" (the focus of this review) irrespective of whether the difference involves structural damage to neurons.

The dopamine receptor family: a primary focus on D2 and D3 receptors in human addiction.

Because of the known dopamine-releasing action of METH on dopamine neurons, there has been interest in establishing whether changes in concentration of several of the different dopamine receptors occur in the dopamine rich striatum and cell body areas of drug users. Whereas most investigation has historically involved a nondiscriminatory focus on the dopamine D2/D3 receptor, a new specific focus of interest is the D3 receptor, in large part because, unlike the D2 and D1 receptors which are distributed uniformly throughout the striatum, the D3 receptor is preferentially localized in limbic areas of the striatum considered to be involved in motivation and reward (Sokoloff et al., 1990; Murray et al., 1994). In addition, there is now an emerging literature showing that D3 selective antagonists can decrease drug-seeking behavior in animal models (reviewed by LeFoll et al., 2005; Heidbreder and Newman, 2010).

At present, in brain PET imaging studies, binding to the striatal D2 receptor is often assessed using radioligands such as ^{11}C-raclopride, which bind to both D2 and D3 dopamine receptors (Malmberg et al., 1993), whereas the radioligand [11]C-(+)-propyl-hexahydro-naptho-oxazin ([11]C-(+)-PHNO) is employed to measure binding to the D3 receptor as it has preferential affinity for the D3 (vs D2) receptor (see Boileau et al., 2009, 2012). Neither probe, however, is absolutely specific for the D2 or D3 receptor. Whereas ^{11}C-raclopride binding is assumed to measure D2 receptor levels in the striatum (as the density of D2 is greater than that of D3), ^{11}C-PHNO binding can be interpreted in a region-dependent manner in which binding in dorsal striatum (high D2 and low D3) most likely reflects D2 receptor binding whereas binding in the substantia nigra and hypothalamus (predominantly D3) mostly reflects D3 receptor levels. ^{11}C-PHNO binding in the ventral pallidum and globus pallidus, areas of mixed D2/D3 binding, has been estimated to represent 75% and 65% of the D3 fraction respectively (Tziortzi et al., 2011; see Boileau et al., 2012).

Four brain imaging investigations report slightly decreased binding to the striatal dopamine D2 receptor in METH users: A change common to users of all abused drugs?

In three brain imaging investigations mean dopamine D2 binding was reported to be decreased in striatum of METH users, but only slightly (Volkow et al., 2001a: −16%, −19% [see Figure 15]; Lee et al., 2009: −8 to −16%; Wang et al., 2011: −11% [caudate only]).

In our own investigation of polydrug methampethamine users employing ^{11}C-PHNO (which as mentioned above binds preferentially to D3 but also to D2 receptors) binding was slightly lower (by 11%) in the D2-rich dorsal striatum of heavier dose METH users (Boileau et al., 2012) consistent with the above observations. The results of our postmortem investigation of D2 receptor protein in METH users are somewhat consistent with the imaging findings as a trend for reduced D2 protein levels (by 24%) was observed in the nucleus accumbens striatal subdivision, but was normal in caudate and putamen (Worsley et al., 2000).

As noted by Volkow and colleagues (2001a), subnormal striatal D2 might not be specific to METH users as similar findings have been reported in users of cocaine, alcohol, and opiates (Volkow et al., 1996; see also Martinez et al., 2004, 2011, 2012). Similarly, we observed modestly lower dopamine D2 protein levels in postmortem nucleus accumbens of users of cocaine (−25%, nonsignificant trend) and of heroin (−37%), (Worsley et al., 2000).

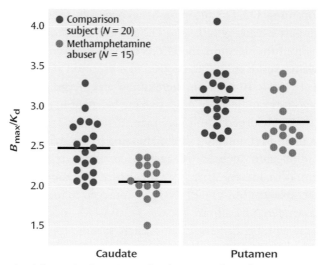

Figure 15 Lower striatal dopamine D2 receptor binding in methamphetamine users: A brain imaging study. This replicated investigation shows decreased striatal binding of [¹¹C]raclopride to the dopamine D2/D3 receptor in human methamphetamine users. The finding is also observed in rats exposed chronically to methamphetamine (Figure 5). What explains low D2 binding? Receptor downregulation? Neurotoxicity-related loss of D2-containing striatal neurones? *Figure reproduced with permission of the American Psychiatric Association, Volkow et al. (2001a).*

Can the slight D2 reduction have any functional consequence?

Assuming that the magnitude of the binding difference directly translates into the identical percentage loss of dopamine receptors, is it likely that a slight 15–25% loss of striatal dopamine D2 receptors will have any functional consequence? In this regard, it is not unreasonable to suggest that this magnitude of reduction may in fact be too small to have any consequence. Thus, results of a brain imaging occupancy study (Kapur et al., 2000) suggest that the threshold of dopamine D2 inactivation for some undesired effects of D2 receptor blocking drugs (e.g. Parkinsonism, prolactin elevation) in the human is approximately 50–60% receptor occupancy. However, subjects of the occupancy study were patients with schizophrenia, who may have had a hyperdopaminergic condition. The threshold for more subtle effects (e.g. related to motivation, hedonic status) does not appear to be known and may be less marked.

A suggestion of functional relevance of the D2 receptor differences is derived from the different imaging studies of potentially meaningful correlations between extent of binding reduction and different outcome measures: Thus, Volkow reports a positive correlation between D2 receptor binding and glucose metabolism in orbitofrontal cortex of the drug users and suggests that dopamine D2 receptor-mediated dysregulation of orbitofrontal cortex might underlie an aspect of compulsive drug taking (Volkow et al., 2001a). London finds a positive correlation between binding and a measure of impulsivity (Lee et al., 2009) and Fowler reports that the subgroup of METH users who relapsed following detoxification showed lower binding than those who did not relapse (Wang et al., 2011). Similarly, Martinez and coworkers (2011) find that cocaine users having lower striatal dopamine receptor binding respond less well to treatment.

All of the above associations do suggest that a D2 receptor reduction (as inferred from the binding differences between METH users and control subjects) might indeed have functional consequence in the drug users. However, the argument still has to be raised that a D2 reduction of such a slight (e.g. 20%) magnitude might not itself have functional consequence but be reflective only of the extent of drug exposure—with the latter correlating with a variety of more substantial brain changes that do have clinical consequence. Countering this argument is the possibility that the actual difference in binding to the radioligands employed (which bind to both D2 and D3 dopamine receptors) and which are susceptible to changes in dopamine levels could be underestimated because of increased D3 receptor expression (see Section 2.1.7) or decreased synaptic dopamine levels (see 2.1.8). However, our (preliminary) postmortem brain investigation of METH users, like the imaging studies, found only a modest reduction in a direct measure of D2 receptor protein. There will continue to be uncertainty on this question until the relationship between striatal binding changes in the in vivo imaging studies and actual receptor concentration is clarified.

Why is striatal D2 low in METH users: Downregulation vs loss of striatal neurones vs preexisting trait?

In principle, lower dopamine D2 receptor binding in the brain imaging studies (in which a single tracer dose concentration of the probe is employed) could be explained by actual differences in receptor protein in intact neurones, loss of striatal neurons containing the D2 receptor, changes in dopamine levels competing with the probe for the binding site, or by differences in receptor affinity for the probe.

Animal findings show that (intravenous) METH exposure can cause a modest reduction of striatal dopamine D2 receptor binding that recovers in about 1 month (Segal et al., 2005), an observation that could be explained by either receptor downregulation or striatal cell loss. The animal proof of concept observation suggests that at least part of the low D2 in the human drug users might be explained by METH exposure. However, low binding to the D2 receptor could also be explained, in principle, by a preexisting trait that might predispose some to develop compulsive drug wanting. In this regard, Lee and colleagues (2009) report a negative correlation between dopamine D2 receptor binding and cumulative METH exposure, but also a negative relationship between impulsivity and D2 binding in the (nonmethamphetamine using) control group. This allows for the possibility that lower D2 receptor number might be related, in part, to a preexisting trait that predisposes (impulsive) drug use, and, in part, by drug exposure.

No data are available in the human that would convincingly indicate that low striatal D2 receptor is more likely to be explained by a receptor downregulation (e.g. consequent to excessive METH-induced dopaminergic stimulation) or to a drug-induced loss of D2 receptor-containing striatal cells. However, findings of enlarged striatal volume in some drug users are consistent with the presence of a pathological process in striatum (see Section 2.5.4). In this regard, the human METH literature needs to consider much more the possibility that the slightly reduced dopamine D2 receptor binding might be explained by slight loss of D2-receptor-containing striatal neurons, in view of some animal findings suggesting that METH toxicity in mammalian brain might not be limited only to dopamine nerve endings and axons (Ryan et al., 1990; see Krasnova and Cadet, 2009) (see also Section 2.2.3).

2.1.7. Preliminary Observations on Striatal Dopamine D1 and D3 Receptors in Methamphetamine Users: Could Increased D3 Receptor Number Explain Drug Wanting?

Lower striatal dopamine D1 receptor-stimulated activity—related to drug tolerance? Only limited, preliminary information is available on the status of the brain dopamine D1 and D3 receptors in METH users.

In a postmortem study we found striatal dopamine D1 protein levels to be normal in caudate and putamen of METH users, but elevated (by 44%) in the nucleus accumbens (Worsley et al., 2000). However, in the same samples (maximal) dopamine-stimulated adenylyl cyclase, a potentially more relevant functional index of D1 receptor activity, was decreased (−25 to −30%) in all three striatal subdivisions, with about 25% of the METH users having values below the lower limit of the control range (Tong et al., 2003). As this finding has previously been reported in METH animal studies (see Segal et al., 2005; Tong et al., 2003), decreased dopamine D1 receptor-linked function could have been consequent to use of METH and, speculatively, related to aspects of drug tolerance. The discrepancy between estimates of D1 receptor number and a measure of function in postmortem brain of the METH users suggests caution in predicting functional changes based on differences only in receptor concentration.

Higher brain dopamine D3 receptor binding—related to drug wanting?

As mentioned above (Section 2.1.6), the finding of a preferential localization of the dopamine D3 receptor to limbic portions of the striatum, considered to be involved in motivation and reward, has prompted investigation of the behavior of the D3 receptor in stimulant conditions.

In a brain imaging investigation of chronic polydrug METH users, employing the dopamine D3-preferring ligand ^{11}C-PHNO, binding was increased in several D3-rich areas of the basal ganglia (substantia nigra, ventral pallidum, globus pallidus) with the change in the D3-rich substantia nigra (+46%) being statistically significant (Boileau et al., 2012). One half of the METH users had values in substantia nigra higher than the upper limit of the control range (Figure 16). This finding, requiring independent replication, might be a

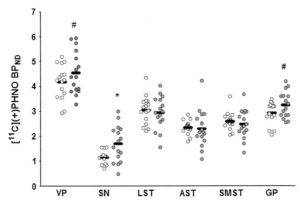

Figure 16 Higher binding of the dopamine D3-preferring ligand [^{11}C]PHNO in methamphetamine users: A brain imaging study. Although [^{11}C]PHNO binds to both dopamine D2 and D3 receptors, the PET imaging probe has preferential affinity for the D3 receptor. Binding was higher in the D3-rich substantia nigra (SN) of the methamphetamine users (gray circles) vs controls (white circles) with a trend for a slight increase in the globus pallidus (GP) and ventral pallidum (VP) *(Reproduced with permission from Boileau et al. (2012).)*. As PHNO binding was related to self-reported "drug wanting" could upregulated dopamine D3 receptor be involved in aspects of drug addiction? (LST, limbic striatum; SMST, sensorimotor striatum; AST, associative striatum; *$P < 0.05$ corrected, #$P < 0.1$ corrected).

feature of use of other abused drugs as increased D3 binding has been previously reported in postmortem brain of human cocaine users (Staley and Mash, 1996; Mash, 1997).

The clinical relevance of higher dopamine D3 binding is speculative as the function of this dopamine receptor subtype in human brain is still unknown. In animals, increased dopamine D3 receptor level has been associated with occurrence of locomotor sensitization to dopamine elevating drugs in dopamine–deficient rats (Bordet et al., 1997, 2000). In this respect, locomotor and dopaminergic sensitization to the behavioral and neurochemical effects of reexposure is one of the most replicated findings in the preclinical addiction literature and has been used to explain increased drug seeking in animals (see Robinson and Berridge, 1993); however, the question whether sensitization (reverse tolerance) actually occurs in human METH users and could explain the ever increasing desire for drug use, or psychosis is still uncertain and debated (see Angrist, 1994; Boileau et al., 2006).

Further, can a D3 increase in METH users be considered a beneficial adaptive response or a pathological consequence of repeated drug-taking (see Boileau et al., 2012 for discussion)? In this regard, Boileau found D3 binding in the drug users to be positively correlated with self-reported "drug wanting" following a priming dose of amphetamine. This question, as well as the issue whether the D3 receptor is in fact one of the substrates for the long-speculated sensitization phenomenon in drug users will be best addressed in clinical investigations of specific D3 agonists and antagonists in stimulant drug users.

2.1.8. Decreased Stimulant-Induced Synaptic Dopamine Release (as Inferred from D2 Receptor Binding Differences) in Methamphetamine Users: Preliminary Observations

Stimulant drugs such as METH (dopamine releaser and uptake blocker) and methylphenidate (dopamine transporter blocker) increase levels of dopamine at the dopamine receptor. By employing a radiotracer that binds to the striatal dopamine D2 receptor (e.g. ^{11}C-raclopride), which competes for dopamine at the D2 receptor, the binding can be compared before and after administration of the stimulant drug, providing a semiquantitative estimate-index of the extent of dopamine increase (Farde et al., 1985; see Cardenas et al., 2004). Thus, stimulant-induced changes in D2 receptor binding are, as expected, decreased in brain of patients with the degenerative dopamine deficiency condition Parkinson's disease (Piccini et al., 2003), whereas the change in binding is greater than normal (presumably reflecting a more marked increase in dopamine release) in some patients with schizophrenia (Laruelle et al., 1996).

In a brain imaging investigation, acute methylphenidate challenge induced a smaller than normal reduction in ^{11}C-raclopride binding in some METH users but with statistically significant differences limited only to the left putamen (Wang et al., 2011). Interestingly, this difference was restricted to the subgroup of drug users who did not complete detoxification, prompting the authors to suggest that low dopamine D2 receptor status might represent a biomarker predictive of relapse during treatment.

This blunting of a stimulant response to changes in striatal D2 receptor binding has also been reported in cocaine users (Volkow et al., 1997; Martinez et al., 2007) allowing for the possibility that the blunted response might be a general consequence of chronic dopaminergic (over-) stimulation or possibly a preexisting trait. However, we might have perhaps expected a more pronounced blunted response in methamphetamine users as compared with that reported for cocaine users based on our postmortem brain finding of a much more marked dopamine depletion in the former group (see Section 2.1.1). The blunted response also suggests that "exaggerated dopaminergic transmission" specifically associated with overactive dopamine release might not be a feature of chronic METH use as could be the case in patients with schizophrenia (Laruelle et al., 1996).

This interesting study, however, is preliminary (small sample size; no scatterplots provided), regionally marginal (as significant changes were limited to only to the left side portion of the putamen) and requires replication. Nevertheless, the findings are interesting as extension of this investigation might ultimately provide a biomarker predictive of treatment relapse.

2.2. NONDOPAMINERGIC CHANGES IN BRAIN OF METHAMPHETAMINE USERS

Animal data show that exposure to METH/amphetamine can cause neuronal damage to nondopaminergic neurones (serotonergic) and alter levels of striatal neuropeptide markers. Do these changes occur in brain of human METH users?

2.2.1. Serotonin Markers: Modestly or Markedly Decreased in Methamphetamine Users?

Variable and generally modest reduction of serotonin markers in postmortem brain of METH users.

As animal data suggest that METH causes release of serotonin (as well as dopamine and noradrenaline) in mammalian brain (Rothman et al., 2001) and might damage serotonin axons and nerve endings (see McCann and Ricaurte, 2004), we measured several serotonin markers (serotonin, metabolite 5-hydroxyindoleacetic acid, rate-limiting biosynthetic enzyme tryptophan hydroxyolase, serotonin transporter) in postmortem brain of 16 METH users who had tested positive for METH but negative for other drugs of abuse in brain, blood (including ethanol) and, where available, scalp hair (Kish et al., 2009). As part of this investigation we first showed that levels of the markers were, as expected, low in postmortem brain of patients with Parkinson's disease, a positive disease control (Kish et al., 2008), and in brain of a heavy dose user of ecstasy (3,4-methylenedioxymethamphetamine, MDMA) a drug shown to cause reduction of brain serotonin markers in experimental animal studies (see Kish et al., 2000, 2010a and references therein).

Unlike the marked reduction of striatal dopamine in the same METH users, only a mild and statistically nonsignificant reduction of striatal serotonin was observed, with a trend for reduction (by 33%) of the serotonin transporter protein in caudate (Kish et al., 2009). Protein levels of tryptophan hydroxylase were slightly (-29%) but significantly decreased in caudate (normal in putamen). In contrast to the modest changes in striatum, levels of serotonin and its metabolite 5-hydroxyindoleacetic acid were more markedly decreased in orbitofrontal cerebral cortices (Brodmann areas 11 and 12) of the drug users. The observation of marked dopamine with only modest serotonin marker differences is consistent with some reports of decreased striatal dopamine, but not serotonin markers, in nonhuman primates exposed to METH or amphetamine (Ricaurte et al., 2005; Yuan et al., 2006; Melega et al., 2008).

The findings in striatum of METH users are similar to those in patients with Parkinson's disease in which changes in serotonin markers are not as severe as those for dopamine (Kish et al., 2008). Serotonergic changes might only be functionally meaningful in brain areas such as the orbitofrontal cerebral cortex, a region previously implicated in METH-related decision-making behavior (Rogers et al., 1999; see Kish et al., 2008), in which loss of serotonin was markedly decreased (-55%, -63%). However, the threshold of loss of serotonin/serotonin neuropil required for functional decompensation in the human is not known. It is also conceivable that the low serotonin transporter finding might be related to the clinical observation of a higher incidence of adverse effects in METH users exposed to a serotonin transporter blocker (Shoptaw et al., 2006).

"Markedly" reduced serotonin transporter binding in brain imaging study of METH users: Findings require confirmation using second generation radioligand.

Sekine and colleagues (2006) provide results of a brain imaging study of serotonin transporter binding using the first generation radioligand ^{11}C-McN-5662 in METH users abstinent for up to approximately 5 years. All drug users had experienced psychosis following use of METH (two with persistent psychosis); however, none had reportedly used "high" doses of METH. Drug hair analysis was employed for some of the subjects although it is not clear whether these analyses were conducted throughout the lengthy abstinence period.

[11]C-McN-5662 binding distribution volume was markedly (by approximately 50%) decreased below normal levels in all examined brain areas, including the cerebellar cortex (a region which contains very low serotonin transporter protein and which is often employed to assess nonspecific binding; Kish et al., 2005). The findings, if valid, are remarkable as the regionally widespread and marked binding reduction in METH users was much more severe than that the regionally specific changes observed in a brain imaging study employing the same radioligand, of users of the serotonin "neurotoxin" MDMA (McCann et al., 2005).

However, there is some question as to the interpretation of the findings as the binding data, unlike those of other studies employing ^{11}C-McN-5662, (e.g. McCann et al., 2005; Parsey et al., 2006; Oquendo et al., 2007) were presented only as distribution volumes, for which the (percentage change) relationship with transporter density is uncertain, and moreover, the radioligand employed is known to have suboptimal characteristics (e.g. high nonspecific binding; see Frankle et al., 2004; Kish et al., 2005). A replication study employing a second-generation serotonin transporter radioligand and more conventional data analyses would help resolve the question whether levels of this brain serotonin marker are likely to be markedly reduced throughout the brain of METH users.

2.2.2. Cholinergic Marker Decrease Limited to Very High Dose Methamphetamine Users: Preliminary Findings

Cholinergic neurons originating in the nucleus basalis-septum area and innervating the cerebral cortex and hippocampus are suspected of being involved in aspects of cognition including executive system function (Mesulam and Geula, 1988). In advanced Alzheimer's disease, activity of the marker acetylcholine synthesizing enzyme for cholinergic neurons, choline acetyltransferase, is decreased in some patients with the dementing disorder (Perry et al., 1977; see Mufson et al., 2003). Because of reports suggestive of cognitive problems in some METH users, we measured activity of choline acetyltransferase in postmortem brain (cerebral cortex, striatum, hippocampus, thalamus) of METH users (all testing positive for METH at autopsy) but found no mean difference in activity of the cholinergic enzyme between the METH and control groups. However, two subjects who tested positive for very high brain/blood METH levels had strikingly severe (up to 94%) enzyme activity reduction (Kish et al., 1999).

We suspect that the two high dose subjects with very low choline acetyltransferase activity may have developed METH-induced hyperthermia, with the high temperature causing some inactivation of the enzyme. This inactivation was not a nonspecific phenomenon as activities of two other "control" enzymes, glyceraldehyde 3-phosphate dehydrogenase and citrate synthetase, were either normal or not substantially decreased.

The finding of very low choline acetyltransferase in subjects suspected of having hyperthermia is consistent with our previous observation of severely decreased activity of the enzyme in autopsied brain of several subjects who died in severe hyperthermia associated with neuroleptic drug use (Kish et al., 1990).

The above findings suggest that cholinergic innervation to cerebral cortex is likely to be generally normal in methamphetamine users, and therefore not explain (suspected) cognitive problems in drug users, but (speculatively) that very high dose methamphetamine exposure associated with severe hyperthermia might cause some inactivation of choline acetyltransferase with possible behavioral consequences.

2.2.3. Neuropeptide Markers: Decreased Striatal Metenkephalin in Methamphetamine Users Reflects of Loss of Striatal Output Neurons?

The function of neuropeptides (dynorphin, met-enkephalin, neuropeptide Y, neurotensin, and substance P) in human striatum is still not understood; however, as the peptides are present in striatal output neurones, it is conceivable that the neuropeptides might subserve aspects of the effects of dopamine on its striatal neurone targets.

The striatum contains a variety of types of neurons including output neurons that utilize GABA as a neurotransmitter and also co-localize neuropeptides including met-enkephalin and dynorphin. In addition, the striatum contains GABAergic interneurons that co-express neuropeptides including neuropeptide Y (see Thiriet et al., 2005). Although experimental animal findings are not entirely consistent, acute exposure of a "high"- dose (but one that presumably is not sufficient to physically damage neurons) of METH can cause substantial increases in met-enkephalin, neurotensin, and substance P levels in some striatal subdivisions (Alburges et al., 2001; see also Frankel et al., 2007) but induce lower levels of neuropeptide Y (Westwood and Hanson, 1999). However, a very high dose regimen of METH can induce histological signs of neuronal apoptosis (programmed cell death) in striatal enkephalin containing neurons, but sparing neuropeptide Y containing neurons, in mice (Thieret et al., 2005).

In postmortem brain of METH users (who had recently used the drug), we found that none of the peptides were increased in striatum, and with the quantitatively most marked change being reduced concentration (by about 40%) of met-enkephalin in the different striatal subdivisions, but with a distinct sparing of neuropeptide Y (Frankel et al., 2007). In contrast, in autopsied brain of cocaine users, striatal met-enkephalin concentrations were normal in striatum with the exception of a nonsignificant trend of for reduction in caudate nucleus (Frankel et al., 2008). The finding of low striatal met-enkephalin in METH users can, in principle, be interpreted as a neuroadapative response to repeated METH-induced dopamine stimulation of striatal output neurons or perhaps to overactivation of the striatal dopamine D1 receptor, resulting in increased release and turnover of the neuropeptide (Alberges et al., 2001).

However, the findings are remarkably consistent with results of the Thiriet investigation (Thieriet et al., 2005) in which a "massive" dose regimen of METH-induced apoptosis in met-enkephalin, but not neuropeptide Y containing striatal neurons. (Thiriet also provides some evidence that neuropeptide Y might protect against METH-induced apoptosis in striatum of mice.) Thus, low striatal met-enkephalin (as a neuronal marker) might possibly reflect some damage to, or loss of striatal output (e.g. striato-pallidal) neurons in human METH. Might the finding of low striatal met-enkephalin in human METH users be related to the structural imaging observations of increased striatal volume in some users of the drug (see Section 2.5.4)?

2.3. DOES METHAMPHETAMINE CAUSE OXIDATIVE STRESS IN HUMAN BRAIN?

Animal data suggest that METH might damage dopamine neurons as a result of oxidative stress consequent to increased formation of dopamine-derived oxyradicals (but see Yuan et al., 2010) in dopamine nerve terminals (for review see Krasnova and Cadet, 2009).

Involvement of oxidative stress is supported by the finding of increased levels of malondialdehyde or malondialdehdye-like lipoperoxidation substances (thiobarbituric acid reactive) in brain of METH-exposed animals (Acikgoz et al., 1998, 2000; Jayanthi et al., 1998; Yamamoto and Zhu, 1998; Kim et al., 1999; Kita et al., 2000; Wan et al., 2000; Gluck et al., 2001; Flora et al., 2002; Iwashita et al., 2004; Horner et al., 2011) and the amelioration of stimulant-induced reduction in brain dopamine markers by antioxidants (De Vito and Wagner, 1989; see Kita et al., 2000). The observation of a regionally widespread increase in malondialdehyde levels in brain of METH-exposed rats supports the possibility that METH might cause oxidative stress in both dopamine rich and poor brain areas (Horner et al., 2011).

As oxidative stress cannot yet be assessed in living human brain, we addressed this question in postmortem brain of METH users (all positive for METH at autopsy) by measuring levels of two "gold standard" lipid peroxidation products, 4-hydroxynonenal and malondialdehyde, that have historically been used to assess oxidative stress (Esterbauer et al., 1991) as well as levels of antioxidant systems that might show a compensatory response to oxidative stress. Drug analyses in brain, blood and where available, scalp hair, disclosed the presence only of METH and/or amphetamine.

2.3.1. Levels of 4-Hydroxynonenal and Malondialdehyde are Markedly Increased in Dopamine-Rich and, to Lesser Extent, Dopamine-Poor Brain Areas of Methamphetamine Users: Postmortem Evidence Suggesting Brain Oxidative Stress

We hypothesized that levels of both lipid peroxidation products would be increased in postmortem brain of METH users, with levels highest in the dopamine-rich striatum, with a much less, if any, increase in cerebral cortex, in which dopaminergic innervation

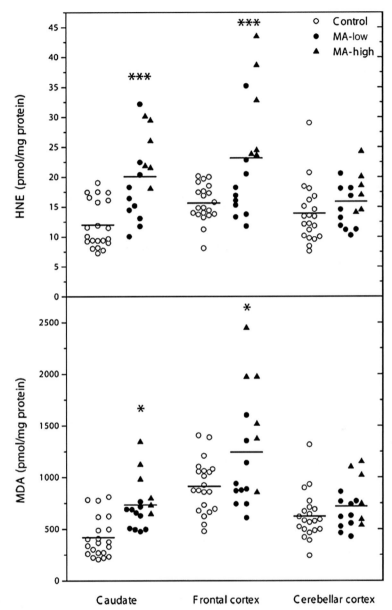

Figure 17 Brain levels of 4-hydroxynonenal (HNE) and malondialdehyde (MDA) are increased in autopsied brain of methamphetamine (MA) users. This finding suggests the possibility of above-normal oxidative stress in brain of methamphetamine users. Consistent with the hypothesis that oxidative stress might be partially dopamine-related, the extent of increase in the two lipid peroxidation products was greatest in the dopamine-rich caudate, followed by the frontal cortex, in which dopamine innervation is relatively "sparse," with only a nonsignificant trend for increase in dopamine-poor cerebellar cortex. Approximately one-half of the drug users had HNE values higher than the upper limit of the control range in the caudate. *Figure reproduced by permission from Fitzmaurice et al. (2006).*

Table 4 Brain Levels of 4-Hydroxynonenal (HNE) and Malondialdehyde (MDA) vs Brain Methamphetamine (MA) Plus Amphetamine Concentration in Autopsied Brain of METH Users

	Dopamine	HNE	MDA
High brain drug level ($n = 6$)			
Caudate	2.3 ± 0.4 (−67%)	$24.5 \pm 2.0^{2,3}$ (+104%)[4]	$935 \pm 108^{2,4}$ (+124%)[4]
Frontal cortex	N.E.	$31.1 \pm 3.5^{2,5}$ (+99%)[4]	$1689 \pm 228^{2,5}$ (+85%)[4]
Cerebeller cortex	N.E.	18.1 ± 1.6 (+30%)	857 ± 109 (+37%)
Low brain drug level ($n = 10$)			
Caudate	3.3 ± 0.5 (−53%)	17.4 ± 2.0^{1} (+45%)	611 ± 34^{1} (+46%)
Frontal cortex	N.E.	18.4 ± 2.2 (+18%)	976 ± 96 (+7%)
Cerebeller cortex	N.E.	14.6 ± 1.1 (+5%)	635 ± 46 (2%)

This table shows that the extent of increase of brain HNE and MDA in METH users is related to levels of drug (MA plus metabolite amphetamine) present in brain at autopsy. Values are mean, SEM; dopamine, ng/mg tissue; HNE and MDA, pmol/mg protein; N.E., not examined.
[1] $p < 0.05$, METH vs control.
[2] $p < 0.001$, METH vs control.
[3] $p < 0.05$ high vs low brain METH cases.
[4] $p < 0.01$ high vs low brain METH cases.
[5] $p < 0.001$ high vs low brain METH cases (one-way ANCOVA with PMI as the covariate, followed by the post hoc Bonferroni's tests).
Source: Reproduced with permission from Fitzmaurice et al. (2006).

is present, but comparatively sparse, and with normal concentrations in the dopamine-poor cerebellar cortex.

The results of our investigation were generally in line with our dopamine-focused working hypothesis: As shown in Figure 17, mean concentrations of both 4-hydroxynonenal and malondialdehyde were markedly increased (by 67% and 75% respectively as compared with control values) in the caudate portion of the striatum whereas the increase in dopamine-poor cerebellar cortex (14–15%) was not statistically significant (Fitzmaurice et al., 2006). In the frontal cortex, in which dopaminergic innervation is not dense, levels were moderately increased (4-hydroxynonenal, +48%; malondialdehyde: +36%). Analysis of individual values in caudate nucleus revealed that 50% of the drug users had values above the upper limit of the control range for 4-hydroxynonenal whereas there was almost complete overlap between the control and drug user range in cerebellar cortex.

Further analysis (see Table 4) involving separation of the METH users into high and low drug groups (based on brain levels of METH plus metabolite amphetamine) showed a dose–response relationship in caudate with the high-dose METH group having higher concentrations for 4-hydroxynonenal in striatum, cerebral cortex, and cerebellar cortex (high dose: +104%, +99%, +30%; low dose: +45%, +18%, +5%) and for malondialdehyde (high dose: +124%, +85%, +37%; low dose, +46%, +7%, +2%).

The regional differences are unlikely to be related to differences in levels of METH in view of our data showing that METH concentrations are homogeneously distributed throughout the postmortem brain of METH users (Kalasinky et al., 2001) (Table 5).

Table 5 Lack of Substantial Regional Distribution Differences of METH in Autopsied Brain of 14 Chronic METH Users

Regional Distribution of Normalized Values of Methamphetamine Plus Amphetamine in 14 Human Methamphetamine Users

Brain region	Normalized "total" methamphetamine concentration
Parietal cortex	87 ± 6
Frontal cortex	95 ± 7
Occipital cortex	103 ± 8
Temporal cortex	108 ± 8
Cingulate cortex	87 ± 6
White matter	120 ± 8
Cerebeller cortex	88 ± 6
Hippocampus	88 ± 5
Putamen	108 ± 11
Caudate	115 ± 13
Globus pallidus internal	109 ± 11
Globus pallidus external	114 ± 11
Hypothalamus	77 ± 5
Medial dorsal thalamus	105 ± 7
Medial pulvinar thalamus	95 ± 6

It is possible that any neurotoxicity of METH might be related to the extent of uptake of the stimulant in different brain regions. However, in this postmortem brain study, "normalized" levels of METH plus its metabolite amphetamine are similar throughout 15 different cerebral cortical and subcortical brain areas of chronic METH users.
Source: Reproduced with permission from Kalasinsky et al. (2001).

To our knowledge, this is the only evidence that METH might cause oxidative stress in human brain, as is the case in animal studies, and points toward a special involvement of dopamine in the process. This finding, requiring replication and extension to METH users in extended abstinence, suggests that METH might cause oxidative stress not only in dopamine-rich brain areas, but also (to a lesser extent) in brain regions that receive a much less dense (cerebral cortex) dopaminergic innervation. Our human data are also consistent with the recent report of regionally widespread increases in malondialdehyde in brain of rats exposed to high doses of METH (Horner et al., 2011).

Although only a limited number of brain regions were examined, the brain drug-aldehyde relationship points toward the possibility that a "low" acute dose of METH might cause oxidative stress perhaps limited to the striatum whereas "higher" doses could result in a regionally more widespread action that might, if sufficiently severe, cause oxidative stress in both dopamine-rich and poor brain areas. Could excessive oxidative stress and damage provide the basis for the reports discussed below (Section 2.5.3) of decreased cerebral cortical gray matter in some METH users?

2.3.2. Levels of Brain Antioxidant Systems in Methamphetamine Users: Preservation of Most Indices but Increased Superoxide Dismutase and Trend for Decreased Glutathione in High Dose Users

A variety of antioxidant systems are present in brain to deal with exposure to harmful oxy-radicals. Several of these brain antioxidant systems are altered in experimental animal studies, for example, striatal glutathione reduction following high-dose METH (Moszczynska et al., 1998), and often show compensatory changes as a consequence of oxidative stress (see Mirecki et al., 2004). To establish whether METH exposure might alter brain levels of key antioxidant defenses, we measured, in postmortem brain of METH users, glutathione and six enzymes of glutathione metabolism, as well as activity of superoxide dismutase (Mirecki et al., 2004).

Brain levels of components of the antioxidant systems were generally preserved but with copper–zinc superoxide dismutase slightly increased (+20%) and a trend for a reduction in glutathione (−35%) in caudate of those METH users having severe (72–97%) dopamine reduction. Both findings have been reported in some METH animal model studies (Moszczynska et al., 1998; Açikgöz et al., 1998, 2000; Kim et al., 1999; but see D'Almeida et al., 1995; Jayanthi et al., 1998).

The human data suggest that oxidative stress might be associated with some modest overutilization of glutathione and a compensatory increase in activity of the enzyme copper–zinc superoxide dismutase, responsible for detoxification of superoxide radical. These changes are also consistent with our observation (Section 2.3.1) of increased levels of lipoperoxidation products in postmortem brain of METH users (Fitzmaurice et al., 2006, Figure 17). Possible functional relevance is suggested by experimental observations showing that inhibition of superoxide dismutase exacerbates the negative effects of METH on dopamine neurones (De Vito and Wagner, 1989) whereas overexpression is neuroprotective (Cadet et al., 1994). Might high (efficacious) dose antioxidant therapy provide protection to any METH-induced neurotoxicity in human brain?

2.4. DOES METHAMPHETAMINE CAUSE GLIOSIS (ACTIVATED MICROGLIOSIS AND REACTIVE ASTROGLIOSIS), A REPUTED INDEX OF NEUROTOXICITY, IN HUMAN BRAIN?

The question whether METH might cause gliosis has been addressed, to date, in two postmortem brain investigations and in a brain imaging study. Parenthetically, I mention also the existence of a very preliminary brain investigation suggesting possibility of abnormal metabolism of glial cells in brain of METH users (Sailasuta et al., 2010a).

2.4.1. Gliosis Is a Common (Obligatory?) Feature of Brain Damage

There is some general agreement in the experimental animal literature involving METH neurotoxicity that morphological criteria indicative of brain neuronal damage include actual neuronal loss, morphological changes such as fiber swelling, neuronal positive silver staining (see Switzer, 2000 for review), and reactive gliosis, involving astrocytes and microglia (Fantegrossi et al., 2008).

Astrogliosis during injury is generally considered to be characterized in part by astrocyte hypertrophy, proliferation, and upregulation of the intermediate filament glial fibrillary acid protein (GFAP) and vimentin (see Zhang et al., 2010; and Figures 18 and 19). However, the question whether astrogliosis involves actual cell proliferation (vs. hypertrophy) is debated (see Norton et al., 1992). Astrocytes are considered to have multiple functions in addition to (somewhat ill-defined) traditional barrier/supportive roles including release of neurotrophic factors, cytokines and involvement in synaptic plasticity. During CNS development and in the adult neurogenic zones, some neural progenitor/stem cells also have the characteristics of astrocytes (GFAP expression) (Kriegstein and Alvarez-Buylla, 2009). Upon brain injury, some GFAP-positive proliferating glial cells have the potential to become pluripotent neural stem cells (Robel et al., 2011).

Following a brain insult microglia can change from a resting to an activated phenotype (assessed using a variety of markers including histocampatibility complex I and II, CD68, CD45, and GLUT5) characterized in part by proliferation, cell hypertrophy, and migration (Oehmichen and Huber, 1976; see Halliday and Stevens, 2011; Streit, 2010). Microglial activation has been considered to have a primary role in removing cellular debris. However, Kettenmann and colleagues (2011) emphasize that "resting" microglia are active in a surveillance mode and that microglial activation is not necessarily a "linear path with a fixed uniform outcome." Indeed, the existence of multiple forms of functionally distinct forms of activated microglia cells is now recognized (Colton, 2009).

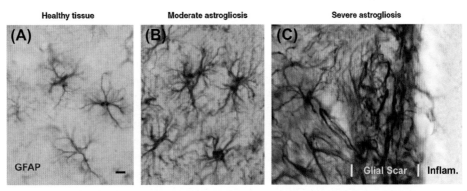

Figure 18 GFAP-staining identifies astrogliosis and glial scar formation following brain injury in the mouse. Reactive astrogliosis occurs following injury in the CNS, although the precise definition of this term continues to be debated. The photomicrograph *(Reproduced with permission from Sofroniew (2009).)* shows GFAP staining of normal astrocytes in healthy mouse cerebral cortex (left), moderately reactive astrogliosis with hypertrophy of cell bodies and processes following brain injury (middle), and severely reactive astrogliosis and glial scar formation following severe injury (right). Does reactive astrogliosis occur in brain of methamphetamine users (see Section 2.4.6)?

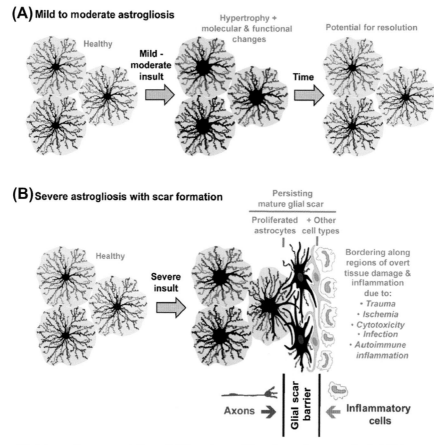

Figure 19 Schematic representation of different severities of reactive astrogliosis. Mild to moderate reactive astrogliosis can potentially resolve with time if the neurotoxic insult is removed (top) whereas severe reactive astrogliosis associated with marked tissue damage can result in a persistent glial scar (bottom) *(Figure reproduced with permission from Sofroniew (2009).)*. Is brain astrogliosis associated with methamphetamine exposure in animal studies short or long-lasting (see Section 8.7.5)?

Some data suggest that microgliosis is an earlier event following brain injury with activated microglia perhaps initiating astrogliosis (see Zhang et al., 2010).

Notwithstanding statements in the literature such as "While it is well accepted that microglial activation contributes to the pathogenesis of several neurodegenerative diseases…" (Venneti et al., 2006), the actual function of glial cells during injury in human brain continues to be debated and is not resolved, with both positive (wound repair, neuroprotection, neuroregeneration) and negative (the "dark side": exacerbation or even cause of injury) possibilities suggested (see Streit et al., 2005; Graeber and Streit, 2010). Streit further cautions against the prevailing assumption that inflammation-like changes, such as microglial activation, are necessarily "…indicative of a chronic and therefore detrimental process that was causative in the development of neurodegeneration" (Streit, 2010).

2.4.2. O'Callaghan Proposes GFAP *as a Biomarker of Brain Damage*: Is Increased GFAP Sufficient to Demonstrate a Neurotoxic Process?

GFAP is an intermediate filament protein in astrocytes forming part of the filament cytoskeleton (see Zhang et al., 2010). Based on histopathological findings in experimental animal studies, O'Callghan has proposed enhanced expression of GFAP as the "ideal biomarker" of all types of nervous system injuries" (O'Callaghan and Sriram, 2005).

Evidence in support of GFAP as a measure of neurotoxicity includes increased brain GFAP consequent to brain trauma (e.g. knife stab) and to exposure of a variety of chemical toxins that cause cell death (with no increase following chemicals that do not cause cell death), (see Norton et al., 1992; O'Callaghan and Sriram, 2005). Although it is difficult to prove a negative, review of the literature suggests that (perhaps) the only experimental animal brain condition that might be associated with increased GFAP without structural neuronal damage is the kindling animal model of epilepsy (Khurgel et al., 1995); however, here it can be argued that methods for demonstration of cell loss might have been too insensitive to detect any change. Or perhaps increased GFAP staining only reflected "dedifferentiation" of some quiescent astrocytes to GFAP-positive neural stem cells in response to kindling?

O'Callaghan suggests that GFAP has the advantage of being a sensitive and specific indicator of neurotoxicity broadly applicable to a variety of toxins/insults, notwithstanding our lack of understanding of the precise role of astrocytosis (repair and recovery vs exacerbation of damage) and of the likely highly variable (and typically uncertain) pathophysiological causes of toxicity associated with different toxins (O'Callaghan and Sriram, 2005).

Although GFAP (measured for example in a tissue homogenate) can be helpful in providing suggestive evidence of a toxic event, *compelling* evidence of *activated* gliosis must, by definition, include histopathological assessment of glial number and morphological evidence of glial activation. Along these lines Graeber and Streit (2010) caution that, despite claims, even after 20 years of research, there "…still is no single molecular marker that allows the unequivocal distinction between resting and activated microglia…" Thus elevated levels of a single glial biochemical marker measured, for example, in a tissue homogenate can be suggestive, but not sufficient, to prove gliosis or neurotoxicity.

I make the above point regarding measurement of glial number and activated state also as Kitamura et al., 2010 (see Section 2.4.6.2) report evidence of increased "resting" but not "activated" or reactive gliosis (microglial and astrocytic) in postmortem brain of METH users—a finding somewhat difficult to interpret.

2.4.3. Gliosis (Microgliosis) is Present in Brain of Patients with Parkinson's disease—but GFAP-Astrogliosis is Minimal: GFAP Might Not always Detect Brain Injury in the Human

If gliosis is a feature of toxic brain injury, there should be evidence of gliosis in brain of patients with progressive degenerative dopamine deficiency conditions that might be relevant to METH as a dopamine neurotoxin. Gliosis has been directly measured in autopsied brain in different dopaminergic disorders and indirectly by employing a reputed biomarker of microgliosis, PK-11195, in brain imaging investigations.

As reviewed by Fellner and colleagues (2011), *microgliosis* is a characteristic of affected brain regions of patients with all examined degenerative dopamine deficiency α-synucleinopathy conditions including Parkinson's disease and multiple system atrophy (a disorder characterized by α-synuclein-positive glial cytoplasmic inclusions in oligodendrocytes), but with continuing debate whether this "pathology" is beneficial or detrimental.

Although extensive, reactive *astrogliosis* is reliably present in affected brain areas of patients with several different degenerative dopamine deficiency disorders (e.g. multiple system atrophy, progressive supranuclear palsy, corticobasal degeneration), reactive astrogliosis positive for GFAP (considered to be the "hallmark" biochemical marker of astrogliosis; Zhang et al., 2010) is "minimal" in Parkinson's disease brain (Song et al., 2009; see Fellner et al., 2011 for review). As a possible explanation for this surprising finding, given that the neuropathology literature considers astrogliosis to be a "pathological hallmark of diseased CNS tissue" (Fellner et al., 2011), Song suggests that the synuclein-involved disease process might have inhibited astrogliosis in Parkinson's disease and in so doing exacerbate the neurodegenerative process. Irrespective, however, of the actual explanation, the dissociation between GFAP-astrogliosis and progressive brain neurodegeneration in Parkinson's disease indicates that caution should be exercised in assuming a one to one correspondence between GFAP astrogliosis and brain injury in human METH studies.

To establish better the relationship between microgliosis and disease progression, a PET radiotracer, ^{11}C-PK11195 was developed, which reputedly binds in living human brain to the microglial mitochondrial protein, now known as the translocator protein 18 kDA (previously peripheral benzodiazepine receptor (Myers et al., 1991; Banati et al., 1997; see Venneti et al., 2006 for review)). The functional meaning of increased translocator protein in activated microglial cells is still unknown. Using a combined autoradiographic-immunohistochemical approach in brain analysis of patients with different neurodegenerative disorders, increased ^{3}H-PK11195 binding "prominently" overlapped with activated microglial cells as compared with GFAP-labeled astrocytes (Venneti et al., 2006) providing some support to the notion that PK11195 might serve as a marker of microgliosis. However, more compelling evidence of PK11195 as a brain imaging marker will require (logistically difficult) correlations between *brain imaging* findings and neuropathological confirmation of actual gliosis in individual subjects.

Consistent with some neuropathological findings (Imamura et al., 2003), ^{11}C-PK11195 binding in a PET imaging study was elevated in both cerebral cortical and subcortical brain areas in patients with Parkinson's disease (Gerhard et al., 2006) and with multiple system atrophy (Gerhard et al., 2003). However, a statistically significant difference in Parkinson's disease brain was not observed in one recent investigation (Bartels et al., 2010) with the authors reporting high variability of values and concluding that the radioligand might not be sufficiently reliable or accurate. In this regard, the

PK11195 radiotracer has a variety of limitations including high nonspecific binding and low brain penetration, prompting a number of investigators to develop a second generation ligand to address these deficiencies (see Rusjan et al., 2011).

2.4.4. Gliosis (Microgliosis and Astrocytosis) is Present in Brain of Some Patients and Experimental Animals with MPTP-Parkinsonism—Long after Exposure to the Neurotoxin: Can Acute Exposure to a Neurotoxin (e.g. Methamphetamine) Cause Persistent, Progressive Neurodegeneration?

As discussed below (Section 2.4.6.4), Sekine has obtained preliminary brain imaging data employing the PK11195 radioligand that his group interprets as suggesting the presence of microgliosis throughout the brain of METH users *reportedly abstinent from the drug for up to 2 years*. This finding, if true, is surprising as gliosis has typically been considered to be only a transient event if the neurotoxic insult is not persistent (O'Callaghan and Sriram, 2005).

What is the evidence for a *persistent* gliosis long after the toxic event in the human brain literature?

In 1983 Langston and colleagues described a group of patients who developed severe Parkinsonism as consequence of self-administration of a new "synthetic heroin" that contained 1-methyl-4-phenyl-1,2,3,6-tetrahydropyridine (MPTP). MPTP was later shown to destroy dopamine neurons in nonhuman primates (Burns et al., 1983). Neuropathological examination of a small number of the subjects exposed to MPTP revealed evidence of both microgliosis and GFAP-astrogliosis (Langston et al., 1999). The striking nature of the finding was that the gliosis was observed *years after exposure to the neurotoxin*, whereas gliosis has generally been considered to be a transient event (but see below).

As Langston notes that it is unlikely that the subjects could have continued to be exposed to MPTP following the original insult (the drug was not known to be available after 1982), he suggests that a single chemical insult might have initiated an active, ongoing neuronal degeneration (see also Vingerhoets et al., 1994). This possibility receives some support from the experimental finding of microgliosis and astrocyte activation in substantia nigra of MPTP-treated monkeys 1 year following exposure to the toxin (Barcia et al., 2004; see also Schintu et al, 2009).

This MPTP-Parkinsonism finding is mentioned in the context of a brain imaging observation suggestive of persistence of microgliosis in METH users withdrawn for many years from the drug (Section 2.4.6.4).

2.4.5. Gliosis (Microgliosis, Astrogliosis, Increased PK11195 Binding) is Present in Some Animal Models of Methamphetamine Exposure: A Transient or Long Lasting Change?

As illustrated in Figure 20, animal data now indicate that METH, at some level of exposure, can cause both microgliosis (Bowyer et al., 1994; LaVoie et al., 2004; Thomas et al., 2004b) and GFAP-related astrogliosis (Hess et al., 1990; Bowyer et al., 1994; O'Callaghan

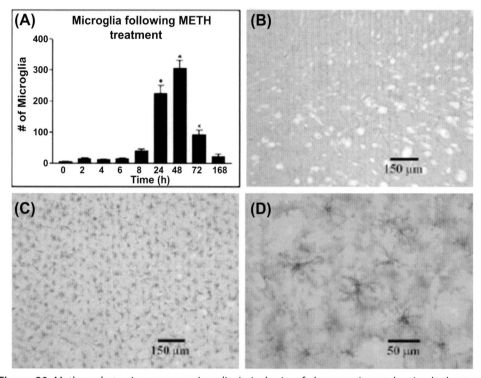

Figure 20 Methamphetamine causes microgliosis in brain of the experimental animal: short- vs longlasting duration? This study provides the time course of increased microglial number (stained with ILB$_4$) in striatum of rats exposed to 4×10 mg/kg methamphetamine *(Figure reproduced with permission from Thomas et al. (2004b).)*. In this investigation microgliosis peaks at about 2 days following methamphetamine exposure (A). The figure also shows a representative section of striatum from a control (B) and a methamphetamine-treated (C) rat and a higher magnification of the field from a methamphetamine-treated rat at the 48 h time point (D). If this animal study is relevant to the human, would the reader not expect that a section of striatum should "light up" with obvious microgliosis ("C" vs "B") if the sample were derived from a human methamphetamine user who died 1–2 days following use of the drug? Further, would the reader really expect to find evidence of microgliosis as long as 1–2 years following last use of the drug? For comparison, see Figures 21 and 22 below for investigations of microgliosis in brain of human methamphetamine users.

and Miller, 1994; Krasnova et al., 2010; Sakoori and Murphy, 2010; McFadden et al., 2012) in striatum and in other brain regions (for review of animal data see Krasnova and Cadet, 2009). Similarly, striatal ^{3}H–PK11195 binding, the reputed microgliosis marker, is increased in striatum of METH–treated rodents (Escubedo et al., 1998; Ladenheim et al., 2000; Pubill et al., 2003; Fantegrossi et al., 2008).

Animal data are not entirely clear, however, on the duration of gliosis following METH exposure. In this regard, microgliosis (6 days; LaVoie et al., 2004; 1 week: Thomas et al., 2004b) and increased GFAP (10 days; Cappon et al., 2000; 21 days; O'Callaghan and Miller, 1994) had not returned to normal levels at the last time point examined in some of the above investigations. Especially noteworthy in this regard is the report of increased GFAP staining in striatum of METH-treated monkeys *1.5 years* following drug exposure (Harvey et al., 2000). This experimental animal finding suggests that METH-induced gliosis might, unexpectedly, be a longlasting event, a possibility relevant to interpretation of human studies of METH users.

Comparison of results of two experimental studies of METH, in which rats self-administered METH (to mimic the human condition) showed that in the first investigation animals allowed to self-administer METH demonstrated low striatal dopamine transporter and increased GFAP (Krasnova et al., 2010) whereas the second study showed low striatal dopamine transporter but only a nonsignificant trend for increased GFAP (McFadden et al., 2012). The investigators reporting elevated GFAP concluded that the neurochemical findings represent "damage" whereas, notwithstanding dopamine transporter reduction lasting for at least 30 days, the study associated with an equivocal GFAP response concluded that the changes represent "neuroadaptation." Similarly, the evidence that brain gliosis or increased PK11195 binding is not observed in rats exposed to the METH derivative MDMA has been used by some to conclude that the drug is not neurotoxic (Pubill et al., 2003). Here it would seem that GFAP is used (rightly or wrongly) as a critical determinant to establish whether neurochemical changes are likely or not to be associated with (structural) damage vs neuroadaptation.

2.4.6. Gliosis Status in Brain of Human Methamphetamine Users: A Clear Picture is Yet to Emerge

To date, there have been three studies aimed at some assessment of gliosis in human METH users: two postmortem (Toronto, Japan) and one brain imaging (Japan) investigation.

2.4.6.1. Toronto Postmortem Brain Study Shows No Qualitative Signs of Obvious, Marked Gliosis

Two postmortem studies have been conducted addressing the possibility of gliosis in brain of humans exposed to METH. Subjects of both investigations were chronic METH users who tested positive for METH at autopsy, indicating recent use of the drug. The first study, based on brain material we examined in our 1996 and 2004 studies, involved routine qualitative-only histopathological analysis whereas the second (Kitamura et al., 2010) employed detailed quantitative analyses of microglial and astroglial markers.

Routine (qualitative only) neuropathological analysis of cerebral cortical and subcortical areas of formalin fixed (half-) brain material used in our Wilson et al., 1996a and Moszczynska et al., 2004 case material (disclosing low striatal levels of dopamine in the frozen half-brain) showed obvious signs of gliosis (in putamen only) in only one of 20 METH users (case 879; see Table 1, Moszczynska et al., 2004) employing GFAP immunohistochemistry; however this subject also had infarcts in the striatum that could well have explained the gliosis. In contrast, the neuropathologist of our study (L Ang) observed marked, obvious, gliosis in putamen of patients with multiple system atrophy, provided over the years as positive controls (Tong et al., 2010). These data, although qualitative only, suggest that marked to severe brain gliosis was unlikely to be a characteristic of the subjects of our investigation.

Work in our laboratory is in progress on the use of Western blot procedures for semiquantitative measurement of biochemical markers of microglial and astroglial cells in postmortem brain homogenates of METH users. However, consistent with an earlier report (Ross et al., 2003), we find GFAP protein levels to be highly variable amongst control subjects in human postmortem brain homogenates.

2.4.6.2. Kitamura Postmortem Brain Study: Increased Striatal Level of Microglial Cells, Trend for Increased GFAP-Astroglial Staining, but No Evidence of Activated or Reactive Glial Cells: Does this Mean No Drug "Neurotoxicity"?

The brain material ($n = 12$) employed in the study of Kitamura and colleagues (2010) largely consisted of the cases previously reported to have decreased striatal dopamine transporter but generally normal VMAT2 protein by immunohistochemistry (Kitamura et al., 2007). Although there was marked overlap in values, median METH blood levels in the Kitamura investigation (4.6 µg/ml) were approximately 10 times those in our Toronto investigation (0.4 µg/ml), suggesting that the Japanese METH users might have used higher doses of the stimulant. Measurement of glial markers was restricted to the striatum (caudate, putamen, nucleus accumbens).

The results of this study first confirm our own ongoing observations of high scatter of levels of glial markers in postmortem human brain (see Figure 21; Kitamura et al., 2010) with large magnitude mean differences between control and drug users for some markers (e.g. CR3/43) not being statistically significant.

Mean GFAP-positive cell density was higher than normal in the different striatal subdivisions of the METH users (by 36%–58%) but the differences were not statistically significant and there was almost complete overlap between drug user and control values. Employing hGLUT5, a member of the glucose transporter family, as a (resting and activated) microglial marker (Horikoshi et al., 2003), a marked increase (+98 to +140%) in hGLUT5-positive cells was observed in striatal subdivisions of the METH users, with about one-third to one-half of the drug users having values above the upper limit of the control range in putamen and nucleus accumbens (Figure 21).

Although the findings above provide some evidence of one aspect of gliosis (apparently (see Norton et al., 1992) increased number of glial cells), Kitamura also reported that the astrocytes generally had normal morphology with only rare detection of *reactive* astrocytes exhibiting hypertrophic cell bodies, shortening of cytoplasmic processes, and nuclear enlargement. Similarly, immunostaining rarely detected *activated* microglia and no correlation was observed between microglial density and levels of the dopamine nerve terminal markers. These negative findings led Kitamura to conclude that the observed glial changes were not part of a METH neurotoxic process.

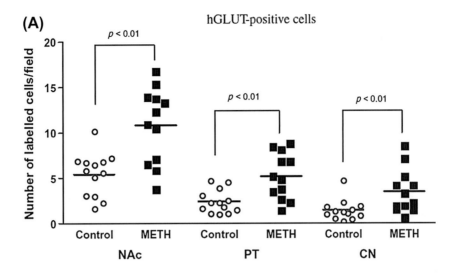

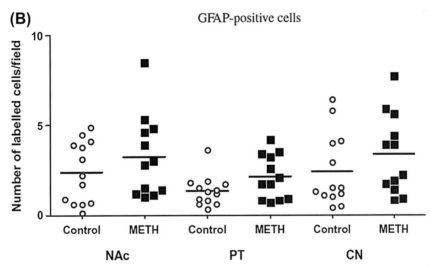

Figure 21 Increased number of hGLUT5-positive microglial cells, but only trend for increased GFAP-positive astrocytes and no glial "activation" in autopsied striatum of methamphetamine users: Conclusion? This study addresses the question whether key indices suggestive of neurotoxicity (microgliosis and astrogliosis) are present in brain of methamphetamine users. Kitamura and colleagues (2010) find, in striatum of methamphetamine users, increased number of hGLUT5-positive microglial cells (A), and a trend for increased number of GFAP-positive astrocytes (B). However, the group finds in the immunohistochemical analysis no evidence of *activated* microglia *or reactive* astrocytes. What should the reader then conclude? Further, what would Kitamura have found if areas outside the dopamine-rich striatum (e.g. cerebral cortex) had been examined?

The Kitamura data suggest possibility of increased number of glial cells in striatum of some chronic (most likely high dose) METH users who had recently taken the drug, but also that striatal glial cells in such subjects might not be in the activated or reactive state expected following acute exposure to a neurotoxin.

This apparent lack of glial activation might simply be explained by absence of neurotoxic action of METH in human brain or could be related to the experimental finding of some tolerance to METH induction of striatal microglial activation and increased GFAP protein following repeated drug exposure (Thomas and Kuhn, 2005; see Mcfadden et al., 2012; Hodges et al., 2011). In this regard, perhaps brain glial activation that had occurred following early exposures to METH is attenuated in the chronic condition (including those subjects of our Toronto study). Alternatively, glial activation could have occurred in brain of the METH users examined by Kitamura, but the interval between last drug use and death might not have provided the appropriate time window to detect changes in the postmortem brain material.

More difficult to explain is the finding of increased density of *resting* microglial cells in striatum of the METH users—a finding not commented on by Kitamura. Could increased number of "resting" microglia examined by Kitamura have been "postactivated" cells undistinguishable by morphology or marker staining from resting cells (see Kettenmann et al., 2011)? Might increased striatal resting microglial number also relate to the brain imaging report (see Section 2.5.4) of enlarged striatal volume in some METH users?

The Kitamura findings raise the question whether increased glial cell number is or is not sufficient to suggest neurotoxicity. Should we require morphological evidence of *activated* microglial cells or *reactive* astrocytes?

2.4.6.3. Little Autopsied Brain Study of Cocaine Users: Evidence of Activated Microglial Cells in Midbrain—Postmortem Literature on Gliosis in Stimulant Drug Users Becomes More Confusing

As animal data suggest that cocaine is not neurotoxic to dopamine neurons, brain of users of cocaine could be considered as a drug control group for studies of METH users. However, as mentioned above, the human brain literature on cocaine provides some reports of lower measures of brain dopamine markers (Wilson et al., 1996c; Little et al., 2003, 2009; Narendran et al., 2012). Although striatum was not apparently assessed, Little and colleagues (2009) report markedly increased (by 108%) number of activated microglia cells in the postmortem midbrain dopamine cell body region (substantia pars compacta) of cocaine users (Table 6), together with a modest reduction in number of melanized midbrain cells (Figure 11). Thus, to date, the postmortem brain evidence for activated microgliosis in brain of human stimulant users is, paradoxically, limited to users of cocaine, but not METH.

This reviewer wonders what would have been found had Kitamura and Little both conducted a regionally extensive assessment of gliosis throughout the brain of METH and cocaine users (see immediately below).

Table 6 Increased Number of RCA-1-Positive Activated Microglial Cells in Autopsied Substantia Nigra of Cocaine Users

Pathological Markers	Controls	Cocaine Users	Statistical Significance
Lewy bodies (H&E, #/50,000 μm^2)	0.94 ± 0.3 N = 17	0.95 ± 0.3 N = 19	$t = 0.013$, df = 34, NS
Free melanin (H&E, # grains/50,000 μm^2)	42 ± 16 N = 15	47 ± 4 N = 17	$t = 0.737$, df = 30, NS
Activated macrophages (cd68 stain +, #/50,000 μm^2)	0.6 ± 0.2 N = 14	2.7 ± 0.6 N = 17	$t = 3.080$, df = 29, $P < 0.01$
Activated microglia (RCA-1 stain +, #/50,000 μm^2)	0.94 ± 0.3 N = 15	0.95 ± 0.3 N = 15	$t = 2.168$, df = 28, $P < 0.05$

If cocaine exposure damages dopamine neurones (see Figure 11), one might expect to find evidence of gliosis in brain of human cocaine users. This preliminary study finds higher than normal (approximately a doubling) density of "activated microglia", as well as activated macrophages, in the substantia nigra pars compacta area of chronic cocaine users vs normal subjects. What would Little have found if the dopamine-rich striatum had been examined?
Source: Reproduced with permission from Little et al. (2009).

2.4.6.4. Sekine Brain Imaging Study: Massive Increase in Binding of "Microglial Marker" PK11196 throughout Brain of Methamphetamine Users Reputedly Abstinent for up to Several Years

In striking contradistinction to the Kitamura postmortem brain finding is the brain PET-imaging report of massively increased binding (by approximately 200–1400%; see Figure 22) of the putative marker of activated microglial cells, ^{11}C-PK11195, in all examined brain areas of METH users abstinent from the drug for 6 months to 4 years (Sekine et al., 2008). The subjects were described as "recreational" only drug users who did not use (presumably high) "toxic" doses.

A key limitation of this study is the use of the radioligand ^{11}C-PK11195 which, as mentioned above, has high nonspecific binding and other deficiencies. As with the earlier PK1195 investigations (Gerhard et al., 2003, 2006), variability of binding was presumably high in control subjects with Sekine employing nonparametric statistics for comparison of group means. No scatterplots were provided making it impossible to establish the extent of overlap between control and drug user values.

Unexplained are differences in regional distribution of binding values in normal brain in the Sekine vs other investigations. For example, Gerhard and colleagues (2006) find in normal brain low binding of PK11195 in striatum, moderate in cerebral cortex, and high in thalamus, similar to the brain regional distribution of binding values our center observes with a second generation radioligand (18F-FEPPA) that binds to the translocator protein (Rusjan et al., 2011). However, Sekine reports similar binding in striatum and thalamus and only trace levels in cerebral cortex, raising some question of the validity of their PET measure. As emphasized by Venneti, in a review of the literature on the use of translocator protein radioligands in imaging, the sensitivity and specificity of PK11195 binding in living human brain still remain to be evaluated (Venneti et al., 2006). Along these lines, I feel that Sekine's article title "Methamphetamine causes microglial activation in the brains of

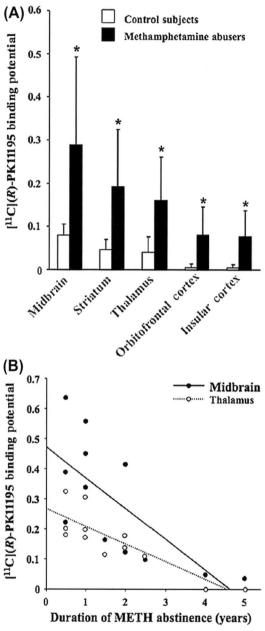

Figure 22 Increased binding of [^{11}C]PK11195 in brain of methamphetamine users: Does methamphetamine cause microglial activation in human brain? This PET brain imaging study shows massively increased binding of the reputed microglial marker [^{11}C]PK11195 (A) throughout the brain of methamphetamine users and (B) as a function of different times during extended abstinence *(Reproduced with permission from Sekine et al. (2008).)*. Does increased PK binding actually equal increased microglial activation? A scatterplot comparing control and subject values is not provided. How do we reconcile this brain imaging study with Kitamura's postmortem brain report of no above-normal microglial activation in (drug-positive) methamphetamine users (see Figure 21) (*, $P < 0.001$).

human abusers" is too strong and as yet unjustified as gliosis, a morphological event, must be confirmed by a postmortem brain finding in the human.

A second limitation of the investigation, critical for interpretation whether the drug affects are recent or persistent, is uncertainty whether the subjects had actually abstained from METH for the entirety of the extended period of time as drug hair data were not provided that would indicate lack of drug exposure during the long (e.g. 2 years) abstinence. On the other hand, the finding of a negative correlation between PK11195 binding and self-reported abstinence, with values "returning" to normal at about 2 years of abstinence, suggests that the abstinence self-report data might have been accurate.

2.4.6.5. Sekine Brain Imaging Study "vs" Kitamura Postmortem Investigation: Long-Term Abstinence Following Methamphetamine Use Activates Persistent Microgliosis in Humans?

In the limited literature on METH and gliosis in human brain, we have, on the one hand, the Kitamura postmortem brain investigation, which reports no evidence of glial activation in likely high dose METH users who had recently taken the drug, whereas the Sekine brain imaging study (assuming PK11195 provided a valid measure of microglial activation) suggests microglial activation lasting years after the last exposure of METH in reportedly lower dose users.

Assuming, very tentatively, that both investigations are valid, these data suggest that, in chronic METH users, the drug might not cause microglial activation (within 1–2 days of last exposure) whereas such activation occurs in later abstinence *and can even persist for years*. Kitamura, attempting to reconcile his findings with those of Sekine, accepts the possibility that the "…long-term survival after METH binges might induce reactive gliosis, as well as the development of neurotoxicity…" (Kitamura et al., 2010).

Although it is generally supposed that gliosis is a transient event following brain injury, one investigation mentioned above reports above-normal GFAP-staining astro-gliosis in striatum of monkeys 1.5 years following METH exposure (Harvey et al., 2000). Similarly, increased astroglial reactivity was observed in brain of rats 4 months following excitotoxic damage by a glutamate receptor agonist (see Oliveira et al., 2003 and references therein). Might the longlasting glial response be related to formation of a "glial scar"? Regarding precedent in human brain for persistent microgliosis following exposure to a chemical neurotoxin, preliminary data, mentioned above, demonstrating active microgliosis and reactive astrocytes years following exposure to the dopamine neurone toxin MPTP suggests that persistent gliosis following acute exposure to a neurotoxin might be possible in the human.

I find the data on brain gliosis in human METH users, although still sparse and difficult to reconcile and interpret, to be highly relevant from a neurotoxicity standpoint as I do suspect that gliosis is an event and sensitive indicator "very close" to a toxic insult. A better understanding of the status of gliosis in METH-exposed human brain will require new studies employing second generation gliosis markers and, for the critical validation of imaging data, postmortem brain investigations assessing actual evidence, in a regionally comprehensive analysis, of glial number and activation state. Further, given the cocaine brain findings mentioned above, a strong case can be made that cocaine users should be included as at least a "drug control" or comparison group in future studies of METH users.

2.5. DOES METHAMPHETAMINE CAUSE HOLES IN HUMAN BRAIN OR A LARGER (GLIAL-FILLED) BRAIN?

If microglial activation, which can accompany neurotoxicity, is in fact present throughout the brain of human chronic METH users (Sekine et al., 2008), it is not unreasonable to expect that some loss of brain neurones might be a feature of METH use in the human.

This issue has yet to be addressed by quantitative neuronal cell counting procedures in postmortem brain investigations of METH users. However, there is now an emerging literature in imaging studies of a putative index of "neuronal integrity" (*N*-acetylaspartate; NAA) as well as of structure measurements (volume, gray matter density).

2.5.1. Postmortem Brain Investigations of Methamphetamine Users: No Disclosure of Obvious, Marked Neuronal Loss

In our postmortem investigation of METH users, routine (qualitative only) neuropathological analysis by a neuropathologist experienced in examination of patients with degenerative dopamine deficiency brain disorders provided no evidence of obvious neuronal loss (e.g. ~35–40%) in any examined cerebral cortical or subcortical brain area (Moszczynska et al., 2004). However, formal neuronal cell counting, which would be necessary to detect slight to moderate neuronal loss, was not conducted. Measurement of neuron-specific enolase protein in brain homogenates also disclosed no reduction in this neuronal marker in striatum, thalamus, and hippocampus, with the exception of a slight (9%) decrease in hippocampal dentate gyrus (Kish et al., 2009). In contrast, measurement of the protein in striatum (the area of degeneration) of patients with multiple system atrophy (as a positive neurodegenerative control) disclosed a 25% reduction (Tong et al., 2010).

Kitamura provides no comment on brain neuronal status (other than absence of ischemic damage) in his postmortem brain report of increased glial cell number in METH users. This most likely suggests that there was no evidence of obvious/marked neuronal cell loss in his case material (Kitamura et al., 2010).

As discussed below, might less severe neuronal cell loss occur in brain of METH users that could only be apparent by employing quantitative measurement procedures?

2.5.2. Brain Imaging Studies Show Decreased *N*-acetylaspartate Levels in Some Methamphetamine Users: Suggestive Neurochemical Evidence of Neuronal Loss and Amelioration by CDP-Choline?

N-acetylaspartate (NAA) is an amino acid derivative, formed by the acetylation of L-aspartic acid, present in very high concentrations (up to 10 mM) in mammalian brain and localized predominantly in neurons (Tallan et al., 1956; for extensive review see Moffett et al., 2007). As decreased brain NAA has been reported in experimental and clinical conditions associated with neuronal cell loss, NAA has been used as a marker of neuronal toxicity (see Benarroch, 2008). However, the still "perplexing" function of NAA in brain continues to be debated with diverse possibilities including involvement in neuronal osmoregulation and mitochondrial energy metabolism (Moffett et al., 2007). Mutations in a gene coding for the NAA catabolic enzyme results in Canavan disease, a fatal leukodystrophy; however the mechanism by which excessive NAA causes toxicity is not understood.

Experimental data suggest that a brain NAA reduction can be associated with neuronal loss or with a reversible state of neuronal dysfunction in the absence of neuronal loss (see Dautry et al., 2000). NAA is considered by some to be an "established marker of neuronal injury or loss." (e.g. Ratai et al., 2011). However, a more conservative reading of the literature suggests that changes in brain NAA *might* reflect differences, to be later proven using validated procedures, in either neuronal number or (mitochondrial?) metabolism within neurones.

The major interest in NAA is related to the observation that this compound can be detected in brain using in vivo proton magnetic resonance spectroscopy (^{1}H-MRS) such that this putative neuronal "viability" marker (NAA or NAA/Creatine ratio) can be used to assess neuronal "integrity" in living brain and follow disease progression (e.g. Cheng et al., 2002; see Benarroch, 2008).

In ^{1}H-MRS investigations of brain of human METH users, NAA and/or NAA/creatine ratio were modestly decreased in cerebral cortical (gray matter) or basal ganglia areas in some (Ernst et al., 2000; Nordhal et al., 2002, 2005; Chang et al., 2005a) but not all (Sekine et al., 2002; Sung et al., 2007; Taylor et al., 2007; Sailasuta et al., 2010b [but decreased in white matter]) studies, suggesting that the general finding might not be robust across investigations. Figure 23 illustrates the NAA reduction (by 3.5%) in a representative number of METH users ($n = 39$) examined by Chang and colleagues (2005a).

Although the data are certainly suggestive of at least a metabolic difference in brain of METH users, the interesting findings still need to corroborated with more accepted measures of brain pathology (e.g. quantitative neuronal cell counting; structural imaging changes on MRI) as described below.

There also should be follow-up on the intriguing, but preliminary, brain imaging finding (in a placebo-controlled study) that 4 weeks oral treatment with cytidine-5′-diphosphate choline (CDP-choline) was associated with increased prefrontal cortical

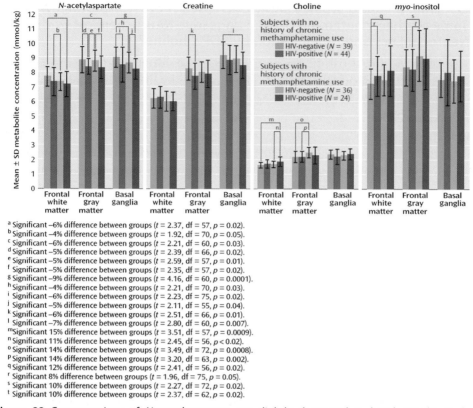

^a Significant –6% difference between groups ($t = 2.37$, df = 57, $p = 0.02$).
^b Significant –4% difference between groups ($t = 1.92$, df = 70, $p = 0.05$).
^c Significant –6% difference between groups ($t = 2.21$, df = 60, $p = 0.03$).
^d Significant –5% difference between groups ($t = 2.39$, df = 66, $p = 0.02$).
^e Significant –5% difference between groups ($t = 2.59$, df = 57, $p = 0.01$).
^f Significant –5% difference between groups ($t = 2.35$, df = 57, $p = 0.02$).
^g Significant –9% difference between groups ($t = 4.16$, df = 60, $p = 0.0001$).
^h Significant –4% difference between groups ($t = 2.21$, df = 70, $p = 0.03$).
ⁱ Significant –6% difference between groups ($t = 2.23$, df = 75, $p = 0.02$).
^j Significant –5% difference between groups ($t = 2.11$, df = 55, $p = 0.04$).
^k Significant –6% difference between groups ($t = 2.51$, df = 66, $p = 0.01$).
^l Significant –7% difference between groups ($t = 2.80$, df = 60, $p = 0.007$).
^mSignificant 15% difference between groups ($t = 3.51$, df = 57, $p = 0.0009$).
ⁿ Significant 11% difference between groups ($t = 2.45$, df = 56, $p < 0.02$).
^o Significant 14% difference between groups ($t = 3.49$, df = 72, $p = 0.0008$).
^p Significant 14% difference between groups ($t = 3.20$, df = 63, $p = 0.002$).
^q Significant 12% difference between groups ($t = 2.41$, df = 56, $p = 0.02$).
^r Significant 8% difference between groups ($t = 1.96$, df = 75, $p = 0.05$).
^s Significant 10% difference between groups ($t = 2.27$, df = 72, $p = 0.02$).
^t Significant 10% difference between groups ($t = 2.37$, df = 62, $p = 0.02$).

Figure 23 Concentrations of *N*-acetylaspartate are slightly decreased in basal ganglia (including caudate and putamen) but not in mid-frontal gray matter in chronic methamphetamine users: A brain imaging study. This figure shows levels of *N*-acetylaspartate, creatine, choline, and myo-inositol in brain of HIV-positive and HIV-negative subjects with and without a history of chronic methamphetamine use. Focus your attention primarily on comparison of *N*-acetylaspartate levels (left side) in basal ganglia and mid-frontal gray matter in HIV-*negative* normal subjects (green bar) vs those in HIV-negative methamphetamine users (blue bar). Note that *N*-acetylaspartate concentrations are slightly decreased in basal ganglia (statistical comparison "h") but not in mid-frontal gray matter. The figure also shows separately that HIV-positive subjects who are not methamphetamine users have slightly lower *N*-acetylaspartate levels in mid-frontal gray matter and basal ganglia (green vs red bar). What do low *N*-acetylaspartate brain concentrations mean? Neuronal loss? Abnormal metabolism only? Both? *Figure reproduced with permission of the American Psychiatric Association, Chang et al. (2005a).*

NAA levels in METH users (Yoon et al., 2010). The finding is especially interesting in view of the previous (also preliminary) report that treatment with this intermediate for membrane phospholipids attenuated measures of craving in cocaine users (Renshaw et al., 1999). Although we did not find activity of the CDP-choline synthesizing enzyme (phosphocholine cytidyl transferase) decreased in postmortem brain of METH users (Ross et al., 2002), could exposure to CDP-choline have stimulated "neuroregeneration" of damaged neurons or perhaps stimulation of neuronal mitochondrial energy production (see Yoon et al., 2010) resulting in normalization of NAA concentration?

2.5.3. Some (but Not All) Brain Imaging Studies Show Lower Cortical Gray Matter in Methamphetamine Users: Are the Differences Caused by Methamphetamine or Other Drugs Used, and Related to Abstinence Time and Gliosis?

Sekine's observation (Sekine et al., 2008) in METH users of a striking (> +1000%) increase in cerebral cortical binding of a putative marker of activated microglial cells in cerebral cortex (if valid), together with reports of decreased cerebral cortical NAA in some drug users, raise the possibility that there might be some cerebral cortical thinning in METH users.

The literature on structural brain changes, employing magnetic resonance imaging (MRI) procedures, in METH users has already been extensively reviewed (Berman et al., 2008, 2009; Chang et al., 2007) and will not be discussed in detail in this review. A key finding in several investigations that employed a representative number of drug users is lower, by up to approximately 10–12% (mean) density of gray matter in cerebral cortex and/or hippocampus in groups of METH users (Thompson et al., 2004; Kim et al., 2006; Schwartz et al., 2010). Figure 24 compares the volume of cerebral cortical brain areas and hippocampus in METH users and normal subjects in the original Thompson investigation. The results of this 2004 study by the London group have been largely replicated in a recent study by the same team that shows smaller than normal gray-matter volumes in different cerebral cortical brain regions of METH dependent users who were also smokers (Morales et al., 2012; see below for further discussion). However, the original finding of smaller hippocampal volume (see Figure 24(A)) was not replicated.

The cerebral cortical brain areas affected are not identical in the different studies and have included cingulate, insular and Brodmann area 10 cortices. A scatterplot provided in the Schwartz investigation disclosed almost complete overlap between drug user and control values, suggesting that the differences in individual studies might not be robust although a group mean difference across studies might be typically observed.

Consistent with the above observations, Nakama and colleagues (2011) recently reported, in a cross-sectional study, a greater than normal age-related "decline" in cerebral cortical gray matter in METH users. In this investigation, scatterplots were provided showing extensive, but not complete, overlap between control and drug user values (Figure 25).

The above results are also consistent with our own observations of polydrug MDMA users in which we observed cerebral cortical thinning primarily restricted to those MDMA users who tested positive for METH in hair (Kish et al., 2010b) (Figure 26)

The structural imaging investigators continue to emphasize the many potential confounds in the studies (Berman et al., 2008; Chang et al., 2007) including use, amongst the investigations, of different measurement procedures, possibility of differences preexisting drug use and the lack of longitudinal studies dealing with the important influence of

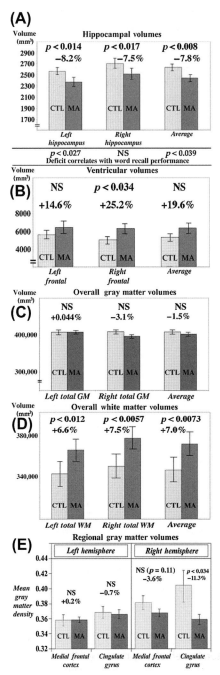

Figure 24 Comparison of brain structure volumes in methamphetamine users (METH, *n* = 22) and control subjects (CTL, *n* = 22): An MRI study. This investigation addresses the question whether methamphetamine exposure might cause loss of brain tissue in the human (*Reproduced with permission from*

abstinence time. In this respect, there is new preliminary evidence suggesting that abstinence time may well influence differences in cerebral cortical volume. Thus, in a recent longitudinal investigation, increased gray matter volume in cerebral cortex in early (4–7 days) to later (1 month) abstinence was observed in a small group of METH users (Morales et al., 2012). Although the basis for the time–dependent increase is not known (gliosis?), the findings, requiring replication, do suggest one explanation for the suspected lack of robust difference in drug user vs control values in the different studies.

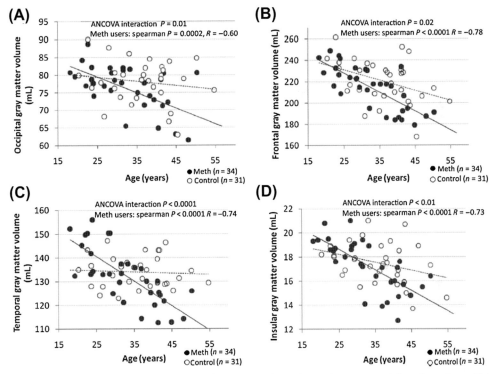

Figure 25 Cerebral cortical gray matter volume in methamphetamine users as a function of age. This is a cross-sectional (not longitudinal) investigation, but it does show a more marked "influence" of advancing age on cerebral cortical gray matter volume in methamphetamine users vs controls. Note however the extensive overlap between individual methamphetamine user and normal groups. *Figure reproduced with permission from Nakama et al. (2011).*

Thompson et al. (2004).). Mean and SEM values are provided for volumes of the hippocampus (A), frontal horn of the ventricles (B), total cerebral gray matter (GM) (C), total cerebral white matter (WM) (D), and gray matter density in medial frontal and cingulate cerebral cortices (E). Note decreased hippocampal (A) volume and gray matter volume in the right cingulate cortex (E) of the drug users. Recently, the same group replicated the finding of decreased gray matter volume in different cerebral cortical brain areas but did not replicate the observation of lower hippocampal volume (Morales et al., 2012).

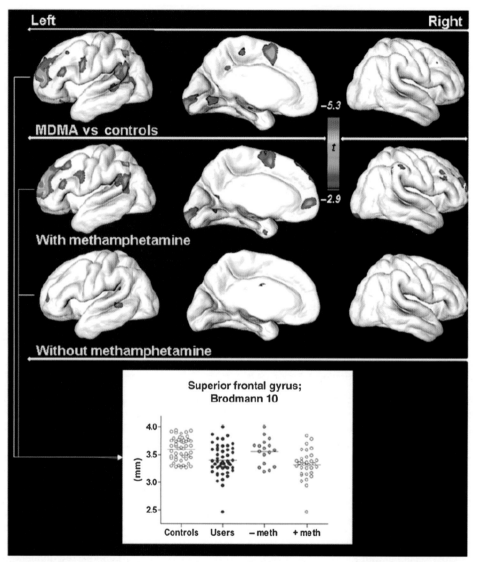

Figure 26 Cerebral cortical "thinning" in ecstasy (MDMA) users who use methamphetamine. This investigation is confounded by inclusion of polydrug users in the study, but it does show that ecstasy users who (knowingly or unknowingly) use methamphetamine have slightly decreased cortical (Brodmann area 10) thickness (see scatterplot and compare third and fourth columns; *t*-statistical map is above) than ecstasy users who do not use the drug. *Figure reproduced with permission of Oxford University Press, Kish et al., 2010b.*

There is also appreciation of the generic confound related to the fact that METH users rarely use only METH and the possibility that any difference in brain size could be related to a co-used drug. Thus, Lawyer and colleagues (2010) report cerebral cortical thinning only in the subgroup of amphetamine users who were also heavy alcohol users, suggesting that alcohol use by METH users might be an important confound.

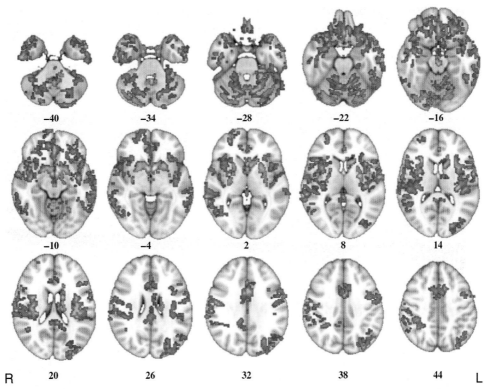

Figure 27 Reduced cerebral cortex but increased striatal volume in cocaine users: Is cerebral cortical thinning a consequence of exposure to most (all?) Abused drugs? This figure *(Reproduced with permission from Ersche et al. (2011).)* was selected to illustrate that structural differences in brain of methamphetamine users (reduced cortical volume; increased striatal volume, see Section 2.5.5) have also been reported in brain of users of other recreational drugs. The figure is a statistical map showing significant differences in grey matter volumes in different brain areas of cocaine users (n=60) and normal controls (n=60) (blue, reduced; red, increased). *Loss* of grey matter volume was observed in orbitofrontal cortex, insula, medial frontal, and anterior cingulate cortex, temporoparietal cortex, and cerebellum whereas an *increase* was reported in striatum. The extent of overlap between controls and drug users is not known as scatterplots were not provided. Since most cocaine users met criteria for nicotine dependence in this study, the argument could also be made that any structural changes might be related to tobacco smoking. Could drug-induced oxidative stress and damage explain the structural differences?

The issue of co-use of tobacco by METH users has also been raised as a potentially significant confound in view of the literature associating smoking with cerebral cortical differences (Kühn et al., 2010; see Schwartz et al., 2010; Morales et al., 2012). The smoking confound will be very difficult to sort out given that most METH users are cigarette smokers (e.g. all of the METH users in the Morales et al., 2012 study). As illustrated in Figure 27, there is also an emerging literature on structural changes in brain of cocaine users reporting decreased cerebral cortical volume in chronic users of this stimulant drug (see Matochik et al., 2003; Ersche et al., 2011 and references therein). Notably, however, in the Ersche investigation of a large ($n = 60$) sample size of cocaine users, most met criteria for nicotine dependence. Might much of the cerebral cortical brain differences in the different drug groups be largely explained by co-use of tobacco?

The literature awaits an explanation for structural "changes" in the relatively (vs striatum) dopamine-poor cerebral cortex of users of drugs of abuse. Could this be explained by drug-induced oxidative stress and damage in brain areas such as the cortex having a relatively sparse dopaminergic innerveration (see Section 2.3)? This is suggested by our report of increased levels of lipoperoxidation products in postmortem frontal cortex (in addition to dopamine-rich striatum) of METH users (Fitzmaurice et al., 2006; see Figure 17 and Table 4), a finding consistent with animal data showing that oxidative stress caused by this stimulant might affect brain areas that do not have high dopamine concentration (Horner et al., 2011). Similarly, results of an animal study investigation suggest that cocaine might also cause oxidative stress and possibly damage in the cerebral cortex of rats (Dietrich et al., 2005; see Figure 28).

Notwithstanding the above interpretational difficulties, the structural imaging reports of subnormal gray matter density in some METH users do provide supporting evidence to the notion that METH might damage nondopaminergic cerebral cortical neurones in

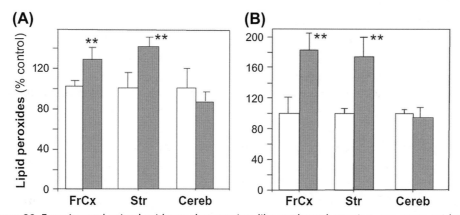

Figure 28 Experimental animal evidence that cocaine, like methamphetamine, can cause oxidative stress in brain. Animal and human data suggest that methamphetamine can cause oxidative stress in brain areas having high or low dopamine concentration (see Section 2.3). This animal study *(Figure reproduced with permission from Dietrich et al. (2005).)* suggests that the same might occur following exposure to cocaine. Could oxidative stress explain human findings of cerebral cortical loss in both methamphetamine and cocaine users (and in tobacco smokers)? The figure shows increased lipid peroxide production in areas of high dopamine (striatum) and lower dopamine (frontal cortex) of rats exposed acutely (A) or 10 days chronically (B) to cocaine (20 mg/kg) and examined 12 h later ($*P < 0.05$; $**P < 0.01$). Similar to results of our study of lipoperoxides in postmortem brain of methamphetamine users (Figure 17), significant differences were observed in striatum and cortex, but not in cerebellum, the latter assumed to have very little dopaminergic innervation.

the human, an event that could be expected to be associated with some gliosis. However, because of the confounds related to use of other drugs that might conceivably damage the cerebral cortex, it is not yet possible to conclude that cerebral cortical thinning is a characteristic of METH exposure in humans.

2.5.4. Some (but Not All) Brain Imaging Studies Show *Increased* Striatal Volume in Methamphetamine Users: Are the Differences Caused by Methamphetamine, Other Drugs Used, and Related to Abstinence Time, Dopamine Status, or Gliosis?

Perhaps the most interesting structural brain imaging finding in METH users is the demonstration in two studies of increased striatal volume in some drug users (Chang et al., 2005b: median 2 months drug abstinence; Jernigan et al., 2005: mean 3 months abstinence), although increased striatal gray matter was not observed by a third group (Schwartz et al., 2010: mean 2 months abstinence.) Interestingly, the London group recently reported *smaller* caudate volumes in METH users (all smokers) in *very early abstinence* (4–7 days) as well as in a control smoking group (Morales et al., 2012). Reconciliation of these data with the findings of increased striatal volume in the Chang and Jernigan studies is suggested by London's preliminary finding of a trend for volume expansion during abstinence in the longitudinal component of the study.

The mean magnitude of the volume increase in the Chang study was approximately 10% (see Figure 29), whereas the percentage change was not disclosed in the Jernigan investigation. Scatterplots comparing individual volume values for drug users and controls were not provided in either Chang or Jernigan studies making it impossible to establish whether the differences were robust.

Consistent with the Chang findings, our group reported a nonsignificant trend for slightly increased (by 4%) striatal volume in the polydrug METH users examined in our VMAT2 PET imaging investigation (Boileau et al., 2008).

Might striatal enlargement in stimulant users be related to the preliminary observations that both acute (Tost et al., 2010; see Figure.30) and chronic (Vernon et al., 2012; see Figure 31) dopamine receptor blockades can induce reversible changes in striatal volume in the human and rodent, respectively. In this regard, Tost reports *decreased* striatal size a few hours following exposure to haloperidol in the human whereas Vernon describes *increased* striatal volume in rats chronically exposed to the same drug. See also Ersche et al., 2011 for a discussion of the literature on influence of dopamine receptor-blocking drugs on brain volume in patients with schizophrenia. Are findings of striatal volume differences following exposure to haloperidol and observations of increased volume reported in human stimulant users related to an action of the drugs on the striatal dopamine system?

If dopamine status in striatum influences striatal size, then one would expect that subjects having the "gold standard" dopamine deficiency condition of Parkinson's disease should have enlarged striatum (assuming the animal investigation employing a dopamine

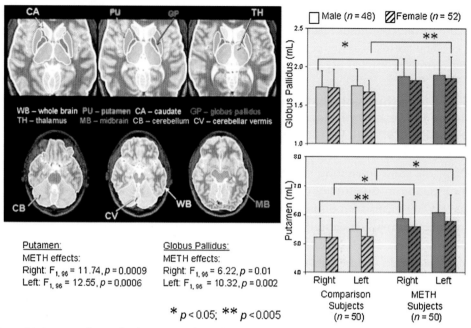

Figure 29 Increased striatal volume in methamphetamine users. Chang and colleagues (2005b) report increased volume of putamen (bottom graph) and globus pallidus (top) in brain of methamphetamine users. Does neurotoxicity-related gliosis, oxidative stress, and/or dopamine deficient state explain increased striatal size? *Figure reproduced with permission from Chang and colleagues (2005b).*

receptor-blocking drug cited in Figure 31 is relevant to the human condition). Obviously, such subjects would need to be assessed early in the disorder and prior to any treatment with dopamine-substitution medication, which might normalize striatal volume. This investigation has, in fact, been conducted in subjects (all presymptomatic) having different genetically determined parkinsonisms (so that the disease can be predicted prior to symptoms) in which increased striatal volume (by approximately 10%) was reported (Reetz et al., 2010).

The authors of the Parkinson's disease study attribute striatal enlargement, not unreasonably, to an "adaptive plasticity" of neurones in striatum consequent to the dopaminergic dysfunction. [Parenthetically, if the adaptive, structural, persistent change is *beneficial*—is this "neurotoxcity"?] Could such dopaminergic dysfunction (deficiency?), changing over time during abstinence, account for observations of striatal enlargement in stimulant users?

2.5.5. Is Striatal Enlargement and Decreased Cerebral Cortical Gray Matter in Stimulant Users a *Preexisting* Condition? MRI Findings Employing Siblings of Drug Users

When I make a finding in brain of a METH user (e.g. low striatal dopamine concentration) I typically acknowledge the generic possibility that the finding might have preexisted use of the drug, yet I strongly feel that this and probably most brain differences

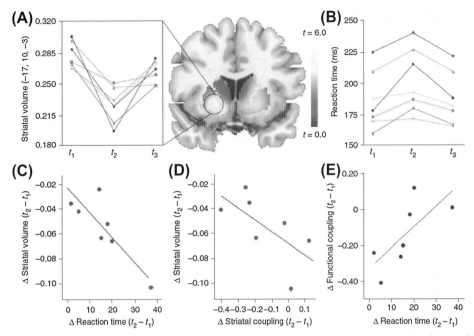

Figure 30 Acute treatment with a dopamine receptor-blocking drug induces a rapid and reversible *decrease* in striatal volume in the human. This study addresses the general question whether drugs acutely affecting the brain dopamine system could cause a change in size of the striatum in humans. As shown in "A", Tost and colleagues (2010) report that striatal volume is decreased in normal subjects ($n = 7$) 1–2 h t_2 following administration of the neuroleptic haloperidol (5 mg/70 kg body weight) but largely normalizes at 24 h (t_1 *one half-life of the drug*). Panels B–E illustrate correlations between striatal volume changes vs reaction time and functional dyscoupling between striatum and motor cortex (see Tost et al., 2010 for details). The authors acknowledge that the explanation is "challenging", but how does the reader explain this remarkable finding? Is it the dopamine receptor blockade that induces striatal volume change or another feature (osmotic-related)? What would have been the effect of acute administration of methamphetamine or cocaine on striatal size? *Figure reproduced with permission from Tost et al., 2010.*

reported in the drug users were in fact caused by drug exposure. Recent findings from Ersche and colleagues (2012) suggest that some of the above-mentioned structural differences in stimulant users might actually have been prior to drug use.

This MRI investigation involved polydrug stimulant drug users (94% cocaine users; 6% amphetamines), *siblings* of the drug users having no history of substance dependence (except nicotine), and unrelated healthy volunteers. Drug users were probably scanned in very early abstinence or while on drug (this is not explicitly stated). A key finding of the study was the identification of some brain areas that were decreased (several cerebral cortical regions) and increased (by 12% in putamen portion of striatum) in gray matter density of both stimulant-dependent subjects *and* their nondependent siblings (see Figure 32). Although there are unresolved issues in the study (e.g. uncertain influence of smoking) the findings do suggest that some structural differences

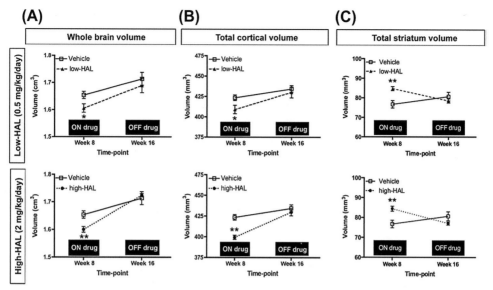

Figure 31 Chronic treatment with a dopamine receptor blocking drug induces a reversible *increase* in striatal volume (and decreased cortical volume) in the rodent. This study addresses, in the experimental animal, the question whether *chronic* administration of drugs affecting the brain dopamine system could cause a change in striatal volume. Vernon and colleagues (2012) show that striatal volume (right column) is increased in rats (*n* = 7–9 per group) chronically treated with haloperidol, with the volume normalizing 2 weeks later. (Note also in the middle column the accompanying decrease in cortical volume, as reported in some human stimulant users; see Section 8.8.4). The authors (like this reviewer) struggle somewhat to explain the interesting finding (e.g. authors' suggestion of "idiosyncratic mechanisms"). What is the readers' preferred speculation? Are changes somehow related to the dopamine receptor-blocking effect of the drug? What would have been the effect of chronic administration of methamphetamine, cocaine, or dopamine agonists on striatal (and cerebral cortical) volume? *Figure reproduced with permission from Vernon and colleagues (2012).*

in stimulant users might not have been consequent to any toxic effect of the drug and were genetically based. Nevertheless, as noted by Volkow and Baler (2012) in the accompanying commentary, some brain areas such as the orbitofrontal cortex showed differences (from normals) in the drug users that were not shared by the siblings, for which a case can be made that the low gray matter density was caused by drug use.

Clarification of some of these issues (e.g. drug exposure vs preexisting difference; possible involvement of abstinence time and gliosis) will be helped by longitudinal brain studies of combined PET (microglial marker) and MRI (structural imaging) investigations in METH users scanned in early and late abstinence. Ultimately, however, there will be need for confirmation of any striatal glial changes by postmortem brain histopathological analyses.

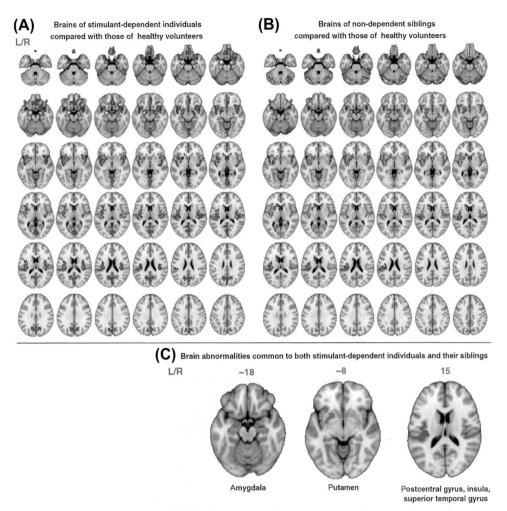

(A) Brains of stimulant-dependent individuals compared with those of healthy volunteers

(B) Brains of non-dependent siblings compared with those of healthy volunteers

(C) Brain abnormalities common to both stimulant-dependent individuals and their siblings

L/R   −18   −8   15

Amygdala   Putamen   Postcentral gyrus, insula, superior temporal gyrus

Figure 32 Gray matter volume in stimulant users vs their siblings: Evidence for shared abnormalities. This structural imaging study addresses the question whether volume differences in brain of stimulant drug users might have preexisted use of the drug. Statistical map (A) compares abnormalities in gray matter volume in brain of polydrug stimulant users ($n=47$) with healthy unrelated subjects ($n=50$) whereas (B) makes the comparison between normals and the nondrug using siblings ($n=49$) of the drug users *(Reproduced with permission from Ersche et al. (2012).)*. Brains areas of abnormally increased (red) and decreased (blue) volume are identified. Focus your attention especially on (C) showing brain differences common to both drug users and siblings in which left amygdala and putamen are enlarged and left postcentral gyrus, superior temporal gyrus, and posterior insula are reduced as compared to normals. The preliminary study provides evidence that some structural differences in brain of stimulant users might have a genetic basis and also in part, be consequent to drug use (i.e. the areas of brain *not* affected in *both* drug users and siblings; see Volkow and Baler (2012) for critical commentary).

2.5.6. What is the Readers' Opinion on the Question of Physical Brain Damage and Methamphetamine Use?

Assume, only for the sake of argument, that I have provided you in this chapter with the entirety of the available relevant findings dealing with the question of METH use and physical brain damage (or differences) in the human and that, with your full understanding of literature evidence, you will provide a reasoned, balanced opinion, as a qualified expert witness, to the trial judge at a sentencing hearing for an accused recently found guilty of possession of METH for the purpose of trafficking.

Assume also that there will be a perfect positive correlation between length of sentence and the judge's appreciation (per your opinion) of the extent of "brain damage" harm caused by the drug. You may be cross-examined by the attorney for the accused on your testimony.

You are asked by the judge specifically to "provide an opinion on the question whether methamphetamine has a harmful effect, if any, on the structure of the human brain (i.e. the public's perception of 'brain damage')." This opinion, using terms that will be easily understood by a lay audience, must be identical to that you would provide to the public and to your scientific colleagues.

What is your bottom line opinion?

2.6. DOES METHAMPHETAMINE CAUSE PARKINSON'S DISEASE OR PERSISTENT PSYCHOSIS PATHOLOGIES?: EPIDEMIOLOGICAL FINDINGS

A different approach to the study of brain pathology of METH users is to establish whether use of the drug is associated with increase risk of developing specific brain disorders.

2.6.1. Epidemiological Study, Requiring Replication, Suggests that High Dose Methamphetamine Might Increase Risk of Developing Parkinson's Disease

Ricaurte and colleagues (2005) suggest, on the basis of their animal findings, that "low" therapeutic doses of amphetamine might actually damage brain dopamine neurones in adults receiving medication for the treatment of ADHD. This question is now being addressed by Ricaurte in a brain imaging study of the dopamine transporter in amphetamine-treated patients with ADHD (Ricaurte, 2011). However, should *high* doses of METH, as some of the animal data mentioned above suggest, damage nigrostriatal dopamine neurons, then it is not unreasonable to suspect that high dose METH use could also damage dopamine neurons in humans and, if the damage is sufficiently severe, cause Parkinsonism. [My position is that it would be highly *unlikely*, based on animal data, if some "very high" dose of METH did not cause at least temporary damage to dopamine neurons in human brain.] In principle, this could arise acutely following a high dose of the drug, or perhaps later in life when age-related loss of dopamine neurons (Kish et al., 1992) in METH-compromised individuals results in dopamine neuron damage exceeding the symptomatic threshold.

I recall well my presentation some years ago at a symposium on methamphetamine at which I made the bold statement that "If methamphetamine causes Parkinson's disease, we really would know this by now." My statement succeeded well in inflaming the audience (primarily) of scientists who had devoted much of their research lives to the demonstration of METH toxicity in experimental animals and who provided a strong pointed response to my position. However, I did feel that my position was not unreasonable in view of the apparent absence of any literature indicating that Parkinsonism can develop acutely following high dose METH, and my discussions with METH rehabilitation centers advising me that Parkinsonism—or even neuroleptic-induced Parkinsonism as treatment of psychosis in a METH user—was not a characteristic feature of drug users.

In other words, if there really were an "epidemic" of Parkinsonism caused by METH, I felt that this would already have become apparent over the past 50 years.

If however, METH does cause persistent, but only subclinical, damage that becomes symptomatic many years later because of age-related dopamine neuron loss and functional decompensation, a longitudinal investigation requiring large number of subjects would be required to address this issue since both METH use and Parkinson's disease are relatively low incidence conditions.

To address this question, Russ Callaghan, an epidemiologist, employed a hospital-record linkage approach in which hospitalized METH users ($n = 40,472$) and a matched proxy group (appendicitis condition; $n = 207,831$) in the US state of California (where METH use is high) were followed for up to 16 years. Cocaine users were selected as a drug control group because cocaine and METH users can be expected to experience, to some extent, similar life style events (e.g. tobacco use, nutritional problems) and both are dopaminergic stimulants. As mentioned above, animal findings suggest that METH, but not cocaine is toxic to brain dopamine neurons although postmortem human brain findings are somewhat equivocal.

We found that the METH cohort had modestly/moderately increased risk of developing Parkinson's disease as compared to the proxy control group (Hazard Ratio = 1.76; 95% confidence interval: 1,12–2.76) and to the matched drug control group of cocaine users (Hazard Ratio = 2.44; 95% confidence interval 1.32–4.41) (Callaghan et al., 2012a).

There are many confounds to this investigation (discussed in detail in Callaghan et al., 2012a) in which results could be explained by inaccuracies in hospital diagnoses of Parkinson's disease and by critical lifestyle factors specifically associated with METH use influencing risk of Parkinson's disease. Nevertheless, these epidemiological data, which must be considered preliminary and require much independent replication, do provide some evidence that METH use sufficiently heavy to warrant a hospital diagnosis of abuse or dependence might increase somewhat risk of developing later in life a degenerative dopamine deficiency disorder. It must also be emphasized that the findings do not necessarily apply to those who use much lower doses of amphetamines for therapeutic purposes (e.g. ADHD, narcolepsy, obesity).

2.6.2. Epidemiological Study, Requiring Replication, Suggests that High-Dose Methamphetamine Might Increase Risk of Developing Persistent Psychosis

The clinically relevant brain pathology of schizophrenia, a persistent psychosis condition, is unknown. However, it has long been speculated that sensitization to the dopamine system might underlie some aspects of the disorder (Lieberman et al., 1997) with the brain imaging finding of enhanced amphetamine-induced release of dopamine in patients with schizophrenia (Laruelle et al., 1996) providing support for this hypothesis.

Japanese investigators have long suspected that repeated exposure to METH, a drug proven to cause psychosis in the human (Griffith et al., 1972), might damage the brain with the consequence of *persistent* psychosis in some drug users (Sato et al., 1992). However, the position of Western psychiatry is that any persistent psychosis (i.e. schizophrenia) must have occurred independently of drug use, either prior to or coincident with METH use (see Angrist, 1994). The unresolved issue is clearly of continuing interest to the psychiatric community as exemplified by two different Clinical Conference articles devoted to this topic in the *American Journal of Psychiatry* spanning 14 years (Flaum and Schulz, 1996 ["When does amphetamine-induced pyschosis become schizophrenia?"]; Grelotti et al., 2010).

Employing, as above (Section 2.6.1), a population-based cohort study from California inpatient hospital discharge records, we compared the risk of developing schizophrenia in hospitalized users of METH (not previously reported to have had a persistent psychotic condition) to a matched proxy control group (appendicitis condition) and to users of other recreational drugs (cocaine, opiates, alcohol, cannabis). Somewhat to our surprise, we found that *all* of the recreational drug groups demonstrated higher risk of subsequent development of schizophrenia than did the proxy control group; however, the METH cohort had a significantly higher risk than the cocaine, opiate, and alcohol groups (hazard ratios of 1.46–2.81), (Callaghan et al., 2012b; See Figure 32). The risk in the METH group was not significantly higher than that in the cannabis group, for which previous epidemiological investigations have reported increased (but still unexplained) risk of developing schizophrenia (see Minozzi et al., 2010) (Figure 33).

There are many generic confounds in this investigation including uncertainty as to the validity of the hospital diagnosis of schizophrenia and absence of schizophrenia prior to admission to hospital with a METH abuse/dependence disorder, and whether the clinicians correctly distinguished between an acute psychotic disorder caused by METH vs persistent psychosis in the drug-free state. To partly address this issue, supplemental

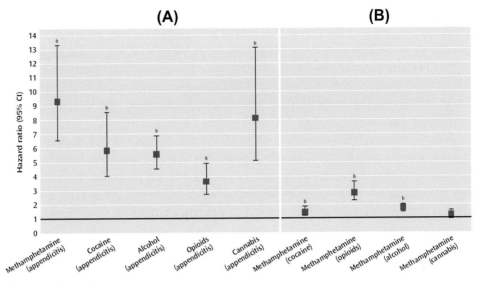

ᵃ Reference group in parentheses.
ᵇ Statistically significant difference, $p < 0.005$.

Figure 33 Does methamphetamine use increase subsequent risk of developing schizophrenia? In an epidemiological study, Japanese clinicians and dopamine-focused neuroscientists (who embrace stimulant-induced sensitization) suggest that exposure to methamphetamine might increase risk of developing permanent psychosis (schizophrenia) later in life. This is a preliminary population-based longitudinal cohort study using California hospital discharge records comparing the hazard ratios (risk) of developing schizophrenia in methamphetamine users, users of other drugs (cocaine, cannabis, alcohol, and opioids) and in a matched proxy control group (individuals hospitalized with appendicitis). Hospital data provided no evidence that subjects had schizophrenia prior to drug use but this possibility cannot be excluded. The left side (A) shows that all drug groups have elevated risk of developing schizophrenia, as compared with the control group, with the methamphetamine and cannabis groups having the highest risk. The right side (B) shows that risk of developing schizophrenia in methamphetamine users relative to the other drug cohort groups is higher than that for cocaine, opioids, and alcohol users, but not for cannabis users. The findings, requiring replication, suggest that there might be a special risk in methamphetamine and cannabis users. However, what explains our surprising observation that risk of schizophrenia is higher in *all* drug groups? To what extent did hospital records miss a diagnosis of schizophrenia preexisting drug use? *Figure reproduced with permission of the American Psychiatric Association, Callaghan et al. (2012b).*

analyses were conducted, which showed similar hazard ratios when using patients who had at least a two-year period of never having received a psychosis-related diagnosis before the start date.

Further, what is the explanation for the finding that risk of schizophrenia was increased in *all* drug cohorts? – Is this a valid finding? Might the data be explained in the context of stress–vulnerability theory of schizophrenia in which exposure to drugs of abuse, all being stressors, might interact with genetic/environmental factors to increase vulnerability to development of persistent psychosis? In a recent review of our finding,

Carter and Hall (2012) who also note the high hazard ratio in all of the drug-using groups find it not surprising that risk should be elevated in users of cocaine, cannabis, and (heavy use) alcohol, as these drugs have been implicated in causing psychosis, and suggest that the hazard ratio for those with opioid diagnoses "be taken as a better estimate of the baseline risk of schizophrenia among substance abusers." However, in so doing, the highest risk of developing schizophrenia (relative to opioid users) remains in the METH and cannabis groups.

Is persistent psychosis associated with METH use different in any respect from than that associated with schizophrenia (see Rounsaville, 2007)? Regarding brain pathophysiological similarities between schizophrenia and METH use, preliminary findings suggest that several possible substrates for the postulated dopaminergic sensitization of schizophrenia (Lieberman et al., 1997) might be different in the two conditions: dopamine release (increased in schizophrenia; Laruelle et al., 1996; blunted in METH users, Wang et al., 2011; brain dopamine D3 receptor binding: increased in METH users, Boileau et al., 2012; normal in schizophrenia, Graff-Guerrero et al., 2009).

Nevertheless, the epidemiological data, requiring replication and explanation, do suggest that exposure to METH, at doses sufficient to result in a hospital diagnosis of a METH condition, might cause a "pathological" change in brain that could increase the risk of schizophrenia more than that associated with use of several other recreational drugs, with the exception of cannabis.

3. CONCLUSIONS

The many generic confounds and lack of documentation in the METH literature make critical review of data difficult.

In studies of METH users, there are many generic difficulties that the investigator has little control over such as the possibility that the difference observed in brain of drug user might have preexisted use of METH and uncertainty whether a reduction in a brain biochemical marker equals loss of whole or part of a neuron. However, there are also interpretational confounds that could be addressed such as use of other drugs (e.g. hair testing, repeated urine drug testing), drug specificity (are the changes observed in METH users also found in other drug users employing the same measurement procedure?), and documentation of extent of overlap between drug user and normal values by disclosure of scatterplots.

A characteristic, defining brain pathology of recreational METH users has yet to be discovered; however, group mean differences in some markers might be characteristic of high dose use of the drug.

Perhaps not unexpectedly, I was unable to discover from the literature any difference in a brain marker that provided an absolute, many-times replicated, separation between normal subjects and chronic METH users. The only study in which the differences between drug users and normal subjects could be considered robust was our postmortem brain investigation of dopamine levels in METH users, in which almost all dopamine values in caudate nucleus fell below the lower limit of the control range, with some drug users having near-total dopamine "loss." However, this investigation is still awaiting independent replication.

Nevertheless, there are differences reported that I believe have a reasonable chance of "typically" being replicated in future investigations and can be considered as potentially characteristic group mean differences in METH users. These are reduced levels of the striatal dopamine transporter (reported in two postmortem and six independent brain imaging studies) and decreased gray matter density in cerebral cortex. Neither of the changes are robust in individual studies, as there is extensive overlap between drug user and control values, and the functional significance and question of drug specificity (in the case of the structural imaging finding) is still uncertain.

Dopamine neuron loss is not yet an established characteristic of METH users and will be difficult to prove.

Experimental animal findings raise the possibility that "high" doses of METH might damage dopamine neurons in the human. Since the available markers of dopamine neurons are not, or might not be, stable measures of dopamine neuronal integrity, dopamine neuron loss will be very difficult to prove in the human. Brain imaging reports of reduced striatal dopamine transporter binding in extended (years) abstinence (assuming the subjects were in fact abstinent) are consistent with a loss of part of the dopamine neuron, but do not constitute proof.

Also relevant to the question of dopamine neuronal damage is the finding that not all brain dopamine markers are convincingly or markedly decreased in human METH users (especially VMAT2 and dopa decarboxylase), as is the case in Parkinson's disease. This suggests that any actual loss of dopaminergic innervation in METH users is not likely to be substantial.

Brain differences in direct/indirect measures related to toxicity: Findings need to be strengthened and replicated with confounds better addressed.

Gliosis and brain volume differences are more accepted indirect and direct measures of toxicity. However, examination of the available literature does not yet provide a clear picture regarding the status of these indices in METH users.

The *gliosis* investigations are primarily limited to only one part of the brain (striatum) in a single postmortem study of users recently exposed to METH (no microglial activation) and a PET imaging investigation of long-abstinent drug users (inferred "microglial activation") that employed a somewhat uncertain first generation microgliosis marker. It is somewhat of a stretch of logic to reconcile the results of the two investigations. The data are still too fragmentary to provide a conclusion.

The *structural brain imaging* findings in cerebral cortex (lower gray matter density) and striatum (increased volume) are also suggestive, but are difficult to interpret because of uncertain influence of other recreational drugs and abstinence time on the outcome measures.

The public should not yet be advised, on the basis of the available evidence, that "physical brain damage" is a characteristic pathology of chronic METH use, as this has not yet been established. However, as my definition of brain damage is broad, I will argue that the minority of METH users who have become addicted to the drug must have in fact suffered some "damage" to the brain (e.g. overexpression of dopamine D3 receptors).

Studies of other measures possibly related to neurotoxicity in METH users are suggestive, but not definitive, including our own postmortem brain investigation finding of markedly increased brain levels of lipid peroxidation products (awaiting replication) and the imaging observations (not always replicated) of decreased brain levels of N-acetylaspartate.

The two epidemiological investigations cited above, associating high-dose METH use with increased risk of developing Parkinson's disease and schizophrenia, are also suggestive, but are quite preliminary and require replication. In this regard, one wonders why such large-scale population-based investigations of METH users are only now being conducted.

RECOMMENDATIONS FOR FUTURE STUDIES

Make drug hair and repeated urine testing mandatory.

It should be mandatory that for all studies of human METH users that drug hair analysis data be provided for those investigations in which hair samples can be obtained. This information is needed because it is obvious that investigators cannot rely only on information from self-reporting and there needs to be proof (more than a single drug urine test) whether subjects used METH chronically and used or did not other recreational drugs. Along these lines, Kim and colleagues (2006) advise that *repeated* urine testing be accomplished in brain scan investigations of drug users in abstinence. The need for better confirmation of drugs actually used is emphasized by recent reports suggesting that some findings in brain of METH users could be explained by co-use of alcohol or tobacco. My experience unfortunately indicates that scientific journals do not typically consider such information to be essential.

Provide scatterplots so readers can appreciate whether changes are robust.

It is obvious that the reader of results of METH investigations must be given the opportunity to establish whether drug user values do or do not fall within the normal control range. For this reason, scatter-plots of key outcome measure values should be included in such publications (at least as attached supplementary information), notwithstanding desire of journal editors and reviewers to economize on space.

More focus on regionally extensive studies of gliosis as a marker "probably" close to physical brain damage.

The reader might notice that I devoted many pages of this review to a lengthy discussion of the question of gliosis in METH users, despite mostly anecdotal data limited to an anatomically restricted postmortem brain investigation and a brain imaging study that employed a probe for which I do not have high confidence. Here I fully recognize that although gliosis, or activated gliosis cannot necessarily equal neurotoxicity, gliosis does, in my opinion, constitute an indirect marker certainly suggestive of, and close to, a damaging process. I further suspect that gliosis is likely to be a more "sensitive" index of toxic insult than the more difficult measures of neuronal number.

Should there continue to be interest in the question of METH and physical brain damage I suggest that there be extension and replication of the gliosis studies of Kitamura and Sekine but with a reduction of overly excessive focus on toxicity limited to the dopamine-rich striatum. In this regard, one wonders whether Kitamura would have discovered activated microgliosis or reactive astrocytes in brain areas outside of the single area (striatum) examined in his postmortem brain study. Future brain imaging investigations should employ a second generation gliosis probe that survives the validation process.

The ideal study will be a longitudinal brain imaging investigation that employs both glial and structural imaging measurements and spans early and late abstinence.

There is still a need for postmortem brain validation studies.

There will also need to be confirmation of brain imaging "gliosis" findings by demonstrated proof of microgliosis and reactive astrogliosis in postmortem brain of METH users employing accepted histopathological procedures. Ideally such studies should also document changes in neuronal morphology (e.g. swollen axons) and number.

Further to this point, investigators also need to be more cautious when making statements regarding the meaning of changes in gliosis markers in the absence of postmortem brain validation. For example, the title of the Sekine brain imaging study employing the PK11195 radioligand: "Methamphetamine causes microglial activation in the brains of human abusers" is unjustified (in part) because microglial activation in human brain can only be established by morphometric histological procedures.

More population-based epidemiological studies are needed.

Finally, notwithstanding the confounds inherent in epidemiological investigations, there need to be more large-scale investigations employing population based data that can be used to address questions whether METH use increases risk of subsequent development of different human brain conditions.

Include a "drug control" group (e.g. cocaine users) in all future METH studies.

We all now recognize that most METH users use other drugs that also might affect the brain. There is also the generic question whether a finding made in brain of METH users would extend to users of all drug groups examined by the investigator, as the answer will bear significantly on the interpretation of the findings. Sorting out the question of drug specificity is important on scientific and public health grounds and should be addressed by inclusion of at least a single drug control group (e.g. users of cocaine) in all future investigations of METH users. Journal editors have the opportunity to make this requirement mandatory. Drug specificity (or lack thereof) will also enhance the credibility of the observation.

ACKNOWLEDGMENTS

Dr. Kish receives research funding from the US NIH National Institute on Drug Abuse (DA 07182, 17301, and 25096) and the Canadian Institutes for Health Research and a salary from the Centre for Addiction and Mental Health, a teaching hospital of the University of Toronto.

REFERENCES

Açikgöz, O., Gönenç, S., Kayatekin, B.M., Uysal, N., Pekçetin, C., Semin, I., Güre, A., 1998. Methamphetamine causes lipid peroxidation and an increase in superoxide dismutase activity in the rat striatum. Brain Res. 813, 200–202.

Açikgöz, O., Gönenç, S., Kayatekin, B.M., Pekçetin, C., Uysal, N., Dayi, A., Semin, I., Güre, A., 2000. The effects of single dose of methamphetamine on lipid peroxidation levels in the rat striatum and prefrontal cortex. Eur. Neuropsychopharmacol. 10, 415–418.

Alburges, M.E., Keefe, K.A., Hanson, G.R., 2001. Contrasting responses by basal ganglia met-enkephalin systems to low and high doses of methamphetamine in a rat model. J.Neurochem. 76, 721–729.

Alexander, G.E., DeLong, M.R., Strick, P.L., 1986. Parallel organization of functionally segregated circuits linking basal ganglia and cortex. Annu. Rev. Neurosci. 9, 357–381.

Angrist, B., 1994. Amphetamine psychosis: clinical; variations of the syndrome. In: Cho, A.K., Segal, D.S. (Eds.), Amphetamine and Its Analogs: Psychopharmacology, Toxicology and Abuse. Academic Press, New York, NY, pp. 387–414.

Banati, R.B., Myers, R., Kreutzberg, G.W., 1997. PK ('peripheral benzodiazepine')—binding sites in the CNS indicate early and discrete brain lesions: microautoradiographic detection of [3H]PK11195 binding to activated microglia. J. Neurocytol. 26, 77–82.

Barcia, C., Sánchez Bahillo, A., Fernández-Villalba, E., Bautista, V., Poza, Y., Poza, M., Fernández-Barreiro, A., Hirsch, E.C., Herrero, M.T., 2004. Evidence of active microglia in substantia nigra pars compacta of parkinsonian monkeys 1 year after MPTP exposure. Glia 46, 402–409.

Bartels, A.L., Willemsen, A.T., Doorduin, J., de Vries, E.F., Dierckx, R.A., Leenders, K.L., 2010. [11C]-PK11195 PET: quantification of neuroinflammation and a monitor of anti-inflammatory treatment in Parkinson's disease? Parkinsonism Relat. Disord. 16, 57–59.

Baselt, R.C., 2004. Disposition of Toxic Drugs and Chemicals in Man, Seventh ed. Biomedical Publications, Foster City, CA.

Benarroch, E.E., 2008. N-Acetylaspartate and N-acetylaspartylglutamate: neurobiology and clinical significance. Neurology 70, 1353–1357.

Berger, M.A., 2005. What has a decade of Daubert wrought? Am. J. Public Health 95, S59–S65.

Berman, S., O'Neill, J., Fears, S., Bartzokis, G., London, E.D., 2008. Abuse of amphetamines and structural abnormalities in the brain. Ann. N.Y. Acad. Sci. 1141, 195–220.

Berman, S.M., Kuczenski, R., McCracken, J.T., London, E.D., 2009. Potential adverse effects of amphetamine treatment on brain and behavior: a review. Mol. Psychiatry 14, 123–142.

Berridge, K.C., Kringelbach, M.L., 2008. Affective neuroscience of pleasure: reward in humans and animals. Psychopharmacology (Berl.) 199, 457–480.

Boileau, I., Dagher, A., Leyton, M., Gunn, R.N., Baker, G.B., Diksic, M., Benkelfat, C., 2006. Modeling sensitization to stimulants in humans: an [11C]raclopride/positron emission tomography study in healthy men. Arch. Gen. Psychiatry 63, 1386–1395.

Boileau, I., Rusjan, P., Houle, S., Wilkins, D., Tong, J., Selby, P., Guttman, M., Saint-Cyr, J.A., Wilson, A.A., Kish, S.J., 2008. Increased vesicular monoamine transporter binding during early abstinence in human methamphetamine users: Is VMAT2 a stable dopamine neuron biomarker? J. Neurosci. 28, 9850–9856.

Boileau, I., Guttman, M., Rusjan, P., Adams, J.R., Houle, S., Tong, J., Hornykiewicz, O., Furukawa, Y., Wilson, A.A., Kapur, S., Kish, S.J., 2009. Decreased binding of the D3 dopamine receptor-preferring ligand [11C]-(+)-PHNO in drug-naive Parkinson's disease. Brain 132, 1366–1375.

Boileau, I., Payer, D., Houle, S., Behzadi, A., Rusjan, P.M., Tong, J., Wilkins, D., Selby, P., George, T.P., Zack, M., Furukawa, Y., McCluskey, T., Wilson, A.A., Kish, S.J., 2012. Higher binding of the dopamine D3 receptor-preferring ligand [11C]-(+)-propyl-hexahydro-naphtho-oxazin in methamphetamine poly-drug users: a positron emission tomography study. J. Neurosci. 32, 1353–1359.

Booij, J., Bergmans, P., Winogrodzka, A., Speelman, J.D., Wolters, E.C., 2001. Imaging of dopamine trans-porters with [123I]FP-CIT SPECT does not suggest a significant effect of age on the symptomatic threshold of disease in Parkinson's disease. Synapse 39, 101–108.

Bordet, R., Ridray, S., Carboni, S., Diaz, J., Sokoloff, P., Schwartz, J.C., 1997. Induction of dopamine D3 receptor expression as a mechanism of behavioral sensitization to levodopa. Proc. Natl. Acad. Sci. U.S.A. 94, 3363–3367.

Bordet, R., Ridray, S., Schwartz, J.C., Sokoloff, P., 2000. Involvement of the direct striatonigral pathway in levodopa-induced sensitization in 6-hydroxydopamine-lesioned rats. Eur. J. Neurosci. 12, 2117–2123.

Bowyer, J.F., Davies, D.L., Schmued, L., Broening, H.W., Newport, G.D., Slikker Jr., W., Holson, R.R., 1994. Further studies of the role of hyperthermia in methamphetamine neurotoxicity. J. Pharmacol. Exp. Ther. 268, 1571–1580.

Burns, R.S., Chiueh, C.C., Markey, S.P., Ebert, M.H., Jacobowitz, D.M., Kopin, I.J., 1983. A primate model of parkinsonism: selective destruction of dopaminergic neurons in the pars compacta of the substantia nigra by N-methyl-4-phenyl-1,2,3,6-tetrahydropyridine. Proc. Natl. Acad. Sci. U.S.A. 80, 4546–4550.

Cadet, J.L., Sheng, P., Ali, S., Rothman, R., Carlson, E., Epstein, C., 1994. Attenuation of methamphetamine-induced neurotoxicity in copper/zinc superoxide dismutase transgenic mice. J. Neurochem. 62, 380–383.

Callaghan, R.C., Cunningham, J.K., Sykes, J., Kish, S.J., 2012a. Increased risk of Parkinson's disease in indi-viduals hospitalized with conditions related to the use of methamphetamine or other amphetamine-type drugs. Drug Alcohol Depend. 120, 35–40.

Callaghan, R.C., Cunningham, J.K., Allebeck, P., Arenovich, T., Sajeev, G., Remington, G., Boileau, I., Kish, S.J., 2012b. Methamphetamine use and schizophrenia: a population-based cohort study in California. Am. J. Psychiatry 169, 389–396.

Cappon, G.D., Pu, C., Vorhees, C.V., 2000. Time-course of methamphetamine-induced neurotoxicity in rat caudate-putamen after single-dose treatment. Brain Res. 863, 106–111.

Cárdenas, L., Houle, S., Kapur, S., Busto, U.E., 2004. Oral D-amphetamine causes prolonged displacement of [11C]raclopride as measured by PET. Synapse 51, 27–31.

Carter, A., Hall, W. 2012. "This large Californian record linkage study provides the first prospective epide-miological evidence to support the…" Evaluation of: Callaghan, R.C., et al. Methamphetamine use and schizophrenia: a population-based cohort study in California. Am. J. Psychiatry. Faculty of 1000, Febru-ary 01, 2012. F1000.com/13515959#eval14892062.

Chang, L., Cloak, C., Patterson, K., Grob, C., Miller, E.N., Ernst, T., 2005b. Enlarged striatum in abstinent methamphetamine abusers: a possible compensatory response. Biol. Psychiatry 57, 967–974.

Chang, L., Ernst, T., Speck, O., Grob, C.S., 2005a. Additive effects of HIV and chronic methamphetamine use on brain metabolite abnormalities. Am. J. Psychiatry 162, 361–369.

Chang, L., Alicata, D., Ernst, T., Volkow, N., 2007. Structural and metabolic brain changes in the striatum associated with methamphetamine abuse. Addiction 102, 16–32.

Cheng, L.L., Newell, K., Mallory, A.E., Hyman, B.T., Gonzalez, R.G., 2002. Quantification of neurons in Alzheimer and control brains with ex vivo high resolution magic angle spinning proton magnetic reso-nance spectroscopy and stereology. Magn. Reson. Imaging 20, 527–533.

Cho, Y.H., Segal, D.S., 1994. In: Cho, A.K., Segal, D.S. (Eds.), Amphetamine and Its Analogs: Psychopharma-cology, Toxicology and Abuse. Academic Press, New York.

Chou, Y.H., Huang, W.S., Su, T.P., Lu, R.B., Wan, F.J., Fu, Y.K., 2007. Dopamine transporters and cognitive function in methamphetamine abuser after a short abstinence: a SPECT study. Eur. Neuropsychophar-macol. 17, 46–52.

Colton, C.A., 2009. Heterogeneity of microglial activation in the innate immune response in the brain. J. Neuroimmune. Pharmacol. 4, 399–418.

Crombag, H.S., Gorny, G., Li, Y., Kolb, B., Robinson, T.E., 2005. Opposite effects of amphetamine self-administration experience on dendritic spines in the medial and orbital prefrontal cortex. Cereb. Cortex 15, 341–348.

D'Almeida, V., Camarini, R., Azzalis, L.A., Mattei, R., Junqueira, V.B., Carlini, E.A., 1995. Antioxidant defense in rat brain after chronic treatment with anorectic drugs. Toxicol. Lett. 81, 101–105.

Daubert v. Merrell Dow Pharmaceuticals, Inc. 1993. United States Supreme Court No. 92–102, 509 U.S. 579.

Dautry, C., Vaufrey, F., Brouillet, E., Bizat, N., Henry, P.G., Condé, F., Bloch, G., Hantraye, P., 2000. Early N-acetylaspartate depletion is a marker of neuronal dysfunction in rats and primates chronically treated with the mitochondrial toxin 3-nitropropionic acid. J. Cereb. Blood Flow Metab. 20, 789–799.

De La Fuente-Fernandez, R., Furtado, S., Guttman, M., Furukawa, Y., Lee, C.S., Calne, D.B., Ruth, T.J., Stoessl, A.J., 2003. VMAT2 binding is elevated in dopa-responsive dystonia: visualizing empty vesicles by PET. Synapse 49, 20–28.

De Vito, M.J., Wagner, G.C., 1989. Methamphetamine-induced neuronal damage: a possible role for free radicals. Neuropharmacology 28, 1145–1150.

Dietrich, J.B., Mangeol, A., Revel, M.O., Burgun, C., Aunis, D., Zwiller, J., 2005. Acute or repeated cocaine administration generates reactive oxygen species and induces antioxidant enzyme activity in dopaminergic rat brain structures. Neuropharmacology 48, 965–974.

Ellison, G., Eison, M.S., Huberman, H.S., Daniel, F., 1978. Long-term changes in dopaminergic innervation of caudate nucleus after continuous amphetamine administration. Science 201, 276–278.

Ersche, K.D., Barnes, A., Jones, P.S., Morein-Zamir, S., Robbins, T.W., Bullmore, E.T., 2011. Abnormal structure of frontostriatal brain systems is associated with aspects of impulsivity and compulsivity in cocaine dependence. Brain 134, 2013–2024.

Ersche, K.D., Jones, P.S., Williams, G.B., Turton, A.J., Robbins, T.W., Bullmore, E.T., 2012. Abnormal brain structure implicated in stimulant drug addiction. Science 335, 601–604.

Ernst, T., Chang, L., Leonido-Yee, M., Speck, O., 2000. Evidence for long-term neurotoxicity associated with methamphetamine abuse: a 1H MRS study. Neurology 54, 1344–1349.

Escubedo, E., Guitart, L., Sureda, F.X., Jiménez, A., Pubill, D., Pallàs, M., Camins, A., Camarasa, J., 1998. Microgliosis and down-regulation of adenosine transporter induced by methamphetamine in rats. Brain Res. 814, 120–126.

Esterbauer, H., Schaur, R.J., Zollner, H., 1991. Chemistry and biochemistry of 4-hydroxynonenal, malonaldehyde and related aldehydes. Free Radic. Biol. Med. 11, 81–128.

Fantegrossi, W.E., Ciullo, J.R., Wakabayashi, K.T., De La Garza 2nd, R., Traynor, J.R., Woods, J.H., 2008. A comparison of the physiological, behavioral, neurochemical and microglial effects of methamphetamine and 3,4-methylenedioxymethamphetamine in the mouse. Neuroscience 151, 533–543.

Farde, L., Ehrin, E., Eriksson, L., Greitz, T., Hall, H., Hedström, C.G., Litton, J.E., Sedvall, G., 1985. Substituted benzamides as ligands for visualization of dopamine receptor binding in the human brain by positron emission tomography. Proc. Natl. Acad. Sci. U.S.A. 82, 3863–3867.

Fellner, L., Jellinger, K.A., Wenning, G.K., Stefanova, N., 2011. Glial dysfunction in the pathogenesis of α-synucleinopathies: emerging concepts. Acta Neuropathol. 121, 675–693.

Fitzmaurice, P.S., Tong, J., Yazdanpanah, M., Liu, P.P., Kalasinsky, K.S., Kish, S.J., 2006. Levels of 4-hydroxynonenal and malondialdehyde are increased in brain of human chronic users of methamphetamine. J. Pharmacol. Exp. Ther. 319, 703–709.

Flaum, M., Schultz, S.K., 1996. When does amphetamine-induced psychosis become schizophrenia? Am. J. Psychiatry 153, 812–815.

Flora, G., Lee, Y.W., Nath, A., Maragos, W., Hennig, B., Toborek, M., 2002. Methamphetamine-induced TNF-alpha gene expression and activation of AP-1 in discrete regions of mouse brain: potential role of reactive oxygen intermediates and lipid peroxidation. Neuromolecular. Med. 2, 71–85.

Frankel, P.S., Alburges, M.E., Bush, L., Hanson, G.R., Kish, S.J., 2007. Brain levels of neuropeptides in human chronic methamphetamine users. Neuropharmacology 53, 447–454.

Frankel, P.S., Alburges, M.E., Bush, L., Hanson, G.R., Kish, S.J., 2008. Striatal and ventral pallidum dynorphin concentrations are markedly increased in human chronic cocaine users. Neuropharmacology 55, 41–46.

Frankle, W.G., Huang, Y., Hwang, D.R., Talbot, P.S., Slifstein, M., Van Heertum, R., Abi-Dargham, A., Laruelle, M., 2004. Comparative evaluation of serotonin transporter radioligands 11C-DASB and 11C-McN 5652 in healthy humans. J. Nucl. Med. 45, 682–694.

Frey, K.A., Koeppe, R.A., Kilbourn, M.R., Vander Borght, T.M., Albin, R.L., Gilman, S., Kuhl, D.E., 1996. Presynaptic monoaminergic vesicles in Parkinson's disease and normal aging. Ann. Neurol. 40, 873–884.

Frey, K., Kilbourn, M., Robinson, T., 1997. Reduced striatal vesicular monoamine transporters after neurotoxic but not after behaviorally-sensitizing doses of methamphetamine. Eur. J. Pharmacol. 334, 273–279.

Frey, K.A., Koeppe, R.A., Kilbourn, M.R., 2001. Imaging the vesicular monoamine transporter. Adv. Neurol. 86, 237–247.

Gerhard, A., Banati, R.B., Goerres, G.B., Cagnin, A., Myers, R., Gunn, R.N., Turkheimer, F., Good, C.D., Mathias, C.J., Quinn, N., Schwarz, J., Brooks, D.J., 2003. [11C](R)-PK11195 PET imaging of microglial activation in multiple system atrophy. Neurology 61, 686–689.

Gerhard, A., Pavese, N., Hotton, G., Turkheimer, F., Es, M., Hammers, A., Eggert, K., Oertel, W., Banati, R.B., Brooks, D.J., 2006. In vivo imaging of microglial activation with [11C](R)-PK11195 PET in idiopathic Parkinson's disease. Neurobiol. Dis. 21, 404–412.

Gluck, M.R., Moy, L.Y., Jayatilleke, E., Hogan, K.A., Manzino, L., Sonsalla, P.K., 2001. Parallel increases in lipid and protein oxidative markers in several mouse brain regions after methamphetamine treatment. J. Neurochem. 79, 152–160.

Graeber, M.B., Streit, W.J., 2010. Microglia: biology and pathology. Acta Neuropathol. 119, 89–105.

Graff-Guerrero, A., Mizrahi, R., Agid, O., Marcon, H., Barsoum, P., Rusjan, P., Wilson, A.A., Zipursky, R., Kapur, S., 2009. The dopamine D2 receptors in high-affinity state and D3 receptors in schizophrenia: a clinical [11C]-(+)-PHNO PET study. Neuropsychopharmacology 34, 1078–1086.

Grahn, J.A., Parkinson, J.A., Owen, A.M., 2008. The cognitive functions of the caudate nucleus. Prog. Neurobiol. 86, 141–155.

Grelotti, D.J., Kanayama, G., Pope, H.G., 2010. Remission of persistent methamphetamine-induced psychosis after electroconvulsive therapy: presentation of a case and review of the literature. Am. J. Psychiatry 167, 17–23.

Griffith, J.D., Cavanaugh, J., Held, J., Oates, J.A., 1972. Dextroamphetamine. Evaluation of psychomimetic properties in man. Arch. Gen. Psychiatry 26, 97–100.

Guttman, M., Burkholder, J., Kish, S.J., Hussey, D., Wilson, A., DaSilva, J., Houle, S., 1997. [11C]RTI-32 PET studies of the dopamine transporter in early dopa-naïve Parkinson's disease: implications for the symptomatic threshold. Neurology 48, 1578–1583.

Halliday, G.M., Stevens, C.H., 2011. Glia: initiators and progressors of pathology in Parkinson's disease. Mov. Disord. 26, 6–17.

Hart, C.L., Marvin, C.B., Silver, R., Smith, E.E., 2012. Is cognitive functioning impaired in methamphetamine users? A critical review. Neuropsychopharmacology 37, 586–608.

Harvey, D.C., Lacan, G., Tanious, S.P., Melega, W.P., 2000. Recovery from methamphetamine induced long-term nigrostriatal dopaminergic deficits without substantia nigra cell loss. Brain Res. 871, 259–270.

Heidbreder, C.A., Newman, A.H., 2010. Current perspectives on selective dopamine D(3) receptor antagonists as pharmacotherapeutics for addictions and related disorders. Ann. N.Y. Acad. Sci. 1187, 4–34.

Hess, A., Desiderio, C., McAuliffe, W.G., 1990. Acute neuropathological changes in the caudate nucleus caused by MPTP and methamphetamine: immunohistochemical studies. J. Neurocytol. 19, 338–342.

Hodges, A.B., Ladenheim, B., McCoy, M.T., Beauvais, G., Cai, N., Krasnova, I.N., Cadet, J.L., 2011. Long-term protective effects of methamphetamine preconditioning against single-day methamphetamine toxic challenges. Curr. Neuropharmacol. 9, 35–39.

Horikoshi, Y., Sasaki, A., Taguchi, N., Maeda, M., Tsukagoshi, H., Sato, K., Yamaguchi, H., 2003. Human GLUT5 immunolabeling is useful for evaluating microglial status in neuropathological study using paraffin sections. Acta Neuropathol. 105, 157–162.

Horner, K.A., Gilbert, Y.E., Cline, S.D., 2011. Widespread increases in malondialdehyde immunoreactivity in dopamine-rich and dopamine-poor regions of rat brain following multiple, high doses of methamphetamine. Front. Syst. Neurosci. 5, 27.

Hornykiewicz, O., Kish, S.J., 1987. Biochemical pathophysiology of Parkinson's disease. Adv. Neurol. 45, 19–34.

Hurd, Y.L., Herkenham, M., 1993. Molecular alterations in the neostriatum of human cocaine addicts. Synapse 13, 357–369.

Imamura, K., Hishikawa, N., Sawada, M., Nagatsu, T., Yoshida, M., Hashizume, Y., 2003. Distribution of major histocompatibility complex class II-positive microglia and cytokine profile of Parkinson's disease brains. Acta Neuropathol. 106, 518–526.

Iwashita, A., Mihara, K., Yamazaki, S., Matsuura, S., Ishida, J., Yamamoto, H., Hattori, K., Matsuoka, N., Mutoh, S., 2004. A new poly(ADP-ribose) polymerase inhibitor, FR261529 [2-(4-chlorophenyl)-5-quinoxalinecarboxamide], ameliorates methamphetamine-induced dopaminergic neurotoxicity in mice. J. Pharmacol. Exp. Ther. 310, 1114–1124.

Jayanthi, S., Ladenheim, B., Cadet, J.L., 1998. Methamphetamine-induced changes in antioxidant enzymes and lipid peroxidation in copper/zinc-superoxide dismutase transgenic mice. Ann. N.Y. Acad. Sci. 844, 92–102.

Jernigan, T.L., Gamst, A.C., Archibald, S.L., Fennema-Notestine, C., Mindt, M.R., Marcotte, T.D., Heaton, R.K., Ellis, R.J., Grant, I., 2005. Effects of methamphetamine dependence and HIV infection on cerebral morphology. Am. J. Psychiatry 162, 1461–1472.

Johanson, C.E., Frey, K.A., Lundahl, L.H., Keenan, P., Lockhart, N., Roll, J., Galloway, G.P., Koeppe, R.A., Kilbourn, M.R., Robbins, T., Schuster, C.R., 2006. Cognitive function and nigrostriatal markers in abstinent methamphetamine abusers. Psychopharmacology (Berl.) 185, 327–338.

Kalasinsky, K.S., Bosy, T.Z., Schmunk, G.A., Reiber, G., Anthony, R.M., Furukawa, Y., Guttman, M., Kish, S.J., 2001. Regional distribution of methamphetamine in autopsied brain of chronic human methamphetamine users. Forensic Sci. Int. 116, 163–169.

Kalasinsky, K.S., Hugel, J., Kish, S.J., 2004. Use of MDA (the "love drug") and methamphetamine in Toronto by unsuspecting users of ecstasy (MDMA). J. Forensic Sci. 49, 1106–1112.

Kapur, S., Zipursky, R., Jones, C., Remington, G., Houle, S., 2000. Relationship between dopamine D(2) occupancy, clinical response, and side effects: a double-blind PET study of first-episode schizophrenia. Am. J. Psychiatry 157, 514–520.

Kettenmann, H., Hanisch, U.K., Noda, M., Verkhratsky, A., 2011. Physiology of microglia. Physiol. Rev. 91, 461–553.

Khurgel, M., Switzer 3rd, R.C., Teskey, G.C., Spiller, A.E., Racine, R.J., Ivy, G.O., 1995. Activation of astrocytes during epileptogenesis in the absence of neuronal degeneration. Neurobiol. Dis. 2, 23–35.

Kilbourn, M.R., Frey, K.A., Vander Borght, T., Sherman, P.S., 1996. Effects of dopaminergic drug treatments on in vivo radioligand binding to brain vesicular monoamine transporters. Nucl. Med. Biol. 23, 467–471.

Kim, H.C., Jhoo, W.K., Choi, D.Y., Im, D.H., Shin, E.J., Suh, J.H., Floyd, R.A., Bing, G., 1999. Protection of methamphetamine nigrostriatal toxicity by dietary selenium. Brain Res. 851, 76–86.

Kim, S.J., Lyoo, I.K., Hwang, J., Chung, A., Hoon Sung, Y., Kim, J., Kwon, D.H., Chang, K.H., Renshaw, P.F., 2006. Prefrontal grey-matter changes in short-term and long-term abstinent methamphetamine abusers. Int. J. Neuropsychopharmacol. 9, 221–228.

Kirkpatrick, M.G., Gunderson, E.W., Johanson, C.E., Levin, F.R., Foltin, R.W., Hart, C.L., 2012. Comparison of intranasal methamphetamine and d-amphetamine self-administration by humans. Addiction 107, 783–791.

Kish, S.J., 2003. What is the evidence that ecstasy (MDMA) can cause Parkinson's disease? Mov. Disord. 18, 1219–1223.

Kish, S.J., 2008. Pharmacologic mechanisms of crystal meth. CMAJ 178, 1679–1682.

Kish, S.J., Shannak, K., Hornykiewicz, O., 1988. Uneven pattern of dopamine loss in the striatum of patients with idiopathic Parkinson's disease. Pathophysiologic and clinical implications. N. Engl. J. Med. 318, 876–880.

Kish, S.J., Kleinert, R., Minauf, M., Gilbert, J., Walter, G.F., Slimovitch, C., Maurer, E., Rezvani, Y., Myers, R., Hornykiewicz, O., 1990. Brain neurotransmitter changes in three patients who had a fatal hyperthermia syndrome. Am. J. Psychiatry 147, 1358–1363.

Kish, S.J., Shannak, K., Rajput, A., Deck, J.H., Hornykiewicz, O., 1992. Aging produces a specific pattern of striatal dopamine loss: implications for the etiology of idiopathic Parkinson's disease. J. Neurochem. 58, 642–648.

Kish, S.J., Zhong, X.H., Hornykiewicz, O., Haycock, J.W., 1995. Striatal 3,4-dihydroxyphenylalanine decarboxylase in aging: disparity between postmortem and positron emission tomography studies? Ann. Neurol. 38, 260–264.

Kish, S.J., Kalasinsky, K.S., Furukawa, Y., Guttman, M., Ang, L., Li, L., Adams, V., Reiber, G., Anthony, R.A., Anderson, W., Smialek, J., DiStefano, L., 1999. Brain choline acetyltransferase activity in chronic, human users of cocaine, methamphetamine, and heroin. Mol. Psychiatry 4, 26–32.

Kish, S.J., Furukawa, Y., Ang, L., Vorce, S.P., Kalasinsky, K.S., 2000. Striatal serotonin is depleted in brain of a human MDMA (ecstasy) user. Neurology 55, 294–296.

Kish, S.J., Kalasinsky, K.S., Derkach, P., Schmunk, G.A., Guttman, M., Ang, L., Adams, V., Furukawa, Y., Haycock, J.W., 2001. Striatal dopaminergic and serotonergic markers in human heroin users. Neuropsychopharmacology 24, 561–567.

Kish, S.J., Furukawa, Y., Chang, L.J., Tong, J., Ginovart, N., Wilson, A., Houle, S., Meyer, J.H., 2005. Regional distribution of serotonin transporter protein in postmortem human brain: is the cerebellum a SERT-free brain region? Nucl. Med. Biol. 32, 123–128.

Kish, S.J., Tong, J., Hornykiewicz, O., Rajput, A., Chang, L.J., Guttman, M., Furukawa, Y., 2008. Preferential loss of serotonin markers in caudate versus putamen in Parkinson's disease. Brain 131, 120–131.

Kish, S.J., Fitzmaurice, P.S., Boileau, I., Schmunk, G.A., Ang, L.C., Furukawa, Y., Chang, L.J., Wickham, D.J., Sherwin, A., Tong, J., 2009. Brain serotonin transporter in human methamphetamine users. Psychopharmacology (Berl.) 202, 649–661.

Kish, S.J., Fitzmaurice, P.S., Chang, L.J., Furukawa, Y., Tong, J., 2010a. Low striatal serotonin transporter protein in a human polydrug MDMA (ecstasy) user: a case study. J. Psychopharmacol. 24, 281–284.

Kish, S.J., Lerch, J., Furukawa, Y., Tong, J., McCluskey, T., Wilkins, D., Houle, S., Meyer, J., Mundo, E., Wilson, A.A., Rusjan, P.M., Saint-Cyr, J.A., Guttman, M., Collins, D.L., Shapiro, C., Warsh, J.J., Boileau, I., 2010b. Decreased cerebral cortical serotonin transporter binding in ecstasy users: a positron emission tomography/[(11)C]DASB and structural brain imaging study. Brain 133, 1779–1797.

Kita, T., Shimada, K., Mastunari, Y., Wagner, G.C., Kubo, K., Nakashima, T., 2000. Methamphetamine-induced striatal dopamine neurotoxicity and cyclooxygenase-2 protein expression in BALB/c mice. Neuropharmacology 39, 399–406.

Kitamura, O., Tokunaga, I., Gotohda, T., Kubo, S., 2007. Immunohistochemical investigation of dopaminergic terminal markers and caspase-3 activation in the striatum of human methamphetamine users. Int. J. Leg. Med. 121, 163–168.

Kitamura, O., Takeichi, T., Wang, E.L., Tokunaga, I., Ishigami, A., Kubo, S., 2010. Microglial and astrocytic changes in the striatum of methamphetamine abusers. Leg. Med. (Tokyo) 12, 57–62.

Kogan, F.J., Nichols, W.K., Gibb, J.W., 1976. Influence of methamphetamine on nigral and striatal tyrosine hydroxylase activity and on striatal dopamine levels. Eur. J. Pharmacol. 36, 363–371.

Krasnova, I.N., Cadet, J.L., 2009. Methamphetamine toxicity and messengers of death. Brain Res. Rev. 60, 379–407.

Krasnova, I.N., Justinova, Z., Ladenheim, B., Jayanthi, S., McCoy, M.T., Barnes, C., Warner, J.E., Goldberg, S.R., Cadet, J.L., 2010. Methamphetamine self-administration is associated with persistent biochemical alterations in striatal and cortical dopaminergic terminals in the rat. PLoS One 5, e8790.

Kriegstein, A., Alvarez-Buylla, A., 2009. The glial nature of embryonic and adult neural stem cells. Annu. Rev. Neurosci. 32, 149–184.

Kühn, S., Schubert, F., Gallinat, J., 2010. Reduced thickness of medial orbitofrontal cortex in smokers. Biol. Psychiatry 68, 1061–1065.

Ladenheim, B., Krasnova, I.N., Deng, X., Oyler, J.M., Polettini, A., Moran, T.H., Huestis, M.A., Cadet, J.L., 2000. Methamphetamine-induced neurotoxicity is attenuated in transgenic mice with a null mutation for interleukin-6. Mol. Pharmacol. 58, 1247–1256.

Lamb, R.J., Henningfield, J.E., 1994. Human d-amphetamine drug discrimination: methamphetamine and hydromorphone. J. Exp. Anal. Behav. 61, 169–180.

Langston, J.W., Ballard, P., Tetrud, J.W., Irwin, I., 1983. Chronic Parkinsonism in humans due to a product of meperidine-analog synthesis. Science 219, 979–980.

Langston, J.W., Forno, L.S., Tetrud, J., Reeves, A.G., Kaplan, J.A., Karluk, D., 1999. Evidence of active nerve cell degeneration in the substantia nigra of humans years after 1-methyl-4-phenyl-1,2,3,6-tetrahydropyridine exposure. Ann. Neurol. 46, 598–605.

Laruelle, M., Abi-Dargham, A., van Dyck, C.H., Rosenblatt, W., Zea-Ponce, Y., Zoghbi, S.S., Baldwin, R.M., Charney, D.S., Hoffer, P.B., Kung, H.F., Innis, R.B., 1995. SPECT imaging of striatal dopamine release after amphetamine challenge. J. Nucl. Med. 36, 1182–1190.

Laruelle, M., Abi-Dargham, A., van Dyck, C.H., Gil, R., D'Souza, C.D., Erdos, J., McCance, E., Rosenblatt, W., Fingado, C., Zoghbi, S.S., Baldwin, R.M., Seibyl, J.P., Krystal, J.H., Charney, D.S., Innis, R.B., 1996. Single photon emission computerized tomography imaging of amphetamine-induced dopamine release in drug-free schizophrenic subjects. Proc. Natl. Acad. Sci. U.S.A. 93, 9235–9240.

LaVoie, M.J., Card, J.P., Hastings, T.G., 2004. Microglial activation precedes dopamine terminal pathology in methamphetamine-induced neurotoxicity. Exp. Neurol. 187, 47–57.

Lawyer, G., Bjerkan, P.S., Hammarberg, A., Jayaram-Lindström, N., Franck, J., Agartz, I., 2010. Amphetamine dependence and co-morbid alcohol abuse: associations to brain cortical thickness. BMC Pharmacol. 10, 5.

Le Foll, B., Goldberg, S.R., Sokoloff, P., 2005. The dopamine D3 receptor and drug dependence: effects on reward or beyond? Neuropharmacology 49, 525–541.

Lee, B., London, E.D., Poldrack, R.A., Farahi, J., Nacca, A., Monterosso, J.R., Mumford, J.A., Bokarius, A.V., Dahlbom, M., Mukherjee, J., Bilder, R.M., Brody, A.L., Mandelkern, M.A., 2009. Striatal dopamine d2/d3 receptor availability is reduced in methamphetamine dependence and is linked to impulsivity. J. Neurosci. 29, 14734–14740.

Lee, C.S., Samii, A., Sossi, V., Ruth, T.J., Schulzer, M., Holden, J.E., Wudel, J., Pal, P.K., de la Fuente-Fernandez, R., Calne, D.B., Stoessl, A.J., 2000. In vivo positron emission tomographic evidence for compensatory changes in presynaptic dopaminergic nerve terminals in Parkinson's disease. Ann. Neurol. 47, 493–503.

Lieberman, J.A., Sheitman, B.B., Kinon, B.J., 1997. Neurochemical sensitization in the pathophysiology of schizophrenia: deficits and dysfunction in neuronal regulation and plasticity. Neuropsychopharmacology 17, 205–229.

Little, K.Y., Kirkman, J.A., Carroll, F.I., Clark, T.B., Duncan, G.E., 1993. Cocaine use increases [3H]WIN 35428 binding sites in human striatum. Brain Res. 628, 17–25.

Little, K.Y., Krolewski, D.M., Zhang, L., Cassin, B.J., 2003. Loss of striatal vesicular monoamine transporter protein (VMAT2) in human cocaine users. Am. J. Psychiatry 160, 47–55.

Little, K.Y., Ramssen, E., Welchko, R., Volberg, V., Roland, C.J., Cassin, B., 2009. Decreased brain dopamine cell numbers in human cocaine users. Psychiatry Res. 168, 173–180.

Lorez, H., 1981. Fluorescence histochemistry indicates damage of striatal dopamine nerve terminals in rats after multiple doses of methamphetamine. Life Sci. 28, 911–916.

Malison, R.T., Best, S.E., van Dyck, C.H., McCance, E.F., Wallace, E.A., Laruelle, M., Baldwin, R.M., Seibyl, J.P., Price, L.H., Kosten, T.R., Innis, R.B., 1998. Elevated striatal dopamine transporters during acute cocaine abstinence as measured by [123I] beta-CIT SPECT. Am. J. Psychiatry 155, 832–834.

Malmberg, A., Jackson, D.M., Eriksson, A., Mohell, N., 1993. Unique binding characteristics of antipsychotic agents interacting with human dopamine D2A, D2B, and D3 receptors. Mol. Pharmacol. 43, 749–754.

Martin, W.R., Sloan, J.W., Sapira, J.D., Jasinski, D.R., 1971. Physiologic, subjective, and behavioral effects of amphetamine, methamphetamine, ephedrine, phenmetrazine, and methylphenidate in man. Clin. Pharmacol. Ther. 12, 245–258.

Martinez, D., Broft, A., Foltin, R.W., Slifstein, M., Hwang, D.R., Huang, Y., Perez, A., Frankel, W.G., Cooper, T., Kieber, H.D., Fischman, M.W., Laruelle, M., 2004. Cocaine dependence and D2 receptor availability in the functional subdivisions of the striatum: relationship with cocaine-seeking behavior. Neuropharmacology 29, 1190–1202.

Martinez, D., Narendran, R., Foltin, R.W., Slifstein, M., Hwang, D.R., Broft, A., Huang, Y., Cooper, T.B., Fischman, M.W., Kleber, H.D., Laruelle, M., 2007. Amphetamine-induced dopamine release: markedly blunted in cocaine dependence and predictive of the choice to self-administer cocaine. Am. J. Psychiatry 164, 622–629.

Martinez, D., Carpenter, K.M., Liu, F., Slifstein, M., Broft, A., Friedman, A.C., Kumar, D., Van Heertum, R., Kleber, H.D., Nunes, E., 2011. Imaging dopamine transmission in cocaine dependence: link between neurochemistry and response to treatment. Am. J. Psychiatry 168, 634–641.

Martinez, D., Saccone, P.A., Liu, F., Slifstein, M., Orlowska, D., Grassetti, A., Cook, S., Broft, A., Van Heertum, R., Comer, S.D., 2012. Deficits in dopamine D(2) receptors and presynaptic dopamine in heroin dependence: commonalities and differences with other types of addiction. Biol. Psychiatry 71, 192–198.

Mash, D.C., 1997. D3 receptor binding in human brain during cocaine overdose. Mol. Psychiatry 2, 5–6.

Matochik, J.A., London, E.D., Eldreth, D.A., Cadet, J.L., Bollam, K.I., 2003. Frontal cortical tissue composition in abstinent cocaine abusers: a magnetic resonance imaging study. Neuroimage 19, 1095–1102.

McCann, U.D., Ricaurte, G.A., 2004. Amphetamine neurotoxicity: accomplishments and remaining challenges. Neurosci. Biobehav. Rev. 27, 821–826.

McCann, U.D., Wong, D.F., Yokoi, F., Villemagne, V., Dannals, R.F., Ricaurte, G.A., 1998. Reduced striatal dopamine transporter density in abstinent methamphetamine and methcathinone users: evidence from positron emission tomography studies with [11C]WIN-35,428. J. Neurosci. 18, 8417–8422.

McCann, U.D., Szabo, Z., Seckin, E., Rosenblatt, P., Mathews, W.B., Ravert, H.T., Dannals, R.F., Ricaurte, G.A., 2005. Quantitative PET studies of the serotonin transporter in MDMA users and controls using [11C]McN5652 and [11C]DASB. Neuropsychopharmacology 30, 1741–1750.

McCann, U.D., Kuwabara, H., Kumar, A., Palermo, M., Abbey, R., Brasic, J., Ye, W., Alexander, M., Dannals, R.F., Wong, D.F., Ricaurte, G.A., 2008. Persistent cognitive and dopamine transporter deficits in abstinent methamphetamine users. Synapse 62, 91–100.

McFadden, L., Hadlock, G.C., Allen, S.C., Vieira-Brock, P.L., Stout, K.A., Ellis, J.D., Hoonakker, A.J., Anderyak, D.M., Neilson, S.M., Wilkins, D.G., Hanson, G.R., Fleckenstein, A.E., 2012. Methamphetamine self-administration causes persistent striatal dopaminergic alterations and mitigates the deficits caused by a subsequent methamphetamine exposure. J. Pharmacol. Exp. Ther. 340, 295–303.

Melega, W.P., Jorgensen, M.J., Laćan, G., Way, B.M., Pham, J., Morton, G., Cho, A.K., Fairbanks, L.A., 2008. Long-term methamphetamine administration in the vervet monkey models aspects of a human exposure: brain neurotoxicity and behavioral profiles. Neuropsychopharmacology 33, 1441–1452.

Mesulam, M.M., Geula, C., 1988. Nucleus basalis (Ch4) and cortical cholinergic innervation in the human brain:observations based on the distribution of acetylcholinesterase and choline acetyltransferase. J. Comp. Neurol. 275, 216–240.

Minozzi, S., Davoli, M., Bargagli, A.M., Amato, L., Vecchi, S., Perucci, C.A., 2010. An overview of systematic reviews on cannabis and psychosis: discussing apparently conflicting results. Drug Alcohol Rev. 29, 304–317.

Mirecki, A., Fitzmaurice, P., Ang, L., Kalasinsky, K.S., Peretti, F.J., Aiken, S.S., Wickham, D.J., Sherwin, A., Nobrega, J.N., Forman, H.J., Kish, S.J., 2004. Brain antioxidant systems in human methamphetamine users. J. Neurochem. 89, 1396–1408.

Moffett, J.R., Ross, B., Arun, P., Madhavarao, C.N., Namboodiri, A.M., 2007. N-Acetylaspartate in the CNS: from neurodiagnostics to neurobiology. Prog. Neurobiol. 81, 89–131.

Morales, A.M., Lee, B., Helleman, G., O'Neill, J., London, E., 2012. Gray-matter volume in methamphetamine dependence: cigarette smoking and changes with abstinence from methamphetamine. Drug Alcohol. Depend. 125, 230–238.

Moszczynska, A., Turenne, S., Kish, S.J., 1998. Rat striatal levels of the antioxidant glutathione are decreased following binge administration of methamphetamine. Neurosci. Lett. 255, 49–52.

Moszczynska, A., Fitzmaurice, P., Ang, L., Kalasinsky, K.S., Schmunk, G.A., Peretti, F.J., Aiken, S.S., Wickham, D.J., Kish, S.J., 2004. Why is parkinsonism not a feature of human methamphetamine users? Brain 127, 363–370.

Mufson, E.J., Ginsberg, S.D., Ikonomovic, M.D., DeKosky, S.T., 2003. Human cholinergic basal forebrain: chemoanatomy and neurologic dysfunction. J. Chem. Neuroanat. 26, 233–242.

Murray, A.M., Ryoo, H.L., Gurevich, E., Joyce, J.N., 1994. Localization of dopamine D3 receptors to mesolimbic and D2 receptors to mesostriatal regions of human forebrain. Proc. Natl. Acad. Sci. U.S.A. 91, 11271–11275.

Myers, R., Manjil, L.G., Cullen, B.M., Price, G.W., Frackowiak, R.S., Cremer, J.E., 1991. Macrophage and astrocyte populations in relation to [3H]PK 11195 binding in rat cerebral cortex following a local ischaemic lesion. J. Cereb. Blood Flow Metab. 11, 314–322.

Nakama, H., Chang, L., Fein, G., Shimotsu, R., Jiang, C.S., Ernst, T., 2011. Methamphetamine users show greater than normal age-related cortical gray matter loss. Addiction 106, 1474–1483.

Narendran, R., Lopresti, B.J., Martinez, D., Mason, N.S., Himes, M., May, M.A., Daley, D.C., Price, J.C., Mathis, C.A., Frankle, W.G., 2012. In vivo evidence for low striatal vesicular monoamine transporter 2 (VMAT2) availability in cocaine abusers. Am. J. Psychiatry 169, 55–63.

Naudon, L., Leroux-Nicollet, I., Costentin, J., 1994. Short-term treatments with haloperidol or bromocriptine do not alter the density of the monoamine vesicular transporter in the substantia nigra. Neurosci. Lett. 173, 1–4.

Nordahl, T.E., Salo, R., Possin, K., Gibson, D.R., Flynn, N., Leamon, M., Galloway, G.P., Pfefferbaum, A., Spielman, D.M., Adalsteinsson, E., Sullivan, E.V., 2002. Low N-acetyl-aspartate and high choline in the anterior cingulum of recently abstinent methamphetamine-dependent subjects: a preliminary proton MRS study. Magnetic resonance spectroscopy. Psychiatry Res. 116, 43–52.

Nordahl, T.E., Salo, R., Natsuaki, Y., Galloway, G.P., Waters, C., Moore, C.D., Kile, S., Buonocore, M.H., 2005. Methamphetamine users in sustained abstinence: a proton magnetic resonance spectroscopy study. Arch. Gen. Psychiatry 62, 444–452.

Norton, W.T., Aquino, D.A., Hozumi, I., Chiu, F.C., Brosnan, C.F., 1992. Quantitative aspects of reactive gliosis: a review. Neurochem. Res. 17, 877–885.

O'Callaghan, J.P., Miller, D.B., 1994. Neurotoxicity profiles of substituted amphetamines in the C57BL/6J mouse. J. Pharmacol. Exp. Ther. 270, 741–751.

O'Callaghan, J.P., Sriram, K., 2005. Glial fibrillary acidic protein and related glial proteins as biomarkers of neurotoxicity. Expert Opin. Drug Saf. 4, 433–442.

Oehmichen, M., Huber, H., 1976. Reactive microglia with membrane features of mononuclear phagocytes. J. Neuropathol. Exp. Neurol. 35, 30–39.

Oliveira, A., Hodges, H., Rezaie, P., 2003. Excitotoxic lesioning of the rat basal forebrain with S-AMPA: consequent mineralization and associated glial response. Exp. Neurol. 179, 127–138.

Oquendo, M.A., Hastings, R.S., Huang, Y.Y., Simpson, N., Ogden, R.T., Hu, X.Z., Goldman, D., Arango, V., Van Heertum, R.L., Mann, J.J., Parsey, R.V., 2007. Brain serotonin transporter binding in depressed patients with bipolar disorder using positron emission tomography. Arch. Gen. Psychiatry 64, 201–208.

Parsey, R.V., Hastings, R.S., Oquendo, M.A., Huang, Y.Y., Simpson, N., Arcement, J., Huang, Y., Ogden, R.T., Van Heertum, R.L., Arango, V., Mann, J.J., 2006. Lower serotonin transporter binding potential in the human brain during major depressive episodes. Am. J. Psychiatry 163, 52–58.

Partilla, J.S., Dempsey, A.G., Nagpal, A.S., Blough, B.E., Baumann, M.H., Rothman, R.B., 2006. Interaction of amphetamines and related compounds at the vesicular monoamine transporter. J. Pharmacol. Exp. Ther. 319, 237–246.

Payer, D.E., Dean, A.C., Boileau, I., 2012. Commentary: what matters in measuring methamphetamine-related cognitive impairments: 'abnormality diction' versus 'everyday import'? Neuropsychopharmacology 37, 1081–1082.

Perry, E.K., Gibson, P.H., Blessed, G., Perry, R.H., Tomlinson, B.E., 1977. Neurotransmitter enzyme abnormalities in senile dementia. Choline acetyltransferase and glutamic acid decarboxylase activities in necropsy brain tissue. J. Neurol. Sci. 34, 247–265.

Piccini, P., Pavese, N., Brooks, D.J., 2003. Endogenous dopamine release after pharmacological challenges in Parkinson's disease. Ann. Neurol. 53, 647–653.

Pifl, C., Drobny, H., Reither, H., Hornykiewicz, O., Singer, E.A., 1995. Mechanism of the dopamine-releasing actions of amphetamine and cocaine: plasmalemmal dopamine transporter versus vesicular monoamine transporter. Mol. Pharmacol. 47, 368–373.

Pubill, D., Canudas, A.M., Pallàs, M., Camins, A., Camarasa, J., Escubedo, E., 2003. Different glial response to methamphetamine- and methylenedioxymethamphetamine-induced neurotoxicity. Naunyn Schmiedebergs Arch. Pharmacol. 367, 490–499.

Ratai, E.M., Annamalai, L., Burdo, T., Joo, C.G., Bombardier, J.P., Fell, R., Hakimelahi, R., He, J., Lentz, M.R., Campbell, J., Curran, E., Halpern, E.F., Masliah, E., Westmoreland, S.V., Williams, K.C., González, R.G., 2011. Brain creatine elevation and N-acetylaspartate reduction indicates neuronal dysfunction in the setting of enhanced glial energy metabolism in a macaque model of neuroAIDS. Magn. Reson. Med. 66, 625–634.

Reetz, K., Tadic, V., Kasten, M., Brüggemann, N., Schmidt, A., Hagenah, J., Pramstaller, P.P., Ramirez, A., Behrens, M.I., Siebner, H.R., Klein, C., Binkofski, F., 2010. Structural imaging in the presymptomatic stage of genetically determined parkinsonism. Neurobiol. Dis. 39, 402–408.

Reneman, L., Booij, J., Lavalaye, J., de Bruin, K., Reitsma, J.B., Gunning, B., den Heeten, G.J., van Den Brink, W., 2002. Use of amphetamine by recreational users of ecstasy (MDMA) is associated with reduced striatal dopamine transporter densities: a [123I]beta-CIT SPECT study-preliminary report. Psychopharmacology (Berl.) 159, 335–340.

Renshaw, P.F., Daniels, S., Lundahl, L.H., Rogers, V., Lukas, S.E., 1999. Short-term treatment with citicoline (CDP-choline) attenuates some measures of craving in cocaine-dependent subjects: a preliminary report. Psychopharmacology (Berl.) 142, 132–138.

Ricuarte, G.A., 2011. PET Studies of Amphetamine Treatment of ADHD. US NIH 5R01MH083967 (accessed on NIH report 01.01.12.).

Ricaurte, G.A., Guillery, R.W., Seiden, L.S., Schuster, C.R., Moore, R.Y., 1982. Dopamine nerve terminal degeneration produced by high doses of methylamphetamine in the rat brain. Brain Res. 235, 93–103.

Ricaurte, G.A., Seiden, L.S., Schuster, C.R., 1984. Further evidence that amphetamines produce long-lasting dopamine neurochemical deficits by destroying dopamine nerve fibers. Brain Res. 303, 359–364.

Ricaurte, G.A., Mechan, A.O., Yuan, J., Hatzidimitriou, G., Xie, T., Mayne, A.H., McCann, U.D., 2005. Amphetamine treatment similar to that used in the treatment of adult attention – deficit/hyperactivity disorder damages dopaminergic nerve endings in the striatum of adult nonhuman primates. J. Pharmacol. Exp. Ther. 315, 91–98.

Robel, S., Berninger, B., Götz, M., 2011. The stem cell potential of glia: lessons from reactive gliosis. Nat. Rev. Neurosci. 12, 88–104.

Robinson, T.E., Berridge, K.C., 1993. The neural basis of drug craving: an incentive-sensitization theory of addiction. Brain Res. Rev. 18, 247.

Robinson, T.E., Kolb, B., 1997. Persistent structural modifications in nucleus accumbens and prefrontal cortex neurons produced by previous experience with amphetamine. J. Neurosci. 17, 8491–8497.

Robinson, T.E., Kolb, B., 1999. Morphine alters the structure of neurons in the nucleus accumbens and neocortex of rats. Synapse 33, 160–162.

Rogers, R.D., Everitt, B.J., Baldacchino, A., Blackshaw, A.J., Swainson, R., Wynne, K., Baker, N.B., Hunter, J., Carthy, T., Booker, E., London, M., Deakin, J.F., Sahakian, B.J., Robbins, T.W., 1999. Dissociable deficits in the decision-making cognition of chronic amphetamine abusers, opiate abusers, patients with focal damage to prefrontal cortex, and tryptophan-depleted normal volunteers: evidence for monoaminergic mechanisms. Neuropsychopharmacology 20, 322–339.

Ross, B.M., Moszczynska, A., Peretti, F.J., Adams, V., Schmunk, G.A., Kalasinsky, K.S., Ang, L., Mamalias, N., Turenne, S.D., Kish, S.J., 2002. Decreased activity of brain phospholipid metabolic enzymes in human users of cocaine and methamphetamine. Drug Alcohol Depend. 67, 73–79.

Ross, G.W., O'Callaghan, J.P., Sharp, D.S., Petrovitch, H., Miller, D.B., Abbott, R.D., Nelson, J., Launer, L.J., Foley, D.J., Burchfiel, C.M., Hardman, J., White, L.R., 2003. Quantification of regional glial fibrillary acidic protein levels in Alzheimer's disease. Acta Neurol. Scand. 107, 318–323.

Rothman, R.B., Baumann, M.H., 2003. Monoamine transporters and psychostimulant drugs. Eur. J. Pharmacol. 479, 23–40.

Rothman, R.B., Baumann, M.H., Dersch, C.M., Romero, D.V., Rice, K.C., Carroll, F.I., Partilla, J.S., 2001. Amphetamine-type central nervous system stimulants release norepinephrine more potently than they release dopamine and serotonin. Synapse 39, 32–41.

Rothman, R.B., Blough, B.E., Baumann, M.H., 2008. Dopamine/serotonin releasers as medications for stimulant addictions. In: Di Giovanni, G., Di Matteo, V., Esposito, E. (Eds.), Progress in Brain Research, Vol. 172. Elsevier, pp. 385–406.

Rounsaville, B.J., 2007. DSM-V research agenda: substance abuse/psychosis comorbidity. Schizophr. Bull. 33, 947–952.

Rusjan, P.M., Wilson, A.A., Bloomfield, P.M., Vitcu, I., Meyer, J.H., Houle, S., Mizrahi, R., 2011. Quantitation of translocator protein binding in human brain with the novel radioligand [18F]-FEPPA and positron emission tomography. J. Cereb. Blood Flow Metab. 31, 1807–1816.

Ryan, L.J., Martone, M.E., Linder, J.C., Groves, P.M., 1988. Cocaine, in contrast to D-amphetamine, does not cause axonal terminal degeneration in neostriatum and agranular frontal cortex of long-evans rats. Life Sci. 43, 1403–1409.

Ryan, L.J., Linder, J.C., Martone, M.E., Groves, P.M., 1990. Histological and ultrastructural evidence that D-amphetamine causes degeneration in neostriatum and frontal cortex of rats. Brain Res. 518, 67–77.

Sailasuta, N., Abulseoud, O., Harris, K.C., Ross, B.D., 2010a. Glial dysfunction in abstinent methamphetamine abusers. J. Cereb. Blood Flow Metab. 30, 950–960.

Sailasuta, N., Abulseoud, O., Hernandez, M., Haghani, P., Ross, B.D., 2010b. Metabolic abnormalities in abstinent methamphetamine dependent subjects. Subst. Abuse, 9–20.

Sakoori, K., Murphy, N.P., 2010. Reduced degeneration of dopaminergic terminals and accentuated astrocyte activation by high dose methamphetamine administration in nociceptin receptor knock out mice. Neurosci. Lett. 469, 309–313.

Sato, M., Numachi, Y., Hamamura, T., 1992. Relapse of paranoid psychotic state in methamphetamine model of schizophrenia. Schizophr. Bull. 18, 115–122.

Schintu, N., Frau, L., Ibba, M., Garau, A., Carboni, E., Carta, A.R., 2009. Progressive dopaminergic degeneration in the chronic MPTP mouse model of Parkinson's disease. Neurotox. Res. 16, 127–139.

Schwartz, D.L., Mitchell, A.D., Lahna, D.L., Luber, H.S., Huckans, M.S., Mitchell, S.H., Hoffman, W.F., 2010. Global and local morphometric differences in recently abstinent methamphetamine-dependent individuals. Neuroimage 50, 1392–1401.

Segal, D.S., Kuczenski, R., O'Neil, M.L., Melega, W.P., Cho, A.K., 2005. Prolonged exposure of rats to intravenous methamphetamine: behavioral and neurochemical characterization. Psychopharmacology (Berl.) 180, 501–512.

Seiden, L.S., Fischman, M.W., Schuster, C.R., 1976. Long-term methamphetamine induced changes in brain catecholamines in tolerant rhesus monkeys. Drug Alcohol Depend. 1, 215–219.

Sekine, Y., Iyo, M., Ouchi, Y., Matsunaga, T., Tsukada, H., Okada, H., Yoshikawa, E., Futatsubashi, M., Takei, N., Mori, N., 2001. Methamphetamine-related psychiatric symptoms and reduced brain dopamine transporters studied with PET. Am. J. Psychiatry 158, 1206–1214.

Sekine, Y., Minabe, Y., Kawai, M., Suzuki, K., Iyo, M., Isoda, H., Sakahara, H., Ashby Jr., C.R., Takei, N., Mori, N., 2002. Metabolite alterations in basal ganglia associated with methamphetamine-related psychiatric symptoms. A proton MRS study. Neuropsychopharmacology 27, 453–461.

Sekine, Y., Ouchi, Y., Takei, N., Yoshikawa, E., Nakamura, K., Futatsubashi, M., Okada, H., Minabe, Y., Suzuki, K., Iwata, Y., Tsuchiya, K.J., Tsukada, H., Iyo, M., Mori, N., 2006. Brain serotonin transporter density and aggression in abstinent methamphetamine abusers. Arch. Gen. Psychiatry 63, 90–100.

Sekine, Y., Ouchi, Y., Sugihara, G., Takei, N., Yoshikawa, E., Nakamura, K., Iwata, Y., Tsuchiya, K.J., Suda, S., Suzuki, K., Kawai, M., Takebayashi, K., Yamamoto, S., Matsuzaki, H., Ueki, T., Mori, N., Gold, M.S., Cadet, J.L., 2008. Methamphetamine causes microglial activation in the brains of human abusers. J. Neurosci. 28, 5756–5761.

Sevak, R.J., Stoops, W.W., Hays, L.R., Rush, C.R., 2009. Discriminative stimulus and subject-rated effects of methamphetamine, d-amphetamine, methylphenidate, and triazolam in methamphetamine-trained humans. J. Pharmacol. Exp. Ther. 328, 1007–1018.

Shoptaw, S., Huber, A., Peck, J., Yang, X., Liu, J., Dang, J., Roll, J., Shapiro, B., Rotheram-Fuller, E., Ling, W., 2006. Randomized, placebo-controlled trial of sertraline and contingency management for the treatment of methamphetamine dependence. Drug Alcohol Depend. 85, 12–18.

Sofroniew, M.V., 2009. Molecular dissection of reactive astrogliosis and glial scar formation. Trends Neurosci. 32, 638–647.

Sokoloff, P., Giros, B., Martres, M.P., Bouthenet, M.L., Schwartz, J.C., 1990. Molecular cloning and characterization of a novel dopamine receptor (D3) as a target for neuroleptics. Nature 347, 146–151.

Song, Y.J., Halliday, G.M., Holton, J.L., Lashley, T., O'Sullivan, S.S., McCann, H., Lees, A.J., Ozawa, T., Williams, D.R., Lockhart, P.J., Revesz, T.R., 2009. Degeneration in different parkinsonian syndromes relates to astrocyte type and astrocyte protein expression. J. Neuropathol. Exp. Neurol. 68, 1073–1083.

Staley, J.K., Mash, D.C., 1996. Adaptive increase in D3 dopamine receptors in the brain reward circuits of human cocaine fatalities. J. Neurosci. 16, 6100–6106.

Staley, J., Basile, M., Wetli, C.V., Hearn, W.L., Flynn, D.D., Ruttenber, A.J., Mash, D.C., 1993. Differential Regulation of the Dopamine Transporter in Cocaine Overdose Deaths vol. 2. National Institute on Drug Abuse Research Monograph Series. Problems of Drug Dependence, Proceedings of the 55th Annual Meeting, pp. 32.

Staley, J.K., Hearn, W.L., Ruttenber, A.J., Wetli, C.V., Mash, D.C., 1994. High affinity cocaine recognition sites on the dopamine transporter are elevated in fatal cocaine overdose victims. J. Pharmacol. Exp. Ther. 271, 1678–1685.

Staley, J.K., Talbot, J.Z., Ciliax, B.J., Miller, G.W., Levey, A.I., Kung, M.P., Kung, H.F., Mash, D.C., 1997. Radioligand binding and immunoautoradiographic evidence for a lack of toxicity to dopaminergic nerve terminals in human cocaine overdose victims. Brain Res. 747, 219–229.

Streit, W.J., 2010. Microglial activation and neuroinflammation in Alzheimer's disease: a critical examination of recent history. Front. Aging Neurosci. 2, 22.

Streit, W.J., Conde, J.R., Fendrick, S.E., Flanary, B.E., Mariani, C.L., 2005. Role of microglia in the central nervous system's immune response. Neurol. Res. 27, 685–691.

Sulzer, D., Sonders, M.S., Poulsen, N.W., Galli, A., 2005. Mechanisms of neurotransmitter release by amphetamines:a review. Prog. Neurobiol. 75, 406–433.

Sung,Y.H., Cho, S.C., Hwang, J., Kim, S.J., Kim, H., Bae, S., Kim, N., Chang, K.H., Daniels, M., Renshaw, P.F., Lyoo, I.K., 2007. Relationship between N-acetyl-aspartate in gray and white matter of abstinent methamphetamine abusers and their history of drug abuse: a proton magnetic resonance spectroscopy study. Drug Alcohol Depend. 88, 28–35.

Switzer 3rd, R.C., 2000. Application of silver degeneration stains for neurotoxicity testing. Toxicol. Pathol. 28, 70–83.

Szalavitz, M., 2011. Why the myth of the meth-damaged brain may hinder recovery. Time CNN November 21. http://healthland.time.com/2011/11/21/why-the-myth-of-the-meth-damaged-brain-may-hinder-recovery.

Tallan, H.H., Moore, S., Stein, W.H., 1956. N-Acetyl-L-aspartic acid in brain. J. Biol. Chem. 219, 257–264.

Taylor, M.J., Schweinsburg, B.C., Alhassoon, O.M., . Gongvatana, A., . Brown, G.G., . Young-Casey, C., . Letendre, S.L., . Grant, I.. HNRC Group, 2007. Effects of human immunodeficiency virus and methamphetamine on cerebral metabolites measured with magnetic resonance spectroscopy. J. Neurovirol. 13, 150–159.

Thiriet, N., Deng, X., Solinas, M., Ladenheim, B., Curtis, W., Goldberg, S.R., Palmiter, R.D., Cadet, J.L., 2005. Neuropeptide Y protects against methamphetamine-induced neuronal apoptosis in the mouse striatum. J. Neurosci. 25, 5273–5279.

Thomas, D.M., Kuhn, D.M., 2005. Attenuated microglial activation mediates tolerance to the neurotoxic effects of methamphetamine. J. Neurochem. 92, 790–797.

Thomas, D.M., Dowgiert, J., Geddes,T.J., Francescutti-Verbeem, D., Liu, X., Kuhn, D.M., 2004a. Microglial activation is a pharmacologically specific marker for the neurotoxic amphetamines. Neurosci. Lett. 367, 349–354.

Thomas, D.M.,Walker, P.D., Benjamins, J.A., Geddes,T.J., Kuhn, D.M., 2004b. Methamphetamine neurotoxicity in dopamine nerve endings of the striatum is associated with microglial activation. J. Pharmacol. Exp. Ther. 311, 1–7.

Thompson, P.M., Hayashi, K.M., Simon, S.L., Geaga, J.A., Hong, M.S., Sui,Y., Lee, J.Y., Toga, A.W., Ling, W., London, E.D., 2004. Structural abnormalities in the brains of human subjects who use methamphetamine. J. Neurosci. 24, 6028–6036.

Tong, J., Ross, B.M., Schmunk, G.A., Peretti, F.J., Kalasinsky, K.S., Furukawa,Y., Ang, L.C., Aiken, S.S., Wickham, D.J., Kish, S.J., 2003. Decreased striatal dopamine D1 receptor-stimulated adenylyl cyclase activity in human methamphetamine users. Am. J. Psychiatry 160, 896–903.

Tong, J.,Wilson, A.A., Boileau, I., Houle, S., Kish, S.J., 2008. Dopamine modulating drugs influence striatal (+)-[11C]DTBZ binding in rats:VMAT2 binding is sensitive to changes in vesicular dopamine concentration. Synapse 62, 873–876.

Tong, J.,Wong, H., Guttman, M.,Ang, L.C., Forno, L.S., Shimadzu, M., Rajput,A.H., Muenter, M.D., Kish, S.J., Hornykiewicz, O., Furukawa, Y., 2010. Brain alpha-synuclein accumulation in multiple system atrophy, Parkinson's disease and progressive supranuclear palsy: a comparative investigation. Brain 133, 172–188.

Tost, H., Braus, D.F., Hakimi, S., Ruf, M.,Vollmert, C., Hohn, F., Meyer-Lindenberg, A., 2010. Acute D2 receptor blockade induces rapid, reversible remodeling in human cortical-striatal circuits. Nat. Neurosci. 13, 920–922.

Tziortzi, A.C., Searle, G.E.,Tzimopoulou, S., Salinas, C., Beaver, J.D., Jenkinson, M., Laruelle, M., Rabiner, E.A., Gunn, R.N., 2011. Imaging dopamine receptors in humans with [11C]-(+)-PHNO: dissection of D3 signal and anatomy. Neuroimage 54, 264–277.

Vander Borght, T., Kilbourn, M., Desmond, T., Kuhl, D., Frey, K., 1995. The vesicular monoamine transporter is not regulated by dopaminergic drug treatments. Eur. J. Pharmacol. 294, 577–583.

Venneti, S., Lopresti, B.J.,Wiley, C.A., 2006.The peripheral benzodiazepine receptor (Translocator protein 18kDa) in microglia: from pathology to imaging. Prog. Neurobiol. 80, 308–322.

Vernon, A.C., Natesan, S., Crum,W.R., Cooper, J.D., Modo, M.,Williams, S.C., Kapur, S., 2012. Contrasting effects of haloperidol and lithium on rodent brain structure: a magnetic resonance imaging study with postmortem confirmation. Biol. Psychiatry (Epub ahead of print).

Vingerhoets, F.J., Snow, B.J., Tetrud, J.W., Langston, J.W., Schulzer, M., Calne, D.B., 1994. Positron emission tomographic evidence for progression of human MPTP-induced dopaminergic lesions. Ann. Neurol. 36, 765–770.

Volkow, N.D., Wang, G.J., Fowler, J.S., Logan, J., Hitzemannn, R., Gatley, S.J., MacGregor, R.R., Wolf, A.P., 1996. Cocaine uptake is decreased in the brain of detoxified cocaine abusers. Neuropsychopharmacology 14, 159–168.

Volkow, N.D., Wang, G.J., Fowler, J.S., Logan, J., Gatley, S.J., Hitzemann, R., Chen, A.D., Dewey, S.L., Pappas, N., 1997. Decreased striatal dopaminergic responsiveness in detoxified cocaine-dependent subjects. Nature 386, 830–833.

Volkow, N.D., Chang, L., Wang, G.J., Fowler, J.S., Ding, Y.S., Sedler, M., Logan, J., Franceschi, D., Gatley, J., Hitzemann, R., Gifford, A., Wong, C., Pappas, N., 2001a. Low level of brain dopamine D2 receptors in methamphetamine abusers:association with metabolism in the orbitofrontal cortex. Am. J. Psychiatry 158, 2015–2021.

Volkow, N.D., Chang, L., Wang, G.J., Fowler, J.S., Franceschi, D., Sedler, M., Gatley, S.J., Miller, E., Hitzemann, R., Ding, Y.S., Logan, J., 2001b. Loss of dopamine transporters in methamphetamine abusers recovers with protracted abstinence. J. Neurosci. 21, 9414–9418.

Volkow, N.D., Chang, L., Wang, G.J., Fowler, J.S., Leonido-Yee, M., Franceschi, D., Sedler, M.J., Gatley, S.J., Hitzemann, R., Ding, Y.S., Logan, J., Wong, C., Miller, E.N., 2001c. Association of dopamine transporter reduction with psychomotor impairment in methamphetamine abusers. Am. J. Psychiatry 158, 377–382.

Volkow, N.D., Wang, G.J., Fowler, J.S., Tomasi, D., Telang, F., 2011. Addiction: beyond dopamine reward circuitry. Proc. Natl. Acad. Sci. U. S. A. 108, 15037–15042.

Volkow, N.D., Baler, R.D., 2012. To stop or not to stop? Science 335, 546–548.

Wan, F.J., Lin, H.C., Huang, K.L., Tseng, C.J., Wong, C.S., 2000. Systemic administration of d-amphetamine induces long-lasting oxidative stress in the rat striatum. Life Sci. 66, PL205–PL212.

Wang, G.J., Smith, L., Volkow, N.D., Telang, F., Logan, J., Tomasi, D., Wong, C.T., Hoffman, W., Jayne, M., Alia-Klein, N., Thanos, P., Fowler, J.S., 2011. Decreased dopamine activity predicts relapse in methamphetamine abusers. Mol. Psychiatry July 12. (Epub ahead of print).

Westwood, S.C., Hanson, G.R., 1999. Effects of stimulants of abuse on extrapyramidal and limbic neuropeptide Y systems. J. Pharmacol. Exp. Ther. 288, 1160–1166.

Wilson, J.M., Kish, S.J., 1996. The vesicular monoamine transporter, in contrast to the dopamine transporter, is not altered by chronic cocaine self-administration in the rat. J. Neurosci. 16, 3507–3510.

Wilson, J.M., Nobrega, J.N., Carroll, M.E., Niznik, H.B., Shannak, K., Lac, S.T., Pristupa, Z.B., Dixon, L.M., Kish, S.J., 1994. Heterogeneous subregional binding patterns of 3H-WIN 35,428 and 3H-GBR 12,935 are differentially regulated by chronic cocaine self-administration. J. Neurosci. 14, 2966–2979.

Wilson, J.M., Kalasinsky, K.S., Levey, A.I., Bergeron, C., Reiber, G., Anthony, R.M., Schmunk, G.A., Shannak, K., Haycock, J.W., Kish, S.J., 1996a. Striatal dopamine nerve terminal markers in human, chronic methamphetamine users. Nat. Med. 2, 699–703.

Wilson, J.M., Levey, A.I., Rajput, A., Ang, L., Guttman, M., Shannak, K., Niznik, H.B., Hornykiewicz, O., Pifl, C., Kish, S.J., 1996b. Differential changes in neurochemical markers of striatal dopamine nerve terminals in idiopathic Parkinson's disease. Neurology 47, 718–726.

Wilson, J.M., Levey, A.I., Bergeron, C., Kalasinsky, K., Ang, L., Peretti, F., Adams, V.I., Smialek, J., Anderson, W.R., Shannak, K., Deck, J., Niznik, H.B., Kish, S.J., 1996c. Striatal dopamine, dopamine transporter, and vesicular monoamine transporter in chronic cocaine users. Ann. Neurol. 40, 428–439.

Worsley, J.N., Moszczynska, A., Falardeau, P., Kalasinsky, K.S., Schmunk, G., Guttman, M., Furukawa, Y., Ang, L., Adams, V., Reiber, G., Anthony, R.A., Wickham, D., Kish, S.J., 2000. Dopamine D1 receptor protein is elevated in nucleus accumbens of human, chronic methamphetamine users. Mol. Psychiatry 5, 664–672.

Yamamoto, B.K., Zhu, W., 1998. The effects of methamphetamine on the production of free radicals and oxidative stress. J. Pharmacol. Exp. Ther. 287, 107–114.

Yoon, S.J., Lyoo, I.K., Kim, H.J., Kim, T.S., Sung, Y.H., Kim, N., Lukas, S.E., Renshaw, P.F., 2010. Neurochemical alterations in methamphetamine-dependent patients treated with cytidine-5'-diphosphate choline: a longitudinal proton magnetic resonance spectroscopy study. Neuropsychopharmacology 35, 1165–1173.

Yuan, J., Hatzidimitriou, G., Suthar, P., Mueller, M., McCann, U., Ricaurte, G., 2006. Relationship between temperature, dopaminergic neurotoxicity, and plasma drug concentrations in methamphetamine-treated squirrel monkeys. J. Pharmacol. Exp. Ther. 316, 1210–1218.

Yuan, J., Darvas, M., Sotak, B., Hatzidimitriou, G., McCann, U.D., Palmiter, R.D., Ricaurte, G.A., 2010. Dopamine is not essential for the development of methamphetamine-induced neurotoxicity. J. Neurochem. 114, 1135–1142.

Zhang, D., Hu, X., Qian, L., O'Callaghan, J.P., Hong, J.S., 2010. Astrogliosis in CNS pathologies: is there a role for microglia? Mol. Neurobiol. 41, 232–241.

Zhong, X.H., Haycock, J.W., Shannak, K., Robitaille, Y., Fratkin, J., Koeppen, A.H., Hornykiewicz, O., Kish, S.J., 1995. Striatal dihydroxyphenylalanine decarboxylase and tyrosine hydroxylase protein in idiopathic Parkinson's disease and dominantly inherited olivopontocerebellar atrophy. Mov. Disord. 10, 10–17.

CHAPTER NINE

The Effects of Alcohol on the Human Nervous System

Kathleen T. Brady[1,2]
[1]Mental Health Service, Ralph H. Johnson VA Medical Center, Charleston, SC, USA
[2]Department of Psychiatry and Behavioral Sciences, Medical University of South Carolina, Charleston, SC, USA

1. INTRODUCTION

Each year in the United States, there are approximately 85,000 deaths and substantial disability from medical and psychiatric consequences of alcohol misuse (Harwood, 2000; Mokdad et al., 2004). Health problems created by excessive alcohol consumption have been rapidly growing (Lieber, 1995; Room et al., 2005). Current statistics indicate that 20–30% of all hospital admissions and health-care costs may be attributable to alcohol use. The prevalence of unhealthy alcohol use is estimated to be 7–20% or more among primary care outpatients, 30–40% among patients in emergency departments, and 50% among patients with trauma (D'Onofrio et al., 1998; Fiellin et al., 2000). There has been a rapid increase in total per capita alcohol consumption, which could lead to lower life expectancies in countries with the highest incidence of heavy alcohol intake (Leon and McCambridge, 2006; Shkolnikov et al., 2001).

Hazardous or harmful alcohol intake covers a spectrum of use associated with varying degrees of risk to health (Table 1 and Figure 1). While both the definition of unhealthy alcohol use and prevalence estimates vary, observational studies indicate that for men under the age of 34 and women under the age of 45, the lowest mortality is associated with no alcohol intake. Above these age cutoffs, weekly intake of no more than five drinks for men or two drinks for women is associated with the lowest mortality (White et al., 2004). While moderate use of alcohol may have beneficial cardiovascular effects, what constitutes the definition of "moderate" depends on age, sex, genetic characteristics, coexisting illnesses, and other factors. The National Institute on Alcohol Abuse and Alcoholism (NIAAA) recommendations for "low risk" alcohol consumption are no more than 4 standard drinks on any one day and no more than 14 standard drinks per week for men and no more than 3 standard drinks on any one day and no more than 7 standard drinks per week for women.

Alcohol dependence is defined by loss of control over drinking behavior and is marked by significant physiologic (withdrawal, tolerance), medical, and psychosocial sequelae. Alcohol dependence is best understood as a chronic and relapsing disorder.

The Effects of Drug Abuse on the Human Nervous System
http://dx.doi.org/10.1016/B978-0-12-418679-8.00009-5

Table 1 Alcohol Definitions

Standard Drink	A drink with 8 g of ethanol—for example, 12 oz. of beer, a small (125 ml) glass of wine, 1.5 oz. of 80 proof spirits
Hazardous drinking	"At-risk" alcohol consumption refers to no more than 4 standard drinks per day and no more than 14 standard drinks per week for men and no more than 3 standard drinks per day and no more than 7 standard drinks per week for women.
Harmful drinking	Pattern of drinking that cause damage to physical and mental health

Alcohol dependence	A cluster of symptoms in which alcohol come to dominate an individual's life with features such as:
	 • A strong desire or compulsion to drink
	 • Difficulty in controlling alcohol use
	 • Physiological withdrawal when drinking reduces
	     • Tolerance, where increasing doses of alcohol are required
	     • Neglect of other aspects of life
	     • Persisting with alcohol use despite evidence of harm

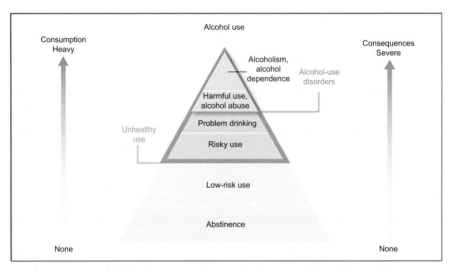

Figure 1 The spectrum of alcohol use extends from abstinence and low-risk use (the most common patterns of alcohol use) to risky use, problem drinking, harmful use and alcohol abuse, and the less common but more severe alcoholism and alcohol dependence. Consumption and the severity of consequences increase from low-risk use through dependence. The areas of the pyramid reflect the approximate prevalence of each category. Clinicians and public health practitioners should be most concerned with the categories in the shaded upper portions of the pyramid (representing unhealthy alcohol use). *(The Spectrum of Alcohol Use (Saitz, 2005)).*

In this article, the effects of alcohol on the human nervous system will be reviewed. In general, many of the effects described are associated with problematic or hazardous use of alcohol or alcohol dependence.

2.1. THE NEUROBIOLOGY OF ALCOHOL

Alcohol's central nervous system (CNS) effects are mediated through a complex interplay between excitatory and inhibitory neural systems. The impact of alcohol on these systems differs with acute versus chronic alcohol use. Alcohol's acute reinforcing effects are produced through interactions with neurotransmitter systems in reward and stress neural circuits. Following chronic exposure, changes in neuronal function in these circuits underlie the development of sensitization, tolerance, withdrawal, and dependence.

Dopamine is the primary neurotransmitter in the mesolimbic system, the major reward pathway in the brain, and is involved in the positive subjective effects associated with acute alcohol intoxication. Alcohol ingestion is associated with dopamine release in the nucleus accumbens, a primary area involved in the reward pathways of the brain (Weiss et al., 1993). However, dopamine is not the only neurotransmitter system involved in the positive subjective effects of alcohol because lesions of the mesolimbic dopamine system in animal models do not completely abolish alcohol-reinforced behavior (Rassnick et al., 1993). Chronic alcohol use and alcohol withdrawal produces decreases in dopamine function, which may play a role in withdrawal symptoms and alcohol relapse (Melis et al., 2005; Volkow et al., 2007).

The endogenous opioid system is also implicated in the actions of alcohol. It has been hypothesized that alcohol reinforcement is, in part, mediated by the release of endogenous opioids in the reward pathways of the brain. Opioid systems influence alcohol-drinking behaviors through interactions with the mesolimbic dopamine system and directly through alcohol-induced increases in extracellular endorphins in the nucleus accumbens (Olive et al., 2001). Opioid receptor antagonists interfere with the rewarding effects of alcohol through actions in the ventral tegmental area, nucleus accumbens, and central nucleus of the amygdala (Koob, 2003). Opioid receptor antagonists, such as naltrexone, have demonstrated efficacy in the treatment of alcohol dependence by decreasing alcohol intake, decreasing craving, and decreasing the reward associated with acute alcohol consumption (Anton, 2008).

Gaba-aminobutyric acid (GABA) is the major inhibitory neurotransmitter in the brain and acts via GABA-A and GABA-B receptors. Alcohol increases GABA activity by increasing its release from neurons and facilitating activity at postsynaptic receptors. Alcohol drinking is suppressed by compounds that interfere with the actions of the GABA-A receptor and compounds that stimulate the GABA-B receptor in the nucleus accumbens, ventral pallidum, bed nucleus of the stria terminalis, and amygdala (Koob, 2004). In the normal brain, GABA and glutamate (described below) work together to reach a healthy balance of neuronal excitability. Chronic alcohol exposure leads to changes in both the GABA and glutamatergic systems. In general, there is a compensatory downregulation of the GABA-ergic system with chronic alcohol use, which contributes to the excess neuronal activation that is characteristic of alcohol withdrawal.

Glutamate is the major excitatory neurotransmitter in the brain and exerts its effects through several receptor subtypes, including the N-methyl-D-aspartate (NMDA) receptor. Glutamate systems are involved in the acute reinforcing actions of alcohol. In animal models, alcohol effects can be mimicked by NMDA receptor antagonists (Colombo and Grant, 1992). In contrast to its effects on GABA, alcohol inhibits glutamatergic activity in the brain. For example, acute alcohol exposure reduces extracellular glutamate levels in reward pathways in the brain including the striatum and the nucleus accumbens (Carboni et al., 1993). NMDA receptors also play a role in neuroplasticity, a process characterized by neural reorganization that likely contributes to hyperexcitability and craving during alcohol withdrawal (Pulvirenti and Diana, 2001). Glutamate systems are also a target in the development of medications for the treatment of alcohol dependence. For example, acamprosate, an agent with FDA approval for the treatment of alcohol dependence, modulates glutamate transmission by acting on NMDA and/or metabotropic glutamate receptors (Littleton, 2007). It has been hypothesized that acamprosate blocks alcohol consumption by dampening excessive glutamatergic activity.

Serotonin (also known as 5-hydroxy-tryptamine or 5-HT) has long been a target of interest for potential pharmacotherapies for alcoholism because of the well-established link between serotonin depletion, impulsivity, and alcohol-drinking behavior in rats and humans (Myers and Veale, 1968; Virkkunen and Linnoila, 1990). Serotonin reuptake inhibitors as well as agents that block specific serotonin receptor subtypes suppress alcohol-reinforced behavior in rodents; however, the impact of serotonin reuptake inhibitors on alcohol consumption in humans has been inconsistent (Johnson, 2008).

Recent research has led to the hypothesis that the transition to alcohol dependence involves the dysregulation of both the neural circuits involved in reward and the neural circuits that mediate behavioral responses to stress. Alcohol-induced perturbation of the brain's stress systems contributes to the negative emotional state seen during alcohol withdrawal. The hypothalamic–pituitary–adrenal (HPA) axis is one of the body's main stress-responsive systems. Release of corticotropin-releasing factor (CRF) from the hypothalamus signals activation of the HPA axis response to stress. In addition, activation of extrahypothalamic CRF systems produces anxiety-like states in animals. A number of studies suggest that extrahypothalamic CRF contributes to the development of alcohol dependence. For example, alcohol-dependent rats exhibit increased extracellular CRF content in the central nucleus of the amygdala (Merlo Pich et al., 1995). Additionally, in rodent studies, CRF antagonists injected directly into the central nucleus of the amygdala suppress both the anxiety-like behavior (Rassnick et al., 1993) and the increase in alcohol drinking (Funk et al., 2006) associated with alcohol dependence.

Neuropeptide-Y (NPY) is also involved in regulating the body's stress response. Unlike CRF, NPY has powerful anxiety-reducing effects in animals and alcohol-dependent rats exhibit decreased NPY content in the central nucleus of the amygdala during withdrawal (Roy and Pandey, 2002). In addition, stimulation of NPY activity in the central nucleus of the amygdala suppresses anxiety-like behavior (Thorsell et al., 2007) and increases in alcohol drinking (Gilpin et al., 2008). The anatomical distributions of CRF and NPY are highly overlapping, suggesting that these systems serve to balance each other in regulating the response to stress.

2.2. ACUTE INTOXICATION

As blood alcohol levels increase in humans, the impact of alcohol on cognitive abilities, psychomotor performance, and vital physiologic functions increases (Naranjo and Bremner, 1993, Table 2).

With chronic use, tolerance to the effects of alcohol develops, so the functional impact of a specific amount of alcohol is dependent on a number of factors including degree of tolerance, rate of intake, body weight, percentage of fat and gender.

Alcohol intoxication initially impacts the frontal lobe region of the brain, causing disinhibition, impaired judgment, and cognitive and problem-solving difficulties. At blood alcohol concentrations between 20 mg% and 99 mg%, along with increasing mood and behavioral changes, the effects of alcohol on the cerebellum can cause motor-coordination problems. With blood alcohol levels of 100–199 mg%, there is neurologic impairment with prolonged reaction time, ataxia, and incoordination. Blood alcohol levels of 200–399 mg% are associated with nausea, vomiting, marked ataxia and hypothermia. Between 400 mg% and 799 mg% blood alcohol level, the onset of alcohol coma can

Table 2 Clinical Manifestations of Blood Alcohol Concentration

Blood Alcohol Level mg%	Clinical Manifestations
20–99	Loss of muscular coordination
	Changes in mood, personality, and behavior
100–99	Neurologic impairment with prolonged reaction time, ataxia, incoordination, and mental impairment
200–299	Very obvious intoxication, except in those with marked tolerance nausea, vomiting, marl-zed ataxia
300–399	Hypothermia, severe dysarthria, amnesia, Stage 1 anesthesia
400–799	Onset of alcoholic coma, with precise level depending on degree of tolerance
	Progressive obtundation, decreases in respiration, blood pressure and body temperature
	Urinary incontinence or retention, reflexes markedly decreased or absent
600–800	Often fatal because of loss of airway protective reflexes from airway obstruction by flaccid tongue, from pulmonary aspiration of gastric contents, or from respiratory arrest from profound central nervous system obstruction

Source: Reproduced with permission (Mayo-Smith, 2009)

occur. Serum levels of alcohol between 600 mg% and 800 mg% are often fatal. Progressive obtundation develops with decreases in blood pressure, respiration, and body temperature. Death may be caused by the loss of protective airway reflexes, aspiration of gastric contents or respiratory/cardiac arrest through depressant effects of alcohol on the medulla oblongata and the pons (Table 2, and Mayo-Smith, 2009).

Severely intoxicated individuals may require admission to the hospital for management in specialized units with close monitoring and respiratory support. In individuals with coma, alternative causes must always be investigated, such as head injury, other drug use, hypoglycemia, or meningitis.

2.3. ALCOHOL WITHDRAWAL

Individuals who drink daily or use alcohol at high levels may become physically dependent on alcohol. It is difficult to predict which patients will require medication-assisted detoxification, but daily use of over 15 standard drinks for men and 10 standard drinks for women are often quoted as levels that are likely to lead to withdrawal symptoms severe enough to require medications (McIntosh and Chick, 2004). However, there is a great deal of individual variability in the relationship between the amount of alcohol consumption and withdrawal symptoms, so careful monitoring after abrupt discontinuation of alcohol use in all individuals who are frequent or heavy drinkers is indicated.

Patients who stop drinking abruptly can experience a spectrum of symptoms ranging from mild sleep disturbance to frank delirium tremens. Withdrawal symptoms are highly individualized in both duration and severity. For the vast majority of patients, withdrawal does not progress beyond mild to moderate symptoms, peaking between 24 and 36 h and gradually subsiding. The severity of withdrawal symptoms relates to a number of factors, but the abruptness of alcohol discontinuation, level of alcohol intake, and the general nutritional/medical status are important determinants of severity. Clinical manifestations of alcohol withdrawal as described in DSM-IV-TR can be found in Table 3 (Grant et al., 2004a). In hospital inpatients, it is important to consider alcohol withdrawal in anyone who becomes confused within a day or two of hospital admission. Early signs are generally those of autonomic hyperactivity and tremor, which typically peak 6–24 h after last drink. Transient hallucinatory states are generally a sign of more severe withdrawal and an indication that more medication is required. The peak incidence for seizures is approximately 36 h after last alcohol intake (usually 12–48 h) and for delirium approximately 72 h (Driessen et al., 2005; Mayo-Smith, 2009).

Delirium tremens (DTs) is most severe form of alcohol withdrawal. Distinguishing features of DTs, include coarse tremor, agitation, tachycardia, and extreme autonomic over activity, complicated by disorientation, hallucinations, or delusions. DTs may develop from milder episodes of withdrawal, usually 72–96 h after the last drink, and last for 2–3 days on average. Hyperpyrexia, ketoacidosis, and circulatory collapse may

Table 3 DSM-IV-TR Diagnostic Criteria for Alcohol Withdrawal

1. Cessation of (or reduction in) alcohol use that has been heavy and prolonged.
2. Two (or more) of the following, developing within several hours to a few days after Criterion A;
   a. autonomic hyperactivity (e.g. sweating or pulse rate>100)
   b. increased hand tremor
   c. insomnia
   d. nausea or vomiting
   e. transient visual, tactile, or auditory hallucinations, or illusions
   f. psychomotor agitation
   g. anxiety
   h. grand mal seizures
3. The symptoms in Criterion B cause clinically significant distress or impairment in social, occupational, or other important areas of functioning.
4. The symptoms are not due to a general medical condition and are not better accounted for by another mental disorder.

The following specifier may be applied to a diagnosis of alcohol withdrawal.

With Perceptual Disturbances. This specifier may be noted in the rare instance when hallucinations with intact reality testing or auditory visual, or tactile illusions occur in the absence of a delirium. *Intact reality testing* means that the person knows that the hallucinations are induced by the substance and do not represent external reality. When hallucinations occur in the absence of intact reality testing, a diagnosis of substance-induced psychotic disorder, with hallucinations, should be considered.

DSM-IV-TR Diagnostic Criteria for Alcohol Withdrawal Delirium

1. Disturbance of consciousness (i.e. reduced clarity of awareness of the environment) with reduced ability to focus, sustain, or shift attention.
2. A change in cognition (such as memory deficit, disorientation, and language disturbance) or the development of a perceptual disturbance that is not better accounted for by a preexisting, established, or evolving dementia.
3. The disturbance develops over a short period (usually hours to days) and tends to fluctuate during the course of the day.
4. There is evidence from the history, physical examination, or laboratory findings that they Symptoms in Criteria A and B developed during, or shortly after, an alcohol withdrawal syndrome.

develop. Before the development of effective therapy for DTs, the mortality rate was substantial. Fortunately, DTs occurs in less than 5% of individuals withdrawing from alcohol and with the development of greater awareness and better treatment, death is now unusual (Mayo-Smith et al., 2004; Saitz et al., 1994).

At a cellular level daily alcohol intake induces brain adaptations as detailed above (Gilpin et al., 2008), leaving the brain with a functional increase in NMDA receptor levels. When alcohol is stopped, these excess receptors combine to cause a large calcium flux into cells, hyperexcitability, and cell death. There is also removal of alcohol-mediated inhibitory actions via GABA and the magnesium controlled inhibitory part of the NMDA receptor. The increase in excitatory glutamate combined with a sudden drop in the brain's inhibitory GABA systems combine to produce noradrenergic "overdrive" and the increase in sympathetic activity which is characteristic of the alcohol withdrawal state.

2.4. ALCOHOL AND SEIZURES

The relation between alcohol and seizures is complex. It has been estimated that the prevalence of epilepsy in alcohol-dependent individuals is three times that of the general population, although the prevalence of alcoholism is only slightly higher in patients with epilepsy than in the general population (Hillbom et al., 2003). Partial seizures and EEG abnormalities are fairly frequently seen in individuals with alcohol use disorders. These are typical of posttraumatic epilepsy, correlating with the common occurrence of brain injury in alcohol misusers. Common causes of occult traumatic brain injuries seen in individuals with alcohol use disorders include parenchymal contusions, subdural hematoma, and subarachnoid hemorrhage (McIntosh and Chick, 2004). Brain imaging should be performed in those presenting with their first alcohol-related seizure to rule out brain trauma. There should also be evaluation of potential metabolic cause for seizures such as hypoglycemia or illicit drug use.

As discussed above, abrupt cessation of alcohol intake after prolonged heavy drinking may trigger alcohol withdrawal seizures, thought to be caused by abrupt declines of the brain alcohol levels and compensatory increases in excitatory neurotransmitter activity. Seizures may occur before the blood alcohol content returns to zero and can occur after short bouts of drinking (1–6 days). The diagnosis of an alcohol withdrawal seizure is usually made because of the presence of other symptoms of alcohol withdrawal, and a history of recent alcohol misuse. There may be a genetic susceptibility to alcohol withdrawal seizures (McIntosh and Chick, 2004).

It has been suggested that alcohol causes between 9% and 25% of cases of status epilepticus, and this may be the first presentation of alcohol-related seizures (Alldredge and Lowenstein, 1993). The outcome of patients with alcohol-related status epilepticus is favorable, but recovery may be compounded by an unduly prolonged postictal state. Repeated cycles of alcohol exposure and withdrawal may lead to a process termed "kindling", or neuronal sensitization, which could precipitate seizures (Ballenger and Post, 1978). This hypothesis has some experimental support (Brown et al., 1988), but one fairly large study showed no correlation between the number of withdrawal episodes and seizures (Hillbom et al., 2003). Clinical experience suggests that in repeated withdrawal episodes, symptoms tend to progress in severity, culminating in seizures or serious psychiatric sequelae. This may also reflect the fact that larger amounts of alcohol are being systematically consumed as the disease progresses.

2.5. WERNICKE–KORSAKOFF'S SYNDROME

Thiamine deficiency is one of the primary causes of Wernicke–Korsakoff syndrome. Wernicke's encephalopathy is characterized as a classical "triad" of ataxia, confusion, and ophthalmoplegia; however, many patients do not present with the classical triad (Harper, 1979). Korsakoff's psychosis is described as an amnestic syndrome with impaired recent memory and relatively intact intellectual function. However, patients rarely have a discrete deficit in forming new memories and often present with more global deficits. In addition, Korsakoff's psychosis is not always preceded by a clear episode of Wernicke's encephalopathy. Comorbid head injury, abnormal gait, memory disturbance, nystagmus, hypothermia, hypotension, acute confusional state, coma, and alcohol withdrawal symptoms can all complicate the presentation, making diagnosis difficult. In addition, diagnosis and treatment are hampered by a lack of accessible means to measure blood and brain thiamine. Because of these factors, unfortunately, the majority of Wernicke–Korsakoff's syndrome diagnoses are made postmortem (McIntosh and Chick, 2004).

As such, it is essential to assess nutritional status in all individuals with alcohol use disorders (Mayo-Smith, 2009). Frequently alcohol-dependent individuals have poor diets, deriving much of their caloric needs from alcohol leading to inadequate oral intake of thiamine. In addition, vomiting, diarrhea, and the actions of alcohol on the gut impair the availability of thiamine to the body. Thiamine (along with other B vitamins) is a coenzyme in glucose metabolism, lipid metabolism, amino acid production, and neurotransmitter synthesis. The body stores a relatively small amount of thiamine and deficiency may present within 2–3 weeks of intake ceasing. The brain is particularly sensitive to a breakdown in vitamin B-dependent glucose metabolism (McIntosh and Chick, 2004).

Improvements in understanding brain neurochemistry and imaging technology have shed light on an association between alcohol, thiamine, and memory. It has been proposed that most organic brain syndromes in alcoholic patients are variants of the Wernicke–Korsakoff syndrome, and that there is no need to consider a separate category of "alcoholic dementia". Alcohol has a direct neurotoxic effect on cortical neurons, but considerable damage is also caused by thiamine deficiency. While it was previously believed that shrinkage of the mammillary bodies was pathognomonic of Wernicke–Korsakoff syndrome, MRI studies have found that mammillary body damage is as common in patients without amnesia as it is on those with amnesia. Scanning studies also demonstrate that Korsakoff patients have widespread cerebral and subcortical atrophy, which is greater than that found in the alcoholic patients without amnesia. The subcortical atrophy is particularly pronounced (Chick, 1997).

Where Wernicke's encephalopathy is suspected, prompt high dose parenteral Vitamin B treatment is required. Oral vitamin supplements are not useful because in alcohol-dependent individuals, thiamine absorption is very variable, with some patients showing little or no absorption due to a reduction in the sodium-dependent transport mechanism caused by alcohol use and malnutrition. The outcome from an episode of Wernicke–Korsakoff syndrome falls roughly into quarters, 25% showing no recovery, 25% slight, 25% significant, and 25% complete recovery of memory (Smith and Hillman, 1999).

2.6. NEUROIMAGING AND ALCOHOL-INDUCED BRAIN CHANGES

Neuroimaging allows direct visualization of the impact of alcohol on brain structure and function. Alcohol-related brain changes include primary or "direct" toxic effects and secondary effects related to the sequelae of liver cirrhosis and problems related to the impairment of vitamin uptake as described above (Mann et al., 2001; Rovira et al., 2008). The primary direct effect of alcohol is volume loss due to cell death as a result of the direct toxic effects of alcohol on neurons. This is often mediated by a compromise of neurotransmitters, receptors, and electrolytes (Spampinato et al., 2005). In some clinical conditions related to alcohol, it remains unclear whether primary effects, secondary effects or a combination of primary and secondary effects are primarily responsible.

Direct brain toxicity is caused by upregulation of NMDA receptors resulting in increased susceptibility to excitatory and cytotoxic effects of glutamate. In addition, acetaldehyde and related metabolites of alcohol can initiate an immune-mediated response resulting in neuronal and white matter damage. Chronic alcohol consumption also leads to decreased neurotropic factors, which results in interference with the normal brain function, dysregulation of neuronal synaptic connectivity and apoptosis. Decreased gene expressions of myelin protein-encoding genes in the glia cells lead to further volume loss in the superior frontal cortex and hippocampus (Harper, 2007; Spampinato et al., 2005). Diffusion tensor imaging has shown that excessive alcohol use causes macro structural tissue shrinkage and microstructure compromise with white matter degradation in multiple fiber systems including the corpus callosum (De Bellis et al., 2008; Pfefferbaum et al., 2009). Of interest, there is regression of brain atrophy and metabolic recovery with abstinence from chronic alcohol use (Bartsch et al., 2007).

Wernicke's encephalopathy is associated with a number of characteristic neuroimaging findings. Demyelination, glial cell proliferation, capillary endothelium hyperplasia, and proliferation with petechial hemorrhage can be seen. Changes occur predominantly in the periventricular region, around the third ventricle in the dorsomedial and pulvinar nuclei of the thalamus and hypothalamus, mammillary bodies, pineal regions and periaqueductal gray region of the midbrain (Victor et al., 1971).

An acquired cerebellar syndrome with anteriosuperior vermal atrophy revealed by CT or MRI is well recognized as related to excessive alcohol use. The clinical presentation is typically a broad-based and unstable gait with upper limbs being rarely involved. Osmotic myelinolysis (osmotic demyelination syndrome, central pontine myelinolysis) is a serious neurological condition classically seen in hyponatremic patients when sodium levels are corrected too rapidly (Yoon et al., 2008). This condition can also be seen in chronic alcoholic patients unrelated to changes in the serum sodium level due to a direct toxic effect of the alcohol on the pontine fibers (Hagiwara et al., 2008).

Marchiafava–Bignami disease is a rare complication of chronic alcohol consumption characterized by primary demyelination and necrosis of the central portion of the corpus callosum (Helenius et al., 2001). Clinical presentations include cognitive impairment, gait disturbance, limb hypertonia, dysarthria, and signs of interhemispheric disconnection. Pathological features include layered necrosis, degeneration of the corpus callosum, predominantly at the body followed by the genu and splenium (Chang et al., 1992). Other white matter tracts such as the anterior and posterior commissures, the corticospinal tract, the external capsule, the hemispheric white matter, and middle cerebellar peduncles may be involved.

2.7. ALCOHOL-RELATED NEUROPATHY

Alcohol misusers may develop "Saturday night palsy", a focal peripheral nerve palsy as a result of nerve compression when heavily sleeping or stuporous. Recovery is usually complete. Chronic alcohol use is also associated with a symmetrical, bilateral mixed sensory and motor peripheral neuropathy, usually of the lower limbs. Individuals may be asymptomatic or present with pain, numbness, burning feet, and hyperalgesia. There may also be muscle weakness and diminished tendon reflexes. These neuropathies are associated with thiamine deficiency and may show some recovery with abstinence from alcohol and thiamine supplementation (McIntosh and Chick, 2004).

2.8. PSYCHIATRIC SEQUALAE

Mood and anxiety disorders are commonly seen in individuals with alcohol use disorders. Recent epidemiologic data suggest that individuals with alcohol use disorders are at least two times more likely to have a mood or anxiety disorder at some time in their lives as compared to the general population (Grant et al., 2004b). Because acute intoxication and withdrawal from alcohol can cause depression and anxiety, it is important to distinguish transient alcohol-related symptoms from enduring symptoms, which require independent treatment. The best approach in individuals complaining of mood/anxiety symptoms while intoxicated or in early withdrawal is to reassess symptoms when patients are not in acute withdrawal and have been abstinent for at least 7–10 days (Brady, 2007).

3. CONCLUSIONS

Alcohol has profound and damaging effects on the human nervous system. While safe levels of alcohol consumption have generally been defined, the impact of alcohol on an individual is variable. Acute alcohol intoxication is characterized by behavioral disinhibition at low blood alcohol levels and life-threatening CNS depression at high blood alcohol levels. Chronic alcohol use is associated with transient effects such as the alcohol withdrawal syndrome, and more permanent effects such as alcohol-related dementia. Education about safe drinking levels is critical to prevention of alcohol-related problems. Early detection and treatment of alcohol dependence is critical to minimizing negative consequences of use.

REFERENCES

Alldredge, B.K., Lowenstein, D.H., 1993. Status epilepticus related to alcohol abuse. Epilepsia 34, 1033–1037.

Anton, R.F., 2008. Naltrexone for the management of alcohol dependence. N. Engl. J. Med. 359, 715–721.

Ballenger, J.C., Post, R.M., 1978. Kindling as a model for alcohol withdrawal syndromes. Br. J. Psychiatry 133, 1–14.

Bartsch, A.J., Homola, G., Biller, A., Smith, S.M., Weijers, H.G., Wiesbeck, G.A., Jenkinson, M., De Stefano, N., Solymosi, L., Bendszus, M., 2007. Manifestations of early brain recovery associated with abstinence from alcoholism. Brain 130, 36–47.

Brady, K.T., 2007. Advances in the treatment of alcohol dependence. J. Clin. Psychiatry 28, 1117–1128.

Brown, M.E., Anton, R.F., Malcolm, R., Ballenger, J.C., 1988. Alcohol detoxification and withdrawal seizures: clinical support for a kindling hypothesis. Biol. Psychiatry 23, 507–514.

Carboni, S., Isola, R., Gessa, G.L., Rossetti, Z.L., 1993. Ethanol prevents the glutamate release induced by N-methyl-D-aspartate in the rat striatum. Neurosci. Lett. 152, 133–136.

Chang, K.H., Cha, S.H., Han, M.H., Park, S.H., Nah, D.L., Hong, J.H., 1992. Marchiafava–Bignami disease: serial changes in corpus callosum on MRI. Neuroradiology 34, 480–482.

Chick, J., 1997. Alcohol and the brain. Curr. Opin. Psychiatry 10, 205–210.

Colombo, G., Grant, K.A., 1992. NMDA receptor complex antagonists have ethanol-like discriminative stimulus effects. Ann. N.Y. Acad. Sci. 654, 421–423.

D'Onofrio, G., Bernstein, E., Bernstein, J., Woolard, R.H., Brewer, P.A., Craig, S.A., Zink, B.J., 1998. Patients with alcohol problems in the emergency department, part 1: improving detection. SAEM Substance Abuse Task Force. Society for Academic Emergency Medicine. Acad. Emerg. Med. 5, 1200–1209.

De Bellis, M.D., Van Voorhees, E., Hooper, S.R., Gibler, N., Nelson, L., Hege, S.G., Payne, M.E., MacFall, J., 2008. Diffusion tensor measures of the corpus callosum in adolescents with adolescent onset alcohol use disorders. Alcohol. Clin. Exp. Res. 32, 395–404.

Driessen, M., Lange, W., Junghanns, K., Wetterling, T., 2005. Proposal of a comprehensive clinical typology of alcohol withdrawal–a cluster analysis approach. Alcohol 40, 308–313.

Fiellin, D.A., Reid, M.C., O'Connor, P.G., 2000. Screening for alcohol problems in primary care: a systematic review. Arch. Intern. Med. 160, 1977–1989.

Funk, C.K., O'Dell, L.E., Crawford, E.F., Koob, G.F., 2006. Corticotropin-releasing factor within the central nucleus of the amygdala mediates enhanced ethanol self-administration in withdrawn, ethanol-dependent rats. J. Neurosci. 26, 11324–11332.

Gilpin, N.W., Misra, K., Koob, G.F., 2008. Neuropeptide Y in the central nucleus of the amygdala suppresses dependence-induced increases in alcohol drinking. Pharmacol. Biochem. Behav. 90, 475–480.

Grant, B.F., Dawson, D.A., Stinson, F.S., Chou, S.P., Dufour, M.C., Pickering, R.P., 2004a. The 12-month prevalence and trends in DSM-IV alcohol abuse and dependence: United States, 1991–1992 and 2001–2002. Drug Alcohol Depend. 74, 223–234.

Grant, B.F., Stinson, F.S., Dawson, D.A., Chou, S.P., Ruan, W.J., Pickering, R.P., 2004b. Co-occurrence of 12-month alcohol and drug use disorders and personality disorders in the United States: results from the National Epidemiologic Survey on Alcohol and Related Conditions. Arch. Gen. Psychiatry 61, 361–368.

Hagiwara, K., Okada, Y., Shida, N., Yamashita, Y., 2008. Extensive central and extrapontine myelinolysis in a case of chronic alcoholism without hyponatremia: a case report with analysis of serial MR findings. Intern. Med. 47, 431–435.

Harper, C., 1979. Wernicke's encephalopathy: a more common disease than realised. A neuropathological study of 51 cases. J. Neurol. Neurosurg. Psychiatry 42, 226–231.

Harper, C., 2007. The neurotoxicity of alcohol. Hum. Exp. Toxicol. 26, 251–257.

Harwood, H.J., 2000. Updating Estimates of the Economic Costs of Alcohol Abuse in the United States: Estimates, Update Methods, and Data. National Institute on Alcohol Abuse and Alcoholism, Bethesda, MD.

Helenius, J., Tatlisumak, T., Soinne, L., Valanne, L., Kaste, M., 2001. Marchiafava–Bignami disease: two cases with favourable outcome. Eur. J. Neurol. 8, 269–272.

Hillbom, M., Pieninkeroinen, I., Leone, M., 2003. Seizures in alcohol-dependent patients: epidemiology, pathophysiology and management. CNS Drugs 17, 1013–1030.

Johnson, B.A., 2008. Update on neuropharmacological treatments for alcoholism: scientific basis and clinical findings. Biochem. Pharmacol. 75, 34–56.

Koob, G.F., 2003. Alcoholism: allostasis and beyond. Alcohol. Clin. Exp. Res. 27, 232–243.

Koob, G.F., 2004. A role for GABA mechanisms in the motivational effects of alcohol. Biochem. Pharmacol. 68, 1515–1525.

Leon, D.A., McCambridge, J., 2006. Liver cirrhosis mortality rates in Britain from 1950 to 2002: an analysis of routine data. Lancet 367, 52–56.

Lieber, C.S., 1995. Medical disorders of alcoholism. N. Engl. J. Med. 333, 1058–1065.

Littleton, J.M., 2007. Acamprosate in alcohol dependence: Implications of a unique mechanism of action. J. Addict. Med. 1, 115–125.

Mann, K., Agartz, I., Harper, C., Shoaf, S., Rawlings, R.R., Momenan, R., Hommer, D.W., Pfefferbaum, A., Sullivan, E.V., Anton, R.F., Drobes, D.J., George, M.S., Bares, R., Machulla, H.J., Mundle, G., Reimold, M., Heinz, A., 2001. Neuroimaging in alcoholism: ethanol and brain damage. Alcohol. Clin. Exp. Res. 25, 104S–109S.

Mayo-Smith, M.F., 2009. Management of alcohol intoxication and withdrawal. In: Ries, R.K., Shannon, M.D., Miller, S.C., Fiellin, D., Saitz, R. (Eds.), Principles of Addiction Medicine. fourth ed. Lippincott Williams & Wilkins, Philadelphia, PA, pp. 559–572.

Mayo-Smith, M.F., Beecher, L.H., Fischer, T.L., Gorelick, D.A., Guillaume, J.L., Hill, A., Jara, G., Kasser, C., Melbourne, J., 2004. Management of alcohol withdrawal delirium. An evidence-based practice guideline. Arch. Intern. Med. 164, 1405–1412.

McIntosh, C., Chick, J., 2004. Alcohol and the nervous system. J. Neurol. Neurosurg. Psychiatry 75 (Suppl. 3), iii16–iii21.

Melis, M., Spiga, S., Diana, M., 2005. The dopamine hypothesis of drug addiction: hypodopaminergic state. Int. Rev. Neurobiol. 63, 101–154.

Merlo Pich, E., Lorang, M., Yeganeh, M., Rodriguez de Fonseca, F., Raber, J., Koob, G.F., Weiss, F., 1995. Increase of extracellular corticotropin-releasing factor-like immunoreactivity levels in the amygdala of awake rats during restraint stress and ethanol withdrawal as measured by microdialysis. J. Neurosci. 15, 5439–5447.

Mokdad, A.H., Marks, J.S., Stroup, D.F., Gerberding, J.L., 2004. Actual causes of death in the United States, 2000. JAMA 291, 1238–1245.

Myers, R.D., Veale, W.L., 1968. Alcohol preference in the rat: reduction following depletion of brain serotonin. Science 160, 1469–1471.

Naranjo, C.A., Bremner, K.E., 1993. Behavioural correlates of alcohol intoxication. Addiction 88, 25–35.

Olive, M.F., Koenig, H.N., Nannini, M.A., Hodge, C.W., 2001. Stimulation of endorphin neurotransmission in the nucleus accumbens by ethanol, cocaine, and amphetamine. J. Neurosci. 21, RC184.

Pfefferbaum, A., Rosenbloom, M., Rohlfing, T., Sullivan, E.V., 2009. Degradation of association and projection white matter systems in alcoholism detected with quantitative fiber tracking. Biol. Psychiatry 65, 680–690.

Pulvirenti, L., Diana, M., 2001. Drug dependence as a disorder of neural plasticity: focus on dopamine and glutamate. Rev. Neurosci. 12, 141–158.

Rassnick, S., Stinus, L., Koob, G.F., 1993. The effects of 6-hydroxydopamine lesions of the nucleus accumbens and the mesolimbic dopamine system on oral self-administration of ethanol in the rat. Brain Res. 623, 16–24.

Room, R., Babor, T., Rehm, J., 2005. Alcohol and public health. Lancet 365, 519–530.

Rovira, A., Alonso, J., Cordoba, J., 2008. MR imaging findings in hepatic encephalopathy. AJNR Am. J. Neuroradiol. 29, 1612–1621.

Roy, A., Pandey, S.C., 2002. The decreased cellular expression of neuropeptide Y protein in rat brain structures during ethanol withdrawal after chronic ethanol exposure. Alcohol. Clin. Exp. Res. 26, 796–803.

Saitz, R., 2005. Clinical practice. Unhealthy alcohol use. N. Engl. J. Med. 352, 596–607.

Saitz, R., Mayo-Smith, M.F., Roberts, M.S., Redmond, H.A., Bernard, D.R., Calkins, D.R., 1994. Individualized treatment for alcohol withdrawal. A randomized double-blind controlled trial. JAMA 272, 519–523.

Shkolnikov, V., McKee, M., Leon, D.A., 2001. Changes in life expectancy in Russia in the mid-1990s. Lancet 357, 917–921.

Smith, I., Hillman, A., 1999. Management of alcohol Korsakoff syndrome. APT 5, 271–278.

Spampinato, M.V., Castillo, M., Rojas, R., Palacios, E., Frascheri, L., Descartes, F., 2005. Top. Magn. Reson. Imaging 16, 223–230.

Thorsell, A., Repunte-Canonigo, V., O'Dell, L.E., Chen, S.A., King, A.R., Lekic, D., Koob, G.F., Sanna, P.P., 2007. Viral vector-induced amygdala NPY overexpression reverses increased alcohol intake caused by repeated deprivations in Wistar rats. Brain 130, 1330–1337.

Victor, M., Adams, R.D., Collins, G.H., 1971. The Wernicke–Korsakoff syndrome. A clinical and pathological study of 245 patients, 82 with post-mortem examinations. Contemp. Neurol. Ser. 7, 1–206.

Virkkunen, M., Linnoila, M., 1990. Serotonin in early onset, male alcoholics with violent behaviour. Ann. Med. 22, 327–331.

Volkow, N.D., Wang, G.J., Telang, F., Fowler, J.S., Logan, J., Jayne, M., Ma, Y., Pradhan, K., Wong, C., 2007. Profound decreases in dopamine release in striatum in detoxified alcoholics: possible orbitofrontal involvement. J. Neurosci. 27, 12700–12706.

Weiss, F., Lorang, M.T., Bloom, F.E., Koob, G.F., 1993. Oral alcohol self-administration stimulates dopamine release in the rat nucleus accumbens: genetic and motivational determinants. J. Pharmacol. Exp. Ther. 267, 250–258.

White, I.R., Altmann, D.R., Nanchahal, K., 2004. Mortality in England and Wales attributable to any drinking, drinking above sensible limits and drinking above lowest-risk level. Addiction 99, 749–756.

Yoon, B., Shim, Y.S., Chung, S.W., 2008. Central pontine and extrapontine myelinolysis after alcohol withdrawal. Alcohol Alcohol. 43, 647–649.

CHAPTER TEN

The Nicotine Hypothesis

James Robert Brašić[1] and Jongho Kim[2]

[1]Section of High Resolution Brain Positron Emission Tomography Imaging, Division of Nuclear Medicine, The Russell H. Morgan Department of Radiology and Radiological Science, The Johns Hopkins University School of Medicine, Johns Hopkins Outpatient Center, Baltimore, MD, USA
[2]Department of Diagnostic Radiology and Nuclear Medicine University of Maryland Medical Center, Baltimore, MD, USA

1. INTRODUCTION

1.1. Nicotine

The effects of acetylcholine (ACh), the neurotransmitter, are separated into two broad categories, muscarinic and nicotinic. Both muscarinic and nicotinic acetylcholine receptors (AChRs) are widely distributed throughout the human body including the nervous system. The manifold manifestations of the effects of ACh on the central nervous system (CNS) are mediated through stimulation of both muscarinic and nicotinic AChRs. This review focuses on the role of dysfunction of central neuronal nicotinic acetylcholine receptors (nAChRs) in the pathogenesis and the pathophysiology of schizophrenia and other neuropsychiatric disorders.

Nicotine, an exogenous substance available in a vast spectrum of forms for human consumption, exerts profound effects throughout the human body. There are several subtypes of neuronal nAChRs in the human body. Ethnic, familial, cultural, environmental, and social influences alter the propensity of a particular person to utilize nicotine in its many forms. Nicotine is extensively metabolized, primarily in the liver (Schroeder, 2005), but also in the lung (Turner et al., 1975), resulting in a variety of metabolites. Cytochrome P450 2A6 (CYP 2A6), a member of the cytochrome P450 mixed-function oxidase system, is the primary enzyme responsible for the oxidation of nicotine and cotinine. The rate of metabolism of nicotine is influenced by genetic, environmental, and other influences. The likelihood that a person will develop nicotine dependence is directly proportional to the persons's rate of metabolism of nicotine (Benowitz, 2008).

Tobacco consumption by smoking cigarettes is a popular form of obtaining nicotine. Cigarette smoking and secondhand exposure to cigarette smoke constitute major public health problems with significant morbidity and mortality (Brašić et al., 2009; Eisner et al., 2007). Facilitating the smoking cessation of smokers may be the most beneficial intervention of a physician (Schroeder, 2005). Cigarette smoking

The Effects of Drug Abuse on the Human Nervous System
http://dx.doi.org/10.1016/B978-0-12-418679-8.00010-1

constitutes a lethal preventable health liability. Cigarette smoking results in approximate annual death rates of $440, 000$ in the United States and $5,000,000$ in the world (Schroeder, 2004). Cigarette smoking has resulted in disabilities, including chronic obstructive pulmonary disease and pulmonary carcinoma (Anonymous, 2003). Total fatal outcomes of cigarette smoking exceed the total fatal outcomes due to human immunodeficiency virus, alcohol and other substances, motor vehicle accidents, and suicide (Centers for Disease Control and Prevention, 2011). Overall nonsmokers live 10 years longer than smokers (Doll et al., 2004). Fatalities due to cigarette smoking are particularly high in people with mental disorders including the abuse of other substances; a third of the deaths in this population result from cigarette smoking (Williams, et al., 2005; Lasser et al., 2000). Even among nonsmokers nine percent of deaths are attributable to inhalation of secondhand cigarette smoke (Anoymous, 2002). Infertility, unfavorable outcomes of pregnancy, carcinomal of the breast, and poor vision due to cataracts and macular degeneration are several of the disorders resulting at least in part from cigarette smoking (Office on Smoking and Health, National Center for Chronic Disease Prevention and Health Promotion, 2004). Many of the adverse health effects of cigarette smoking can be avoided by the consumption of nicotine in a variety of alternative forms including gum, patch, lozenge, nasal spray, and inhaler (Schroeder, 2005). The inhaled particles of smoke pass through the pulmonary circulation into the arterial blood supply to the brain. In the brain the nicotine passes the blood–brain barrier to bind to neuronal nAChRs, sites on post-synaptic neurons (Benowitz, 2008).

1.2. Acetylcholine Receptors

ACh acts widely throughout the organs of the body including the nervous system. Neurons, nerve cells, communicate with each other by secreting a chemical compound across a synapse, the gap between neurons. The receiving postsynaptic neuron contains receptors, structures that bind to the substance emitted by the initiating neuron. Neuronal AChRs exist in two broad groups, muscarinic and nicotinic.

1.3. Muscarinic Acetylcholine Receptors

There exist at least five neuronal muscarinic acetylcholine receptors (mAChRs) sub-types categorized as (A) the group that couples through $G_{q/11}$ proteins to utilize phopholipase-C to mobilize calcium, M_1, M_3, and M_5, and (B) the group that utilizes $G_{i/0}$ proteins to diminish cyclic adenosine monophosphate (cAMP) in cells via reductions of adenylate cyclase, M_2 and M_4 (Langmead et al., 2008). In the CNS, the M_1 subtype predominates primarily in the cortex, the hippocampus, the striatum, and the thalamus (Langmead et al., 2008). The M_1 mAChR is a target of pharmacotherapeutic interventions to improve the cognitive deficits of Alzheimer's disease (AD) and other dementias (Langmead et al., 2008).

Several radiotracers have been developed to delineate the muscarinic system. *RS 3-quinuclidinyl-4-[^{123}I]iodobenzilate (RS [^{123}I]IQNB)* is a radiotracer used for clinical studies in humans to quantify changes in the concentrations of mAChRs (Eckelman, 2006). Since decrements in the uptake of regions of interest identified through imaging with *RS [^{123}I]IQNB* can be due to either (A) a reduced number of mAChRs per cell or (B) a reduced concentration of cells with mAChRs, this radiotracer has not gained widespread clinical use (Eckelman, 2006). *3-[[4-(3-[^{18}F]fluoropropylsulfanyl)-1,2,5-thiadia-zol-3-yl]]-1-methyl-1,2,5,6-tetrahydropyridine. ([^{18}F]FP-TZTP),* a radiotracer to quantitatively estimate the function of the muscarinic system in humans, is a promising agent to monitor this system (Eckelman, 2006). The development of novel techniques to characterize the muscarinic system in health and disease promises to enhance the diagnosis and treatment of neurologic and psychiatric disorders, as well as other diseases.

1.4. Nicotinic Acetylcholine Receptors

nAChRs are channels operated by contact with the ligand. When opened, the channels allow the passage of sodium and calcium ions. Each nAChR is made up of five components. Twelve subunits are categorized as alpha (α) isoforms, $α_2$ to $α_{10}$, and beta (β) isoforms, $β_2$ to $β_4$. The $α_4β_2$ constitute the predominant high-affinity nAChR subtype in the rat and human brain. The $α_4$ subunit participates in the sensitivity to nicotine. The $β_2$ subunit plays a role in dopamine release. The $β_2$ subunit also mediates the behavioral effects of nicotine. An asterisk (⋆) indicates the preceding subunit may occur in multiples in the site on the postsynaptic receptor membrane. Thus, $α_4β_2$⋆ indicates that multiple $β_2$ isoforms are combined with a single $α_4$ isoform. The $α_7$ subunit likely mediates synaptic transmission and sensory gating (Benowitz, 2008).

High-affinity nAChRs are those nAChRs that bind *[^{3}H]nicotine* and *[^{3}H]cytisine*. These high-affinity sites are thought to be composed of $α_4$ and $β_2$ subunits. In the human brain the $α_4β_2$⋆ nAChR subtype is distributed with high density in the thalamus, the basal ganglia, other central structures, and the dentate gyrus, and with moderate density in the cortex, the hippocampal pyramidal layer, the limbic system, and other structures (Gotti et al., 1997).

nAChRs differentiate cigarette smokers from nonsmokers. While increased stimulation by agonists typically cause downregulation, a reduction in the number of cell surface receptors, chronic exposure to high levels of tobacco through cigarette smoking paradoxically causes upregulation, an increment in the number (B_{max}) of high-affinity $α_4β_2$⋆ nAChRs in the cortex, the hippocampus, the cerebellum, the striatum (Court et al., 1998; Gotti et al., 1997; Perry et al., 1999), the nucleus accumbens, and the ventral tegmental areas of the midbrain (Schroeder, 2005). The principal pharmacological site of action of nicotine appears to be the high-affinity $α_4β_2$⋆ nAChRs. Furthermore, human cigarette smokers exhibit a dose-dependent increase in high-affinity $α_4/β_2$⋆ nAChR *[^{3}H]nicotine* binding in the hippocampus and the thalamus (Breese et al., 1997).

High-affinity nAChR binding appears to differentiate people with schizophrenia from people without schizophrenia (Leonard et al., 1998). Both decreased inhibition of the P50 auditory evoked potential to repeated stimulation (Leonard et al., 1998) and the phenotype of schizophrenia have been associated with 15q13-q14 in the region of the α_7 nAChR (Craddock et al., 1999). The absence of a radiotracer for the α_7 nAChR hinders its investigation. However, dysfunction of the nAChR system in schizophrenia is not exclusively of the low-affinity α_7 subtype. For example, an increase in in vitro high-affinity nAChR ([^{3}H]nicotine) binding in the striatum, the hippocampus, and some thalamic nuclei in people with schizophrenia who smoke cigarettes has been reported by some researchers (Court et al., 1998). On the other hand, others have reported reductions in in vitro nAChR (*[^{3}H]nicotine*) binding in the hippocampus, the cortex, and the caudate of people with schizophrenia who smoke cigarettes in contrast to people without schizophrenia who smoke cigarettes (Leonard et al., 1999). Additionally, in vitro $\alpha_4\beta_2\star$ nAChR *[^{3}H]cytisine* density is decreased in the postmortem striata of people with both schizophrenia and Parkinson's disease in contrast to normal control subjects (Durany et al., 2000). Furthermore α_7 nAChR are reduced in the postmortem brains of people with schizophrenia (Kucinski et al., 2010).

Preparations of nicotine may be salutatory for some individuals with schizophrenia just as for some people with Parkingson's disease (Durany et al., 2000). The neuroprotective effects of nicotine observed in vitro in laboratory animals (Costa et al., 2001) likely hold in some humans.

Future research is needed to compare and contrast the number of high-affinity $\alpha_4/\beta_2\star$ neuronal nAChRs in the hippocampus, amygdala, cingulate, and other limbic structures, as well as the hypothalamus, striatum, and cortex of people with and without schizophrenia, using with positron emission tomography (PET).

A goal of future research is to demonstrate an objective measure to quantify high-affinity neuronal nAChR binding in the living human brain. If there exist marked reductions in high-affinity neuronal nAChR binding in the living brains of people with schizophrenia, then we shall have a tool to identify the deficits in people with schizophrenia and other neuropsychiatric disorders as well as a means to quantify the effects of pharmacological and other interventions to treat the deficits. Thus, there will be a means to assess people with schizophrenia and to monitor their progress during novel therapies. The needed technique will provide the means to quantitatively measure the occupancy by novel therapeutic agents of nicotinic receptors in the brains to facilitate the determination of optimal therapeutic doses. This procedure will thus facilitate the development of innovative pharmacological agents for schizophrenia, nicotine dependence, and other neuropsychiatric disorders.

The needed procedure likely will contribute to the knowledge about schizophrenia and other neuropsychiatric disorders, as well as nicotine dependence and other addictions.

2.1. SCHIZOPHRENIA

The high prevalence of cigarette smoking in individuals with schizophrenia (Gotti et al., 1997; Leonard et al., 1998) suggests that dysfunction of the nAChRs may (1) play a greater role in the relentless progressive deterioration common in people with schizophrenia who smoke cigarettes, and (2) provide clues to the development of novel effective treatments. The results of this review are crucial to determine biological markers for subtypes of people with schizophrenia, nicotine dependence, and other neuropsychiatric disorders.

The dopamine hypothesis of schizophrenia attributes positive symptoms to increased dopaminergic subcortical neurotransmission and negative symptoms to decreased dopaminergic cortical neurotransmission (Kucinski et al., 2010).

If a brain imaging procedure can determine alterations in the density and distributions of neuronal nAChRs in the living human brain of people with schizophrenia who do and do not smoke cigarettes, then those who likely will benefit from the administration of nicotine and nicotinic agonists can possibly be identified. Additionally, a subclass of people with schizophrenia who will not benefit from nicotinic agonists can likely be identified. They can then be spared the adverse effects of fruitless administrations of nicotine and nicotinic agonists.

The demonstration that a positive allosteric modulator of the α7 neuronal nAChR enhanced the current flow through α7 neuronal nAChR in the presence of ACh in cell cultures suggests that this agent exhibits the potential to alleviate the cognitive impairments associated with auditory gating inhibition in people with schizophrenia (Timmermann et al., 2007). This research suggests that positive allosteric modulators of neuronal nAChR may be fruitful agents to ameliorate the auditory gating inhibitions characteristic of people with schizophrenia. Additionally, other novel therapeutic interventions for schizophrenia and related neuropsychiatric disorders, as well as nicotine dependence and other addictions will likely be based on the results of this protocol.

Schizophrenia is a devastating disorder without effective treatments presenting in adolescence or young adulthood and afflicting approximately one percent of the population (American Psychiatric Association, 2000). Both genetic and environmental items likely play a role in the pathogenesis of schizophrenia (Tandon et al., 2008), a disorder likely resulting from alterations in the brain development during the prenatal period (Lafargue and Brasic, 2000).

The observation that many people with schizophrenia smoke cigarettes (Herrán et al., 2000; Williams and Ziedonis, 2004) suggests that the brains of these individuals may respond abnormally to nicotine. In contrast to control smokers without schizophrenia, smokers with schizophrenia have greater levels of nicotine and cotinine due to an increased nicotine intake per cigarette. Thus, people with schizophrenia are able to obtain large amounts of nicotine from cigarettes (Williams et al., 2005). Greater positive symptoms and fewer negative symptoms are seen in people with schizophrenia who smoke cigarettes heavily (Williams and Ziedonis, 2004). Although smoking may reduce negative symptoms in people with schizophrenia (Gotti and Clementi, 2004), people with schizophrenia with nicotine dependence have poor outcomes (Aguilar et al., 2005).

Anecdotal studies suggest that nicotine and nicotine agonists are effective to treat some people with neurobiological disorders, e.g. Tourette syndrome (Dursun and Reveley, 1997; Sanberg et al., 1997). This review may (1) identify biological subgroups of people with schizophrenia who may be helped by nicotine preparations and related compounds and (2) provide the means to document the effects of therapies on the underlying deficits in high-affinity $\alpha 4\beta 2\star$ neuronal nAChRs of the affected portions of the brains of individuals with schizophrenia and related conditions (Brašić et al., 2009, 2010; Wong et al., 2002; Zhou et al., 2002, 2004).

Research on autopsy material supports the hypothesis that nAChRs in the brain are dysfunctional in people with schizophrenia and nicotine dependence. In postmortem studies the neuronal nAChRs normally stimulated by nicotine are reduced in density in the brains of people with schizophrenia (Breese et al., 2000; Leonard et al., 2000). While postmortem studies suggest that binding to high-affinity neuronal nAChRs in the brains of people with schizophrenia who smoke cigarettes is half that of people without schizophrenia who smoke cigarettes (Breese et al., 2000; De Luca et al., 2006; Leonard et al., 1998), the difference is likely reduced by the postmortem changes.

Likely, the apparent alterations in neuronal nAChRs of people with schizophrenia represent an interaction between an inherent genetic effect and a drug effect of nicotine (De Luca et al., 2006). Since the cholinergic genes, CHRNA4 and CHRNB2, confer susceptibility to schizophrenia, some individuals with polymorphisms in CHRNA4 and CHRNB2 may be vulnerable to develop both nicotine dependence and schizophrenia (De Luca et al., 2006). Furthermore, among people with schizophrenia, heavy cigarette smoking is associated with the CHRNA4 rs3746372 allele 1 and CHRNA4 rs3787116 and rs3746372 suggesting an interaction between schizophrenia and the number of cigarettes smoked (Voineskos et al., 2007). At postmortem the density of neuronal nAChRs in the cortex exists in a continuum from highest density to lowest density as follows: (1) people without schizophrenia who smoke cigarettes; (2) people with schizophrenia who smoke cigarettes; (3) healthy people without schizophrenia who do not smoke cigarettes, i.e. normal control subjects; and (4) people with schizophrenia who do not smoke cigarettes (Breese et al., 2000).

In schizophrenia dysfunction of neuronal nAChRs may play an important role in the cognitive and attention deficits and relentless progressive deterioration (Sacco et al., 2005). Abnormalities in the transmission of auditory sensory information through the amygdala (Cromwell et al., 2005) likely plays a role in the symptoms of schizophrenia. People with schizophrenia exhibit alterations in P50 (Fresnán et al., 2007), N100 (Turetsky et al., 2008), N200 (Groom et al., 2008), and P300 (Begré et al., 2008; Groom et al., 2008) auditory evoked potentials. The $\alpha 7$ neuronal nAChR subtype is likely associated with the P50 auditory inhibition marker on human chromosome 15 (Leonard et al., 1998; Leonard and Freedman, 2006; Li et al., 2004; Williams and Ziedonis, 2004). In contrast to healthy controls increased memory load led to increases in event-related desynchronization and synchronization in people with schizophrenia and, to a lesser extent, their co-twins discordant for schizophrenia (Bachman et al., 2008). Nicotine may reduce the disturbed information sensory gating of people with schizophrenia.

Exposure to smoking and related scenes produces activation in the hippocampus of smokers. Smoking-related cues result in activation of associative learning portions of the brain including the hippocampus. After rats addicted to nicotine experience withdrawal, the activation of high-affinity nicotinic receptors returns to baseline levels within 24 h. However, CA1 pyramidal neurons may exhibit an increased sensitivity to nicotine resulting in persistent susceptibility to utilize nicotine long after the last use of nicotine (Penton et al., 2011).

Utilizing kinetic modeling (Fujita et al., 1999, 2000; Yokoi et al., 1999), we demonstrated that *(S)-5-[^{123}I]iodo-3-(2-azetidinylmethoxy)pyridine (5-[^{123}I]IA)*, a novel potent radioligand prepared by radioiododestannylation (Musachio et al., 2001) for high-affinity $\alpha_4\beta_2\star$ neuronal nAChRs (Musachio et al., 1999), provides a means to evaluate the density and the distribution of neuronal nAChRs in the living human brain of healthy adults (Brašić et al., 2009; Wong et al., 2001). Dynamic single photon emission computed tomography (SPECT) on a Trionix TriadXLT collected images over 6 h with 20 acquisitions (Brašić et al., 2009; Wong et al., 2001). A 2-compartmental (plasma and brain tissue) model was used for the kinetic analysis and parametric imaging of volumes of interest (VOIs) drawn on co-registered magnetic resonance imaging (MRI) scans (Brašić et al., 2009; Wong et al., 2001). The plasma radioactivity of centrifuged whole arterial blood obtained during the scan was measured with a gamma counter (Brašić et al., 2009). Plasma radiometabolites were also assayed with high performance lipid chromatography (HPLC) (Brašić et al., 2009; Hilton et al., 2000). The metabolite-corrected plasma input function was used for kinetic modeling (Brašić et al., 2009; Fujita et al., 1999, 2000). Binding potentials were calculated by the mathematical modeling utilizing the procedures for other radioligands of this series (Brašić et al., 2009; Yokoi et al., 1999). High plasma nicotine level was significantly associated with low *5-[^{123}I]IA* binding in the caudate head, the cerebellum, the cortex, the fusiform gyrus, the hippocampus, the parahippocampus, the pons, the putamen, and the thalamus. These findings confirm that *5-[^{123}I]IA* competes with nicotine to occupy nAChRs. We conclude that *5-[^{123}I]IA* is a safe, well-tolerated, and effective pharmacologic agent for human subjects to estimate high-affinity α_4/β_2 neuronal in the living human brain (Brašić et al., 2001, 2004, 2009; Wong et al., 2001; Zhou et al., 2001).

Since the high-resolution research tomography (HRRT), a brain dedicated PET, has superior resolution of 2 mm, 3–5 times smaller than SPECT, we now seek to perform HRRT, in place of SPECT, using PET [^{18}F] radiotracers, with greater binding to nAChRs developed by (Horti and Wong, 2009) in place of $5\text{-}[^{123}I]IA$, an SPECT radiotracer.

We have demonstrated that $2\text{-}[^{18}F]fluoro\text{-}3\text{-}(2(S)\text{-}azetidinylmethoxy)pyridine$ $(2\text{-}[^{18}F]FA)$, (Schildan et al., 2007; Sorger et al., 2007) a compound with the same chemical moiety to bind to the high–affinity $\alpha_4\beta_2$ neuronal nAChRs as $5\text{-}[^{123}I]IA$, has approximately twice the uptake in the thalami of healthy nonsmoking control adult volunteers than smoking adults with schizophrenia who smoked cigarettes the day of the scan (Brašić et al., 2010). Utilizing $2\text{-}[^{18}F]FA$, Brody et al. (2006) demonstrated that in human smokers inhalation of one or two mouthfuls of cigarette smoke led to uptake in half the $\alpha_4\beta_2\star$ neuronal nAChRs and that uptake of half the $\alpha_4\beta_2\star$ neuronal nAChRs occurred with plasma nicotine levels of 0.87 ng/mL (Brody et al., 2006). Additionally, PET with $2\text{-}[^{18}F]FA$ demonstrated greater regional uptake in the brainstem and the cerebellum in smokers than in nonsmokers (Wüllner et al., 2008).

2.2. NACHR RADIOTRACERS

Several agents have been developed to image nAChRs (Brašić et al., 2009). $5\text{-}[^{123}I]$ IA, $2\text{-}[^{18}F]FA$, and $6\text{-}[^{18}F]fluoro\text{-}3\text{-}(2(S)\text{-}azetidinylmethoxy)pyridine$ $(6\text{-}[^{18}F]FA)$ are agents available for nuclear neuroimaging in humans with relatively slow uptake requiring several hours of administration of radiotracer to attain equilibrium (Horti et al., 2010). While some healthy adults can tolerate the hours required for the administration and the uptake of $5\text{-}[^{123}I]IA$, $2\text{-}[^{18}F]FA$, and $6\text{-}[^{18}F]FA$, many patients with schizophrenia and other neuropsychiatric disorders cannot tolerate the discomforts and the inconveniences of these scans. Another drawback to the utilization of $5\text{-}[^{123}I]IA$ (Brašić et al., 2009), $2\text{-}[^{18}F]FA$ (Brašić et al., 2010), and $6\text{-}[^{18}F]FA$ is the relatively limited uptake of these compounds outside the thalamus. This is a particular drawback for the evaluation of schizophrenia and other neuropsychiatric disorders with likely deficits in the dorsolateral prefrontal cortex and other extrathalamic regions of the brain.

2.3. ALZHEIMER'S DISEASE

2.3.1. Overview

Decrements in ACh transmitting neurons are hypothesized to play a role in the pathogenesis and pathophysiology of mild cognitive impairment (MCI) and AD (Kendziorra et al., 2010). Quantitative measurements of the density and the distribution of the high–affinity $\alpha_4\beta_2\star$ neuronal nAChRs provides a tool to guage the possible deficits present in MCI and progressive in AD. Additionally estimation of the extent of dysfunction of high–affinity $\alpha_4\beta_2\star$ neuronal nAChRs in MCI, AD, and related disorders may provide a tool to monitor the beneficial and adverse effects of therapeutic interventions for these disorders.

$2\text{-}[^{18}F]FA$ (370 MBq) was administered by continuous infusion in a vein in an antecubital fossa over 90 s to 24 nonsmoking participants including 8 with MCI and 9 with AD as well as 7 age-matched healthy control subjects (Kendziorra et al., 2010). PET (ECAT EXACT HR+, CTI/Siemens, Knoxville, TN, USA) was performed in the 2-D acquisition mode to obtain 63 slices with a resolution of 4.7 mm full-width half maximum (Kendziorra et al., 2010), in the first 120 min (4×15 s, 4×1 min, 5×2 min, 5×5 min, and 8×6 min) as well as 6 h after the radiotracer administration (4×15 min) (Kendziorra et al., 2010). Input function arterial blood sampling was obtained to determine the counts of radioactivity with a Cobra gamma counter (Packard Instrument Company, Meriden, CT, USA) (16 in 3 min, then samples at 3, 4, 5, 6, 8, 10, 12, 14, 18, 25, 35, 50, 60 min, and then every 30 min until 7 h). Samples of arterial blood were obtained to measure the nonmetabolized fraction at 14, 60, 120, 240, 330, and 420 min (Kendziorra et al., 2010). Utilizing coregistered PET and MRI visualizations, the non-displaceable binding potential (BP_{ND}) (Innis et al., 2007) was calculated with the corpus callosum or the cerebellum as the reference region (Kendziorra et al., 2010). Since both patients with MCI and AD demonstrated reduced BP_{ND}s in the regions affected in AD, the corpus callosum was utilized as the reference region. Progression to AD was observed solely in those people with MCI who had reductions BP_{ND}s. The cognitive impairment was inversely proportional to the BP_{ND}.

2.3.2. Sample Size Determination

Pilot studies of schizophrenia are often hindered by limited knowledge of the study variables. While the sample sizes of this study are so small that detailed statistical analysis is inappropriate, a strategy can be developed to assess small numbers of subjects in research studies. Utilizing a hypothesized proportional difference in the variable of interest between the experimental group and healthy volunteers, the minimal sample size to detect the hypothesized variations in the study variable in the experimental (P_1) and control (P_2) groups can be estimated. Proportions are estimated to range between 0 and 1 ($0 < P_1 < 1$, and $0 < P_2 < 1$). For example, suppose that based on prior research with 20 subjects in each group, we hypothesize that typical values of the study variable are exhibited by (1) a low proportion of subjects with schizophrenia, say, $P_1 = 0.05$, and (2) a high proportion of healthy adults, say, $P_2 = 0.75$. We estimate the power for a two-sample comparison of proportions to test the null hypothesis, H_0: $P_1 = P_2$, where the proportion of typical values of the variable of interest is P_1 in subjects with schizophrenia and P_2 in healthy volunteers. We assume that alpha (α) = 0.0500 (two-tailed), that there are 20 participants with schizophrenia, and that there are 20 control participants. The power resulting from the expected and slightly different proportions are listed in Table 1 (StataCorp, 1999). Even if the proportions obtained in the study deviate slightly from the hypothesized proportions, reasonable power is obtained with a sample size of 20 in each of the two groups as demonstrated by Table 1. For a pilot study of participants with schizophrenia or another rare disorder, these are reasonable powers with a sample size of 20 in each group (Brašić et al., 2003, StataCorp, 1999). The power resulting from the expected and slightly different proportions are listed in Table 1 (StataCorp, 1999).

Table 1 Power for Variable Proportions in 20 Participants with Study Condition (P_1) and 20 Healthy Control Participants (P_2) (StataCorp, 1999)

P_1	P_2	Power
0.05	0.65	0.6894
0.05	0.70	0.7832
0.05	0.75	0.8671
0.05	0.80	0.9338
0.10	0.65	0.5572
0.10	0.70	0.6582
0.10	0.75	0.7582
0.10	0.80	0.8501

Even if the proportions obtained in the study deviate slightly from the hypothesized proportions, reasonable power is obtained with the sample size of 10 in each of the two groups as demonstrated by Table 1. For a pilot study these are reasonable powers with a sample size of 10 in each of two groups (StataCorp, 1999).

3. CONCLUSIONS

The pathways for ACh, dopamine, serotonin, and other neurotransmitters are intimately related. Exposure to nicotine, nicotinic agonists, cocaine and other dopaminergic agents, and other substances affects the other neurotransmitter systems. For this reason, quantitative records of exposure to nicotine and other substances need to be obtained at baseline as well as during and after scans and other measurements of nicotinic receptors and other neurotransmitters. For this reason, we include tools to capture the key information for exposure to cigarettes and other forms of nicotine, various forms of cocaine (Please refer to Table A.1 in the appendix).

Characterization of people with schizophrenia remains a challenge for clinicians and researchers. Several procedures have been developed to standardize the criteria for schizophrenia (American Psychiatric Association, 2000; Brašić et al., 2009, 2010). Tools to measure the nuances of both positive (Andreasen, 1984b) and negative (Andreasen, 1984a) symptoms of schizophrenia will be invaluable to the identification of subtypes of people with schizophrenia. The effort required to accurately structured tools to identify the subtle differences between groups of individuals with schizophrenia (Andreasen, 1984a,b) will be rewarded by the precise characterization of subclasses of the disorder.

The Cocaine Administration Preference Questionnaire (Table A.1 in the Appendix) is a twelve-item form to be completed by the patient repeatedly before, during, and after treatments. Scores of individual items are tabulated for each section—A. Nasal inhalation, B. Smoking a joint, and C. Intravenous injection. For each section, a positive score indicates preference for the indicated route of administration, while a negative score indicates a dislike of the indicated route of administration. A score of zero for a section indicates indifference to the suggested route of administration (Table A.1 in the Appendix).

The density and the distribution of nicotinic receptors likely are influenced by the phase of the menstrual cycle. By analogy the binding potential for dopamine $D_{2/3}$ receptors ($D_{2/3}$ Rs) in increased in the periovulatory and luteal phases and is decreased for the follicular phase (Wong et al., 1988). Like nicotinic receptors and other types of receptors throughout the brain exhibit similar waxing and waning with the menstrual cycle. For this reason, capturing the salient data to determine the phase of the menstrual cycle when the patient enters studies is a key foundation for meaningful data analysis. For this reason, all women are asked to complete both the Menstrual Cycle Questionnaire (Table A.2 in the Appendix) to obtain clinical relevant clinical data and the Menstrual Cycle Diary (Table A.3 in the Appendix) to obtain retrospective data to determine the specific phase of the menstrual cycle. The Menstrual Cycle Questionnaire (Table A.2 in the Appendix) is completed retrospectively by each woman one the date of screening for each study. The Menstrual Cycle Diary (Table A.3 in the Appendix) is given to each woman on the date of screening for each study to be completed prospectively to document the start and stop dates for all menstrual cycles from the start of the study until the completion of the study. If the woman records the start and the stop dates for at least three menstrual cycles, then a reasonable estimation of the phase of the menstrual cycle on the study date. Additionally, plasma is obtained from each woman on the study date for measurement of prolactin and progesterone levels to confirm the data recorded by the patient.

While identification of the phase of the menstrual cycle is crucial to the interpretation of scans and other quantitative measurements of nicotinic receptors and other neurotransmitters, caution is required to conduct research on pregnant women. The effects of exposures to radiation, chemicals, toxins, and other experimental agents on fetuses may be deleterious. For this reason participants who are pregnant and who may be pregnant are prudently excluded from research studies with experimental interventions. Women who are likely to become pregnant during the course of a research study are wisely excluded from imaging studies of nAChRs in the adult brain. Women may be told that they are welcome to return to participate in studies once they are not pregnant.

Characterization of the smoking preferences of all participants in research studies of nAChRs in the brain is needed to identify the subgroups for analyses. For this reason the Smoking Preferences Questionnaire (Table A.4 in Appendix 1) is administered to all potential participants as part of the screening procedure at baseline, during, and after protocols to visualize nAChRs.

Additionally researchers who administer nicotine patches will benefit from the administration of the Nicotine Patch Adverse Events Scale (Table A.5 in Appendix 2) to each patient before, during, and after each administration an a nicotine patch.

GLOSSARY OF ABBREVIATIONS

2-[^{18}F]FA, 2-[^{18}F]fluoro-3-(2(S)-azetidinylmethoxy)pyridine a radiotracer to visualize the high-affinity α4β2★ neuronal nicotinic acetylcholine receptors (nAChRs) by means of positron emission tomography (PET).

5-[^{123}I]IA, (S)-5-[^{123}I]iodo-3-(2-azetidinylmethoxy)pyridine a radiotracer to visualize the high-affinity α4β2★ neuronal nicotinic acetylcholine receptors (nAChRs) by means of single-photon computed tomography (SPECT).

6-[^{18}F]FA, 6-[^{18}F]fluoro-3-(2(S)-azetidinylmethoxy)pyridine a radiotracer to visualize the high-affinity α4β2★ neuronal nicotinic acetylcholine receptors (nAChRs) by means of positron emission tomography (PET).

[^{18}F]AZAN, (−)-2-(6-[^{18}F]fluoro-2,3′-bipyridin-5′-yl)-7-methyl-7-aza-bicyclo[2.2.1]heptane ([^{18}F]JHU87522) a radiotracer to visualize the high-affinity α4β2★ neuronal nicotinic acetylcholine receptors (nAChRs) by means of positron emission tomography (PET).

[^{18}F]FP-TZTP, 3-[[4-(3-[^{18}F]fluoropropylsulfanyl)-1,2,5-thiadiazol-3-yl]]-1-methyl-1,2,5, 6-tetrahydropyridine a radiotracer to quantitatively estimate the function of the muscarinic system in humans.

ACh acetylcholine, an excitatory neurotransmitter

AChRs acetylcholine receptors, structures on postsynaptic neurons that send excitatory impulses on the neurons when activated by binding to acetylcholine (ACh) transmitted from the presynaptic neuron across the synapse.

AD Alzheimer's disease, a disorder typically presenting in people aged more than 65 years characterized by the progressive decline of cognitive functions.

BP$_{ND}$ binding potential

CHRNA4 a cholinergic gene conferring susceptibility to schizophrenia.

CHRNB2 a cholinergic gene conferring susceptibility to schizophrenia.

CNS central nervous system, the brain and the spinal cord.

cAMP cyclic adenosine monophosphate, a compound to provide energy for cellular activities.

HPLC high performance lipid chromatography, a technique to separate proteins present in fluids

HRRT high resolution research tomography, a positron emission tomography (PET) tool with resolution approaching 2 mm.

mAChRs mucarinic acetylcholine receptors, acetylcholine receptors (AChRs) activated by muscarine.

nAChRs nicotinic acetylcholine receptors, acetylcholine receptors (AChRs) activated by nicotine.

MCI mild cognitive impairment, slight decline in intellectual function.

PET positron emission tomography, an imaging technique to visualize the physiology of organs through the detection of positrons released through the decay of radiotracers administered to the patient.

RS [^{123}I]IQNB, RS 3-quinuclidinyl-4-[^{123}I]iodobenzilate a radiotracer for clinical studies in humans to quantify changes in the concentrations of mAChRs.

SPECT single-photon computed tomography, an imaging technique to visualize the physiology of organs through the detection of photons released through the decay of radiotracers administered to the patient.

VOIs volumes of interest, three-dimensional structures to be represented by means of imaging techniques.

Table A.1 Cocaine Administration Preference Questionnaire. This form is completed repeatedly by the patient before, during, and after each protocol. For each section a positive score indicates a preference, a negative score indicates a dislike, and a score of zero indicates indifference.

Name_____

Date_____

Instructions: Please answer all items based on how you feel right now. Please encircle your response.

A. Nasal administration preference items

| 1. I like to snort cocaine (through the nose). | No = 0 | Yes = 1 |

| 2. I like to see others snort cocaine (through the nose). | No = 0 | Yes = 1 |

| 3. Seeing others snort cocaine (through the nose) turns me on. | No = 0 | Yes = 1 |

| 4. Seeing other people snort cocaine (through the nose) turns me off. | No = 0 | Yes = −1 |

Nasal administration preference score (Total of scores 1–4)   _____

B. Smoking administration preference items

| 5. I like to smoke cocaine. | No = 0 | Yes = 1 |

| 6. I like to see others smoke cocaine. | No = 0 | Yes = 1 |

| 7. Seeing other people smoke cocaine turns me on. | No = 0 | Yes = 1 |

| 8. Seeing other people smoke cocaine turns me off. | No = 0 | Yes = −1 |

Smoking administration preference score (Total of scores 5–8)   _____

C. Intravenous administration preference items

| 9. I like to shoot up (inject intravenously) cocaine. | No = 0 | Yes = 1 |

| 10. I like to see others shoot up (inject intravenously) cocaine. | No = 0 | Yes = 1 |

| 11. Seeing other people shoot up (inject intravenously) cocaine turns me on. | No = 0 | Yes = 1 |

| 12. Seeing other people shoot up (inject intravenously) cocaine turns me off. | No = 0 | Yes = −1 |

Intravenous administration preference score (Total of scores 9–12)   _____

Table A.2 Menstrual Cycle Questionnaire. To obtain the relevant data to determine the phase of the menstrual cycle on the date of the study, please ask the patient to complete this form for every assessment.

Instructions: Please complete this form as accurately as possible. You may write any additional information on the reverse page.

Subject Name: _____

History Number: _____

Date: _____/ _____/ _____

1. Age of menarche (when you started your first period): _____

2. Do you use birth control?   (circle one)   Yes   No

      If "Yes," please describe._____

      If "Yes," how long have you used this form of birth control?:

_____

3. Average number of days between periods of menstruation: _____

4. Have you ever had problems associated with menstruation? Please describe.
_____

5. What was the first day of last month's menstruation? _____

6. What was the first day of your last menstruation? _____

7. Have you ever been pregnant? (circle one)   Yes   No

      If "Yes", please describe.
_____
_____

Table A.3 Menstrual Cycle. To obtain the relevant data to determine the phase of the menstrual cycle on the date of the study, please ask the patient to complete this form on screening for every study. Please ask the patient to return to the study physician on completion of the study.

Instructions: Please record the start date and the stop date for every menstrual cycle until the completion of the study. Please give this diary to the study physician on the date you complete the study. Please record additional information on the back of this page.

Subject Name: _____

Date: _____/ _____/ _____

Menstrual cycle 1

Start Date: _____/ _____/ _____

Stop Date: _____/ _____/ _____

Menstrual cycle 2

Start Date: _____/ _____/ _____

Stop Date: _____/ _____/ _____

Menstrual cycle 3

Start Date: _____/ _____/ _____

Stop Date: _____/ _____/ _____

Menstrual cycle 4

Start Date: _____/ _____/ _____

Stop Date: _____/ _____/ _____

Menstrual cycle 4

Start Date: _____/ _____/ _____

Stop Date: _____/ _____/ _____

Table A.4 Smoking Preferences Questionnaire. To obtain the relevant data to determine the smoking preferences of all subjects on the date of the study, please ask the patient to complete this form on screening, during, and after every study. Please ask the patient to return to the study physician on completion of the study.

Name_____

Date_____

Time_____

Please answer all items based on your personal lifetime experience. You may record additional information on the back of this page. Thank you.

1. How many cigarettes do you smoke daily?_____

2. How many years in your entire lifetime have you smoked cigarettes?_____

3. What brand of cigarettes do you smoke today?_____

4. What brand of cigarettes have you smoked most in your lifetime?_____

_____

_____

5. How many times in your lifetime have you quit smoking cigarettes?_____

6. What is the longest period of time you quit smoking cigarettes?_____

7. What is the length of the most recent period you quit smoking cigarettes?_____

Table A.5 Nicotine Patch Advere Events Checkelist. To obtain the relevant data to determine the adverse events experienced by each subject resulting from exposure to nicotine patches on the date of the study, please ask the patient to complete this form on screening, during, and after every administration of nicotine patches during the course of the study. Please ask the patient to return to the study physician on completion of the study.

Nicotine Patch Adverse Events Checklist

Name_____

Rater_____

Date_____

Please rate your experience with each item in the week after the most recent nicotine patch application.

	A	Q	M	MO	MS	VS
1. Skin redness	0	1	2	3	4	5
2. Itching	0	1	2	3	4	5
3. Burning	0	1	2	3	4	5
4. Insomnia	0	1	2	3	4	5
5. Abnormal dreams	0	1	2	3	4	5
6. Nervousness	0	1	2	3	4	5
7. Muscle cramps	0	1	2	3	4	5
8. Upset stomach	0	1	2	3	4	5
9. Diarrhea	0	1	2	3	4	5
10. Incoordination	0	1	2	3	4	5
11. Fever	0	1	2	3	4	5

A = absent; Q = questionable; M = mild; MO = moderate; MS = moderately severe; VS = very severe

REFERENCES

Aguilar, M.C., Gurpegui, M., Diaz, F.J., de Leon, J., 2005. Br. J. Psychiatry 186, 215.

American Psychiatric Association, 2000. Diagnostic and Statistical Manual of Mental Disorders, fourth ed., Text Revision (DSM-IV-TR™), American Psychiatric Association, Washington, DC.

Andreasen, N.C., 1984a. The Schedule for the Assessment of Negative Symptoms (SANS). The University of Iowa, Iowa City, Iowa.

Andreasen, N.C., 1984b. The Schedule for the Assessment of Positive Symptoms (SAPS). The University of Iowa, Iowa City, Iowa.

Anonymous, 2002. MMWR Morb. Mortal. Wkly. Rep. 51, 300.

Anonymous, 2003. MMWR Morb. Mortal. Wkly. Rep. 52, 842.

Bachman, P., Kim, J., Yee, C.M., Therman, S., Manninen, M., Lönnqvist, J., Kaprio, J., Huttunen, M.O., Näätänen, R., Cannon, T.D., 2008. Schizophr. Res. 103, 293.

Begré, S., Kleinlogel, H., Kiefer, C., Strik, W., Dierks, T., Federspiel, A., 2008. Neurobiol. Dis. 30, 270.

Benowitz, N.L., 2008. Clin. Pharmacol. Ther. 83, 531.

Brašić, J.R., 2003. Treatment of movement disorders in autism spectrum disorders. In: Hollander, E. (Ed.), Autism Spectrum Disorders. Medical Psychiatry Series, vol. 24. Marcel Dekker, Inc., New York, pp. 273–346. ISBN 0-8247-0715-X. http://www.dekker.com.

Brašić, J.R., Cascella, N., Hussain, B., Bisuna, B., Kumar, A., Raymont, V., Guevara, M.R., Horti, A., Wong, D.F., 2010. J. Nucl. Med. 51 (Suppl. 2), 388, (abstract).

Brašić, J.R., Musachio, J.L., Zhou, Y., Cascella, N., Nestadt, G., Osman, M., Gay, O., Hilton, J., Kuwabara, H., Crabbe, A., Rousset, O., Fan, H., Al-Humadi, M.A., Al-Humadi, B.A., Arkles, J.S., Thompson, T., Kalaff, A., Smith, J., Reinhardt, M.J., Dogan, A.S., Ford, B., Wong, D.F., 2001. Mov. Disord. 16 (Suppl. 1), S54, (abstract).

Brašić, J.R., Rohde, C.A., Maris, M.A., Wong, D.F., 2003. In: 4th Annual Rett Syndrome Symposium. Rett Syndrome Research Foundation, Baltimore, Maryland. 23–25 June 2003, 46, (abstract). www.rsrf.org.

Brašić, J.R., Zhou, Y., Musachio, J.L., Hilton, J., Fan, H., Crabb, A., Endres, C.J., Reinhardt, M.J., Dogan, A.S., Alexander, M., Rousset, O., Maris, M.A., Galecki, J., Nandi, A., Wong, D.F., 2009. Synapse 63 (4), 339.

Brašić, J.R., Zhou, Y., Musachio, J., Hilton, J., Fan, H., Crabb, A., Endres, C.J., Reinhardt, M.J., Dogan, A.S., Alexander, M., Rousset, O., Maris, M.A., Kuwabara, H., Wong, D.F., 2004. Baltimore, MarylandIn: 5th Annual Rett Syndrome Symposium, June 28–30, 2004, Inn at the Colonnade. Rett Syndrome Research Foundation (RSRF), Cincinnati, Ohio. 44, (abstract). www.rsrf.org.

Breese, C.R., Lee, M.J., Adams, C.E., Sullivan, B., Logel, J., Gillen, K.M., Marks, M.J., Collins, A.C., Leonard, S., 2000. Neuropsychopharmacology 23, 351.

Breese, C.R., Marks, M.J., Logel, L., Adams, C.E., Sullivan, B., Collins, A.C., Leonard, S., 1997. J. Pharmacol. Exp. Ther. 282, 7.

Brody, A.L., Mandelkern, M.A., London, E.D., Olmstead, R.E., Farahi, J., Scheibal, D., Jou, J., Allen, V., Tiongson, E., Chefer, S.I., Koren, A.O., Mukhin, A.G., 2006. Arch. Gen. Psychiatry 63, 907.

Centers for Disease Control and Prevention, 2011. Smoking & Tobacco Use. Tobacco-Related Mortality. . (accessed 04.02.11.). http://www.cdc.gov/tobacco/data_statistics/fact_sheets/health_effects/tobacco_related_mortality/index.htm.

Costa, G., Abin-Carriquiry, J.A., Dajas, F., 2001. Brain Res. 888, 336.

Court, J.A., Lloyd, S., Thomas, N., Piggott, M.A., Marshall, E.F., Morris, C.M., Lamb, H., Perry, R.H., Johnson, M., Perry, E.K., 1998. Neuroscience 87, 63.

Craddock, N., Lendon, C., Cichon, S., Culverhouse, R., Detera-Wadleigh, S., Devon, R., Faraone, S., Foroud, S., Gejman, P., Leonard, S., McInnis, M., Owen, M.J., Riley, B., Armstrong, C., Barden, N., van Broeckhoven, C., Ewald, H., Folstein, S., Gerhard, D., Goldman, D., Gurling, H., Kelsoe, J., Levinson, D., Muir, W., Philippe, A., Pulver, A., Wildenauer, D., 1999. Am. J. Med. Genet. B Neuropsychiatr. Genet. 88, 244.

Cromwell, H.C., Anstrom, K., Azarov, A., Woodward, D.J., 2005. Brain Res. 1043, 12.

De Luca, V., Voineskos, S., Wong, G., Kennedy, J.L., 2006. Exp. Brain Res. 174, 292.

Doll, R., Peto, R., Boreham, J., Sutherland, I., 2004. BMJ 328, 1519.

Durany, N., Zöchling, R., Boissl, K.W., Paulus, W., Ransmayr, G., Tatschner, T., Danielczyk, W., Jellinger, K., Deckert, J., Riederer, P., 2000. Neurosci. Lett. 287, 109.

Dursun, S.M., Reveley, M.A., 1997. Psychol. Med. 27, 483.

Eckelman, W.C., 2006. Curr. Pharm. Des. 12, 3901.

Eisner, M.D., Wang, Y., Haight, T.J., Balmes, J., Hammond, S.K., Tager, I.B., 2007. Ann. Epidemiol. 17, 364.

Fresán, A., Apiquian, R., García-Anaya, M., de la Fuente-Sandoval, C., Nicolini, H., Graff-Guerrero, A., 2007. Schizophr. Res. 97, 128.

Fujita, M., Tamagnan, G., Zoghbi, S.S., Al-Tikriti, M.S., Baldwin, R.M., Seibyl, J.P., Innis, R.B., 2000. J. Nucl. Med. 41, 1552.

Fujita, M., Tamagnan, G., Zoghbi, S.S., Innis, R.B., 1999. ProCeeding of the Society for Neurosciencevol. 25. 285, (abstract).

Gotti, C., Clementi, F., 2004. Prog. Neurobiol. 74, 363.

Gotti, C., Fornasari, D., Clementi, F., 1997. Prog. Neurobiol. 53, 199.

Groom, M.J., Bates, A.T., Jackson, G.M., Calton, T.G., Liddle, P.F., Hollis, C., 2008. Biol. Psychiatry 63, 784.

Herrán, A., de Santiago, A., Sandoya, M., Fernández, M.J., Diez-Manrique, J.F., Vázquez-Barquero, J.L., 2000. Schizophr. Res. 41, 373.

Hilton, J., Yokoi, F., Dannals, R.F., Ravert, H.T., Szabo, Z., Wong, D.F., 2000. Nucl. Med. Biol. 27 (6), 627.

Horti, A.G., Gao, Y., Kuwabara, H., Dannals, R.F., 2010. Life Sci. 86, 575.

Horti, A.G., Wong, D.F., 2009. P. E. T. Clin. 4 (1), 89.

Innis, R.B., Cunningham, V.J., Delforge, J., Fujita, M., Gjedde, A., Gunn, R.N., Holden, J., Houle, S., Huang, S.C, Ichise, M., Iida, H., Ito, H., Kimura, Y., Koeppe, R.A., Knudsen, G.M., Knuuti, J., Lammertsma, A.A., Laruelle, M., Logan, J., Maguire, R.P., Mintun, M.A., Morris, E.D., Parsey, R., Price, J.C., Slifstein, M., Sossi, V., Suhara, T., Votaw, J.R., Wong, D.F., and Carson, R.E., 2007. Consensus nomenclature for in vivo imaging of reversibly binding radioligands. J. Cereb. Blood Flow Metab. 27, 1533–1539.

Kendziorra, K., Wolf, H., Meyer, P.M., Barthel, H., Hesse, S., Becker, G.A., Luthardt, J., Schildan, A., Patt, M., Sorger, D., Seese, A., Gertz, H.-J., Sabri, O., 2010. Eur. J. Nucl. Med. Mol. Imaging. http://dx.doi.org/10.1007/s00259-010-1644-5.

Kucinski, A.J., Stachowiak, M.K., Wersinger, S.R., Lippiello, P.M., Bencherif, M., December 6, 2010. Curr. Pharm. Biotechnol. (Epub ahead of print).

Langmead, C.J., Watson, J., Reavill, C., 2008. Pharmacol. Ther. 117, 232.

Lafargue, T., Brasic, J., 2000. Med. Hypotheses 55 (4), 314.

Lasser, K., Boyd, J.W., Woolhandler, S., Himmelstein, D.U., McCormick, D., Bor, D.H., 2000. JAMA 284, 2606.

Leonard, S., Breese, C.R., Lee, M.J., Logel, J., Adams, C.E., Freedman, R., 1999. Schizophr. Res. 36 (1–3), 74, (abstract, special issue).

Leonard, S., Breese, C., Adams, C., Benhammou, K., Gault, J., Stevens, K., Lee, M., Adler, L., Olincy, A., Ross, R., Freedman, R., 2000. Eur. J. Pharmacol. 393, 237.

Leonard, S., Freedman, R., 2006. Biol. Psychiatry 60, 115.

Leonard, S., Gault, J., Adams, C., Breese, C.R., Rollins, Y., Adler, L.E., Olincy, A., Freedman, R., 1998. Restor. Neurol. Neurosci. 12, 195.

Li, C.-H., Liao, H.-M., Chen, C.-H., 2004. Neurosci. Lett. 372, 1.

Musachio, J.L., Brasic, J.R., Scheffel, U.A., Rauseo, P.A., Fan, H., Osman, M., Kellar, K.J., Xiao, Y., Hilton, J., Zhou, Y., Wong, D.F., 2001. Abstracts of Papers – American Chemical Society. In: American Chemical Society Meeting. American Chemical Society, Washington, District of Columbia. April 1, 2001, 221(Part 2), 183–NUCL (abstract).

Musachio, J.L., Villemagne, V.L., Scheffel, U.A., Dannals, R.F., Dogan, A.S., Yokoi, F., Wong, D.F., 1999. Nucl. Med. Biol. 26, 201.

Office on Smoking and Health, National Center for Chronic Disease Prevention and Health Promotion, 2004. The Health Consequences of Smoking: A Report of the Surgeon General. Office on Smoking and Health, National Center for Chronic Disease Prevention and Health Promotion, Washington, DC.

Perry, D.C., Dávila-García, M.I., Stockmeier, C.A., Kellar, K.J., 1999. J. Pharmacol. Exp. Ther. 289, 1545.

Penton, R.E., Quick, M.W., Lester, R.A.J., 2011. J. Neurosci. 31 (7), 2584.

Sacco, K.A., Termine, A., Seyal, A., Dudas, M.M., Vessicchio, J.C., Krishnan-Sarin, S., Jatlow, P.I., Wexler, B.E., George, T.P., 2005. Arch. Gen. Psychiatry 62, 649.

Sanberg, P.R., Silver, A.A., Shytle, R.D., Philipp, M.K., Cahill, D.W., Fogelson, H.M., McConville, B.J., 1997. Pharmacol. Ther. 4, 21.

Schildan, A., Patt, M., Sabri, O., 2007. Synthesis procedure for routine production of 2-[^{18}F]fluoro-3-(2(S)-azetidinylmethoxy)pyridine (2-[^{18}F]F-A-85380). Appl. Radiat. Isot. 65, 1244–1248.

Schroeder, S.A., 2005. JAMA 294, 482.

Schroeder, S.A., 2004. N. Engl. J. Med. 350, 293.

Sorger, D., Becker, G.A., Patt, M., Schildan, A., Grossmann, U., Schliebs, R., Seese, A., Kendziorra, K., Kluge, M., Brust, P., Mukhin, A.G., Sabri, O., 2007. Measurement of the α4β2★ nicotinic acetylcholine receptor ligand 2-[^{18}F]fluoro-A-85380 and its metabolites in human blood during PET investigation: a methodological study. Nucl. Med. Biol. 34, 331–342.

StataCorp, 1999. Stata Statistical Software: Release 6.0. Stata Corporation, College Station, TX.

Tandon, R., Keshavan, M.S., Nasrallah, H.A., 2008. Schizophr. Res. 102, 1.

Timmermann, D.B., Grønlien, J.H., Kohlhaas, K.L., Nielsen, E.Ø., Dam, E., Jørgensen, T.D., Ahring, P.K., Peters, D., Holst, D., Christensen, J.K., Malysz, J., Briggs, C.A., Gopalakrishnan, M., Olsen, G.M., 2007. J. Pharmacol. Exp. Ther. 323, 294.

Turetsky, B.I., Greenwood, T.A., Olincy, A., Radant, A.D., Braff, D.L., Cadenhead, K.S., Dobie, D.J., Freedman, R., Green, M.F., Gur, R.E., Gur, R.C., Light, G.A., Mintz, J., Nuechterlein, K.H., Schork, N.J., Seidman, L.J., Siever, L.J., Silverman, J.M., Stone, W.S., Swerdlow, N.R., Tsuang, D.W., Tsuang, M.T., Calkins, M.E., 2008. Biol. Psychiatry 64 (12), 1051.

Turner, D.M., Armitage, A.K., Briant, R.H. and Dollery, C.T., 1975. Metabolism of nicotine by the isolated perfused dog lung. Xenobiotica 5, 539–551.

Voineskos, S., De Luca, V., Mensah, A., Vincent, J.B., Potapova, N., Kennedy, J.L., 2007. J. Psychiatry Neurosci. 32 (6), 412.

Williams, J.M., Ziedonis, D., 2004. Addict. Behav. 29, 1067.

Williams, J.M., Ziedonis, D.M., Abanyie, F., Steinberg, M.L., Foulds, J., Benowitz, N.L., 2005. Schizophr. Res. 79, 323.

Wong, D.F., Brasic, J.R., Zhou, Y., Gay, O., Crabb, A.H., Kuwabara, H., Hilton, J., Osman, M., Scheffel, U.A., Roursset, O., Fan, H., Dannals, R.F., Musachio, J.L., 2001. J. Nucl. Med. 42 (5 Suppl.), 142 (abstract).

Wong, D.F., Broussolle, E.P., Wand, G., Villemagne, V., Dannals, R.F., Links, J.M., Zacur, H.A., Harris, J., Naidu, S., Braestrup, C., Wagner Jr., H.N., Gjedde, A., 1988. Ann. N.Y. Acad. Sci. 515, 203.

Wong Jr., D.F., Maini, A., Alexander, M., Zhou, Y., Brasic, J., Scheffel, U., Fan, H., Hilton, J., Dannals, R., Musachio, J., 2002. J. Nucl. Med. 43 (5 Suppl.), 109 (abstract).

Wüllner, U., Gündisch, D., Herzog, H., Minnerop, M., Joe, A., Warnecke, M., Jessen, F., Schütz, C., Reinhardt, M., Eschner, W., Klockgether, T., Schmaljohann, J., 2008. Neurosci. Lett. 430, 34.

Yokoi, F., Musachio, J., Hilton, J., Kassiou, M., Ravert, H.T., Mathews, W.B., Dannals, R.F., Stephane, M., Wong, D.F., 1999. J. Nucl. Med. 40 (5 Suppl.), 263 (abstract).

Zhou, Y., Brašić, J.R., Fan, H., Maini, A., Wong, D.F., 2002. In: 3rd Annual Rett Syndrome Symposium, June 17–19, 2002, Inn at the Colonnade. Rett Syndrome Research Foundation, Baltimore, Maryland. Cincinnati, Ohio, 31 (abstract).

Zhou, Y., Brašić, J.R., Musachio, J.L., Crabb, A.H., Fan, H., Hilton, J., Wong, D.F., 2004. In: 5th Annual Rett Syndrome Symposium, June 28–30, 2004, Inn at the Colonnade. Rett Syndrome Research Foundation (RSRF), Baltimore, Maryland. Cincinnati, Ohio, 45 (abstract).

Zhou, Y., Brasic, J.R., Musachio, J.L., Zukin, S.R., Kuwabara, H., Crabb, A.H., Endres, C.J., Hilton, J., Fan, H., Wong, D.F., 2001. In: Nuclear Science Symposium Conference Record, 2001 Institute of Electrical and Electronics Engineers (IEEE), Incorporated. . 3, 1340.

CHAPTER ELEVEN

Smoking Effects in the Human Nervous System

Zev Schuman-Olivier[1], Luke E. Stoeckel[2], Erika Weisz[3] and A. Eden Evins[4]

[1]Department of Psychiatry, Cambridge Hospital, Cambridge, MA; Cambridge Health Alliance, Somerville, MA; Harvard Medical School, Boston, MA
[2]Department of Psychiatry, Massachusetts General Hospital, Boston, MA; Harvard Medical School, Boston, MA
[3]Department of Psychiatry, Massachusetts General Hospital, Boston, MA
[4]Department of Psychiatry, Massachusetts General Hospital, Boston, MA; Harvard Medical School, Boston, MA

1. INTRODUCTION

Tobacco smoking is the leading cause of preventable mortality worldwide (WHO, 2009). There are more than a billion smokers worldwide, an estimated 700 million children are exposed to second-hand smoke at home, and about five million people die from tobacco-related illnesses each year (WHO, 2009). The prevalence of smoking in the US decreased to 20.6% in 2008 (Cigarette smoking among adults and trends in smoking cessation – United States, 2008, 2009). Nonetheless, tobacco use is still the most common cause of preventable death and disease in the US, accounting for approximately 438,000 premature deaths annually, representing 18% of total yearly deaths (Mokdad et al., 2004). Smoking reduces the median survival of smokers by an average of 10 years, but by stopping cigarette smoking, a patient improves survival even among those who stop after age 50 (Doll et al., 2004). Nonetheless, habitual smokers find it extremely difficult to successfully stop smoking. Although more than two-thirds of smokers would like to stop, and 40% make at least one cessation attempt per year, only 3–5% of smokers per year are successful in stopping long term on their own (Messer et al., 2007). While most of the morbidity and mortality from smoking is due to toxic compounds and carcinogens ingested during tobacco smoke inhalation, nicotine is the active ingredient in the development of addiction and the substance responsible for its central nervous system (CNS) effects.

This chapter is focused primarily on the effects of inhaled, tobacco-derived nicotine on nicotinic acetylcholine receptors (nAChRs) and the resulting effect in the human CNS. An average smoker delivers small pulses of nicotine via the lungs into the arterial blood that reaches an arterial concentration in the range of 50 ng/mL within 5 min of smoking tobacco (Henningfield et al., 1993), activating nAChRs. Generally, in chronic tobacco smokers, a level of 10–50 ng/mL of nicotine is maintained in the brain, potentially resulting in cycles of activation and densensitization throughout the day (Picciotto et al., 2008). Smokers in special clinical populations (e.g. schizophrenia) may extract more nicotine

The Effects of Drug Abuse on the Human Nervous System
http://dx.doi.org/10.1016/B978-0-12-418679-8.00011-3

per cigarette (Williams et al.). With repeated smoking episodes throughout the day, the peak arterial nicotine concentration associated with each cigarette is superimposed onto a steady-state nicotine level that may increase with repeated cigarette consumption during the day, depending on the intercigarette interval (nicotine has a half-life of about 2 h) (Dani and Heinemann, 1996). The metabolism of nicotine can vary based on individual genetic differences (e.g. CYP 2A6 activity) (Hukkanen et al., 2005) and on the presence of other substances that may influence nicotine metabolism (e.g. the inclusion of additives such as menthol, which slows hepatic metabolism of nicotine) (Benowitz et al., 2004). In the following sections, we will describe the current state of literature regarding nAChR structure, subunit diversity, and function in human brain, and then describe the way nicotine's actions at the nAChR effects systems throughout the human brain.

2.1. NICOTINIC ACETYLCHOLINE RECEPTORS

2.1.1. Nicotine Receptor Structure and Subunit Diversity

nAChRs belong to the superfamily of ligand-gated ion channels that includes g-aminobutyric acid A (GABA-A), glycine, and serotonin 3 (5-HT3) receptors (Dani, 2001; Karlin, 2002). nAChRs belong to the evolutionary Cys-loop family of receptors (Ortells and Lunt, 1995). The structure of the nicotinic receptor–channel complex arises from five polypeptide subunits assembled like sides of a barrel around a central water-filled, cation-selective pore (Cooper et al., 1991). Mammalian nAChR subunits have four trans-membrane loops and are cation selective, permeable to small monovalent and divalent cations. See Figure 1.

Most of these subunits assemble to form heteropentameric ion-channel protein structures with various combinations of α and β subunits (Karlin, 2002). α and β subunits jointly form binding sites for agonists and control functional properties of nAChRs, including affinities of activated and desensitized nAChRs (Karlin, 2002; Wang and Sun, 2005). In heteromeric nAChRs, both α and β subunits contribute to the pharmacology of the binding site. For example, in $\alpha 4 \beta 2$ receptors, ACh binds in a small pocket formed between the $\alpha 4$ and its adjacent $\beta 2$ subunit. Some subunits, particularly $\alpha 7$, are able to form homopentamers (Couturier et al., 1990). In homomeric nAChRs, the ligand binding site is defined by the chirality of the adjacent $\alpha 7$ subunits (Dani and Bertrand, 2007).

There are muscle-type and neuronal-type nAChRs. We will focus here on neuronal nAChRs. While there is tremendous potential for nAChR diversity within the CNS due to the possibility for high levels of variation in the combinations of nAChR subunits, a limited number of subunit combinations are favored. Nicotinic AChRs containing $\alpha 4$ and $\beta 2$ subunits (Wada et al., 1989), together with $\alpha 7$ nAChRs, are the most abundant and widely expressed nAChRs in the brain (See Sher et al., 2004 and Dani, 2001 for reviews). In the next section, several notable variations that are functionally involved in human smoking will be highlighted.

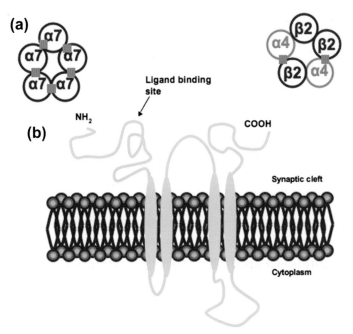

Figure 1 Structure of nAChRs. Nicotinic AChRs are formed by five subunits, which can be homomeric (α) or heteromeric (α/β). (a) Organization of subunits in neuronal homomeric α7-nAChRs and heteromeric α4β2-nAChRs. (b) One subunit of the nAChR contains (1) a large N- and a small C-terminal extracellular domains, (2) four transmembrane domains (M1-M4), and (3) a long cytoplasmic loop between M3 and M4. *(Adapted with permission from Yang et al., 2009).*

Based on their evolutionary development (Le Novere and Changeux, 1995), as well as pharmacologic and physiologic properties (Dani, 2001), two general functional classes of nAChRs exist in the human CNS:

1. Standard neuronal subunits (α2–α6 and β2–β4) that form nAChRs in αβ combinations,
2. Subunits (α7–α9) capable of forming homomeric nAChRs that are inhibited by α-bungarotoxin.

The α7 subunit is widely distributed in the mammalian CNS, while α9-nAChRs are present in inner ear (Elgoyhen et al., 1994).

The nAChRs can also be categorized by genes and chromosomal location. Alpha genes 2–7, and beta genes 2–4 are expressed in human CNS, while alpha 1 and beta 1, as well as delta, epsilon and gamma are expressed in muscle (Albuquerque et al., 2009; Millar and Harkness, 2008). While α8 has appeared in chick tissue (Schoepfer et al., 1990), it has not been found in mammalian tissue (Millar and Harkness, 2008). Notably, α5 and β3 are auxiliary subunits that often coassemble with other α and β combinations, similarly there is evidence that α10 subunits can coassemble with α9 (Plazas et al., 2005).

2.1.2. Locations of Nicotinic AChRs

Nicotinic AChRs are distributed throughout the human brain in regions involved in the endogenous cholinergic system. Cholinergic cell bodies are located in a loosely contiguous axis running from the cranial nerve nuclei of the brain stem to the medullary tegmentum and pontomesencephalic tegmentum, continuing rostrally through the diencephalon to the telencephalon (Dani and Bertrand, 2007; Woolf, 1991). There are four major cholinergic subsystems above the brain stem that innervate nearly every neural area. One cholinergic system arises from neurons mainly in the pedunculopontine tegmentum and the laterodorsal pontine tegmentum, providing widespread innervation to the thalamus and midbrain dopaminergic areas and also descending innervation to the caudal pons and brain stem. The second major cholinergic system arises from various basal forebrain nuclei, including the medial septum, the vertical and horizontal diagonal band of Broca, the substantia innominata, and the nucleus basalis of Meynert, and makes broad projections throughout the cortex and hippocampus. In general, these first two cholinergic projection systems provide broad, diffuse, and generally sparse innervation to wide areas of the brain (Woolf, 1991; Kasa, 1986). The other major cholinergic subsystems are exceptions to this principle of broad cholinergic innervation. The third subsystem arises from a collection of cholinergic interneurons located in the striatum. Unlike many broadly projecting cholinergic neurons throughout the brain, these cholinergic interneurons make up approximately 2% of the striatal neurons, and they provide very rich local innervation throughout the striatum and the olfactory tubercle (Zhou et al., 2002). A fourth cholinergic system originates in the medial habenula and terminates in the interpeduncular nucleus (Mulle et al., 1991).

2.1.2.1. Cholinergic Projections

Cholinergic systems provide diffuse innervations to practically all of the brain, but a relatively small number of cholinergic neurons make sparse projections that reach broad areas. Thus, the activity of a rather few cholinergic neurons can influence relatively large neuronal structures (see Figure 2). Despite the sparse innervations, cholinergic activity drives or modulates a wide variety of behaviors. It is thought that cholinergic systems particularly affect discriminatory processes by increasing the signal to-noise ratio and by helping to evaluate the significance and relevance of stimuli. Recent studies suggest the presence of multiple brain regions sending and/or receiving cholinergic projections that are relevant to the effects of smoking, including the following: striatum, ventral tegmental area (VTA)/substantia nigra (SN), medial habenula (MHb)/interpeduncular nucleus (IPN), hippocampus, cerebellum, periaqueductal grey (PAG), supraoptic nucleus, spinal cord, and laterodorsal tegmental nucleus (LDT)/pedunclopontine tegmental nucleus (PPT).

2.1.3. Selected nAChR Subtypes: CNS Distribution and Function

While there is a high level of variation in nAChR combinations, this chapter focuses on selected nAChR combinations most relevant to the effects of tobacco smoking on the human brain.

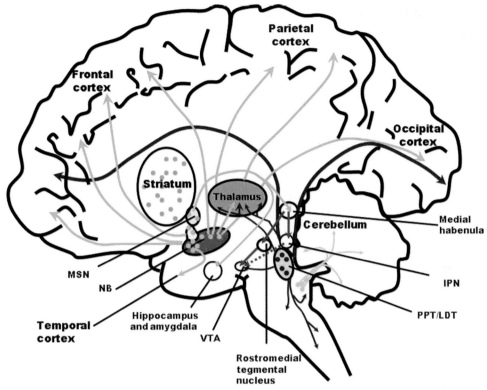

Figure 2 Major cholinergic systems in the human brain (Perry et al., 1999; Baldwin et al.). Two major pathways project widely to different brain areas: (1) basal-forebrain cholinergic neurons (blue, including the nucleus basalis (NB) and medial septal nucleus (MSN)) and (2) pedunculopontine (PPT)–lateral dorsal tegmental (LDT) neurons (purple). Other cholinergic neurons include striatal interneurons (orange) and vestibular nuclei (green). The putative relationship of the habenula–interpeduncular (IPN) pathway to reward circuitry is shown in red. This circuitry is likely dominated by GABA/glutamate neurons, with a hypothesized projection from the IPN to the ventral tegmental area (VTA) shown by a red dotted line.

2.1.3.1. α4β2

Heteromeric $(\alpha 4)_2(\beta 2)_3$ receptors are found in the brain in both pre- and postsynaptic locations and are excitatory by means of Na^+ and K^+ permeability in a central cation channel. $\alpha 4$ and $\beta 2$ subunits are highly expressed in many areas of the brain but are very richly expressed in thalamus, cortex, and hippocampus (Sher et al., 2004). These receptors bind nicotinic agonists (nicotine, epibatidine, acetylcholine, cytisine, varenicline) and antagonists (mecamyline, α-conotoxin). While often found in $(\alpha 4)_2(\beta 2)_3$ combinations, these receptors also frequently combine with an additional subtype, altering the overall receptors function. While these $(\alpha 4)_2(\beta 2)_3$ forms have a high sensitivity to agonists, manipulation in *Xenopus* oocytes have demonstrated an alternate stoichiometric form $(\alpha 4)3(\beta 2)_2$ that has a relatively lower sensitivity to agonists and higher Ca^{2+} permeability (Moroni and Bermudez, 2006; Tapia et al., 2007); however, this stoichiometry has yet to be found in nature. The α4β2★ containing subtypes of nAChRs play a critical role in mediating the acute reinforcing properties of nicotine (Pons et al., 2008), especially those including an α5 and/or α6 subunit. These receptors can be found in presynaptic (Hill et al., 1993), postsynaptic (Charpantier et al., 1998; Klink et al., 2001), and somatic (Sorenson et al., 1998), locations on dopaminergic neurons. There are also β2 subunits containing nAChRs in the medial habenula (Quick et al., 1999).

2.1.3.2. α7

Homomeric α7 subunits are abundant throughout the mammalian brain (Dominguez del Toro et al., 1994). While α7 receptors are generally homomeric, heteromeric combinations of α7 and α8 exist uncommonly in chick tissue; however, no reports of heteromeric α7 nAChR exist in mammalian tissue (Millar and Harkness, 2008). They are present with particularly high density in the hippocampus (Alkondon and Albuquerque, 2004) where they are found on interneurons (Arnaiz-Cot et al., 2008) and astrocytes, (Sharma and Vijayaraghavan, 2001) and their activation stimulates calcium influx and calcium-induced calcium release from intracellular stores. The α7 nAChR-mediated activation of such interneurons can result in either inhibition or disinhibition of principal pyramidal neurons in cortex (Alkondon et al., 2000) and hippocampus (Ji and Dani, 2000). These receptors are excitatory mainly through the mechanism of increased Ca^{2+} permeability in the central cation channel. These receptors bind nicotinic agonists (nicotine, epibatidine, choline, dimethylphenylpiperazinium) and antagonists (mecamyline, memantine, α-bungarotoxin). Functional α7 nAChR subunits have been demonstrated on hippocampal interneurons where they are associated with fast, strong excitatory effects compared with non-α7 nAChRs (McQuiston and Madison, 1999). In the hippocampus, the timing of cholinergic inputs, acting through α7-nAChRs, results in the promotion of long-term potentiation (LTP) (Berg, 2011; Gu and Yakel, 2011). The high expression and heavy regulation of α7 subunits in perinatal ages and evidence showing α7 supports neural plasticity and opposes apoptosis has led to the suggestion that α7 could play a major role in brain development (Broide and Leslie, 1999). α7-nAChRs on adult-born neurons appear to be critical for normal development and integration of the neurons into circuits and are essential for normal dendritic development of the neurons in hippocampus dentate gyrus (Campbell et al., 2010, 2011). Homomeric α7 subunits are also widely distributed in the cerebellar cortex (Graham et al., 2002). α7-nAChRs have been reported to be expressed in reduced numbers and with suboptimal function in the CNS of people with schizophrenia (Court et al., 1999), a factor thought to underlie some of the cognitive deficits associated with the disorder and possibly the high rate of nicotine dependence among this patient population (Leonard et al., 2000).

2.1.3.3. α3β4

Heteromeric $(\alpha3)_2(\beta4)_3$ are common in central and peripheral nervous systems. Within the PNS they are referred to as *ganglionic* receptors because of their presence in sympathetic and parasympathetic ganglia. These receptors result in excitatory postsynaptic potentials mainly through increased Na^+ and K^+ permeability. These receptors bind nicotinic agonists (nicotine, epibatidine, acetylcholine, dimethylphenylpiperazinium, carbachol) and antagonists (mecamylamine). Within the CNS they are prevalent in both the MHb and locus coeruleus (LC) neurons (Mulle et al., 1991; Quick et al., 1999) and possibly the IPN (Sher et al., 2004). They affect presynaptic and postsynaptic excitations. These receptors bind nicotinic agonists (nicotine, epibatidine, acetylcholine, cytisine) and antagonists (mecamylamine). The α3 is present on VTA and SN dopaminergic neurons (Charpantier et al., 1998). The peripheral α3β4 receptor co-locates with α5, a relationship that is increasingly attracting attention after genomic studies found clinical implications from variants of this combination as described below.

2.1.3.4. α5

The α5 subunit needs to be co-expressed as an accessory subunit with another α and β combination, such as α3β4 (Ramirez-Latorre et al., 1996) or α4β2 (Bailey et al.). The variant CHRNA5 (D398N) greatly increases the risk for nicotine dependence, lung cancer, and chronic obstructive pulmonary disease. The lower mRNA expression of CHRNA5 along with the nonrisk allele of rs16969968 is protective for nicotine dependence and lung cancer (Bierut, 2010). A relationship with *CHRNA3* SNP rs1051730 and SNP rs578776 may also influence this genetic relationship (Ware et al., 2011). The addition of an accessory α5 subunit affects nicotine upregulation by α4β2★ (Mao et al., 2011). Nicotine activates the habenulo-interpeduncular pathway through α5-containing nAChRs, triggering an inhibitory motivational signal that acts to limit nicotine intake. The MHb projects mainly to the IPN, which, in turn, appears to inhibit the motivational response to nicotine intake. Thus, inactivation of the MHb and IPN both result in increased intake of nicotine (Fowler et al.). Consistent with these studies, overexpression of β4 results in enhanced activity of the MHb, resulting in the opposing effect, e.g. aversion to nicotine. Reversal of nicotine aversion in engineered mice overexpressing β4 is achieved by expression of the α5 D397N in MHb neurons. Similarly, α5 reexpression in the MHb of α5 KO mice normalizes their nicotine intake (Fowler et al.). Therefore, it has been proposed that the MHb acts as a gatekeeper in the control of nicotine consumption and that the balanced contribution of β4 and α5 subunits is critical for this function (Frahm et al.). Also, when α5 is expressed together with α4β2 or α3β4, it accelerates desensitization (see Wang and Sun, 2005).

2.1.3.5. α6 (α4/α6/β2; α6/β3; α4/α6/β2/β3)

The α6 subunit is a ligand-binding subunit that is the site of action of a–conotoxin MII (Azam et al., 2008), which is often co-expressed with α4β2, though β3 and α6 may co-assemble to form heteromeric channels (Le Novere et al., 1996) especially in LC (Lena et al., 1999). α6★-nAChRs dominate in the effects of nicotine in the nucleus accumbens (NAc) (Exley et al., 2008), playing a role in relating the effects of α4 subunits in nicotine self-administration and its long-term maintenance (Exley et al., 2011). Nicotine selectively activates dopaminergic neurons within the posterior VTA through $\alpha4\alpha6(\star)$ nAChRs(Zhao-Shea et al., 2011). The effects of $\alpha6\beta2\star$ nAChRs expressed in the VTA are necessary for the effects of systemic nicotine on dopamine (DA) neuron activity and DA-dependent behaviors such as locomotion and reinforcement (Gotti et al., 2010). It has been shown in animal models that $\alpha6\beta2\star$ nAChRs in the VTA and/or in its projections are necessary and sufficient to establish nicotine self-administration behavior (Pons et al., 2008). This effect is likely due to the dominant role of $\alpha6\star$-nAChRs in dynamic filtering (frequency-sensitive regulation of DA neuronal activity and terminal DA release) of action potential-dependent DA release in the NAc (Yang et al., 2009). $\alpha6$ is expressed by ST and VTA catecholaminergic neurons (Charpantier et al., 1998). When co-assembling with the $\beta3$ subunit, $\beta3$ can be important for the correct assembly, stability and/or transport of $\alpha6\star$-nAChRs in nigrostriatal dopaminergic neurons and influences their subunit composition. However, $\beta3$ subunit expression is not essential for the expression of $\alpha6\star$ (Gotti et al., 2005) and there is ample evidence of $\alpha6\beta2$ nAChRs without $\beta3$. There was a higher expression of mRNA in SN than VTA for $\alpha5$, $\alpha6$ (biggest difference), and $\beta3$ mRNAs during postnatal development (days 20–40) (Azam et al., 2007). One recent study found functional $\alpha6\star$-nAChRs are naturally expressed on GABAergic presynaptic boutons, in which they mediate cholinergic modulation of GABA release onto DA neurons in the VTA. Acute exposure to smoking-relevant concentrations of nicotine desensitizes, rather than activates, these presynaptic $\alpha6\star$-nAChRs and eliminates cholinergic enhancement of GABA release (Yang et al., 2011).

2.1.4. Mechanisms and Localization of nAChR Activity

Many nAChRs are in presynaptic locations; however, many are postsynaptic, and there is evidence of nAChRs being distributed to preterminal, axonal, dendritic, and somatic locations (Lena et al., 1993; Zarei et al., 1999). The localization of nicotinic AChRs, which can cause intracellular changes through multiple mechanisms, can affect the way they modulate cellular activity and signal transmission (see review by Penton and Lester, 2009).

2.1.4.1. Presynaptic nAChR Activity

The activation of presynaptic nAChRs initiates direct and indirect intracellular Ca^{2+} signals that potentiate neurotransmitter release through the following mechanisms: (1) a small, direct Ca^{2+} influx via nAChR activity (Vernino et al., 1992; Seguela et al., 1993; Castro and Albuquerque, 1995) that (2) may trigger Ca^{2+}-induced Ca^{2+} release from intracellular Ca^{2+} stores (Sharma and Vijayaraghavan, 2003) and (3) the activation of nAChRs further causes membrane depolarization that activates voltage-gated Ca^{2+} channels in presynaptic terminals (Tredway et al., 1999). The overall effect is that presynaptic nAChR activity elevates Ca^{2+} levels in presynaptic terminals, in turn leading to an increase in neurotransmitter release.

Presynaptic α7 and β2★ nAChRs modulate excitatory amino acid release through different cellular mechanisms. Stimulation of presynaptic α7 nAChRs leads to Ca^{2+}-induced Calcium Release (CICR) (Le Magueresse and Cherubini, 2007). In contrast, stimulation of non-α7 nAChRs recruits voltage-operated calcium channels (VOCC) in PC12 cell culture (Dickinson et al., 2007), while β2★ nAChRs have been shown to recruit VOCCs in murine brain cells (Dickinson et al., 2008). The overall effect is that presynaptic nAChR activity elevates intraterminal calcium and contributes to the increased neurotransmitter release.

Chronic nicotine exposure differentially affects the function of presynaptic nAChR subtypes, which modulate the release of noradrenaline (NA), glutamate (GLU), DA, and ACh (Grilli et al., 2005), as well as the activity of nonnicotinic receptors such as the ionotropic N-methyl-D-aspartate (NMDA) and amino-3- hydroxy-5-methylisoxazole-4-propionic acid (AMPA) receptors (Grilli et al., 2009). The changes that occurred after chronic nicotine treatment may therefore produce important modifications both of the integrated responses of nicotine itself, and also of the normal physiological responses of ACh (Marchi and Grilli, 2010).

2.1.4.2. Postsynaptic

α7 nACRs are present in pyramidal neurons in the hippocampus, and therefore, are not only involved in presynaptic regulation of transmitter release, but also play a critical role postsynaptically (Ji et al., 2001). There are postsynaptic nAChRs in striatum and VTA; however, it is thought that the presynaptic modulators of GABA and GLU signaling play a larger role in development of reinforcement and conditioning related to tobacco smoking (Koob and LeMoal, 2006).

2.1.4.3. Preterminal Activity

Preterminal nAChRs located before the presynaptic terminal bouton indirectly affect neurotransmitter release by activating voltage-gated channels and, potentially by initiating action potentials (Lena et al., 1993). The evidence for preterminal nAChR influences is strongest at some GABAergic synapses. Preterminal nAChR activation depolarizes the membrane locally, thereby activating voltage-gated channels that directly mediate the presynaptic calcium influx underlying enhanced GABA release (Tredway et al., 1999). In other cases, the stimulation of neurotransmitter release has been shown to be independent from action potential generation, thus suggesting a strictly presynaptic localization of these receptors (Gray et al., 1996; McGehee and Role, 1995). On chick ciliary ganglia, specific sorting and targeting mechanisms may be in place to cluster the α3★ nAChR, postsynaptically on the cell soma, and α7★ nAChRs peri-synaptically on somatic spines (Shoop et al., 1999).

2.2. THE EFFECTS OF NICOTINE ON nAChRS IN THE HUMAN BRAIN

In the CNS, ACh acts as a broad modulator of cortical activity, with major projection to both the thalamus and the cortex (Woolf, 1991; Kasa, 1986). ACh plays a major role in arousal, attention, and detection of salient environmental cues (Perry et al., 1999).

2.2.1. Basic Function of Nicotinic Receptors

Upon binding ACh, the nAChR ion channel is stabilized in the open conformation for several milliseconds. Then the open pore of the receptor/channel closes to a resting state or closes to a desensitized state that is unresponsive to ACh or other agonists for many milliseconds or more. While open, nAChRs conduct cations, which can cause a local depolarization of the membrane and produce an intracellular ionic signal. Although Na^+ and K^+ carry most of the nAChR current, Ca^{2+} can also make a significant contribution (Vernino et al., 1992; Seguela et al. 1993; Castro and Albuquerque, 1995; Decker and Dani, 1990; Dani and Mayer, 1995). The marked variability in Ca^{2+} permeability among various subtypes results in substantially different properties for each subtype, which will be discussed in detail later. The most basic conformational states of nAChRs are the closed state at rest, the open state, and the desensitized state.

The heteromeric receptors have a higher affinity for nicotine in the brain relative to the neuromuscular junction. A recent study demonstrated this is a consequence of enhanced interactions with a specific tryptophan residue 149 (TrpB). A cation–π interaction that is absent in the muscle-type receptor is quite strong in α4β2. In addition, a hydrogen bond to a backbone carbonyl that is weak in the muscle type is enhanced in α4β2. Both effects are substantial, and in combination they account fully for the differential sensitivity to nicotine of the two receptors. The side chain of residue 153 in loop B distinguishes the two receptor types and influences the shape of the binding site aromatic box, allowing a stronger interaction between nicotine and TrpB in high-affinity receptors (Xiu et al., 2009).

2.2.2. nAChR Activation

The calcium permeability of neuronal nAChRs is also significant. While muscle nAChRs have a pCa^{2+}/pNa^+ of only 0.2, most heteromeric neuronal nAChRs have pCa^{2+}/pNa^+ ratios in the range of 1–1.5 (McGehee and Role, 1995). The notable exception is α7, which with a pCa^{2+}/pNa^+ of ~20 is even more permeable than the prototypic ligand-gated calcium-permeable channel, the NMDA receptor (McGehee and Role, 1995). A particularly important feature of nAChRs, which is different from other calcium entry pathways such as NMDA receptors or voltage–dependent calcium channels, is that they allow calcium influx without requiring cell depolarization. Activation of nAChRs allows Ca^{2+} influx into "resting" cells, influencing their metabolic state as well as their responsiveness to incoming synaptic stimuli. Therefore, nAChRs have an important role not only in synaptic transmission, but also in the regulation of calcium-dependent cellular events that include activation and modulation of other ion channels, excitability, secretion, motility and migration, gene expression, cell differentiation, and survival (Rathouz et al., 1996). The presynaptic influx of Ca^{2+} needed to activate the release machinery could result either from the secondary activation of voltage-dependent calcium channels (triggered by nAChRs-induced depolarization) or from direct influx through the nAChRs channels themselves (Lena and Changeux, 1997).

The kinetics for nAChR activation, closure, and desensitization are influenced by the amino acid sequence of the subunits. Nicotinic AChR activation is best described by its dose–response profile, with receptor activation as a function of agonist concentration. The $\alpha7$ nAChR has a relatively low affinity for ACh activation, with an effective dose for half-activation (EC50) at approximately $200\,\mu M$ ACh. In a transgenic cell line (HEK-293) that expresses both $\alpha4$ and $\beta2$ AChRs, the $\alpha4\beta2$ nAChR had a higher affinity, with activation best described by a biphasic activation curve, with both high-affinity ($\alpha4\beta2$) [EC50 = $1.6\,\mu M$] and low-affinity ($\alpha4\beta2$) [EC50 = $62\,\mu M$] components, and the high/low ratio is approximately 25/75% (Buisson and Bertrand, 2001). Homomeric $\alpha7$ receptors have a low affinity for ACh, yet undergo rapid activation and desensitization, display high calcium permeability, and are irreversibly blocked by α-bungarotoxin (Castro and Albuquerque, 1995) (see Table 1). These receptors have been found in the hippocampus (Castro and Albuquerque, 1995) and hypothalamus (Uteshev et al., 1996) and are selectively eliminated in *CHRNA7* KO mice (Orr-Urtreger et al., 1997). However, native $\alpha7$-containing receptors with different biophysical properties (slow desensitization) and pharmacology (reversible block by α-bungarotoxin) have also been described (Cuevas et al., 2000). This diversity may be due to co-assembly with different subunits in heteropentameric structures (Khiroug et al., 2002) or to the presence of an $\alpha7$-2 splice variant (Severance and Cuevas, 2004).

Most heteromeric nAChRs display relatively slow activation and desensitization kinetics. A major determinant of these properties is the β subunit, with $\beta2$ containing

Table 1 Relationship between Activation & Desensitization-Induced Inhibition of nAChRs

nAChR Subtype	Agonist	EC_{50} (μM)	IC_{50} (μM)	Ratio EC_{50}/IC_{50}
Rat $\alpha4\beta2$	Nicotine	14	<0.01	>1400
Rat $\alpha7$	Nicotine	90	1.3	69

Activation is expressed as the EC50 value, which is the concentration of agonist producing half-maximal response amplitude. Since $\alpha4\beta2$ receptors may exist in both the high sensitivity and low sensitivity forms, the EC50 represents an average of the two forms. Inhibition is expressed as the IC50 value, which is the concentration of agonist reducing the amplitude of the test response by 50%. Notably, more agonist is required to desensitize $\alpha7$ receptors than $\alpha4\beta2$ receptors in rats. Yet, $\alpha7$ receptors desensitize rapidly after activation (in milliseconds) compared to $\alpha4\beta2$ receptors that desensitize slowly after activation (in seconds). The ratio of activation to inhibition for $\alpha7$ is much lower than $\alpha4\beta2$, likely representing a difference in the process of desensitization between $\alpha7$-containing nAChRs and those without $\alpha7$. *Source:* Adapted from with permission from Giniatullin et al., 2005.

receptors desensitizing faster than β4-containing receptors. The β2 subunit is also a major determinant of the pharmacological sensitivity of nAChRs (Sher et al., 2004).

2.2.3. Cellular Regulation of nAChR by Chronic Nicotine

Chronic nicotine exposure causes densensitization of nAChRs in the short term and upregulation of nAChR number in the long term. The combination of desensitization and upregulation of nAChRs has been thought to represent the basis of nicotine tolerance and dependence (Wang and Sun, 2005); however, the mechanistic contribution of upregulation to the development of nicotine dependence has more recently come under scrutiny.

2.2.3.1. nAChR Subunit Desensitization Characteristics

Desensitization of nAChRs causes alterations in receptor affinity for agonists (Fenster et al., 1997). There are two interconverting states of desensitization, with one state (D1-"shallow") having a relatively rapid entry and recovery rate, and the other (D2-"deep") undergoing significantly slower transitions. Upon shorter applications of agonist, desensitization is dominated by the fast on/off kinetics of D1, whereas longer applications of agonist favor the transition (or stabilization) to the slowly recovering D2 conformation (Marszalec et al., 2005).

Characteristics of receptor subunits cause desensitization to occur at differing speeds and at various concentrations of agonist. Although α7 receptors exhibit pronounced and rapid desensitization, much higher concentrations of nicotine are required than with other nAChRs. Heteromeric nAChRs can be ordered from fast to slow in terms of the onset of desensitization as follows: α3β2>α4β2>α3β4>α4β4. For heteromeric nAChRs, the α subunit makes a significant contribution in determining the apparent nicotine affinity of the active and desensitized states of a nAChR, while the β subunit makes a significant contribution in determining the overall time course of the development of desensitization of a nAChR. In addition, even though external calcium ions produce subtle effects on the kinetics and apparent affinities of nicotine for activation and desensitization of the αβ nAChRs, they do not alter the pattern of contributions of these various subunits. In contrast, for the homomeric α7 nAChR, which displays faster kinetics and generally lower affinities for nicotine than the αβ pairs, calcium increases the apparent affinity of nicotine for both the active and desensitized states (Fenster et al., 1997; Hsu et al., 1996). While a study with oocytes suggested that the insertion of an α5 accessory subunit into α3β4 or α4β2 nAChRs causes the nAChRs to desensitize faster (Ramirez-Latorre et al., 1996), a more recent study in murine brain suggest α4β2-nAChRs that include α5 subunits, as compared to those without α5 subunits, show faster rates of recovery from desensitization and less desensitization by brief exposures to low concentrations of nicotine. These studies demonstrate the importance of considering findings related to the mechanisms of desensitization within the context of differential results based on receptor subtypes and tissue type.

2.2.3.2. Mechanisms of nAChR Desensitization

Considering the differences just described, it is important to note that the following description of mechanisms of nAChR desensitization is based primarily on investigations conducted with the α4β2 subunit in various expression systems.

Nicotinic AChRs have an intrinsically slow rate of recovery from desensitization after a 30–60 min treatment with levels of nicotine related to use of tobacco (Fenster et al., 1997). Desensitization of nAChRs induced by prolonged exposure to nicotine may result in a reduced Ca^{2+} influx, thereby promoting the dephosphorylated state of α4β2 receptors. Recovery from the "deep" desensitized conformation would be markedly slowed, and receptors would become "trapped" in a chronically desensitized/deactivated state (Lukas, 1991; Peng et al., 1994). While nicotine accumulation during a smoking day is enough to produce nAChR desensitization (Ortells and Arias, 2010), low agonist concentrations can induce desensitization even without nAChR activation, a process named "high-affinity desensitization" (Giniatullin et al., 2005).

In general, the rate of recovery from desensitization is governed by the balance between phosphatase and kinase activity (see Figure 3). Additionally, an increase in intracellular Ca^{2+} enhances the onset of nAChR desensitization (Wang and Sun, 2005).

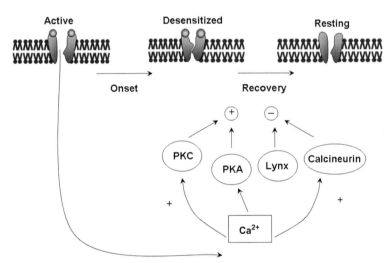

Figure 3 Densensitization and recovery of nAChRs. Recovery from desensitization of α4β2 nAChRs in adrenal chromaffin cells is modulated by phosphorylation. After the application of nAChR agonists at high concentrations causes Ca^{2+} influx, receptors proceed into the desensitized state, followed by transition into the resting state. Note that two agonist molecules remain bound to the desensitized receptor. Recovery from the desensitized resting state can be accelerated by phosphorylation involving protein kinase (C) (PKC) or c-AMP-dependent protein kinase A (PKA), whereas dephosphorylation caused by calcineurin delays recovery. Intracellular Ca^{2+} thus fine tunes the recovery process by shifting the balance between PKC and/or calcineurin activity. Lynx is a newly identified CNS protein that may also modulate desensitization. (*Adapted with permission from Giniatullin et al., 2005*).

Activation of protein kinase A (PKA) accelerates recovery from desensitization and its inhibition slows α4β2 recovery from desensitization (Nishizaki and Sumikawa, 1998). Inhibiting phosphatase or activating protein kinase C (PKC) also accelerates recovery from desensitization (Wang and Sun, 2005). PKC-dependent phosphorylation of α4 subunits changes the rates governing the transitions from "deep" to "shallow" desensitized conformations and effectively increases the overall rate of recovery from desensitization. Long-lasting dephosphorylation may underlie the "permanent" inactivation of α4β2 receptors observed after chronic nicotine (Fenster et al., 1999a). Indeed, it has been reported that prolonged treatment with PKC inhibitors will also drive α4β2 nAChRs to a functionally inactive conformation (Eilers et al., 1997). Chronic phorbol 12-myristate 13-acetate (PMA) treatment, which downregulates PKC activity (Favaron et al., 1990) promotes an increase in the number of α4β2 receptors (Gopalakrishnan et al., 1997), consistent with the suggestion that the dephosphorylated state of the receptor could either directly or indirectly serve as a signal for preventing receptor turnover (Peng et al., 1994). Wang suggested two mechanisms for the effects of phosphorylation on desensitization: one being a direct pathway via interaction between ligand and the nAChR, with another being induced by activation of PKA or PKC. There are multiple sequences for phosphorylation by both PKA and PKC in the intracellular loop between TMIII and TMIV of the α4 subunit (Wang and Sun, 2005).

Researchers continue to investigate other modulators of densensitization. For instance, lynx-1 is a newly identified CNS protein (Miwa et al., 1999; Ibanez-Tallon et al., 2002), also referred to as a "prototoxin", based on the finding that it adopts the three-fingered toxin fold characteristic of α-neurotoxins, such as α-bungarotoxin, which bind and block nAChRs. Lynx-1 co-localizes with α4β2 and α7 nAChRs in the somatodendritic compartment of neurons in many brain areas (the cerebral cortex, thalamus, SN, and cerebellum), and some claim it may modulate desensitization (Ibanez-Tallon et al., 2002).

2.2.3.3. Differential Desensitization and the Development of Tolerance and Reinforcement to Nicotine

As described earlier, the rate of desensitization and recovery from desensitization can vary depending on which types of nAChRs are expressed on a particular neuron or in different areas of the brain. Thus, multiple phases of nAChR desensitization and recovery could underlie aspects of tolerance such that a second dose of nicotine following immediately after an initial dose does not elicit the same effects (Pidoplichko et al., 1997).

Differential desensitization has been postulated as a cause of nicotine addiction: desensitization may be required for activation of mesolimbic dopamine and development of dependence, because of differential rates of desensitization. If presynaptic α7-nAChRs on GLU terminals desensitize more slowly than presynaptic α4β2-nAChRs on GABA terminals in VTA, then this may result in net activation of VTA DA neurons and increased DA firing in NAc (see detailed discussion in review, Koob and LeMoal, 2006). It is supposed that by acting on presynaptic α7-nAChRs (desensitized less than non-α7-nAChRs) located on GLU terminals, nicotine at concentrations experienced by smokers can produce long-term enhancement of glutamatergic transmission in the VTA (Schilstrom et al., 1998), whereas activation of presynaptic α4β2-nAChRs can only cause transient enhancement of GABAergic transmission. These presynaptic α4β2-nAChRs become significantly and quickly desensitized during long-term exposure to low concentrations of nicotine (Mansvelder et al., 2002). As a result, GABAergic terminals, rather than glutamatergic terminals, become insensitive to tonically released ACh from cholinergic afferents from the PPT and LDT (Oakman et al., 1995; Bolam et al., 1991), which will in turn lead to long-term activation of glutamatergic input accompanied with depression of GABAergic input to VTA DAergic neurons that is experienced by tobacco smokers. Collectively, the differential desensitization properties of these two nicotinic receptor subtypes may explain why low concentrations of nicotine tends to drive the activity of VTA DAergic neurons toward long-term excitation that underlie the course of nicotine addiction process (Hogg et al., 2003) (see Figure 4). Thus, in vivo experiments

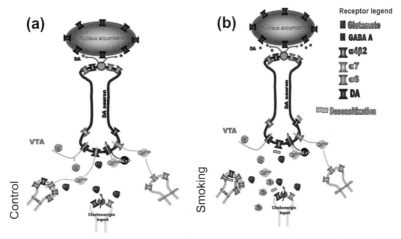

Figure 4 Role of nAChRs in nicotine addiction process. In the VTA, α6*- and α4β2-nAChRs are located on GABAergic terminals and provide inhibitory inputs onto DAergic neuons, while α7-nAChRs are located on glutamatergic terminals and activation of these receptors enhances GLU release and increases excitability of DAergic neurons. Endogenous ACh released from cholinergic terminals projected from PPT and LDT can modulate the excitability of both GABAergic and glutamatergic terminals. (a): Under control conditions, endogenous ACh can activate α6*- and α4β2-nAChRs on GABAergic terminals and α7-nAChRs on glutamatergic terminals. Thus postsynaptic DAergic neurons will receive balanced inhibitory and excitatory inputs. (b): In smoking conditions, α6*- and α4β2-nAChRs, rather than α7-nAChRs, are desensitized rapidly after chronic exposure to low concentrations of nicotine, thus inhibiting GABAergic inhibitory inputs (disinhibition). But endogenous ACh can still significantly enhance glutamatergic inputs onto the DAergic neurons. As a result, the increased excitation of DAergic neurons will result in a net increase in DAergic neuron firing and more DA release in NAc. *(Adapted with permission from Yang et al., 2009).*

observed that a single exposure to nicotine increases DA release in NAc from VTA for more than 1 h (Schilstrom et al., 1998; Di Chiara and Imperato, 1988).

2.2.4. nAChR Upregulation

The generally accepted view is that overstimulation induced by agonists leads to a reduction in the number of receptors. However, desensitization by long-term exposure to nicotine triggers an increase in the number of nAChRs with high affinity, and a decrease in the number of nAChRs with low affinity (Buisson and Bertrand, 2001). As Wang and Sun (2005) points out, chronic nicotine treatment does not elicit any change in mRNA, which suggests that upregulation is not due to transcription, and that a post-transcriptional mechanism is responsible for this phenomenon. Similarly, blocking protein synthesis by anisomycin or inhibiting protein glycosylation by tunicamycin both prevented receptor upregulation on cortical neurons, suggesting that upregulation in receptor number is not due to the preexisting intracellular pool that supplies receptors to surface, but due to additional synthesis and transport to the cell surface (Wang and Sun, 2005). Nicotine-induced upregulation of nicotinic ligand binding has been shown to arise from increases in mature assembled nAChR protein (Marks et al., 2011).

The cause of this upregulation phenomenon has not been fully explained; however, the literature describes several models to explain the mechanism of functional upregulation of α4β2-subtype nAChRs and some of the most promising hypotheses are described below:

2.2.4.1. Densensitization and Endocystosis

Upregulation was hypothesized to be initiated by the desensitization of surface nAChRs (Fenster et al., 1999b). Upregulation possibly relies on the fact that brain nAChRs undergo rapid desensitization and consequent inactivation after prolonged exposure to agonist (Fenster et al., 1999b). α4β2-subtype receptors might be recycled rapidly from the cell membrane. Chronic exposure to nicotine would slow receptor endocytosis from the membrane and increase their membrane density by inserting additional receptors from a submembraneous pool.

2.2.4.2. The "Maturational Enhancer Model"

The hypothesis that nicotine, penetrating into the cell, facilitates the maturation of nAChRs precursors in the ER is supported by direct investigations of the biosynthesis, trafficking, and degradation of nAChRs within the cells (Sallette et al., 2005; Corringer et al., 2006).

2.2.4.3. Pharmacological Chaperone of nAChRs

Nicotine and other nicotinic ligands can act on an assembly of precursors as pharmacological chaperones to cause upregulation by promoting the assembly of mature AChR pentamers (Kuryatov et al., 2005). A clear aspect of the chaperone hypothesis is that the basic molecular interaction that causes upregulation takes place within the endoplasmic reticulum rather than at the cell surface (Lester et al., 2009). The $(\alpha4)_2(\beta2)_3$ stoichiometry would languish in the endoplasmic reticulum (ER), but nicotine acts as a pharmacological chaperone to stabilize $(\alpha4)_2(\beta2)_3$ nAChRs and presumably, as a consequence, ER exit sites. The result: nAChRs reach the plasma membrane in greater numbers (Srinivasan et al.). Interaction of ubiquilin-1 with the $\alpha3$ subunit draws the receptor subunit and proteosome into a complex. These data suggest that ubiquilin-1 limits the availability of unassembled nAChR subunits in neurons by drawing them to the proteosome, thus regulating nicotine-induced upregulation (Ficklin et al., 2005).

2.2.4.4. State Affinity

It is proposed that $\alpha4\beta2$-subtype receptors exist in two interconvertible states, one with high affinity for nicotine and the other with low affinity for nicotine, and that chronic exposure to nicotine (or a nAChR ligand) stabilizes a larger fraction of receptors in the high-affinity, large-conductance state. In this model, it is assumed that the receptor undergoes a transition from a desensitized, nonfunctional state to an alternative state that can be activated in the continuous presence of nicotine (Buisson and Bertrand, 2002). Long-term exposure to agonist alters the ratio of high- to low-affinity nAChRs. For example, overnight nicotine exposure at concentrations comparable to those experienced by a smoker's brain increases the high- versus low-affinity ratio (Buisson and Bertrand, 2001).

In a comprehensive review, Penton and Lester conclude that upregulation of $\alpha4\beta2\star$ nAChRs does not represent a direct homeostatic response to persistent receptor desensitization, but may in fact be a fortuitous event resulting from the ability of nicotine to cross-membranes, accumulate in cells, and bind to nAChR assembly intermediates, suggesting that the upregulation of nAChRs is an unfortunate epiphenomenon with no causal role in addiction to nicotine. This interesting hypothesis is notable and compels further research and investigation. Importantly, it remains possible that several of these putative mechanisms may act together in a way that is not yet fully understood.

2.3. EFFECT OF nAChr ACTIVATION ON OTHER NEUROTRANSMITTERS IN HUMAN CNS

Presynaptic nAChRs have been found to increase the release of nearly every neurotransmitter that has been examined (Gray et al., 1996; McGehee and Role, 1995; Albuquerque et al., 1997; Alkondon et al., 1997; Radcliffe and Dani, 1998; Guo et al., 1998; Jones et al., 1999; Li et al., 1998; Role and Berg, 1996; Wonnacott, 1997). In this next section, we review the evidence supporting the role of nAChR activation on the release of other neurotransmitters in the human CNS.

2.3.1. Acetylcholine Release

Heteromeric nAChRs facilitate or potentiate ACh release in many different brain areas (Rowell and Winkler, 1984; Beani et al., 1985). ACh release from the IPN is an example where β4★ nAChRs have been shown to play a role in facilitating ACh release (Grady et al., 2001). There is minimal evidence for an α7 auto-receptor role; however, α7 plays a facilitatory presynaptic role in the cholinergic synapses of the chick ciliary ganglion (Coggan et al., 1997).

Chronic nicotine has dramatic effects on the function of nACh autoreceptors, probably through downregulation of nACh autoreceptor function. In hippocampal synaptosomes, acute doses of nicotine no longer evoke ACh release in rats chronically administered nicotine (Grilli et al., 2005). This finding is in keeping with results reported in a hippocampal slice model (Lapchak et al., 1989). Because the K^+-evoked release remained unchanged, chronic nicotine should not have negatively affected the ACh exocytotic machinery. The results are more consistent with downregulation of nACh autoreceptor function.

2.3.2. Glutamate Release

Many studies have described nAChR modulation of GLU release in the brain, with α7 nAChRs reported to play a major role (Gray et al., 1996; McGehee and Role, 1995). The most significant body of research on the mechanisms through which nAChRs modulate GLU release at various presynaptic locations focuses on the hippocampus; nAChRs have also been shown to be involved in GLU release in the VTA, striatum, as well as in the developing sensory cortex and cerebellum.

Nicotine, acting via α7 nAChRs, typically increased the frequency, but not the amplitude, of miniature excitatory postsynaptic currents in both hippocampal slices and cell cultures, suggesting a major presynaptic effect (Gray et al., 1996). Within the hippocampus, nicotinic stimulation enhances GLU release on time scales extending from seconds to a few minutes (Radcliffe and Dani, 1998), and contributes to the induction of synaptic plasticity (Ge and Dani, 2005; Mansvelder and McGehee, 2000). In some but not all cases, highly calcium-permeable α7 nAChRs mediate GLU release within seconds directly by the rapid increase in Ca^{2+} concentration near the active zones. The forms of enhancement lasting on a time scale of minutes require elevated concentrations of intraterminal calcium acting as a second messenger to indirectly modify glutamatergic synaptic transmission through Ca^{2+}-induced long-term changes in the secretory machinery. Somatic or postsynaptic nAChRs can initiate a Ca^{2+} signal that can act via calmodulin to reduce the responsiveness of NMDA receptors (Fisher and Dani, 2000). Properly localized calcium signals mediated by nAChRs initiate enzymatic activity (such as protein kinases and phosphatases) that is known to modify glutamatergic synapses (Hu et al., 2002).

Nicotinic receptors modulate glutamatergic synaptic plasticity, affecting the crucial hippocampal processes of LTP, long-term depression (LTD), and short-term depression (STD). Nicotinic AChRs directly influence GLU transmission via presynaptic sites. Cholinergic input, through either an ion channel receptor (α7 nAChR) or the G protein-coupled receptor (mAChR) (muscarinic, which is unrelated to nicotine), can directly induce hippocampal synaptic plasticity in a timing- and context-dependent manner. With timing shifts in the millisecond range, different types of synaptic plasticity are induced through different AChR subtypes with different mechanisms (presynaptic or postsynaptic) (Gu and Yakel, 2011). When nAChR activity at presynaptic GLU receptors precedes or matches the arrival of an action potential at the presynaptic terminal, the intraterminal Ca^{2+} signal initiated by nAChR activity adds to that from voltage-gated Ca^{2+} channels to enhance the probability of GLU release (Tredway et al., 1999; Gray et al., 1996). Properly timed presynaptic nAChR activity within a critical time window, arriving just before electrical stimulation of glutamatergic afferents, boosts the release of GLU and enhances induction of long-term synaptic potentiation (Berg, 2011; Gu and Yakel, 2011; Sharma and Vijayaraghavan, 2003; Ji et al., 2001; Drever et al., 2011). The depolarization contributed by nAChRs helps to relieve the Mg^{2+} block of postsynaptic NMDA receptors. In addition, the Ca^{2+} signal contributed by nAChRs supplements that of the NMDA receptor to enhance the probability of LTP induction. However, if that postsynaptic nAChR-initiated signal precedes presynaptic activity, a mismatch between nicotinic-induced activity and the postsynaptic response develops. In that case, synaptic LTD can be observed (Ge and Dani, 2005). In addition, indirect influences arise from GABAergic activity, inhibiting pyramidal neurons and preventing induction of short-term potentiation (STP) or LTP (Ji and Dani, 2000; Ji et al., 2001). α7 nAChR-dependent LTP is likely due to a postsynaptic effect that requires NMDAR activation and prolongation of NMDAR-mediated calcium transients in the spines and GluR2-containing AMPAR synaptic insertion. The α7 nAChR-dependent STD appears to be mediated primarily through presynaptic inhibition of GLU release (Gu and Yakel, 2011).

Most were presynaptic α7 nAChRs receptors and were located on glutamatergic axon terminals. The majority of presynaptic α7 nAChRs were found in perisynaptic (61%) loci, placing them in an ideal location for modulating glutamatergic transmission via local depolarization and/or increases in intracellular calcium through α7 nAChRs. The absence of direct cholinergic synaptic input to presynaptic α7 nAChRs indicates that these receptors are likely to be activated by endogenous ACh or choline diffusing from local cholinergic terminals (Jones and Wonnacott, 2004). In this process, choline acts as a selective agonist for α7 nAChRs, while it does not seem to be an agonist for any other subtypes (Albuquerque et al., 1998). These cholinergic afferents into the midbrain act via presynaptic (mainly) α7* nAChRs on GLU terminals to boost GLU transmission (Schilstrom et al., 1998), in a process that leads to increased dopamine release and development of behavioral reinforcement. In addition, work by Wonnacott and Kaiser has implicated α7 in the stimulation of GLU release from striatal slices, which in turn translates as an increase in DA release (Kaiser and Wonnacott, 2000) (see section below of DA release for a detailed description).

Finally, α7★ nAChRs likely play a synergistic role with glutamatergic NMDA receptors in activity-dependent plasticity in early pos-natal life. α7★ nAChRs potentiate GLU transmission between mossy fibers and granule cells in the developing cerebellum (De Filippi et al., 2001). Similar findings were reported for other synapses in the developing rat cerebellum but not in adult cerebellum (Kawa, 2002; Campbell et al., 2010). In rat auditory cortex, presynaptic α7 nAChRs have been shown to stimulate GLU release and selectively potentiate NMDA receptor-mediated synaptic transmission (Aramakis and Metherate, 1998).

2.3.3. GABA Release

Nicotinic AChR modulation of presynaptic GABA occurs at different sites on GABA neurons and in multiple brain regions, most notably within the hippocampus, the VTA, and the IPN. There is a strong evidence for preterminal nAChR influences at GABAergic synapses. Preterminal nAChR activation depolarizes the membrane locally, activating voltage-gated channels that directly mediate presynaptic calcium influx, enhancing GABA release (Lena et al., 1993). Functional α7- and α4β2-like nAChRs are present on somato-dendritic and/or preterminal/terminal regions of interneurons in the CA1 field of the rat hippocampus and in the human cerebral cortex. Activation of the different nAChR subtypes present in the preterminal/terminal areas of the interneurons triggers release of GABA (Albuquerque et al., 2000).

Within the hippocampus, activation of nAChRs on GABAergic interneurons can evoke inhibitory activity in CA1 pyramidal neurons, contributing to the modulation of information processing (Alkondon et al., 1997). Because nAChRs (particularly the α7 subtype) are elevated during development, they can be important for tuning the number and strength of developing synaptic connections in the hippocampus through regulation of GABA release (Broide and Leslie, 1999; Maggi et al., 2001). Within CA1 interneurons, GABA mediated postsynaptic inhibitory currents can be stimulated by both α4β2 and α7 nAChRs receptors. Responses mediated by α7 nAChRs appear to be short-lived, whereas those mediated by α4β2 nAChRs are long-lasting (Alkondon et al., 1999). These GABAergic interneurons seem to determine if inhibition or disinhibition of hippocampal pyramidal cells will occur (Alkondon et al., 2000; Ji and Dani, 2000).

The VTA is a very important region where nicotine modulates GABA release (Tapper et al., 2004). Nicotine activates and then desensitizes α4β2-like receptors on GABAergic neurons, which normally innervate the DA cells of the VTA. The desensitization of this GABAergic, inhibitory drive onto DA cells is suggested to represent a major determinant for nicotine-induced increased excitability of the VTA and of the persistent enhancement of DA release in the NAc found during smoking, a mechanism potentially underlying nicotine addiction. (See Effects of nicotine on DA section for more details). In addition, nicotine induces GABA release in the IPN where multiple receptor populations seem to be present (Mulle et al., 1991). Finally, experiments within mouse thalamic nuclei, such as dorsal lateral geniculate or ventrobasal complex, have demonstrated presynaptic nAChRs can increase the release of GABA by producing an influx of calcium in the presynaptic compartment, possibly contributing to increase in signal-to-noise ratio observed in thalamus in vivo during arousal (Lena and Changeux, 1997).

2.3.4. Dopamine Release

Nicotine and endogenous ACh act on dopaminergic neurons in the midbrain, at their source, and in the striatum where many DA fibers terminate, exerting multiple modulatory influences over DA signaling. Modulatory effects include (1) presynaptic effects due to differential desensitization of GABA and GLU afferents, (2) increased LTP from glutamatergic synapses, as well as (3) direct cholinergic transmission via postsynaptic nAChRs. These effects are likely critical for nicotine reinforcement.

Cholinergic afferents into the midbrain acting via presynaptic (mainly) $\alpha 7^\star$ nAChRs on GLU terminals boost GLU transmission (Schilstrom et al., 1998). Nicotinic activity also supplies some excitatory drive to midbrain GABA projection neurons and interneurons via (mainly) $\beta 2^\star$ nAChRs (Mansvelder et al., 2002; Pidoplichko et al., 2004). The DA neurons express a variety of nAChR subunits: $\alpha 4$–$\alpha 7$ and $\beta 2$, with $\beta 2^\star$ nAChRs predominating (Klink et al., 2001; Pidoplichko et al., 1997; Wooltorton et al., 2003). Acting through these excitatory and inhibitory inputs and nAChRs located on the DA neurons, nicotinic receptors influence the firing modes and firing frequency of DA neurons (Schilstrom et al., 2000; Schilstrom et al., 2003). Specifically, DAergic neurons within the posterior VTA seem to be essential for nicotine reinforcement. Rats will self-administer nicotine directly into the posterior, but not into the anterior VTA (Ikemoto et al., 2006). Nicotine increases burst firing of dopaminergic cells. This effect of nicotine is due to $\alpha 7$ nicotinic receptor-mediated presynaptic facilitation of GLU release in the VTA, whereas other nicotinic receptors seem to induce an increase in firing frequency (Schilstrom et al., 2003).

The development of nicotine reinforcement through release of DA in NAc is thought to involve a theory of differential desensitization (Pidoplichko et al., 1997; Mansvelder et al., 2002; Tapper et al., 2004; Pidoplichko et al., 2004) (see Figures 4 and 5). This theory suggests that nicotine initially activates nAChRs on DA neurons, causing an increase in burst firing and overall firing rate. The nAChRs on the VTA DA neurons, which are mainly $\alpha 4\beta 2^\star$, largely desensitize over the next few minutes. Simultaneously, nicotine activates presynaptic $\alpha 7^\star$ nAChRs, boosting glutamatergic synaptic transmission onto DA neurons. Because $\alpha 7^\star$ nAChRs have a relatively low affinity for nicotine, the low concentrations of nicotine achieved by smokers do not significantly desensitize the $\alpha 7^\star$ nAChRs (Wooltorton et al., 2003). Postsynaptic $\beta 2^\star$ nAChRs initially

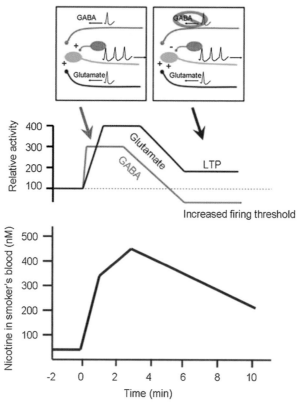

Figure 5 Different time course of inputs to VTA DA neurons in response to nicotine. The upper diagram depicts the changes in activity of the different VTA cell types following nicotine exposure. The middle panel demonstrates the relative activity of GABAergic and glutamatergic inputs to VTA DA neurons that correspond temporally to changes in arterial nicotine levels due to nicotine administration. *(Adapted with permission from Mansvelder et al., 2002).*

depolarize DA neurons, causing them to fire action potentials while presynaptic α7★ nAChRs boost GLU release. The combination of enhanced glutamatergic release and strong postsynaptic response produces LTP of the glutamatergic afferents with shift in AMPA/NMDA ratio (Mao et al., 2011). The shift in AMPA/NMDA ratio is larger in young animals, suggesting a stronger role in plasticity during adolescence (Placzek et al., 2009).

Within the DA target area of the striatum, cholinergic interneurons contribute only approximately 2% of the striatal neurons, but they are large and have extremely dense axonal arbors that provide denser cholinergic innervation than is seen anywhere else in the brain (Dani and Bertrand, 2007). The action of cholinergic transmission in the striatum is via direct synaptic transmission and volume transmission (de Rover et al., 2002; Koos and Tepper, 2002). Dopaminergic nerve terminals in the striatum express as many as five different nAChR subtypes (α4β2, α4α5β2, α4α6β2β3, α6β2β3, and α6β2 Grady et al., 2007). There appears to be a major role for α6*-nAChRs in regulating both DA neuron firing and synaptic release of DA in the striatum (Drenan et al., 2008) (see Figure 4). Interestingly, α6* are increased in adolescents compared to adults, potentially conferring a role in the increased plasticity noted after exposures in adolescents (Azam et al., 2007; Placzek et al., 2009). Nicotine decreases tonic DA release in the striatum that is evoked by single-action potentials (Zhou et al., 2001), and nicotine also alters the frequency dependence of DA release that is electrically evoked by stimulus trains (Rice and Cragg, 2004; Zhang and Sulzer, 2004). While desensitization of nAChRs indeed curbs dopamine released by stimuli emulating tonic firing, it allows a rapid rise in DA from stimuli emulating phasic firing patterns associated with incentive/salience paradigms. Nicotine may thus enhance the contrast of dopamine signals associated with behavioral cues (Zhang and Sulzer, 2004).

2.3.5. Noradrenaline Release

Nicotine effects NA release from LC neurons to hippocampus and paraventricular nucleus. A single dose of nicotine is associated with an increase in terminal tyrosine hydroxylase (TH) activity 2–6 days later and a subsequent increase in NA release; chronic nicotine exposure is associated with induction of TH in the hippocampus (Mitchell et al., 1993). Chronic exposure to nicotine enhances the function of ionotropic GLU receptors mediating NA release in the hippocampus (Risso et al., 2004). TH expression is induced by stress or nicotine in both the adrenal medulla and LC. The induction in the adrenal medulla is dependent primarily on transcriptional mechanisms, whereas that in the LC is apparently dependent primarily on posttranscriptional mechanisms (Osterhout et al., 2005). Blockade of α3β4 receptors blocks NA release in the hippocampus (Luo et al., 1998). Expression of both α3 and β2 has been implicated as a mediator of NA release in LC nuclei projecting to paraventricular nucleus (Okada et al., 2008). Yet, there is a predominance of putative α6β2β3-nAChRs in the LC, suggesting a limited role of α3β4-containing nAChRs in NA release (Le Novere and Changeux, 1995). The presence of multiple subtypes of nAChRs in the LC may reflect the complexity and possibly the physiological importance of the cholinergic control of the noradrenergic system in the brain (Lena et al., 1999).

2.3.6. Serotonin Release

Nicotinic modulation of the serotonergic system is also well documented. Nicotinic receptors are expressed in the major serotonergic nuclei (Picciotto et al., 2001). Systemic nicotine administration stimulates serotonin (5-HT) release in frontal cortex (Ribeiro et al., 1993) as well as from the striatum (Reuben and Clarke, 2000). Nicotine stimulates 5-HT release from dorsal raphe neurons (DRN) while also influencing cell firing. In one study, nicotine (10–300 μM) induced a concentration-dependent two- to sevenfold increase of serotonin release (Mihailescu et al., 1998). Intracellular recordings from DRN slices also demonstrated that nicotine modulates NA and 5-HT release, causing cell excitation and inhibition, respectively. Interestingly, the nAChRs mediating the release of the two neurotransmitters are pharmacologically different (Li et al., 1998).

Within adolescent mouse brain models of depression, the antidepressant effects of nicotine have been shown to vary based on the presence of monoamine oxidase-inhibiting (MAO-I) constituents in tobacco smoke. Specifically, through putative activation of serotonergic mechanisms, adolescent mice were less depressed and more active when an MAO-I and nicotine were administered simultaneously, compared with either substance alone (Villegier et al., 2010).

The effects of nicotine exposure on pain and sleep may be mediated through its effect on serotonin. Nicotinic modulation of 5-HT release underlies nicotine-induced analgesia with tonic modulation of serotonergic modulation in the spinal cord (Cordero-Erausquin et al., 2004). Nicotine suppresses the ponto-geniculo-occipital (PGO) spikes of rapid eye movement (REM) sleep in cats, and nicotine inhibits the activity of LDT and PPT neurons and consequently the generation of PGO spikes through stimulation of DRN serotoninergic neurons (Mihailescu et al., 1998).

2.3.7. Endogenous Opioids

The endogenous opioid system modulates the brain's reward circuitry. Acute exposure to nicotine leads to a release of opioid peptides in specific brain regions, resulting in an activation of their corresponding receptors. For example, an increased concentration of β-endorphin has been found in the hypothalamus after acute nicotine administration in rodents (Marty et al., 1985). With chronic exposure to nicotine, as well as the other prototypical drugs of abuse, significant adaptive changes in the level and expression of endogenous opioid peptides and receptors occur. These adaptive changes are thought to contribute to the homeostatic or allostatic adaptations of the brain, which have been associated with drug dependence (Trigo et al.). Endogenous opioid peptides derived from preproenkephalin are involved in the antinociceptive and rewarding properties of nicotine and participate in the expression of physical nicotine dependence. Nicotine-induced antinociception, rewarding effects, and physical dependence are decreased in mice lacking the preproenkephalin gene (Berrendero et al., 2005).

There are often conflicting clinical reports of the effect of nicotine on anxiety. This is reflected in a complex experimental literature. For example, anxiolytic-like effects of nicotine are blocked by the μ-opioid antagonist β-funaltrexamine, while its anxiogenic-like effects are enhanced by the δ-opioid antagonist naltrindol (Balerio et al., 2005). Interestingly, the anxiogenic and rewarding effects of nicotine are attenuated in β endorphin knockout (KO) mice while nicotine-induced physical dependence was maintained in these mutants. This suggests that β-endorphin, together with other endogenous opioid peptides, may selectively influence the addiction process by mediating the anxiolytic and rewarding properties of nicotine (Trigo et al., 2009).

Endogenous enkephalins and β-endorphins acting on μ-opioid receptors are involved in nicotine-rewarding effects, whereas opioid peptides derived from prodynorphin participate in nicotine aversive responses. An upregulation of μ-opioid receptors that could counteract the development of nicotine tolerance has been reported after chronic nicotine treatment, whereas the downregulation induced on kappa-opioid receptors seems to facilitate nicotine tolerance. Endogenous enkephalins acting on μ-opioid receptors also play a role in the development of physical dependence to nicotine (Berrendero et al., 2010). A review summarizing the role of specific opioid peptides and receptors in various stages of the addiction process is beyond the scope of this chapter (Drews and Zimmer).

3. CONCLUSIONS

Tobacco smoking has dramatic and widespread effects on multiple neurotransmitter systems through the actions of several nAChR subunit combinations. The diversity of subunit arrangements and wide distribution of receptor locations complicates simple description of nicotine's effects on the human brain. In its role as presynaptic modulator, nicotine appears to modulate the signal-to-noise ratio of multiple neurotransmitters. Through the process of differential desensitization in the VTA, nicotine activates reward circuitry with consequent development of tolerance and withdrawal.

Several areas of promising research remain beyond the scope of this chapter, including epigenetics of nicotine and its effects on chromatin and histones (Volkow, 2011), effects of nAChRs on neurogenesis (Campbell et al., 2011), effects of nicotine on feeding behaviors (Audrain-McGovern and Benowitz, 2011), and differential developmental effects of smoking during fetal life (Dwyer et al., 2008) and adolescence (Placzek et al., 2009). Further research is needed if we are to reduce the prevalence of nicotine dependence and the societal burden of addiction to tobacco smoking.

ACKNOWLEDGMENTS

We would like to thank Michael Marks for his substantive editorial contribution by reviewing a draft of this chapter before publication.

REFERENCES

Albuquerque, E.X., et al., 1997. Properties of neuronal nicotinic acetylcholine receptors: pharmacological characterization and modulation of synaptic function. J. Pharmacol. Exp. Ther. 280 (3), 1117–1136.

Albuquerque, E.X., et al., 1998. Contribution of nicotinic receptors to the function of synapses in the central nervous system: the action of choline as a selective agonist of alpha 7 receptors. J. Physiol. Paris 92 (3–4), 309–316.

Albuquerque, E.X., et al., 2000. Neuronal nicotinic receptors in synaptic functions in humans and rats: physiological and clinical relevance. Behav. Brain Res. 113 (1–2), 131–141.

Albuquerque, E.X., et al., 2009. Mammalian nicotinic acetylcholine receptors: from structure to function. Physiol. Rev. 89 (1), 73–120.

Alkondon, M., Albuquerque, E.X., 2004. The nicotinic acetylcholine receptor subtypes and their function in the hippocampus and cerebral cortex. Prog. Brain Res. 145, 109–120.

Alkondon, M., et al., 1997. Neuronal nicotinic acetylcholine receptor activation modulates gamma-aminobutyric acid release from CA1 neurons of rat hippocampal slices. J. Pharmacol. Exp. Ther. 283 (3), 1396–1411.

Alkondon, M., et al., 1999. Choline and selective antagonists identify two subtypes of nicotinic acetylcholine receptors that modulate GABA release from CA1 interneurons in rat hippocampal slices. J. Neurosci. 19 (7), 2693–2705.

Alkondon, M., et al., 2000. Nicotinic receptor activation in human cerebral cortical interneurons: a mechanism for inhibition and disinhibition of neuronal networks. J. Neurosci. 20 (1), 66–75.

Aramakis, V.B., Metherate, R., 1998. Nicotine selectively enhances NMDA receptor-mediated synaptic transmission during postnatal development in sensory neocortex. J. Neurosci. 18 (20), 8485–8495.

Arnaiz-Cot, J.J., et al., 2008. Allosteric modulation of alpha 7 nicotinic receptors selectively depolarizes hippocampal interneurons, enhancing spontaneous GABAergic transmission. Eur. J. Neurosci. 27 (5), 1097–1110.

Audrain-McGovern, J., Benowitz, N.L., 2011. Cigarette smoking, nicotine, and body weight. Clin. Pharmacol. Ther. 90 (1), 164–168.

Azam, L., Chen, Y., Leslie, F.M., 2007. Developmental regulation of nicotinic acetylcholine receptors within midbrain dopamine neurons. Neuroscience 144 (4), 1347–1360.

Azam, L., Yoshikami, D., McIntosh, J.M., 2008. Amino acid residues that confer high selectivity of the alpha6 nicotinic acetylcholine receptor subunit to alpha-conotoxin MII[S4A, E11A, L15A]. J. Biol. Chem. 283 (17), 11625–11632.

Bailey, C.D., et al. The nicotinic acetylcholine receptor alpha5 subunit plays a key role in attention circuitry and accuracy. J. Neurosci. 30 (27), 9241–9252.

Baldwin, P.R., Alanis, R., Salas, R. The role of the habenula in nicotine addiction. J. Addict Res. Ther. S1 (2).

Balerio, G.N., Aso, E., Maldonado, R., 2005. Involvement of the opioid system in the effects induced by nicotine on anxiety-like behaviour in mice. Psychopharmacology (Berl) 181 (2), 260–269.

Beani, L., et al., 1985. The effect of nicotine and cytisine on 3H-acetylcholine release from cortical slices of guinea-pig brain. Naunyn Schmiedebergs Arch. Pharmacol. 331 (2–3), 293–296.

Benowitz, N.L., Herrera, B., Jacob 3rd, P., 2004. Mentholated cigarette smoking inhibits nicotine metabolism. J. Pharmacol. Exp. Ther. 310 (3), 1208–1215.

Berg, D.K., 2011. Timing is everything, even for cholinergic control. Neuron 71 (1), 6–8.

Berrendero, F., et al., 2005. Nicotine-induced antinociception, rewarding effects, and physical dependence are decreased in mice lacking the preproenkephalin gene. J. Neurosci. 25 (5), 1103–1112.

Berrendero, F., et al., 2010. Neurobiological mechanisms involved in nicotine dependence and reward: participation of the endogenous opioid system. Neurosci. Biobehav. Rev. 35 (2), 220–231.

Bierut, L.J., 2010. Convergence of genetic findings for nicotine dependence and smoking related diseases with chromosome 15q24-25. Trends Pharmacol. Sci. 31 (1), 46–51.

Bolam, J.P., Francis, C.M., Henderson, Z., 1991. Cholinergic input to dopaminergic neurons in the substantia nigra: a double immunocytochemical study. Neuroscience 41 (2–3), 483–494.

Broide, R.S., Leslie, F.M., 1999. The alpha7 nicotinic acetylcholine receptor in neuronal plasticity. Mol. Neurobiol. 20 (1), 1–16.

Buisson, B., Bertrand, D., 2001. Chronic exposure to nicotine upregulates the human (alpha)4((beta)2 nicotinic acetylcholine receptor function. J. Neurosci. 21 (6), 1819–1829.

Buisson, B., Bertrand, D., 2002. Nicotine addiction: the possible role of functional upregulation. Trends Pharmacol. Sci. 23 (3), 130–136.

Campbell, N.R., et al., 2010. Endogenous signaling through alpha7-containing nicotinic receptors promotes maturation and integration of adult-born neurons in the hippocampus. J. Neurosci. 30 (26), 8734–8744.

Campbell, N.R., et al., 2011. Nicotinic control of adult-born neuron fate. Biochem. Pharmacol. 82 (8), 820–827.

Castro, N.G., Albuquerque, E.X., 1995. alpha-Bungarotoxin-sensitive hippocampal nicotinic receptor channel has a high calcium permeability. Biophys. J. 68 (2), 516–524.

Charpantier, E., et al., 1998. Nicotinic acetylcholine subunit mRNA expression in dopaminergic neurons of the rat substantia nigra and ventral tegmental area. Neuroreport 9 (13), 3097–3101.

Cigarette smoking among adults and trends in smoking cessation – United States, 2008. 2009. MMWR Morb. Mortal. Wkly. Rep. 58 (44), 1227–1232.

Coggan, J.S., et al., 1997. Direct recording of nicotinic responses in presynaptic nerve terminals. J. Neurosci. 17 (15), 5798–5806.

Cooper, E., Couturier, S., Ballivet, M., 1991. Pentameric structure and subunit stoichiometry of a neuronal nicotinic acetylcholine receptor. Nature 350 (6315), 235–238.

Cordero-Erausquin, M., et al., 2004. Nicotine differentially activates inhibitory and excitatory neurons in the dorsal spinal cord. Pain 109 (3), 308–318.

Corringer, P.J., Sallette, J., Changeux, J.P., 2006. Nicotine enhances intracellular nicotinic receptor maturation: a novel mechanism of neural plasticity? J. Physiol. Paris 99 (2–3), 162–171.

Court, J., et al., 1999. Neuronal nicotinic receptors in dementia with Lewy bodies and schizophrenia: alpha-bungarotoxin and nicotine binding in the thalamus. J. Neurochem. 73 (4), 1590–1597.

Couturier, S., et al., 1990. A neuronal nicotinic acetylcholine receptor subunit (alpha 7) is developmentally regulated and forms a homo-oligomeric channel blocked by alpha-BTX. Neuron 5 (6), 847–856.

Cuevas, J., Roth, A.L., Berg, D.K., 2000. Two distinct classes of functional 7-containing nicotinic receptor on rat superior cervical ganglion neurons. J. Physiol. 525 (Pt 3), 735–746.

Dani, J.A., Bertrand, D., 2007. Nicotinic acetylcholine receptors and nicotinic cholinergic mechanisms of the central nervous system. Annu. Rev. Pharmacol. Toxicol. 47, 699–729.

Dani, J.A., Heinemann, S., 1996. Molecular and cellular aspects of nicotine abuse. Neuron 16 (5), 905–908.

Dani, J.A., Mayer, M.L., 1995. Structure and function of glutamate and nicotinic acetylcholine receptors. Curr. Opin. Neurobiol. 5 (3), 310–317.

Dani, J.A., 2001. Overview of nicotinic receptors and their roles in the central nervous system. Biol. Psychiatry 49 (3), 166–174.

De Filippi, G., Baldwinson, T., Sher, E., 2001. Evidence for nicotinic acetylcholine receptor activation in rat cerebellar slices. Pharmacol. Biochem. Behav. 70 (4), 447–455.

de Rover, M., et al., 2002. Cholinergic modulation of nucleus accumbens medium spiny neurons. Eur. J. Neurosci. 16 (12), 2279–2290.

Decker, E.R., Dani, J.A., 1990. Calcium permeability of the nicotinic acetylcholine receptor: the single-channel calcium influx is significant. J. Neurosci. 10 (10), 3413–3420.

Di Chiara, G., Imperato, A., 1988. Drugs abused by humans preferentially increase synaptic dopamine concentrations in the mesolimbic system of freely moving rats. Proc. Natl. Acad. Sci. U.S.A. 85 (14), 5274–5278.

Dickinson, J.A., et al., 2007. Differential coupling of alpha7 and non-alpha7 nicotinic acetylcholine receptors to calcium-induced calcium release and voltage-operated calcium channels in PC12 cells. J. Neurochem. 100 (4), 1089–1096.

Dickinson, J.A., Kew, J.N., Wonnacott, S., 2008. Presynaptic alpha 7- and beta 2-containing nicotinic acetylcholine receptors modulate excitatory amino acid release from rat prefrontal cortex nerve terminals via distinct cellular mechanisms. Mol. Pharmacol. 74 (2), 348–359.

Doll, R., et al., 2004. Mortality in relation to smoking: 50 years' observations on male British doctors. BMJ 328 (7455), 1519.

Dominguez del Toro, E., et al., 1994. Immunocytochemical localization of the alpha 7 subunit of the nicotinic acetylcholine receptor in the rat central nervous system. J. Comp. Neurol. 349 (3), 325–342.

Drenan, R.M., et al., 2008. In vivo activation of midbrain dopamine neurons via sensitized, high-affinity alpha 6 nicotinic acetylcholine receptors. Neuron 60 (1), 123–136.

Drever, B.D., Riedel, G., Platt, B., 2011. The cholinergic system and hippocampal plasticity. Behav. Brain Res. 221 (2), 505–514.

Drews, E., Zimmer, A. Modulation of alcohol and nicotine responses through the endogenous opioid system. Prog. Neurobiol. 90 (1), 1–15.

Dwyer, J.B., Broide, R.S., Leslie, F.M., 2008. Nicotine and brain development. Birth Defects Res. C Embryo Today 84 (1), 30–44.

Eilers, H., et al., 1997. Functional deactivation of the major neuronal nicotinic receptor caused by nicotine and a protein kinase C-dependent mechanism. Mol. Pharmacol. 52 (6), 1105–1112.

Elgoyhen, A.B., et al., 1994. Alpha 9: an acetylcholine receptor with novel pharmacological properties expressed in rat cochlear hair cells. Cell 79 (4), 705–715.

Exley, R., et al., 2008. Alpha6-containing nicotinic acetylcholine receptors dominate the nicotine control of dopamine neurotransmission in nucleus accumbens. Neuropsychopharmacology 33 (9), 2158–2166.

Exley, R., et al., 2011. Distinct contributions of nicotinic acetylcholine receptor subunit alpha4 and subunit alpha6 to the reinforcing effects of nicotine. Proc. Natl. Acad. Sci. U.S.A. 108 (18), 7577–7582.

Favaron, M., et al., 1990. Down-regulation of protein kinase C protects cerebellar granule neurons in primary culture from glutamate-induced neuronal death. Proc. Natl. Acad. Sci. U.S.A. 87 (5), 1983–1987.

Fenster, C.P., et al., 1997. Influence of subunit composition on desensitization of neuronal acetylcholine receptors at low concentrations of nicotine. J. Neurosci. 17 (15), 5747–5759.

Fenster, C.P., et al., 1999a. Regulation of alpha4beta2 nicotinic receptor desensitization by calcium and protein kinase C. Mol. Pharmacol. 55 (3), 432–443.

Fenster, C.P., et al., 1999b. Upregulation of surface alpha4beta2 nicotinic receptors is initiated by receptor desensitization after chronic exposure to nicotine. J. Neurosci. 19 (12), 4804–4814.

Ficklin, M.B., Zhao, S., Feng, G., 2005. Ubiquilin-1 regulates nicotine-induced up-regulation of neuronal nicotinic acetylcholine receptors. J. Biol. Chem. 280 (40), 34088–34095.

Fisher, J.L., Dani, J.A., 2000. Nicotinic receptors on hippocampal cultures can increase synaptic glutamate currents while decreasing the NMDA-receptor component. Neuropharmacology 39 (13), 2756–2769.

Fowler, C.D., et al. Habenular alpha5 nicotinic receptor subunit signalling controls nicotine intake. Nature. 471 (7340), 597–601.

Frahm, S., et al. Aversion to nicotine is regulated by the balanced activity of beta4 and alpha5 nicotinic receptor subunits in the medial habenula. Neuron. 70 (3), 522–535.

Ge, S., Dani, J.A., 2005. Nicotinic acetylcholine receptors at glutamate synapses facilitate long-term depression or potentiation. J. Neurosci. 25 (26), 6084–6091.

Giniatullin, R., Nistri, A., Yakel, J.L., 2005. Desensitization of nicotinic ACh receptors: shaping cholinergic signaling. Trends Neurosci. 28 (7), 371–378.

Gopalakrishnan, M., Molinari, E.J., Sullivan, J.P., 1997. Regulation of human alpha4beta2 neuronal nicotinic acetylcholine receptors by cholinergic channel ligands and second messenger pathways. Mol. Pharmacol. 52 (3), 524–534.

Gotti, C., et al., 2005. Expression of nigrostriatal alpha 6-containing nicotinic acetylcholine receptors is selectively reduced, but not eliminated, by beta 3 subunit gene deletion. Mol. Pharmacol. 67 (6), 2007–2015.

Gotti, C., et al., 2010. Nicotinic acetylcholine receptors in the mesolimbic pathway: primary role of ventral tegmental area alpha6beta2* receptors in mediating systemic nicotine effects on dopamine release, locomotion, and reinforcement. J. Neurosci. 30 (15), 5311–5325.

Grady, S.R., et al., 2001. Nicotinic agonists stimulate acetylcholine release from mouse interpeduncular nucleus: a function mediated by a different nAChR than dopamine release from striatum. J. Neurochem. 76 (1), 258–268.

Grady, S.R., et al., 2007. The subtypes of nicotinic acetylcholine receptors on dopaminergic terminals of mouse striatum. Biochem. Pharmacol. 74 (8), 1235–1246.

Graham, A., et al., 2002. Immunohistochemical localisation of nicotinic acetylcholine receptor subunits in human cerebellum. Neuroscience 113 (3), 493–507.

Gray, R., et al., 1996. Hippocampal synaptic transmission enhanced by low concentrations of nicotine. Nature 383 (6602), 713–716.

Grilli, M., et al., 2005. Chronic nicotine differentially affects the function of nicotinic receptor subtypes regulating neurotransmitter release. J. Neurochem. 93 (5), 1353–1360.

Grilli, M., et al., 2009. NMDA-mediated modulation of dopamine release is modified in rat prefrontal cortex and nucleus accumbens after chronic nicotine treatment. J. Neurochem. 108 (2), 408–416.

Gu, Z., Yakel, J.L., 2011. Timing-dependent septal cholinergic induction of dynamic hippocampal synaptic plasticity. Neuron 71 (1), 155–165.

Guo, J.Z., Tredway, T.L., Chiappinelli, V.A., 1998. Glutamate and GABA release are enhanced by different subtypes of presynaptic nicotinic receptors in the lateral geniculate nucleus. J. Neurosci. 18 (6), 1963–1969.

Henningfield, J.E., et al., 1993. Higher levels of nicotine in arterial than in venous blood after cigarette smoking. Drug Alcohol Depend. 33 (1), 23–29.

Hill Jr., J.A., et al., 1993. Immunocytochemical localization of a neuronal nicotinic receptor: the beta 2-subunit. J. Neurosci. 13 (4), 1551–1568.

Hogg, R.C., Raggenbass, M., Bertrand, D., 2003. Nicotinic acetylcholine receptors: from structure to brain function. Rev. Physiol. Biochem. Pharmacol. 147, 1–46.

Hsu, Y.N., et al., 1996. Sustained nicotine exposure differentially affects alpha 3 beta 2 and alpha 4 beta 2 neuronal nicotinic receptors expressed in *Xenopus oocytes*. J. Neurochem. 66 (2), 667–675.

Hu, M., et al., 2002. Nicotinic regulation of CREB activation in hippocampal neurons by glutamatergic and nonglutamatergic pathways. Mol. Cell Neurosci. 21 (4), 616–625.

Hukkanen, J., Jacob 3rd, P., Benowitz, N.L., 2005. Metabolism and disposition kinetics of nicotine. Pharmacol. Rev. 57 (1), 79–115.

Ibanez-Tallon, I., et al., 2002. Novel modulation of neuronal nicotinic acetylcholine receptors by association with the endogenous prototoxin lynx1. Neuron 33 (6), 893–903.

Ikemoto, S., Qin, M., Liu, Z.H., 2006. Primary reinforcing effects of nicotine are triggered from multiple regions both inside and outside the ventral tegmental area. J. Neurosci. 26 (3), 723–730.

Ji, D., Dani, J.A., 2000. Inhibition and disinhibition of pyramidal neurons by activation of nicotinic receptors on hippocampal interneurons. J. Neurophysiol. 83 (5), 2682–2690.

Ji, D., Lape, R., Dani, J.A., 2001. Timing and location of nicotinic activity enhances or depresses hippocampal synaptic plasticity. Neuron 31 (1), 131–141.

Jones, I.W., Wonnacott, S., 2004. Precise localization of alpha7 nicotinic acetylcholine receptors on glutamatergic axon terminals in the rat ventral tegmental area. J. Neurosci. 24 (50), 11244–11252.

Jones, S., Sudweeks, S., Yakel, J.L., 1999. Nicotinic receptors in the brain: correlating physiology with function. Trends Neurosci. 22 (12), 555–561.

Kaiser, S., Wonnacott, S., 2000. alpha-bungarotoxin-sensitive nicotinic receptors indirectly modulate [(3)H] dopamine release in rat striatal slices via glutamate release. Mol. Pharmacol. 58 (2), 312–318.

Karlin, A., 2002. Emerging structure of the nicotinic acetylcholine receptors. Nat. Rev. Neurosci. 3 (2), 102–114.

Kasa, P., 1986. The cholinergic systems in brain and spinal cord. Prog. Neurobiol. 26 (3), 211–272.

Kawa, K., 2002. Acute synaptic modulation by nicotinic agonists in developing cerebellar Purkinje cells of the rat. J. Physiol. 538 (Pt 1), 87–102.

Khiroug, S.S., et al., 2002. Rat nicotinic ACh receptor alpha7 and beta2 subunits co-assemble to form functional heteromeric nicotinic receptor channels. J. Physiol. 540 (Pt 2), 425–434.

Klink, R., et al., 2001. Molecular and physiological diversity of nicotinic acetylcholine receptors in the midbrain dopaminergic nuclei. J. Neurosci. 21 (5), 1452–1463.

Koob, G.F., Le Moal, M., 2006. Neurobiology of Addiction. Elsevier Inc, London, UK, p. 490.

Koos, T., Tepper, J.M., 2002. Dual cholinergic control of fast-spiking interneurons in the neostriatum. J. Neurosci. 22 (2), 529–535.

Kuryatov, A., et al., 2005. Nicotine acts as a pharmacological chaperone to up-regulate human alpha4beta2 acetylcholine receptors. Mol. Pharmacol. 68 (6), 1839–1851.

Lapchak, P.A., et al., 1989. Presynaptic cholinergic mechanisms in the rat cerebellum: evidence for nicotinic, but not muscarinic autoreceptors. J. Neurochem. 53 (6), 1843–1851.

Le Magueresse, C., Cherubini, E., 2007. Presynaptic calcium stores contribute to nicotine-elicited potentiation of evoked synaptic transmission at CA3-CA1 connections in the neonatal rat hippocampus. Hippocampus 17 (4), 316–325.

Le Novere, N., Changeux, J.P., 1995. Molecular evolution of the nicotinic acetylcholine receptor: an example of multigene family in excitable cells. J. Mol. Evol. 40 (2), 155–172.

Le Novere, N., Zoli, M., Changeux, J.P., 1996. Neuronal nicotinic receptor alpha 6 subunit mRNA is selectively concentrated in catecholaminergic nuclei of the rat brain. Eur. J. Neurosci. 8 (11), 2428–2439.

Lena, C., Changeux, J.P., 1997. Role of Ca^{2+} ions in nicotinic facilitation of GABA release in mouse thalamus. J. Neurosci. 17 (2), 576–585.

Lena, C., et al., 1999. Diversity and distribution of nicotinic acetylcholine receptors in the locus ceruleus neurons. Proc. Natl. Acad. Sci. U.S.A. 96 (21), 12126–12131.

Lena, C., Changeux, J.P., Mulle, C., 1993. Evidence for "preterminal" nicotinic receptors on GABAergic axons in the rat interpeduncular nucleus. J. Neurosci. 13 (6), 2680–2688.

Leonard, S., et al., 2000. Smoking and schizophrenia: abnormal nicotinic receptor expression. Eur. J. Pharmacol. 393 (1–3), 237–242.

Lester, H.A., et al., 2009. Nicotine is a selective pharmacological chaperone of acetylcholine receptor number and stoichiometry. Implications for drug discovery. AAPS J. 11 (1), 167–177.

Li, X., et al., 1998. Presynaptic nicotinic receptors facilitate monoaminergic transmission. J. Neurosci. 18 (5), 1904–1912.

Lukas, R.J., 1991. Effects of chronic nicotinic ligand exposure on functional activity of nicotinic acetylcholine receptors expressed by cells of the PC12 rat pheochromocytoma or the TE671/RD human clonal line. J. Neurochem. 56 (4), 1134–1145.

Luo, S., et al., 1998. alpha-conotoxin AuIB selectively blocks alpha3 beta4 nicotinic acetylcholine receptors and nicotine-evoked norepinephrine release. J. Neurosci. 18 (21), 8571–8579.

Maggi, L., Sher, E., Cherubini, E., 2001. Regulation of GABA release by nicotinic acetylcholine receptors in the neonatal rat hippocampus. J. Physiol. 536 (Pt 1), 89–100.

Mansvelder, H.D., McGehee, D.S., 2000. Long-term potentiation of excitatory inputs to brain reward areas by nicotine. Neuron 27 (2), 349–357.

Mansvelder, H.D., Keath, J.R., McGehee, D.S., 2002. Synaptic mechanisms underlie nicotine-induced excitability of brain reward areas. Neuron 33 (6), 905–919.

Mao, D., Gallagher, K., McGehee, D.S., 2011. Nicotine potentiation of excitatory inputs to ventral tegmental area dopamine neurons. J. Neurosci. 31 (18), 6710–6720.

Marchi, M., Grilli, M., 2010. Presynaptic nicotinic receptors modulating neurotransmitter release in the central nervous system: functional interactions with other coexisting receptors. Prog. Neurobiol. 92 (2), 105–111.

Marks, M.J., et al., 2011. Increased nicotinic acetylcholine receptor protein underlies chronic nicotine-induced up-regulation of nicotinic agonist binding sites in mouse brain. J. Pharmacol. Exp. Ther. 337 (1), 187–200.

Marszalec, W., Yeh, J.Z., Narahashi, T., 2005. Desensitization of nicotine acetylcholine receptors: modulation by kinase activation and phosphatase inhibition. Eur. J. Pharmacol. 514 (2–3), 83–90.

Marty, M.A., et al., 1985. Effects of nicotine on beta-endorphin, alpha MSH, and ACTH secretion by isolated perfused mouse brains and pituitary glands, in vitro. Pharmacol. Biochem. Behav. 22 (2), 317–325.

McGehee, D.S., Role, L.W., 1995. Physiological diversity of nicotinic acetylcholine receptors expressed by vertebrate neurons. Annu. Rev. Physiol. 57, 521–546.

McQuiston, A.R., Madison, D.V., 1999. Nicotinic receptor activation excites distinct subtypes of interneurons in the rat hippocampus. J. Neurosci. 19 (8), 2887–2896.

Messer, K., et al., 2007. The California Tobacco Control Program's effect on adult smokers: (1) Smoking cessation. Tob. Control 16 (2), 85–90.

Mihailescu, S., et al., 1998. Effects of nicotine and mecamylamine on rat dorsal raphe neurons. Eur. J. Pharmacol. 360 (1), 31–36.

Millar, N.S., Harkness, P.C., 2008. Assembly and trafficking of nicotinic acetylcholine receptors (Review). Mol. Membr. Biol. 25 (4), 279–292.

Mitchell, S.N., et al., 1993. Increases in tyrosine hydroxylase messenger RNA in the locus coeruleus after a single dose of nicotine are followed by time-dependent increases in enzyme activity and noradrenaline release. Neuroscience 56 (4), 989–997.

Miwa, J.M., et al., 1999. lynx1, an endogenous toxin-like modulator of nicotinic acetylcholine receptors in the mammalian CNS. Neuron 23 (1), 105–114.

Mokdad, A.H., et al., 2004. Actual causes of death in the United States, 2000. JAMA 291 (10), 1238–1245.

Moroni, M., Bermudez, I., 2006. Stoichiometry and pharmacology of two human alpha4beta2 nicotinic receptor types. J. Mol. Neurosci. 30 (1–2), 95–96.

Mulle, C., et al., 1991. Existence of different subtypes of nicotinic acetylcholine receptors in the rat habenulo-interpeduncular system. J. Neurosci. 11 (8), 2588–2597.

Nishizaki, T., Sumikawa, K., 1998. Effects of PKC and PKA phosphorylation on desensitization of nicotinic acetylcholine receptors. Brain Res. 812 (1–2), 242–245.

Oakman, S.A., et al., 1995. Distribution of pontomesencephalic cholinergic neurons projecting to substantia nigra differs significantly from those projecting to ventral tegmental area. J. Neurosci. 15 (9), 5859–5869.

Okada, S., et al., 2008. Role of brain nicotinic acetylcholine receptor in centrally administered corticotropin-releasing factor-induced elevation of plasma corticosterone in rats. Eur. J. Pharmacol. 587 (1–3), 322–329.

Orr-Urtreger, A., et al., 1997. Mice deficient in the alpha7 neuronal nicotinic acetylcholine receptor lack alpha-bungarotoxin binding sites and hippocampal fast nicotinic currents. J. Neurosci. 17 (23), 9165–9171.

Ortells, M.O., Arias, H.R., 2010. Neuronal networks of nicotine addiction. Int. J. Biochem. Cell Biol. 42 (12), 1931–1935.

Ortells, M.O., Lunt, G.G., 1995. Evolutionary history of the ligand-gated ion-channel superfamily of receptors. Trends Neurosci. 18 (3), 121–127.

Osterhout, C.A., et al., 2005. Induction of tyrosine hydroxylase in the locus coeruleus of transgenic mice in response to stress or nicotine treatment: lack of activation of tyrosine hydroxylase promoter activity. J. Neurochem. 94 (3), 731–741.

Peng, X., et al., 1994. Nicotine-induced increase in neuronal nicotinic receptors results from a decrease in the rate of receptor turnover. Mol. Pharmacol. 46 (3), 523–530.

Penton, R.E., Lester, R.A., 2009. Cellular events in nicotine addiction. Semin. Cell Dev. Biol. 20 (4), 418–431.

Perry, E., et al., 1999. Acetylcholine in mind: a neurotransmitter correlate of consciousness? Trends Neurosci. 22 (6), 273–280.

Picciotto, M.R., et al., 2001. Neuronal nicotinic acetylcholine receptor subunit knockout mice: physiological and behavioral phenotypes and possible clinical implications. Pharmacol. Ther. 92 (2–3), 89–108.

Picciotto, M.R., et al., 2008. It is not "either/or": activation and desensitization of nicotinic acetylcholine receptors both contribute to behaviors related to nicotine addiction and mood. Prog. Neurobiol. 84 (4), 329–342.

Pidoplichko, V.I., et al., 1997. Nicotine activates and desensitizes midbrain dopamine neurons. Nature 390 (6658), 401–404.

Pidoplichko, V.I., et al., 2004. Nicotinic cholinergic synaptic mechanisms in the ventral tegmental area contribute to nicotine addiction. Learn. Mem. 11 (1), 60–69.

Placzek, A.N., Zhang, T.A., Dani, J.A., 2009. Age dependent nicotinic influences over dopamine neuron synaptic plasticity. Biochem. Pharmacol. 78 (7), 686–692.

Plazas, P.V., et al., 2005. Stoichiometry of the alpha9alpha10 nicotinic cholinergic receptor. J. Neurosci. 25 (47), 10905–10912.

Pons, S., et al., 2008. Crucial role of alpha4 and alpha6 nicotinic acetylcholine receptor subunits from ventral tegmental area in systemic nicotine self-administration. J. Neurosci. 28 (47), 12318–12327.

Quick, M.W., et al., 1999. Alpha3beta4 subunit-containing nicotinic receptors dominate function in rat medial habenula neurons. Neuropharmacology 38 (6), 769–783.

Radcliffe, K.A., Dani, J.A., 1998. Nicotinic stimulation produces multiple forms of increased glutamatergic synaptic transmission. J. Neurosci. 18 (18), 7075–7083.

Ramirez-Latorre, J., et al., 1996. Functional contributions of alpha5 subunit to neuronal acetylcholine receptor channels. Nature 380 (6572), 347–351.

Rathouz, M.M., Vijayaraghavan, S., Berg, D.K., 1996. Elevation of intracellular calcium levels in neurons by nicotinic acetylcholine receptors. Mol. Neurobiol. 12 (2), 117–131.

Reuben, M., Clarke, P.B., 2000. Nicotine-evoked [^{3}H]5-hydroxytryptamine release from rat striatal synaptosomes. Neuropharmacology 39 (2), 290–299.

Ribeiro, E.B., et al., 1993. Effects of systemic nicotine on serotonin release in rat brain. Brain Res. 621 (2), 311–318.

Rice, M.E., Cragg, S.J., 2004. Nicotine amplifies reward-related dopamine signals in striatum. Nat. Neurosci. 7 (6), 583–584.

Risso, F., et al., 2004. Chronic nicotine causes functional upregulation of ionotropic glutamate receptors mediating hippocampal noradrenaline and striatal dopamine release. Neurochem. Int. 44 (5), 293–301.

Role, L.W., Berg, D.K., 1996. Nicotinic receptors in the development and modulation of CNS synapses. Neuron 16 (6), 1077–1085.

Rowell, P.P., Winkler, D.L., 1984. Nicotinic stimulation of [^{3}H]acetylcholine release from mouse cerebral cortical synaptosomes. J. Neurochem. 43 (6), 1593–1598.

Sallette, J., et al., 2005. Nicotine upregulates its own receptors through enhanced intracellular maturation. Neuron 46 (4), 595–607.

Schilstrom, B., et al., 1998. N-methyl-D-aspartate receptor antagonism in the ventral tegmental area diminishes the systemic nicotine-induced dopamine release in the nucleus accumbens. Neuroscience 82 (3), 781–789.

Schilstrom, B., et al., 2000. Putative role of presynaptic alpha7* nicotinic receptors in nicotine stimulated increases of extracellular levels of glutamate and aspartate in the ventral tegmental area. Synapse 38 (4), 375–383.

Schilstrom, B., et al., 2003. Dual effects of nicotine on dopamine neurons mediated by different nicotinic receptor subtypes. Int. J. Neuropsychopharmacol. 6 (1), 1–11.

Schoepfer, R., et al., 1990. Brain alpha-bungarotoxin binding protein cDNAs and MAbs reveal subtypes of this branch of the ligand-gated ion channel gene superfamily. Neuron 5 (1), 35–48.

Seguela, P., et al., 1993. Molecular cloning, functional properties, and distribution of rat brain alpha 7: a nicotinic cation channel highly permeable to calcium. J. Neurosci. 13 (2), 596–604.

Severance, E.G., Cuevas, J., 2004. Distribution and synaptic localization of nicotinic acetylcholine receptors containing a novel alpha7 subunit isoform in embryonic rat cortical neurons. Neurosci. Lett. 372 (1–2), 104–109.

Sharma, G., Vijayaraghavan, S., 2001. Nicotinic cholinergic signaling in hippocampal astrocytes involves calcium-induced calcium release from intracellular stores. Proc. Natl. Acad. Sci. U.S.A. 98 (7), 4148–4153.

Sharma, G., Vijayaraghavan, S., 2003. Modulation of presynaptic store calcium induces release of glutamate and postsynaptic firing. Neuron 38 (6), 929–939.

Sher, E., et al., 2004. Physiological roles of neuronal nicotinic receptor subtypes: new insights on the nicotinic modulation of neurotransmitter release, synaptic transmission and plasticity. Curr. Top. Med. Chem. 4 (3), 283–297.

Shoop, R.D., et al., 1999. Neuronal acetylcholine receptors with alpha7 subunits are concentrated on somatic spines for synaptic signaling in embryonic chick ciliary ganglia. J. Neurosci. 19 (2), 692–704.

Sorenson, E.M., Shiroyama, T., Kitai, S.T., 1998. Postsynaptic nicotinic receptors on dopaminergic neurons in the substantia nigra pars compacta of the rat. Neuroscience 87 (3), 659–673.

Srinivasan, R., et al. Nicotine up-regulates alpha4beta2 nicotinic receptors and ER exit sites via stoichiometry-dependent chaperoning. J. Gen. Physiol. 137 (1), 59–79.

Tapia, L., Kuryatov, A., Lindstrom, J., 2007. Ca^{2+} permeability of the (alpha4)3(beta2)2 stoichiometry greatly exceeds that of (alpha4)2(beta2)3 human acetylcholine receptors. Mol. Pharmacol. 71 (3), 769–776.

Tapper, A.R., et al., 2004. Nicotine activation of alpha4* receptors: sufficient for reward, tolerance, and sensitization. Science 306 (5698), 1029–1032.

Tredway, T.L., Guo, J.Z., Chiappinelli, V.A., 1999. N-type voltage-dependent calcium channels mediate the nicotinic enhancement of GABA release in chick brain. J. Neurophysiol. 81 (2), 447–454.

Trigo, J.M., et al. The endogenous opioid system: a common substrate in drug addiction. Drug Alcohol Depend. 108 (3), 183–194.

Trigo, J.M., Zimmer, A., Maldonado, R., 2009. Nicotine anxiogenic and rewarding effects are decreased in mice lacking beta-endorphin. Neuropharmacology 56 (8), 1147–1153.

Uteshev, V.V., Stevens, D.R., Haas, H.L., 1996. Alpha-bungarotoxin-sensitive nicotinic responses in rat tuberomammillary neurons. Pflugers Arch. 432 (4), 607–613.

Vernino, S., et al., 1992. Calcium modulation and high calcium permeability of neuronal nicotinic acetylcholine receptors. Neuron 8 (1), 127–134.

Villegier, A.S., et al., 2010. Age influences the effects of nicotine and monoamine oxidase inhibition on mood-related behaviors in rats. Psychopharmacology (Berl) 208 (4), 593–601.

Volkow, N.D., 2011. Epigenetics of nicotine: another nail in the coughing. Sci. Transl. Med. 3 (107), 107ps43.

Wada, E., et al., 1989. Distribution of alpha 2, alpha 3, alpha 4, and beta 2 neuronal nicotinic receptor subunit mRNAs in the central nervous system: a hybridization histochemical study in the rat. J. Comp. Neurol. 284 (2), 314–335.

Wang, H., Sun, X., 2005. Desensitized nicotinic receptors in brain. Brain Res. Brain Res. Rev. 48 (3), 420–437.

Ware, J.J., van den Bree, M.B., Munafo, M.R., 2011. Association of the CHRNA5-A3-B4 gene cluster with heaviness of smoking: a meta-analysis. Nicotine Tob. Res. 13 (12), 1167–1175.

WHO, 2009. WHO REPORT on the global TOBACCO epidemic. Implementing Smoke-Free Environments, WHO, Editor 2009. WHO Press, Geneva.

Williams, J.M., et al. Shorter interpuff interval is associated with higher nicotine intake in smokers with schizophrenia. Drug Alcohol Depend. 118 (2–3), 313–319.

Wonnacott, S., 1997. Presynaptic nicotinic ACh receptors. Trends Neurosci. 20 (2), 92–98.

Woolf, N.J., 1991. Cholinergic systems in mammalian brain and spinal cord. Prog. Neurobiol. 37 (6), 475–524.

Wooltorton, J.R., et al., 2003. Differential desensitization and distribution of nicotinic acetylcholine receptor subtypes in midbrain dopamine areas. J. Neurosci. 23 (8), 3176–3185.

Xiu, X., et al., 2009. Nicotine binding to brain receptors requires a strong cation-pi interaction. Nature 458 (7237), 534–537.

Yang, K., et al., 2011. Functional nicotinic acetylcholine receptors containing alpha6 subunits are on GABAergic neuronal boutons adherent to ventral tegmental area dopamine neurons. J. Neurosci. 31 (7), 2537–2548.

Yang, K.C., Jin, G.Z., Wu, J., 2009. Mysterious alpha6-containing nAChRs: function, pharmacology, and pathophysiology. Acta Pharmacol. Sin. 30 (6), 740–751.

Zarei, M.M., et al., 1999. Distributions of nicotinic acetylcholine receptor alpha7 and beta2 subunits on cultured hippocampal neurons. Neuroscience 88 (3), 755–764.

Zhang, H., Sulzer, D., 2004. Frequency-dependent modulation of dopamine release by nicotine. Nat. Neurosci. 7 (6), 581–582.

Zhao-Shea, R., et al., 2011. Nicotine-mediated activation of dopaminergic neurons in distinct regions of the ventral tegmental area. Neuropsychopharmacology 36 (5), 1021–1032.

Zhou, F.M., Liang, Y., Dani, J.A., 2001. Endogenous nicotinic cholinergic activity regulates dopamine release in the striatum. Nat. Neurosci. 4 (12), 1224–1229.

Zhou, F.M., Wilson, C.J., Dani, J.A., 2002. Cholinergic interneuron characteristics and nicotinic properties in the striatum. J. Neurobiol. 53 (4), 590–605.

CHAPTER TWELVE

Cognitive Effects of Nicotine

Mehmet Sofuoglu[1], Aryeh I. Herman[1], Cendrine Robinson[2] and Andrew J. Waters[2]

[1]Department of Psychiatry and VA Connecticut Healthcare System, School of Medicine, Yale University, West Haven, CT, USA
[2]Department of Medical and Clinical Psychology, Uniformed Services University of the Health Science, Bethesda, MD, USA

1. INTRODUCTION

Cigarette smoking is the primary cause of preventable death in developed countries. An estimated 435,000 premature deaths in the U.S. and 5.5 million deaths worldwide are caused by smoking each year (CDC, 2008). Approximately half of all cigarette smokers will die as a result of smoking-related diseases, including lung cancer, coronary heart disease, stroke, and chronic obstructive pulmonary disease In the United States, it is estimated that 30% of the deaths caused by cancer each year result from cigarette smoking. Lung cancer results in approximately 1.2 million deaths worldwide, and over 90% of those cases are caused by cigarette smoking (Jemal et al., 2008). To put these figures in perspective, it is estimated that more individuals in the United States die from smoking-related causes than from alcohol-related causes, car accidents, suicide, AIDS, homicide, and illegal drug use combined. The estimated total economic and healthcare cost of cigarette smoking in the United States is $193 billion per year (CDC, 2008).

Over the past 50 years, the rate of smoking in the United States has decreased from 40% to 20%, but there has been less of a decline in the smoking rate among people with low incomes, low educational levels, psychiatric disorders and/or other addictions (CDC, 2011). Quitting smoking is associated with immediate health benefits regardless of age or the presence of smoking-related diseases (Menzin et al., 2009; Godtfredsen and Prescott, 2011), but even when smokers utilize evidence-based cessation treatments, only 15–25% of those who quit succeed in avoiding tobacco use for at least one year (Fiore et al., 2008; Herman and Sofuoglu, 2010). Thus, it is necessary to develop more effective treatments for nicotine addiction. The development of new treatments requires a better understanding of the individual factors that contribute to the initiation and maintenance of nicotine addiction.

A large body of evidence from animal and human studies supports the notion that nicotine has cognitive-enhancing effects. Smokers report that smoking has beneficial effects on concentration and memory (Piper et al., 2004; Russell et al., 1974; Wesnes and Warburton, 1983), and abstinence from smoking is associated with decreases in cognitive function such as difficulty concentrating, impaired attention, and reductions in the

The Effects of Drug Abuse on the Human Nervous System
http://dx.doi.org/10.1016/B978-0-12-418679-8.00012-5

efficiency of working memory (Harrison et al., 2009; Hatsukami et al., 1984; Hughes and Hatsukami, 1986; Jacobsen et al., 2005; McClernon et al., 2008; Xu et al., 2005). However, nicotine use enhances performance in several domains of cognitive functioning, including attention, working memory, and complex task performance in satiated smokers and nonsmokers (Baschnagel and Hawk, 2008; Ernst et al., 2001; Foulds et al., 1996; Heishman, 1998; Lawrence et al., 2002; Meinke et al., 2006; Mumenthaler et al., 1998; Trimmel and Wittberger, 2004).

Over the past decade, there have been great advances in the understanding of the neurobiology of the nicotinic acetylcholine receptor (nAChR) as it relates to cognitive function and the reward (Changeux, 2010; Dos Santos Coura and Granon, 2012). Furthermore, functional neuroimaging studies provide essential information regarding the brain regions that mediate the rewarding and cognitive effects of nicotine (Newhouse et al., 2011; Sharma and Brody, 2009). As a result of these advances, the cognitive-enhancing effects of nicotine are increasingly recognized as important factors that contribute to the initiation and maintenance of smoking (Levin et al., 2006). Nicotine may positively reinforce smoking behaviors by enhancing cognitive function, especially among individuals in whom normal cognitive functioning is impaired. A high prevalence of smoking is observed among individuals with schizophrenia (de Leon and Diaz, 2005) and attention deficit hyperactivity disorder (ADHD) (Milberger et al., 1997). These psychiatric disorders are associated with cognitive impairments (Chamberlain et al., 2011). Medications that target the $\alpha7$ and $\alpha4\beta2$ nAChRs have also emerged as cognitive-enhancers for the treatment of neuropsychiatric disorders (Wallace and Porter, 2011).

The goal of this chapter is to provide a brief overview of the cognitive effects of nicotine. The first section of this review focuses on the cognitive effects of nicotine in humans. We then review the neurobiological mechanisms of the cognitive effects of nicotine with a focus on the nicotinic acetylcholine (ACh) and dopamine (DA) receptors. Finally, we address the potential treatment implications of this area of research. The chapter will primarily focus on the acute effects that nicotine has on cognitive performance; the long-term (chronic) cognitive effects of smoking will not be covered (Swan and Lessov-Schlaggar, 2007). For more details, several recent reviews provide excellent overviews of behavioral pharmacology (Heishman et al., 2010), neuroimaging (Newhouse et al., 2011; Sharma and Brody, 2009), and preclinical studies (Dos Santos Coura and Granon, 2012; Mansvelder et al., 2006; Poorthuis et al., 2009) of this broad topic.

2.1. COGNITIVE EFFECTS OF NICOTINE IN HUMANS

In their meta-analysis, Heishman et al. (2010) found that there was little consistency in the dose–response functions of nicotine both within and across domains. They concluded that nicotine improves performance on tasks requiring motor abilities, attention, and memory functions even in the absence of the confounding effects of withdrawal relief (Heishman et al., 2010).

Given the availability of several excellent reviews of the cognitive effects of nicotine in humans, we have chosen to focus on a few studies that illustrate the methodology used in and the typical results obtained from human studies that examine the effects that nicotine administration has on cognitive performance.

Myers et al. (2008) conducted a placebo-controlled double-blind study that examined the dose-dependent effects of nicotine that was administered via a nasal spray (placebo, 1 or 2 mg) in 28 smokers. The participants in this study were tested twice: once after overnight abstinence and once under ad libitum smoking conditions. At each session, the smokers received nasal sprays that contained a placebo, 1 or 2 mg of nicotine in a random order at 90-min intervals. After each dose was administered, various tests of cognitive function, including the continuous performance test (CPT), an arithmetic test, and the N-back test, were administered. In the CPT, the participants were shown a series of letters in rapid succession, and they were asked to press a button when the target letter (X) appeared. In the arithmetic test, the participants were asked to determine whether the solutions to single-digit addition or subtraction problems were correct. In the N-back test, the participants were asked to remember a series of letters that were presented individually on a computer screen, and they were asked to identify whether a letter was repeated with one intervening letter. In the ad libitum smoking condition, nicotine enhanced performance on both the CPT and the arithmetic test in a dose-related manner, but it did not affect working memory performance, which had been assessed using the N-back test. Smokers showed more prominent cognitive impairment in the smoking abstinent condition, and nicotine administration improved cognitive function. This study was well designed, and it demonstrates that nicotine has cognitive-enhancing effects on attentional and computational task performance while controlling for both the nicotine dose and the abstinence interval (Myers et al., 2008).

Another study by Poltavski and Petros (2006) addressed the question of whether the cognitive-enhancing effects of nicotine were moderated by the baseline attention level of an individual. A total of 62 nonsmokers with low- and high-attention levels were recruited for their study. The participants were treated with either a placebo or 7 mg nicotine patch, and each of them completed the Wisconsin Card Sorting Test (WCST), the classic Stroop task, and the CPT. In the Stroop task, the participants were asked to press a button on the basis of the color of the word that appeared on a computer screen while ignoring its meaning. In the WCST, the participants were instructed to place cards sequentially below four key cards, but they were not informed of the rule by which the cards were to be sorted. Instead, the participants received positive verbal reinforcement if they arrived at the correct sorting strategy. After every 10 consecutive cards, the rule was changed, and the participant was tasked with finding the next correct strategy. Participants in the low attention group who were treated with nicotine performed better on the CPT test compared with participants who were treated with the placebo. However, nicotine significantly impaired the performance of participants in the high attention group on the WCST. These results suggest that nicotine optimizes performance on cognitive tasks instead of improving it, and baseline cognitive function is important in modulating the effects of nicotine (Poltavski and Petros, 2006).

It is worth noting that the acute cognitive-enhancing effects of nicotine noted above may mediate some of the acute mood-enhancing or mood-stabilizing effects of nicotine (Waters and Sutton, 2000). For example, by improving attentional focus on a benign distracter stimulus, nicotine may alleviate the negative consequences of a stressor (Kassel and Shiffman, 1997).

Recently, implicit cognition researchers (Wiers and Stacy, 2006) have assessed the impact of smoking cues (vs. control cues) on cognition (Waters and Sayette, 2006). For example, the smoking Stroop task assesses attentional bias to smoking cues, and the Implicit Association Test assesses automatic (implicit) memory associations. Few studies have examined the acute effect of nicotine on task performance; however one study reported that memory associations to smoking cues became less positive after smoking (vs. not smoking) a cigarette (Waters et al., 2007). In addition, attentional bias was reduced by smoking (vs. not smoking) a cigarette (Waters et al., 2009). Acute smoking may reduce the distracting influence of cigarette cues on cognitive performance.

The brain regions that are activated by nicotine administration have been studied by functional neuroimaging studies in humans. In one of the earliest pharmacological functional magnetic resonance imaging (fMRI) studies, Stein et al. (1998) administered saline followed by three doses of nicotine (0.75, 1.50, and 2.25 mg/70 kg) intravenously. Their study found that nicotine activated several brain regions, including the nucleus accumbens, amygdala, cingulate, and frontal cortex, in a dose-dependent manner. These brain regions are known to be involved in the reward and cognitive functions. Another study (Rose et al., 2003) used positron emission tomography imaging to examine the changes in regional cerebral blood flow (rCBF). That study found that in cigarette smokers, nicotine increased or normalized the amount of rCBF in the left frontal region and decreased the amount of rCBF in the left amygdala, which concurs with the results of the Stein et al. study. In several other fMRI studies, the administration of nicotine via nicotine gum enhanced neuronal activity in prefrontal and parietal brain regions (Giessing et al., 2006; Thiel and Fink, 2008; Vossel et al., 2008). Together, these results support the notion that nicotine-induced activation of the prefrontal cortex plays a role in the cognitive-enhancing effects of nicotine. The results of the neuroimaging studies are consistent with the well-established observations that the prefrontal cortex plays a role in a number of cognitive functions including attention, working memory, response inhibition, affective processing, decision making, and goal-directed behavior (Miller and Cohen, 2001). As will be summarized below, nAChRs in the prefrontal cortex modulate the functions of many other neurotransmitters, including glutamate, DA, GABA, serotonin, norepinephrine, and ACh.

2.2. NEUROBIOLOGY OF THE COGNITIVE EFFECTS OF NICOTINE

Nicotine, which is the main addictive chemical in tobacco smoke, is essential in continued and compulsive tobacco use (Benowitz, 2009). Nicotine enters cerebral circulation within 10–60 s after a cigarette puff, and it binds to the nAChRs that are normally activated by ACh (Rose et al., 1999). nAChRs are ligand-gated ion channels that are permeable to sodium, potassium, and calcium ions. These receptors are excitatory and show relatively fast responses; their response times are of the order of milliseconds (Clader and Wang, 2005; Dani and Bertrand, 2007). It is important to note that ACh is hydrolyzed by the enzyme acetylcholinesterase within milliseconds of its release into the synaptic cleft; in contrast, no such rapid breakdown mechanism exists to remove nicotine from the synaptic cleft, so it activates the nAChR longer than ACh (Penton and Lester, 2009). This prolonged activation of the nAChR by nicotine results in the desensitization of the receptor and in its temporary inability to be activated by subsequent agonist activity. The desensitization and tolerance of the nAChR are thought to be crucial in the development of nicotine addiction (Picciotto et al., 2008; Quick and Lester, 2002).

Most nAChRs in the CNS are located presynaptically, and they modulate the release of several neurotransmitters, such as ACh, DA, serotonin, glutamate, GABA, and norepinephrine (Dani and Bertrand, 2007). Some nAChRs, such as those on the dopaminergic neurons in the ventral tegmental area, are also located postsynaptically. nAChRs can either be heteromeric channels that are formed by a combination of α and β subunits (e.g. $\alpha4\beta2$, $\alpha3\beta4$) or homomeric channels that are formed by a group of α subunits (e.g. $\alpha6$ or $\alpha7$). The two most commonly expressed nAChRs in the brain are $\alpha4\beta2$ and $\alpha7$ nAChRs (Dani and Bertrand, 2007). Activation of nAChRs increases extracellular levels of DA in the nucleus accumbens and the prefrontal cortex; these brain areas are thought to be critical in mediating the rewarding and cognitive effects of nicotine, respectively (Balfour, 2009; Corrigall et al., 1992; Dos Santos Coura and Granon, 2012; Rahman et al., 2008).

The cellular mechanisms of nicotine-induced cognitive enhancement are not well characterized, but both the prefrontal cortex and hippocampal brain regions have been implicated in this effect (Leiser et al., 2009; Sarter et al., 2009). Electrophysiological data suggest that nicotine results in cognitive enhancement by improving the signal-to-noise ratio in the prefrontal cortex, and other evidence suggests that nicotine facilitates synaptic plasticity in the prefrontal cortex (Couey et al., 2007). The nAChR subunits that mediate the cognitive effects of nicotine may include $\alpha2$, $\alpha3$, $\alpha4$, $\alpha5$, $\alpha7$, $\beta2$, and $\beta4$ (Changeux, 2010). As will be summarized below, most of the studies that have been conducted to date have focused on the $\alpha7$ and $\beta2$ subunits (Kenney and Gould, 2008).

2.2.1. Nicotinic Acetylcholine Receptors

2.2.1.1. *a7nAChR*

α7 nAChRs are abundant in many brain regions that are associated with cognitive functions, including the hippocampus and prefrontal cortex (Gotti et al., 2007; Leiser et al., 2009). Like the NMDA type of glutamate receptor, α7 nAChRs are highly permeable to calcium, which allows them to enhance the release of neurotransmitters (e.g. glutamate) and to modulate synaptic plasticity (Gray et al., 1996; Quik et al., 1997; Seguela et al., 1993). Relative to α4β2 nAChRs, a7nAChRs have a low affinity for nicotine and do not become desensitized at low nicotine concentrations (Quick and Lester, 2002; Wooltorton et al., 2003). This delayed desensitization of the a7nAChRs may be a mechanism that allows the release of several neurotransmitters, including DA, to be maintained after the α4β2 nAChRs have been desensitized (Giniatullin et al., 2005).

a7nAChR knock-out mice show impairment in attention and working memory tasks (Fernandes et al., 2006; Hoyle et al., 2006). In a study by Young et al. (2004), a7nAChR knock-out mice showed more errors of commission in a sustained attention task than the wild-type. It is also possible that both the distribution and density of various nAChR subtypes differ significantly between wild-type and a7nAChR knock-out mice due to compensatory changes during development (Young et al., 2004).

In humans, a7nAChRs may play a key role in the relationship between smoking and sensory gating sensitivity in individuals with schizophrenia (Adler et al., 1993; Nomikos et al., 2000; Taiminen et al., 1998). Between 75% and 85% of individuals with schizophrenia smoke cigarettes (de Leon and Diaz, 2005), and as many as 90% of them have cognitive deficits in at least one domain (e.g. attention, memory, or executive functioning) (Palmer et al., 1997; Leonard et al., 2001; Medalia et al., 2008; Poirier et al., 2002; Reichenberg et al., 2006). Postmortem examinations of the brains of schizophrenic patients revealed reductions in the density of a7nAChRs in the hippocampus (Breese et al., 2000; Freedman et al., 1995; Guan et al., 1999; Martin-Ruiz et al., 2003), which has been linked to the sensory gating dysfunction that occurs in schizophrenia (Potter et al., 2006). Sensory gating is a process by which irrelevant stimuli are separated from meaningful ones, and it may underlie both sensory overload and the cognitive deficits that are observed in schizophrenic patients. Sensory gating dysfunction is measured as a reduced response to the middle latency (50 ms) component of an auditory event-related potential (Croft et al., 2001). Both nicotine and GTS-21 (DMXB-A), which is a partial a7nAChR agonist, have been shown to reverse auditory gating deficits in a number of animal models and in schizophrenic patients (Martin and Freedman, 2007), and several a7nAChR agonists are under investigation to reduce the cognitive deficits in individuals with schizophrenia, ADHD, and/or Alzheimer's disease (Wallace and Porter, 2011).

2.2.1.2. *α4β2 nAChR*

Compared with a7nAChRs, α4β2 nAChRs have a high affinity for nicotine and become desensitized at low concentrations of nicotine that are within the range of nicotine concentrations that is generally found in the blood of smokers (Gotti et al., 1997). The α4β2 receptor subtype has a high affinity for a number of agonists including nicotine, ACh, varenicline, and cytisine. The activation of α4β2 nAChRs that are located in DAergic cell bodies and presynaptic terminals increases DA release in both the nucleus accumbens and the prefrontal cortex (Chen et al., 2003), which, in turn, may contribute to the rewarding and cognitive-enhancing effects of nicotine, respectively.

The β2 subunit, which is found in over 90% of nAChR pentamers, is highly expressed in the basal ganglia, the thalamus, and the hippocampus (Perry et al., 1992, 1995; Spurden et al., 1997). Mice that lack the β2 subunit of the nAChR demonstrate deficits in attention, working memory, and behavioral flexibility (Granon and Changeux, 2006; Granon et al., 2003; Guillem et al., 2011). It was reported that nicotine did not enhance associative memory performance in β2 knock-out mice, whereas associative memory performance was the expected response to nicotine administration in wild-type mice. In a more recent study, β2 knock-out mice displayed deficits in exploratory behavior that could be partially alleviated by nicotine treatment (Besson et al., 2008).

Pharmacological studies that used partial agonists of the α4β2 nAChR to study its role in cognitive functioning support their role in cognitive functions in a manner that is consistent with the aforementioned findings. One of these partial agonists, AZD3480 enhanced both attention and episodic memory function in healthy volunteers (Dunbar et al., 2007). Similarly, varenicline, which is another partial agonist for the α4β2 nAChR and which is marketed as a treatment for smoking cessation (Rollema et al., 2007), alleviated learning deficits in mice that had been induced by either alcohol administration (Gulick and Gould, 2008) or nicotine withdrawal (Raybuck et al., 2008). In a recent study of cigarette smokers, 10 days of varenicline treatment improved working memory and attention deficits that were induced by nicotine withdrawal (Patterson et al., 2009b). The partial agonists of the α4β2 nAChR may potentially be used as cognitive-enhancing agents for the treatment neuropsychiatric disorders with cognitive deficits as cognitive-enhancing agents.

2.2.1.3. Other nAChR

In addition to α4β2 and α7 subtypes, α2, α5, α3, and β4 subunits may also participate in cognitive-enhancing effect of nicotine (Changeux, 2010). For example, the a5 subunit is widely expressed both in the central and peripheral nervous systems as part of α4β2, α3β2, and α3β4 nAChRs. Although this subunit lacks key residues that could be involved in the binding of either nicotine or ACh, its inclusion changes the function of the nAChR in which it is included. Mice that lack the α5 subunit demonstrate increased nicotine reward responses and reduced aversion to high doses of nicotine (Jackson et al., 2010). Further, these mice have reduced cognitive performance in attention tasks relative to the performance of wild-type mice (Bailey et al., 2010).

In a recent study, Winterer et al. (2010) found a significant association between a functional variant, rs16969968, of the gene that encodes the a5 subunit of nAChRs

(CHRNA5) and performance on the N-back working memory task that is consistent with the results of the aforementioned preclinical studies. The rs16969968 SNP has been associated with the age at which an individual initiates cigarette smoking, the severity of the nicotine dependence as measured by the Fagerström Test for Nicotine Dependence score, and the number of cigarettes that an individual smokes each day (Berrettini et al., 2008; Bierut et al., 2007; Saccone et al., 2007; Winterer et al., 2010). These findings suggest that the vulnerability to nicotine dependence that is associated with this particular SNP may be mediated by a reduction in the cognitive performance of the individual. Thus, individuals with baseline cognitive impairments may be more vulnerable to nicotine dependence for the cognitive-enhancing effects of nicotine (Winterer et al., 2010).

2.2.2. Dopamine

DA is implicated in a number of cognitive functions, including working memory, attention, and response inhibition (Colzato et al., 2009; Nieoullon, 2002; Tanila et al., 1998). DA dysfunction has also been implicated in psychiatric disorders that are associated with poor attention and working memory function such as ADHD and schizophrenia (Cheon et al., 2003; Seeman and Kapur, 2000). As will be summarized below, studies are beginning to shed light on the role of DA in nicotine-induced cognitive enhancement.

DA acts via five receptor subtypes (D1-D5) (D1-D5) (Sealfon and Olanow, 2000; Sokoloff and Schwartz, 1995; Zhu et al., 2008). The DA receptors are also classified into two main receptor families: the D1-like family (which includes the D1 and D5 receptors) and the D2-like family (which includes the D2, D3 and D4 receptors). The D2 receptor family also functions as an autoreceptor that acts to reduce the release of DA (Missale et al., 1998). Among the DA receptors, D2 and D4 are the primary receptors that have been examined in relation to the cognitive effects of nicotine. The D2 receptor family is of particular interest, and it has been implicated in set shifting and cognitive flexibility (van Holstein et al., 2011). Blocking the D2 receptors in the prefrontal cortices of rats has been shown to impair their set shifting abilities without changing their abilities to perform working memory tasks (Floresco et al., 2006).

2.2.2.1. D2 Receptor

Several studies have shown that genetic variation in the human D2 receptor gene modulates abstinence-induced changes in cognitive measures (Evans et al., 2009; Gilbert et al., 2004) and nicotine's effects on cognitive performance.

Jacobsen et al. (2006) reported that following the administration of a nicotine patch, smokers who carried the 957T allele of the gene for the D2 receptor experienced some impairment in their working memory abilities during a task that involved a high verbal working memory load. This particular 957T allele increases the binding availability of the D2 receptor (Hirvonen et al., 2004), which suggests that the reduced working memory function may be due to excess baseline levels of DA in carriers of the 957T allele. Alternatively, the working memory performance of individuals who were homo-zygous for the 957C allele was not appreciably different between placebo and nicotine patch conditions. Thus, the authors suggested that individuals who carry two copies of the 957C allele may not be able to further increase DA activity during the performance of tasks that involve a high working memory load (Jacobsen et al., 2006). This study illus-trates the way in which genetic variation that controls D2 receptor levels may influence the cognitive responses to nicotine.

2.2.2.2. D4 Receptor

Both the structure and pharmacology of the D4 receptor are similar to those of the D2 receptor (Van Tol et al., 1991). One study found evidence that the D4 receptor gene may modulate the attentional bias for smoking-related words that was observed in ex-smokers using a modified Stroop task (Munafo and Johnstone, 2008). Ex-smokers who carried at least one allele with 7 (long) or more repeats had significantly increased levels of color naming interference (Stroop effect) when tested using smoking-related words compared with ex-smokers who carried six or fewer repeats on both alleles, but this difference was not observed among current smokers. The DRD4 7-repeat (long) allele is associated with reduced DA activity in comparison with the 2- or 4-repeat variants (short) (Asghari et al., 1995). These findings suggest that the long allele of the *DRD4* gene predicts that abstinent smokers will experience greater attentional bias for smoking cues in abstinent smokers possibly through reduced DA activity (Asghari et al., 1995).

2.2.2.3. Catechol-O-methyltransferase

Catechol-O-methyltransferase (COMT) is an enzyme that inactivates DA, and it is associated with DA regulation, cognitive processes, and the cognitive effects of nico-tine. COMT contains a well-studied single nucleotide polymorphism that results in the presence of either a methionine (Met) or valine (Val) (val158met) in the enzyme (Sengupta et al., 2008). The COMT enzyme that contains Met is one-fourth as active as the COMT enzyme that contains Val. Therefore, because the Val allele results in a form of COMT that has increased enzymatic efficiency compared with the Met allele, lower levels of DA occur in the prefrontal cortex (Guo et al., 2007). In a pioneer-ing study, Loughead et al. (2008) studied the influence of variations in COMT on cognitive deficits and brain function during abstinence from smoking. Smokers were tested under two conditions: normal smoking and overnight abstinence (the total duration of which was 14h). In each condition, the working memory performance of the smokers was tested using the visual N-back task. During abstinence, the smokers who carried two copies of the Val allele exhibited decreased fMRI BOLD signals in both the bilateral dorsal lateral prefrontal cortex and the dorsal cingulate/medial PFC. They also exhibited slower reaction times in the N-back task compared with their

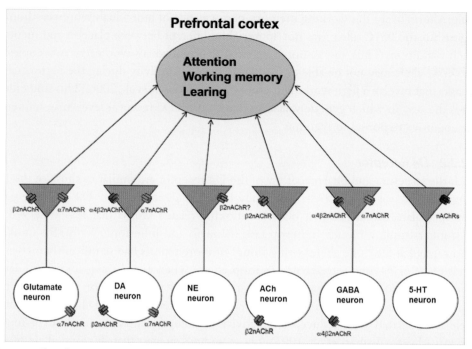

Figure 1 This illustrates the hypothesized effects of nicotinic acetylcholine receptors (nAChRs) on the regulation of dopamine (DA), glutamate, norepinephrine (NE), serotonin (5-HT), GABA, and acetylcholine (ACh) release in the prefrontal cortex. Activation of nAChRs enhances the release of neurotransmitters, but the exact types and locations of these nAChRs must still be determined. See Dos Santos Coura and Granon (2012) for details.

performances under normal smoking conditions (Loughead et al., 2008). These differences were not observed in smokers who carried at least one copy of the Met allele (Loughead et al., 2008).

In a recent study, we investigated the role of the COMT (val158met) polymorphism in acute responses to nicotine that was administered intravenously in a sample of African-American (n = 56) and European-American smokers (Herman et al., 2013). The study included a single laboratory session in which smokers were challenged with saline that was followed by the administration of 0.5 and 1.0 mg/70 kg doses of nicotine that were given at 30-min intervals following overnight abstinence from smoking. The cognitive measures that we investigated included the Mathematical Processing, CPT, and the Stroop Test, all of which were administered twice: once at the beginning of the session and once after the last nicotine administration. In African-Americans, the Val/Val genotype was associated with poorer performance on the CPT and the Stroop Test, but this was not the case for European-American smokers. The reduced transmission of DA during abstinence from smoking might have enhanced the way in which the COMT polymorphism affected cognitive performance such that a tighter control of synaptic DA levels in smokers with the Val/Val genotype resulted in poorer performance on the Stroop Test and the CPT (Zhang et al., 2012).

2.3.3. Other Neurotransmitters

In addition to ACh and DA, other neurotransmitters participate in mediating the cognitive-enhancing effects of nicotine in both the hippocampus and the prefrontal cortex (Dos Santos Coura and Granon, 2012; Parikh et al., 2008; Sarter et al., 2009). The connections between the prefrontal cortex and many other cortical and subcortical areas, including the limbic system and the hippocampus, create a functional circuit that serves many cognitive functions, including attention, working memory, response inhibition, and decision making.

In the prefrontal cortex, the key neurotransmitters that are involved in cognitive functions include glutamate, DA, norepinephrine, serotonin, GABA, and ACh. The precise mechanisms for the cognitive-enhancing effects of nicotine have not yet been determined, but a working hypothesis for the interaction between nAChRs and neurotransmitter release in the prefrontal cortex is shown in Figure 1.

3. CONCLUSIONS

To summarize, human studies have demonstrated that nicotine has cognitive-enhancing effects in both nonsmokers and minimally deprived smokers. Cognitive functions in humans that are particularly improved by nicotine administration include fine motor functions, attentional functions, working memory, and episodic memory. These findings are consistent with human neuroimaging studies that have demonstrated activation in the prefrontal and parietal cortices following nicotine administration. Preclinical studies have implicated both α4β2 and α7 nAChRs in the cognitive-enhancing effects of nicotine. The α7 subunit appears to modulate a sensory filtering function, and it may play an important role in the cognitive deficits that are associated with schizophrenia. Further, the β2 subunit appears to be essential in mediating the cognitive functions that are associated with attention, working memory, and behavioral flexibility. The mechanisms of the cognitive-enhancing effects of nicotine may be mediated via the modulation of the release of various neurotransmitters by α7β2 and α7 nAChRs in the prefrontal cortex. These neurotransmitters include DA, glutamate, serotonin, norepinephrine, GABA, and ACh, all of which contribute to the cognitive functions that take place in the prefrontal cortex (See Figure 12.1).

3.1. Treatment Implications

3.1.1. Targeting Cognitive Function as a Treatment for Smoking Cessation

Mounting evidence suggests that individuals with cognitive deficits may be more vulnerable to nicotine addiction (Yakir et al., 2007). In population-based studies, smokers were found to have deficits in the cognitive functions that are related to attention, working memory, and impulse control (Wagner et al., 2012). These deficits were not correlated with lifetime nicotine use, and even smokers with low amounts of nicotine exposure display these deficits. These findings suggest that these deficits existed in some individuals before they began smoking. Further, Winterer et al. (2010) found that a genetic variation in the α5 nAChR may increase the degree to which an individual is vulnerable to nicotine dependence and may be associated with reduced performance on a working memory task. Presumably, individuals with genetic vulnerability may derive particular benefit from the cognitive-enhancing effects of nicotine (Winterer et al., 2010). Cognitive deficits are also common among patients with psychiatric disorders; for example, 75–90% of schizophrenia patients show evidence of cognitive deficits. Similarly, individuals with ADHD have impairment in their attention function, and ADHD is associated with elevated rates of smoking. Among smokers who were trying to quit smoking, it was found that poorer performance on the N-back test (a working memory task) predicted relapses (Patterson et al., 2009a).

We have also found that abstinence-induced deterioration in the performance of an individual on the Rapid Visual Information Processing Task, which assesses sustained attention and working memory, predicted whether smokers would relapse to smoking at the end of the study (Kang et al., 2012). Still more studies have reported that attentional biases to smoking cues predict relapses in smokers who are attempting to stop smoking (Janes et al., 2010; Powell et al., 2010; Waters et al., 2003). These findings suggest that cognitive-enhancement or cognitive-retraining may be an effective strategy for enabling people with nicotine addictions to quit smoking, especially smokers with cognitive deficits. Several behavioral and pharmacological cognitive-enhancement approaches have been under investigation, including approaches that use nAChR agonists (see below).

3.1.2. Subtype-Selective nAChR Agonists as Cognitive-Enhancers

Agonists that are selective for nAChRs may provide more effective cognitive enhancement than nicotine. Although nicotine produces cognitive enhancement, its therapeutic effects are limited due to rapid desensitization that temporarily renders the receptor inactive. As a result, nicotine also acts as an nAChR antagonist. It is important to note that desensitization of the nAChR is specific to both the agonist and the nicotinic receptor subtype. Conceivably, agonists that are selective for the various nAChR subtypes may be more effective cognitive-enhancers than nicotine. One such subtype-specific group of agonists that is under development as a cognitive enhancer is a group of α7 nAChR agonists. A promising group of medications are those that are selective for the α7 receptors (Wallace and Porter, 2011). These medications are undergoing clinical trials as cognitive-enhancers for patients with schizophrenia, Alzheimer's disease, and ADHD.

ACKNOWLEDGMENTS

This research was supported by the Veterans Administration Mental Illness Research, Education and Clinical Center (MIRECC) and National Institute on Drug Abuse (NIDA) grants R01 DA020752, K02-DA021304 (MS), and K12-DA-019446 (AIH). Dr Sofuoglu serves as an expert witness on behalf of Pfizer in lawsuits related to varenicline.

REFERENCES

Adler, L.E., Hoffer, L.D., Wiser, A., Freedman, R., 1993. Normalization of auditory physiology by cigarette smoking in schizophrenic patients. Am. J. Psychiatry 150, 1856–1861.

Asghari, V., Sanyal, S., Buchwaldt, S., Paterson, A., Jovanovic, V., Van Tol, H.H., 1995. Modulation of intracellular cyclic AMP levels by different human dopamine D4 receptor variants. J. Neurochem. 65, 1157–1165.

Bailey, C.D., De Biasi, M., Fletcher, P.J., Lambe, E.K., 2010. The nicotinic acetylcholine receptor alpha5 subunit plays a key role in attention circuitry and accuracy. J. Neurosci. 30, 9241–9252.

Balfour, D.J., 2009. The neuronal pathways mediating the behavioral and addictive properties of nicotine. Handb. Exp. Pharmacol. 192, 209–233.

Baschnagel, J.S., Hawk Jr., L.W., 2008. The effects of nicotine on the attentional modification of the acoustic startle response in nonsmokers. Psychopharmacology (Berl) 198, 93–101.

Benowitz, N.L., 2009. Pharmacology of nicotine: addiction, smoking-induced disease, and therapeutics. Annu. Rev. Pharmacol. Toxicol. 49, 57–71.

Berrettini, W., Yuan, X., Tozzi, F., Song, K., Francks, C., Chilcoat, H., Waterworth, D., Muglia, P., Mooser, V., 2008. Alpha-5/alpha-3 nicotinic receptor subunit alleles increase risk for heavy smoking. Mol. Psychiatry 13, 368–373.

Besson, M., Suarez, S., Cormier, A., Changeux, J.P., Granon, S., 2008. Chronic nicotine exposure has dissociable behavioural effects on control and beta2-/- mice. Behav. Genet. 38, 503–514.

Bierut, L.J., Madden, P.A., Breslau, N., Johnson, E.O., Hatsukami, D., Pomerleau, O.F., Swan, G.E., Rutter, J., Bertelsen, S., Fox, L., Fugman, D., Goate, A.M., Hinrichs, A.L., Konvicka, K., Martin, N.G., Montgomery, G.W., Saccone, N.L., Saccone, S.F., Wang, J.C., Chase, G.A., Rice, J.P., Ballinger, D.G., 2007. Novel genes identified in a high-density genome wide association study for nicotine dependence. Hum. Mol. Genet. 16, 24–35.

Breese, C.R., Lee, M.J., Adams, C.E., Sullivan, B., Logel, J., Gillen, K.M., Marks, M.J., Collins, A.C., Leonard, S., 2000. Abnormal regulation of high affinity nicotinic receptors in subjects with schizophrenia. Neuropsychopharmacology 23, 351–364.

Center for Disease Control, 2008. Smoking-attributable mortality, years of potential life lost, and productivity losses–United States, 2000-2004. MMWR Morb. Mortal. Wkly. Rep. 57, 1226–1228.

Center for Disease Control, 2011. Current cigarette smoking prevalence among working adults–United States, 2004-2010. MMWR Morb. Mortal. Wkly. Rep. 60, 1305–1309.

Chamberlain, S.R., Robbins, T.W., Winder-Rhodes, S., Muller, U., Sahakian, B.J., Blackwell, A.D., Barnett, J.H., 2011. Translational approaches to frontostriatal dysfunction in attention-deficit/hyperactivity disorder using a computerized neuropsychological battery. Biol. Psychiatry 69, 1192–1203.

Changeux, J.P., 2010. Nicotine addiction and nicotinic receptors: lessons from genetically modified mice. Nat. Rev. Neurosci. 11, 389–401.

Chen, Y., Sharples, T.J., Phillips, K.G., Benedetti, G., Broad, L.M., Zwart, R., Sher, E., 2003. The nicotinic alpha 4 beta 2 receptor selective agonist, TC-2559, increases dopamine neuronal activity in the ventral tegmental area of rat midbrain slices. Neuropharmacology 45, 334–344.

Cheon, K.A., Ryu, Y.H., Kim, Y.K., Namkoong, K., Kim, C.H., Lee, J.D., 2003. Dopamine transporter density in the basal ganglia assessed with [^{123}I]IPT SPET in children with attention deficit hyperactivity disorder. Eur. J. Nucl. Med. Mol. Imaging 30, 306–311.

Clader, J.W., Wang, Y., 2005. Muscarinic receptor agonists and antagonists in the treatment of Alzheimer's disease. Curr. Pharm. Des. 11, 3353–3361.

Colzato, L.S., van den Wildenberg, W.P., van Wouwe, N.C., Pannebakker, M.M., Hommel, B., 2009. Dopamine and inhibitory action control: evidence from spontaneous eye blink rates. Exp. Brain Res. 196, 467–474.

Corrigall, W.A., Franklin, K.B., Coen, K.M., Clarke, P.B., 1992. The mesolimbic dopaminergic system is implicated in the reinforcing effects of nicotine. Psychopharmacology (Berl) 107, 285–289.

Couey, J.J., Meredith, R.M., Spijker, S., Poorthuis, R.B., Smit, A.B., Brussaard, A.B., Mansvelder, H.D., 2007. Distributed network actions by nicotine increase the threshold for spike-timing-dependent plasticity in prefrontal cortex. Neuron 54, 73–87.

Croft, R.J., Lee, A., Bertolot, J., Gruzelier, J.H., 2001. Associations of P50 suppression and desensitization with perceptual and cognitive features of "unreality" in schizotypy. Biol. Psychiatry 50, 441–446.

Dani, J.A., Bertrand, D., 2007. Nicotinic acetylcholine receptors and nicotinic cholinergic mechanisms of the central nervous system. Annu. Rev. Pharmacol. Toxicol. 47, 699–729.

de Leon, J., Diaz, F.J., 2005. A meta-analysis of worldwide studies demonstrates an association between schizophrenia and tobacco smoking behaviors. Schizophr. Res. 76, 135–157.

Dos Santos Coura, R., Granon, S., 2012. Prefrontal neuromodulation by nicotinic receptors for cognitive processes. Psychopharmacology. http://dx.doi.org/10.1007/s00213-011-2596-6.

Dunbar, G., Boeijinga, P.H., Demazieres, A., Cisterni, C., Kuchibhatla, R., Wesnes, K., Luthringer, R., 2007. Effects of TC-1734 (AZD3480), a selective neuronal nicotinic receptor agonist, on cognitive performance and the EEG of young healthy male volunteers. Psychopharmacology (Berl) 191, 919–929.

Ernst, M., Heishman, S.J., Spurgeon, L., London, E.D., 2001. Smoking history and nicotine effects on cognitive performance. Neuropsychopharmacology 25, 313–319.

Evans, D.E., Park, J.Y., Maxfield, N., Drobes, D.J., 2009. Neurocognitive variation in smoking behavior and withdrawal: genetic and affective moderators. Genes Brain Behav. 8, 86–96.

Fernandes, C., Hoyle, E., Dempster, E., Schalkwyk, L.C., Collier, D.A., 2006. Performance deficit of alpha7 nicotinic receptor knockout mice in a delayed matching-to-place task suggests a mild impairment of working/episodic-like memory. Genes Brain Behav. 5, 433–440.

Fiore, M.C., Jaen, C.R., Baker, T.B., 2008. Treating Tobacco Use and Dependence: 2008 Update. Clinical Practice Guideline. U.S. Department of Health and Human Services, Rockville, MD.

Floresco, S.B., Magyar, O., Ghods-Sharifi, S., Vexelman, C., Tse, M.T., 2006. Multiple dopamine receptor subtypes in the medial prefrontal cortex of the rat regulate set-shifting. Neuropsychopharmacology 31, 297–309.

Foulds, J., Stapleton, J., Swettenham, J., Bell, N., McSorley, K., Russell, M.A., 1996. Cognitive performance effects of subcutaneous nicotine in smokers and never-smokers. Psychopharmacology (Berl) 127, 31–38.

Freedman, R., Hall, M., Adler, L.E., Leonard, S., 1995. Evidence in postmortem brain tissue for decreased numbers of hippocampal nicotinic receptors in schizophrenia. Biol. Psychiatry 38, 22–33.

Giessing, C., Thiel, C.M., Rosler, F., Fink, G.R., 2006. The modulatory effects of nicotine on parietal cortex activity in a cued target detection task depend on cue reliability. Neuroscience 137, 853–864.

Gilbert, D., McClernon, J., Rabinovich, N., Sugai, C., Plath, L., Asgaard, G., Zuo, Y., Huggenvik, J., Botros, N., 2004. Effects of quitting smoking on EEG activation and attention last for more than 31 days and are more severe with stress, dependence, DRD2 A1 allele, and depressive traits. Nicotine Tob. Res. 6, 249–267.

Giniatullin, R., Nistri, A., Yakel, J.L., 2005. Desensitization of nicotinic ACh receptors: shaping cholinergic signaling. Trends Neurosci. 28, 371–378.

Godtfredsen, N.S., Prescott, E., 2011. Benefits of smoking cessation with focus on cardiovascular and respiratory comorbidities. Clin. Respir. J. 5, 187–194.

Gotti, C., Fornasari, D., Clementi, F., 1997. Human neuronal nicotinic receptors. Prog. Neurobiol. 53, 199–237.

Gotti, C., Moretti, M., Gaimarri, A., Zanardi, A., Clementi, F., Zoli, M., 2007. Heterogeneity and complexity of native brain nicotinic receptors. Biochem. Pharmacol. 74, 1102–1111.

Granon, S., Changeux, J.P., 2006. Attention-deficit/hyperactivity disorder: a plausible mouse model? Acta Paediatr. 95, 645–649.

Granon, S., Faure, P., Changeux, J.P., 2003. Executive and social behaviors under nicotinic receptor regulation. Proc. Natl. Acad. Sci. U.S.A. 100, 9596–9601.

Gray, R., Rajan, A.S., Radcliffe, K.A., Yakehiro, M., Dani, J.A., 1996. Hippocampal synaptic transmission enhanced by low concentrations of nicotine. Nature 383, 713–716.

Guan, Z.Z., Zhang, X., Blennow, K., Nordberg, A., 1999. Decreased protein level of nicotinic receptor alpha7 subunit in the frontal cortex from schizophrenic brain. Neuroreport 10, 1779–1782.

Guillem, K., Bloem, B., Poorthuis, R.B., Loos, M., Smit, A.B., Maskos, U., Spijker, S., Mansvelder, H.D., 2011. Nicotinic acetylcholine receptor beta2 subunits in the medial prefrontal cortex control attention. Science 333, 888–891.

Gulick, D., Gould, T.J., 2008. Varenicline ameliorates ethanol-induced deficits in learning in C57BL/6 mice. Neurobiol. Learn. Mem. 90, 230–236.

Guo, S., Chen da, F., Zhou, D.F., Sun, H.Q., Wu, G.Y., Haile, C.N., Kosten, T.A., Kosten, T.R., Zhang, X.Y., 2007. Association of functional catechol O-methyl transferase (COMT) Val108Met polymorphism with smoking severity and age of smoking initiation in Chinese male smokers. Psychopharmacology (Berl) 190, 449–456.

Harrison, E.L., Coppola, S., McKee, S.A., 2009. Nicotine deprivation and trait impulsivity affect smokers' performance on cognitive tasks of inhibition and attention. Exp. Clin. Psychopharmacol. 17, 91–98.

Hatsukami, D.K., Hughes, J.R., Pickens, R.W., Svikis, D., 1984. Tobacco withdrawal symptoms: an experimental analysis. Psychopharmacology (Berl) 84, 231–236.

Heishman, S.J., 1998. What aspects of human performance are truly enhanced by nicotine? Addiction 93, 317–320.

Heishman, S.J., Kleykamp, B.A., Singleton, E.G., 2010. Meta-analysis of the acute effects of nicotine and smoking on human performance. Psychopharmacology 210, 453–469.

Herman, A.I., Sofuoglu, M., 2010. Comparison of available treatments for tobacco addiction. Curr. Psychiatry Rep. 12, 433–440.

Herman, A.I., Jatlow, P.I., Gelernter, J., Listman, J.B., Sofuoglu, M., , 2013. COMT Val158Met modulates subjective responses to intravenous nicotine and cognitive performance in abstinent smokers. Pharmacogenomics J , http://dx.doi.org./10.1038/tpj.2013.1. [Epub ahead of print] PubMed PMID: 23459442; PubMed Central PMCID: PMC3675163.

Hirvonen, M., Laakso, A., Nagren, K., Rinne, J.O., Pohjalainen, T., Hietala, J., 2004. C957T polymorphism of the dopamine D2 receptor (DRD2) gene affects striatal DRD2 availability in vivo. Mol. Psychiatry 9, 1060–1061.

Hoyle, E., Genn, R.F., Fernandes, C., Stolerman, I.P., 2006. Impaired performance of alpha7 nicotinic receptor knockout mice in the five-choice serial reaction time task. Psychopharmacology (Berl) 189, 211–223.

Hughes, J.R., Hatsukami, D., 1986. Signs and symptoms of tobacco withdrawal. Arch. Gen. Psychiatry 43, 289–294.

Jackson, K.J., Marks, M.J., Vann, R.E., Chen, X., Gamage, T.F., Warner, J.A., Damaj, M.I., 2010. Role of alpha5 nicotinic acetylcholine receptors in pharmacological and behavioral effects of nicotine in mice. J. Pharmacol. Exp. Ther. 334, 137–146.

Jacobsen, L.K., Krystal, J.H., Mencl, W.E., Westerveld, M., Frost, S.J., Pugh, K.R., 2005. Effects of smoking and smoking abstinence on cognition in adolescent tobacco smokers. Biol. Psychiatry 57, 56–66.

Jacobsen, L.K., Pugh, K.R., Mencl, W.E., Gelernter, J., 2006. C957T polymorphism of the dopamine D2 receptor gene modulates the effect of nicotine on working memory performance and cortical processing efficiency. Psychopharmacology (Berl) 188, 530–540.

Janes, A.C., Pizzagalli, D.A., Richardt, S., deB Frederick, B., Chuzi, S., Pachas, G., Culhane, M.A., Holmes, A.J., Fava, M., Evins, A.E., Kaufman, M.J., 2010. Brain reactivity to smoking cues prior to smoking cessation predicts ability to maintain tobacco abstinence. Biol. Psychiatry 67, 722–729.

Jemal, A., Thun, M.J., Ries, L.A., Howe, H.L., Weir, H.K., Center, M.M., Ward, E., Wu, X.C., Eheman, C., Anderson, R., Ajani, U.A., Kohler, B., Edwards, B.K., 2008. Annual report to the nation on the status of cancer, 1975–2005, featuring trends in lung cancer, tobacco use, and tobacco control. J. Natl. Cancer Inst. 100, 1672–1694.

Kang, N., Robinson, C., Wetter, D.W., Cinciripini, P., Yisheng Li, Y., Waters, A.J., 2012. Abstinence-induced decrements in sustained attention predict relapse in smoking cessation. Presentation at the Forthcoming American Psychological Association Conference, Orlando, FL.

Kassel, J.D., Shiffman, S., 1997. Attentional mediation of cigarette smoking's effect on anxiety. Health Psychol. 16, 359–368.

Kenney, J.W., Gould, T.J., 2008. Modulation of hippocampus-dependent learning and synaptic plasticity by nicotine. Mol. Neurobiol. 38, 101–121.

Lawrence, N.S., Ross, T.J., Stein, E.A., 2002. Cognitive mechanisms of nicotine on visual attention. Neuron 36, 539–548.

Leiser, S.C., Bowlby, M.R., Comery, T.A., Dunlop, J., 2009. A cog in cognition: how the alpha7 nicotinic acetylcholine receptor is geared towards improving cognitive deficits. Pharmacol. Ther. 122, 302–311.

Leonard, S., Adler, L.E., Benhammou, K., Berger, R., Breese, C.R., Drebing, C., Gault, J., Lee, M.J., Logel, J., Olincy, A., Ross, R.G., Stevens, K., Sullivan, B., Vianzon, R., Virnich, D.E., Waldo, M., Walton, K., Freedman, R., 2001. Smoking and mental illness. Pharmacol. Biochem. Behav. 70, 561–570.

Levin, E.D., McClernon, F.J., Rezvani, A.H., 2006. Nicotinic effects on cognitive function: behavioral characterization, pharmacological specification, and anatomic localization. Psychopharmacology (Berl) 184, 523–539.

Loughead, J., Wileyto, E.P., Valdez, J.N., Sanborn, P., Tang, K., Strasser, A.A., Ruparel, K., Ray, R., Gur, R.C., Lerman, C., 2008. Effect of abstinence challenge on brain function and cognition in smokers differs by COMT genotype. Mol. Psychiatry 14, 820–826.

Mansvelder, H.D., van Aerde, K.I., Couey, J.J., Brussaard, A.B., 2006. Nicotinic modulation of neuronal networks: from receptors to cognition. Psychopharmacology (Berl) 184, 292–305.

Martin-Ruiz, C.M., Haroutunian, V.H., Long, P., Young, A.H., Davis, K.L., Perry, E.K., Court, J.A., 2003. Dementia rating and nicotinic receptor expression in the prefrontal cortex in schizophrenia. Biol. Psychiatry 54, 1222–1233.

Martin, L.F., Freedman, R., 2007. Schizophrenia and the alpha7 nicotinic acetylcholine receptor. Int. Rev. Neurobiol. 78, 225–246.

McClernon, F.J., Kollins, S.H., Lutz, A.M., Fitzgerald, D.P., Murray, D.W., Redman, C., Rose, J.E., 2008. Effects of smoking abstinence on adult smokers with and without attention deficit hyperactivity disorder: results of a preliminary study. Psychopharmacology (Berl) 197, 95–105.

Medalia, A., Thysen, J., Freilich, B., 2008. Do people with schizophrenia who have objective cognitive impairment identify cognitive deficits on a self report measure? Schizophr. Res. 105, 156–164.

Meinke, A., Thiel, C.M., Fink, G.R., 2006. Effects of nicotine on visuo-spatial selective attention as indexed by event-related potentials. Neuroscience 141, 201–212.

Menzin, J., Lines, L.M., Marton, J., 2009. Estimating the short-term clinical and economic benefits of smoking cessation: do we have it right? Expert Rev. Pharmacoecon. Outcomes Res. 9, 257–264.

Milberger, S., Biederman, J., Faraone, S.V., Chen, L., Jones, J., 1997. Further evidence of an association between attention-deficit/hyperactivity disorder and cigarette smoking. Findings from a high-risk sample of siblings. Am. J. Addict. 6, 205–217.

Miller, E.K., Cohen, J.D., 2001. An integrative theory of prefrontal cortex function. Annu. Rev. Neurosci. 24, 167–202.

Missale, C., Nash, S.R., Robinson, S.W., Jaber, M., Caron, M.G., 1998. Dopamine receptors: from structure to function. Physiol. Rev. 78, 189–225.

Mumenthaler, M.S., Taylor, J.L., O'Hara, R., Yesavage, J.A., 1998. Influence of nicotine on simulator flight performance in non-smokers. Psychopharmacology (Berl) 140, 38–41.

Munafo, M.R., Johnstone, E.C., 2008. Smoking status moderates the association of the dopamine D4 receptor (DRD4) gene VNTR polymorphism with selective processing of smoking-related cues. Addict. Biol. 13, 435–439.

Myers, C.S., Taylor, R.C., Moolchan, E.T., Heishman, S.J., 2008. Dose-related enhancement of mood and cognition in smokers administered nicotine nasal spray. Neuropsychopharmacology 33, 588–598.

Newhouse, P.A., Potter, A.S., Dumas, J.A., Thiel, C.M., 2011. Functional brain imaging of nicotinic effects on higher cognitive processes. Biochem. Pharmacol. 82 (8), 943–951.

Nieoullon, A., 2002. Dopamine and the regulation of cognition and attention. Prog. Neurobiol. 67, 53–83.

Nomikos, G.G., Schilstrom, B., Hildebrand, B.E., Panagis, G., Grenhoff, J., Svensson, T.H., 2000. Role of alpha7 nicotinic receptors in nicotine dependence and implications for psychiatric illness. Behav. Brain Res. 113, 97–103.

Palmer, B.W., Heaton, R.K., Paulsen, J.S., Kuck, J., Braff, D., Harris, M.J., Zisook, S., Jeste, D.V., 1997. Is it possible to be schizophrenic yet neuropsychologically normal? Neuropsychology 11, 437–446.

Parikh, V., Man, K., Decker, M.W., Sarter, M., 2008. Glutamatergic contributions to nicotinic acetylcholine receptor agonist-evoked cholinergic transients in the prefrontal cortex. J. Neurosci. 28, 3769–3780.

Patterson, F., Jepson, C., Loughead, J., Perkins, K., Strasser, A.A., Siegel, S., Frey, J., Gur, R., Lerman, C., 2009a. Working memory deficits predict short-term smoking resumption following brief abstinence. Drug Alcohol Depend. 106, 61–64.

Patterson, F., Jepson, C., Strasser, A.A., Loughead, J., Perkins, K.A., Gur, R.C., Frey, J.M., Siegel, S., Lerman, C., 2009b. Varenicline improves mood and cognition during smoking abstinence. Biol. Psychiatry 65, 144–149.

Penton, R.E., Lester, R.A., 2009. Cellular events in nicotine addiction. Semin. Cell Dev. Biol. 20, 418–431.

Perry, E.K., Court, J.A., Johnson, M., Piggott, M.A., Perry, R.H., 1992. Autoradiographic distribution of [^{3}H]nicotine binding in human cortex: relative abundance in subicular complex. J. Chem. Neuroanat. 5, 399–405.

Perry, E.K., Morris, C.M., Court, J.A., Cheng, A., Fairbairn, A.F., McKeith, I.G., Irving, D., Brown, A., Perry, R.H., 1995. Alteration in nicotine binding sites in Parkinson's disease, Lewy body dementia and Alzheimer's disease: possible index of early neuropathology. Neuroscience 64, 385–395.

Picciotto, M.R., Addy, N.A., Mineur, Y.S., Brunzell, D.H., 2008. It is not "either/or": activation and desensitization of nicotinic acetylcholine receptors both contribute to behaviors related to nicotine addiction and mood. Prog. Neurobiol. 84, 329–342.

Piper, M.E., Piasecki, T.M., Federman, E.B., Bolt, D.M., Smith, S.S., Fiore, M.C., Baker, T.B., 2004. A multiple motives approach to tobacco dependence: the Wisconsin Inventory of Smoking Dependence Motives (WISDM-68). J. Consult. Clin. Psychol. 72, 139–154.

Poirier, M.F., Canceil, O., Bayle, F., Millet, B., Bourdel, M.C., Moatti, C., Olie, J.P., Attar-Levy, D., 2002. Prevalence of smoking in psychiatric patients. Prog. Neuropsychopharmacol. Biol. Psychiatry 26, 529–537.

Poltavski, D.V., Petros, T., 2006. Effects of transdermal nicotine on attention in adult non-smokers with and without attentional deficits. Physiol. Behav. 87, 614–624.

Poorthuis, R.B., Goriounova, N.A., Couey, J.J., Mansvelder, H.D., 2009. Nicotinic actions on neuronal networks for cognition: general principles and long-term consequences. Biochem. Pharmacol. 78, 668–676.

Potter, D., Summerfelt, A., Gold, J., Buchanan, R.W., 2006. Review of clinical correlates of P50 sensory gating abnormalities in patients with schizophrenia. Schizophr. Bull. 32, 692–700.

Powell, J., Dawkins, L., West, R., Pickering, A., 2010. Relapse to smoking during unaided cessation: clinical, cognitive and motivational predictors. Psychopharmacology 212, 537–549.

Quick, M.W., Lester, R.A., 2002. Desensitization of neuronal nicotinic receptors. J. Neurobiol. 53, 457–478.

Quik, M., Philie, J., Choremis, J., 1997. Modulation of alpha7 nicotinic receptor-mediated calcium influx by nicotinic agonists. Mol. Pharmacol. 51, 499–506.

Rahman, S., Zhang, Z., Papke, R.L., Crooks, P.A., Dwoskin, L.P., Bardo, M.T., 2008. Region-specific effects of N, N'-dodecane-1,12-diyl-bis-3-picolinium dibromide on nicotine-induced increase in extracellular dopamine in vivo. Br. J. Pharmacol. 153, 792–804.

Raybuck, J.D., Portugal, G.S., Lerman, C., Gould, T.J., 2008. Varenicline ameliorates nicotine withdrawal-induced learning deficits in C57BL/6 mice. Behav. Neurosci. 122, 1166–1171.

Reichenberg, A., Weiser, M., Caspi, A., Knobler, H.Y., Lubin, G., Harvey, P.D., Rabinowitz, J., Davidson, M., 2006. Premorbid intellectual functioning and risk of schizophrenia and spectrum disorders. J. Clin. Exp. Neuropsychol. 28, 193–207.

Rollema, H., Chambers, L.K., Coe, J.W., Glowa, J., Hurst, R.S., Lebel, L.A., Lu, Y., Mansbach, R.S., Mather, R.J., Rovetti, C.C., Sands, S.B., Schaeffer, E., Schulz, D.W., Tingley 3rd, F.D., Williams, K.E., 2007. Pharmacological profile of the alpha4beta2 nicotinic acetylcholine receptor partial agonist varenicline, an effective smoking cessation aid. Neuropharmacology 52, 985–994.

Rose, J.E., Behm, F.M., Westman, E.C., Coleman, R.E., 1999. Arterial nicotine kinetics during cigarette smoking and intravenous nicotine administration: implications for addiction. Drug Alcohol Depend. 56, 99–107.

Rose, J.E., Behm, F.M., Westman, E.C., Mathew, R.J., London, E.D., Hawk, T.C., Turkington, T.G., Coleman, R.E., 2003. PET studies of the influences of nicotine on neural systems in cigarette smokers. Am. J. Psychiatry 160, 323–333.

Russell, M.A.H., Peto, J., Patel, U.A., 1974. Classification of smoking by factorial structure of motives. J. R. Stat. Soc. 137, 313–346.

Saccone, S.F., Hinrichs, A.L., Saccone, N.L., Chase, G.A., Konvicka, K., Madden, P.A., Breslau, N., Johnson, E.O., Hatsukami, D., Pomerleau, O., Swan, G.E., Goate, A.M., Rutter, J., Bertelsen, S., Fox, L., Fugman,

D., Martin, N.G., Montgomery, G.W., Wang, J.C., Ballinger, D.G., Rice, J.P., Bierut, L.J., 2007. Cholinergic nicotinic receptor genes implicated in a nicotine dependence association study targeting 348 candidate genes with 3713 SNPs. Hum. Mol. Genet. 16, 36–49.

Sarter, M., Parikh, V., Howe, W.M., 2009. nAChR agonist-induced cognition enhancement: integration of cognitive and neuronal mechanisms. Biochem. Pharmacol. 78, 658–667.

Sealfon, S.C., Olanow, C.W., 2000. Dopamine receptors: from structure to behavior. Trends Neurosci. 23, S34–S40.

Seeman, P., Kapur, S., 2000. Schizophrenia: more dopamine, more D2 receptors. Proc. Natl. Acad. Sci. U.S.A. 97, 7673–7675.

Seguela, P., Wadiche, J., Dineley-Miller, K., Dani, J.A., Patrick, J.W., 1993. Molecular cloning, functional properties, and distribution of rat brain alpha 7: a nicotinic cation channel highly permeable to calcium. J. Neurosci. 13, 596–604.

Sengupta, S., Grizenko, N., Schmitz, N., Schwartz, G., Bellingham, J., Polotskaia, A., Stepanian, M.T., Goto, Y., Grace, A.A., Joober, R., 2008. COMT Val108/158Met polymorphism and the modulation of task-oriented behavior in children with ADHD. Neuropsychopharmacology 33, 3069–3077.

Sharma, A., Brody, A.L., 2009. In vivo brain imaging of human exposure to nicotine and tobacco. Handb. Exp. Pharmacol. 192, 145–171.

Sokoloff, P., Schwartz, J.C., 1995. Novel dopamine receptors half a decade later. Trends Pharmacol. Sci. 16, 270–275.

Spurden, D.P., Court, J.A., Lloyd, S., Oakley, A., Perry, R., Pearson, C., Pullen, R.G., Perry, E.K., 1997. Nicotinic receptor distribution in the human thalamus: autoradiographical localization of [^{3}H]nicotine and [^{125}I] alpha-bungarotoxin binding. J. Chem. Neuroanat. 13, 105–113.

Stein, E.A., Pankiewicz, J., Harsch, H.H., Cho, J.K., Fuller, S.A., Hoffmann, R.G., Hawkins, M., Rao, S.M., Bandettini, P.A., Bloom, A.S., 1998. Nicotine-induced limbic cortical activation in the human brain: a functional MRI study. Am. J. Psychiatry 155, 1009–1015.

Swan, G.E., Lessov-Schlaggar, C.N., 2007. The effects of tobacco smoke and nicotine on cognition and the brain. Neuropsychol. Rev. 17, 259–273.

Taiminen, T.J., Salokangas, R.K., Saarijarvi, S., Niemi, H., Lehto, H., Ahola, V., Syvalahti, E., 1998. Smoking and cognitive deficits in schizophrenia: a pilot study. Addict. Behav. 23, 263–266.

Tanila, H., Bjorklund, M., Riekkinen Jr., P., 1998. Cognitive changes in mice following moderate MPTP exposure. Brain Res. Bull. 45, 577–582.

Thiel, C.M., Fink, G.R., 2008. Effects of the cholinergic agonist nicotine on reorienting of visual spatial attention and top-down attentional control. Neuroscience 152, 381–390.

Trimmel, M., Wittberger, S., 2004. Effects of transdermally administered nicotine on aspects of attention, task load, and mood in women and men. Pharmacol. Biochem. Behav. 78, 639–645.

van Holstein, M., Aarts, E., van der Schaaf, M.E., Geurts, D.E., Verkes, R.J., Franke, B., van Schouwenburg, M.R., Cools, R., 2011. Human cognitive flexibility depends on dopamine D2 receptor signaling. Psychopharmacology 218, 567–578.

Van Tol, H.H., Bunzow, J.R., Guan, H.C., Sunahara, R.K., Seeman, P., Niznik, H.B., Civelli, O., 1991. Cloning of the gene for a human dopamine D4 receptor with high affinity for the antipsychotic clozapine. Nature 350, 610–614.

Vossel, S., Thiel, C.M., Fink, G.R., 2008. Behavioral and neural effects of nicotine on visuospatial attentional reorienting in non-smoking subjects. Neuropsychopharmacology 33, 731–738.

Wagner, M., Schulze-Rauschenbach, S., Petrovsky, N., Brinkmeyer, J., von der Goltz, C., Grunder, G., Spreckelmeyer, K.N., Wienker, T., Diaz-Lacava, A., Mobascher, A., Dahmen, N., Clepce, M., Thuerauf, N., Kiefer, F., de Millas, J.W., Gallinat, J., Winterer, G., 2012. Neurocognitive impairments in non-deprived smokers-results from a population-based multi-center study on smoking-related behavior. Addict. Biol. http://dx.doi.org/10.1111/j.1369–1600.2011.00429.x/.

Wallace, T.L., Porter, R.H., 2011. Targeting the nicotinic alpha7 acetylcholine receptor to enhance cognition in disease. Biochem. Pharmacol. 82, 891–903.

Waters, A.J., Carter, B.L., Robinson, J.D., Wetter, D.W., Lam, C.Y., Cinciripini, P.M., 2007. Implicit attitudes to smoking are associated with craving and dependence. Drug Alcohol Depend. 91, 178–186.

Waters, A.J., Carter, B.L., Robinson, J.D., Wetter, D.W., Lam, C.Y., Kerst, W., Cinciripini, P.M., 2009. Attentional bias is associated with incentive-related physiological and subjective measures. Exp. Clin. Psychopharmacol. 17, 247–257.

Waters, A.J., Sayette, M.A., 2006. Implicit cognition and tobacco addiction. In: Wiers, R.W., Stacy, A.W. (Eds.), Handbook on Implicit Cognition and Addiction. Sage, Thousand Oaks, CA, pp. 309–338.

Waters, A.J., Shiffman, S., Sayette, M.A., Paty, J.A., Gwaltney, C.J., Balabanis, M.H., 2003. Attentional bias predicts outcome in smoking cessation. Health Psychol. 22, 378–387.

Waters, A.J., Sutton, S.R., 2000. Direct and indirect effects of nicotine/smoking on cognition in humans. Addict. Behav. 25, 29–43.

Wesnes, K., Warburton, D.M., 1983. Effects of smoking on rapid information-processing performance. Neuropsychobiology 9, 223–229.

Wiers, R.W.H.J., Stacy, A.W., 2006. Handbook of Implicit Cognition and Addiction. Sage Publications, Thousand Oaks.

Winterer, G., Mittelstrass, K., Giegling, I., Lamina, C., Fehr, C., Brenner, H., Breitling, L.P., Nitz, B., Raum, E., Muller, H., Gallinat, J., Gal, A., Heim, K., Prokisch, H., Meitinger, T., Hartmann, A.M., Moller, H.J., Gieger, C., Wichmann, H.E., Illig, T., Dahmen, N., Rujescu, D., 2010. Risk gene variants for nicotine dependence in the CHRNA5-CHRNA3-CHRNB4 cluster are associated with cognitive performance. Am. J. Med. Genet. B Neuropsychiatr. Genet. 153B, 1448–1458.

Wooltorton, J.R., Pidoplichko, V.I., Broide, R.S., Dani, J.A., 2003. Differential desensitization and distribution of nicotinic acetylcholine receptor subtypes in midbrain dopamine areas. J. Neurosci. 23, 3176–3185.

Xu, J., Mendrek, A., Cohen, M.S., Monterosso, J., Rodriguez, P., Simon, S.L., Brody, A., Jarvik, M., Domier, C.P., Olmstead, R., Ernst, M., London, E.D., 2005. Brain activity in cigarette smokers performing a working memory task: effect of smoking abstinence. Biol. Psychiatry 58, 143–150.

Yakir, A., Rigbi, A., Kanyas, K., Pollak, Y., Kahana, G., Karni, O., Eitan, R., Kertzman, S., Lerer, B., 2007. Why do young women smoke? III. Attention and impulsivity as neurocognitive predisposing factors. Eur. Neuropsychopharmacol. 7 (5), 339–351.

Young, J.W., Finlayson, K., Spratt, C., Marston, H.M., Crawford, N., Kelly, J.S., Sharkey, J., 2004. Nicotine improves sustained attention in mice: evidence for involvement of the alpha7 nicotinic acetylcholine receptor. Neuropsychopharmacology 29, 891–900.

Zhang, L., Dong, Y., Doyon, W.M., Dani, J.A., 2012. Withdrawal from chronic nicotine exposure alters dopamine signaling dynamics in the nucleus accumbens. Biol. Psychiatry 71, 184–191.

Zhu, H., Clemens, S., Sawchuk, M., Hochman, S., 2008. Unaltered D1, D2, D4, and D5 dopamine receptor mRNA expression and distribution in the spinal cord of the D3 receptor knockout mouse. J. Comp. Physiol. A Neuroethol. Sens. Neural Behav. Physiol. 194, 957–962.

CHAPTER THIRTEEN

Effects of Cannabis and Cannabinoids in the Human Nervous System

Harold Kalant[1,2]
[1]Department of Pharmacology & Toxicology, University of Toronto, ON, Canada
[2]Centre for Addiction and Mental Health, Toronto, ON, Canada

1. INTRODUCTION

1.1. Definitions and Concepts

The term *cannabis*, when used without any further qualifying term, is generally understood to mean a crude preparation of the dried upper leaves and flowering tops of the plant *Cannabis sativa* or *Cannabis indica*. Such preparations, known by a variety of names such as marijuana, bhang, maconha in different countries and languages, have been used virtually worldwide for centuries or possibly millennia, for medicinal, ceremonial, and recreational purposes (Kalant, 1972; Russo, 2007). All of these uses are based upon the pharmacological actions of a group of substances called *cannabinoids*, of which the most important is Δ^9-tetrahydrocannabinol (THC), that are also responsible for most of the undesired or adverse effects of cannabis, regardless of the intended purpose for which it is used.

The definition of a cannabinoid was originally a chemical designation applied to a group of constituents of the cannabis plant that shared a common molecular structure thought to be characteristic of the cannabis plant, or were derived from such a structure (Joyce and Curry, 1970; Mechoulam, 1970). More recent evidence suggests that similar chemical structures may also be found in other plants than cannabis (Gertsch et al., 2010). It was formerly thought that the biological activity of cannabinoids depended upon their lipophilicity, which would enable them to exert a nonspecific action in the lipid phase of cell membranes, comparable to that of general anesthetics (Thomas et al., 1990).

However, in the late 1980s, two membrane-bound receptors were identified in mammalian tissues, through which the pharmacologically active cannabinoids exerted their actions, and which were therefore named *CB1 and CB2 cannabinoid receptors* (Devane et al., 1988; Bidaut-Russell et al., 1990; Svizenska et al., 2008). Endogenous ligands for those receptors were found shortly afterward, that were named *endocannabinoids*, and that do not share the same ring structure as the *phytocannabinoids* isolated from the cannabis plant. The two best-studied endocannabinoids are arachidonoyl ethanolamine (AEA, anandamide) and 2-arachidonylglycerol (2-AG), although many analogous N-acylethanolamides, including the palmitoyl, oleoyl, and stearoyl analogs (PEA, OEA, and SEA respectively), and arachidonoyldopamine are present in even larger amounts in the brain (Hansen, 2010;

The Effects of Drug Abuse on the Human Nervous System
http://dx.doi.org/10.1016/B978-0-12-418679-8.00013-7

De Petrocellis and Di Marzo, 2009). In addition, virodhamine, an O-acyl analog of AEA in which the ethanolamine hydroxyl group is joined to the arachidonic carbonyl group in an ester linkage, was identified in rat brain and a number of peripheral tissues (Porter et al., 2002; Ottria et al., 2012). These endocannabinoids are synthesized enzymatically by specific lipases from lipoproteins in the cell membranes, act locally upon the receptors, and are then degraded by other enzymes in the immediate cellular vicinity. The whole complex of synthesizing enzymes, endocannabinoids, endocannabinoid receptors, and degradative enzymes is referred to as the *endocannabinoid system*. Finally, many entirely synthetic molecules have now been produced that are capable of binding to the cannabinoid receptors, or to the enzymes that produce or degrade the endocannabinoids.

As a result of these discoveries, the term *cannabinoid* is now defined functionally rather than chemically. It is applied to any molecule—phytocannabinoid, endocannabinoid, or synthetic analog—that can interact with components of the endocannabinoid system, whether as a receptor agonist, partial agonist or antagonist, or as an enzyme substrate or inhibitor. The term will be used in the latter sense in this review.

Although the actions of phytocannabinoids are exerted through the endocannabinoid system, the pattern of effects of cannabis as usually used, though similar to that of the endocannabinoids, is not identical and in some particulars even appears to be opposite. The possible reasons for the differences will be discussed in the course of this review. The shared basic mechanisms of action, as well as the differences, also underlie the actual and proposed therapeutic uses of cannabis and cannabinoids (Pertwee, 2009, 2012a), and the adverse effects that are commonly encountered in both medical and nonmedical use.

1.2 Structure and Orientation of the Chapter

The chapter begins with a brief overview of the endocannabinoid system and its roles in different functions of the nervous system. In relation to each of these functions, differences between the effects of the endocannabinoids and of exogenous phytocannabinoids will be indicated where knowledge permits. A proposed explanation for the differences will then be offered. The principal part of the chapter will be devoted to a review of the therapeutic and adverse effects of cannabis and pure cannabinoids in the human nervous system that will not include therapeutic recommendations or advocacy of specific policies.

2.1. ACTIONS AND EFFECTS OF ENDOCANNABINOIDS AND CANNABIS

2.1.1. Location and Functions of the Endocannabinoid System

Some of the components of the endocannabinoid system appeared early in the course of evolution, being found in a number of unrelated invertebrates (McPartland, 2004), but the presence together of all the components of the functional system may be limited to vertebrates (Elphick and Egertova, 2001).

Within the mammalian nervous system, CB1 receptors are very widely distributed, including most prominently the cerebellar cortex, hippocampus, basal ganglia, hypothalamus, amygdala, cerebral cortex, and spinal cord pain pathways (Pertwee, 2001; Egerton et al., 2006; McPartland et al., 2007). CB2 receptors are found in peripheral sensory neurons and some brainstem neurons, as well as in microglia, which are derived from macrophages (Kaur et al., 2001) but can influence the activity of adjacent neurons (Malan et al., 2002).

By far, the greatest part of the research on the endocannabinoid system has been done in rodents, but the regional and cellular distribution of CB1 receptors and of fatty acid amide hydrolase (FAAH), the enzyme which degrades anandamide, in human brain is in good general agreement with those in rat and monkey brain (Romero et al., 2002; Ong and Mackie, 1999). Therefore, it is a reasonable assumption that the endocannabinoid system in the human brain functions in much the same way as in the brains of rodents and other mammals.

More recently, other receptors that also bind endocannabinoids and mediate some of their physiological effects have been identified in mammalian brain and other tissues. A G-protein-linked enzyme designated as GPR55, and previously considered an "orphan receptor" found in substantial amounts in various parts of the brain, adrenals, small intestine and spleen, has been identified as a cannabinoid receptor (Pertwee, 2007; Rydberg et al., 2007). AEA, PEA, OEA and virodhamine bind to it as agonists at nanomolar concentrations, and even some phytocannabinoid and synthetic cannabinoid CB1 agonists bind more strongly to the GPR55 as agonists than they do to the CB1 receptor (Pertwee, 2007; Rydberg et al., 2007; Kreitzer and Stella, 2009). Perhaps even more importantly, cannabidiol (CBD) was found to bind strongly as an inhibitor to GPR55 (Rydberg et al., 2007). Another G protein-linked "orphan" receptor, designated GPR18, also appears to function as a cannabinoid receptor for the CBD analog o-1602 (Ashton, 2012; Console-Bram et al., 2012; McHugh, 2012), although the evidence is not uniformly consistent with this view (Lu et al., 2013).

Anandamide and N-arachidonoyldopamine also act as agonists at a transient receptor potential vanilloid-type one channel known as TRPV1 or vanilloid receptor 1 (De Petrocellis and Di Marzo, 2009). This is a calcium channel that is distributed very widely throughout the body, including sensory neurons, brain, epithelium, muscle, liver, bone, blood vessels, lymphocytes, and other cell types. In many of these cell types, the TRPV1 channels and CB1 receptors are found together, and interact in the regulation of cytosolic Ca^2 concentration.

The constituents of the endocannabinoid system have been shown repeatedly to be localized in synapses, in which the CB receptors are located in the presynaptic terminal. Impulse activation of the presynaptic terminal, with release of its neurotransmitter, activates the postsynaptic neuron, and Ca^{2+} influx together with membrane depolarization stimulate the production of endocannabinoids by postsynaptic enzyme action on membrane phospholipoproteins (Wang and Ueda, 2009). Anandamide and 2-AG are released into the synaptic cleft and diffuse back to the presynaptic CB receptors. Combination with the receptors produces an inhibitory effect on the further presynaptic release of neurotransmitter. This occurs only in presynaptic terminals that have been activated; application of endocannabinoid to a resting terminal does not affect its subsequent release of neurotransmitter (Heifets and Castillo, 2009). The remaining free endocannabinoid is carried into the presynaptic terminal by a specific transporter, where anandamide is degraded by an FAAH and 2-AG by a monoacylglycerol lipase (Bisogno et al., 2001; Kano et al., 2009; Saturnino et al., 2010). The wide distribution of FAAH in the human central nervous system is similar to that of the cannabinoid receptors (Romero et al., 2002).

There is some evidence that cannabinoids such as CBD, which have biological activity but little or no affinity for either CB1 or CB2 receptors (though it is now known to act through the GPR55 and GPR18 receptors), as well as some of the newer synthetic analogs, may act on the enzymes that produce or degrade the endocannabinoids and thus alter the amounts available to act on the CB1 or CB2 receptors (Bisogno et al., 2001; Massi et al., 2008; Saturnino et al., 2010). However, Massi et al. reported that CBD stimulated FAAH activity and reduced anandamide content in glioma tumor tissue, whereas Watanabe et al. and Bisogno et al. reported that CBD inhibited FAAH activity in mouse brain microsomes and neuroblastoma cell membranes; the CBD concentrations used in these studies were fairly similar, but the tumor cell lines were different. THC and cannabinol, which do act on CB receptors, were also reported to inhibit FAAH (Watanabe et al., 1996), thus raising the possibility that the more active cannabinoids might act directly on their receptors as well as indirectly by altering the levels of free endocannabinoids.

The common functional effect of endocannabinoid action at most synapses is inhibition of release of neurotransmitter from the presynaptic terminal. This has been found in a variety of neurons that release different transmitters, some excitatory and others inhibitory. Among the transmitters found to be affected in the same manner are glutamate, gamma-aminobutyrate (GABA), glycine, acetylcholine, norepinephrine, dopamine, and serotonin (Pazos et al., 2005; Egerton et al., 2006; Katona et al., 2006; Moreira and Lutz, 2008; Kano et al., 2009). Therefore, the endocannabinoids function as modulators of other modulators, and the net effect upon any specific neuronal pathway may depend upon the balance of other modulatory inputs to that pathway that can be altered by the endocannabinoids. If 2-AG, for example, is acting primarily to inhibit release of glutamate, the result will be a depression of activity of the glutamate-stimulated neuron, whereas if it is acting primarily on GABA release, the net effect will be a disinhibition of the GABA-inhibited neuron (Freund et al., 2003; Moreira and Lutz, 2008). The process differs in various respects, however, in excitatory and inhibitory synapses (Heifets and Castillo, 2009). For example, anandamide is involved in synaptic plasticity of corticostriatal dopaminergic neurons, while 2-AG is involved in hippocampal glutaminergic neurons. The time requirement for co-occurrence of presynaptic activation and endocannabinoid receptor occupation to produce LTD varies from milliseconds in somatosensory cortex to several minutes in hippocampus and dorsal striatum. Regardless of these and other mechanistic differences, the general role of endocannabinoids is essentially similar in synaptic plasticity changes in almost all parts of the brain.

These endocannabinoid effects on excitatory and inhibitory synaptic transmission constitute an important form of synaptic plasticity, defined as the ability of a synapse to exhibit adaptive modification of the postsynaptic response as a result of changes in presynaptic impulse frequency and in external stimuli (Alger, 2009). Synaptic plasticity is involved in such widely different functions as associative learning, control of motor function, tolerance, and sensory adaptation. Such adaptive changes can be of short or long duration. Depolarization-induced suppression of inhibitory transmission (DSI) and depolarization-induced suppression of excitatory transmission (DSE) are very rapid and short-lasting forms (Diana and Marty, 2004; Sheinin et al., 2008) that are induced by very brief presynaptic impulses, and appear to be mediated entirely by the retrograde endocannabinoid signaling within the synapse. Repeated or tetanic presynaptic neuronal excitation, on the other hand, gives rise to long-term potentiation (LTP) or long-term depression (LTD) that probably involves additional mechanisms such as altered expression or activation of the enzymes that produce or degrade the endocannabinoids, and altered responsiveness of the CB receptors (Zhu, 2006; Heifets and Castillo, 2009).

Another potential source of endocannabinoid response variation is the existence of extrasynaptic receptors. GABA activation of $GABA_A$ receptors located intrasynaptically gives rise to short-lasting phasic inhibition, whereas action on extrasynaptic receptors gives rise to tonic inhibition (Farrant and Nusser, 2005). Sigel et al. (2011) have reported a direct effect of 2-AG upon certain extrasynaptic GABA receptors that results in potentiation of the inhibitory response to GABA. However, the concentration of 2-AG required for this effect was in the low micromolar range, and a similar concentration of THC did not produce GABA potentiation. Mice lacking CB1 and CB2 receptors showed hypomotility when injected with 2-AG (10 mg/kg intra-venous), but mice lacking these extrasynaptic-type GABA receptors showed hypermotility. These 2-AG concentrations and doses are very high in comparison with concentrations and dosages of THC used by humans. It is therefore not yet clear whether direct action of endocannabinoids on GABA receptors plays a role in the effects experienced by humans.

An additional level of modulation of endocannabinoid action is provided by the presence of allosteric binding sites on the cannabinoid receptors (Console-Bram et al., 2012). These are sites at which molecules other than endocannabinoids may bind and, in so doing, may produce conformational changes in the receptor that alter the binding affinity and efficacy of the principal or orthosteric binding site for the endocannabinoids. Allosteric modulators may be purely synthetic (Price et al., 2005) or endogenous molecules (Bauer et al., 2012; Pamplona et al., 2012; Pertwee, 2012b). Conversely, classical endocannabinoids can act as allosteric modulators of receptors for other transmitters or modulators. For example, 2-AG has been reported to be a negative allosteric modulator of the A3 adenosine receptor (Lane et al., 2010). The effects of allosteric modulators on the cannabinoid receptors are specific for both the receptor type and the intracellular signaling cascade through which the actions of the receptor are mediated (Anavi-Goffer et al., 2012). Therefore, they raise the possibility that this approach may yield highly selective agents for therapeutic modulation of very specific effects of cannabinoid therapy while minimizing the risk of unwanted side effects.

Three further points are important for any comparison of the endocannabinoid system with the effects of cannabis and phytocannabinoids in humans. The first is the fact that release of endocannabinoids is an "on-demand" process that occurs when the postsynaptic neuron is activated by the binding of the neurotransmitter with its postsynaptic receptor. The resulting depolarization, together with Ca^{2+} entry, mainly triggers the rapid synthesis and release of newly formed endocannabinoid, although there is also some evidence that "on-demand" release of preformed endocannabinoid from some unidentified pool may occur (Basavarajappa, 2007; Bisogno and Di Marzo, 2007; Min et al., 2010; Alger and Kim, 2011). In either case, this type of "on-demand" release is rapid, short-lasting, and optimally suited for immediate response to temporary environmental and physiological stimuli (Moreira and Lutz, 2008). Long-lasting changes in endocannabinoid tonic function that are involved in such processes as learning and memory appear to be mediated by changes in receptor response and in expression and activity of endocannabinoid synthesizing and degrading enzymes (Alger and Kim, 2011).

The second, and related, point is that rapid on-demand synthesis and release implies that the amounts of endocannabinoid involved in normal regulation of synaptic function are relatively small. If the CB receptors are exposed to much larger amounts over longer periods, a state that has been referred to as "overload" (Lichtman et al., 2010), it results in desensitization and downregulation of CB1 receptors. Acute inhibition of monoacylglycerol lipase greatly increased the brain levels of 2-AG and resulted in exaggerated CB1-induced behavioral and physiological effects, including analgesia, hypothermia, and hypomotility (Kinsey et al., 2009). In contrast, genetic knockout of the enzyme in mice produced chronically elevated levels of 2-AG but unchanged or decreased behavioral responses, because of CB1 desensitization (Chanda et al., 2010). Since cannabis and its contained THC produce longer-lasting effects than the very short effects of on-demand endocannabinoid release, it has been suggested that they are more likely to result in overload phenomena that may be related to some of their adverse effects. For example, Sarne et al. (2011) suggested that dose-related overload may convert neuroprotective effects of THC into neurotoxic effects. Similarly, the facilitation of extinction learning by the CB1 agonist WIN 55,212-22, involving LTP of synaptic transmission in the prefrontal cortex, hippocampus, and amygdala, changes to an impairment of extinction learning when the same agonist is given chronically (Kaplan and Moore, 2011). This point will be discussed further below.

The third point is that the vast bulk of research literature on cannabinoids and synaptic plasticity deals with the endocannabinoids. There is remarkably little literature on the effects of cannabis, phytocannabinoids and other exogenous CB ligands in relation to synaptic plasticity. Liu et al. (2010) examined LTP in glutamate synapses on dopaminergic neurons in the rat brain ventral tegmental area (VTA). A single exposure to THC or to HU210, a synthetic CB1 agonist, did not produce LTD in those synapses, but after five daily injections there was a transient depression of the excitatory neurotransmission. Robbe et al. (2003) similarly studied LTD in tetanically stimulated glutamate synapses of afferent fibers from the prefrontal cortex to the nucleus accumbens of the mouse, as well as in GABA synapses. Both the excitatory and the inhibitory synapses showed LTD development that was mediated by endocannabinoids. However, application of an exogenous CB1 agonist, WIN-51,221, markedly decreased excitatory transmission, but also blocked the development of endocannabinoid-mediated LTD.

Koch et al. (2009) studied the effect of Sativex®, a standardized cannabis extract containing equal concentrations of THC and CBD, on LTP and LTD as represented by sustained increase or decrease, respectively, in motor cortex evoked potentials in human subjects. Transcranial theta burst stimulation (TBS) was used to produce the LTP and LTD: continuous TBS produced LTD, whereas intermittent TBS produced LTP, in the absence of any cannabinoid treatment. After four weeks of daily Sativex treatment, the LTP phenomenon was unaltered, but the LTD was replaced by increase in the amplitude of the evoked potentials, i.e. by an apparent LTP. The authors referred to this as a "shift in the polarity of synaptic plasticity", but correctly pointed out that it was not certain that the change was caused by the drug treatment, or that the changes in evoked potential amplitude produced by TBS really are the same phenomenon as endocannabinoid-dependent synaptic plasticity. It should be noted that once again the effect of exogenous cannabinoids was seen only after chronic treatment.

2.1.2. Neuronal Proliferation and Maturation

Synaptic plasticity, though an important mechanism of adaptation in the nervous system, is not the only one. Important long-term adaptations may involve the production of new neurons from neural stem cells that are found in the ventricular ependyma and subependyma and in the hippocampus of the adult brain (Morshead and Van der Kooy, 2001; Wolf and Ullrich, 2008; Oudin et al., 2011). Proliferation, differentiation, and migration of these cells to their ultimate locations in other sites in the brain are believed to play an important role in the formation of new memory circuits and in repair of neuronal damage. Endocannabinoids have been shown to initiate and stimulate proliferation and migration of neural stem cells (Molina-Holgado et al., 2007; Rubio-Araiz et al., 2008; Wolf and Ullrich, 2008; Galve-Roperh et al., 2009; Oudin et al., 2011). In contrast, anandamide was found to inhibit the differentiation of the stem cells into mature neuronal phenotypes (Rueda et al., 2002), but differentiation did occur when the diacylglycerol lipases that synthesize

2-AG were sharply downregulated (Walker et al., 2010). Thus, endocannabinoids stimulate proliferation of neural stem cells and their migration to end locations, but differentiation does not occur until a different controlling factor, possibly glutamatergic, takes over.

No comparable studies of the effects of exogenous cannabinoids on neural stem cell proliferation, migration, or differentiation could be found. However, there are numerous publications concerning the effects of cannabis or cannabinoids on other aspects of neural maturation in the fetus and the adolescent, both human and nonhuman. The effects of intrauterine exposure to cannabis or single cannabinoids on morphological and functional development of the rat brain from infancy to adulthood have been described in several comprehensive reviews (Dalterio, 1986; Campolongo et al., 2009a,b). Prenatal exposure gave rise to impairments in cognitive and locomotor functions and emotionality that were evident when the offspring reached adulthood.

For obvious reasons, it is not possible to study the effects of in utero exposure on brain morphology in humans in the same detail as is possible in laboratory animals, and alterations of brain development in the humans must to a considerable extent be inferred from functional alterations. Developmental studies in humans exposed to cannabis in utero have been in continuous progress for many years (Fried et al., 2003; Goldschmidt et al., 2000; Richardson et al., 2002; Smith et al., 2010). These studies have shown mild but persistent changes in memory, problem solving, attention, and impulsivity that are suggestive of maturational defects in the brain.

It has been known for many years that adolescent rats are more sensitive to brain developmental disturbances by exposure to cannabis than adult rats are. Prepubertal rats exposed to a limited period of cannabis administration, and then studied after prolonged abstinence, showed long-lasting deficits in learning and memory, as well as increased anxiety behaviors, that persisted into adulthood (Fehr et al., 1976; Stiglick and Kalant, 1982), whereas rats receiving the same exposure as young adults did not develop these alterations (Stiglick and Kalant, 1985; O'Shea et al., 2004). These findings raise the possibility that during brain maturation in adolescence, heavy exposure to cannabis might prevent the growth of axons and the establishment of large numbers of synaptic connections that normally accompany experience and learning. It is necessary to avoid untested generalizations, however. After the same drug exposure and testing schedule as in the circular maze experiments, rats that were exposed to cannabis as adults showed faster shuttle-box avoidance learning than placebo-treated controls (Stiglick and Kalant, 1985), but it is not possible to say whether this represented improved learning or hypersensitivity to footshock.

Possible support for this interpretation is provided by the results of MRI studies of the brains of late teen-aged males who had used cannabis heavily throughout adolescence. Wilson et al. (2000) reported smaller overall brain size and thinner cortex in early heavy users than in age-matched users who did not begin until after age 17. Lopez-Larson et al. (2011) similarly found significant cortical thinness in frontal cortex and insula, compared to matched nonusers, and the degree of thinning appeared to be greater, the earlier the age of onset of heavy use. In similarly matched groups of late teenagers, Ashtari et al. (2009) found altered frontal–temporal connections (arcuate tract) in early-onset users compared to nonusers; this tract was among areas that they had previously found to undergo substantial maturational change during adolescence. Hermann et al. (2007) compared a group of young adult cannabis users (mean age 20 ± 2 years) who had begun cannabis use at about 16 years of age with a matched group of nonusers, with the respective status confirmed by hair analysis for THC and CBD. Tests of attention, memory and short-term auditory memory were significantly impaired in the user group, and spectroscopic magnet resonance imaging of the brain revealed evidence of a reduced neuronal and axonal integrity (reduced ratio of N-acetylaspartate to total creatinine) in the prefrontal cortex of the users. Unfortunately, the age of first use of cannabis is low enough that it is not possible to be sure whether the observed effects were due to impaired maturation or to chronic toxicity that might occur at any age. A similar relation between age of onset of cannabis use and alteration of cerebral circulation has been observed.

It has been suggested by various authors (e.g. Galve-Roperh et al., 2009; Bossong and Niesink, 2010; Caballero and Tseng, 2012) that impaired brain maturation caused by cannabis use during early adolescence may constitute the neurological basis for later onset of schizophrenia. This point will be pursued below in relation to specific disturbances of neuronal function.

2.1.3. Neuroprotection

One of the widely studied effects of cannabinoids is their neuroprotective effect, i.e. the ability to prevent or decrease the severity of brain damage caused by a variety of circulatory, mechanical, metabolic, or toxic injuries. Both CB1 and CB2 receptors contribute to this protective effect. For example, CB1 receptor knock-out mice show increased severity, both anatomical and functional, of ischemic stroke than wild type littermates (Parmentier-Batteur et al., 2002). There is a very large body of recent literature, including numerous excellent reviews, concerning the nature and mechanisms of the neuroprotective actions of the endocannabinoid system in both human brain diseases and experimental animal models. Some reviews have concentrated on specific groups of brain diseases in which endocannabinoids have been reported to be potentially useful, including cerebrovascular occlusion and reperfusion (Zhang et al., 2009; Alonso-Alconada et al., 2010, 2011), traumatic brain damage (Shohami et al., 2011), neuroinflammatory diseases (Eljaschewitsch et al., 2006; Zhang et al., 2009; Kubajewska and Constantinescu, 2010) and neurodegenerative conditions such as Parkinson, Huntington and Alzheimer diseases (Palazuelos et al., 2009; Scotter et al., 2010; Fernandez-Ruiz et al., 2011). Other reviews have concentrated principally on the molecular mechanisms by which endocannabinoids can exert neuroprotective action, such as prevention of excitotoxicity (Marsicano et al., 2003), immunomodulation, (Bisogno and Di Marzo, 2010) and combinations of multiple mechanisms (Eljaschewitsch et al., 2006).

Despite their somewhat different orientations, these reviews give a very consistent picture of the neuroprotective action of endocannabinoids, and of the numerous elements of which it consists. These include, among others:

- inhibition of glutamatergic afferents, with resulting prevention of excitotoxicity
- inhibition of MAP kinase-induced and tumor necrosis factor α-induced inflammatory responses by microglia and macrophages to brain tissue injury that result in aggravation of the damage
- antioxidant action of phenolic groups against reactive oxygen species (ROS), and decreased formation of ROS by monoamine oxidase, cyclooxygenase, etc., under influence of nitric oxide, Ca^{2+} and other factors in injured brain tissue
- decreased neuronal activity, produced in part by hypothermia (Leker et al., 2003), resulting in decreased metabolic requirements, with consequent reduction of damage during hypoxia and glucose deficiency.

The changes have been described in brain tissue of humans dying of the various diseases mentioned, as well as of animals with experimental models of these diseases (Scotter et al., 2010). Endocannabinoid levels are increased in injured brain tissue, as well as in animals subjected to intense excitatory brain stimulation by kainic acid (Marsicano et al., 2003), possibly as a defensive reaction to the injury. Both CB1 and CB2 receptors affect residual damage in different ways. There is less information about the role of CB2 receptors in neuroprotection, and the existing information is somewhat confusing. The protective action of CB2 agonists appears to consist mainly of anti-inflammatory activity, whereas that of CB1 agonists is to a considerable extent due to hypothermia. In keeping with that interpretation, maximum protection against ischemia/reperfusion damage was reported to occur with a combination of CB2 agonist and CB1 antagonist (Zhang et al., 2008). A further problem with CB1 agonists is that at doses giving maximum neuroprotection they produce psychoactive effects that would be undesirable side effects in the presence of brain injury. Inhibition of FAAH has been proposed as a more selective manner of achieving the same protective action without the undesired psychoactivity (Hwang et al., 2010).

THC, CBD, HU-210, dexanabinol (HU-211), ajulemic acid (HU-239) and numerous other synthetic cannabinoids also exert neuroprotective activity similar to that of the endocannabinoids in experimental models. The protective effect of CBD is not exerted through either CB1 or CB2 receptors, but a $5-HT_{1A}$ receptor blocker prevented the protective effect of CBD on infract size and cerebral blood flow (Mishima et al., 2005). This raises the possibility, but does not prove, that CBD acts as a 5-HT receptor agonist. A further point of interest about the neuroprotective effect of CBD is that the intensity of effect at first increased with increasing dose, reached a maximum, and then decreased and finally disappeared with further increases of dose. This is reminiscent of the endocannabinoid "overload" described in Section 2.1 above.

The duration of neuroprotective activity with repeated administration of cannabinoids is a matter of potential importance with respect to the possibility of clinical use. In the mouse model of ischemic stroke produced by middle cerebral artery occlusion, daily administration of THC for 14 days resulted in desensitization and downregulation of CB1 receptors, and the beneficial effects of THC on body temperature, cerebral blood flow, and infarct size were markedly reduced; this did not happen with CBD (Hayakawa et al., 2007).

Crude cannabis preparations do not appear to have been tested for their neuroprotective effect. Moreover, despite the promising results with individual cannabinoids in preclinical studies, there is virtually no literature on clinical trials. One study with dexanabinol (Knoller et al., 2002), consisting of a Phase II trial in patients with severe closed head injury, showed promising results in that intracranial pressure was promptly reduced in the dexanabinol group but not the placebo group, the cerebral perfusion pressure was significantly better, and the 3- and 6-month total outcome was superior to that in the placebo group. However, no Phase III trial has so far been reported.

In contrast, there is also evidence of neurotoxic effects of exogenous cannabinoids such as THC and WIN55,212-2 when added to neuronal cultures or injected into the cerebral ventricles, or when administered together with ethanol (Fowler et al., 2010). The toxic effects were produced at higher cannabinoid concentrations than those required to activate CB receptors in their physiological functions. The question of possible neurotoxic effects of cannabis in heavy users will be taken up further below.

2.1.4. Sensory Pathways

2.1.4.1. Olfactory

Sensory detection of odors occurs via G-protein-linked receptors found in the nasal mucosa, as well as in other parts of the olfactory tract. CB1 receptors and TRPV1 receptors, as well as other components of the endocannabinoid system, are also found in a variety of locations in peripheral and central sensory pathways, including the olfactory pathway. Therefore, it seems likely that cannabinoids affect olfactory sensation (Lötsch et al, 2012). Indeed, Walter et al. (2011) did report that THC increased odor threshold for vanillin and decreased odor discrimination in healthy normosmic humans. However, effects on sensory impulse transmission are varied. In the olfactory pathway, for example, 2-AG has been found to influence odor detection differently in hungry vs sated *Xenopus* larvae. In the hungry larvae, the activity of 2-AG synthesizing enzymes is increased, and 2-AG levels in the sensory epithelium are increased. The result is a reduced threshold for olfactory stimulus detection, which is thought to help the larvae detect and locate food (Breunig et al., 2010). In contrast, $1\,\mu M$ THC inhibited excitatory evoked potentials in the rat olfactory cortex, whereas cannabis extract with the THC removed had a stimulatory effect (Whalley et al., 2004). However, neither THC (doses up to $10\,mg/kg$) nor the FAAH inhibitor URB-597 affected the acquisition or the performance of an olfactory discrimination task in intact rats (Sokolic et al., 2011). Both drugs did impair performance after reversal of the cues; this is interpreted as an impairment of learning rather than of the sensory modality itself.

2.1.4.2. Auditory

The endocannabinoid 2-AG appears to play a similar role in the avian and murine auditory systems as in the other neuronal pathways already discussed. Released postsynaptically, it acts on presynaptic CB1 receptors located primarily on activated glutamatergic terminals to reduce the release of glutamate (Penzo and Pena, 2009; Zhao et al., 2009; Sedlacek et al., 2011). Since the postsynaptic cells involved are inhibitory, the result of the endocannabinoid effect on them is disinhibition of the principal cells of the cochlear nucleus; the end result is not related to auditory acuity, but to perceptual localization of the source of a sound.

Auditory gating is the process by which irrelevant stimuli are filtered out, permitting attention to be focused on the relevant ones in a given situation. This can be studied in humans and in rodents by electroencephalographic recording of the evoked responses to a conditioned tone paired with a confounding tone. The ratio of the two responses is a measure of the efficacy of gating. The synthetic cannabinoid agonist WIN55,212-2 disrupted gating, and its effect was prevented by prior administration of the CB1 antagonist SR141716A (Dissanayake et al., 2008). Impairment of sensory gating has been found repeatedly in schizophrenic patients, and it has been suggested that this action of cannabinoids might be a mechanism contributing to the precipitation of schizophrenia (Hajós et al., 2008).

A possibly related phenomenon is the auditory evoked mismatch negativity, which is a negative wave in the EEG evoked by the unexpected introduction of a discordant auditory stimulus into a train of repeated stimuli. Impairment of mismatch negativity is interpreted as a deficit in attention to one's environmental stimulus input, and is also seen in schizophrenic subjects. In healthy human subjects, the mismatch negativity was impaired by rimonabant, a CB1 antagonist (Roser et al., 2011), which suggests that the endocannabinoid system normally contributes to the production of the negativity. In chronic heavy cannabis users, in contrast, the mismatch negativity recorded from frontal electrodes was significantly reduced in comparison to occasional light users and nonusers (Roser et al., 2010; Rentzsch et al., 2011). In contrast, among schizophrenic patients, those who used cannabis regularly had greater mismatch negativity than those who did not use (Rentzsch et al., 2011); this might be indicative of cannabis use as self-medication by some schizophrenic patients.

In contrast to the olfactory studies mentioned in Section 2.4.1, a similarly designed auditory discrimination task, once learned, was clearly impaired by THC and by URB-597, but unaffected by the CB1 antagonist rimonabant (Sokolic et al., 2011). THC and URB-597 also impaired performance of the auditory task after cue reversal. It is not clear whether the adverse effect of THC and of endocannabinoid on performance of the learned discrimination represents an effect on auditory perception or on choice of the appropriate response. The absence of an effect of rimonabant alone also suggests that the endocannabinoid system is not essential for maintenance of normal test performance, and therefore raises the possibility that the adverse effect of THC and URB-597 is again an "overload" phenomenon.

2.1.4.3. Pain

As noted earlier, the constituents of the endocannabinoid system are found in abundance along the pain pathways in the dorsal horns of the spinal cord, as well as in the thalamus, dorsal raphe nucleus, and the periaqueductal grey matter (Pertwee, 2001; Bushlin et al., 2010). In experimental models of neuropathic pain, osteoarthritis, endometriosis, and other causes of chronic pain, the concentrations of CB1 receptors as well as of endocannabinoids and their synthesizing enzymes in the spinal and supraspinal pain pathways are increased (Petrosino et al., 2007; Dmitrieva et al., 2010; Sagar et al., 2010). Administration of CB1 agonists, or inhibition of FAAH, reduced pain responses more in animals with chronic pain than in healthy controls (Sagar et al., 2010). CB2 receptors, which were formerly thought to exist essentially in peripheral nonneural tissues, are now known to be present also in peripheral sensory nerves and in parts of the central pain pathways (Anand et al., 2009). CB2-selective agonists have produced analgesia or reduction of hyperalgesia in models of neuropathic and inflammatory pain. There is extensive anatomical overlap of the opioid and cannabinoid receptor systems, and it appears probable that functional interactions between them occur in the production of analgesia.

The analgesic effect of THC and other exogenous cannabinoids, as well as of standardized cannabis extract containing both THC and CBD, has been well demonstrated in humans suffering from chronic neuropathic pain, cancer pain, and chronic musculoskeletal pain. A very thorough review of all clinical trials involving smoked cannabis, oral cannabinoids of CB1 and CB2 specificity, and modulators of endocannabinoid metabolism in the treatment of neuropathic pain concluded that all forms of cannabinoid therapy appear to provide some degree of relief of pain, though CB1 agonists produce significant side effects due to their psychoactivity (Rahn and Hohmann, 2009). Ware et al. (2010a,b) have also reported modest relief of chronic neuropathic pain by low-dose smoked cannabis, which also provided improved sleep and only minor side effects.

Cannabinoids have not shown satisfactory analgesic effect against acute pain such as postsurgical pain, and in this respect they are less versatile as analgesics than morphine and other strong opioids. However, there is some clinical support for the utility of combined therapy with cannabinoids and opioids in the treatment of chronic pain (Narang et al., 2008; Elikottil et al., 2009) as a way of obtaining superior analgesia with reduced side effects of each drug. The comparability of analgesic effects of cannabinoids in animal models of chronic pain and human patients gives no evidence for endocannabinoid overload in this action.

2.1.5. Nausea

Nausea and vomiting are initiated by the coordinated interaction of a number of sub-nuclei of the nucleus of the tractus solitarius in the medulla oblongata, where afferent information from brainstem chemoreceptors, vagal afferent fibers from the stomach and intestine, gustatory and vestibular inputs, and descending fibers from the cerebral cortex is integrated (Hornby, 2001). The dorsal vagal complex contains neurons with CB1 receptors, which are inhibitory and 5-HT_3 receptors, which are excitatory to the vagal afferent fibers that activate the integrated emetic process. Thus, it is a site at which CB1 agonists and 5-HT_3 antagonists can exert antiemetic and antinauseant effects.

An animal model of chemically induced nausea in the rat, which lacks the neural organization for vomiting, is a mouth-gaping response and posture that is produced by stimuli that evoke nausea in humans. This response, produced by cisplatin or lithium in the rat, is alleviated both by 5-HT_3 antagonists such as ondansetron and granisetron, and by CB1 agonists such as THC, anadamide and WIN55,212-2 (Parker and Limebeer, 2006). GPR55 receptors are also found in this region, but there is not yet sufficient evidence to permit a conclusion as to whether they also participate in control of the nausea response (Schicho and Storr, 2012). Both in this rat model and in humans, however, nausea and vomiting due to cancer chemotherapy or to HIV drugs is controlled by lower doses of ondansetron and related drugs than of THC, which in turn is better than the older phenothiazine antinauseants (Söderpalm et al., 2001; Machado Rocha et al., 2008). The effect of THC cannot be improved by raising the dose because the psychoactive side effects are too troubling to many patients, especially older patients who have not previously used cannabis for nonmedical purposes. However, the combination of lower doses of ondansetron and a CB1 agonist might give a superior effect with fewer side effects (Parker and Limebeer, 2006).

2.1.6. Appetite and Food Intake

Perhaps not surprisingly, regulation of appetite and food intake involves many of the same nervous system structures that regulate nausea and vomiting. Excellent reviews are available (Fride et al., 2005; Capasso and Izzo, 2008). As in the case of nausea, endocannabinoids participate in all the peripheral and central components that converge on the dorsal vagal nucleus. Vagal afferents from the gastrointestinal tract carry information indicating the presence of food, which tends to inhibit further food intake. Endocannabinoids inhibit these vagal fibers and thus tend to promote further eating. Centers in the hypothalamus similarly convey information to the hindbrain about the chemical composition of the blood as affected by the ingested food. The limbic system provides information about the sensory qualities of the food ("pleasing" or "displeasing"), and cerebral cortical inputs convey information about environmental context that may affect the act of consuming food. All of these separate inputs are integrated in the hindbrain dorsal vagal nucleus and the hypothalamus. The net effect of CB1 agonists is typically to promote food intake, and of CB1 antagonists to decrease or inhibit food intake.

There is a selective interaction with different dietary constituents. In sham-feeding experiments, in which the ingested food exits via a gastric cannula so that only its orosensory effects are in play, administration of a fatty diet stimulated endocannabinoid production in the rat small intestine, whereas carbohydrate or protein meals did not (DiPatrizio et al., 2011). Conversely, infusion of the CB1 antagonist rimonabant into the jejunum reduced the sham-ingestion of a fatty meal but not of normal chow. These effects were shown to involve only the local CB1 receptors on neurons of the enteric plexus. In both rats and humans, administration of THC, anandamide, or 2-AG causes hyperphagia and weight gain, which is blocked by rimonabant, and fasting increases the levels of anandamide and 2-AG in the Nucleus accumbens and hypothalamus as well as in the small intestine (Capasso and Izzo, 2008). Obese humans have presumably chronically increased activity of the intestinal endocannabinoid system, and administration of CB1 antagonists reduces food consumption and body weight (Bermudez-Silva et al., 2010). The availability of selective agents that act only in the periphery may make future therapeutic use possible in the treatment of obesity, without the dangerous depression and suicide risk that resulted in the termination of such use of rimonabant.

Fride et al. (2005) make a fascinating connection between the endocannabinoid actions in the newborn and in the adult. In the newborn, the binding of endocannabinoids to CB1 receptors promotes an early postnatal commencement of suckling. This is inhibited by CB1 antagonists such as rimonabant, if given within the first 24 h after birth. The only food the suckling response provides is maternal milk, which has a complete balance of the necessary nutritional factors for infant survival and growth. Thus, the "pleasing" gustatory quality of the food works in harmony with the intestinal appetite stimulus and the motor response of suckling to maximize the protective effect on the infant. In contrast, in the adult, the pleasing sensory qualities of experiencing the effects of the food cause the ingestion-stimulating effect of cannabis and CB1-linked cannabinoids to be directed toward high-sugar and high-fat snack foods. Thus, the weight gain produced by cannabis or THC in cachectic patients leads to weight gain associated with fat deposition in adipose tissue, rather than to protein synthesis and tissue regeneration, and superior effect has been reported with megestrol acetate than with dronabinol (Bruera, 1992; Tchekmedyian et al., 1992; Jatoi et al., 2002). However, these comparisons were carried out before the advent of newer more selective cannabinoid agents that might permit the use of higher dosage without the adverse psychoactive effects of CB1 agonists.

2.1.7. Sleep

Cannabis has a long history of use as a sedative and hypnotic medication. The earlier literature concerning the effects of cannabis, THC, and other exogenous cannabinoids on sleep patterns in the rat and human has been reviewed recently by Arias-Carrión et al. (2011). The major effects observed were increased slow-wave sleep and REM sleep and reduced waking time. These were later shown to be essentially the same as the effects of anandamide (Mechoulam et al., 1997; Murillo-Rodríguez et al., 1998) and opposite to those of the CB1 antagonist SR141716A (Santucci et al., 1996). Later research has shown the effects of activation and inhibition of the synthesizing and degrading enzymes and the membrane transporter of anandamide, as well as its downstream intracellular signaling mechanisms, to be fully consistent with the cannabinoid effects mentioned above.

There appears to be a reciprocal interaction between the endocannabinoid system and sleep (Chen and Bazan, 2005). Anandamide and 2-AG undergo reciprocal cyclic changes in concentration in rat brain during the light and dark phases of the diurnal cycle (Valenti et al., 2004). This pattern was disrupted by prolonged sleep deprivation, especially of REM sleep, which resulted in increased levels of 2-AG in the hippocampus but no change in anandamide. These changes were accompanied by disruptions of synaptic plasticity and hippocampal memory processes. Such changes in function as a result of sleep deprivation serve to confirm the role of the endocannabinoid system as a regulator of these functions in the normal state.

Other cannabinoids such as nabilone and Sativex® also improve the subjective qualities of sleep in patients with painful or other sleep-disturbing illnesses (Russo et al., 2007; Ware et al., 2010a,b). Tolerance to these effects on sleep is said not to occur with prolonged use of cannabis preparations; this is somewhat surprising, given the "overload" phenomena, receptor desensitization and response alterations seen with other effects of cannabinoids noted above, and deserves further study.

2.1.8. Affective Responses

Cannabinoid effects on affective responses have been studied most extensively in relation to depression and to anxiety and stress (Ashton and Moore, 2011). Numerous studies have indicated a mood-elevating and antidepressant activity of the endocannabinoid system. FAAH knockout mice, which have elevated brain levels of endocannabinoids, also had higher baseline serotonergic activity in the frontal cortex and showed less anxiety in the open-field test and greater social interaction than the normal controls (Cassano et al., 2011). The CB1 antagonist rimonabant abolished the difference in behavioral patterns, and also abolished the serotonin response to stimulation of the cortex and hippocampus, which appears to be an important locus of endocannabinoid modulation in affective responses (Haj-Dahmane and Shen, 2011). Inhibition of endocannabinoid synthesis in rats produces immobility in the forced swim test, which is regarded as depression-like behavior. In contrast, the CB1 agonist WIN55,212-2 exerts antidepressant-like behavior, which is abolished both by rimonabant and by serotonin depletion before the test (Bambico et al., 2007; Mangieri and Piomelli, 2007; Gorzalka and Hill, 2011). These changes are to some extent paralleled in humans by low circulating levels of endocannabinoids and reduced CB1 receptor activity in depressed patients, and the production of depression and risk of suicide in human patients receiving rimonabant for the treatment of obesity (Lazary et al., 2011).

In contrast to these antidepressant effects of the finely adjusted actions of the endo-cannabinoid system, high doses of cannabis in humans appear to increase the risk of depression, especially in the young (Ashton and Moore, 2011). Daily administration of WIN55,212-2 to adolescent rats for 20 days produced depression-like effects in behavioral tests and electrphysiological evidence of decreased serotonergic activity, whereas the same treatment did not produce such results in adult rats (Bambico et al., 2010). Population studies in several countries have yielded similar observations of increased risk of later onset of depression in adolescents who began regular use of cannabis before the age of 14–15 (Kalant, 2004), though the nature of the link is still not wholly clear (Degenhardt et al., 2003, 2007; Hall and Degenhardt, 2009). Despite this uncertainty, the fact that cannabis use did not exert an *anti*depressant action in these heavy users implies some difference from the endocannabinoid actions described above, such as receptor desensitization, for example.

The effects of cannabinoids on anxiety, fear, and stress-related affective reactions have also received much attention in the clinical and research literature. As is well known, the acute application of a stressor to a normal organism elicits sequential release of hypothalamic corticotrophin-releasing factor (CRF), pituitary adrenocorticotropic hormone (ACTH), adrenal corticosteroid that produces metabolic adaptations throughout the body, assisting in the response to stress. This sequential response is in part initiated by the endocannabinoid system, through a rapid increase in FAAH activity, thus reducing the amount of anandamide acting on glutamatergic terminals. The resulting disinhibition of glutamatergic stimulus greatly increases the tonic activity of the hypothalamic–pituitary axis (Hill et al., 2010) The released corticosteroid completes a feedback circuit by activating corticosteroid receptors both in the hypothalamus and in the medial prefrontal cortex (mPFC) that sharply reduce ACTH secretion when the stressor has been removed. The response to activation of these glucocorticoid receptors also involves the endocannabinoid system. Local application of a CB1 receptor antagonist to the rat mPFC delays the termination of ACTH secretion after the end of the stress, and mice lacking CB1 receptors also show a prolonged secretory response (Hill et al., 2011).

Activation of the endocannabinoid system also influences the affective components of the stress response, and stress-associated memory and conditioning, as well as initiating stress-associated analgesia (Finn, 2010). Stress-induced analgesia is mediated by activation of a descending inhibitory pathway that modulates transmission in the ascending pain pathway (Vaughan, 2006; Butler and Finn, 2009). This process involves endocannabinoid receptors on both glutamatergic and GABAergic synapses, at various levels in the brain and spinal cord, as well as interactions between the endocannabinoid and the endogenous opioid systems. It seems probable that stress-induced analgesia is generated at the same points as the cannabinoid-induced analgesia described in Section 2.4.3, and that stress is simply one of a number of factors that can activate the same pathways.

Other parts of the stress response include fear as an acute reaction (Marsicano et al., 2002), and anxiety and depression as components of chronic or repeated stress reactions. CB1 receptors are present in high concentration in the amygdala, hippocampus, and cortex, and many studies have observed altered emotionality as a result of activation or blockade of these receptors. Mutant mice lacking them, and rats subject to pharmacological blockade of them, show anxiety-like and depression-like patterns of behavior on a variety of tests, together with reduced basal secretion and reduced responsiveness of the adrenal cortex (Viveros et al., 2005). CBD, though not a CB1 agonist, can affect anxiety through other sites of action on the endocannabinoid system. Crippa et al. (2004, 2011) attempted to localize the site of the anxiolytic action of CBD in the human brain by studying changes in regional cerebral blood flow in normal subjects and in patients with generalized social anxiety disorder. In a double-blind placebo-controlled cross-over study, CBD produced a reduction in anticipatory anxiety in the healthy subjects and in subjective anxiety ratings in the patients (relative to placebo), and also reduced blood flow in the hippocampus, left parahippocampal gyrus and inferior temporal gyrus. However, there was no correlation between the anxiety ratings and the magnitude of changes in blood flow.

Localization at such a macrostructural level is difficult to interpret when one considers the functional differences of endocannabinoid activity in substructures only millimeters apart. For example, the acquisition of conditioned fear responses in rats and mice depends on the level of endocannabinoid signaling in different parts of the amygdala: acute blockade of CB1 receptors in the central amygdala greatly increased conditioned fear responses (Kamprath et al., 2011), while CB1 blockade in the basolateral amygdala prevented the formation of conditioned fear responses (Tan et al., 2011). These findings suggest that the effects of cannabinoids on fear-conditioned responses in the intact organism are highly variable, depending on the relative degrees of cannabinoid modulation in different parts of the brain under the influence of various external factors.

That such is indeed the case in both experimental animals and humans is amply documented in several excellent reviews (Viveros et al., 2005; Moreira and Lutz, 2008; Akirav, 2011). There is a clear and strong pattern of biphasic dose–response relations, with anxiolytic effects predominating at low doses of exogenous cannabinoid agonists or low levels of endocannabinoid activation, and anxiogenic effects predominating at high doses. Moreover, the effects at a given dose can vary greatly as a function of altered external circumstances or concurrent physiologic influences, combined action with other drugs, and of individual experience with cannabis or cannabinoids. An additional source of variability that has received very little attention until recently is sex difference in the responses of the endocannabinoid system, especially at times of sexual maturation (Krebs-Kraft et al., 2010; Viveros et al., 2011). This is an important topic that merits much more research than it has had in the past.

2.1.9. Seizure Susceptibility

In keeping with its inhibitory effect on excitatory impulse transmission, the endogenous cannabinoid system also exerts an inhibitory influence on seizure susceptibility. It has been known for many years that cannabis has antiseizure activity in humans (Kalant, 1972; Russo, 2007). In rat models of epilepsy, both THC and CBD exert antiseizure activity similar to that of phenytoin (Sofia et al., 1976; Corcoran et al., 1978; Consroe et al., 1982).

More recent research demonstrates the role of the endocannabinoid system in this action by various types of evidence. For example, the CB1 agonist WIN55,252-2 delayed the development of kindled seizures in rats, whereas the FAAH inhibitor URB597 failed to do so because it also inhibited neuronal proliferation (Wendt et al., 2011). Conversely, the CB1 receptor blocker SR141716 facilitated the reacquisition of audiogenic epilepsy in rats that had lost their seizures through repeated testing (Vinogradova et al., 2011). In rats subjected to a single period of pilocarpine-induced status epilepticus, the subsequent development of epilepsy took place over a period of about a month, but was preceded by a marked loss of hippocampal CB1 receptors (Falenski et al., 2009). The FAAH inhibitor URB597 protected mice against electroshock-induced seizures, and the effect was synergistic or additive with the effect of diazepam, depending on the dose ratios of the two drugs (Naderi et al., 2008). Another FAAH inhibitor, AM374, also increased anandamide levels in the rat hippocampus and protected against kainite-induced seizures as well as excitotoxic neuronal damage (Karanian et al., 2007). In the fetus and the newborn rat brain, glutamatergic excitatory activity is not yet highly developed and GABAergic synapses are excitatory rather than inhibitory (Bernard et al., 2005). In this situation, blockade of CB1 receptors also led to epileptic discharges, but through a different locus of action than in the mature brain.

One study provides apparently conflicting findings, in that the ketogenic diet prevented induction of pentylenetetrazole-induced seizures in mice without affecting either CB1 receptor expression or brain levels of endocannabinoids (Hansen et al., 2009). The levels of a number of other acylethanolamines, especially oleoylethanolamide, were actually reduced in the animals on ketogenic diet. However, these animals grew much more slowly than their normal diet controls, and their plasma contained more than twice as much betahydroxybutyrate, so that the mechanism of the antiepileptic effect of the ketogenic diet may be quite different from, and independent of, that of the endocannabinoid effect. This should not be surprising, since the endocannabinoid system is obviously not the only factor affecting neuronal excitability.

Unlike normal mice, mice totally lacking FAAH had greatly increased levels of endogenous anandamide in the hippocampus, and reacted to exogenous anandamide administration with a greatly increased susceptibility to kainate- or bicuculline-induced seizures (Clement et al., 2003). Perhaps analogously, prolonged exposure to WIN55,212-2 in a hippocampal culture model led to downregulation of CB1 receptors in both glutamatergic and GABAergic synapses, and the progressive appearance of low-Mg^{2+}-induced seizure discharges despite the continued presence of the WIN55,212-2 (Blair et al., 2009). Both of these examples appear to represent "overload" effects comparable to those noted above with other actions of the endocannabinoids.

However, another possible explanation of enhanced seizure susceptibility caused by either endocannabinoid overactivity or exogenous cannabinoid administration is based on the balance of glutamatergic and GABAergic activity in the animal model employed. If the model is one that involves primarily GABA-dependent control of neuronal excitability, and sufficiently high doses of cannabinoid are used that strongly inhibit GABAergic synaptic transmission, the result could be increased seizure susceptibility (Lutz, 2004).

Despite the abundance of experimental literature on the subject, there is almost nothing on clinical evaluation of cannabis or cannabinoids in the treatment of epilepsy in humans. There are a number of survey reports, such as that by concerning 136 patients seen in an epilepsy clinic, of whom 48% reported use of cannabis at some time and 21% were current users. Approximately two-thirds of the current users reported beneficial effects on seizure severity and half reported lower seizure frequency; the others reported no effect on their epilepsy. There are also very rare case reports (e.g. Mortati et al., 2007) of marked improvement after initiation of therapy with cannabis extract. However, there is only one double-blind placebo-controlled study of epileptic patients who received supplementary treatment with CBD after conventional medications failed to provide adequate control. Each patient remained on the previous medication but received in addition oral capsules of CBD or placebo in a cross-over design. The addition of CBD resulted in significant reduction of seizure frequency (Cunha et al., 1980). The lack of further studies is surprising, but the availability of a variety of endocannabinoid agents without unwanted psychoactive effects may encourage well-designed clinical trials.

2.1.10. Motor Control

The extensive and complex role of the endocannabinoid system in the functional control of the motor pathways has been thoroughly described in detail in several comprehensive reviews (van der Stelt and Di Marzo, 2003; Fernández-Ruiz, 2009; El Manira and Kyriakatos, 2010). The topic has interested many investigators because of the hope that knowledge of the mechanisms of endocannabinoid modulation of motor control might yield cannabinoid-derived medications capable of providing symptomatic relief, and possibly correction of the underlying pathology, in motor control disorders such as Parkinson's disease (PD), Huntington's disease (HD), Tourette's syndrome and others. Several points in these recent reviews suggest a basis for that hope: (1) There is a very heavy concentration of CB1 receptors, and of the enzymes for synthesis and degradation of the endocannabinoids, in the basal ganglia, cerebellum, and other brain structures involved in motor coordination and control. Indeed, the acquisition of a conditioned delayed eyeblink response, which involves LTD of synapses on cerebellar Purkinje cells, has been proposed as a quantitative measure of endocannabinoid function (Edwards and Skosnik, 2007). (2) The receptors are located on the presynaptic terminals of all the main fiber types involved in motor control, including glutamatergic, GABAergic, and dopaminergic axons (Morera-Herreras et al., 2012). (3) Alterations of the receptors have been found in tissue from the brains of patients with PD and HD, and the alterations have regressed in those who have responded well to treatment.

In the presymptomatic stages of both PD and HD, the CB1 receptors in the basal ganglia are downregulated and desensitized both in human patients and in animal models of PD and HD, and anandamide levels in the cerebrospinal fluid are elevated, possibly as a compensatory response (Pisani et al., 2010). This would lead to increased excitotoxicity, oxidative stress, and inflammatory reaction, as described in Section 2.3, and thus add to the toll of neuronal loss until the symptomatic stage is reached. In the latter stage the patterns of alteration in the two diseases differ. In PD, the CB1 receptors become upregulated and the resulting block of excitatory transmission gives rise to the characteristic motor inhibition and bradykinesia. In HD, in contrast, the receptor downregulation is increased, leading to hyperkinesia. These receptor changes have been found both in tissue from patients and in animal models (Fernández-Ruiz, 2009).

Theoretically, a CB1 receptor blocker should be helpful in late stage PD to alleviate bradykinesia. Trials in animal models have yielded some suggestive results, but the only controlled, double-blind cross-over study in 19 human PD patients showed neither benefit nor harm from the oral administration of an alcoholic extract of cannabis with a THC:CBD ratio of 2:1 (Carroll et al., 2004; Health Canada, 2010). Similarly, in theory the hyperkinesia of advanced HD might benefit from the inhibitory action of CB1 agonists, but this possibility has not yet been tested in a well-designed clinical trial. Such a trial of CBD in patients with symptomatic HD showed no significant response to treatment with even a fairly high dose (Consroe et al., 1991; Health Canada, 2010), but this is not really relevant because CBD is neither a CB1 nor a CB2 ligand, and it is an inhibitor of the GPR55. A small number of uncontrolled case reports yielded conflicting findings. However, a later trial of nabilone, an agonist at both CB1 and CB2 receptors, in 44 patients with advanced HD found no improvement in overall motor score, but there was a small but significant improvement of chorea (Curtis et al., 2009b). The latter study provides justification for further study with a wider dose range, and possibly more selectively acting drugs. Overall, however, it seems likely that the changes in endocannabinoid activity in both Huntington's and Parkinson's diseases affect too many different control mechanisms to make single-drug therapy likely to be very successful.

Tourette's syndrome offers a different problem from that seen in PD and HD, in that it does not involve malfunction of the regulation of motor pathways, but rather, the sudden initiation of contextually irrelevant movements. It is therefore considered a neuropsychiatric disorder rather than a purely neurological problem. Claims have been made for the success of THC treatment of this disorder. However, a fairly recent

Cochrane review found only two methodologically sound published trials (Curtis et al., 2009a) and concluded that THC treatment had yielded only minimal improvement in tic frequency and severity, insufficient to warrant the use of cannabinoid therapy.

Perhaps the greatest current interest in relation to cannabinoids and motor control attaches to its effects on spasticity in patients with multiple sclerosis. An antispasticity action was described in the Indian Hemp Drugs Commission Report (Kalant, 1972), and many modern publications have dealt with the effects of cannabis or individual cannabinoids. However, the reported results have been rather contradictory (Kalant and Porath-Waller, 2012). A review by Zajicek and Apostu (2011) concluded that in the best-designed trials, cannabinoids relieved the pain and the subjective sensation of spasm, but that objective measurements of spasm failed to demonstrate a clear effect. However, in a recent randomized, double-blind, placebo-controlled clinical trial, 30 patients with poorly controlled spasticity, and who were not using cannabis or other illicit drugs, were tested with smoked cannabis in a carefully controlled dosage. This resulted in significant reduction of both subjective and objective measures of spasticity after cannabis treatment, versus no change after placebo. However, the dose of cannabis that reduced the score on the modified Ashworth scale of spasticity also resulted in a significant increase in unwanted psychoactive effects (Corey-Bloom et al., 2012). These results suggest that the earlier contradictory results may have been due to insufficiently high dosage of cannabinoid, but that effective doses may exceed the tolerable limit in many or most patients.

2.1.11. Cognitive Functions

2.1.11.1. Cognition-Related Brain Functions in Animals

Given the important roles of synaptic plasticity and of axonal sprouting and new synapse formation in learning, laying down of new memory traces, extinction or active unlearning, tolerance and physiological adaptations of various kinds, it can readily be anticipated that the endocannabinoid system must be involved as part of the neuronal mechanisms of many cognitive functions. Surprisingly, however, there is very little published work on a facilitatory role of endocannabinoid signaling in learning and memory. Most of the support for such a role is inferred from the finding of negative effects in knockout mice lacking CB1 receptors, which show inability to extinguish a conditioned fear response, or to develop sensitization of the rewarding effects of cocaine (Marsicano et al., 2002; Heifets and Castillo, 2009). Conversely, the endocannabinoid transporter AM404, which facilitates endocannabinoid binding to the receptors, and the CB1 agonist WIN55,212-2, both enhanced extinction of conditioned fear and water maze spatial learned responses (Chhatwal et al., 2005; Pamplona et al., 2006). More recently, Campolongo et al. (2009) reported that infusion of WIN55,212-2 directly into the basolateral amygdala of rats immediately after inhibitory avoidance training facilitated the consolidation of the learned response, and a CB1 receptor blocker inhibite consolidation.

Davies et al. (2002) suggested that a possible reason for failure to find more examples of facilitation of learning by endocannabinoid or cannabinoid action might be that almost all experimenters have used systemic administration of the cannabinoid, which would affect allcannabinoid binding sites in the brain. Since the actions of endocannabinoids, in particular, are highly variable at different brain sites, different neuron types, and different behavioral tests, the resulting medley of actions would make it very difficult to recognize facilitatory actions; the results cited above are consistent with this concept of variability of endocannabinoid function. Biphasic dose–response curves can also be seen with CB1 receptor blockers. AM251, a CB1 blocker, given at low doses to rats improved recognition memory, but at high doses it had no effect (Bialuk and Winnicka, 2011). This adds to the complexity of the picture but at present no comprehensive explanation is apparent.

Davies et al. argued that the proper type of experiment would involve local injection of the cannabinoid or endocannabinoid directly into specific sites in the hippocampus. Nevertheless, a number of such experiments have been carried out (Akirav, 2011) and the consistent pattern has been that cannabinoid agonists inhibit learning both when given systemically and when injected intracerebrally (Akirav, 2011). Working memory appears to be the most sensitive to disruption by very low doses of THC (Varvel et al., 2001; Egerton et al., 2006). These findings suggest that if there is a more general facilitatory role of endocannabinoids on a greater variety of cognitive tests, the reversal point at which inhibitory effects predominate must be very low relative to those observed with anxiety, neuroprotection, and other functions described above.

2.1.11.2. Cerebral Blood Flow Studies

In contrast to these studies of the cellular anatomy of cannabinoid actions, others have approached the subject by studies in the intact organism. One such approach has been functional imaging of the brain, both in rodents and in humans. Acute intravenous administration of THC or its active metabolite 11-hydroxy-THC produced a variable pattern of changes in regional blood flow in the rat brain, with increases in about half of the areas measured, and decreases in the others (Bloom et al., 1997). After administration of anandamide, the picture was more prominently one of decreased regional flow, with more regions being affected for longer after higher doses (Stein et al., 1998). The assumption underlying these studies is that regional blood flow is a reflection of neuronal activity, but in the light of the extreme diversity of different neuron types within the same small region, and the complex interactions among them as illustrated by the mechanism of postsynaptically formed endocannabinoid acting at inhibitory presynaptic receptors, it is difficult to see what functional interpretation can be placed on the changes observed. The difference in cannabinoid effects within areas in the amygdala that are only millimeters apart serves to illustrate this point.

Essentially, the same problem is seen in the blood flow studies in humans. Acute administration resulted in global increases in cerebral blood flow, and regional increases in the frontal lobes, that were accompanied by subjective changes ("high", anxiety, anger, confusion) and increases in heart rate, that were correlated in magnitude with the peak plasma THC level (Mathew and Wilson, 1993). However, neither the blood flow measures nor the physiological and emotional measures are sufficiently fine-grained to permit any mechanistic interpretation. The same limitation of interpretability applies to another acute study in which a cognitive performance task was also included (O'Leary et al., 2002). In addition, the trials were not truly acute because, for ethical reasons, only experienced marijuana users were allowed to take part, so that the effects may well have been modified by varying degrees of tolerance or abstinence. In another study in which the participants had to perform two versions of a decision-making task, one version that allowed for reasonable calculation and the other that was more probabilistic in nature (Vaidya et al., 2012). Somewhat more selective changes in regional cerebral blood flow were seen according to the nature of the task, but in both tasks the chronic users had greater increases in blood flow than the nonuser controls. This again illustrates the difficulties of interpretation: did the greater increase in blood flow signify an adaptive response to the drug, or a need for greater cerebral energy expenditure because of poorer task efficiency? Two other studies (Tunving et al., 1986; Lundqvist et al., 2001) in chronic users who were withdrawn for only 1–2 days before blood flow measurement, and who were not required to perform any mental task during the study, found either reduced hemispheric blood flow or no difference from controls. Those who had initially low blood flow showed significant improvement when reexamined after 9–60 days of detoxication.

In view of the many individual and intergroup differences in these studies, and the multiple differences in the results of the studies, it is not surprising that a systematic review of 41 human brain imaging studies found such heterogeneity that no meta-analysis was possible (Martin-Santos et al., 2010).

2.1.11.3. Human Cognitive Functions

There have been many studies of the effects of cannabis on a variety of cognitive functions in human users of cannabis, both acute and chronic. Two extensive literature reviews give very representative pictures of the most consistent findings. Ranganathan and D'Souza (2006) dealt primarily with effects on memory, and found dose-related impairments of immediate and delayed recall of information presented while the subjects were under the influence of the drug. Encoding, consolidation and retrieval of memory were all affected, despite many sources of variation in different subject selection methods, function tests used, sample sizes and other aspects of design.

More recent evidence continues to yield similar findings. For example, Solowij et al. (2011), in a study of 181 subjects aged 16–20 years, found that cannabis users (mean duration of use 2.4 years) found significantly worse measures of verbal learning and memory than in alcohol users and in nonusers of any drug. The impairment was worse, the earlier the age of onset of regular use of cannabis and the greater the extent of use. Essentially, similar results have been obtained by others as well (Wagner et al., 2010; Takagi et al., 2011).

Dose-dependent acute inhibitory effects of THC on attention and information processing in human subjects have been studied by measurement of P300 EEG responses during performance of a monitoring task (D'Souza et al., 2012). This is clearly a cannabinoid effect, because a triplet repeat polymorphism in the gene encoding the CB1 receptor caused a marked change in sensitivity to the effect of THC on the P300 response (Stadelmann et al., 2011). However, fMRI studies have shown different effects on auditory and visual information processing (Winton-Brown et al., 2011). A 10 mg dose of THC that produced anxiety, intoxication and psychotic symptoms in healthy volunteers decreased activation in both temporal cortices during auditory processing, but variably increased or decreased activation in different areas of the visual cortex during visual processing. CBD, which did not produce cannabis-like subjective symptoms, had effects opposite to those of THC in the temporal cortex.

Crean et al. (2011) reviewed a broader spectrum of cognitive functions designated as executive functions. These included attention, concentration, decision-making, impulsivity, inhibition (self-control of responses), reaction time, risk taking, verbal fluency, and working memory. All these functions have been found to be impaired acutely in a dose-dependent manner, but there is no unanimity because on each of the functions described there are some studies that found no difference between users and nonusers. Once again, variations in experimental design and execution, differences in definition of *acute* versus *chronic* and *withdrawal*, and length of time elapsed since last use undoubtedly account for much of the differences in conclusions.

In long-term users seen after periods of withdrawal of at least 21 days, which the authors presumably consider sufficient for essentially complete detoxication, so that persistent intoxication effects do not confuse the assessment of residual brain changes, there is again a considerable measure of disagreement among the various studies, due to the same sources of variation mentioned above. Crean et al. conclude that while some components of executive function usually recover completely with the passage of time after cessation of cannabis use, those in which deficits are most likely to persist for very long periods of time are decision making, concept formation, and planning. Just as with effects on maturation, residual long-term effect on executive functions are most likely to occur in those who began regular cannabis use early in adolescence (or during intrauterine development), used cannabis most heavily and for the longest time. There is generally good agreement between the conclusions based on these studies and the clinical impressions based on population studies (e.g. Kalant, 2004; Hall and Degenhardt, 2009).

A growing body of evidence attests to the ability of cannabis to impair psychomotor functions involved in complex tasks such as operation of motor vehicles or aircraft (Kalant, 2004; Ramaekers et al., 2004; Kelly et al., 2004; Sewell et al., 2009; Calabria et al., 2010). It is beyond the scope of the present review to examine this evidence in greater detail, but it represents a significant health and social problem resulting from the actions of cannabis on complex mental functions.

3. CONCLUSIONS

- There has been enormous progress made in recent years, in analyzing the structure and functions of the endocannabinoid system in the nervous system, and it is clear that it is involved in modulating a very broad range of nervous system activities, including synaptic transmission, neuronal growth and maturation, integration of incoming information from many different sources within the body and outside, and linkage between information and responses. However, there are still large gaps in knowledge at the cellular level, and knowledge of the basic workings of the endocannabinoid system still do not offer comprehensive understanding of the effects at the level of the whole organism, including the subjective effects underlying nonmedical use, and the functional effects that may be therapeutically useful or harmful.

- In theory, such a broad range of functions offers the possibility of many therapeutic applications, and it seems probable that in the future many clinically useful developments will become available. They will be useful to the degree that they offer highly selective and controllable degrees of intervention in the workings of the endocannabinoid system, which crude cannabis (marijuana) does not offer, and to the extent that they are therefore free of the undesirable psychoactive effects of currently available cannabinoids.

- There are major differences between the low-dose effects of endogenous cannabinoids and the higher-dose and more prolonged effects of phytocannabinoids. As a result, most of the potentially therapeutically useful effects of cannabinoids are likely to come from the development of endocannabinoid derivatives and analogs, while the higher-dose effects are mainly related to the adverse effects of crude cannabis preparations. The major adverse effects in the nervous system are related to alterations in neuronal growth and maturation, and to impairments of cognitive function including memory and executive functions.

- As with most drugs, the adverse effects are closely related to the age at which use begins, the amount and duration of use, and the length of time after cessation of use.

REFERENCES

Akirav, I., 2011. The role of cannabinoids in modulating emotional and non-emotional memory processes in the hippocampus. Front. Behav. Neurosci. 5, 1–11 Article 34.

Alger, B.E., 2009. Endocannabinoid signaling in neural plasticity. Curr. Top. Behav. Neurosci. 1, 141–172.

Alger, B.E., Kim, J., 2011. Supply and demand for endocannabinoids. Trends Neurosci. 34, 304–315.

Alonso-Alconada, D., Alvarez, F.J., Alvarez, A., Mielgo, V.E., Goni-de-Cerio, F., et al., 2010. The cannabinoid receptor agonist WIN55,212-2 reduces the initial cerebral damage after hypoxic-ischemic injury in fetal lambs. Brain Res. 1362, 150–159.

Alonso-Alconada, D., Alvarez, A., Hilario, E., 2011. Cannabinoid as a neuroprotective strategy in perinatal hypoxic-ischemic injury. Neurosci. Bull. 27, 275–285.

Anand, P., Whiteside, G., Fowler, C.J., Hohmann, A.G., 2009. Targeting CB2 receptors and the endocannabinoid system for the treatment of pain. Brain Res. Rev. 60, 255–266.

Anavi-Goffer, S., Baillie, G., Irving, A.J., Gertsch, J., Greig, I.R., et al., 2012. Modulation of L-α-lysophosphatidylinositol/GPR55 mitogen-activated protein kinase (MAPK) signaling by cannabinoids. J. Biol. Chem. 287, 91–104.

Arias-Carrión, O., Huitrón-Reséndiz, S., Arankowsky-Sandoval, G., Murillo-Rodríguez, E., 2011. Biochemical modulation of the sleep-wake cycle: endogenous sleep-inducing factors. J. Neurosci. Res. 89, 1143–1149.

Ashtari, M., Cervellione, K., Cottone, J., Ardekani, B.A., Kumra, S., 2009. Diffusion abnormalities in adolescents and young adults with a history of heavy cannabis use. J. Psychiatr. Res. 43, 189–204.

Ashton, C.H., Moore, P.B., 2011. Endocannabinoid system dysfunction in mood and related disorders. Acta Psychiatr. Scand. 124, 250–261.

Ashton, J.C., 2012. The atypical cannabinoid o-1602: targets, actions, and the central nervous system. CNS Agents Med. Chem. 12, 233–239.

Bambico, F.R., Katz, N., Debonnel, G., Gobbi, G., 2007. Cannabinoids elicit antidepressant-like behavior and activate serotonergic neurons through the medial prefrontal cortex. J. Neurosci. 27, 11700–11711.

Bambico, F.R., Nguyen, N.T., Katz, N., Gobbi, G., 2010. Chronic exposure to cannabinoids during adolescence but not during adulthood impairs emotional behaviour and monoaminergic neurotransmission. Neurobiol. Dis. 37, 641–655.

Basavarajappa, B.S., 2007. Neuropharmacology of the endocannabinoid signaling system – molecular mechanisms, biological actions and synaptic plasticity. Curr. Neuropharmacol. 5, 81–97.

Bauer, M., Chicca, A., Tamborrini, M., Eisen, D., Lerner, R., et al., 2012. Identification and quantification of a new family of peptide endocannabinoids (Pepcans) showing negative allosteric modulation at CB1 receptors. J. Biol. Chem. 287, 36944–36967.

Bermudez-Silva, F.J., Viveros, M.P., McPartland, J.M., Rodriguez de Fonseca, F., 2010. The endocannabinoid system, eating behavior and energy homeostasis: the end or a new beginning. Pharmacol. Biochem. Behav. 95, 375–382.

Bernard, C., Milh, M., Morozov, Y.M., Ben-Ari, Y., Freund, T.F., et al., 2005. Altering cannabinoid signaling during devgelopment disrupts neuronal activity. Proc. Natl. Acad. Sci. U.S.A 102, 9388–9393.

Bialuk, I., Winnicka, M.M., 2011. AM251, cannabinoids receptor ligand, improves recognition memory in rats. Pharmacol. Rep. 63, 670–679.

Bidaut-Russell, M., Devane, W.A., Howlett, A.C., 1990. Cannabinoid receptors and modulation of cyclic AMP accumulation in rat brain. J. Neurochem. 55, 21–26.

Bisogno, T., Di Marzo, V., 2007. Short- and long-term plasticity of the endocannabinoid system in neuropsychiatric and neurological disorders. Pharmacol. Res. 56, 428–442.

Bisogno, T., Di Marzo, V., 2010. Cannabinoid receptors and endocannabinoids: role in neuroinflammatory and neurodegenerative disorders. CNS Neurol. Disord.: Drug Targets 9, 564–573.

Bisogno, T., Hanus, L., De Petrocellis, L., Tchilibon, S., Ponde, D.E., et al., 2001. Molecular targets for cannabidiol and its synthetic analogues: effect on vanilloid VR1 receptors and on the cellular uptake and enzymatic hydrolysis of anandamide. Br. J. Pharmacol. 134, 845–852.

Blair, R.E., Deshpande, L.S., Sombati, S., Elphick, M.R., Martin, B.R., et al., 2009. Prolonged exposure to WIN55,212-2 causes downregulation of the CB1 receptor and the development of tolerance to its anticonvulsant effects in the hippocampal neuronal culture model of acquired epilepsy. Neuropharmacology 57, 208–218.

Bloom, A.S., Tershner, S., Fuller, S.A., Stein, E.A., 1997. Cannabinoid-induced alterations in regional cerebral blood flow in the rat. Pharmacol. Biochem. Behav. 57, 625–631.

Bossong, M.G., Niesink, R.J., 2010. Adolescent brain maturation: the endogenous cannabinoid system and the neurobiology of cannabis-induced schizophrenia. Prog. Neurobiol. 92, 370–385.

Breunig, E., Czesnik, D., Piscitelli, F., Di Marzo, V., Manzini, I., et al., 2010. Endocannabinoid modulation in the olfactory epithelium. Results Probl. Cell Differ. 52, 139–145.

Bruera, E., 1992. Clinical management of anorexia and cachexia in patients with advanced cancer. Oncology 49 (Suppl. 2), 35–42.

Bushlin, I., Rozenfeld, R., Devi, L.A., 2010. Cannabinoid-opioid interactions during neuropathic pain and analgesia. Curr. Opin. Pharmacol. 10, 80–86.

Butler, R.K., Finn, D.P., 2009. Stress-induced analgesia. Prog. Neurobiol. 88, 184–202.

Caballero, A., Tseng, K.Y., 2012. Association of cannabis use during adolescence, prefrontal CB1 receptor signaling, and schizophrenia. Article 101 Front. Pharmacol. 3, 1–6. http://dx.doi.org/10.3389/fphar.2012.00101 (accessed 08.11.12.).

Calabria, B., Degenhardt, L., Hall, W., Lynskey, M., 2010. Does cannabis use increase the risk of death? Systematic review of epidemiological evidence on adverse effects of cannabis use. Drug Alcohol Rev. 29, 318–330.

Campolongo, P., Trezza, V., Palmery, M., Trabace, L., Cuomo, V., 2009a. Developmental exposure to cannabinoids causes subtle and enduring neurofunctional alterations. Int. Rev. Neurobiol. 85, 117–133.

Campolongo, P., Trezza, V., Ratano, P., Palmery, M., Cuomo, V., 2009b. Developmental consequences of perinatal cannabis exposure: behavioral and neuroendocrine effects in adult rodents. Psychopharmacology 214, 5–15.

Campolongo, P., Roozendaal, B., Trezza, V., Hauer, D., Schelling, G., et al., 2009. Endocannabinoids in the rat basolateral amygdala enhance memory consolidation and enable glucocorticoid modulation of memory. Proc. Natl. Acad. Sci. U.S.A 106, 4888–4893.

Capasso, R., Izzo, A.A., 2008. Gastrointestinal regulation of food intake: general aspects and focus on anandamide and oleoylethanolamide. J. Neuroendocrinol. 20 (Suppl. 1), 39–46.

Carroll, C.B., Bain, P.G., Teare, L., Liu, X., Joint, C., et al., 2004. Cannabis for dyskinesia in Parkinson disease: a randomized double-blind crossover study. Neurology 63, 1245–1250.

Cassano, T., Gaetani, S., Macheda, T., Laconca, L., Romano, A., et al., 2011. Evaluation of the emotional phenotype and serotonergic neurotransmission of fatty acid amide hydrolase-deficient mice. Psychopharmacology 214, 465–476.

Chanda, P.K., Gao, Y., Mark, L., Btesh, J., Strassle, B.W., et al., 2010. Monoacylglycerol lipase activity is a critical modulator of the tone and integrity of the endocannabinoid system. Mol. Pharmacol. 78, 996–1003.

Chhatwal, J.P., Myers, K.M., Ressler, K.J., Davis, M., 2005. Regulation of gephyrin and $GABA_A$ receptor binding within the amygdala after fear acquisition and extinction. J. Neurosci. 25, 502–506.

Chen, C., Bazan, N.G., 2005. Lipid signaling: sleep, synaptic plasticity, and neuroprotection. Prostaglandins Other Lipid Mediators 77, 65–76.

Clement, A.B., Hawkins, E.G., Lichtman, A.H., Cravatt, B.F., 2003. Increased seizure susceptibility and proconvulsant activity of anadamide in mice lacking fatty acid amide hydrolase. J. Neurosci. 23, 3916–3923.

Console-Bram, L., Marcu, J., Abood, M.E., 2012. Cannabinoid receptors: nomenclature and pharmacological principles. Prog. Neuro-Psychopharmacol. Biol. Psychiatry 38, 4–15.

Consroe, P., Benedito, M.A., Leite, J.R., Carlini, E.A., Mechoulam, R., 1982. Effects of cannabidiol on behavioral seizures cause by convulsant drugs or current in mice. Eur. J. Pharmacol. 83, 293–298.

Consroe, P., Laguna, J., Allender, J., Snider, S., Stern, L., et al., 1991. Controlled clinical trial of cannabidiol in Huntington's disease. Pharmacol. Biochem. Behav. 40, 701–708.

Corcoran, M.E., McCaughran Jr., J.A., Wada, J.A., 1978. Antiepileptic and prophylactic effects of tetrahydrocannabinols in amygdaloid kindled rats. Epilepsia 19, 47–55.

Corey-Bloom, J., Wolfson, T., Gamst, A., Jin, S., Marcotte, T.D., et al., 2012. Smoked cannabis for spasticity in multiple sclerosis: a randomized, placebo-controlled trial. Can. Med. Assoc. J. 184, 1143–1150.

Crean, R.D., Crane, N.A., Mason, B.J., 2011. An evidence-based review of acute and long-teerm effects of cannabis use on executive cognitive functions. J. Addict. Med. 5, 1–8.

Crippa, J.A., Derenusson, G.N., Ferrari, T.B., Wichert-Ana, L., Duran, F.L., et al., 2011. Neural basis of anxiolytic effects of cannabidiol (CBD) in generalized social anxiety disorder. J. Psychopharmacol. 25, 121–130.

Crippa, J.A., Zuardi, A.W., Garrido, G.E., Wichert-Ana, L., Guarnieri, R., et al., 2004. Effects of cannabidiol (CBD) on regional cerebral blood flow. Neuropsychopharmacology 29, 417–426.

Cunha, J.M., Carlini, E.A., Pereira, A.E., Ramos, O.L., Pimentel, C., et al., 1980. Chronic administration of cannabidiol to healthy volunteers and epileptic patients. Pharmacology 21, 175–185.

Curtis, A., Clarke, C.E., Rickards, H., 2009a. Cannabinoids for Tourette's Syndrome. Published online 7.10.09. The Cochrane Library. http://dx.doi.org/10.1002/14651858.CD006565.pub2.

Curtis, A., Mitchell, I., Patel, S., Ives, N., Rickards, H., 2009b. A pilot study using nabilone for symptomatic treatment in Huntington's disease. Mov. Disord. 24, 2254–2259.

Dalterio, S.L., 1986. Cannabinoid exposure: effects on development. Neurobehav. Toxicol. Teratol. 8, 345–352.

Davies, S.N., Pertwee, R.G., Riedel, G., 2002. Functions of cannabinoids in the hippocampus. Neuropharmacology 42, 993–1007.

Degenhardt, L., Hall, W., Lynskey, M., 2003. Exploring the association between cannabis use and depression. Addiction 98, 1493–1504.

Degenhardt, L., Tennant, C., Gilmour, S., Schofield, D., Nash, L., et al., 2007. The temporal dynamics of relationships between cannabis, psychosis and depression among young adults with psychotic disorders: findings from a 10-month prospective study. Psychol. Med. 37, 927–934.

De Petrocellis, L., Di Marzo, V., 2009. Role of endocannabinoids and endovanilloids in Ca^{2+} signalling. Cell Calcium 45, 611–624.

Devane, W.A., Dysarz 3rd, F.A., Johnson, M.R., Melvin, L.S., Howlett, A.C., 1988. Determination and characterization of a cannabinoid receptor in rat brain. Mol. Pharmacol. 34, 605–613.

Diana, M.A., Marty, A., 2004. Endocannabinoid-mediated short-term synaptic plasticity: depolarization-induced suppression of inhibition (DSI) and depolarization-induced suppression of excitation (DSE). Br. J. Pharmacol. 142, 9–19.

DiPatrizio, N.V., Astarita, G., Schwartz, G., Li, X., Piomelli, D., 2011. Endocannabinoid signal in the gut controls dietary fat intake. Proc. Natl. Acad. Sci. U.S.A 108, 12904–12908.

Dissanayake, D.W., Zachariou, M., Marsden, C.A., Mason, R., 2008. Auditory gating in rat hippocampus and medial prefrontal cortex: effect of the cannabinoid agonist WIN55,212-2. Neuropharmacology 55, 1397–1404.

Dmitrieva, N., Nagabukuro, H., Resuehr, D., Zhang, G., McAllister, S.L., et al., 2010. Endocannabinoid involvement in endometriosis. Pain 151, 703–710.

D'Souza, D.C., Fridberg, D.J., Skosnik, P.D., Williams, A., Roach, B., et al., 2012. Dose-related modulation of event-related potentials to novel and target stimuli by intravenous 9-THC in humans. Neuropsychopharmacology 37, 1632–1646.

Edwards, C.R., Skosnik, P.D., 2007. Cerebellar-dependent learning as a neurobehavioral index of the cannabinoid system. Crit. Rev. Neurobiol. 19, 29–57.

Egerton, A., Allison, C., Brett, R.R., Pratt, J.A., 2006. Cannabinoids and prefrontal cortical function: insights from preclinical studies. Neurosci. Biobehav. Rev. 30, 680–695.

El Manira, A., Kyriakatos, A., 2010. The role of endocannabinoid signaling in motor control. Physiology 25, 230–238.

Elikottil, J., Gupta, P., Gupta, K., 2009. The analgesic potential of cannabinoids. J. Opioid Manag. 5, 341–357.

Eljaschewitsch, E., Witting, A., Mawrin, C., Lee, T., Schmidt, P.M., et al., 2006. The endocannabinoid anandamide protects neuronsduring CNS inflammation by induction of MKP-1 in microglial cells. Neuron 49, 67–79.

Elphick, M.R., Egertova, M., 2001. The neurobiology and evolution of cannabinoid signaling. Philos. Trans. R. Soc., B 356, 381–408.

Falenski, K.W., Carter, D.S., Harrison, A.J., Martin, B.R., Blair, R.E., et al., 2009. Temporal characterization of changes in hippocampal cannabinoid CB_1 receptor expression following pilocarpine-induced status epilepticus. Brain Res. 1262, 64–72.

Farrant, M., Nusser, Z., 2005. Variations on an inhibitory theme: phasic and tonic activation of $GABA_A$ receptors. Nat. Rev. Neurosci. 6, 215–229.

Fehr, K.A., Kalant, H., LeBlanc, A.E., 1976. Residual learning deficit after heavy exposure to cannabis or alcohol in rats. Science 192, 1249–1251.

Fernández-Ruiz, J., 2009. The endocannabinoid system as a target for the treatment of motor dysfunction. Br. J. Pharmacol. 156, 1029–1040.

Fernández-Ruiz, J., Moreno-Martet, M., Rodriguez-Cueto, C., Palomo-Garo, C., Gomez-Canas, M., et al., 2011. Prospects for cannabinoid therapies in basal ganglia disorders. Br. J. Pharmacol. 163, 1365–1378.

Finn, D.P., 2010. Endocannabinoid-mediated modulation of stress responses: physiological and pathophysiological significance. Immunobiology 215, 629–646.

Fowler, C.J., Rojo, M.L., Rodriguez-Gaztelumendi, A., 2010. Modulation of the endocannabinoid system: neuroprotection or neurotoxicity? Exp. Neurol. 224, 37–47.

Freund, T.F., Katona, I., Piomelli, D., 2003. Role of endogenous cannabinoids in synaptic signaling. Physiol. Rev. 83, 1017–1066.

Fride, E., Bregman, T., Kirkham, T.C., 2005. Endocannabinoids and food intake: newborn suckling and appetite regulation in adulthood. Exp. Biol. Med. 230, 225–234.

Fried, P., Watkinson, B., Gray, R., 2003. Differential effects on cognitive functioning in 13- to 16-year-olds prenatally exposed to cigarettes and marihuana. Neurotoxicol. Teratol. 25, 427–436.

Galve-Roperh, I., Palazuelos, J., Aguado, T., Guzman, M., 2009. The endocannabinoid system and the regulation of neural development: potential implications in psychiatric disorders. Eur. Arch. Psychiatry Clin. Neurosci. 259, 371–382.

Gertsch, J., Pertwee, R.G., Di Marzo, V., 2010. Phytocannabinoids beyond the cannabis plant – do they exist? Br. J. Pharmacol. 160, 523–529.

Goldschmidt, L., Day, N.L., Richardson, G.A., 2000. Effects of prenatal marijuana exposure on child behavior problems at age 10. Neurotoxicol. Teratol. 22, 325–336.

Gorzalka, B.B., Hill, M.N., 2011. Putative role of endocannabinoid signaling in the etiology of depression and actions of antidepressants. Prog. Neuro-Psychopharmacol. Biol. Psychiatry 35, 1575–1585.

Haj-Dahmane, S., Shen, R.Y., 2011. Modulation of the serotonin system by endocannabinoid signaling. Neuropharmacology 61, 414–420.

Hajós, M., Hoffmann, W.E., Kocsis, B., 2008. Activation of cannabinoid-1 receptors disrupts sensory gating and neuronal oscillation: relevance to schizophrenia. Biol. Psychiatry 63, 1075–1083.

Hall, W., Degenhardt, L., 2009. Adverse health effects of non-medical cannabis use. Lancet 374, 1383–1391.

Hansen, H.S., 2010. Palmitoylethanolamide and other anandamide congeners. Proposed role in the diseased brain. Exp. Neurol. 224, 48–55.

Hansen, S.L., Nielsen, A.H., Knudsen, K.E., Artmann, A., Petersen, G., et al., 2009. Ketogenic diet is anti-epileptogenic in pentylenetetrazole kindled mice and decrease [sic] levels of N-acylethanolamines in hippocampus. Neurochem. Int. 54, 199–204.

Hayakawa, K., Mishima, K., Nozako, M., Ogata, A., Hazekawa, M., et al., 2007. Repeated treatment with cannabidiol but not Δ^9-tetrahydrocannabinol has a neuroprotective effect without the development of tolerance. Neuropharmacology 52, 1079–1087.

Health Canada, 2010. Information for Health Care Professionals: Marihuana [Monograph]. Retrieved from: http://www.hc-sc.gc.ca/dhp-mps/marihuana/how-comment/medpract/infoprof/index-eng.php.

Heifets, B.D., Castillo, P.E., 2009. Endocannabinoid signaling and long-term synaptic plasticity. Ann. Rev. Physiol. 71, 283–306.

Hermann, D., Sartorius, A., Welzel, H., Walter, S., Skopp, G., et al., 2007. Dorsolateral prefrontal cortex N-acetylaspartate/total creatinine (NAA/tCr) loss in male recreational cannabis users. Biol. Psychiatry 61, 1281–1289.

Hill, M.N., McLaughlin, R.J., Pan, B., Fitzgerald, M.L., Roberts, C.J., et al., 2011. Recruitment of prefrontal cortical endocannabinoid signaling by glucocorticoids contributes to termination of the stress response. J. Neurosci. 31, 10506–10515.

Hill, M.N., Patel, S., Campolongo, P., Tasker, J.G., Wotjak, C.T., et al., 2010. Functional interactions between stress and the endocannabinoid system: from synaptic signaling to behavioral output. J. Neurosci., 14980–14986.

Hornby, P.J., 2001. Central neurocircuitry associated with emesis. Am. J. Med. 1118, (Suppl. 1), 106S–112S.

Hwang, J., Adamson, C., Butler, D., Janero, D.R., Makriyannis, A., et al., 2010. Enhancement of endocannabinoid signaling by fatty acid amide hydrolase inhibition: a neuroprotective therapeutic modality. Life Sci. 86, 615–623.

Jatoi, A., Windschitl, H.E., Loprinzi, C.L., Sloan, J.A., Dakhil, S.R., et al., 2002. Dronabinol versus megestrol acetate versus combination therapy for cancer-associated anorexia: a North Central Cancer Treatment Group study. J. Clin. Oncol. 20, 2912–2913.

Joyce, C.R.B., Curry, S.H. (Eds.), 1970. The Botany and Chemistry of Cannabis. J&A Churchill, London.

Kalant, H., 2004. Adverse effects of cannabis on health: an update of the literature since 1996. Prog. Neuro-Psychopharmacol. Biol. Psychiatry 28, 849–863.

Kalant, H., Porath-Waller, A.J., 2012. Clearing the Smoke on Cannabis. Series No. 5: Medical Use of Cannabis and Cannabinoids. Canadian Centre on Substance Abuse, Ottawa. ISBN 978-1-926705-75-0.

Kalant, O.J., 1972. Report of the Indian Hemp Drugs Commission, 1893-94: a critical review. Int. J. Addict. 7, 77–96.

Kamprath, K., Romo-Parra, H., Haring, M., Gaburro, S., Doengi, M., et al., 2011. Short-term adaptation of conditioned fear responses through endocannabinoid signaling in the central amygdale. Neuropsychopharmacology 36, 652–663.

Kano, M., Ohno-Shosaku, T., Hashimotodani, Y., Uchigashima, M., Watanabe, M., 2009. Endocannabinoid-mediated control of synaptic transmission. Physiol. Rev. 89, 309–380.

Kaplan, G.B., Moore, K.A., 2011. The use of cognitive enhancers in animal models of fear extinction. Pharmacol. Biochem. Behav. 99, 217–228.

Karanian, D.A., Karim, S.L., Wood, J.T., Williams, J.S., Lin, S., et al., 2007. Endocannabinoid enhamcement protects against kainic acid-induced seizures and associated brain damage. J. Pharmacol. Exp. Ther. 322, 1059–1066.

Katona, I., Urban, G.M., Wallace, M., Ledent, C., Jung, K.M., et al., 2006. Molecular composition of the endocannabinoid system at glutamatergic synapses. J. Neurosci. 26, 5628–5637.

Kaur, C., Hao, A.J., Wu, C.H., Ling, E.A., 2001. Origin of microglia. Microsc. Res. Tech. 54, 2–9.

Kelly, E., Darke, S., Ross, J., 2004. A review of drug use and driving: epidemiology, impairment, risk factors and risk perceptions. Drug Alcohol Rev. 23, 319–344.

Kinsey, S.G., Long, J.Z., O'Neal, S.T., Abdullah, R.A., Poklis, J.L., et al., 2009. Blockade of endocannabinoid-degrading enzymes attenuates neuropathic pain. J. Pharmacol. Exp. Ther. 330, 902–910.

Knoller, N., Levi, L., Shoshan, I., Reichenthal, E., Razon, N., et al., 2002. Dexanabinol (HU-211) in the treatment of severe closed head injury: a randomized, placebo-controlled, phase II clinical trial. Crit. Care Med. 30, 548–554.

Koch, G., Mori, F., Codeca, C., Kusayanagi, H., Monteleone, F., et al., 2009. Cannabis-based treatment induces polarity-reversing plasticity assessed by theta burst stimulation in humans. Brain Stim. 2, 229–233.

Krebs-Kraft, D.L., Hill, M.N., Hillard, C.J., McCarthy, M.M., 2010. Sex difference in cell proliferation in developing rat amygdala mediated by endocannabinoids has implication for social behavior. Proc. Natl. Acad. Sci. U.S.A 107, 20535–20540.

Kreitzer, F.R., Stella, N., 2009. The therapeutic potential of novel cannabinoid receptors. Pharmacol. Ther. 122, 83–96.

Kubajewska, I., Constantinescu, C.S., 2010. Cannabinoidds and experimental models of multiple sclerosis. Immunobiology 215, 647–657.

Lane, J.R., Beukers, M.W., Mulder-Krieger, T., Ijzerman, A.P., 2010. The endocannabinoid 2-arachidonyl-glycerol is a negative allosteric modulator of the hum, an A3 adenosine receptor. Biochem. Pharmacol. 79, 48–56.

Lazary, J., Juhasz, G., Hunyady, L., Bagdy, G., 2011. Personalized medicine can pave the way for the safe use of CB1 receptor antagonists. Trends Pharmacol. Sci. 32, 270–280.

Leker, R.R., Gai, N., Mechoulam, R., Ovadia, H., 2003. Drud-induced hypothermia reduces ischemic damage: effects of the cannabinoid HU-21. Stroke 34, 2000–2006.

Lichtman, A.H., Blankman, J.L., Cravatt, B.F., 2010. Endocannabinoid overload. Mol. Pharmacol. 78, 993–995.

Liu, Z., Han, J., Jia, L., Maillet, J.C., Bai, G., et al., 2010. Synaptic neurotransmission depression in ventral tegmental dopamine neurons and cannabinoid-associated addictive learning. PLoS One 5, e15634.

Lopez-Larson, M.P., Bogorodzki, P., Rogowska, J., McGlade, E., King, J.B., et al., 2011. Altered prefrontal and insular cortical thickness in adolescent marijuana users. Behav. Brain Res. 220, 164–172.

Lötsch, J., Geisslinger, G., Hummel, T., 2012. Sniffing out pharmacology: interactions of drugs with human olfaction. Trends Pharmacol. Sci. 33, 193–199.

Lu, V.B., Puhl 3rd, H.L., Ikeda, S.R., 2013. N-Arachidonyl glycine does not activate G protein-coupled receptor 18 signaling via canonical pathways. Mol. Pharmacol. 83, 267–282.

Lundqvist, T., Jonsson, S., Warkentin, S., 2001. Frontal lobe dysfunction in long-term cannabis users. Neurotoxicol. Teratol. 23, 437–443.

Lutz, B., 2004. On-demand activation of the endocannabinoid system in the control of neuronal excitability and epileptiform seizures. Biochem. Pharmacol. 68, 1691–1698.

Machado Rocha, F.C., Stefano, S.C., De Cassia Haiek, R., Rosa Oliveira, L.M., Da Silveira, D.X., 2008. Therapeutic use of *Cannabis sativa* on chemotherapy-induced nausea and vomiting among cancer patients: systematic review and meta-analysis. Eur. J. Cancer Care 17, 431–443.

Malan Jr., T.P., Ibrahim, M.M., Vanderah, T.W., Makriyannis, A., Porreca, F., 2002. Inhibition of pain responses by activation of CB2 cannabinoid receptors. Chem. Phys. Lipids 121, 191–200.

Mangieri, R.A., Piomelli, D., 2007. Enhancement of endocannabinoid signaling and the pharmacotherapy of depression. Pharmacol. Res. 56, 360–366.

Marsicano, G., Goodenough, S., Monory, K., Hermann, H., Eder, M., et al., 2003. CB1 cannabinoid receptors and on-demand defense against excitotoxicity. Science 302, 84–88.

Marsicano, G., Wotjak, C.T., Azad, S.C., Bisogno, T., Rammes, G., et al., 2002. The endogenous cannabinoid system controls extinction of aversive memories. Nature 418, 530–534.

Martin-Santos, R., Fagundo, A.B., Crippa, J.A., Atakan, Z., Bhattacharyya, S., et al., 2010. Neuroimaging in cannabis use: a systematic review of the literature. Psychol. Med. 40, 383–398.

Massi, P., Valenti, M., Vaccani, A., Gasperi, V., Perletti, G., et al., 2008. 5-Lipoxygenase and anandamide hydrolase (FAAH) mediate the antitumor activity of cannabidiol, a non-psychoactive cannabinoid. J. Neurochem. 104, 1091–1100.

Mathew, R.J., Wilson, W.H., 1993. Acute changes in cerebral blood flow after smoking marijuana. Life Sci. 52, 757–767.

McHugh, D., 2012. GPR18 in microglia: impliations for the CNS and endocannabinoid system signaling. Br. J. Pharmacol. 167, 1575–1582.

McPartland, J.M., 2004. Phylogenomic and chemotaxonomic analysis of the endocannabinoid system. Brain Res. Brain Res. Rev. 45, 18–29.

McPartland, J.M., Glass, M., Pertwee, R.G., 2007. Meta-analysis of cannabinoid ligand binding affinity and receptor distribution: interspecies differences. Br. J. Pharmacol. 152, 583–593.

Mechoulam, R., 1970. Marihuana chemistry: recent advances in cannabinoid chemistry open the area to more sophisticated biological research. Science 168, 1159–1166.

Mechoulam, R., Fride, E., Hanuš, L., Sheskin, T., Bisogno, T., et al., 1997. Anandamide may mediate sleep induction. Nature 89, 25–26.

Min, R., Di Marzo, V., Mansvelder, H.D., 2010. DAG lipase involvement in depolarization-induced suppression of inhibition: does endocannabinoid biosynthesis always meet the demand? Neuroscientist 16, 608–613.

Mishima, K., Hayakawa, K., Abe, K., Ikeda, T., Egashira, N., et al., 2005. Cannabidiol prevents cerebral infarction via a serotonergic 5-hydroxytryptamine 1A receptor-dependent mechanism. Stroke 36, 1077–1082.

Molina-Holgado, F., Rubio-Araiz, A., Garcia-Ovejero, D., Williams, R.J., Moore, J.D., et al., 2007. CB2 cannabinoid receptors promote mouse neural stem cell proliferation. Eur. J. Neurosci. 25, 629–634.

Moreira, F.A., Lutz, B., 2008. The endocannabinoid system: emotion, learning and addiction. Addict. Biol. 13, 196–212.

Morera-Herreras, T., Miguelez, C., Aristieta, A., Ruiz-Ortega, J.A., Ugedo, L., 2012. Endocannabinoid modulation of dopaminergic motor circuits. Front. Pharmacol. 3, 110. http://dx.doi.org/10.3389/fphar.2012.00110.

Morshead, C.M., van der Kooy, D., 2001. A new "spin" on neural stem cells? Curr. Opin. Neurobiol. 11, 59–65.

Mortati, K., Dworetzky, B., Devinsky, O., 2007. Marijuana: an effective antiepileptic treatment in partial epilepsy? A case report and review of the literature. Rev. Neurol. Dis. 4, 103–106.

Murillo-Rodríguez, E., Sanchez-Alavez, M., Navarro, L., Martínez-González, D., Drucker-Colín, R., et al., 1998. Anandamide modulates sleep and memory in rats. Brain Res. 812, 270–274.

Naderi, N., Aziz Ahari, F., Shafaghi, B., Najarkolaei, A.H., Motamedi, F., 2008. Evaluation of interactions between cannabinoid compounds and diazepam in electroshock-induced seizure model in mice. J. Neural Transm. 115, 1501–1511.

Narang, S., Gibson, D., Wasan, A.D., Ross, E.L., Michna, E., et al., 2008. Efficacy of dronabinol as an adjuvant treatment for chronic pain patients on opioid therapy. J. Pain 9, 254–264.

O'Leary, D.S., Block, R.I., Koeppel, J.A., Flaum, M., Schultz, S.K., et al., 2002. Neuropsychopharmacology 26, 802–816.

Ong, W.Y., Mackie, K., 1999. A light and electron microscopic study of the CB1 cannabinoid receptor in primate brain. Neuroscience 92, 1177–1191.

O'Shea, M., Singh, M.E., McGregor, I.S., Mallet, P.E., 2004. Chronic cannabinoid exposure produces lasting memory impairment and increased anxiety in adolescent but not adult rats. J. Psychopharmacol. 18, 502–508.

Ottria, R., Casati, S., Ciuffreda, P., 2012. Optimized synthesis and characterization of N-acylethanolamines and O-acylethanolamines, important family of lipid-signalling molecules. Chem. Phys. Lipids 165, 705–711.

Oudin, M.J., Gajendra, S., Williams, G., Hobbs, C., Lalli, G., et al., 2011. Endocannabinoids regulate the migration of subventricular zone-derived neuroblasts in the postnatal brain.

Palazuelos, J., Aguado, T., Pazos, M.R., Julien, B., Carrasco, C., et al., 2009. Microglial CB2 cannabinoid receptors are neuroprotective in Huntington's disease excitotoxicity. Brain 132, 3152–3164.

Pamplona, F.A., Ferreira, J., Menezes de Lima Jr., O., Duarte, F.S., Bento, A.F., et al., 2012. Anti-inflammatory lipoxin A4 is an endogenous allosteric enhancer of CB1 cannabinoid receptor. Proc. Natl. Acad. Sci. U.S.A 109, 21134–21139.

Pamplona, F.A., Prediger, R.D.S., Pandolfo, P., Takahashi, R.N., 2006. The cannabinoid receptor agonist WIN55,212-2 facilitates the extinction of contextual fear memory and sspatial memory in rats. Psychopharmacology 188, 641–649.

Parker, L.A., Limebeer, C.L., 2006. Conditioned gaping in rats: selective measure of nausea. Autonom. Neurosci. 129, 36–41.

Parmentier-Batteur, S., Jin, K., Mao, X.O., Xie, L., Greenberg, D.A., 2002. Increased severity of stroke in CB1 cannabinoid receptor knock-out mice. J. Neurosci. 22, 9771–9775.

Pazos, M.R., Nuñez, E., Benito, C., Tolon, R.M., Romero, J., 2005. Functional neuroanatomy of the endocannabinoid system. Pharmacol. Biochem. Behav. 81, 239–247.

Penzo, M.A., Peña, J.L., 2009. Endocannabinoid-mediated long-term depression in the avian midbrain expressed presynaptically and postsynaptically. J. Neurosci. 29, 4131–4139.

Pertwee, R.G., 2001. Cannabinoid receptors and pain. Prog. Neurobiol. 63, 569–611.

Pertwee, R.G., 2007. GPR55: a new member of the cannabinoid receptor clan? Br. J. Pharmacol. 152, 984–986.

Pertwee, R.G., 2009. Emerging strategies for exploiting cannabinoid receptor agonists as medicines. Br. J. Pharmacol. 156, 397–411.

Pertwee, R.G., 2012a. Targeting the endocannabinoid system with cannabinoid receptor agonists: pharmacological strategies and therapeutic possibilities. Philos. Trans. R. Soc., B 367, 3353–3363.

Pertwee, R.G., 2012b. Lipoxin A4 is an allosteric endocannabinoid that strengthens anandamide-induced CB1 receptor activation. Proc. Natl. Acad. Sci. U.S.A 109, 20781–20782.

Petrosino, S., Palazzo, E., de Novellis, V., Bisogno, T., Rossi, F., et al., 2007. Changes in spinal and supraspinal endocannabinoid levels in neuropathic rats. Neuropharmacology 52, 415–422.

Pisani, V., Moschella, V., Bari, M., Fezza, F., Galati, S., et al., 2010. Dynamic changes of anandamide in the cerebrospinal fluid of Parkinson's disease patients. Mov. Disord. 25, 920–924.

Porter, A.C., Sauer, J.M., Knierman, M.D., Becker, G.W., Berna, M.J., et al., 2002. Characterization of a novel endocannabinoid, virodhamine, with antagonist activity at the CB1 receptor. J. Pharmacol. Exp. Ther. 301, 1020–1024.

Price, M.R., Baillie, G.L., Thomas, A., Stevenson, L.A., Easson, M., et al., 2005. Allosteric modulation of the cannabinoid CB1 receptor. Mol. Pharmacol. 68, 1484–1495.

Rahn, E.J., Hohmann, A.G., 2009. Cannabinoids as pharmacotherapies for neuropathic pain: from the bench to the bedside. Neurotherapeutics 6, 713–737.

Ramaekers, J.G., Berghaus, G., van Laar, M., Drummer, O.H., 2004. Dose related risk of motor vehicle crashes after cannabis use. Drug Alcohol Depend. 73, 109–119.

Ranganathan, M., D'Souza, D.C., 2006. The acute effects of cannabinoids on memory in humans: a review. Psychopharmacology 188, 425–444.

Rentzsch, J., Buntebart, E., Stadelmeier, A., Gallinat, J., Jockers-Scherubl, M.C., 2011. Differential effects of chronic cannabis use on preattentional cognitive functioning in abstinent schizophrenic patients and healthy subjects. Schizophrenia Res. 130, 222–227.

Richardson, G.A., Ryan, C., Willford, J., Day, N.L., Goldschmidt, L., 2002. Prenatal alcohol and marijuana exposure: effects on neuropsychological outcomes at 10 years. Neurotoxicol. Teratol. 24, 309–320.

Robbe, D., Alonso, G., Manzoni, O.J., 2003. Exogenous and endogenous cannabinoids control synaptic transmission in mice nucleus accumbens. Ann. N.Y. Acad. Sci. 1003, 212–225.

Romero, J., Hillard, C.J., Calero, M., Rabano, A., 2002. Fatty acid amide hydrolase localization in the human central nervous system: an immunohistochemical study. Brain Res. Mol. Brain Res. 100, 85–93.

Roser, P., Della, B., Norra, C., Uhl, I., Brune, M., et al., 2010. Auditory mismatch negativity deficits in long-term heavy cannabis users. Eur. Arch. Psychiatry Clin. Neurosci. 260, 491–498.

Roser, P., Haussleiter, I.S., Chong, H.J., Maier, C., Kawohl, W., et al., 2011. Inhibition of cerebral type 1 cannabinoid receptors is associated with impaired auditory mismatch negativity generation in the ketamine model of schizophrenia. Psychopharmacology 218, 611–620.

Rubio-Araiz, A., Arevalo-Martin, A., Gomez-Torres, O., Navarro-Galve, B., Garcia-Ovejero, D., et al., 2008. The endocannabinoid system modulates a transient TNF pathway that induces neural stem cell proliferation. Mol. Cell. Neurosci. 38, 374–380.

Rueda, D., Navarro, B., Martinez-Serrano, A., Guzman, M., Galve-Roperh, I., 2002. The endocannabinoid anandamide inhibits neuronal progenitor cell differentiation through attenuation of the Rap-1/B-Raf/ERK pathway. J. Biol. Chem. 277, 46645–46650.

Russo, E.B., 2007. History of cannabis and its preparations in saga, science, and sobriquet. Chem. Biodiversity 4, 1614–1648.

Russo, E.B., Guy, G.W., Robson, P.J., 2007. Cannabis, pain, and sleep: lessons from therapeutic clinical trials of Sativex, a cannabis-based medicine. Chem. Biodiversity 4, 1729–1743.

Rydberg, E., Larsson, N., Sjöberg, S., Hjorth, S., Hermansson, N.-O., et al., 2007. The orphan receptor GPR55 is a novel cannabinoid receptor. Br. J. Pharmacol. 152, 1092–1101.

Sagar, D.R., Staniaszek, L.E., Okine, B.N., Woodhams, S., Norris, L.M., et al., 2010. Tonic modulation of spinal hyperexcitability by the endocannabinoid receptor system in a rat model of osteoarthritis pain. Arthritis Rheum. 62, 3666–3676.

Santucci, V., Storme, J.J., Soubrié, P., Le Fur, G., 1996. Arousal-enhancing properties of the CB1 cannabinoid receptor antagonist SR141716A in rats as assessed by electroencephalographic spectral and sleep-waking analysis. Life Sci. 58, PL103–PL110.

Sarne, Y., Asaf, F., Fishbein, M., Gafni, M., Keren, O., 2011. The dual neuroprotective-neurotoxic profile of cannabinoid drugs. Br. J. Pharmacol. 163, 1391–1401.

Saturnino, C., Petrosino, S., Ligresti, A., Palladino, C., De Martino, G., et al., 2010. Synthesis and biological evaluation of new potential inhibitors of N-acylethanolamine hydrolyzing acid amidase. Bioorg. Med. Chem. Lett. 20, 1210–1213.

Schicho, R., Storr, M., 2012. A potential role for GPR55 in gastrointestinal functions. Curr. Opin. Pharmacol. 12, 653–658.

Scotter, E.L., Abood, M.E., Glass, M., 2010. The endocannabinoid system as a target for the treatment of neurodegenerative disease. Br. J. Pharmacol. 160, 480–498.

Sedlacek, M., Tipton, P.W., Brenowitz, S.D., 2011. Sustained firing of cartwheel cells in the dorsal cochlear nucleus evokes endocannabinoid release and retrograde suppression of parallel fiber synapses. J. Neurosci. 31, 15807–15817.

Sewell, R.A., Poling, J., Sofuoglu, M., 2009. The effect of cannabis compared with alcohol on driving. Am. J. Addict. 18, 185–193.

Sheinin, A., Talani, G., Davis, M.I., Lovinger, D.M., 2008. Endocannabinoid- and mGluR5-dependent short-term synaptic depression in an isolated neuron/bouton preparation from the hippocampal CA1 region. J. Neurophysiol. 100, 1041–1052.

Shohami, E., Cohen-Yeshurun, A., Magid, L., Algali, M., Mechoulam, R., 2011. Endocannabinoids and traumatic brain injury. Br. J. Pharmacol. 163, 1402–1410.

Sigel, E., Baur, R., Rácz, I., Marazzi, J., Smart, T.G., et al., 2011. The major central endocannabinoid directl acts at GABA$_A$ receptors. Proc. Natl. Acad. Sci. U.S.A 108, 18150–18155.

Smith, A.M., Longo, C.A., Fried, P.A., Hogan, M.J., Cameron, I., 2010. Effects of marijuana on visuospatial working memory: an fMRI study in young adults. Psychopharmacology 210, 429–438.

Söderpalm, A.H.V., Schuster, A., Dew Wit, H., 2001. Antiemetic efficacy of smoked marijuana. Subjective and behavioral effects on nausea induced by syrup of ipecac. Pharmacol. Biochem. Behav. 69, 343–350.

Sofia, R.D., Solomon, T.A., Barry III, H., 1976. Anticonvulsant activity of Δ^9-tetrahydrocannabinol compared with three other drugs. Eur. J. Pharmacol. 35, 7–16.

Sokolic, L., Long, L.E., Hunt, G.E., Arnold, J.C., McGregor, I.S., 2011. Disruptive effects of the prototypical cannabinoid Δ^9-tetrahydrocannabinol and the fatty acid amide inhibitor [sic] URB-597 on go/no-go auditory discrimination performance and olfactory reversal learning in rats. Behav. Pharmacol. 22, 191–202.

Solowij, N., Jones, K.A., Rozman, M.E., Davis, S.M., Ciarrochi, J., et al., 2011. Verbal learning and memory in adolescent cannabis users, alcohol users and non-users. Psychopharmacology 216, 131–144.

Stadelmann, A.M., Juckel, G., Arning, L., Gallinat, G., Epplen, J.T., et al., 2011. Association between a cannabinoid receptor gene (CNR1) polymorphism and and cannabinoid-induced alterations of the auditory event-related P300 potential. Neurosci. Lett. 496, 60–64.

Stein, E.A., Fuller, S.A., Edgemond, W.S., Campbell, W.B., 1998. Selective effects of the endogenous cannabinoid arachidonylethanolamide (anandamide) on regional cerebral blood flow in the rat. Neuropsychopharmacology 19, 481–491.

Stiglick, A., Kalant, H., 1982. Learning impairment in the radial-arm maze following prolonged cannabis ntreatment in rats. Psychopharmacology 77, 117–123.

Stiglick, A., Kalant, H., 1985. Residual effects of chronic cannabis treatment on behavior in mature rats. Psychopharmacology 85, 436–439.

Svizenska, I., Dubovy, P., Sulcova, A., 2008. Cannabinoid receptors 1 and 2 (CB1 and CB2), their distribution, ligands and functional involvement in nervous system structures – a short review. Pharmacol. Biochem. Behav. 90, 501–511.

Takagi, M., Yucel, M., Cotton, S.M., Baliz, Y., Tucker, A., et al., 2011. Verbal memory, learning, and executive functioning among adolescent inhalant and cannabis users. J. Stud. Alcohol Drugs 72, 96–105.

Tan, H., Lauzon, N.M., Bishop, S.F., Chi, N., Bechard, M., et al., 2011. Cannabinoid transmission in the basolateral amygdala mediates fear memory formation via functional inputs to the prelimbic cortex. J. Neurosci. 31, 5300–5312.

Tchekmedyian, N.S., Halpert, C., Heber, D., 1992. Nutrition in advanced cancer: anorexia as an outcome variable and target of therapy. JPEN, J. Parenter. Enteral Nutr. 16 (Suppl. 6), 88S–92S.

Thomas, B.F., Compton, D.R., Martin, B.R., 1990. Characterization of the lipophilicity of natural and synthetic analogs of delta 9-tetrahydrocannabinol and its relationship to pharmacological potency. J. Pharmacol. Exp. Ther. 255, 624–630.

Tunving, K., Thulin, S.O., Risberg, J., Warkentin, S., 1986. Regional cerebral blood flow in long-term heavy cannabis use. Psychiatr. Res. 17, 15–21.

Vaidya, J.G., Block, R.I., O'Leary, D.S., Ponto, L.B., Ghoneim, M.M., et al., 2012. Effects of chronic marijuana use on brain activity during monetary decision-making. Neuropsychopharmacology 37, 618–629.

Valenti, M., Viganò, D., Casico, M.G., Rubino, T., Steardo, L., et al., 2004. Differential diurnal variations in anandamide and 2-rachidonoyl-glycerol levels in rat brain. Cell. Mol. Life Sci. 61, 945–950.

Van der Stelt, M., Di Marzo, V., 2003. The endocannabinoid system in the basal ganglia and in the mesolimbic reward system: implications for neurological and psychiatric disorders. Eur. J. Pharmacol. 480, 133–150.

Varvel, S.A., Hamm, R.J., Martin, B.R., Lichtman, A.H., 2001. Differential effects of Δ^9THC on spatial reference and working memory in mice. Psychopharmacology 157, 142–150.

Vaughan, C.W., 2006. Stressed-out endogenous cannabinoids relieve pain. Trends Pharmacol. Sci. 27, 69–71.

Vinogradova, L.V., Shatskova, A.B., van Rijn, C.M., 2011. Pro-epileptic effects of the cannabinoid receptor antagonist SR141716 in a model of audiogenic epilepsy. Epilepsy Res. 96, 250–256.

Viveros, M.P., Marco, E.M., File, S.E., 2005. Endocannabinoid system and stress and anxiety responses. Pharmacol. Biochem. Behav. 81, 331–342.

Viveros, M.P., Marco, E.M., López-Gallardo, M., García-Segura, L.M., Wagner, E.J., 2011. Framework for sex differences in adolescent neurobiology: a focus on cannabinoids. Neurosci. Biobehav. Rev. 35, 1740–1751.

Wagner, D., Becker, B., Gouzoulis-Mayfrank, E., Daumann, J., 2010. Interactions between specific parameters of cannabis use and verbal memory. Prog. Neuro-Psychopharmacol. Biol. Psychiatry 34, 871–876.

Walker, D.J., Suetterlin, P., Reisenberg, M., Williams, G., Doherty, P., 2010. Down-regulation of diacylglycerol lipase-alpha during neural stem cell differentiation: identification of elements that regulate transcription. J. Neurosci. Res. 88, 735–745.

Walter, C., Oertel, B.G., Felden, L., Nöth, U., Deichmann, R., et al., 2011. The effects of delta-9-tetrahydrocannabinol on nasal chemosensitivity: a pharmacological fMRI study in healthy volunteers [Abstract]. Naunyn-Schmiedeberg's Arch. Pharmacol. 383 (Suppl. 1), 75.

Wang, J., Ueda, N., 2009. Biology of endocannabinoid synthesis system. Prostaglandins Other Lipid Mediators 89, 112–119.

Ware, M.A., Fitzcharles, M.A., Joseph, J., Shir, Y., 2010a. The effects of nabilone on sleep in fibromyalgia: results of a randomized controlled trial. Anesth. Analg. 110, 604–610.

Ware, M.A., Wang, T., Shapiro, S., Robinson, A., Ducruet, T., et al., 2010b. Smoked cannabis for chronic neuropathic pain: a randomized controlled trial. Can. Med. Assoc. J. 182, E694–E701.

Watanabe, K., Kayano, Y., Matsunaga, T., Yamamoto, I., Yoshimura, H., 1996. Inhibition of anandamide amidase activity in mouse brain microsomes by cannabinoids. Biol. Pharm. Bull. 19, 1109–1111.

Wendt, H., Soerensen, J., Wotjak, C.T., Potschka, H., 2011. Targeting the endocannabinoid system in the amygdala kindling model of temporal lobe epilepsy in mice. Epilepsia 52, e62–65.

Whalley, B.J., Wilkinson, J.D., Williamson, E.M., Constanti, A., 2004. A novel component of cannabis extract potentiates excitatory synaptic transmission in rat olfactory cortex in vitro. Neurosci. Lett. 365, 58–63.

Wilson, W., Mathew, R., Turkington, T., Hawk, T., Coleman, R.E., et al., 2000. Brain morphological changes and early marijuana use: a magnetic resonance and positron emission tomography study. J. Addict. Dis. 19, 1–22.

Winton-Brown, T.T., Allen, P., Bhattacharyya, S., Borgwardt, S.J., Fusar-Poli, P., et al., 2011. Modulation of auditory and visual processing by delta-9-tetrahydrocannabinol and cannabidiol: an FMRI study. Neuropsychopharmacology 36, 1340–1348.

Wolf, S.A., Ullrich, O., 2008. Endocannabinoids and the brain immune system: new neurons at the horizon? J. Neuroendocrinol. 20 (Suppl. 1), 15–19.

Zajicek, J.P., Apostu, V.I., 2011. Role of cannabinoids in multiple sclerosis. CNS Drugs 25, 187–201.

Zhang, M., Martin, B.R., Adler, M.W., Razdan, R.K., Ganea, D., et al., 2008. Modulation of the balance between cannabinoid CB_1 and CB_2 receptor activation during cerebral ischemic/reperfusion injury. Neuroscience 152, 753–760.

Zhang, M., Martin, B.R., Adler, M.W., Razdan, R.K., Kong, W., et al., 2009. Modulation of cannabinoid receptor activation as a neuroprotective strategy for EAE and stroke. J. Neuroimmune Pharmacol. 4, 249–259.

Zhao, Y., Rubio, M., Tzounopoulos, T., 2009. Distinct functional and anatomical architecture of the endocannabinoid system in the auditory brainstem. J. Neurophysiol. 101, 2434–2446.

Zhu, P.J., 2006. Endocannabinoid signaling and synaptic plasticity in the brain. Crit. Rev. Neurobiol. 18, 113–124.

CHAPTER FOURTEEN

Cannabis, Cannabinoids, and the Association with Psychosis

Rajiv Radhakrishnan[1,3], Peter H. Addy[1], R. Andrew Sewell[1,2,3], Patrick D. Skosnik[1,2,3], Mohini Ranganathan[1,2,3] and Deepak Cyril D'Souza[1,2,3]

[1]Psychiatry Service, VA Connecticut Healthcare System, West Haven, CT
[2]Abraham Ribicoff Research Facilities, Connecticut Mental Health Center, New Haven, CT
[3]Department of Psychiatry, Yale University School of Medicine, New Haven, CT

1. INTRODUCTION

An association between cannabis and psychosis has long been suspected. The increasing rates of cannabis use, the use of cannabis for medical purposes, the increasing potency of cannabis, the earlier age of onset of cannabis use by adolescents, the greater availability of synthetic cannabinoids, and recent advances in our understanding of the role of endogenous cannabinoids in the brain have both, renewed and reinvigorated interest in the link between cannabis use and psychosis.

In this chapter, we review the literature pertaining to the association between cannabinoids and psychosis. At the outset, it is important to highlight a few distinctions. Firstly, the distinction between 'psychotic symptoms' and a 'psychotic disorder' is an important one. Psychotic symptoms refer to alterations in perception, hallucinations, delusions and disorganized thinking and speech. In experimental studies with cannabinoids, these symptoms are transient and abate spontaneously. A psychotic disorder, such as schizophrenia, is characterized by, not only positive psychotic symptoms but also negative symptoms (such as amotivation, social withdrawal and blunted affect) and cognitive deficits (such as impairments in attention, memory, working memory and executive functioning). Furthermore, these symptoms are persistent and result in significant socio-occupational dysfunction. Secondly, a distinction must be made between 'cannabinoids' and 'cannabis'. Much of the experimental evidence is derived from studies that use Δ^9-tetrahydrahydrocannabinol (Δ^9-THC), the main psychoactive ingredient of cannabis. Cannabis, however contains more than 70 different cannabinoids apart from Δ^9-THC, including cannabidiol (CBD) and cannabinol (CBN). Therefore, cannabis is more than just Δ^9-THC. Synthetic cannabinoids are compounds derived from analogues of cannabinoid receptor (CB-1R) agonists and bind to CB-1R with greater affinity, but are not known to be found in cannabis. Synthetic cannabinoids are hence not cannabis, and are more than just Δ^9-THC.

The Effects of Drug Abuse on the Human Nervous System
http://dx.doi.org/10.1016/B978-0-12-418679-8.00014-9

2.1. TRANSIENT BEHAVIORAL AND COGNITIVE EFFECTS OF CANNABINOIDS: NONEXPERIMENTAL EVIDENCE

2.1.1. Anecdotal Reports of Cannabis Effects and Surveys of Cannabis Users

Cannabis use was suggested as a cause of "moral degeneracy" probably as early as the Eleventh century A.D. as is illustrated in the story of Hasan ibn al-Sabbah. His fanatical group of followers, referred to as the "Hashishiyans", (a term from which the English word "assassin" is derived), allegedly earned the moniker because of their use of "hashish" (*Cannabis indica*) (Nahas, 1982). The earliest report of an association between cannabis and altered behavior can be traced back to 1235A.D. when Ibn Beitar, an Arabian physician, made the link between insanity and the consumption of "hashish" (Warnock, 1903). In the Nineteenth century, the French psychiatrist Jacques-Joseph Moreau a member of the famous Parisian Club des Hashischins, a club dedicated to the exploration of drug-induced experiences, was the first to describe the psychotomimetic effects of hashish in a systematic manner. In his 1845 book, Hashish and Mental Illness, he reported that hashish intoxication could precipitate "acute psychotic reactions, generally lasting but a few hours, but occasionally as long as weeks" (Moreau, 1973). Since then, there have been numerous anecdotal reports describing psychotic symptoms in the context of cannabis intoxication (Brook, 1984; Chopra and Smith, 1974; D'Souza, 2007; Grossman, 1969; Keeler and Moore, 1974; Marshall, 1897; Moreau, 1973; Smith, 1968; Spencer, 1971; Talbott and Teague, 1969; Thacore, 1973). The symptoms commonly reported include depersonalization, derealization, paranoia, ideas of reference, flight of ideas, pressured thought, disorganized thinking, persecutory delusions, grandiose delusions, auditory and visual hallucinations, and impairments in attention and memory in an otherwise clear consciousness. The subjective experience with cannabis however varies to a great degree and depends on various extrinsic and intrinsic factors, including personality traits, such as schizotypy (Barkus et al., 2006), individual set of beliefs/expectations and the setting in which cannabis is used (Adamec et al., 1976). While rich in detail, individual accounts are fraught with some confounds and are difficult to generalize.

Some of the limitations of anecdotal accounts can be addressed in population-based surveys. The Indian Hemp Drugs Commission Report (1894), an exceptionally detailed report comprising some seven volumes and 3,281 pages, documents that cannabis use may have been responsible for psychotic reactions in 222 of 2344 cases (i.e. 9.5%) admitted in asylums in India. More recent studies suggest that between 20% and 50% of individuals report paranoia, persecutory ideas, and hallucinations while under the influence of cannabis (Green et al., 2003; Reilly et al., 1998).

2.1.2. Effects of Medicinal Cannabinoids

Treatment with synthetic cannabinoids has also been linked to transient psychotic symptoms with. Dronabinol (Δ^9-THC), nabilone (9-trans-ketocannabinoid) and levonantradol have been used to manage pain syndromes, chemotherapy-induced nausea and spasticity from multiple sclerosis. Patients being treated with these synthetic cannabinoids have reported to experience transient psychotic symptoms, such as "loss of control", thought disturbances, feelings of unreality, apprehension, fear and paranoia, anxiety and panic, dissociation, depersonalization, dysphoria, difficulty concentrating, hallucinations, perceptual alterations, amnesia and anxiety (Citron et al., 1985; Cronin et al., 1981; Diasio et al., 1981; Heim et al., 1984; Heim et al., 1981; Jain et al., 1981; Kenny and Wilkinson, 1982; Laszlo et al., 1981; Leweke et al., 1999; 1998; Sheidler et al., 1984; Stambaugh et al., 1984; Stuart-Harris et al., 1983; Volkow et al., 1991; Wesnes et al., 2009). The severity of these symptoms appears to be dose-dependent and dependent on the affinity of the compound for the CB-1R receptor. In fact, Levonantradol, which has 30 times higher affinity for the CB-1R receptor than Δ^9-THC and was being developed as an analgesic agent had to be abandoned from further drug development because of the high incidence of intolerable behavioral side effects.

In two systematic reviews of randomized clinical trials that examined the efficacy of synthetics cannabinoids for chemotherapy-induced nausea and vomiting, the use of these synthetic cannabinoids led to hallucinations in 6% of patients, paranoia in 5% of patients, and was responsible for nearly 30% of dropouts, primarily because of intolerable behavioral effects. (Machado Rocha et al., 2008; Tramer et al., 2001). Importantly, in these trials, hallucinations and paranoia were seen exclusively with cannabinoids, and not with other antiemetic agents. These effects appear to increase both with increasing potency, dose and with repeated dosing.

2.1.3. Synthetic Cannabinoids (Spice, K2)

The phenomenon of "Spice" provides another very recent source of nonexperimental evidence linking cannabinoids to psychosis. "Spice" refers to synthetic cannabinoids that have been sprayed onto a herbal substrate of dried leaves, flowers and stems from a number of plants. These products are marketed under a number of suggestive names, like "Spice", K2, Yucatan Fire, Skunk, Moon Rocks, and others. For convenience, we heretofore refer to these products as "Spice". It is sold for $10–50 in attractive and colorful packages, weighing about 3 gm, via the internet, headshops, and gas stations. "Spice" does not contain tobacco or phytocannabinoids (such as Δ^9-THC or CBD). Instead it contains one or more synthetic cannabinoids from the Pfizer (CP) group (CP-47,497 and CP-47,497-C8), the aminoalkylindole group (JWH-018, JWH-073, JWH-081, JWH-122, JWH-210 JWH-250) the HU group (HU-211) or the benzoylindole group (RCS-4) (Auwarter et al., 2009; Dresen et al., 2010; Lapoint et al., 2011; RTL, 2012; Schneir et al., 2011; Uchiyama et al., 2011; Uchiyama et al., 2009; Uchiyama et al., 2010). These compounds are high affinity, full agonists of brain cannabinoid receptors and are between 10–200 times more potent than THC. While the product labels list a number of herbs (e.g. alfalfa, blue violet, nettle leaf, etc.) often, these cannot be detected or are not relevant to the effects of "Spice". Similarly, the absence of cannabinoid-2 receptor

(CB2R) agonists, which are generally not psychoactive, in "Spice" suggests that it is sold for its psychoactive effects. Even though "Spice" is advertised as "natural herbs" or "harmless incense," "not for human consumption" or "for aromatherapy only," it is typically smoked. This labeling places the burden of responsibility on the user.

"Spice", has become the second most frequently used illicit substance after herbal cannabis in the U.S. (NIDA, 2011). The promise of a stronger high than cannabis, easy access, affordability, a perception that the products are legal, a perception that the products are safe "like marijuana", and the difficulty in detecting these compounds in standard urine toxicology tests have likely contributed to "Spice"'s growing popularity and use. The use of "Spice" is increasing amongst youth in the US. In the most recent Monitoring the Future survey, 11% of high school seniors nationwide admitted using "Spice" in the previous year, compared to 36% who admitted using cannabis in the past year (NIDA, 2011). "Spice" is typically smoked using a pipe resulting in rapid (10–20 minutes) onset of effects (Every-Palmer, 2011; Vandrey et al., 2012).

2.1.3.1. *The Acute Effects of "Spice"*

There are no published controlled data on the effects of "Spice". The synthetic cannabinoids present in "Spice" have not undergone the rigorous testing to which most other drugs are subjected. Most of the information about "Spice" effects consists of retrospective case reports from ER visits (Lapoint et al., 2011; Muller et al., 2010a; Schneir et al., 2011) and Poison Control Centers (AAPCC, 2011), media and law-enforcement reports of catastrophic events associated with "Spice" use (Businessweek, 2011; CBS, 2011; CNN, 2011; News, 2011; NPR, 2011; Post, 2011) and survey data (Every-Palmer, 2011; Hu et al., 2011; Vandrey et al., 2012). Reported effects reported to Poison Control Centers include agitation, drowsiness, and hallucinations (62% of calls). Emergency room reports from the US document more severe effects including anxiety, agitation, disorientation, hallucinations, and paranoia (Benford and Caplan, 2011; Lapoint et al., 2011; Mir et al., 2011; Schneir et al., 2011; Vearrier and Osterhoudt, 2010). Users in an internet survey most commonly endorsed feeling the following effects "most of the time" or "every time" they used "Spice": hallucinated (3%), felt paranoid (11%), produced a dream-like state (26%) (Vandrey et al., 2012).

Psychotic relapses may occur following the use of "Spice" in patients with preexisting psychotic disorders (Every-Palmer, 2010, 2011; Muller et al., 2010a). Müller (Muller et al., 2010b) reported a 25-year-old man with a family history of schizophrenia and a personal history of psychotic episodes precipitated by cannabis use. The patient had been stable for 2 years until he smoked "Spice" 3 times over the course of a month. He was hospitalized for exhibiting anxiety, paranoid delusions, and hallucinations. Every-Palmer described five forensic patients with sudden agitation, disorganization, and delusions after using "Spice" containing JWH-018 and/or CP47,497 (Every-Palmer, 2010). Interestingly, only one of the five patients displayed insight into the possible psychotogenic nature of "Spice" (Every-Palmer, 2010). Every-Palmer conducted a follow-up survey of 15 inpatients with Serious Mental Illness in a forensic psychiatric facility (Every-Palmer, 2011). Patients exhibited psychotic symptoms and anxiety after smoking "Spice"; however, few reported tolerance (23%) and none reported withdrawal symptoms.

The adverse clinical events reported in case reports with the use of "Spice" involve altered consciousness, confusion, anxiety, irritability, agitation, paranoia, hallucinations, and psychosis (Every-Palmer, 2010; Hurst et al., 2011; Muller et al., 2010b; Schneir et al., 2011; Vearrier and Osterhoudt, 2010). Psychotic symptoms are reported in patients with no previous history of psychotic symptoms. However, the majority of case reports to date discuss people 25 years or younger (youngest cases reported age 16 (Cohen et al., 2012; Mir et al., 2011)). Therefore it is possible that "Spice" exacerbated a preexisting prodrome syndrome. Case reports and cross-sectional surveys are only able to show correlation and cannot elucidate causation.

The sparse literature on "Spice" effects reviewed above has a number of limitations, including selection bias, a reliance on the accuracy of written record or subject recall, the uncontrolled nature of the evidence, the inadequate characterization of cases, the lack of standardized assessments, the confounding effects of other drugs used concomitantly with "Spice", variable dose and routes of administration, and variable set and setting. Cases reported by the media and law enforcement may represent extremes that might not be generalizable. The temporal profile, range, and intensity of "Spice" effects, and whether the effects are dose-related and biphasic, are not known. Furthermore, the relationship between dose, effects, and blood/urine levels of the parent compound and metabolites is not known.

Analysis of the content of "Spice" suggest that adverse events following consumption of synthetic cannabinoid preparations are unlikely to be due to impurities or residue from the manufacturing process, but rather to effects of the active drug or interactions with other psychoactive chemicals from herbs blended into products marketed as cannabis alternatives (Ginsburg et al., 2012).

2.1.3.2. The Changing Composition of "Spice"

One unique aspect of the "Spice" phenomenon is that the constituents of "Spice" are changing over time. This might be related to manufacturers staying one step ahead of legislation. The first generation of "Spice" contained CP-47,497-C8, JWH-018 and JWH-073. As first observed in the UK (Dargan et al., 2011), Germany (Lindigkeit et al., 2009), and Japan (Kikura-Hanajiri et al., 2011), as these compounds were banned, they were replaced by structurally similar and pharmacologically active compounds that skirted the regulations. In March 2011, the U.S. Drug Enforcement Agency (DEA) placed JWH-018, JWH-073, JWH-200, CP-47,497, and (C8)-CP47,497 present in "Spice", on the Schedule 1 list under 21 U.S.C. 811(h) of the Controlled Substances Act (21 CFR Part 1308) (NIDA, 2011). The banned compounds have been replaced in the U.S. by structurally related ones including JWH-081, JWH-250, JWH-122, and JWH-210 (Today, 2011). Furthermore, some "Spice" samples are explicitly labeled not to contain banned cannabinoids.

While the specific combination of synthetic cannabinoids in "Spice" may change over time, the primary site and mechanism of action of its components remains unchanged. However, unlike Δ9-THC—the principal active constituent of cannabis, which is a low efficacy, partial agonist at CB-1RRs—the cannabinoids in "Spice" are high potency, high efficacy, high affinity, full agonists at CB-1RR. (Atwood et al., 2010; Atwood et al., 2011; Huffman and Padgett, 2005; Huffman et al., 2005; Lindigkeit et al., 2009; Marriott and Huffman, 2008). This may explain the higher rates of adverse events associated with "Spice" use relative to cannabis use. Furthermore, people who do not appreciate the differences between cannabis and these high potency, full agonist synthetic cannabinoids may be lulled into a false sense of security and therefore, be more likely to use "Spice" and experience adverse events.

2.2. TRANSIENT BEHAVIORAL AND COGNITIVE EFFECTS OF CANNABINOIDS: EXPERIMENTAL EVIDENCE

2.2.1 Gendral Studies

Studies conducted in a scientific manner provide an opportunity to control variables such as dose and setting, and to objectively measure positive symptoms, negative symptoms, cognitive and psychophysiological effects. In one of the earliest semi-experimental studies, Siler et al. (1933) examined soldiers in the Panama after they were given cannabis and reported that it led to "mild intoxication" but not psychotic symptoms (Siler et al., 1933). The cannabis was of a strain that was grown in the Isthmus of the Panama, and the content of Δ^9-THC may have been variable. One of the earliest studies using a method of standardization is contained in the so-called "Mayor's Report" or "LaGuardia Committee Report" of 1944. The mayor of New York City, Mayor Fiorello LaGuardia was concerned about the widespread use of cannabis and appointed a Committee on Marihuana in 1939 in order to assess the nature and extent of physiological and psychological effects of cannabis on humans. The clinical portion of the study was conducted in 72 inmates from penitentiaries in New York who were given about 2–5 ml of cannabis concentrate (equivalent to 30–75 mg oral tetrahydrocannabinol and 830 mg smoked tertahydrocannabinol). Nine (12.5%) of the subjects developed psychotic reactions. The symptoms resolved within 34 h in 6 subjects, persisted for days to months in 2 subjects, while one subject who was given numerous, large doses of cannabis concentrate, went on to develop a persistent psychotic illness similar to schizophrenia. Cannabis was also found to produce significant dose-related impairments in memory, problem-solving ability, hand steadiness and balance (Mayor's Committee on Marijuana, 1944). Although the study has serious limitations in terms of the mental health status of the subjects given that they were prison inmates and that they had been exposed to various other drugs over their lifetime, the study drew attention to the negative consequences of cannabis use. Ames (1958) examined the effects of unassayed oral doses of cannabis extract in 10 medically trained individuals. Subjects were given between 240420 mg of cannabis extract (50 to 70 mg Δ^9-THC) and their subjective experiences were documented. Despite interindividual variations in experiences, the most frequently reported psychotic symptoms were perceptual changes, fragmented thinking, dissociation between thoughts and action, visual illusions and hallucinations, derealization and depersonalization. Some subjects experienced paranoia, which took on delusional proportions. The experiences ranged from a conviction that there were hidden recorders in the room, to fears that the individual was going to be hypnotized or subjected to electroconvulsive therapy. One subject refused to answer questions altogether for fear of being certified as insane. Other similar quasi-experimental studies of cannabis have reported a range of dose-related psychotic symptoms with cannabis (Isbell et al., 1967; Isbell and Jasinski, 1969; Renault et al., 1974; Siler, 1933; Williams et al., 1946).

The first controlled double-blind study was conducted in 1968 in the Behavioral Pharmacology Laboratory of Boston University. Weil et al (1968) administered cannabis extract containing 0.9% Δ^9-THC in a double blind manner and did not observe any psychotic effects. They however, reported that experienced users were more susceptible to perceptual effects compared to naïve users, suggesting that there may be some form of "pharmacological sensitization" with continued use (Weil et al., 1968). This was further tested by Jones and Stone, (1970) who examined the effects of smoked and oral cannabis extract (0.5 g containing 0.9% Δ^9-THC) in 10 heavy cannabis users, but found that they had lesser effects suggestive of "tolerance" (Jones and Stone, 1970). They also noticed significant differences between the oral and smoked routes. Melges et al, 1970 administered oral Δ^9-THC in three doses (20 mg, 40 mg, 60 mg) in a randomized, double-blind placebo-controlled study in 8 healthy individuals (Melges et al., 1970). Subjects were noticed to experience significant difficulty integrating their thoughts and manifested loose-associations and a lack of goal-directedness, apart from deficits in short-term memory. Renault et al wondered whether the route of administration made a difference to the effects observed. They developed a system to deliver cannabis smoke assayed to contain 1.5% Δ^9-THC (Renault et al., 1971) and subsequently used their apparatus to examine the effects of cannabis smoke in humans (Renault et al., 1974). Psychosis was observed in one of the 7 subjects who participated in a 10 day, multiple dosing paradigm (Renault et al., 1974). The method was however not foolproof and an unmeasurable amount of Δ^9-THC was lost in the process of combustion and delivery. Kaufmann et al, (2010) examined the effects of oral cannabis extract on a sample consisting of only female subjects. Cannabis, but not diazepam, increased scores on the Brief Psychiatric Rating Scale (BPRS) a scale that is commonly used to measure symptoms in schizophrenia. One of the subjects experienced "severe" psychotic symptoms that lasted for several hours (Kaufmann et al., 2010). Another aspect that has come to light is that different cannabinoids may have different neurophysiological and behavioral effects, although these studies await replication. Fusar-Poli et al (2009) used oral doses of Δ^9-THC (10 mg) and cannabidiol (CBD) (600 mg), the other main psychoactive ingredient of cannabis in a functional magnetic resonance (fMRI) paradigm that examined brain responses to emotional expression of faces (Fusar-Poli et al., 2009). Δ^9-THC was found to result in increased psychotic symptoms and increased skin conductance responses while processing fearful faces. CBD, on the other hand led to a reduction in anxiety and a decrease in skin conductance response. Δ^9-THC and CBD were also found to have opposite effects on blood oxygen-level dependent (BOLD) responses in tasks of verbal recall, response-inhibition, processing fearful facial expressions, auditory processing and visual processing (Bhattacharyya et al., 2010). The relatively small sample size, the route of drug administration and lack of assessment of drug availability and the fact that the authors' selected areas where activation was in opposite directions post hoc, are significant limitations of this study.

Other studies have used the intravenous route to examine the psychotomimetic properties of Δ^9-THC in order to ensure standardization of drug-delivery and adequate plasma levels, both in healthy subjects (D'Souza et al., 2004) (D'Souza et al., 2008b) (Morrison et al., 2009) (Morrison et al., 2011) and in patients with schizophrenia (D'Souza et al., 2005). D'Souza et al. (2004), administered intravenous Δ^9-THC in two doses (2.5 mg, and 5 mg), in a double blind, randomized, placebo-controlled study in healthy adults ($n = 22$). Subjects were screened to rule out significant psychiatric disorder or family history of Axis I disorders (D'Souza et al., 2004). The study found that Δ^9-THC produced transient positive psychotic symptoms (Figure 1) including

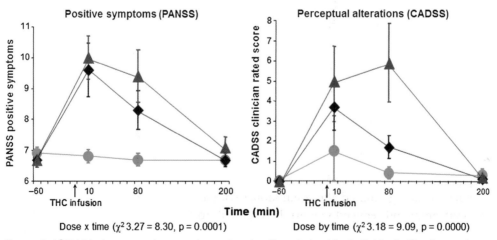

Figure 1 Δ^9-THC induces transient psychotomimetic effects in healthy individuals. The figure shows the effects of Δ^9-THC on the seven-item positive symptom subscales of the Positive and Negative Syndrome Scale (PANSS) (left panel) and on the eight-item clinician-rated subscale of the Clinician Administered Dissociative Symptoms Scale (CADSS) (right panel). (Green line = placebo, Red line = 2.5 mg Δ^9-THC, Blue line = 5 mg Δ^9-THC.)

perceptual alterations, negative symptoms, mood symptoms such as euphoria and anxiety, and cognitive deficits, especially in attention, working memory and verbal recall, without altering general orientation. In a similar study in healthy individuals, using almost identical methods, Morrison et al (2009), showed that intravenous Δ^9-THC (2.5 mg) produced similar effects on positive psychotic symptoms, mood, and cognition.

We provide a brief summary of the effects observed with cannabis, Δ^9-THC and other synthetic cannabinoids below. It is interesting to note that these compounds produce the full range of positive psychotic symptoms, negative symptoms and cognitive deficits seen in schizophrenia.

2.2.2. Positive Symptoms

Cannabis, Δ^9-THC and other cannabinoids induce a range of transient, positive psychotic symptoms which are qualitatively similar to that seen in schizophrenia. These symptoms include suspiciousness, paranoid and grandiose delusions, conceptual disorganization, fragmented thinking, and perceptual alterations. Additionally, Δ^9-THC results in depersonalization, derealization, distorted sensory perceptions, altered body perception, and feelings of unreality. Time perception abnormalities are known to occur in schizophrenia, but have received little attention (Andreasen et al., 1999; Carroll et al., 2009; Davalos et al., 2003; Tysk, 1983). Cannabinoids have been shown to alter time perception in both preclinical (Conrad et al., 1972; Han and Robinson, 2001; McClure and McMillan, 1997; Schulze et al., 1988) and clinical studies (Hicks et al., 1984; Mathew et al., 1998; McDonald et al., 2003; Sewell and D'Souza, April 3rd, 2011; Stone et al., 2010; Tinklenberg et al., 1972). Cannabinoids have also been found to disrupt performance on the binocular depth inversion task, a potential surrogate marker for psychosis seen in patients with acute paranoid schizophrenic or schizophreniform psychosis (Koethe et al., 2006). This effect has been observed with cannabis resin (Emrich et al., 1991), nabilone (a synthetic analogue of Δ^9-THC) (Leweke et al., 2000), dronabinol (a synthetic analogue of THC) (Koethe et al., 2006), and in chronic cannabis users (Semple et al., 2003).

2.2.3. Negative Symptoms

Δ^9-THC also produces a range of effects similar to the negative symptoms of schizophrenia, including blunted affect, emotional withdrawal, psychomotor retardation, lack of spontaneity and reduced rapport. It is difficult to tease out whether these "negative symptoms" were primary or were a consequence of the sedating and cataleptic effects of cannabinoids observed in animal studies. It is also unclear if the negative symptoms were an external manifestation of internal preoccupation with positive psychotic experiences. Furthermore, acute pharmacological studies may be limited in their capacity to "model" negative symptoms.

2.2.4. Cognitive Deficits

Cannabis, Δ^9-THC and other synthetic cannabinoids also produce transient, dose-related cognitive impairments, especially in the domains of verbal learning, short-term memory, working memory, executive function, abstract ability, decision making, and attention (Hart et al., 2001; Heishman et al., 1990; Hooker and Jones, 1987; Leweke et al., 1998; Marks and MacAvoy, 1989; Miller et al., 1977; Ranganathan and D'Souza, 2006). These effects are not limited to humans and are also seen in rodents and nonhuman primates (reviewed in Lichtman et al., 2002; Wilson and Nicoll, 2002). Interestingly, the profile of impairment observed in different cognitive domains is similar to that observed in schizophrenia (Heinrichs and Zakzanis, 1998). The cognitive impairment produced by Δ^9-THC is most pronounced in the domain of verbal learning and memory (Ranganathan and D'Souza, 2006), which is also one of the domains of significant impairment in schizophrenia (Heinrichs and Zakzanis, 1998). Figure 2 illustrates the effects of Δ^9-THC on the Hopkins Verbal Learning Test (HVLT) in healthy subjects (D'Souza et al., 2005). Δ^9-THC produced robust dose-dependent impairments on both, immediate and delayed (30-minute) verbal recall. Δ^9-THC also increased the number of "false positives" and "intrusions" on the HVLT. Similar findings have been recently reported by Henquet et al., (Henquet et al., 2006) and Morrison et al (Morrison et al., 2009).

The acute effects of cannabinoids are likely modulated by genetic and personality factors. This would explain why only a small minority of people experience the

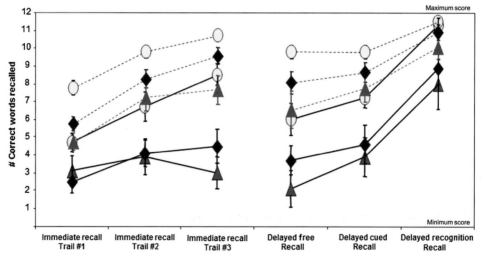

Figure 2 Δ^9-THC impairs verbal memory. The figure shows the effects of Δ^9-THC on verbal learning and recall (means $\pm$ SEM) as measured by the Hopkins Verbal Learning Test in healthy subjects (dashed lines) and patients with schizophrenia (solid lines). (Green line = placebo, Red line = 2.5 mg Δ^9-THC, Blue line = 5 mg Δ^9-THC.)

psychotomimetic effects of cannabinoids. Henquet et al (2006) examined the effects of the interaction of Catechol-O-methyl transferase (COMT) polymorphism and a trait index of psychosis liability on smoked Δ^9-THC (0.3 mg/kg) on cognitive performance and psychosis in 30 healthy individuals (Henquet et al., 2006). They found that individuals with the Val/Val polymorphism and high scores on psychosis liability had higher Δ^9-THC-induced psychotic symptoms.

2.3. EFFECTS OF CANNABINOIDS ON SCHIZOPHRENIA PATIENTS

The three drugs most commonly misused by schizophrenia patients are tobacco, alcohol, and cannabis (Margolese et al., 2004; McCreadie, 2002). Of illicit drugs, cannabis is the most frequently used (Bersani et al., 2002; Buhler et al., 2002; Cuffel et al., 1993; Farrell et al., 1998; Fowler et al., 1998; Green et al., 2005; Hambrecht and Hafner, 1996; Jablensky et al., 2000; Kessler et al., 1995; Linszen et al., 1994; McCreadie, 2002; Mueser et al., 1992). A recent meta-analysis showed that about 25% of schizophrenia patients in clinical samples have been diagnosed with cannabis abuse or dependence (Koskinen et al., 2009); in contrast, the rate of cannabis abuse and dependence in the general population has been estimated to be 1.13% and 0.32%, respectively (Compton et al., 2004). Information about the acute effects of cannabis and cannabinoids on schizophrenia patients comes from subjective reports, retrospective self-assessments, longitudinal and observational studies, and experimental studies.

2.3.1. Nonexperimental

2.3.1.1. Retrospective Self-assessments

Drug users' stated beliefs about the effects that they experience from drugs are often inaccurate but nevertheless frequently drive drug-taking behavior (Dixon et al., 1991). Behavior is often based on attitudes that spring from beliefs; moreover, beliefs that are based on personal experience have a stronger influence in the formation of attitudes than information gained in other ways, and better predict later behavior. Thus, what cannabis-using schizophrenia patients believe are the effects that cannabis has on them can be a crucial determinant in their use of cannabis, including whether they continue to use or relapse.

Many qualitative studies have been conducted in North America, Australia, and the United Kingdom investigating self-reported reasons for drug use in patients with psychotic disorders. Some protocols ask open-ended questions (Baigent et al., 1995; Fowler et al., 1998); in others patients select reasons for use from predetermined lists (Addington and Duchak, 1997; Dixon et al., 1991; Test et al., 1989; Warner et al., 1994). These studies reveal three main beliefs that motivate use of cannabis:

1. Cannabis use causes relaxation and enhances positive mood (Dixon et al., 1991; Fowler et al., 1998).

2. Cannabis use decreases negative emotions (such as depression) (Addington and Duchak, 1997; Baker et al., 2002; Dixon et al., 1991; Fowler et al., 1998; Gearon et al., 2001; Goswami et al., 2004; Green et al., 2004c; Schofield et al., 2006; Spencer et al., 2002).

3. Cannabis use enhances social occasions and makes it easier to cope with people (Fowler et al., 1998; Test et al., 1989).

Cannabis is also reported to reverse neuroleptic effects. In one study, half of 45 participants reported that they were using drugs (including cannabis) or alcohol to cope with or reduce auditory hallucinations feelings of suspiciousness or paranoia, and two out of five were using drugs when they were experiencing medication side effects (Gregg et al., 2009). However, a recent review of 14 studies specifically addressing self-reported reasons for cannabis use in patients with psychotic disorders concluded that the three reasons listed above were the most cited, with only a minority reporting that cannabis use relieved medication side effects or symptoms of psychosis such as hallucinations and suspiciousness (Dekker et al., 2009). Patients with psychosis appear more likely to use cannabis for mood elevation than for relaxation; less likely to use out of habit or for social reasons, and more likely to use in order to cope with boredom or other negative affective states (Green et al., 2004a).

Despite these expectations, however, the actual subjective effects reported by patients following cannabis use are often negative. Although patients initially report feeling "inspired, relaxed, energized, or active" following cannabis use, they also frequently describe an exacerbation of positive psychotic symptoms and increases in dysphoria and aggression following the initial positive effects (Knudsen and Vilmar, 1984). Some patients report cannabis to be unpleasant and to cause adverse psychic effects (Negrete et al., 1986). Severe exacerbations of psychosis and functional deterioration sometimes follow periods of moderate cannabis use (Treffert, 1978), and more than 50% of subjects in one study reported that cannabis increased their positive symptoms (Addington and Duchak, 1997). Although some schizophrenia patients experience reductions in anxiety, depression and negative symptoms, others report increased suspiciousness, derealization, and variable effects on hallucinations (Arndt et al., 1992; Dixon et al., 1990; Hekimian and Gershon, 1968; Peralta and Cuesta, 1992; Weil, 1970). Recently diagnosed schizophrenia patients have also reported increased visual and auditory hallucinations and confusion after cannabis use, as well as longer-term social problems, depression, and less control over thoughts (Peters et al., 2009a).

It is possible that from the patients' viewpoint cannabis use is beneficial because it decreases their preoccupation with psychotic symptoms, even as it increases objectively measured symptoms. Alternatively, while cannabis may increase psychotic symptoms, it may also reduce symptom-related distress so that schizophrenia patients experience the overall effects of cannabis as beneficial. However, retrospective self-report data are subject to distortion. Individuals who misuse drugs typically use denial and rationalization to justify their use. In addition, cannabis alters perception and has amnestic effects that may influence the interpretation of events and therefore interfere with the accurate recall of drug effects. Longitudinal studies that follow cohorts of schizophrenia patients over time and observational studies that assess cognitive and behavioral functioning are better able to measure objective data such as hospital admissions or changes in cognition.

2.3.1.2. Longitudinal and Observational Studies

Most longitudinal studies find that the more cannabis schizophrenia patients use, the more problems they have (Andreasson et al., 1987; Linszen et al., 1994; Rais et al., 2008; Yucel et al., 2008). Cannabis use interferes with the therapeutic alliance (Dixon, 1999; Wilk et al., 2006), is associated with less access to nonpharmacological interventions (Regier et al., 1990; Wilk et al., 2006), and increased chance of prematurely early discharge from the hospital (Brunette et al., 1997), and an increased need for interventions and longer hospitalizations (Kivlahan et al., 1991). Cannabis-using schizophrenia patients have worse psychopathology and higher relapse rates (Johns, 2001; Linszen et al., 1994; Swartz et al., 2008), as well as an increased need for interventions and longer hospitalizations (Kivlahan et al., 1991). A recent prospective cohort study of 145 male schizophrenia patients followed for 12 months found that those who used cannabis were more frequently hospitalized than noncannabis-using patients although they did not differ with respect to psychopathology (van Dijk et al., 2012).

Although most studies concur that cannabis use has a negative effect of cannabis use on the course of schizophrenia, some studies find positive or equivocal effects as well. One study found no baseline differences in psychopathology between cannabis users and nonusers (Zisook et al., 1992). In the CATIE cohort, cannabis use was even associated with equal or higher psychosocial functioning compared to abstinent patients (Swartz et al., 2006). One group of cannabis-using schizophrenia patients had better cross-sectional scores on affective measurements compared with their noncannabis-using counterparts (Peralta and Cuesta, 1992), and a subgroup of schizophrenia subjects that used cannabis less than once daily in a Dutch cohort reported similar positive effects of cannabis on anxiety and depressive symptoms (Linszen et al., 1994). Other authors have also reported less anxiety and fewer depressive symptoms in schizophrenic patients using cannabis (Johns, 2001). Cannabis-using schizophrenia patients have lower negative-symptom scores (Peralta and Cuesta, 1992), and cannabis-using adolescents with first-episode psychosis have lower negative symptom scores and a better prognosis that those who do not use cannabis (Baeza et al., 2009).

The literature on the effect of cannabis use on cognitive functioning in patients with schizophrenia is also mixed. While acutely, cannabinoids appear to worsen cognitive test performance in schizophrenia patients, cross-sectional studies suggest that schizophrenia patients who abuse cannabis have better cognitive performance (Joyal et al., 2003; Løberg and Hugdahl, 2009; Potvin et al., 2005; Ringen et al., 2010; Scholes and Martin-Iverson, 2010; Stirling et al., 2005). A cross-sectional study of nearly a thousand schizophrenia patients compared with nearly a thousand sibling and over 500 controls in a battery of cognitive tasks found poorer performance on immediate verbal learning, processing speed, and working memory in the current cannabis users, but better performance on acquired knowledge, facial affect recognition, and face recognition identity in lifetime users (Meijer et al., 2011). More recently, DeRosse et al (2010) showed that patients with schizophrenia who also abused cannabis had better performance on measures of processing speed, verbal fluency and verbal learning and memory that patients with schizophrenia who did not use cannabis (DeRosse et al., 2010). The study was subsequently critiqued for inadequate methods of analysis (Suckling, 2011). However, a review of 23 studies on cannabis, schizophrenia, and cognition by Løberg and Hugdahl, (2009) also found that 14 studies reported *better* cognition in the cannabis-using groups (Løberg and Hugdahl, 2009).

Furthermore, two recent meta-analysis, one with 8 studies on the effects of cannabis on neurocognitive functioning in schizophrenia with a total sample size of 942 (Rabin et al., 2011), and the other with 10 studies (total sample size = 572) (Yucel et al., 2012) concluded that patients with schizophrenia who use cannabis (Rabin et al., 2011) and patients with first-episode psychosis who use cannabis (Yucel et al., 2012) have better neurocognitive performance. The studies are cross-sectional in design and hence cannot ascribe causality to the association between cannabis use and better neurocognitive functioning. While it is possible that cannabis use improves cognition, it is more likely that cannabis using patients have higher baseline neurocognitive functioning which would be essential in order for them to be able to procure and use an illegal substance, while at the same time avoid the legal system. Hence, although there is a deterioration in cognitive performance with the pathological processes associated with schizophrenia, the performance could appear to be better when examined in cross-sectional studies. This notion is also supported by the fact that the improved cognitive performance is not specific to cannabis but is also seen in patients with schizophrenia who abuse other illicit substances (Potvin et al., 2008).

It is difficult to attribute consequences solely to cannabis in naturalistic studies, because cannabis is often used in combination with nicotine, alcohol, and other illicit drugs. Moreover, positive and negative effects of cannabis are likely to be dose-related, and dose-response relationships are all but impossible to assess in naturalistic studies because of uncertainty over dose and potency of the cannabis. Some of these limitations have been addressed in experimental studies.

2.3.2. Experimental Studies

Only three studies have directly recorded the effects of Δ^9-THC or cannabis administration to schizophrenia patients in order to study acute effects on psychosis outcomes, cognition, and side effects. In the first study, in 1934, Lindeman and Malamud administered unassayed doses of hashish to a group of schizophrenia patients, who experienced an exacerbation of their symptoms (Lindemann and Malamud, 1934).

The second study, seven decades later, was a randomized, double-blind, placebo-controlled study that examined whether schizophrenia patients were more vulnerable than healthy controls to the cognitive and propsychotic effects of Δ^9-THC (D'Souza et al., 2005). Thirteen stable antipsychotic-medicated schizophrenia patients and 22 healthy controls were given 0 mg, 2.5 mg, and 5 mg Δ^9-THC intravenously in a three-day, double-blind, randomized, counterbalanced study, with test days were separated by at least a week (more than three times the elimination half-life of Δ^9-THC) in order to minimize carryover effects. Subjects were abstinent from caffeine, alcohol, and illicit drugs for two weeks before testing until study completion, confirmed through urinary testing. A three-point or greater increase on the Positive and Negative Syndrome Scale (PANSS)(Kay et al., 1986) positive symptom subscale was considered a clinically significant response; the PANSS was administered several times after Δ^9-THC or placebo administration. Acute Δ^9-THC effects on neuropsychological functioning were also tested using a verbal fluency test (Corkin et al., 1964), the Hopkins Verbal Learning Test (Brandt, 1991) for learning and immediate and delayed recall, and a continuous performance test (Gordon, 1986) to measure attention. Motor side effects were measured using the Abnormal Involuntary Movement Scale (AIMS) for dyskinesias, the Barnes Akathisia Scale for akathisia, and the Simpson Angus Scale (SAS) for parkinsonism.

There were similarities between the schizophrenia patients and healthy controls, as well as differences. Plasma Δ^9-THC and 11-nor-Δ^9-carboxy-THC (metabolite) levels were the same in both patient and control groups. Δ^9-THC transiently increased positive symptoms in both schizophrenia patients and matched healthy controls. These effects were dose-related, occurred 10 to 20 minutes after Δ^9-THC administration, and resolved within four hours. Eighty percent of the schizophrenia patients but only 35% of control subjects had a clinically significant increase in psychosis in response to 2.5 mg Δ^9-THC, and 75% of schizophrenia patients but only 50% of control subjects had a suprathreshold response to 5 mg (Figure 3). Δ^9-THC acutely impaired immediate recall, delayed free recall, and delayed cued recall in a dose-dependent fashion, as well as increasing omission errors in the attention task. Schizophrenia patients were more sensitive to the cognitive effects of cannabis, particularly impairment of memory and attention (Figure 2), although there were no significant group-by-dose interactive effects. Δ^9-THC had no effects on verbal fluency, nor did it make subjects more calm and relaxed. There was no interaction between group, dose, and time, nor was there a difference between groups in Δ^9-THC's effects on other measures, such as feeling high (VAS) or perceptual alterations (on the Clinician-Administered Dissociative States Scale). Δ^9-THC increased total scores on the AIMS (dyskinesia) as well as on the SAS (rigidity).

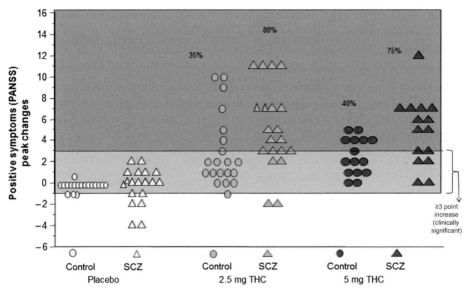

Figure 3 Enhanced sensitivity to the psychotomimetic effects of Δ^9-THC in schizophrenia. The figure shows the peak increase in on the positive symptom subscale of the Positive and Negative Symptoms Scale (PANNS) (group means ± 1 SD) in healthy subjects (controls) and in patients with schizophrenia.

The third study was a *momentary assessment study* that investigated the complicated dynamics of cannabis use and its varied effects in psychotic patients in the context of daily life by using the Experience Sampling Method (ESM) (Henquet et al., 2006). The study tracked the acute effects of cannabis on mood and psychotic symptoms in the daily life of 80 cannabis users; 42 psychotic patients and 38 healthy controls. In ESM, subjects received a digital wristwatch and a recording booklet. Twelve times a day on six consecutive days, the watch beeped randomly about once every 90 minutes, signaling the subjects to complete Likert scales on affect, thoughts, symptom severity, and activity at the moment of the beep. This systematic observation of recreational cannabis use in daily life revealed that not only was the frequency of cannabis use significantly higher in patients than in controls, but that neither positive nor negative affect predicted cannabis use at the next sampling point, indicating that there was no evidence for "self-medication". Similarly, no associations were found between delusions or hallucinations and subsequent cannabis use. Cannabis acutely induced hallucinatory experiences in patients but not healthy controls, and decreases in negative affect were observed after cannabis use in patients but not in controls, indicating that patients were more sensitive to the positive mood-enhancing effects of cannabis as well as the negative propsychotic effects. Patients were also became more sociable with cannabis, whereas no such effect was observed in the control group. *Post hoc* analysis suggested that cannabis may have immediate positive effects on mood, followed by decreases in mood and increases in hallucinations several hours later. The authors concluded that this delay between immediate reward and negative consequences might explain why schizophrenia patients continue to use cannabis.

In summary, cannabis is one of the drugs most commonly misused by schizophrenia patients. Information about the acute effects of cannabis and cannabinoids on schizophrenia patients comes mostly from subjective reports, retrospective self-assessments, longitudinal studies, with few experimental studies. Subjectively, schizophrenia patients are motivated to smoke cannabis because they feel that it causes relaxation and enhances positive mood, decreases negative emotions (such as depression) and enhances social occasions and makes it easier to cope with people, with some also reporting relief of medication side effects or psychotic symptoms such as hallucinations and suspiciousness. The actual subjective effects reported by patients following cannabis use are often negative, however.

Most longitudinal studies find that the more cannabis schizophrenia patients use, the more problems they have, although some do find positive or equivocal effects as well. Only two modern studies have directly recorded the effects of Δ^9-THC or cannabis administration to schizophrenia patients. Schizophrenia patients are more sensitive to the cognitive effects of cannabis, particularly impairment of memory and attention, and it did not make patients more calm and relaxed. Cannabis may have immediate positive effects on mood, followed by decreases in mood and increases in hallucinations several hours later. This delay between immediate reward and negative consequences might explain why schizophrenia patients continue to use cannabis. It is also possible that from the patients' viewpoint, cannabis use is beneficial because it decreases their preoccupation with psychotic symptoms (making them more bearable) and elevates their mood, even considering the cost of increased symptoms.

2.4. EFFECTS OF CANNABINOIDS ON BRAIN STRUCTURE AND FUNCTION

2.4.1. Effects of Cannabinoids on Brain Structure

The effects of chronic exposure to cannabinoids on brain structure have been studied in animals and humans. In general, chronic administration of cannabinoids to rats and nonhuman primate produces dose-dependent neurotoxic changes in areas with a high density of CB-1RRs. However, it should be noted that few of these studies replicate the pattern of cannabis exposure typical in humans. Brian regions showing evidence of neurotoxic effects include the hippocampus, amygdala, and cortex (Chan et al., 1998; Downer et al., 2007; Harper et al., 1977; Heath et al., 1980; Landfield et al., 1988; Lawston et al., 2000; Scallet, 1991; Scallet et al., 1987). For example, treatment of hippocampal slices and neuronal cultures with Δ^9-THC resulted in shrinkage of neuronal cell bodies and nuclei, breakages in DNA strands and apoptosis (Chan et al., 1998). In a study that used a pattern of chronic Δ^9-THC administration to rats (5 times weekly for 8 months, i.e approximately 30% of the life-span) similar to cannabis exposure in humans, Landfield et al (1988) noted significant morphological changes in the hippocampus which included decreased neuronal density and increased glial cell reactivity (Landfield et al., 1988). Hippocampus specific toxic changes were also noticed with chronic administration of the synthetic cannabinoid WIN 55,212-2 administered twice daily for 21 days (Lawston et al., 2000).

Studies that have examined the effect of cannabis use on brain structure in humans have yielded contradictory results with some studies suggesting changes in specific brain regions while others showing no changes. In one of the earliest case-series of 10 chronic cannabis abusers, Campbell et al (1971) found significant cerebral atrophy and ventricular enlargement among abusers when compared with control subjects (Campbell et al., 1971). However, the technique of pneumoencephalography and the choice of controls are limitations of that study. Subsequent studies using computerized tomography failed to detect significant structural abnormalities (Co et al., 1977; Hannerz and Hindmarsh, 1983; Kuehnle et al., 1977).

More recent studies using magnetic resonance imaging (MRI) have reported mixed results. While several studies did not find any changes (Block et al., 2000; Jager et al., 2007; Medina et al., 2007a; Tzilos et al., 2005), others reported global changes in both gray and white matter density (Wilson et al., 2000), focal changes in the hippocampal and parahippocampal areas (Matochik et al., 2005; Medina et al., 2007c), cerebellum (Cousijn et al., 2012; Solowij et al., 2011) and alterations in cortical gyrification (Mata et al., 2010). The mixed findings are likely due to the variability in cannabis exposure in the samples studied. Thus, while studies with lower frequency and/or amount of cannabis exposure show no change (Block et al., 2000; Jager et al., 2007; Medina et al., 2007a; Tzilos et al., 2005) or an increase (Medina et al., 2007b) in hippocampal and parahippocampal volumes, studies of samples with heavier cannabis exposure show reductions (Matochik et al., 2005).

Studies that have investigated the effect of cannabis use on structural brain abnormalities in patients with psychotic disorders (Table 1) have also yielded mixed results. The lack of consistent findings in these studies may be because of intrinsic differences in the samples, differences in the amount of cannabis exposure (current vs lifetime history), concomitant use of other drug (ranging from cannabis only to polydrug users), duration of illness and the effects of antipsychotic medication. Taken together, it appears that if exposure to cannabis alters brain structure, the effects are small. Furthermore, whether these structural changes have any functional consequences is not clear.

2.4.2. Effects of Cannabinoids on Brain Function (Psychophysiology)

Psychophysiological measures may be particularly informative in elucidating the mechanisms mediating the link between cannabinoids and psychosis. EEG is one of the few available methodologies that can directly measure neural events (postsynaptic potentials) with high temporal precision in humans (Luck et al., 2010). Psychophysiological indices using event-related potentials (ERPs) have been shown to one of the most consistently

Table 1 Cannabis Effects on Brain Structure in Schizophrenia

Study	Method	Participants	Results
Cahn et al. (2004)	MRI	27 S + C, 20 S − CB (naïve)	No difference in total brain, GM, WM or caudate nucleus volumes
Szeszko et al. (2007)	MRI	20 S + C, 31 S − C, 56 HC	Anterior cingulate GM volume: S + C < S − C, HC
Potvin et al. (2007)	MRI	12 S + SM; 5 SC + 2; 2 S + EtOH; 5 S + EtOH; 11 S − C, 15 HC	Ventral striatal GM density: S + SM > 2
Rais et al. (2008)	MRI	19 S + C; 32 S − C; 51 HC	Over five years: Loss of GM volume: S + C > S − C > HC; LV enlargement: S + C > S − C, HC; TV enlargement: S + C > S − C, HC
Bangalore et al. (2008)	MRI	Untreated first episode psychosis: 15 S + C; 24 S − C	Right posterior cingulate GM density: S + C > S − C
Wobrock et al. (2009)	MRI	20 S + SM (primarily C); 21 S − SM	No change in volume of amygdala, hippocampus, superior temporal gyrus and cingulate cortex.
Peters et al. (2009b)	DTI	24 S + C (onset < 17 y); 11 S − C	FA in frontal WM, uncinate fasciculus and anterior internal capsule: S + C > S − C
Dekker et al.	DTI	10 S + C (onset < 15 years); 8 S + C (≥17 years); 8 S − C	FA density in splenium: S − C < S + C (<15 years); WM demsity in splenium, right occipital lobe and left temporal lobe: S − C < S + C (<15 years)
Solowij et al. (2011)	MRI	17 S (47% S + C) 31 HC (48% SM)	Cerebeller WM reduction in CB using healthy subjects as well as S + C
James et al. (2011)	MRIDTI	16 S + C, 16 S − C 28 HC	GM density in temporal fusiform gyrus, parahhippocampal gyrus, ventral striatum, right middle temporal gyrus, insular cortex, precuneus, right paracingulate gyrus, dorsolateral prefrontal cortex, left postcentral gyrus, lateral occipital cortex and cerebellum: S + C < S − C. FA IN brain stem, internal capsule, corona radiata, superior and inferior longitudinal fasciculus: S + C < S − C

S + C = patients with psychotic illness and cannabis use; S − C = patients with psychotic illness without cannabis use; GM = gray matter; WM = white matter; HC = healthy controls; SM = substance misuse (abuse or dependence); EtOH = alcohol; LV = lateral ventricle; TV = third ventricle; DTI = diffusion tensor imaging; FA = fractional anisotropy.

replicated psychophysiological abnormalities in schizophrenia, and have yielded some of the largest effect sizes in the neurobiological study of schizophrenia (Heinrichs, 2004; Turetsky et al., 2007). ERPs are averaged electroencephalographic (EEG) responses time-locked to particular stimuli or events (Figure 4).

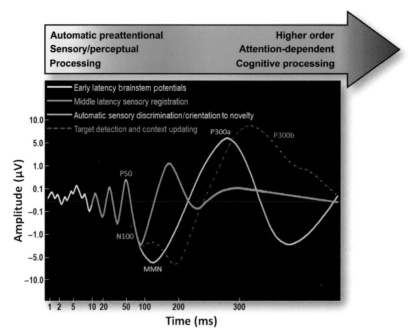

Figure 4 Schematic depiction of event-related potentials. Illustration demonstrating prototypical event-related potentials (ERPs) thought to be sensitive to both cannabinoids and schizophrenia. Time is shown using a logarithmic scale. Stages of information processing from automatic preattentional to higher order attentional processes are shown from left to right, respectively. *Adapted from Picton et al. (1974) and Rissling et al. (2010).*

As described by Solowij and Michie (2007), ERPs from several paradigms have been shown to exhibit parallel disruptions in both psychotic disorders and following the administration of exogenous cannabinoids. These paradigms include auditory sensory gating (P50) and the P300 (Solowij and Michie, 2007). A brief review of recent studies utilizing these psychophysiological outcomes in relation to psychosis and cannabinoids follows. Moreover, descriptions of several other putative biomarkers are included, namely mismatch negativity (MMN) and the N100 response. For a historical review of EEG studies on cannabis prior to 1990, see Struve and Straumanis (1990) (Struve and Straumanis, 1990). See Figure 4 for a schematic representation of ERPs thought to altered both in relation to cannabinoids and schizophrenia.

2.4.2.1. Auditory Sensory Gating (P50)
The P50 is a positive voltage, midlatency (~50 ms), preattentive ERP component elicited by discrete auditory stimuli (e.g. brief white noise clicks). In the P50 sensory gating or "dual-click" procedure, the amplitude of the P50 to a second paired click (S2) is reduced relative to the P50 amplitude to the first click (S1; 500 ms interstimulus

interval). While the P50 to S1 is related to the capacity of the central nervous system to register salient stimuli, the attenuation of the P50 amplitude to S2 (compared to S1) is associated with the automatic suppression or inhibitory gating of redundant and irrelevant stimuli. P50 suppression deficits in schizophrenia were first observed in the early 1980s (for recent reviews, see Braff and Light, 2004; Patterson et al., 2008; Potter et al., 2006; Thaker, 2008; Turetsky et al., 2007), and have also been demonstrated in clinically unaffected relatives, and individuals with schizotypal personality disorder (Cadenhead et al., 2000; de Wilde et al., 2007). Recent meta-analyses of over 20 sensory gating studies in schizophrenia have resulted in large effect sizes (Bramon et al., 2004; de Wilde et al., 2007; Patterson et al., 2008). Clinically, alterations in sensory gating may represent an inability to filter out redundant and irrelevant sensory information, thus leading to perceptual overload, which could contribute to positive symptomatology (Boutros et al., 1999; Thaker, 2008).

The brain regions thought to mediate auditory sensory gating include the hippocampus, the temporoparietal region, and the prefrontal cortex (Grunwald et al., 2003; Luntz-Leybman et al., 1992). Interestingly, in both humans and nonhuman primates, it has been shown that each of these areas have a high density of CB-1RRs (Eggan and Lewis, 2007). This fact, along with reports that THC intoxication can induce severe perceptual distortions (D'Souza et al., 2004), suggests that cannabinoids may contribute to psychosis by altering sensory gating. While no studies to date have examined the effects of acute cannabinoids on sensory gating, preclinical studies in rats suggest that cannabinoid agonists (CP-55940 and WIN 55,212-2) disrupt sensory gating in animal analogs of this paradigm (N40) (Dissanayake et al., 2008; Zachariou et al., 2008).

In contrast to the lack of any data on the acute effects of cannabinoids on P50 gating in humans several preliminary studies have demonstrated disruptions in P50 suppression in chronic cannabis users abstinent for at least 24 h (Patrick et al., 1999; Patrick and Struve, 2000). Furthermore, Edwards et al. (2009) assessed heavy cannabis users after 24 h of abstinence and replicated the findings of abnormal P50 gating in the cannabis group (larger S2/S1 amplitude ratios), which was positively correlated with the estimated number of joints smoked in the previous 6 months (Edwards et al., 2009). Finally, Rentzsch et al. (2007) assessed sensory gating in schizophrenia patients and healthy controls with and without comorbid cannabis abuse. It was demonstrated that cannabis-using controls (assessed after 28 days of abstinence) had significant P50 gating deficits, which correlated with the number of years of cannabis consumption (Rentzsch et al., 2007). However, cannabis using schizophrenia patients did not differ from cannabis-free schizophrenia patients, suggesting normalization and protection via antipsychotic medication. In sum, a common information processing deficit could link psychosis and cannabinoids at the level of inhibitory sensory gating, possibly through disruptions in GABAergic modulation of inhibitory neural circuits.

2.4.2.2. Target Detection (P300)

One of the most consistent psychophysiological findings in SZ, particularly in the auditory modality, is the P300 response (Braff, 1993; Bramon et al., 2005; Bramon et al., 2004; Duncan, 1988; Jeon and Polich, 2003; Roth and Cannon, 1972; Solowij and Michie, 2007; Turetsky et al., 2007; Turetsky et al., 1998). The P300 is a late positive, postattentional ERP component thought to be related to directed attention, contextual updating of working memory, and the attribution of salience to deviant or novel stimuli (Polich and Criado, 2006). This response is typically elicited utilizing standard "oddball" paradigms in which low probability target tones (~20%) are embedded within a repeating sequence of high probability standard tones (~80%) differing in some physical dimension (e.g. frequency or duration).

Rather than being generated by a single neural source, it this thought that the P300, particularly the P300b component, reflects activity from a distributed neural ensemble including such areas as the thalamus, hippocampus, inferior parietal lobe, superior temporal gyrus, and frontal cortex (Kiehl et al., 2001). Abnormal reductions in P300 amplitude and increased latencies haves been reliably shown in both schizophrenia patients and unaffected relatives, with moderate effect sizes reported (Bramon et al., 2005; Bramon et al., 2004; Jeon and Polich, 2003). While P300 deficits are one of the most robust findings in the neurobiology of SZ, it should also be noted that abnormalities of the P300 are not specific to psychosis, and have been reported in Alzheimer's disease, bipolar disorder (O'Donnell et al., 2004; Thaker, 2008), and especially alcoholism and substance abuse (Polich et al., 1994; Singh and Basu, 2009).

Several studies have reported on the effect of acute cannabinoids on the P300 response I humans. Using standardized active THC cigarettes, Ilan et al. (2005) reported a reduction in P300 amplitude during a visuospatial N-back working memory task (Ilan et al., 2005). In another study, oral THC (10 mg) was shown to reduce the amplitude of

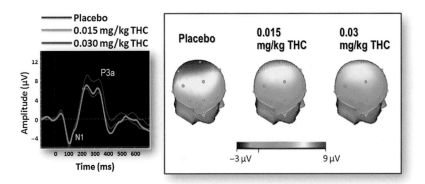

Figure 5 Effect of Δ^9–THC on P300a in healthy subjects. (Left). Grand-averaged novelty P300a waveforms at electrode Cz across THC dose conditions. (Right). Topographic voltage maps from the peak grand-averaged P300a across THC dose conditions.

the P300 measured during an auditory choice reaction task in healthy humans (Roser et al., 2008). More recently, D'Souza et al. (2012) examined the effect of several doses of intravenous (IV) THC on the novelty P300a and target P300b in healthy participants They showed that THC decreased of the amplitude of both the P300a (Figure 5) and P300b (Figure 6) in a dose-dependent manner (D'Souza et al., 2012). Further, these doses of THC also produced concomitant psychotomimetic effects. No effects on P300a or P300b latency were observed. These results suggest that acute THC can disrupt cortical processes responsible for context updating (P300b) and the automatic orientation of attention (P300a), while leaving processing speed intact. Furthermore, no effects of THC on the N100 (Figures 5 and 6) was observed suggesting that cannabinoids do not have significant effects on early sensory registration.

The results of studies assessing the effect of chronic cannabis use on the P300 response have been equivocal. Solowij et al. (1991) reported that a small sample of current cannabis users (12 h of abstinence) exhibited decreased P300 amplitudes during an auditory selective attention task (Solowij et al., 1991). While a follow up study with a larger sample size was unable to replicate the amplitude results, Solowij et al. (1995) demonstrated increased P300 latencies in heavy cannabis users utilizing the same task. Interestingly, it was reported that frequency of cannabis use was correlated with P300 latency (Solowij et al., 1995). More recently, decreased P300 amplitudes during a complex auditory selective attention task in cannabis users were reported, which was more pronounced in early onset users (Kempel et al., 2003).

By contrast, using a simple dual-stimulus oddball task (both auditory and visual), Patrick et al. (1999) were unable to detect P300 amplitude differences in age-matched, psychiatrically screened heavy cannabis users (tested after 24 h of abstinence). It was actually demonstrated increased P300 amplitudes in a small sample of cannabis users

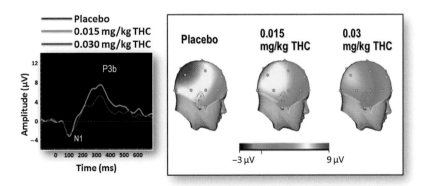

Figure 6 Effect of Δ⁹–THC on P300b in healthy subjects. Left: Grand-averaged target P300b waveforms at electrode Pz across THC dose conditions. Right: Topographic voltage maps from the peak grand-averaged P300b across THC dose conditions.

during a dual-stimulus visual discrimination task (Skosnik et al., 2008b). Further, de Sola et al. (2008) reported that cannabis users exhibited normal P300 responses in a simple auditory P300 task, and that higher cannabis use was correlated with shorter latency (de Sola et al., 2008). The reasons for these discrepant results are unclear. The initial pattern that seems to have emerged is that chronic cannabis users are impaired during more cognitively challenging selective attention tasks (Kempel et al., 2003; Solowij et al., 1991; Solowij et al., 1995), but retain normal ERP responses when simple dual-stimulus discrimination tasks are utilized (de Sola et al., 2008; Skosnik et al., 2008b).

To summarize the data on the P300 ERP, the majority of studies to date examining the P300 response in the context of both chronic and acute cannabinoids are suggestive of impairments in novelty detection, selective attention and working memory. These findings indicate some level of neurobiological overlap between the ERP deficits observed in psychosis and in the context of chronic and acute cannabinoids.

2.4.2.3. Mismatch Negativity (MMN)

MMN is an automatic, preattentive, negative voltage ERP component that is generated primarily in the superior temporal and prefrontal cortex (Naatanen and Alho, 1995; Rinne et al., 2000). It occurs approximately 100200 ms after an auditory stimulus that deviates in frequency or duration from a sequence of standard auditory stimuli. It is thought to index basic auditory processing and sensory memory, and is relatively independent of attention. Early studies of MMN in schizophrenia patients found abnormal amplitudes to stimuli deviating in either duration or frequency. Since then, numerous studies have observed alterations in the MMN ERP, which have been recently reviewed by Näätänen and Kahkonen (2009). A metanalysis of over 39 studies of MMN on schizophrenia patients (up to year 2003) reported a large effect size of 0.99 (Umbricht and Krljes, 2005). Interestingly, MMN does not appear to be altered in other psychiatric disorders such as unipolar and bipolar depression (Umbricht et al., 2003). Hence, MMN may be a particularly specific and useful biomarker for auditory disturbances in schizophrenia.

To date, only two studies have examined auditory MMN in relation to cannabinoids. Juckel et al. (2007) determined the acute effects of oral THC or cannabis extract containing both THC and the nonpsychoactive constituent cannabidiol (CBD) on MMN using a frequency deviance paradigm. Surprisingly, oral THC did not alter MMN amplitude as compared to placebo (Juckel et al., 2007). Further, THC + CBD was actually shown to increase the amplitude of the MMN ERP. The authors postulated that cannabis extract, with the addition of CBD, enhanced MMN by virtue of CBD's reputed antipsychotic effects (Zuardi et al., 2006a; Zuardi et al., 2009; Zuardi et al., 2006b; Zuardi et al., 1995; Zuardi et al., 1991). The lack of a MMN effect with pure THC could be related to the dose of THC chosen (10 mg orally), or due to the inter and intraindividual variability inherent in oral routes of administration.

More recently, the effect of chronic cannabis exposure on MMN utilizing frequency and duration deviance was examined. Initial analysis showed that cannabis users exhibited decreased MMN amplitudes at electrode Cz in the frequency deviance condition (Roser et al., 2008). More striking was the fact that both long-term and heavier users of cannabis had significantly lower MMN amplitudes compared to short-term or light users (Roser et al., 2008). Moreover, duration of cannabis exposure negatively correlated with MMN amplitudes in the frequency deviance condition at frontal electrode sites (Fz and F4). In other words, those showing smaller amplitudes had greater overall chronicity of cannabis use. While these data are only preliminary, it appears that chronic, heavy use of cannabis may be associated with MMN ERP deficits in a pattern similar to schizophrenia patients.

2.4.2.4. N100

While less well studied than the EEG biomarkers discussed above, an additional psychophysiological index that may prove useful in the study of psychosis and cannabinoids is the N100 ERP. The N100 component (or N160/N170 in the visual modality) is a large exogenous ERP that exists regardless of task demands (although it can me modulated by attention) (Hillyard et al., 1973). It is thought to be related to basic perceptual processing, and in the auditory domain, is likely generated by auditory and frontal cortices. While initially reported to be aberrant in schizophrenia in the 1970s (Saletu et al., 1971), most studies assaying this response have demonstrated abnormal N100s in both schizophrenia patients and their clinically unaffected relatives (Rosburg et al., 2008; Turetsky et al., 2008). Given the parallel auditory and visual disturbances in both schizophrenia patients and individuals consuming cannabis, this ERP component could prove useful in examining the neural mechanisms linking cannabinoids and psychosis.

The acute administration of cannabinoids (IV THC) failed to affect the N100 component during an auditory oddball paradigm (D'Souza et al., 2012). In contrast, Skosnik et al. (2006) demonstrated robust differences in the visual N160 response in chronic cannabis users tested after 24 h of abstinence (Skosnik et al., 2006) using a repetitive photic stimuli paradigm. No differences in N160 latencies were shown. This effect was further demonstrated in the auditory modality for discrete 1000 Hz tones during an associative learning task (N100) (Skosnik et al., 2008a). However, a subsequent study utilizing the same auditory stimuli with a new sample of cannabis users failed to replicate this finding (Edwards et al., 2008). Furthermore, a recent study of the effect of chronic cannabis use on the gamma-band (40 Hz) auditory steady state response (ASSR) showed no differences in the N100 component. It therefore appears that effect of cannabinoids on the N100 ERP remains equiviocal, and may vary depending on modality and experimental paradigm.

2.5. PERSISTENT BEHAVIORAL AND COGNITIVE EFFECTS OF CANNABINOIDS

The evidence for the persistent effects of cannabinoids in humans, derives primarily from nonexperimental, epidemiological studies of cannabis. The latter have a number of limitations including sampling biases, power of the study (i.e overpowered/underpowered sample sizes), presence of unknown confounders, lifetime exposure to multiple drugs, period-, time-, cohort- effects and direction of causality also applies to this extensive literature.

2.5.1. Positive Symptoms

The question of whether chronic exposure to cannabinoids can "cause" persistent symptoms or a psychotic disorder is an important one.

One of the first large epidemiological studies that attempted to answer this question was a longitudinal, 15-year cohort study of 45,570 Swedish conscripts who had been conscripted between 1969 and 1970 (Andreasson et al., 1987). Since Sweden mandates military service, almost all of the male population of Sweden (97%) aged 1820 years were included. Andreasson et al (1987) observed a dose-response relationship between self-reported cannabis use at conscription (age 18 years) and psychiatric hospitalization for schizophrenia over the ensuing 15 years. Conscripts who reported having used cannabis at least once in their lifetime had a 2.4-fold (95% confidence interval 1.83.3) increased risk of developing schizophrenia over the course of 15 years. This relative risk rose to 6 fold (95% CI = 48.9) in those who had used cannabis more than 50 times in their lifetime. Furthermore, the risk remained significantly high despite adjusting for other factors such as psychiatric illness at the time of conscription, solvent abuse and parental separation.

In a 27-year follow-up study of the same cohort and a re-analysis of the data, Zammit et al (2002) replicated the findings of Andreasson et al, (1987) showing that cannabis use was associated with a linear risk of developing schizophrenia with a relative risk increasing from 2.2 (95% CI 1.72.8) in those who had used cannabis at least once to 6.7% (95% CI 4.510) in those who had used cannabis more than 50 times in their lifetime (Zammit et al., 2002). When potential confounders such as psychiatric diagnosis at conscription, IQ score, degree of social integration, disturbed behavior in childhood, cigarette smoking and place of upbringing were included in the regression analysis, the adjusted relative risk was 1.5 (95% CI = 1.12.0) in those who had used cannabis at least once to 3.1 (95% CI = 1.75.5) from in those who had used cannabis more than 50 times in their lifetime. The relative risk for schizophrenia was significantly higher in those who developed schizophrenia within 5 years of conscription which brings into question the direction of causality, i.e. whether cannabis use led to schizophrenia or whether subjects used cannabis in an attempt to self-medicate incipient symptoms of schizophrenia. The study attempted to answer this question in a secondary analysis that excluded those who developed a diagnosis of schizophrenia within 5 years of conscription. In this secondary analysis the adjusted relative risk remained significant only for those who had used cannabis more than 50 times (adjusted relative risk = 2.5, 95% CI = 1.25.1). The study needs to be interpreted with caution since while 24.3% of the sample had used any drug, a very small percent (3.4%) had used only cannabis. While the analysis controlled for cigarette smoking, it failed to control for the use of stimulants and other drugs. Also, the fact that presumably weak confounders such as "place of upbringing" and "cigarette smoking" contributed substantially, along with other variables in reducing the adjusted relative risk by approximately 50% in the regression analysis highlights the difficulties inherent in interpreting epidemiological data and raises the issue of other unknown confounders. Similar criticism has been raised by other authors (Johnson, 1990; Johnson et al., 1988; Negrete, 1989), including the facts that (1) the use of other drugs was more common in the cannabis-using group, (2) the association between cannabis use and schizophrenia may be mediated by a third, as yet unknown factor, and (3) the follow-up study, a quarter century later, failed to address the issue of confounding due to use of other drugs, many of which are also known to precipitate psychosis.

A longitudinal Dutch study, the Netherlands Mental Health Survey and Incidence Study (NEMESIS), reported that cannabis use at baseline was associated with an increased risk of psychosis (van Os et al., 2002). The study assessed 7,076 subjects at baseline (1996), 5,618 subjects at a first time-point (1997) and 4,848 subjects at a second time-point (1999) via telephonic interviews, and found 10 subjects who developed psychosis, while 38 subjects endorsed individual items on the Brief Psychiatric Rating Scale (BPRS). The findings of the study are limited by the small numbers in the outcome of interest (psychosis) despite the large sample size, also reflected in the wide confidence intervals of the relative risk. The German prospective Early Developmental Stages of Psychopathology study (EDSP) which used in-person interviews in the assessment of 923 individuals from the general population (aged 1424 years) showed that cannabis use was associated with an increased risk of psychotic symptoms and persistent use increased this risk (Kuepper et al., 2011b). The EDSP study importantly points to the directionality of the association between cannabis use and psychosis. This is in contrast with other recent studies (Ferdinand et al., 2005; McGrath et al., 2010), which have shown the relationship to be bidirectional, alluding to the possibility of a phenomenon of "self-medication".

Fergusson et al, (2005) attempted to tease apart the causal link between cannabis use and psychosis in a dataset of a 25-year longitudinal study in New Zealand, the Christchurch Health and Development Study (CHDS) birth cohort comprising 1265 children (635 males, 630 females) (Fergusson et al., 2005). The study showed that daily use of cannabis was associated with 2.33.3-fold higher risk of psychosis that among nonusers. One of the limitations of the study is that the data was derived from 10 items of the Symptom

Checklist-90, the items on which overlap with personality traits such as schizotypy and paranoia and that the study did not attempt to delineate psychotic symptoms due to the acute effects of cannabis use from persistent effects (Mirken and Earleywine, 2005).

This finding of increased psychosis risk has been reported in several other prospective studies (Arseneault et al., 2002; Henquet et al., 2005a; van Os et al., 2002; Weiser et al., 2002). The issue of reverse causality i.e. psychosis leading to cannabis use rather than cannabis use leading to psychosis, can be addressed to some extent in prospective studies by excluding subjects with psychosis at the outset of the study. In a systematic review of longitudinal studies Moore et al. found that any cannabis use (pooled adjusted OR = 1.41, 95% CI 1.20 ± 1.65) was associated with a 40% increased risk of psychotic disorder, and the risk increased in a dose-dependent fashion with greater cannabis exposure (OR = 2.09, 1.54 ± 2.84) (Moore et al., 2007).

Recent studies have attempted to explore other factors that may moderate the increased risk of psychosis outcomes with cannabis use. It is being increasingly recognized that adolescence may be a particular period of increased vulnerability to the effects of cannabis and additional factors such as, schizotypy or other trait measures of liability to psychosis, and childhood abuse may mediate the causation of schizophrenia with prolonged and persistent cannabis use. In a 26-year follow-up study of another birth cohort in New Zealand ($n = 1037$) called the Dunedin Multidisciplinary Health and Development Study, Arseneault et al (2002) reported that individuals using cannabis at age 15 years were 4 times more likely to have a diagnosis of schizophreniform disorder at age 26 years (Arseneault et al., 2002). This risk was higher than the risk with cannabis use at age 18 years, after controlling for the presence of self-reported psychotic symptoms at age 11 years. The study design has significant strengths in that it provided data on psychotic symptoms at age 11 years (i.e. before the onset of cannabis use), on cannabis use at ages 15 years and 18 years and DSM-IV diagnosis at age 26 years. However, psychotic symptoms at age 11 years significantly increased the risk of a diagnosis of schizophreniform psychosis at age 26 years, and when used as a covariate in the analysis the risk of cannabis use was no longer significant, although it remained elevated (Odds Ratio = 3.1). The small sample size is another factor that limits the interpretation of these results.

In a sibling-pair analysis using samples of patients with a psychotic disorder ($n = 1120$), their siblings ($n = 1057$) and community controls ($n = 590$), the Genetic Risk and Outcome in Psychosis (GROUP) investigators found that siblings of patients with psychosis had greater sensitivity to positive and negative schizotypy, strait measures of psychosis liability and had higher psychotic experiences as measured on the CAPE (Community Assessment of Psychic Experiences) (GROUP, 2011). An exploratory analysis showed that familial risk moderated, rather than mediated the increased sensitivity to cannabis (GROUP, 2011).

More recently, the interactive effects of childhood maltreatment and cannabis abuse have been examined. While two cross-sectional and one prospective study show that cannabis use and childhood maltreatment act additively to increase the risk of psychosis (Harley et al., 2010; Houston et al., 2008; Konings et al., 2012), this finding was not replicated in the German EDSP dataset (Kuepper et al., 2011a). Future studies that examine the interaction between genetic liability, trait measures of psychosis liability, cannabis use and other environmental factors may provide greater insights into the complex architecture and mechanisms that lead to the causation of psychosis.

Interestingly, although meta-analytical studies suggest that cannabis might account for between 8% and 14% of schizophrenia cases (Henquet et al., 2005b; Moore et al., 2007), the four fold increase in the rates of cannabis use over the last four decades (Aust et al., 2002; Zammit et al., 2002) has not resulted in a commensurate 40% to 70% increase in prevalence of schizophrenia. Some studies suggest that the rates of schizophrenia may be decreasing (Der et al., 1990), while others suggest the contrary (Ajdacic-Gross et al., 2007; Hickman et al., 2007).

2.5.2. Negative Symptoms

Apart from the link with psychosis, as detailed above, chronic heavy cannabis use has also been associated with an "amotivational syndrome" (Halikas et al., 1982; Hall and Solowij, 1998; Kolansky and Moore, 1971; Millman and Sbriglio, 1986; Tennant and Groesbeck, 1972). This syndrome is characterized by apathy, amotivation, social withdrawal, narrowing of one's personal repertoire of interests, lethargy, impairment in memory and concentration, impaired judgment and decision making, and poor sociooccupational functioning. All these symptoms share similarities with the negative symptoms of schizophrenia. The nosological status of the syndrome is however, debated and the confounding effects of concomitant polydrug abuse, poverty, low socioeconomic status, or preexisting psychiatric disorders cannot be ruled out.

2.5.3. Cognitive Deficits

Several studies suggest that chronic, heavy cannabis use may lead to impairments in memory and attention (Bolla et al., 2002; Lundqvist, 2005; Pope et al., 2001a; Pope et al., 1995; Pope and Yurgelun-Todd, 1996; Solowij and Battisti, 2008). Solowij and Mitchie (2007) have suggested that cognitive dysfunction associated with long-term or heavy cannabis use is a cognitive endophenotype of schizophrenia (Solowij and Michie, 2007). In a comprehensive review, Solowij and Battisti (2008) concluded that chronic heavy cannabis use was associated with impairments in memory (Solowij and Battisti, 2008) that persisted beyond the period of acute intoxication and was related to the frequency, duration, dose and age of onset of cannabis use. Fontes, et al (2011), evaluated the neuropsychological performance of 104 chronic, heavy cannabis users and found that, compared to controls, chronic cannabis users had significant impairment on the cognitive domains of sustained attention, impulse control and executive functioning (Fontes et al., 2011). Additionally, similar to the literature on the risk of psychosis, individuals who used cannabis in adolescence (before the age of 15 years) had greater deficits. The authors however, did not assess whether subjects were in withdrawal or had residual effects from their last use of cannabis at the time of assessment.

While chronic, heavy cannabis users have deficits in cognitive processes, especially memory and attention in the context of ongoing cannabis use, the question of whether these impairments are persistent or a result of withdrawal and residual effects is unclear. While one study demonstrated an absence of persistent neuropsychological deficits in frequent long-term cannabis users after 28 days of abstinence (Pope et al., 2001b), other studies have shown variable durations to full recovery, ranging from a week (Jager et al., 2006), to 28 days (Pope et al., 2001a), to three months of abstinence (Fried et al., 2005), with some studies showing recovery only after an average of two years of abstinence (Hall and Solowij, 1998; Solowij, 1995). A recent review provides a summary of the literature to date (Crean et al., 2011).

Among studies in which neuropsychological assessments were performed 3 weeks or later after the last use of cannabis, five out of seven studies showed no impairment in attention (Lyons et al., 2004; Pope, 2002; Pope et al., 2003; Pope et al., 2001a; Pope et al., 2002; Pope et al., 2001b; Verdejo-Garcia et al., 2005), while two showed persisting impairment (Bolla et al., 2002; Solowij, 1995). One study revealed a trend toward impairment in decision making/risk taking (Verdejo-Garcia et al., 2006). There was, no impairment on response inhibition measured by the Stroop test (Bolla et al., 2002; Lyons et al., 2004; Pope et al., 2003; Pope et al., 2001a; Pope et al., 2002), and on working memory (Verdejo-Garcia et al., 2005) while all (Bolla et al., 2002; Pope et al., 2003; Pope et al., 2001a; Pope et al., 2002), but one (Lyons et al., 2004) found an impairment on the Wisconsin Card Sorting Test, a test of reasoning/problem solving. There was no impairment in verbal memory in two (Pope et al., 2002; Pope et al., 2001b) of the three studies that used the Buschke's Selective Reminding Test (BSRT), a test of memory of word lists. When the data from the third study (Pope et al., 2003) was stratified based on age at onset of cannabis use, significantly greater impairment was noticed in those who had first use cannabis before the age of 17 years again suggesting that like for positive symptoms, earlier age of onset of cannabis use may be associated with greater persistent cognitive deficits. It is important to note that none of these studies were designed to determine whether the cognitive impairments predated cannabis use.

2.6. CANNABINOIDS, PSYCHOSIS, AND CAUSALITY

Does exposure to cannabinoids "cause" psychosis where none would have otherwise existed? The commonly applied criteria to establish disease causality include temporality, strength and direction of the association, biological gradient (dose), consistency, specificity, coherence, experimental evidence, and biologic plausibility.

2.6.1. Temporality

Experimental evidence from laboratory studies clearly demonstrates a robust temporal relationship between exposure to cannabinoids and psychotic *symptoms*. The onset of cannabis use may precede, follow or co-occur with the onset of schizophrenia. Allebeck and colleagues reported that in 69% of a schizophrenic patient sample from a Swedish case registry ($n = 112$), cannabis abuse preceded the onset of psychotic symptoms by at least one year (Allebeck et al., 1993b). Further, in only 11% did the onset of psychotic symptoms precede the onset of cannabis abuse. Similarly, Linszen et al. found that cannabis abuse preceded the onset of psychotic symptoms by at least 1 year in 23 of 24 cannabis-abusing recent onset schizophrenic patients (Linszen et al., 1994). Hambrecht and Hafner in their study of first-episode schizophrenic patients found that 14.2% of the sample had a lifetime history of drug abuse with cannabis being the most frequently abused drug (88%) (Hambrecht and Hafner, 1996; Hambrecht and Hafner, 2000). Furthermore, drug abuse preceded the first sign of schizophrenia by more than a year but typically by more than 5 years in 27.5% of patients. In 37.9% of individuals, drug abuse followed the first sign of schizophrenia, and in 34.6% of individuals, the first sign of schizophrenia and drug abuse started within the same month. Related to the above, some studies suggest that cannabis and other substance use is associated with an earlier age of and more abrupt onset of psychotic symptoms in schizophrenic patients (Addington and Addington, 1998; Allebeck et al., 1993a; Andreasson et al., 1987; Andreasson et al., 1989; Cleghorn et al., 1991; Green et al., 2004b; Hambrecht and Hafner, 1996; Linszen et al., 1994; McGuire et al., 1994; Van Mastrigt et al., 2004; Veen et al., 2004).

However, schizophrenia begins insidiously, and evolves through several identifiable stages with the emergence of psychotic symptoms as the final step in the evolution of the disorder. As a result, while it may be easy to pinpoint the emergence of positive psychotic symptoms in retrospective studies, pinpointing the onset of the less obvious prodromal symptoms is extremely challenging. Further, if as the neurodevelopmental hypothesis posits, that the pathophysiological processes underlying the illness precede the clinical manifestations by years or even decades and that these processes may even begin in utero, then, the argument about a temporal relationship is no longer relevant.

Thus, while there is evidence suggesting a temporal association between cannabis use and the onset of positive psychotic symptoms, the temporal relationship between cannabis use and the less obvious symptoms has not been studied.

Dose: Several studies reviewed here provide evidence of a doseresponse relationship between exposure to cannabinoids and the risk of both psychotic symptoms and disorder.

Direction: The case of reverse causality has been proposed whereby risk for schizophrenia predisposes to cannabis use, rendering the association between cannabis and psychotic illness merely an epiphenomenon of a shared vulnerability for both psychosis and cannabis (Collip et al., 2008; Macleod, 2007). Since several longitudinal studies excluded people with psychosis at baseline, or adjusted for psychotic symptoms in the analysis, the observed association between cannabis and psychosis is unlikely to reflect reverse causation (Moore et al., 2007).

Strength: Cannabis exposure increases the odds of developing schizophrenia modestly (by 40%) even after controlling for many potential confounding variables (Moore et al., 2007).

Specificity: While there is a strong association between cigarette smoking and schizophrenia, there is little evidence to support the notion that cigarette smoking "causes" schizophrenia. Further, the association between cannabis use is weaker for anxiety or affective disorders (Moore et al., 2007).

Biologic Plausibility: The effects of cannabinoids on key neurotransmitters and known to be implicated in psychosis, and also neurodevelopmental processes provide biological plausibility for the association (D'Souza, 2007; D'Souza et al., 2009; Sewell et al., 2009). The acute effects of cannabinoids on DA, GABA and glutamate neurotransmission may explain some of the acute positive, negative and cognitive symptoms of cannabinoids. But these acute effects may be difficult to explain how exposure to cannabinoids causes a persistent psychotic disorder such as schizophrenia. The findings that early exposure to cannabis is associated with a greater risk for psychosis outcome than later exposure may provide some clues toward the underlying mechanism.

2.7. THE EFFECTS OF CANNABINOIDS ON NEURODEVELOPMENT

The acute effects of cannabinoids on DA, GABA and glutamate neurotransmission may explain some of the acute positive, negative and cognitive symptoms of cannabinoids. But these acute effects may be difficult to explain how exposure to cannabinoids contributes to the risk for a persistent psychotic disorder such as schizophrenia. The findings that early exposure to cannabis is associated with a greater risk for psychosis outcome than later exposure may provide some clues toward the underlying mechanism. Emerging evidence suggest that cannabinoids can influence neurodevelopment. The effect of cannabinoids on neurodevelopment is particularly relevant since schizophrenia, one view of schizophrenia is that it is a neurodevelopmental disorder (Rapoport et al., 2005; Weinberger, 1996).

An important feature of the endocannabinoid signaling system is that its different components are present in the embryo even at a preneuronal stage i.e. even before the onset of neurogenesis (Psychoyos et al., 2012). Exposure to the THC analogue, O-2545, has been shown to impair early processes in the differentiation of the primordial brain (Psychoyos et al., 2008). Endocannabinoids also regulate important neurodevelopmental processes such as neurogenesis, neural specification, neural maturation, neuronal migration, axonal elongation, and glia formation (Berghuis et al., 2005; Berghuis et al., 2007; Galve-Roperh et al., 2009; Watson et al., 2008).

The distribution of CB-1R receptors over the time course from the embryonal stage to adulthood is telling in this regard. CB-1R receptors are present on neural proliferator cells and immature neural cells in the embryo (Aguado et al., 2006). In the postnatal period CB1-R are present on radial-type glial cells and B-type cells, (which are thought to constitute neural stem cells) (Berghuis et al., 2005) and subsequently, in the adult brain they are present on the subventricular/ ventricular zone progenitor cells and on both, glutamatergic neurons and GABAergic neurons (Oudin et al., 2011). This temporal distribution is consistent with its role in neurogenesis, neural specification and neural maturation (Aguado et al., 2006; Berghuis et al., 2005; Berghuis et al., 2007; Fernandez-Ruiz et al., 2000; Galve-Roperh et al., 2007; Harkany et al., 2007; Jin et al., 2004; Mulder et al., 2008; Watson et al., 2008). Also, CB-1R receptors are initially abundant in the intermediate zone of the cortical plate region of the neocortical brain, are then found in white matter regions of the embryonic brain and gradually move to occupy layers IVI of the cortex in the adult brain, a pattern consistent with neuronal migration. The presence of CB-1R since early stages of brain development along with its transient presence in atypical locations like the white matter regions is further suggestive of its role in neuronal migration.

Another finding which supports this hypothesis is that the enzyme DAG Lipase which is involved in the synthesis of the endocannabinoid 2-archidonylglycerol (2-AG), appears to localize to axonal tracts in the embryo and transitions to localize onto dendritic fields in the adult brain (Harkany et al., 2007). This is also accompanied by a transition in endocannabinoid concentrations, 2-AG being most abundant in the early embryonal period while anandamide (AEA) peaks in the perinatal period (Galve-Roperh et al., 2009).

Further evidence for the role of endocannabinoids in neurodevelopmental processes comes from studies on knock-out mice. CB-1R receptor knock-out (KO) mice have reduced proliferation of neural progenitor cells in the hippocampus and subventricular zones (Jiang et al., 2005). A similar effect was found with the CB-1R antagonist, rimonabant (Gobbi et al., 2005). On the other hand, both mice deficient in Fatty Acid Amide Hydrolase (FAAH) the enzyme that breaks down endocannabinoids, and pharmacological inhibition of FAAH and DAGL led to increased proliferation of neural progenitor cells in the hippocampus. CB-1R KO and FAAH-1 deficient mice show altered radial migration and layering of pyramidal neurons (Mulder et al., 2008). Consistent with these results, inhibition of CB-1R led to failure of the corticothalamic axons to invade the dorsal striatum (Mulder et al., 2008). CB-1R are also found to play a role in interneuron migration and interneuron migration (Berghuis et al., 2005).

Chronic exposure to Δ^9-THC in early stages of neurodevelopment has been shown to result in altered pattern of cholecystokinin-positive (CCK) densities, a marker of GABAergic interneurons in the hippocampus (Berghuis et al., 2005). AEA is known to stimulate CCK-positive interneuron migration but inhibit brain derived neurotrophic factor (BDNF)-induced dendritic branching (Berghuis et al., 2005). Additionally, in animal studies, Δ^9-THC has been shown to alter BDNF expression (Butovsky et al., 2005; Derkinderen et al., 2003; Maj et al., 2007; Rubino et al., 2006; Valjent et al., 2001). Δ^9-THC has been shown to induce BDNF mRNA transcription via CB-1R receptors (Derkinderen et al., 2003; Rubino et al., 2006; Valjent et al., 2001). D'Souza et al. (2008a,b) showed that a socially relevant dose Δ^9-THC resulted in an increase in serum BDNF levels in healthy adult subjects (D'Souza et al., 2008a). Interestingly, light users of cannabis had lower basal BDNF levels. The lower basal BDNF levels in light users of cannabis suggest that chronic exposure to cannabinoids can lead to a suppression of BDNF release.

The effects of in utero exposure to cannabis on the developing brain is an area of great interest. Studies on a postmortem human fetal brain collection from midgestational mothers with cannabis use is currently underway (reviewed in (Jutras-Aswad et al., 2009)). In utero exposure to cannabis impacts the maturation of the dopamine neurotransmitter system (including tyrosine hydroxylase activity, decreased dopamine D1 and D2 receptor mRNA levels in amygdala), opioid system (decrease proenkephalin mRNA levels, increased mu receptor expression, decreased kappa receptor expression in mediodorsal nucleus of thalamus) and the serotonin system (decreased serotonin in raphe nucleus and ventral hippocampus) (Molina-Holgado et al., 1997). The effects of in utero cannabis exposure on the cannabinoid system are mixed, with some studies showing a change in CB-1R receptor density while other studies show no significant change (Garcia-Gil et al., 1999; Wang et al., 2004). There have been no systematic studies on the effects of in utero exposure on the GABA and glutamate systems.

The behavioral and cognitive effects of in utero exposure to cannabis have been examined in two longitudinal studies, namely the Maternal Health Practices and Child Development Project (MHPCD) (Goldschmidt et al., 2008) and the Ottawa Prenatal Prospective Study (OPPS) (Fried et al., 2003). The MHPCD is a prospective study (since 1982) of the effects of prenatal marijuana and alcohol exposure on the intelligence test performance of 648 children in a low-income, African-American and Caucasian population in Pittsburgh, Pennsylvania. Newborns and infants with prenatal marijuana exposure (PME) demonstrated increased tremors, exaggerated startle response, and poor habituation to novel stimuli (Fried, 1982; Fried and Makin, 1987; Fried and O'Connell, 1987). They showed increased hyperactivity, inattention, and impulsive symptoms by age 10 (Fried et al., 1998; Goldschmidt et al., 2000) and had higher rates of delinquency and externalizing behavioral problems as compared to age-appropriate nonexposed children (Goldschmidt et al., 2000). The study also found that PME predicted lower scores on the verbal reasoning and short-term memory area scores of the Stanford-Binet Intelligence Scale (SBIS) at age 3 and at age 6. PME predicted lower scores on the Bayley Scale of Infant Development, Mental Development Index at 8 months but this was no longer significant at age 18 months. A path analysis suggested that the effects on cognition were mediated by inattention and impulsivity (Goldschmidt et al., 2000). This also appears to be consistent with animal studies in which animals exposed prenatally to Δ^9-THC show a selective disturbance of the frontostriatopallidal proenkephalin/D2 dopamine receptor circuit, a circuit important in inhibitory control behavior (Frank et al., 2007).

In the OPPS, initiated in 1978, the effects of alcohol, cigarette smoking and marijuana use in the prenatal period of 698 women were studied in a birth cohort comprising of 190 children. The population was a low-risk, Caucasian, predominantly middle-class cohort in Canada. While prenatal cannabis use was associated with significantly lower performance on memory and verbal subscales of the McCarthy Scales of Children's Abilities at age 4, these were no longer significant at age 5 and age 6. The question remains as to the validity of these scales in capturing significant differences in specific domains of cognition, rather than a global cognitive functioning.

Adolescence is another critical period of neurodevelopment as brain development continues well into young adulthood (around age 25) (Crews et al., 2007). Animal studies have shown that exposure to cannabinoids in adolescence has more deleterious effects than exposure in adulthood (Cha et al., 2006; O'Shea et al., 2004; Quinn et al., 2008; Schneider, 2008; Schneider et al., 2008). Acute administration of the cannabinoid agonist, WIN 55,212-2 resulted in greater anxiety-related behavior in rats when WIN was administered during puberty compared to when administered in adult rats. Chronic cannabinoid treatment during puberty induced persistent deficits in object recognition and social behavior such as, social play and self-grooming (Schneider et al., 2008). In another study, exposure of adolescent rats to Δ^9-THC during the critical pubertal period led to deficits in spatial memory in adulthood and corresponding changes in hippocampal morphology and NMDA receptor expression (Rubino et al., 2009). Studies in humans show that earlier onset of cannabis use is associated with poorer cognitive performance (Ehrenreich et al., 1999; Kempel et al., 2003), premature alteration in cortical gyrification pattern (Mata et al., 2010) and smaller gray matter volumes (Kumra et al., 2012).

In conclusion, the endocannabinoid system appears to influence neurodevelopment. Perturbation of the endocannabinoid system in the rapidly changing brain, as is the case in adolescence, by excessive or nonphysiological stimulation, as may be the case with exposure to exogenous cannabinoids, may have far reaching consequences. This would be especially so in the presence of already altered neurodevelopmental processes such as the case in individuals with a high risk for schizophrenia. While admittedly speculative, exogenous cannabinoids by disrupting the endocannabinoid system and interfering with neurodevelopmental processes may provide a mechanism by which exposure to cannabinoids during adolescence may increase the risk for the development of schizophrenia. Clearly further work is necessary to understand the consequences of perturbing the endocannabinoid system during critical periods of neurodevelopment.

3. CONCLUSIONS

Cannabinoids can induce transient schizophrenia-like positive, negative and cognitive symptoms in healthy individuals. Cannabinoids also produce some psychophysiological deficits also known to be present in schizophrenia. Also clear is that in individuals with an established psychotic disorder, cannabinoids can exacerbate symptoms, trigger relapse, and have negative consequences on the course of the illness. Schizophrenic patients and others who are psychosis prone may be more likely to experience transient positive, negative and cognitive symptoms following exposure to cannabinoids, and these effects may be greater in magnitude and duration relative to healthy individuals.

Whether cannabinoids "cause" a persistent psychotic disorder such as schizophrenia is not as clear. Increasing evidence suggests that early and heavy cannabis exposure may increase the risk of developing a psychotic disorder such as schizophrenia. The relationship between cannabis exposure and schizophrenia fulfills some, but not all, of the usual criteria for causality. However, most people who use cannabis do not develop schizophrenia, and many people diagnosed with schizophrenia have never used cannabis. Furthermore, the increase in cannabis use, the use of more potent forms of cannabis and the earlier age of first use should be accompanied or followed by a commensurate increase in the rates of schizophrenia or an earlier age of onset of the illness.

However, data on the rates of schizophrenia have been mixed with some studies suggesting a decrease, others suggesting an increase and others suggesting no change. Therefore, exposure to cannabis is neither a necessary nor a sufficient cause of schizophrenia — similar to cigarette smoking being neither necessary nor sufficient to cause lung cancer or the role of dietary sodium and hypertension. More likely, cannabis exposure is a component or contributing cause which interacts with other known (genetic, environmental) and unknown factors, culminating in schizophrenia. In the absence of known causes of schizophrenia, however, and the implications for public health policy should such a link be established (Hall and Pacula, 2003), the role of component causes such as cannabinoid exposure should remain a focus of further study. The increasing availability of high potency cannabis, the increasing use of "medicinal" cannabis, and the increased availability of highly potent, full agonist synthetic cannabinoids, might provide further evidence to either refute or prove the link between cannabinoids and psychosis. Finally, further studies are warranted to determine a biological mechanism by which cannabinoids "cause" psychosis.

ACKNOWLEDGMENTS

The authors wish to acknowledge support from the (1) Department of Veterans Affairs (RAS, IGS, MR, DCD), (2) National Institute of Mental Health (DCD), (3) National Institute of Drug Abuse (PDS, DCD), (4) Stanley Medical Research Institute (MR), and (5) NARSAD (MR, PDS).

REFERENCES

AAPCC, 2011. Synthetic Marijuana. Data Updated December 12, 2011.

Adamec, C., Pihl, R.O., Leiter, L., 1976. An analysis of the subjective marijuana experience. Int. J. Addict. 11, 295–307.

Addington, J., Addington, D., 1998. Effect of substance misuse in early psychosis. Br. J. Psychiatry Suppl. 172, 134–136.

Addington, J., Duchak, V., 1997. Reasons for substance use in schizophrenia. Acta Psychiatr. Scand. 96, 329–333.

Aguado, T., Palazuelos, J., Monory, K., Stella, N., Cravatt, B., Lutz, B., Marsicano, G., Kokaia, Z., Guzman, M., Galve-Roperh, I., 2006. The endocannabinoid system promotes astroglial differentiation by acting on neural progenitor cells. J. Neurosci. 26, 1551–1561.

Ajdacic-Gross, V., Lauber, C., Warnke, I., Haker, H., Murray, R.M., Rossler, W., 2007. Changing incidence of psychotic disorders among the young in Zurich. Schizophr. Res. 95, 9–18.

Allebeck, P., Adamsson, C., Engstrom, A., Rydberg, U., 1993a. Cannabis and schizophrenia: a longitudinal study of cases treated in Stockholm County. Acta Psychiatr. Scand. 88, 21–24.

Allebeck, P., Adamsson, C., Engstrom, A., Rydberg, U., 1993b. Cannabis and schizophrenia: a longitudinal study of cases treated in Stockholm County. (erratum appears in Acta Psychiatr Scand 1993 Oct; 88(4):304), Acta Psychiatr. Scand. 88, 21–24.

Ames, F., 1958. A clinical and metabolic study of acute intoxication with *Cannabis sativa* and its role in the model psychoses. J. Ment. Sci. 104, 972–999.

Andreasen, N.C., Nopoulos, P., O'Leary, D.S., Miller, D.D., Wassink, T., Flaum, M., 1999. Defining the phenotype of schizophrenia: cognitive dysmetria and its neural mechanisms. Biol. Psychiatry 46, 908–920.

Andreasson, S., Allebeck, P., Engstrom, A., Rydberg, U., 1987. Cannabis and schizophrenia. A longitudinal study of Swedish conscripts. Lancet 2, 1483–1486.

Andreasson, S., Allebeck, P., Rydberg, U., 1989. Schizophrenia in users and nonusers of cannabis. A longitudinal study in Stockholm County. Acta Psychiatr. Scand. 79, 505–510.

Arndt, S., Tyrrell, G., Flaum, M., Andreasen, N., 1992. Comorbidity of substance abuse and schizophrenia: the role of pre-morbid adjustment. Psychol. Med. 22, 379–388.

Arseneault, L., Cannon, M., Poulton, R., Murray, R., Caspi, A., Moffitt, T.E., 2002. Cannabis use in adolescence and risk for adult psychosis: longitudinal prospective study. BMJ 325, 1212–1213.

Atwood, B.K., Huffman, J., Straiker, A., Mackie, K., 2010. JWH018, a common constituent of 'Spice' herbal blends, is a potent and efficacious cannabinoid CB receptor agonist. Br. J. Pharmacol. 160, 585–593.

Atwood, B.K., Lee, D., Straiker, A., Widlanski, T.S., Mackie, K., 2011. CP47, 497-C8 and JWH073, commonly found in 'Spice' herbal blends, are potent and efficacious CB(1) cannabinoid receptor agonists. Eur. J. Pharmacol. 659, 139–145.

Aust, R., Sharp, C., Goulden, C., 2002. Prevalence of Drug Use: Key Findings from the 2001/2002 British Crime Survey, Findings 182. Home Office Research, Development and Statistics Directorate, London.

Auwarter, V., Dresen, S., Weinmann, W., Muller, M., Putz, M., Ferreiros, N., 2009. 'Spice' and other herbal blends: harmless incense or cannabinoid designer drugs? J. Mass Spectrom. 44, 832–837.

Baeza, I., Graell, M., Moreno, D., Castro-Fornieles, J., Parellada, M., Gonzalez-Pinto, A., Paya, B., Soutullo, C., de la Serna, E., Arango, C., 2009. Cannabis use in children and adolescents with first episode psychosis: influence on psychopathology and short-term outcome (CAFEPS study). Schizophr. Res. 113, 129–137.

Baigent, M., Holme, G., Hafner, R.J., 1995. Self reports of the interaction between substance abuse and schizophrenia. Aust. N. Z. J. Psychiatry 29, 69–74.

Baker, A., Lewin, T., Reichler, H., Clancy, R., Carr, V., Garrett, R., Sly, K., Devir, H., Terry, M., 2002. Motivational interviewing among psychiatric in-patients with substance use disorders. Acta Psychiatr. Scand. 106, 233–240.

Bangalore, S.S., Prasad, K.M., Montrose, D.M., Goradia, D.D., Diwadkar, V.A., Keshavan, M.S., 2008. Cannabis use and brain structural alterations in first episode schizophrenia–a region of interest, voxel based morphometric study. Schizophr. Res. 99, 1–6.

Barkus, E.J., Stirling, J., Hopkins, R.S., Lewis, S., 2006. Cannabis-induced psychosis-like experiences are associated with high schizotypy. Psychopathology 39, 175–178.

Benford, D.M., Caplan, J.P., 2011. Psychiatric sequelae of spice, K2, and synthetic cannabinoid receptor agonists. Psychosomatics 52, 295.

Berghuis, P., Dobszay, M.B., Wang, X., Spano, S., Ledda, F., Sousa, K.M., Schulte, G., Ernfors, P., Mackie, K., Paratcha, G., Hurd, Y.L., Harkany, T., 2005. Endocannabinoids regulate interneuron migration and morphogenesis by transactivating the TrkB receptor. Proc. Natl. Acad. Sci. U.S.A 102, 19115–19120.

Berghuis, P., Rajnicek, A.M., Morozov, Y.M., Ross, R.A., Mulder, J., Urban, G.M., Monory, K., Marsicano, G., Matteoli, M., Canty, A., Irving, A.J., Katona, I., Yanagawa, Y., Rakic, P., Lutz, B., Mackie, K., Harkany, T., 2007. Hardwiring the brain: endocannabinoids shape neuronal connectivity. Science 316, 1212–1216.

Bersani, G., Orlandi, V., Kotzalidis, G., Pancheri, P., 2002. Cannabis and schizophrenia: impact on onset, course, psychopathology and outcomes. Eur. Arch. Psychiatry Clin. Neurosci. 252, 86–92.

Bhattacharyya, S., Morrison, P.D., Fusar-Poli, P., Martin-Santos, R., Borgwardt, S., Winton-Brown, T., Nosarti, C., CM, O.C., Seal, M., Allen, P., Mehta, M.A., Stone, J.M., Tunstall, N., Giampietro, V., Kapur, S., Murray, R.M., Zuardi, A.W., Crippa, J.A., Atakan, Z., McGuire, P.K., 2010. Opposite effects of delta-9-tetrahydrocannabinol and cannabidiol on human brain function and psychopathology. Neuropsychopharmacology 35, 764–774.

Block, R.I., O'Leary, D.S., Ehrhardt, J.C., Augustinack, J.C., Ghoneim, M.M., Arndt, S., Hall, J.A., 2000. Effects of frequent marijuana use on brain tissue volume and composition. Neuroreport 11, 491–496.

Bolla, K.I., Brown, K., Eldreth, D., Tate, K., Cadet, J.L., 2002. Dose-related neurocognitive effects of marijuana use. Neurology 59, 1337–1343.

Boutros, N.N., Belger, A., Campbell, D., D'Souza, C., Krystal, J., 1999. Comparison of four components of sensory gating in schizophrenia and normal subjects: a preliminary report. Psychiatry Res. 88, 119–130.

Braff, D.L., 1993. Information processing and attention dysfunctions in schizophrenia. Schizophr. Bull. 19, 233–259.

Braff, D.L., Light, G.A., 2004. Preattentional and attentional cognitive deficits as targets for treating schizophrenia. Psychopharmacology (Berl.) 174, 75–85.

Bramon, E., Rabe-Hesketh, S., Sham, P., Murray, R.M., Frangou, S., 2004. Meta-analysis of the P300 and P50 waveforms in schizophrenia. Schizophr. Res. 70, 315–329.

Bramon, E., McDonald, C., Croft, R.J., Landau, S., Filbey, F., Gruzelier, J.H., Sham, P.C., Frangou, S., Murray, R.M., 2005. Is the P300 wave an endophenotype for schizophrenia? A meta-analysis and a family study. NeuroImage 27, 960–968.

Brandt, J., 1991. The hopkins verbal learning test: development of a new memory test with six equivalent forms. Clin. Neuropsychol. 5, 125–142.

Brook, M., 1984. Psychosis after cannabis abuse. BMJ 288, 1381 (Clinical research ed.).

Brunette, M.F., Mueser, K.T., Xie, H., Drake, R.E., 1997. Relationships between symptoms of schizophrenia and substance abuse. J. Nerv. Ment. Dis. 185, 13–20.

Buhler, B., Hambrecht, M., Loffler, W., an der Heiden, W., Hafner, H., 2002. Precipitation and determination of the onset and course of schizophrenia by substance abuse—a retrospective and prospective study of 232 population-based first illness episodes. Schizophr. Res. 54, 243–251.

Businessweek, B, 2011. The Big Business of Synthetic Highs (Bloomberg Businessweek).

Butovsky, E., Juknat, A., Goncharov, I., Elbaz, J., Eilam, R., Zangen, A., Vogel, Z., 2005. In vivo up-regulation of brain-derived neurotrophic factor in specific brain areas by chronic exposure to delta-tetrahydrocannabinol. J. Neurochem. 93, 802–811.

Cadenhead, K.S., Light, G.A., Geyer, M.A., Braff, D.L., 2000. Sensory gating deficits assessed by the P50 event-related potential in subjects with schizotypal personality disorder. Am. J. Psychiatry 157, 55–59.

Cahn, W., Hulshoff Pol, H.E., Caspers, E., van Haren, N.E., Schnack, H.G., Kahn, R.S., 2004. Cannabis and brain morphology in recent-onset schizophrenia. Schizophr. Res. 67, 305–307.

Campbell, A.M., Evans, M., Thomson, J.L., Williams, M.J., 1971. Cerebral atrophy in young cannabis smokers. Lancet 2, 1219–1224.

Carroll, C.A., O'Donnell, B.F., Shekhar, A., Hetrick, W.P., 2009. Timing dysfunctions in schizophrenia span from millisecond to several-second durations. Brain Cogn. 70, 181–190.

CBS, 2011. Synthetic Marijuana: How Dangerous Is It?. (CBS Miami).

Cha, Y.M., White, A.M., Kuhn, C.M., Wilson, W.A., Swartzwelder, H.S., 2006. Differential effects of delta9-THC on learning in adolescent and adult rats. Pharmacol. Biochem. Behav. 83, 448–455.

Chan, G.C., Hinds, T.R., Impey, S., Storm, D.R., 1998. Hippocampal neurotoxicity of delta9-tetrahydrocannabinol. J. Neurosci. 18, 5322–5332.

Chopra, G.S., Smith, J.W., 1974. Psychotic reactions following cannabis use in East Indians. Arch. Gen. Psychiatry 30, 24–27.

Citron, M.L., Herman, T.S., Vreeland, F., Krasnow, S.H., Fossieck Jr., B.E., Harwood, S., Franklin, R., Cohen, M.H., 1985. Antiemetic efficacy of levonantradol compared to delta-9-tetrahydrocannabinol for chemotherapy-induced nausea and vomiting. Cancer Treat Rep. 69, 109–112.

Cleghorn, J.M., Kaplan, R.D., Szechtman, B., Szechtman, H., Brown, G.M., Franco, S., 1991. Substance abuse and schizophrenia: effect on symptoms but not on neurocognitive function. J. Clin. Psychiatry 52, 26–30.

CNN, 2011. Synthetic marijuana a growing trend among teens, authorities say (CNN).

Co, B.T., Goodwin, D.W., Gado, M., Mikhael, M., Hill, S.Y., 1977. Absence of cerebral atrophy in chronic cannabis users. Evaluation by computerized transaxial tomography. JAMA 237, 1229–1230.

Cohen, J., Morrison, S., Greenberg, J., Saidinejad, M., 2012. Clinical presentation of intoxication due to synthetic cannabinoids. Pediatrics.

Collip, D., Myin-Germeys, I., Van Os, J., 2008. Does the concept of "sensitization" provide a plausible mechanism for the putative link between the environment and schizophrenia? Schizophr. Bull. 34, 220–225.

Compton, W.M., Grant, B.F., Colliver, J.D., Glantz, M.D., Stinson, F.S., 2004. Prevalence of marijuana use disorders in the United States: 1991–1992 and 2001–2002. JAMA 291, 2114–2121.

Conrad, D.G., Elsmore, T.F., Sodetz, F.J., 1972. 9-tetrahydrocannabinol: dose-related effects on timing behavior in chimpanzee. Science 175, 547–550 New York, N.Y.

Corkin, S., Milner, B., Rasmussen, T., 1964. Effects of different cortical excisions on sensory thresholds in man. Trans. Am. Neurol. Assoc. 89, 112–116.

Cousijn, J., Wiers, R.W., Ridderinkhof, K.R., van den Brink, W., Veltman, D.J., Goudriaan, A.E., 2012. Grey matter alterations associated with cannabis use: results of a VBM study in heavy cannabis users and healthy controls. NeuroImage 59, 3845–3851.

Crean, R.D., Crane, N.A., Mason, B.J., 2011. An evidence based review of acute and long-term effects of cannabis use on executive cognitive functions. J. Addict. Med. 5, 1–8.

Crews, F., He, J., Hodge, C., 2007. Adolescent cortical development: a critical period of vulnerability for addiction. Pharmacol. Biochem. Behav. 86, 189–199.

Cronin, C.M., Sallan, S.E., Gelber, R., Lucas, V.S., Laszlo, J., 1981. Antiemetic effect of intramuscular levonantradol in patients receiving anticancer chemotherapy. J. Clin. Pharmacol. 21, 43S–50S.

Cuffel, B., Heithoff, K., Lawson, W., 1993. Correlates of patterns of substance abuse among patients with schizophrenia. Hosp. Community Psychiatry 44, 247–251.

D'Souza, D.C., 2007. Cannabinoids and psychosis. Int. Rev. Neurobiol. 78, 289–326.

D'Souza, D.C., Perry, E., MacDougall, L., Ammerman, Y., Cooper, T., Wu, Y.T., Braley, G., Gueorguieva, R., Krystal, J.H., 2004. The psychotomimetic effects of intravenous delta-9-tetrahydrocannabinol in healthy individuals: implications for psychosis. Neuropsychopharmacology 29, 1558–1572.

D'Souza, D.C., Abi-Saab, W.M., Madonick, S., Forselius-Bielen, K., Doersch, A., Braley, G., Gueorguieva, R., Cooper, T.B., Krystal, J.H., 2005. Delta-9-tetrahydrocannabinol effects in schizophrenia: implications for cognition, psychosis, and addiction. Biol. Psychiatry 57, 594–608.

D'Souza, D.C., Pittman, B., Perry, E., Simen, A., 2008a. Preliminary evidence of cannabinoid effects on brain-derived neurotrophic factor (BDNF) levels in humans. Psychopharmacology (Berl.).

D'Souza, D.C., Ranganathan, M., Braley, G., Gueorguieva, R., Zimolo, Z., Cooper, T., Perry, E., Krystal, J., 2008b. Blunted psychotomimetic and amnestic effects of delta-9-tetrahydrocannabinol in frequent users of cannabis. Neuropsychopharmacology 33, 2505–2516.

D'Souza, D.C., Sewell, R.A., Ranganathan, M., 2009. Cannabis and psychosis/schizophrenia: human studies. Eur. Arch. Psychiatry Clin. Neurosci. 259, 413–431.

D'Souza, D.C., Fridberg, D.J., Skosnik, P.D., Williams, A., Roach, B., Singh, N., Carbuto, M., Elander, J., Schnakenberg, A., Pittman, B., Sewell, R.A., Ranganathan, M., Mathalon, D., 2012. Dose-related modulation of event-related potentials to novel and target stimuli by intravenous delta(9)-THC in humans. Neuropsychopharmacology.

Dargan, P.I., Hudson, S., Ramsey, J., Wood, D.M., 2011. The impact of changes in UK classification of the synthetic cannabinoid receptor agonists in 'Spice'. Int. J. Drug Policy 22, 274–277.

Davalos, D.B., Kisley, M.A., Ross, R.G., 2003. Effects of interval duration on temporal processing in schizophrenia. Brain Cogn. 52, 295–301.

de Sola, S., Tarancon, T., Pena-Casanova, J., Espadaler, J.M., Langohr, K., Poudevida, S., Farre, M., Verdejo-Garcia, A., de la Torre, R., 2008. Auditory event-related potentials (P3) and cognitive performance in recreational ecstasy polydrug users: evidence from a 12-month longitudinal study. Psychopharmacology (Berl.) 200, 425–437.

de Wilde, O.M., Bour, L.J., Dingemans, P.M., Koelman, J.H., Linszen, D.H., 2007. A meta-analysis of P50 studies in patients with schizophrenia and relatives: differences in methodology between research groups. Schizophr. Res. 97, 137–151.

Dekker, N., Schmitz, N., Peters, B.D., van Amelsvoort, T.A., Linszen, D.H., de Haan, L. Cannabis use and callosal white matter structure and integrity in recent-onset schizophrenia. Psychiatry Res. 181, 51–56.

Dekker, N., Linszen, D., De Haan, L., 2009. Reasons for cannabis use and effects of cannabis use as reported by patients with psychotic disorders. Psychopathology 42, 350–360.

Der, G., Gupta, S., Murray, R.M., 1990. Is schizophrenia disappearing? Lancet 335, 513–516.

Derkinderen, P., Valjent, E., Toutant, M., Corvol, J.C., Enslen, H., Ledent, C., Trzaskos, J., Caboche, J., Girault, J.A., 2003. Regulation of extracellular signal-regulated kinase by cannabinoids in hippocampus. J. Neurosci. 23, 2371–2382.

DeRosse, P., Kaplan, A., Burdick, K.E., Lencz, T., Malhotra, A.K., 2010. Cannabis use disorders in schizophrenia: effects on cognition and symptoms. Schizophr. Res. 120, 95–100.

Diasio, R.B., Ettinger, D.S., Satterwhite, B.E., 1981. Oral levonantradol in the treatment of chemotherapy-induced emesis: preliminary observations. J. Clin. Pharmacol. 21, 81S–85S.

Dissanayake, D.W., Zachariou, M., Marsden, C.A., Mason, R., 2008. Auditory gating in rat hippocampus and medial prefrontal cortex: effect of the cannabinoid agonist WIN55,212-2. Neuropharmacology 55, 1397–1404.

Dixon, L., 1999. Dual diagnosis of substance abuse in schizophrenia: prevalence and impact on outcomes. Schizophr. Res. 35 (Suppl.), S93–100.

Dixon, L., Haas, G., Weiden, P., Sweeney, J., Frances, A., 1990. Acute effects of drug abuse in schizophrenic patients: clinical observations and patients' self-reports. Schizophr. Bull. 16, 69–79.

Dixon, L., Haas, G., Weiden, P.J., Sweeney, J., Frances, A.J., 1991. Drug abuse in schizophrenic patients: clinical correlates and reasons for use. Am. J. Psychiatry 148, 224–230.

Downer, E.J., Gowran, A., Campbell, V.A., 2007. A comparison of the apoptotic effect of delta(9)-tetrahydrocannabinol in the neonatal and adult rat cerebral cortex. Brain Res. 1175, 39–47.

Dresen, S., Ferreiros, N., Putz, M., Westphal, F., Zimmermann, R., Auwarter, V., 2010. Monitoring of herbal mixtures potentially containing synthetic cannabinoids as psychoactive compounds. J. Mass Spectrom. 45, 1186–1194.

Duncan, C.C., 1988. Event-related brain potentials: a window on information processing in schizophrenia. Schizophr. Bull. 14, 199–203.

Edwards, C.R., Skosnik, P.D., Steinmetz, A.B., Vollmer, J.M., O'Donnell, B.F., Hetrick, W.P., 2008. Assessment of forebrain-dependent trace eyeblink conditioning in chronic cannabis users. Neurosci. Lett. 439, 264–268.

Edwards, C.R., Skosnik, P.D., Steinmetz, A.B., O'Donnell, B.F., Hetrick, W.P., 2009. Sensory gating impairments in heavy cannabis users are associated with altered neural oscillations. Behav. Neurosci. 123, 894–904.

Eggan, S.M., Lewis, D.A., 2007. Immunocytochemical distribution of the cannabinoid CB1 receptor in the primate neocortex: a regional and laminar analysis. Cereb. Cortex 17, 175–191.

Ehrenreich, H., Rinn, T., Kunert, H.J., Moeller, M.R., Poser, W., Schilling, L., Gigerenzer, G., Hoehe, M.R., 1999. Specific attentional dysfunction in adults following early start of cannabis use. Psychopharmacology (Berl.) 142, 295–301.

Emrich, H.M., Weber, M.M., Wendl, A., Zihl, J., von Meyer, L., Hanisch, W., 1991. Reduced binocular depth inversion as an indicator of cannabis-induced censorship impairment. Pharmacol. Biochem. Behav. 40, 689–690.

Every-Palmer, S., 2010. Warning: legal synthetic cannabinoid-receptor agonists such as JWH-018 may precipitate psychosis in vulnerable individuals. Addiction 105, 1859–1860.

Every-Palmer, S., 2011. Synthetic cannabinoid JWH-018 and psychosis: an explorative study. Drug Alcohol Depend. 117, 152–157.

Farrell, M., Howes, S., Taylor, C., Lewis, G., Jenkins, R., Bebbington, P., Jarvis, M., Brugha, T., Gill, B., Meltzer, H., 1998. Substance misuse and psychiatric comorbidity: an overview of the OPCS national psychiatric morbidity survey. Addict. Behav. 23, 909–918.

Ferdinand, R.F., Sondeijker, F., van der Ende, J., Selten, J.P., Huizink, A., Verhulst, F.C., 2005. Cannabis use predicts future psychotic symptoms, and vice versa. Addiction 100, 612–618.

Fergusson, D.M., Horwood, L.J., Ridder, E.M., 2005. Tests of causal linkages between cannabis use and psychotic symptoms. Addiction 100, 354–366.

Fernandez-Ruiz, J., Berrendero, F., Hernandez, M.L., Ramos, J.A., 2000. The endogenous cannabinoid system and brain development. Trends Neurosci. 23, 14–20.

Fontes, M.A., Bolla, K.I., Cunha, P.J., Almeida, P.P., Jungerman, F., Laranjeira, R.R., Bressan, R.A., Lacerda, A.L., 2011. Cannabis use before age 15 and subsequent executive functioning. Br. J. Psychiatry 198, 442–447.

Fowler, I., Carr, V., Carter, N., Lewin, T., 1998. Patterns of current and lifetime substance use in schizophrenia. Schizophr. Bull. 24, 443–455.

Frank, M.J., Moustafa, A.A., Haughey, H.M., Curran, T., Hutchison, K.E., 2007. Genetic triple dissociation reveals multiple roles for dopamine in reinforcement learning. Proc. Natl. Acad. Sci. U.S.A 104, 16311–16316.

Fried, P.A., 1982. Marihuana use by pregnant women and effects on offspring: an update. Neurobehav. Toxicol. Teratol. 4, 451–454.

Fried, P.A., Makin, J.E., 1987. Neonatal behavioural correlates of prenatal exposure to marihuana, cigarettes and alcohol in a low risk population. Neurotoxicol. Teratol. 9, 1–7.

Fried, P.A., O'Connell, C.M., 1987. A comparison of the effects of prenatal exposure to tobacco, alcohol, cannabis and caffeine on birth size and subsequent growth. Neurotoxicol. Teratol. 9, 79–85.

Fried, P.A., Watkinson, B., Gray, R., 1998. Differential effects on cognitive functioning in 9- to 12-year olds prenatally exposed to cigarettes and marihuana. Neurotoxicol. Teratol. 20, 293–306.

Fried, P.A., Watkinson, B., Gray, R., 2003. Differential effects on cognitive functioning in 13- to 16-year-olds prenatally exposed to cigarettes and marihuana. Neurotoxicol. Teratol. 25, 427–436.

Fried, P.A., Watkinson, B., Gray, R., 2005. Neurocognitive consequences of marihuana—a comparison with pre-drug performance. Neurotoxicol. Teratol. 27, 231–239.

Fusar-Poli, P., Crippa, J.A., Bhattacharyya, S., Borgwardt, S.J., Allen, P., Martin-Santos, R., Seal, M., Surguladze, S.A., O'Carrol, C., Atakan, Z., Zuardi, A.W., McGuire, P.K., 2009. Distinct effects of {delta}9-tetrahydrocannabinol and cannabidiol on neural activation during emotional processing. Arch. Gen. Psychiatry 66, 95–105.

Galve-Roperh, I., Aguado, T., Palazuelos, J., Guzman, M., 2007. The endocannabinoid system and neurogenesis in health and disease. Neuroscientist 13, 109–114.

Galve-Roperh, I., Palazuelos, J., Aguado, T., Guzman, M., 2009. The endocannabinoid system and the regulation of neural development: potential implications in psychiatric disorders. Eur. Arch. Psychiatry Clin. Neurosci. 259, 371–382.

Garcia-Gil, L., Romero, J., Ramos, J.A., Fernandez-Ruiz, J.J., 1999. Cannabinoid receptor binding and mRNA levels in several brain regions of adult male and female rats perinatally exposed to delta9-tetrahydrocannabinol. Drug Alcohol Depend. 55, 127–136.

Gearon, J.S., Bellack, A.S., Rachbeisel, J., Dixon, L., 2001. Drug-use behavior and correlates in people with schizophrenia. Addict. Behav. 26, 51–61.

Ginsburg, B.C., McMahon, L.R., Sanchez, J.J., Javors, M.A., 2012. Purity of synthetic cannabinoids sold online for recreational use. J. Anal. Toxicol. 36, 66–68.

Gobbi, G., Bambico, F.R., Mangieri, R., Bortolato, M., Campolongo, P., Solinas, M., Cassano, T., Morgese, M.G., Debonnel, G., Duranti, A., Tontini, A., Tarzia, G., Mor, M., Trezza, V., Goldberg, S.R., Cuomo, V., Piomelli, D., 2005. Antidepressant-like activity and modulation of brain monoaminergic transmission by blockade of anandamide hydrolysis. Proc. Natl. Acad. Sci. U.S.A 102, 18620–18625.

Goldschmidt, L., Day, N.L., Richardson, G.A., 2000. Effects of prenatal marijuana exposure on child behavior problems at age 10. Neurotoxicol. Teratol. 22, 325–336.

Goldschmidt, L., Richardson, G.A., Willford, J., Day, N.L., 2008. Prenatal marijuana exposure and intelligence test performance at age 6. J. Am. Acad. Child Adolesc. Psychiatry 47, 254–263.

Gordon, M., 1986. Microprocessor-based assessment of attention deficit disorders (ADD). Psychopharmacol. Bull. 22, 288–290.

Goswami, S., Mattoo, S.K., Basu, D., Singh, G., 2004. Substance-abusing schizophrenics: do they self-medicate? Am. J. Addict. 13, 139–150.

Green, B., Kavanagh, D., Young, R., 2003. Being stoned: a review of self-reported cannabis effects. Drug Alcohol Rev. 22, 453–460.

Green, A.I., Tohen, M.F., Hamer, R.M., Strakowski, S.M., Lieberman, J.A., Glick, I., Clark, W.S., 2004a. First episode schizophrenia-related psychosis and substance use disorders: acute response to olanzapine and haloperidol. Schizophr. Res. 66, 125–135.

Green, A.I., Tohen, M.F., Hamer, R.M., Strakowski, S.M., Lieberman, J.A., Glick, I., Clark, W.S., Group, H.R., 2004b. First episode schizophrenia-related psychosis and substance use disorders: acute response to olanzapine and haloperidol.[see comment]. Schizophr. Res. 66, 125–135.

Green, B., Kavanagh, D.J., Young, R.M., 2004c. Reasons for cannabis use in men with and without psychosis. Drug Alcohol Rev. 23, 445–453.

Green, B., Young, R., Kavanagh, D., 2005. Cannabis use and misuse prevalence among people with psychosis. Br. J. Psychiatry 187, 306–313.

Gregg, L., Barrowclough, C., Haddock, G., 2009. Development and validation of a scale for assessing reasons for substance use in schizophrenia: the ReSUS scale. Addict. Behav. 34, 830–837.

Grossman, W., 1969. Adverse reactions associated with cannabis products in India. Ann. Intern. Med. 70, 529–533.

GROUP, G.R.a.O.i.P.I., 2011. Evidence that familial liability for psychosis is expressed as differential sensitivity to cannabis: an analysis of patient-sibling and sibling-control pairs. Arch. Gen. Psychiatry 68, 138–147.

Grunwald, T., Boutros, N.N., Pezer, N., von Oertzen, J., Fernandez, G., Schaller, C., Elger, C.E., 2003. Neuronal substrates of sensory gating within the human brain. Biol. Psychiatry 53, 511–519.

Halikas, J., Weller, R., Morse, C., 1982. Effects of regular marijuana use on sexual performance. J. Psychoactive Drugs 14, 59–70.

Hall, W., Pacula, R., 2003. Cannabis Use and Dependence: Public Health and Public Policy. Cambridge University Press, London.

Hall, W., Solowij, N., 1998. Adverse effects of cannabis. Lancet 352, 1611–1616.

Hambrecht, M., Hafner, H., 1996. Substance abuse and the onset of schizophrenia. Biol. Psychiatry 40, 1155–1163.

Hambrecht, M., Hafner, H., 2000. Cannabis, vulnerability, and the onset of schizophrenia: an epidemiological perspective. Aust. N. Z. J. Psychiatry 34, 468–475.

Han, C.J., Robinson, J.K., 2001. Cannabinoid modulation of time estimation in the rat. Behav. Neurosci. 115, 243–246.

Hannerz, J., Hindmarsh, T., 1983. Neurological and neuroradiological examination of chronic cannabis smokers. Ann. Neurol. 13, 207–210.

Harkany, T., Guzman, M., Galve-Roperh, I., Berghuis, P., Devi, L.A., Mackie, K., 2007. The emerging functions of endocannabinoid signaling during CNS development. Trends Pharmacol. Sci. 28, 83–92.

Harley, M., Kelleher, I., Clarke, M., Lynch, F., Arseneault, L., Connor, D., Fitzpatrick, C., Cannon, M., 2010. Cannabis use and childhood trauma interact additively to increase the risk of psychotic symptoms in adolescence. Psychol. Med. 40, 1627–1634.

Harper, J.W., Heath, R.G., Myers, W.A., 1977. Effects of *Cannabis sativa* on ultrastructure of the synapse in monkey brain. J. Neurosci. Res. 3, 87–93.

Hart, C.L., van Gorp, W., Haney, M., Foltin, R.W., Fischman, M.W., 2001. Effects of acute smoked marijuana on complex cognitive performance. Neuropsychopharmacology 25, 757–765.

Heath, R.G., Fitzjarrell, A.T., Fontana, C.J., Garey, R.E., 1980. Cannabis sativa: effects on brain function and ultrastructure in rhesus monkeys. Biol. Psychiatry 15, 657–690.

Heim, M.E., Romer, W., Queisser, W., 1981. Clinical experience with levonantradol hydrochloride in the prevention of cancer chemotherapy-induced nausea and vomiting. J. Clin. Pharmacol. 21, 86S–89S.

Heim, M.E., Queisser, W., Altenburg, H.P., 1984. Randomized crossover study of the antiemetic activity of levonantradol and metoclopramide in cancer patients receiving chemotherapy. Cancer Chemother. Pharmacol. 13, 123–125.

Heinrichs, R.W., 2004. Meta-analysis and the science of schizophrenia: variant evidence or evidence of variants? Neurosci. Biobehav. Rev. 28, 379–394.

Heinrichs, R.W., Zakzanis, K.K., 1998. Neurocognitive deficit in schizophrenia: a quantitative review of the evidence. Neuropsychology 12, 426–445.

Heishman, S.J., Huestis, M.A., Henningfield, J.E., Cone, E.J., 1990. Acute and residual effects of marijuana: profiles of plasma THC levels, physiological, subjective, and performance measures. Pharmacol. Biochem. Behav. 37, 561–565.

Hekimian, L.J., Gershon, S., 1968. Characteristics of drug abusers admitted to a psychiatric hospital. JAMA 205, 125–130.

Henquet, C., Krabbendam, L., Spauwen, J., Kaplan, C., Lieb, R., Wittchen, H.U., van Os, J., 2005a. Prospective cohort study of cannabis use, predisposition for psychosis, and psychotic symptoms in young people. BMJ 330, 11.

Henquet, C., Murray, R., Linszen, D., van Os, J., 2005b. The environment and schizophrenia: the role of cannabis use. Schizophr. Bull. 31, 608–612.

Henquet, C., Rosa, A., Krabbendam, L., Papiol, S., Fananas, L., Drukker, M., Ramaekers, J.G., van Os, J., 2006. An experimental study of catechol-o-methyltransferase Val158Met moderation of delta-9-tetrahydrocannabinol-induced effects on psychosis and cognition. Neuropsychopharmacology 31, 2748–2757.

Hickman, M., Vickerman, P., Macleod, J., Kirkbride, J., Jones, P.B., 2007. Cannabis and schizophrenia: model projections of the impact of the rise in cannabis use on historical and future trends in schizophrenia in England and Wales. Addiction 102, 597–606.

Hicks, R.E., Gualtieri, C.T., Mayo Jr., J.P., Perez-Reyes, M., 1984. Cannabis, atropine, and temporal information processing. Neuropsychobiology 12, 229–237.

Hillyard, S.A., Hink, R.F., Schwent, V.L., Picton, T.W., 1973. Electrical signs of selective attention in the human brain. Science 182, 177–180.

Hooker, W.D., Jones, R.T., 1987. Increased susceptibility to memory intrusions and the Stroop interference effect during acute marijuana intoxication. Psychopharmacology 91, 20–24.

Houston, J.E., Murphy, J., Adamson, G., Stringer, M., Shevlin, M., 2008. Childhood sexual abuse, early cannabis use, and psychosis: testing an interaction model based on the national comorbidity survey. Schizophr. Bull. 34, 580–585.

Hu, X., Primack, B.A., Barnett, T.E., Cook, R.L., 2011. College students and use of K2: an emerging drug of abuse in young persons. Subst. Abuse Treat. Prev. Policy 6, 16.

Huffman, J.W., Padgett, L.W., 2005. Recent developments in the medicinal chemistry of cannabimimetic indoles, pyrroles and indenes. Curr. Med. Chem. 12, 1395–1411.

Huffman, J.W., Zengin, G., Wu, M.J., Lu, J., Hynd, G., Bushell, K., Thompson, A.L., Bushell, S., Tartal, C., Hurst, D.P., Reggio, P.H., Selley, D.E., Cassidy, M.P., Wiley, J.L., Martin, B.R., 2005. Structure–activity relationships for 1-alkyl-3-(1-naphthoyl)indoles at the cannabinoid CB(1) and CB(2) receptors: steric and electronic effects of naphthoyl substituents. New highly selective CB(2) receptor agonists. Bioorg. Med. Chem. 13, 89–112.

Hurst, D., Loeffler, G., McLay, R., 2011. Psychosis associated with synthetic cannabinoid agonists: a case series. Am. J. Psychiatry 168, 1119.

Ilan, A.B., Gevins, A., Coleman, M., Elsohly, M.A., de Wit, H., 2005. Neurophysiological and subjective profile of marijuana with varying concentrations of cannabinoids. Behav. Pharmacol. 16, 487–496.

Isbell, H., Jasinski, D.R., 1969. A comparison of LSD-25 with (-)-delta-9-trans-tetrahydrocannabinol (THC) and attempted cross tolerance between LSD and THC. Psychopharmacologia 14, 115–123.

Isbell, H., Gorodetzsky, C.W., Jasinski, D., Claussen, U., von Spulak, F., Korte, F., 1967. Effects of (–)delta-9-trans-tetrahydrocannabinol in man. Psychopharmacologia 11, 184–188.

Jablensky, A., McGrath, J., Herrman, H., Castle, D., Gureje, O., Evans, M., Carr, V., Morgan, V., Korten, A., Harvey, C., 2000. Psychotic disorders in urban areas: an overview of the study on low prevalence disorders. Aust. N. Z. J. Psychiatry 34, 221–236.

Jager, G., Kahn, R.S., Van Den Brink, W., Van Ree, J.M., Ramsey, N.F., 2006. Long-term effects of frequent cannabis use on working memory and attention: an fMRI study. Psychopharmacology 185, 358–368.

Jager, G., Van Hell, H.H., De Win, M.M., Kahn, R.S., Van Den Brink, W., Van Ree, J.M., Ramsey, N.F., 2007. Effects of frequent cannabis use on hippocampal activity during an associative memory task. Eur. Neuropsychopharmacol. 17, 289–297.

Jain, A.K., Ryan, J.R., McMahon, F.G., Smith, G., 1981. Evaluation of intramuscular levonantradol and placebo in acute postoperative pain. J. Clin. Pharmacol. 21, 320S–326S.

James, A., Hough, M., James, S., Winmill, L., Burge, L., Nijhawan, S., Matthews, P.M., Zarei, M., 2011. Greater white and grey matter changes associated with early cannabis use in adolescent-onset schizophrenia (AOS). Schizophr. Res. 128, 91–97.

Jeon, Y.W., Polich, J., 2003. Meta-analysis of P300 and schizophrenia: patients, paradigms, and practical implications. Psychophysiology 40, 684–701.

Jiang, W., Zhang, Y., Xiao, L., Van Cleemput, J., Ji, S.P., Bai, G., Zhang, X., 2005. Cannabinoids promote embryonic and adult hippocampus neurogenesis and produce anxiolytic- and antidepressant-like effects. J. Clin. Invest. 115, 3104–3116.

Jin, K., Xie, L., Kim, S.H., Parmentier-Batteur, S., Sun, Y., Mao, X.O., Childs, J., Greenberg, D.A., 2004. Defective adult neurogenesis in CB1 cannabinoid receptor knockout mice. Mol. Pharmacol. 66, 204–208.

Johns, A., 2001. Psychiatric effects of cannabis. Br. J. Psychiatry 178, 116–122.

Johnson, B.A., 1990. Psychopharmacological effects of cannabis. Br. J. Hosp. Med. 43 (122), 114-6–118-20.

Johnson, B.A., Smith, B.L., Taylor, P., 1988. Cannabis and schizophrenia. Lancet 1, 592–593.

Jones, R.T., Stone, G.C., 1970. Psychological studies of marijuana and alcohol in man. Psychopharmacologia 18, 108–117.

Joyal, C.C., Halle, P., Lapierre, D., Hodgins, S., 2003. Drug abuse and/or dependence and better neuropsychological performance in patients with schizophrenia. Schizophr. Res. 63, 297–299.

Juckel, G., Roser, P., Nadulski, T., Stadelmann, A.M., Gallinat, J., 2007. Acute effects of delta-9-tetrahydro-cannabinol and standardized cannabis extract on the auditory evoked mismatch negativity. Schizophr. Res. 97, 109–117.

Jutras-Aswad, D., DiNieri, J.A., Harkany, T., Hurd, Y.L., 2009. Neurobiological consequences of maternal cannabis on human fetal development and its neuropsychiatric outcome. Eur. Arch. Psychiatry Clin. Neurosci. 259, 395–412.

Kaufmann, R.M., Kraft, B., Frey, R., Winkler, D., Weiszenbichler, S., Backer, C., Kasper, S., Kress, H.G., 2010. Acute psychotropic effects of oral cannabis extract with a defined content of delta-9-tetrahydrocannabinol (THC) in healthy volunteers. Pharmacopsychiatry 43, 24–32.

Kay, S.R., Opler, L.A., Fiszbein, A., 1986. Significance of positive and negative syndromes in chronic schizophrenia. Br. J. Psychiatry 149, 439–448.

Keeler, M.H., Moore, E., 1974. Paranoid reactions while using marijuana. Dis. Nerv. Syst. 35, 535–536.

Kempel, P., Lampe, K., Parnefjord, R., Hennig, J., Kunert, H.J., 2003. Auditory-evoked potentials and selective attention: different ways of information processing in cannabis users and controls. Neuropsychobiology 48, 95–101.

Kenny, J.B., Wilkinson, P.M., 1982. Levonantradol effectiveness in cancer patients resistant to conventional anti-emetics. Clin. Oncol. 8, 335–339.

Kessler, R., Foster, C., Saunders, W., Stang, P., 1995. Social consequences of psychiatric disorders, I: educational attainment. Am. J. Psychiatry 152, 1026–1032.

Kiehl, K.A., Laurens, K.R., Duty, T.L., Forster, B.B., Liddle, P.F., 2001. Neural sources involved in auditory target detection and novelty processing: an event-related fMRI study. Psychophysiology 38, 133–142.

Kikura-Hanajiri, R., Uchiyama, N., Goda, Y., 2011. Survey of current trends in the abuse of psychotropic substances and plants in Japan. Leg. Med. (Tokyo) 13, 109–115.

Kivlahan, D.R., Heiman, J.R., Wright, R.C., Mundt, J.W., Shupe, J.A., 1991. Treatment cost and rehospitalization rate in schizophrenic outpatients with a history of substance abuse. Hosp. Community Psychiatry 42, 609–614.

Knudsen, P., Vilmar, T., 1984. Cannabis and neuroleptic agents in schizophrenia. Acta Psychiatr. Scand. 69, 162–174.

Koethe, D., Gerth, C.W., Neatby, M.A., Haensel, A., Thies, M., Schneider, U., Emrich, H.M., Klosterkotter, J., Schultze-Lutter, F., Leweke, F.M., 2006. Disturbances of visual information processing in early states of psychosis and experimental delta-9-tetrahydrocannabinol altered states of consciousness. Schizophr. Res. 88, 142–150.

Kolansky, H., Moore, W.T., 1971. Effects of marihuana on adolescents and young adults. J Psychiatr. Nurs. Ment. Health Serv. 9, 9–16.

Konings, M., Stefanis, N., Kuepper, R., de Graaf, R., ten Have, M., van Os, J., Bakoula, C., Henquet, C., 2012. Replication in two independent population-based samples that childhood maltreatment and cannabis use synergistically impact on psychosis risk. Psychol. Med. 42, 149–159.

Koskinen, J., Lohonen, J., Koponen, H., Isohanni, M., Miettunen, J., 2009. Rate of cannabis use disorders in clinical samples of patients with schizophrenia: a meta-analysis. Schizophr. Bull.

Kuehnle, J., Mendelson, J.H., Davis, K.R., New, P.F., 1977. Computed tomographic examination of heavy marijuana smokers. JAMA 237, 1231–1232.

Kuepper, R., Henquet, C., Lieb, R., Wittchen, H.U., van Os, J., 2011a. Non-replication of interaction between cannabis use and trauma in predicting psychosis. Schizophr. Res. 131, 262–263.

Kuepper, R., van Os, J., Lieb, R., Wittchen, H.U., Hofler, M., Henquet, C., 2011b. Continued cannabis use and risk of incidence and persistence of psychotic symptoms: 10 year follow-up cohort study. BMJ 342, d738.

Kumra, S., Robinson, P., Tambyraja, R., Jensen, D., Schimunek, C., Houri, A., Reis, T., Lim, K., 2012. Parietal lobe volume deficits in adolescents with schizophrenia and adolescents with cannabis use disorders. J. Am. Acad. Child Adolesc. Psychiatry 51, 171–180.

Landfield, P.W., Cadwallader, L.B., Vinsant, S., 1988. Quantitative changes in hippocampal structure following long-term exposure to delta 9-tetrahydrocannabinol: possible mediation by glucocorticoid systems. Brain Res. 443, 47–62.

Lapoint, J., James, L.P., Moran, C.L., Nelson, L.S., Hoffman, R.S., Moran, J.H., 2011. Severe toxicity following synthetic cannabinoid ingestion. Clin. Toxicol. 49, 760–764.

Laszlo, J., Lucas Jr., V.S., Hanson, D.C., Cronin, C.M., Sallan, S.E., 1981. Levonantradol for chemotherapy-induced emesis: phase I-II oral administration. J. Clin. Pharmacol. 21, 51S–56S.

Lawston, J., Borella, A., Robinson, J.K., Whitaker-Azmitia, P.M., 2000. Changes in hippocampal morphology following chronic treatment with the synthetic cannabinoid WIN 55,212-2. Brain Res. 877, 407–410.

Leweke, M., Kampmann, C., Radwan, M., Dietrich, D.E., Johannes, S., Emrich, H.M., Munte, T.F., 1998. The effects of tetrahydrocannabinol on the recognition of emotionally charged words: an analysis using event-related brain potentials. Neuropsychobiology 37, 104–111.

Leweke, F.M., Schneider, U., Thies, M., Munte, T.F., Emrich, H.M., 1999. Effects of synthetic delta9-tetrahydrocannabinol on binocular depth inversion of natural and artificial objects in man. Psychopharmacology (Berl.) 142, 230–235.

Leweke, F.M., Schneider, U., Radwan, M., Schmidt, E., Emrich, H.M., 2000. Different effects of nabilone and cannabidiol on binocular depth inversion in man. Pharmacol. Biochem. Behav. 66, 175–181.

Lichtman, A.H., Varvel, S.A., Martin, B.R., 2002. Endocannabinoids in cognition and dependence. Prostaglandins Leukot. Essent. Fatty Acids 66, 269–285.

Lindemann, E., Malamud, W., 1934. Experimental analysis of the psychopathological effects of intoxicating drug. Am. J. Psychiatry 90, 853–881.

Lindigkeit, R., Boehme, A., Eiserloh, I., Luebbecke, M., Wiggermann, M., Ernst, L., Beuerle, T., 2009. Spice: a never ending story? Forensic Sci. Int. 191, 58–63.

Linszen, D.H., Dingemans, P.M., Lenior, M.E., 1994. Cannabis abuse and the course of recent-onset schizophrenic disorders. Arch. Gen. Psychiatry 51, 273–279.

Løberg, E.M., Hugdahl, K., 2009. Cannabis use and cognition in schizophrenia. Front. Hum. Neurosci. 3, 53.

Luck, S.J., Mathalon, D.H., O'Donnell, B.F., Hamalainen, M.S., Spencer, K.M., Javitt, D.C., Uhlhaas, P.J., 2010. A roadmap for the development and validation of event-related potential biomarkers in schizophrenia research. Biol. Psychiatry.

Lundqvist, T., 2005. Cognitive consequences of cannabis use: comparison with abuse of stimulants and heroin with regard to attention, memory and executive functions. Pharmacol. Biochem. Behav. 81, 319–330.

Luntz-Leybman, V., Bickford, P.C., Freedman, R., 1992. Cholinergic gating of response to auditory stimuli in rat hippocampus. Brain Res. 587, 130–136.

Lyons, M.J., Bar, J.L., Panizzon, M.S., Toomey, R., Eisen, S., Xian, H., Tsuang, M.T., 2004. Neuropsychological consequences of regular marijuana use: a twin study. Psychol. Med. 34, 1239–1250.

Løberg, E.M., Hugdahl, K., 2009. Cannabis use and cognition in schizophrenia. Front. Hum. Neurosci. 3, 53.

Machado Rocha, F.C., Stefano, S.C., De Cassia Haiek, R., Rosa Oliveira, L.M., Da Silveira, D.X., 2008. Therapeutic use of Cannabis sativa on chemotherapy-induced nausea and vomiting among cancer patients: systematic review and meta-analysis. Eur. J. Cancer Care 17, 431–443.

Macleod, J., 2007. Cannabis use and symptom experience amongst people with mental illness: a commentary on Degenhardt et al. Psychol. Med. 37, 913–916.

Maj, P.F., Collu, M., Fadda, P., Cattaneo, A., Racagni, G., Riva, M.A., 2007. Long-term reduction of brain-derived neurotrophic factor levels and signaling impairment following prenatal treatment with the cannabinoid receptor 1 receptor agonist (R)-(+)-[2,3-dihydro-5-methyl-3-(4-morpholinyl-methyl) pyrrolo[1,2,3-de]-1,4-benzoxazin-6-yl]-1- naphthalenylmethanone. Eur. J. Neurosci. 25, 3305–3311.

Margolese, H., Malchy, L., Negrete, J., Tempier, R., Gill, K., 2004. Drug and alcohol use among patients with schizophrenia and related psychoses: levels and consequences. Schizophr. Res. 67, 157–166.

Marks, D.F., MacAvoy, M.G., 1989. Divided attention performance in cannabis users and non-users following alcohol and cannabis separately and in combination. Psychopharmacology 99, 397–401.

Marriott, K.S., Huffman, J.W., 2008. Recent advances in the development of selective ligands for the cannabinoid CB(2) receptor. Curr. Top. Med. Chem. 8, 187–204.

Marshall, C., 1897. The active principle of Indian help; a preliminary communication. Lancet 1, 235–238.

Mata, I., Perez-Iglesias, R., Roiz-Santianez, R., Tordesillas-Gutierrez, D., Pazos, A., Gutierrez, A., Vazquez-Barquero, J.L., Crespo-Facorro, B., 2010. Gyrification brain abnormalities associated with adolescence and early-adulthood cannabis use. Brain Res. 1317, 297–304.

Mathew, R.J., Wilson, W.H., Turkington, T.G., Coleman, R.E., 1998. Cerebellar activity and disturbed time sense after THC. Brain Res. 797, 183–189.

Matochik, J.A., Eldreth, D.A., Cadet, J.L., Bolla, K.I., 2005. Altered brain tissue composition in heavy marijuana users. Drug Alcohol Depend. 77, 23–30.

Mayor's Committee on Marijuana, 1944. The LaGuardia Committee Report: The Marihuana Problem in the City of New York. The New York Academy of Medicine, New York.

McClure, G.Y., McMillan, D.E., 1997. Effects of drugs on response duration differentiation. VI: differential effects under differential reinforcement of low rates of responding schedules. J. Pharmacol. Exp. Ther. 281, 1368–1380.

McCreadie, R.G., 2002. Use of drugs, alcohol and tobacco by people with schizophrenia: case-control study. Br. J. Psychiatry 181, 321–325.

McDonald, J., Schleifer, L., Richards, J.B., de Wit, H., 2003. Effects of THC on behavioral measures of impulsivity in humans. Neuropsychopharmacology 28, 1356–1365.

McGrath, J., Welham, J., Scott, J., Varghese, D., Degenhardt, L., Hayatbakhsh, M.R., Alati, R., Williams, G.M., Bor, W., Najman, J.M., 2010. Association between cannabis use and psychosis-related outcomes using sibling pair analysis in a cohort of young adults. Arch. Gen. Psychiatry 67, 440–447.

McGuire, P.K., Jones, P., Harvey, I., Bebbington, P., Toone, B., Lewis, S., Murray, R.M., 1994. Cannabis and acute psychosis. Schizophr. Res. 13, 161–167.

Medina, K.L., Hanson, K.L., Schweinsburg, A.D., Cohen-Zion, M., Nagel, B.J., Tapert, S.F., 2007a. Neuropsychological functioning in adolescent marijuana users: subtle deficits detectable after a month of abstinence. J. Int. Neuropsychol. Soc. 13, 807–820.

Medina, K.L., Nagel, B.J., Park, A., McQueeny, T., Tapert, S.F., 2007b. Depressive symptoms in adolescents: associations with white matter volume and marijuana use. J. Child Psychol. Psychiatry 48, 592–600.

Medina, K.L., Schweinsburg, A.D., Cohen-Zion, M., Nagel, B.J., Tapert, S.F., 2007c. Effects of alcohol and combined marijuana and alcohol use during adolescence on hippocampal volume and asymmetry. Neurotoxicol. Teratol. 29, 141–152.

Meijer, J.H., Dekker, N., Koeter, M.W., Quee, P.J., van Beveren, N.J., Meijer, C.J., 2011. Cannabis and cognitive performance in psychosis: a cross-sectional study in patients with non-affective psychotic illness and their unaffected siblings. Psychol. Med., 1–12.

Melges, F.T., Tinklenberg, J.R., Hollister, L.E., Gillespie, H.K., 1970. Marihuana and temporal disintegration. Science 168, 1118–1120.

Miller, L.L., McFarland, D., Cornett, T.L., Brightwell, D., 1977. Marijuana and memory impairment: effect on free recall and recognition memory. Pharmacol. Biochem. Behav. 7, 99–103.

Millman, R.B., Sbriglio, R., 1986. Patterns of use and psychopathology in chronic marijuana users. Psychiatr. Clin. North Am. 9, 533–545.

Mir, A., Obafemi, A., Young, A., Kane, C., 2011. Myocardial infarction associated with use of the synthetic cannabinoid K2. Pediatrics 128, e1622–e1627.

Mirken, B., Earleywine, M., 2005. The cannabis and psychosis connection questioned: a comment on Fergusson et al. 2005. Addiction 100, 714–715 author reply 715–716.

Molina-Holgado, F., Alvarez, F.J., Gonzalez, I., Antonio, M.T., Leret, M.L., 1997. Maternal exposure to delta 9-tetrahydrocannabinol (delta 9-THC) alters indolamine levels and turnover in adult male and female rat brain regions. Brain Res. Bull. 43, 173–178.

Moore, T.H., Zammit, S., Lingford-Hughes, A., Barnes, T.R., Jones, P.B., Burke, M., Lewis, G., 2007. Cannabis use and risk of psychotic or affective mental health outcomes: a systematic review. Lancet 370, 319–328.

Moreau, J., 1973. Hashish and Mental Illness. Raven, New York.

Morrison, P.D., Zois, V., McKeown, D.A., Lee, T.D., Holt, D.W., Powell, J.F., Kapur, S., Murray, R.M., 2009. The acute effects of synthetic intravenous delta9-tetrahydrocannabinol on psychosis, mood and cognitive functioning. Psychol. Med. 39, 1607–1616.

Morrison, P.D., Nottage, J., Stone, J.M., Bhattacharyya, S., Tunstall, N., Brenneisen, R., Holt, D., Wilson, D., Sumich, A., McGuire, P., Murray, R.M., Kapur, S., Ffytche, D.H., 2011. Disruption of frontal theta coherence by delta9-tetrahydrocannabinol is associated with positive psychotic symptoms. Neuropsychopharmacology 36, 827–836.

Mueser, K., Bellack, A., Blanchard, J., 1992. Comorbidity of schizophrenia and substance abuse: implications for treatment. J. Consult. Clin. Psychol. 60, 845–856.

Mulder, J., Aguado, T., Keimpema, E., Barabas, K., Ballester Rosado, C.J., Nguyen, L., Monory, K., Marsicano, G., Di Marzo, V., Hurd, Y.L., Guillemot, F., Mackie, K., Lutz, B., Guzman, M., Lu, H.C.,

Galve-Roperh, I., Harkany, T., 2008. Endocannabinoid signaling controls pyramidal cell specification and long-range axon patterning. Proc. Natl. Acad. Sci. U.S.A 105, 8760–8765.

Muller, H., Huttner, H.B., Kohrmann, M., Wielopolski, J.E., Kornhuber, J., Sperling, W., 2010a. Panic attack after spice abuse in a patient with ADHD. Pharmacopsychiatry 43, 152–153.

Muller, H., Sperling, W., Kohrmann, M., Huttner, H.B., Kornhuber, J., Maler, J.M., 2010b. The synthetic cannabinoid Spice as a trigger for an acute exacerbation of cannabis induced recurrent psychotic episodes. Schizophr. Res. 118, 309–310.

Naatanen, R., Alho, K., 1995. Generators of electrical and magnetic mismatch responses in humans. Brain Topogr. 7, 315–320.

Naatanen, R., Kahkonen, S., 2009. Central auditory dysfunction in schizophrenia as revealed by the mismatch negativity (MMN) and its magnetic equivalent MMNm: a review. Int. J. Neuropsychopharmacol. 12, 125–135.

Naatanen, R., Picton, T., 1987. The N1 wave of the human electric and magnetic response to sound: a review and an analysis of the component structure. Psychophysiology 24, 375–425.

Nahas, G.G., 1982. Hashish in Islam Ninth to Eighteenth century. Bull. N.Y. Acad. Med. 58, 814–831.

Negrete, J.C., 1989. Cannabis and schizophrenia. Br. J. Addict. 84, 349–351.

Negrete, J.C., Knapp, W.P., Douglas, D.E., Smith, W.B., 1986. Cannabis affects the severity of schizophrenic symptoms: results of a clinical survey. Psychol. Med. 16, 515–520.

News, F., 2011. Fake Weed, Real Drug: K2 Causing Hallucinations in Teens (Fox News).

NIDA, 2011. Info Facts: Spice.

NPR, 2011. Friend Says On 911 Call Demi Moore Was Convulsing (NPR).

O'Donnell, B.F., Vohs, J.L., Hetrick, W.P., Carroll, C.A., Shekhar, A., 2004. Auditory event-related potential abnormalities in bipolar disorder and schizophrenia. Int. J. Psychophysiol. 53, 45–55.

O'Shea, M., Singh, M.E., McGregor, I.S., Mallet, P.E., 2004. Chronic cannabinoid exposure produces lasting memory impairment and increased anxiety in adolescent but not adult rats. J. Psychopharmacol. 18, 502–508.

Oudin, M.J., Gajendra, S., Williams, G., Hobbs, C., Lalli, G., Doherty, P., 2011. Endocannabinoids regulate the migration of subventricular zone-derived neuroblasts in the postnatal brain. J. Neurosci. 31, 4000–4011.

Patrick, G., Struve, F.A., 2000. Reduction of auditory P50 gating response in marihuana users: further supporting data. Clin. Electroencephalogr. 31, 88–93.

Patrick, G., Straumanis, J.J., Struve, F.A., Fitz-Gerald, M.J., Leavitt, J., Manno, J.E., 1999. Reduced P50 auditory gating response in psychiatrically normal chronic marihuana users: a pilot study. Biol. Psychiatry 45, 1307–1312.

Patterson, J.V., Hetrick, W.P., Boutros, N.N., Jin, Y., Sandman, C., Stern, H., Potkin, S., Bunney Jr., W.E., 2008. P50 sensory gating ratios in schizophrenics and controls: a review and data analysis. Psychiatry Res. 158, 226–247.

Peralta, V., Cuesta, M.J., 1992. Influence of cannabis abuse on schizophrenic psychopathology. Acta Psychiatr. Scand. 85, 127–130.

Peters, B., de Koning, P., Dingemans, P., Becker, H., Linszen, D., de Haan, L., 2009a. Subjective effects of cannabis before the first psychotic episode. Aust. N. Z. J. Psychiatry 43, 1155–1162.

Peters, B.D., de Haan, L., Vlieger, E.J., Majoie, C.B., den Heeten, G.J., Linszen, D.H., 2009b. Recent-onset schizophrenia and adolescent cannabis use: MRI evidence for structural hyperconnectivity? Psychopharmacol. Bull. 42, 75–88.

Picton, T.W., Hillyard, S.A., Krausz, H.I., Galambos, R., 1974. Human auditory evoked potentials. I. Evaluation of components. Electroencephalogr. Clin. Neurophysiol. 36, 179–190.

Polich, J., Corey-Bloom, J., 2005. Alzheimer's disease and P300: review and evaluation of task and modality. Curr. Alzheimer Res. 2, 515–525.

Polich, J., Criado, J.R., 2006. Neuropsychology and neuropharmacology of P3a and P3b. Int. J. Psychophysiol. 60, 172–185.

Polich, J., Pollock, V.E., Bloom, F.E., 1994. Meta-analysis of P300 amplitude from males at risk for alcoholism. Psychol. Bull. 115, 55–73.

Pope Jr., H.G., 2002. Cannabis, cognition, and residual confounding. JAMA 287, 1172–1174.

Pope Jr., H.G., Yurgelun-Todd, D., 1996. The residual cognitive effects of heavy marijuana use in college students. JAMA 275, 521–527.

Pope Jr., H.G., Gruber, A.J., Yurgelun-Todd, D., 1995. The residual neuropsychological effects of cannabis: the current status of research. Drug Alcohol Depend. 38, 25–34.

Pope Jr., H.G., Gruber, A.J., Hudson, J.I., Huestis, M.A., Yurgelun-Todd, D., 2001a. Neuropsychological performance in long-term cannabis users. Arch. Gen. Psychiatry 58, 909–915.

Pope Jr., H.G., Gruber, A.J., Yurgelun-Todd, D., 2001b. Residual neuropsychologic effects of cannabis. Curr. Psychiatry Rep. 3, 507–512.

Pope Jr., H.G., Gruber, A.J., Hudson, J.I., Huestis, M.A., Yurgelun-Todd, D., 2002. Cognitive measures in long-term cannabis users. J. Clin. Pharmacol. 42, 41S–47S.

Pope Jr., H.G., Gruber, A.J., Hudson, J.I., Cohane, G., Huestis, M.A., Yurgelun-Todd, D., 2003. Early-onset cannabis use and cognitive deficits: what is the nature of the association? Drug Alcohol Depend. 69, 303–310.

Post, W., 2011. The Growing Buzz on 'Spice' – the Marijuana Alternative (Washington Post).

Potter, D., Summerfelt, A., Gold, J., Buchanan, R.W., 2006. Review of clinical correlates of P50 sensory gating abnormalities in patients with schizophrenia. Schizophr. Bull. 32, 692–700.

Potvin, S., Briand, C., Prouteau, A., Bouchard, R.H., Lipp, O., Lalonde, P., Nicole, L., Lesage, A., Stip, E., 2005. CANTAB explicit memory is less impaired in addicted schizophrenia patients. Brain Cogn. 59, 38–42.

Potvin, S., Mancini-Marie, A., Fahim, C., Mensour, B., Levesque, J., Karama, S., Beauregard, M., Rompre, P.P., Stip, E., 2007. Increased striatal gray matter densities in patients with schizophrenia and substance use disorder: a voxel-based morphometry study. Psychiatry Res. 154, 275–279.

Potvin, S., Joyal, C.C., Pelletier, J., Stip, E., 2008. Contradictory cognitive capacities among substance-abusing patients with schizophrenia: a meta-analysis. Schizophr. Res. 100, 242–251.

Psychoyos, D., Hungund, B., Cooper, T., Finnell, R.H., 2008. A cannabinoid analogue of delta9-tetrahydrocannabinol disrupts neural development in chick. Birth Defects Res. B Dev. Reprod. Toxicol. 83, 477–488.

Psychoyos, D., Vinod, K.Y., Cao, J., Xie, S., Hyson, R.L., Wlodarczyk, B., He, W., Cooper, T.B., Hungund, B.L., Finnell, R.H., 2012. Cannabinoid receptor 1 signaling in embryo neurodevelopment. Birth Defects Res. B Dev. Reprod. Toxicol.

Quinn, H.R., Matsumoto, I., Callaghan, P.D., Long, L.E., Arnold, J.C., Gunasekaran, N., Thompson, M.R., Dawson, B., Mallet, P.E., Kashem, M.A., Matsuda-Matsumoto, H., Iwazaki, T., McGregor, I.S., 2008. Adolescent rats find repeated delta(9)-THC less aversive than adult rats but display greater residual cognitive deficits and changes in hippocampal protein expression following exposure. Neuropsychopharmacology 33, 1113–1126.

Rabin, R.A., Zakzanis, K.K., George, T.P., 2011. The effects of cannabis use on neurocognition in schizophrenia: a meta-analysis. Schizophr. Res. 128, 111–116.

Rais, M., Cahn, W., Van Haren, N., Schnack, H., Caspers, E., Hulshoff Pol, H., Kahn, R., 2008. Excessive brain volume loss over time in cannabis-using first-episode schizophrenia patients. Am. J. Psychiatry 165, 490–496.

Ranganathan, M., D'Souza, D.C., 2006. The acute effects of cannabinoids on memory in humans: a review. Psychopharmacology (Berl.) 188, 425–444.

Rapoport, J.L., Addington, A.M., Frangou, S., Psych, M.R., 2005. The neurodevelopmental model of schizophrenia: update 2005. Mol. Psychiatry 10, 434–449.

Regier, D.A., Farmer, M.E., Rae, D.S., Locke, B.Z., Keith, S.J., Judd, L.L., Goodwin, F.K., 1990. Comorbidity of mental disorders with alcohol and other drug abuse. Results from the Epidemiologic Catchment Area (ECA) study. JAMA 264, 2511–2518.

Reilly, D., Didcott, P., Swift, W., Hall, W., 1998. Long-term cannabis use: characteristics of users in an Australian rural area. Addiction 93, 837–846.

Renault, P.F., Schuster, C.R., Heinrich, R., Freeman, D.X., 1971. Marihuana: standardized smoke administration and dose effect curves on heart rate in humans. Science 174, 589–591.

Renault, P.F., Schuster, C.R., Freedman, D.X., Sikic, B., de Mello, D.N., 1974. Repeat administration of marihuana smoke to humans. Arch. Gen. Psychiatry 31, 95–102.

Rentzsch, J., Penzhorn, A., Kernbichler, K., Plockl, D., Gomez-Carrillo de Castro, A., Gallinat, J., Jockers-Scherubl, M.C., 2007. Differential impact of heavy cannabis use on sensory gating in schizophrenic patients and otherwise healthy controls. Exp. Neurol. 205, 241–249.

Ringen, P.A., Vaskinn, A., Sundet, K., Engh, J.A., Jonsdottir, H., Simonsen, C., Friis, S., Opjordsmoen, S., Melle, I., Andreassen, O.A., 2010. Opposite relationships between cannabis use and neurocognitive functioning in bipolar disorder and schizophrenia. Psychol. Med. 40, 1337–1347.

Rinne, T., Alho, K., Ilmoniemi, R.J., Virtanen, J., Naatanen, R., 2000. Separate time behaviors of the temporal and frontal mismatch negativity sources. NeuroImage 12, 14–19.

Rissling, A.J., Makeig, S., Braff, D.L., Light, G.A., 2010. Neurophysiologic markers of abnormal brain activity in schizophrenia. Curr. Psychiatry Rep. 12, 572–578.

Rosburg, T., Boutros, N.N., Ford, J.M., 2008. Reduced auditory evoked potential component N100 in schizophrenia–a critical review. Psychiatry Res. 161, 259–274.

Roser, P., Della, B., Norra, C., Uhl, I., Brune, M., Juckel, G. 2008. Auditory mismatch negativity deficits in long-term heavy cannabis users. Eur. Arch. Psychiatry Clin. Neurosci.

Roth, W.T., Cannon, E.H., 1972. Some features of the auditory evoked response in schizophrenics. Arch. Gen. Psychiatry 27, 466–471.

RTL, 2012. Synthetic Cannabinoid Urine Test.

Rubino, T., Vigano, D., Premoli, F., Castiglioni, C., Bianchessi, S., Zippel, R., Parolaro, D., 2006. Changes in the expression of G protein-coupled receptor kinases and beta-arrestins in mouse brain during cannabinoid tolerance: a role for RAS-ERK cascade. Mol. Neurobiol. 33, 199–213.

Rubino, T., Realini, N., Braida, D., Guidi, S., Capurro, V., Vigano, D., Guidali, C., Pinter, M., Sala, M., Bartesaghi, R., Parolaro, D., 2009. Changes in hippocampal morphology and neuroplasticity induced by adolescent THC treatment are associated with cognitive impairment in adulthood. Hippocampus 19, 763–772.

Saletu, B., Itil, T.M., Saletu, M., 1971. Auditory evoked response, EEG, and thought process in schizophrenics. Am. J. Psychiatry 128, 336–344.

Scallet, A.C., 1991. Neurotoxicology of cannabis and THC: a review of chronic exposure studies in animals. Pharmacol. Biochem. Behav. 40, 671–676.

Scallet, A.C., Uemura, E., Andrews, A., Ali, S.F., McMillan, D.E., Paule, M.G., Brown, R.M., Slikker Jr., W., 1987. Morphometric studies of the rat hippocampus following chronic delta-9-tetrahydrocannabinol (THC). Brain Res. 436, 193–198.

Schneider, M., 2008. Puberty as a highly vulnerable developmental period for the consequences of cannabis exposure. Addict. Biol. 13, 253–263.

Schneider, M., Schomig, E., Leweke, F.M., 2008. Acute and chronic cannabinoid treatment differentially affects recognition memory and social behavior in pubertal and adult rats. Addict. Biol. 13, 345–357.

Schneir, A.B., Cullen, J., Ly, B.T., 2011. "Spice" girls: synthetic cannabinoid intoxication. J. Emerg. Med. 40, 296–299.

Schofield, D., Tennant, C., Nash, L., Degenhardt, L., Cornish, A., Hobbs, C., Brennan, G., 2006. Reasons for cannabis use in psychosis. Aust. N. Z. J. Psychiatry 40, 570–574.

Scholes, K.E., Martin-Iverson, M.T., 2010. Cannabis use and neuropsychological performance in healthy individuals and patients with schizophrenia. Psychol. Med. 40, 1635–1646.

Schulze, G.E., McMillan, D.E., Bailey, J.R., Scallet, A., Ali, S.F., Slikker Jr., W., Paule, M.G., 1988. Acute effects of delta-9-tetrahydrocannabinol in rhesus monkeys as measured by performance in a battery of complex operant tests. J. Pharmacol. Exp. Ther. 245, 178–186.

Semple, D.M., Ramsden, F., McIntosh, A.M., 2003. Reduced binocular depth inversion in regular cannabis users. Pharmacol. Biochem. Behav. 75, 789–793.

Sewell, A., D'Souza, D.C., 3 April 2011. Impairment of Time Estimation and Reproduction by Δ^9-THC.

Sewell, R.A., Ranganathan, M., D'Souza, D.C., 2009. Cannabinoids and psychosis. Int. Rev. Psychiatry 21, 152–162.

Sheidler, V.R., Ettinger, D.S., Diasio, R.B., Enterline, J.P., Brown, M.D., 1984. Double-blind multiple-dose crossover study of the antiemetic effect of intramuscular levonantradol compared to prochlorperazine. J. Clin. Pharmacol. 24, 155–159.

Siler, J.F., Sheep, W.L., Bates, L.B., Clark, G.F., Cook, G.W., Smith, W.A., 1933. Marihuana smoking in Panama. Mil. Surg. 73, 269–280.

Singh, S.M., Basu, D., 2009. The P300 event-related potential and its possible role as an endophenotype for studying substance use disorders: a review. Addict. Biol. 14, 298–309.

Skosnik, P.D., Krishnan, G.P., Vohs, J.L., O'Donnell, B.F., 2006. The effect of cannabis use and gender on the visual steady state evoked potential. Clin. Neurophysiol. 117, 144–156.

Skosnik, P.D., Edwards, C.R., O'Donnell, B.F., Steffen, A., Steinmetz, J.E., Hetrick, W.P., 2008a. Cannabis use disrupts eyeblink conditioning: evidence for cannabinoid modulation of cerebellar-dependent learning. Neuropsychopharmacology 33, 1432–1440.

Skosnik, P.D., Park, S., Dobbs, L., Gardner, W.L., 2008b. Affect processing and positive syndrome schizotypy in cannabis users. Psychiatry Res. 157, 279–282.

Smith, D.E., 1968. Acute and chronic toxicity of marijuana. J. Psychedelic Drugs 2, 37–47.

Solowij, N., 1995. Do cognitive impairments recover following cessation of cannabis use? Life Sci. 56, 2119–2126.

Solowij, N., Battisti, R., 2008. The chronic effects of cannabis on memory in humans: a review. Curr. Drug Abuse Rev. 1, 81–98.

Solowij, N., Michie, P.T., 2007. Cannabis and cognitive dysfunction: parallels with endophenotypes of schizophrenia? J. Psychiatry Neurosci. 32, 30–52.

Solowij, N., Michie, P.T., Fox, A.M., 1991. Effects of long-term cannabis use on selective attention: an event-related potential study. Pharmacol. Biochem. Behav. 40, 683–688.

Solowij, N., Michie, P.T., Fox, A.M., 1995. Differential impairments of selective attention due to frequency and duration of cannabis use. Biol. Psychiatry 37, 731–739.

Solowij, N., Yucel, M., Respondek, C., Whittle, S., Lindsay, E., Pantelis, C., Lubman, D.I., 2011. Cerebellar white-matter changes in cannabis users with and without schizophrenia. Psychol. Med. 41, 2349–2359.

Spencer, D.J., 1971. Cannabis-induced psychosis. Int. J. Addict. 6, 323–326.

Spencer, C., Castle, D., Michie, P.T., 2002. Motivations that maintain substance use among individuals with psychotic disorders. Schizophr. Bull. 28, 233–247.

Stambaugh Jr., J.E., McAdams, J., Vreeland, F., 1984. Dose ranging evaluation of the antiemetic efficacy and toxicity of intramuscular levonantradol in cancer subjects with chemotherapy-induced emesis. J. Clin. Pharmacol. 24, 480–485.

Stirling, J., Lewis, S., Hopkins, R., White, C., 2005. Cannabis use prior to first onset psychosis predicts spared neurocognition at 10-year follow-up. Schizophr. Res. 75, 135–137.

Stone, J.M., Morrison, P.D., Nottage, J., Bhattacharyya, S., Feilding, A., McGuire, P.K., 2010. Delta-9-tetrahydro-cannabinol disruption of time perception and of self-timed actions. Pharmacopsychiatry 43, 236–237.

Struve, F.A., Straumanis, J.J., 1990. Electroencephalographic and evoked-potential methods in human marijuana research - historical review and future-trends. Drug Dev. Res. 20, 369–388.

Stuart-Harris, R.C., Mooney, C.A., Smith, I.E., 1983. Levonantradol: a synthetic cannabinoid in the treatment of severe chemotherapy-induced nausea and vomiting resistant to conventional anti-emetic therapy. Clin. Oncol. 9, 143–146.

Suckling, J., 2011. Correlated covariates in ANCOVA cannot adjust for pre-existing differences between groups. Schizophr. Res. 126, 310–311.

Swartz, M., Wagner, H., Swanson, J., Stroup, T., McEvoy, J., Canive, J., Miller, D., Reimherr, F., McGee, M., Khan, A., Van Dorn, R., Rosenheck, R., Lieberman, J., 2006. Substance use in persons with schizophrenia: baseline prevalence and correlates from the NIMH CATIE study. J. Nerv. Ment. Dis. 194, 164–172.

Swartz, M.S., Stroup, T.S., McEvoy, J.P., Davis, S.M., Rosenheck, R.A., Keefe, R.S., Hsiao, J.K., Lieberman, J.A., 2008. What CATIE found: results from the schizophrenia trial. Psychiatr. Serv. 59, 500–506 Washington, D.C.

Szeszko, P.R., Robinson, D.G., Sevy, S., Kumra, S., Rupp, C.I., Betensky, J.D., Lencz, T., Ashtari, M., Kane, J.M., Malhotra, A.K., Gunduz-Bruce, H., Napolitano, B., Bilder, R.M., 2007. Anterior cingulate grey-matter deficits and cannabis use in first-episode schizophrenia. Br. J. Psychiatry 190, 230–236.

Talbott, J.A., Teague, J.W., 1969. Marihuana psychosis. Acute toxic psychosis associated with the use of cannabis derivatives. JAMA 210, 299–302.

Tennant Jr., F.S., Groesbeck, C.J., 1972. Psychiatric effects of hashish. Arch. Gen. Psychiatry 27, 133–136.

Test, M.A., Wallisch, L.S., Allness, D.J., Ripp, K., 1989. Substance use in young adults with schizophrenic disorders. Schizophr. Bull. 15, 465–476.

Thacore, V.R., 1973. Bhang psychosis. Br. J. Psychiatry 123, 225–229.

Thaker, G.K., 2008. Neurophysiological endophenotypes across bipolar and schizophrenia psychosis. Schizophr. Bull. 34, 760–773.

Tinklenberg, J.R., Kopell, B.S., Melges, F.T., Hollister, L.E., 1972. Marihuana and alcohol, time production and memory functions. Arch. Gen. Psychiatry 27, 812–815.

Today, U, 2011 'Spice' makers change recipes to sidestep bans.

Tramer, M.R., Carroll, D., Campbell, F.A., Reynolds, D.J., Moore, R.A., McQuay, H.J., 2001. Cannabinoids for control of chemotherapy induced nausea and vomiting: quantitative systematic review. BMJ 323, 16–21.

Treffert, D.A., 1978. Marijuana use in schizophrenia: a clear hazard. Am. J. Psychiatry 135, 1213–1215.

Turetsky, B.I., Colbath, E.A., Gur, R.E., 1998. P300 subcomponent abnormalities in schizophrenia: I. Physiological evidence for gender and subtype specific differences in regional pathology. Biol. Psychiatry 43, 84–96.

Turetsky, B.I., Calkins, M.E., Light, G.A., Olincy, A., Radant, A.D., Swerdlow, N.R., 2007. Neurophysiological endophenotypes of schizophrenia: the viability of selected candidate measures. Schizophr. Bull. 33, 69–94.

Turetsky, B.I., Greenwood, T.A., Olincy, A., Radant, A.D., Braff, D.L., Cadenhead, K.S., Dobie, D.J., Freedman, R., Green, M.F., Gur, R.E., Gur, R.C., Light, G.A., Mintz, J., Nuechterlein, K.H., Schork, N.J., Seidman, L.J., Siever, L.J., Silverman, J.M., Stone, W.S., Swerdlow, N.R., Tsuang, D.W., Tsuang, M.T., Calkins, M.E., 2008. Abnormal auditory N100 amplitude: a heritable endophenotype in first-degree relatives of schizophrenia probands. Biol. Psychiatry 64, 1051–1059.

Tysk, L., 1983. Time estimation by healthy subjects and schizophrenic patients: a methodological study. Percept. Mot. Skills 56, 983–988.

Tzilos, G.K., Cintron, C.B., Wood, J.B., Simpson, N.S., Young, A.D., Pope Jr., H.G., Yurgelun-Todd, D.A., 2005. Lack of hippocampal volume change in long-term heavy cannabis users. Am. J. Addict. 14, 64–72.

Uchiyama, N., Kikura-Hanajiri, R., Kawahara, N., Haishima, Y., Goda, Y., 2009. Identification of a cannabinoid analog as a new type of designer drug in a herbal product. Chem. Pharm. Bull. 57, 439–441.

Uchiyama, N., Kikura-Hanajiri, R., Ogata, J., Goda, Y., 2010. Chemical analysis of synthetic cannabinoids as designer drugs in herbal products. Forensic Sci. Int. 198, 31–38.

Uchiyama, N., Kikura-Hanajiri, R., Goda, Y., 2011. Identification of a novel cannabimimetic phenylacetylindole, cannabipiperidiethanone, as a designer drug in a herbal product and its affinity for cannabinoid CB and CB receptors. Chem. Pharm. Bull. 59, 1203–1205.

Umbricht, D., Krljes, S., 2005. Mismatch negativity in schizophrenia: a meta-analysis. Schizophr. Res. 76, 1–23.

Umbricht, D., Koller, R., Schmid, L., Skrabo, A., Grubel, C., Huber, T., Stassen, H., 2003. How specific are deficits in mismatch negativity generation to schizophrenia? Biol. Psychiatry 53, 1120–1131.

Valjent, E., Pages, C., Rogard, M., Besson, M.J., Maldonado, R., Caboche, J., 2001. Delta 9-tetrahydrocannabinol-induced MAPK/ERK and Elk-1 activation in vivo depends on dopaminergic transmission. Eur. J. Neurosci. 14, 342–352.

van Dijk, D., Koeter, M.W., Hijman, R., Kahn, R.S., van den Brink, W., 2012. Effect of cannabis use on the course of schizophrenia in male patients: aprospective cohort study. Schizophr. Res.

Van Mastrigt, S., Addington, J., Addington, D., 2004. Substance misuse at presentation to an early psychosis program. Soc. Psychiatry Psychiatr. Epidemiol. 39, 69–72.

van Os, J., Bak, M., Hanssen, M., Bijl, R.V., de Graaf, R., Verdoux, H., 2002. Cannabis use and psychosis: a longitudinal population-based study. Am. J. Epidemiol. 156, 319–327.

Vandrey, R., Dunn, K.E., Fry, J.A., Girling, E.R., 2012. A survey study to characterize use of Spice products (synthetic cannabinoids). Drug Alcohol Depend. 120, 238–241.

Vearrier, D., Osterhoudt, K.C., 2010. A teenager with agitation: higher than she should have climbed. Pediatr. Emerg. Care. 26, 462–465.

Veen, N.D., Selten, J.P., van der Tweel, I., Feller, W.G., Hoek, H.W., Kahn, R.S., 2004. Cannabis use and age at onset of schizophrenia. Am. J. Psychiatry 161, 501–506.

Verdejo-Garcia, A.J., Lopez-Torrecillas, F., Aguilar de Arcos, F., Perez-Garcia, M., 2005. Differential effects of MDMA, cocaine, and cannabis use severity on distinctive components of the executive functions in polysubstance users: a multiple regression analysis. Addict. Behav. 30, 89–101.

Verdejo-Garcia, A., Bechara, A., Recknor, E.C., Perez-Garcia, M., 2006. Executive dysfunction in substance dependent individuals during drug use and abstinence: an examination of the behavioral, cognitive and emotional correlates of addiction. J. Int. Neuropsychol. Soc. 12, 405–415.

Volkow, N., Fowler, J., Wolf, A., Gillespi, H., 1991. In: Bethesda, M.D. (Ed.), Metabolic studies of drugs of abuse, National Institute of Drug Abuse, pp. 47–53. Report no. 105.

Wang, X., Dow-Edwards, D., Anderson, V., Minkoff, H., Hurd, Y.L., 2004. In utero marijuana exposure associated with abnormal amygdala dopamine D2 gene expression in the human fetus. Biol. Psychiatry 56, 909–915.

Warner, R., Taylor, D., Wright, J., Sloat, A., Springett, G., Arnold, S., Weinberg, H., 1994. Substance use among the mentally ill: prevalence, reasons for use, and effects on illness. Am. J. Orthopsychiatry 64, 30–39.

Warnock, J., 1903. Insanity from hasheesh. J. Ment. Sci. 49, 96–110.

Watson, S., Chambers, D., Hobbs, C., Doherty, P., Graham, A., 2008. The endocannabinoid receptor, CB1, is required for normal axonal growth and fasciculation. Mol. Cell. Neurosci. 38, 89–97.

Weil, A.T., 1970. Adverse reactions to marihuana. Classification and suggested treatment. N. Engl. J. Med. 282, 997–1000.

Weil, A.T., Zinberg, N.E., Nelsen, J.M., 1968. Clinical and psychological effects of marihuana in man. Science 162, 1234–1242.

Weinberger, D.R., 1996. On the plausibility of "the neurodevelopmental hypothesis" of schizophrenia. Neuropsychopharmacology 14, 1S–11S.

Weiser, M., Knobler, H.Y., Noy, S., Kaplan, Z., 2002. Clinical characteristics of adolescents later hospitalized for schizophrenia. Am. J. Med. Genet. 114, 949–955.

Wesnes, K., Annas, P., Edgar, C., Deeprose, C., Karlsten, R., Philipp, A., Kalliomaki, J., Segerdahl, M., 2009. Nabilone produces marked impairments to cognitive function and changes in subjective state in healthy volunteers. J. Psychopharmacol. (Oxford, England).

Wilk, J., Marcus, S.C., West, J., Countis, L., Hall, R., Regier, D.A., Olfson, M., 2006. Substance abuse and the management of medication nonadherence in schizophrenia. J. Nerv. Ment. Dis. 194, 454–457.

Williams, E.H., Himmelsbach, C.K., Lloyd, B.J., Ruble, D.C., Wikler, A., 1946. Studies on marihuana and parahexyl compound. Public Health Rep. 61, 1059–1083.

Wilson, R.I., Nicoll, R.A., 2002. Endocannabinoid signaling in the brain. Science 296, 678–682.

Wilson, W., Mathew, R., Turkington, T., Hawk, T., Coleman, R.E., Provenzale, J., 2000. Brain morphological changes and early marijuana use: a magnetic resonance and positron emission tomography study. J. Addict. Dis. 19, 1–22.

Wobrock, T., Sittinger, H., Behrendt, B., D'Amelio, R., Falkai, P., 2009. Comorbid substance abuse and brain morphology in recent-onset psychosis. Eur. Arch. Psychiatry Clin. Neurosci. 259, 28–36.

Yucel, M., Solowij, N., Respondek, C., Whittle, S., Fornito, A., Pantelis, C., Lubman, D.I., 2008. Regional brain abnormalities associated with long-term heavy cannabis use. Arch. Gen. Psychiatry 65, 694–701.

Yucel, M., Bora, E., Lubman, D.I., Solowij, N., Brewer, W.J., Cotton, S.M., Conus, P., Takagi, M.J., Fornito, A., Wood, S.J., McGorry, P.D., Pantelis, C., 2012. The impact of cannabis use on cognitive functioning in patients with schizophrenia: a meta-analysis of existing findings and new data in a first-episode sample. Schizophr. Bull. 38, 316–330.

Zachariou, M., Dissanayake, D.W., Coombes, S., Owen, M.R., Mason, R., 2008. Sensory gating and its modulation by cannabinoids: electrophysiological, computational and mathematical analysis. Cogn. Neurodyn. 2, 159–170.

Zammit, S., Allebeck, P., Andreasson, S., Lundberg, I., Lewis, G., 2002. Self reported cannabis use as a risk factor for schizophrenia in Swedish conscripts of 1969: historical cohort study. BMJ 325, 1199.

Zisook, S., Heaton, R., Moranville, J., Kuck, J., Jernigan, T., Braff, D., 1992. Past substance abuse and clinical course of schizophrenia. Am. J. Psychiatry 149, 552–553.

Zuardi, A.W., Rodrigues, J.A., Cunha, J.M., 1991. Effects of cannabidiol in animal models predictive of antipsychotic activity. Psychopharmacology (Berl.) 104, 260–264.

Zuardi, A.W., Morais, S.L., Guimaraes, F.S., Mechoulam, R., 1995. Antipsychotic effect of cannabidiol. J. Clin. Psychiatry 56, 485–486.

Zuardi, A.W., Crippa, J.A., Hallak, J.E., Moreira, F.A., Guimaraes, F.S., 2006a. Cannabidiol, a *Cannabis sativa* constituent, as an antipsychotic drug. Braz. J. Med. Biol. Res. 39, 421–429.

Zuardi, A.W., Hallak, J.E., Dursun, S.M., Morais, S.L., Sanches, R.F., Musty, R.E., Crippa, J.A., 2006b. Cannabidiol monotherapy for treatment-resistant schizophrenia. J. Psychopharmacol. 20, 683–686.

Zuardi, A.W., Crippa, J.A., Hallak, J.E., Pinto, J.P., Chagas, M.H., Rodrigues, G.G., Dursun, S.M., Tumas, V., 2009. Cannabidiol for the treatment of psychosis in Parkinson's disease. J. Psychopharmacol. 23, 979–983.

CHAPTER FIFTEEN

Effects of MDMA on the Human Nervous System

Una D. McCann[1] and George A. Ricaurte[2]
[1]Department of Psychiatry, The Johns Hopkins University School of Medicine, Baltimore, MD, USA
[2]Department of Neurology, The Johns Hopkins University School of Medicine, Baltimore, MD, USA

1. INTRODUCTION

(±)3,4-Methylenedioxymethamphetamine (MDMA) is a ring-substituted amphetamine analog that was originally synthesized by Merck in Darmstadt, Germany. It was included as one of the several structurally related chemical compounds in a patent application that was filed in 1912, and approved in 1914 (Merck, 2012). MDMA received little attention for the next several decades. In 1953, the US Army Chemical Center funded a series of preclinical studies of MDMA, conducted at the University of Michigan, in mice, rats, monkeys, dogs, and guinea pigs (Hardman et al., 1973). Although it is sometimes stated that these studies were an effort to identify a potential "brainwashing" agent (Holland, 2001), these studies actually appear to have been general toxicity studies identifying lethal dosages (LD50s) of a variety of chemical agents (Hardman et al., 1973).

There are reports that MDMA was used by "New Age" thinkers in the 1960s, who appreciated its ability to produce feelings of well-being and "connectedness" (McDowell and Kleber, 1994; Watson and Beck, 1991). MDMA was introduced to psychologists in Northern California in 1976, with therapists on the East Coast following suit, shortly thereafter. In brief, therapists began employing MDMA as a psychotherapeutic adjunct or "catalyst" to enhance and improve results of insight-oriented psychotherapy (Shulgin, 1980). Although the number of patients who were treated with MDMA-induced psychotherapy is not known, its continued use in psychotherapeutic sessions took place until such practice became illegal in 1985 (Beck, 1978).

Contemporaneous with MDMA's use in psychotherapeutic circles was its use in recreational settings. Despite its popularity as a recreational drug, MDMA did not receive much attention from the federal government as a potential drug of abuse until 1985. In particular, an aggressive marketing campaign by a particular MDMA distributor in Texas made MDMA widely available, not only at dance venues, but at bars and convenience stores (McDowell and Kleber, 2004). This practice came to the attention of Senator Lloyd Bentson, who petitioned the Drug Enforcement Administration (DEA). As a

The Effects of Drug Abuse on the Human Nervous System
http://dx.doi.org/10.1016/B978-0-12-418679-8.00015-0

475

result of Senator Bentson's efforts, MDMA was placed on Schedule I, and has remained there, since July 1, 1985 (Lawn, 1985).

In its decision to place MDMA on Schedule I, the DEA raised concerns over increasing recreational MDMA use, and that use of MDMA may pose a threat to public health, since a close structural relative, 3,4-methylenedioxyamphetamine (MDA), had recently been found to produce toxic effects on brain serotonin (5-HT)-containing neurons in rodents (Ricaurte et al., 1985). In addition to concerns about substance abuse and neurotoxicity, the DEA maintained that MDMA had no medical utility. This claim was challenged by a number of mental health specialists who believed that MDMA facilitated psychotherapy (Greer, 1985; Greer and Tolbert, 1986).

Despite its newly illegal status, MDMA became popular on college campuses and in large organized dance parties known as "raves", in spite of emerging evidence that MDMA, like MDA, was toxic toward brain 5-HT neurons in animals (Schmidt et al., 1985; Stone et al., 1986; Commins et al., 1987).

Since the discovery of the 5-HT neurotoxic potential of MDMA in animals, there has been significant research interest in its mechanisms of neurotoxicity, the parameters necessary to produce neurotoxicity, and whether such neurotoxicity extends to humans who use MDMA in recreational (or therapeutic) settings. Over 2001–2010, the Food and Drug Administration in the United States has approved a limited number of human MDMA studies to go forward. Government-sanctioned clinical research of MDMA in Switzerland and other European countries has also been conducted.

This chapter, as suggested by its title, will focus on the effects of MDMA on the human nervous system. To this end, we will begin by reviewing MDMA's mechanisms of action and its effects on the sympathetic nervous system (SNS). The bulk of our discussion will focus on MDMA's effects on the central nervous system (CNS), including its pharmacological and neurotoxic effects. Because a full understanding of MDMA pharmacology and neurotoxicology in humans requires understanding of its effects in preclinical models, these will be discussed as appropriate. Studies that have attempted to identify functional consequences of MDMA in humans, as well as potential neuropsychiatric sequelae of MDMA consumption, will be discussed. We will conclude with a summary of the research directed at determining the effects of MDMA on the human nervous system, limitations of this research, and potential future directions for MDMA studies in clinical populations.

2.1. MECHANISM OF ACTION

Like most amphetamine analogs, MDMA stimulates the CNS and the SNS, primarily by releasing 5-HT, dopamine (DA), and norepinephrine (NE), and blocking their inactivation via reuptake (Steele et al., 1994; Green et al., 2003). Key to these actions is the interaction between MDMA and monoamine transporters. This is known from results of in vitro studies using nerve ending suspensions (synaptosomes) (Nichols et al., 1982; Steele et al., 1987), transfected cells (Rudnick and Wall, 1992; Verrico et al., 2007) and in vivo studies using microdialysis (Gudelsky and Yamamoto, 2008) and transgenic animals (Fox et al., 2007). Of note, results from receptor binding studies suggest that MDMA, in addition to its indirect monoamine-mediated actions, may also produce direct effects on monoamine receptors, particularly 5-HT$_2$ receptors (Battaglia et al., 1989a,b; Teitler et al., 1990; but see Sadzot et al., 1989). In this regard, it is important to recall that MDMA, as typically used on the street and in most experimental and clinical studies, consists of an equal mixture of S(+) and R(−) enantiomers, with the R(−) enantiomer appearing to have the most prominent direct action on 5HT$_2$ receptors (Lyon et al., 1986). Both MDMA enantiomers also appear to have the potential to inhibit monoamine oxidase activity (Leonardi and Azmitia, 1994).

A major, as yet unanswered, question is whether the above mentioned effects of MDMA on monoamine-containing neurons fully account for its spectrum of behavioral effects. While it may well be that the subjective effects of MDMA are the end-product of a mix and blend of effects of MDMA on pre- and postsynaptic elements of 5-HT, DA, and NE neurons, a unique site or mechanism of action of MDMA has not been excluded. Early binding studies identified a novel binding site for MDMA (Gehlert et al., 1985). Although subsequent studies revealed that the binding site in question was located on the glass-fiber filters used to carry out the binding assays, rather than in the brain tissue under study (Wang et al., 1987; but see Zaczek et al., 1989).

That release of 5-HT and other monoamines may not be the sole mechanism by which MDMA produces its behavioral effects is suggested by the observation that fluoxetine, which blocks MDMA-induced 5-HT release (Schmidt et al, 1987), does not block MDMA's reinforcing subjective effects in humans (McCann and Ricaurte, 1993; but see Lietchi et al., 2000). Similar observations have been reported in animals (Piper et al., 2008). Of course, this leaves open the possibility that DA and NE, acting alone or in concert with 5-HT, may mediate MDMA's behavioral effects (Liechti and Vollenweider, 2000, 2001). These considerations notwithstanding a novel mechanism of action for MDMA are conceivable. Indeed, several investigators have postulated that MDMA may be the prototype of a new drug class (Nichols et al., 1986; Vollenweider et al., 1998).

2.2. PHARMACOLOGY

2.2.1. Absorption and Metabolism

MDMA is typically taken orally in the form of tablets, capsules, or pills. The dose generally ranges from 75 to 150 mg, although experienced users often take more (Parrott, 2005). MDMA is well absorbed from the gastrointestinal tract and its onset of action is typically within 30 min of ingestion even though the time to peak plasma concentrations is approximately 2–3 h (Kolbrich et al., 2008a,b; Mueller et al., 2009). The plasma half-life of

MDMA in humans ranges from 6 to 9 h (Chu et al., 1996; de la Torre et al., 2000; Kolbrich et al., 2008a,b). Notably, MDMA has nonlinear pharmacokinetics (Chu et al., 1996; de la Torre et al., 2000; Kolbrich et al., 2008a,b), with increases in plasma MDMA concentrations exceeding those predicted by the increase in dose. Also of note is the fact that nonlinear MDMA pharmacokinetics begins at plasma MDMA concentrations of approximately 125–150 ng/ml (Mueller et al., 2009). This effectively renders every consumer who takes a typical dose of MDMA a "poor metabolizer", because the hepatic cytochrome P_{450} isoenzyme that metabolizes MDMA (CYP2D6) is inhibited and/or saturated at this relatively low plasma MDMA concentration (125–150 ng/ml) (Mueller et al., 2009).

In humans, the major route of degradative metabolism of MDMA involves ring O-demethylenation, with formation of the catechol dihydroxymethamphetamine (HHMA) (Meyer et al., 2008). HHMA is then O-methylated to form HMMA, and/or N-demethylated to form HHA. In contrast to what is seen in rodents, only a small fraction of MDMA is N-demethylated to MDA in humans. In particular, in rodents, approximately 50% of MDMA is metabolized to MDA (Chu et al., 1996), whereas in humans, <5% of MDMA is converted to MDA (de la Torre et al., 2004; Kolbrich et al., 2008a,b). Unlike other MDMA metabolites, MDA is known to be pharmacologically active in humans (Naranjo et al., 1967). In addition to O-demethylenation and N-demethylation, other identified metabolic pathways of MDMA in vivo include deamination and conjugation (O-glucuronidation and O-sulfation) (Meyer et al., 2008).

2.2.2. Sympathomimetic Effects

Sympathomimetic effects of MDMA include mydriasis, tachycardia, increased diastolic and systolic blood pressure, and decreased appetite (Downing, 1986; Vollenweider et al., 1998; Mas et al., 1999; Cami et al., 2000; Kolbich et al., 2008a,b). Effects on blood pressure and heart rhythm are not always predictable or dose-related (see Vollenweider et al., 1998; Hua et al., 2009). Interestingly, the beta-blocker, pindolol, prevents MDMA-induced increases in heart rate but not increases in blood pressure (Hysek et al., 2010). Cardiovascular complications occasionally include cardiac arrhythmias, hypertensive crises, and stroke (hemorrhagic or ischemic) but these are relatively rare (Ricaurte and McCann, 2005). Body temperature is also typically mildly elevated (see Parrot et al., 2012) but significant, sometimes fatal, malignant hyperthermia can occur (Henry et al., 1992), resulting in rhabdomyolysis and multiorgan failure. In these instances, uncoupling or disruption of the balance between heat production and dissipation is suspected (Mills et al., 2004). Recent findings in animals suggest that neuronal activity in the dorsomedial hypothalamus, a brain region known to be involved in stress responses and thermoregulation, is necessary for some of the sympathomimetic effects of MDMA to occur (Rusyniak et al., 2008). MDMA-elicited increases in peripheral sympathetic discharge to the cutaneous vascular bed undoubtedly also contribute to hyperthermia, by reducing heat dissipation through the skin (Blessing et al., 2003).

2.2.3. Subjective Behavioral Effects

Shulgin and Nichols (1978) were the first to report on the subjective effects of MDMA. They indicated that MDMA produced an alteration in consciousness with sensual and emotional overtones. Other commonly reported behavioral effects of MDMA include increased physical energy, subjective sense of closeness to others, heightened sensual awareness, euphoria, loquaciousness, and empathy toward others (Downing, 1986; Kolbrich et al., 2008a). These effects typically last several hours, and it is not uncommon for MDMA consumers to use a "booster" dose to recoup or prolong the above mentioned effects. The days after MDMA ingestion, users typically experience lack of energy, dysphoria, and difficulty concentrating.

2.3. NEUROTOXICOLOGY

2.3.1. Animals

2.3.1.1. Serotonin Neurotoxicity

Over 1975–2000, an extensive body of data has accrued demonstrating that MDMA has neurotoxic potential toward brain 5-HT neurons. Animals treated with MDMA develop lasting depletions of brain of 5-HT, 5-hydroxyindoleacetic acid (5-HIAA), tryptophan hydroxylase (TPH), the serotonin transporter (SERT), and the vesicular monoamine transporter-type 2 (see Steele et al., 1994; Gibb et al., 1994; Lew et al., 1997; Green et al., 2003; Gudelsky and Yamamoto, 2003). Notably, these lasting neurochemical deficits in MDMA-treated animals are accompanied by morphologic/structural signs of damage to 5-HT axons and axon terminals. For example, Commins et al. (1987) found evidence of axon terminal degeneration using the silver degeneration method of Fink and Heimer. Using more specific immunocytochemical methods, Molliver and colleagues demonstrated loss of 5-HT-immunoreactive (IR) axons and axon terminals (O'Hearn et al., 1988; Wilson et al., 1989; Axt et al., 1994). Importantly, at short survival times after MDMA administration (1–3 days), the same investigators found numerous, markedly swollen 5-HT-IR axons, many of which appeared fragmented (Axt et al., 1994; Molliver et al., 1990). These data, coupled with tract-tracing results based on anterograde transport of tritiated proline from the raphe to the forebrain (Callahan, 2001) and evidence of the "pruning" phenomenon after MDMA exposure (Fischer et al, 1995), provide strong evidence that lasting serotonergic neurochemical deficits produced by MDMA are the result of destruction of 5-HT axons and axon terminals. Indeed, virtually identical effects are produced by 5,7-dihydroxytryptamine (5,7-DHT), a well-documented 5-HT neurotoxin (Jonsson, 1980; Baumgarten and Lachenmay, 2004).

Thus, as used in this chapter, the term *serotonin neurotoxicity* refers to a lasting loss of various chemical markers that are unique to 5-HT neurons (5-HT, 5-HIAA, TPH, and

SERT) and are accompanied by anatomic evidence of structural damage to 5-HT axons. The term makes no assumptions about reversibility of the neuronal injury or potential functional consequences.

Serotonin neurotoxicity after MDMA has been shown to have broad species generality. In particular, MDMA-induced 5-HT neurotoxicity has been documented in rats, guinea pigs, squirrel monkeys, cynomolgus monkeys, rhesus monkeys, and baboons (see Steele et al., 1994; Gibb et al., 1994). In all these species, MDMA neurotoxicity is selective for 5-HT neurons. Noradrenergic neurons are invariably spared. DA neurons are also typically unaffected except after unusually high dosages or when MDMA treatment is carried out at a high ambient temperature (Commins et al., 1987; Yuan et al., 2002; but see Sanchez et al., 2004). However, even under these conditions, the toxic effect of MDMA on brain DA neurons is modest, except in the mouse. For reasons that have yet to be elucidated, mice differ from all other animal species in which MDMA has been tested in that the mouse incurs selective DA neural injury, with 5-HT neurons only developing toxicity when the dosage of MDMA is inordinately high (Stone et al., 1987; Logan et al., 1988).

A few investigators have questioned the 5-HT neurotoxic potential of MDMA based on the observation that treatment with MDMA and structurally related drugs (e.g. p-chloroamphetamine (PCA)) does not generally lead to a glial reaction (Rowland et al., 1993; O'Callaghan and Miller, 1994; Pubill et al., 2003; Wang et al., 2004, 2005; Rothman et al., 2004; Thomas et al., 2004; but see Wilson et al., 1993; Aguirre et al., 1999; Orio et al., 2004). However, when considering this line of reasoning, it is important to recognize that established 5-HT neurotoxins (e.g. 5, 7-DHT) also fail to consistently induce a glial reaction in brain regions distant from the site of intracerebral toxin injection (see Hardin et al., 1994; Frankfurt and Azmitia, 1984; Rowland et al., 1993; Bendotti et al., 1994; Straiko et al., 2007). Thus, in our view, rather than casting doubt on the neurotoxic potential of MDMA and other substituted amphetamines (e.g. PCA), these findings indicate that additional research is needed to more fully characterize the determinants, timing, nature, and role of astroglial and microglial responses to neuron-selective lesions involving fine, relatively sparse, axonal projections (e.g. 5-HT projections). Indeed, when much denser axonal projections are damaged by MDMA (e.g. dopaminergic axonal projections to mouse striatum), both a glial reaction and positive silver staining response are consistently observed (O'Callaghan and Miller, 1994; Thomas et al., 2004), and the neurotoxic potential of MDMA toward DA neurons in the mouse goes unquestioned.

More recently, Rothman and colleagues questioned the 5-HT neurotoxic potential of MDMA based on results obtained in Western blot studies revealing no reduction in the abundance of a 70 kDa band that was thought to correspond to the SERT protein (Rothman et al., 2003, 2004; see also Wang et al., 2004, 2005). These investigators postulated that lasting serotonergic effects of MDMA represented neuroadapative changes (involving SERT trafficking), rather than damage to brain 5-HT neurons. To examine this issue further, we studied the in vivo expression of the SERT protein and other 5-HT neuronal markers (5-HT, 5-HIAA, and TPH) after MDMA and related drugs, using the established 5-HT neurotoxin, 5,7-DHT, as a positive control. Results indicated that the 70 kD band in question did not correspond to the SERT protein, as it did not exhibit the known relative regional distribution of brain SERT was resistant to 5,7-DHT treatment, and was present in SERT-KO animals (Xie et al., 2006). In contrast, a more diffuse band at approximately 63–68 kD (which has the expected regional brain distribution of the SERT, was markedly reduced after 5,7-DHT treatment and absent in SERT-KO animals) was decreased after MDMA administration (Xie et al., 2006), indicating that the SERT protein (like other 5-HT neuronal markers) is decreased after MDMA exposure. Recently, these results have been confirmed by others (Bhide et al., 2009; Cunningham et al., 2009; Biezonski and Meyer, 2010).

Still others have questioned the 5-HT neurotoxic potential of MDMA because recovery of axonal 5-HT markers has been shown to occur over time (Battaglia et al., 1991; Scanzello et al., 1993; Hatzidimitriou et al., 1999). However, with regard to recovery, it is important to recall that MDMA neurotoxicity is limited to 5-HT axons and axon terminals and that cell bodies in the brainstem raphe nuclei are invariably spared. Sparing of 5-HT nerve cell bodies leaves open the possibility that, with time, regenerative sprouting can occur. Indeed, after 5, 7-DHT lesions that spare nerve cell bodies, regenerative sprouting of 5-HT axons has been demonstrated (Zhou and Azmitia, 1986; Frankfurt and Beaudet, 1987). Thus, the recovery observed after MDMA exposure should not be taken to indicate that MDMA lacks neurotoxic potential toward 5-HT axons and axon terminals. Rather, the recovery observed reflects a fundamental feature of the MDMA lesion: namely, that it is restricted to axons and axon terminals and that, by sparing 5-HT nerve cell bodies, leaves open the potential for future regenerative sprouting. A key question in this regard is whether the regenerative sprouting that takes place leads to the reestablishment of the original 5-HT innervations patterns. In nonhuman primates that have sustained severe 5-HT lesions, this does not appear to be the case (Fischer et al, 1995; Hatzidimitriou et al., 1999).

2.3.1.2. Relevance of Animal Data to Humans

As discussed in detail elsewhere (Ricaurte et al., 2000; McCann and Ricaurte, 2007) and as recently emphasized by Easton and Marsden (2006), determining the relevance of 5-HT neurotoxicity findings in animals to human MDMA users requires consideration of a number of important factors including possible species differences in drug metabolism, route and schedule of drug administration, and, perhaps most importantly, dosage. Each of these factors is discussed, in turn, below:

Species differences in drug metabolism: 5-HT neurotoxic effects of MDMA have been demonstrated in rats, guinea pigs, squirrel monkeys, cynomolgus monkeys, rhesus monkeys, and baboons. The mouse is the only animal that is relatively resistant to MDMA-induced 5-HT neurotoxicity but, curiously, develops DA neurotoxicity. As there are marked differences in MDMA metabolism among these various species, it seems unlikely that species differences in drug metabolism will render humans insensitive to 5-HT neurotoxicity. Nevertheless, use of animal models in which the metabolism of MDMA resembles that observed in humans seems advisable for future studies. As yet, it is unclear if the neurotoxic effects of MDMA are produced by the parent compound or metabolites. Should the parent compound be the active toxic chemical species, the importance of species differences in MDMA metabolism for issues related to neurotoxicity would be diminished.

Route of drug administration: It have shown that regardless of whether MDMA is administered orally or subcutaneously, it produces comparable 5-HT neurotoxic effects in the rat. Similar results have been obtained in rhesus monkeys (Kleven et al., 1989; Slikker et al., 1988, 1989), although the oral route of administration may afford modest degree of protection in the squirrel monkey (Ricaurte et al., 1988c).

Dose: At first glance, doses of MDMA that produce 5-HT neurotoxicity in rats (10–20 mg/kg (Schmidt et al., 1985; Schmidt, 1987)) appear much higher than those typically used by humans (approximately 1–2 mg/kg). However, when comparing doses across species, it is important to take into account differences in body size, particularly when these are substantial (as is the case between a 0.20 kg rat or a 1 kg squirrel monkey and a 70 kg human). This is because fundamental physiological processes that govern drug responses are a function of body size, and these typically occur faster in smaller animals. Accordingly, interspecies scaling principles dictate that small animals generally require larger doses (on an mg/kg basis) than humans to sustain the same drug effect. Indeed, by taking into account known relationships between body size, physiological processes, and drug responses, allometric equations have been developed to allow for more accurate extrapolation of doses in small laboratory animals to humans. For example, a commonly used allometric equation is

$$D_{\text{human}} = D_{\text{animal}}(W_{\text{human}}/W_{\text{animal}})^{0.67}$$

where D, is the dose of drug in milligrams (mg), W, the weight in kilograms (kg), and 0.67 is an empirically determined exponent. It is to be noted that the foregoing allometric equation incorporates various models for extrapolating animal data to humans, and is widely used experimentally as well as the pharmaceutical industry and the Food and Drug Administration (FDA). When applied to MDMA, this equation indicates that a 10–20 mg/kg dose, which is known to produce 5-HT neurotoxic effects in a 0.2 kg rat (Schmidt et al., 1985; Schmidt, 1987 also Preliminary results) translates to a 1.45–2.89 mg/kg dose in a 70 kg human being. Notably, this is the same dosage range of MDMA required to produce the psychoactive effect of interest in human beings. Thus, the available animal data suggest that the margin of safety for MDMA (with regard to 5-HT neurotoxicity) may be narrow. In other words, there is reason to suspect that even single doses of MDMA may be associated with a significant risk of brain 5-HT neurotoxicity in humans. However, despite the strengths of the allometric approach, it is important to recognize that the method is not perfect, and that it may yield estimates of doses in animals that are not exactly equivalent to those in humans (or vice versa), for a variety of reasons including possible species differences in absorption, metabolism, or elimination.

Mechanisms: The mechanism by which MDMA damages brain 5-HT neurons has yet to be elucidated. An extensive series of pharmacological and toxicological studies has implicated endogenous brain DA and its oxidation products (see Gibb et al., 1994). However, much of the data implicating brain DA in MDMA neurotoxicity is confounded by drug effects on body temperature (Green et al., 2003), which can strongly influence MDMA neurotoxicity. Moreover, brain DA does not appear to be essential for the expression of MDMA neurotoxicity because even animals with near-total depletions of brain DA develop MDMA-induced 5-HT neurotoxic changes (Yuan et al., 2002). Recently, it was presented evidence that tyrosine, after nonenzymatic conversion to d(ihydr)o(xy)p(henyl)a(lanine) (DOPA) (then enzymatic conversion to DA), contributes to MDMA neurotoxicity, particularly in brain regions receiving little or no dopaminergic innervation. However, this finding is at seeming odds with the failure of reserpine to protect against MDMA neurotoxicity (Hekmatpanah and Peroutka, 1990; Yuan et al., 2002), and recent unpublished observations in our laboratory that the DOPA decarboxylase inhibitor, NSD1015, does not protect against MDMA-induced 5-HT neurotoxicity, if temperature changes are controlled. In addition to DA-derived free radicals, glutamate (GLU)-mediated reactive nitrogen species have been proposed to play a role in the neurotoxicity of amphetamine-type stimulants. However, the mechanism by which reactive species injure the 5-HT neuron is not entirely clear, nor is it certain that reactive species are the cause (rather than the consequence) of neurotoxicity. Energy dysregulation, possibly secondary to inhibition of mitochondrial function, may also be involved in the 5-HT neurotoxic action of MDMA. However, glucoprivation with 2-deoxy-D-glucose, a competitive inhibitor of glucose metabolism, fails to potentiate MDMA-induced toxicity to brain 5-HT neurons (Yuan et al., 2002). Based on the well-established neuroprotective effect of SERT inhibitors (Schmidt et al, 1987) and the finding that inhibitors of Na^+/H^+ and/or Na^+/Ca^{++} exchange potentiate METH-induced neurotoxicity, we proposed that ionic dysregulation, possibly secondary to prolonged transporter activation (known to be linked to ion flux), may play a role in substituted amphetamine neurotoxicity (Callahan et al., 2001).

However, more direct evidence in support of this hypothesis is needed. Finally, in recent years, there has been growing interest in the possibility that metabolites of MDMA play a role in the neurotoxic process.

2.3.2. Humans

Like animals, humans appear to be susceptible to MDMA-induced 5-HT neurotoxicity. In particular, there are two validated measures of MDMA-induced 5-HT damage in humans: (1) reductions in cerebrospinal fluid concentrations of 5-HIAA obtained at a time point when acute pharmacological effects have dissipated (i.e. weeks to months) after the last MDMA exposure and; (2) reductions in the SERT, as measured by PET, weeks to months after MDMA exposure. Abstinent MDMA users, like MDMA-treated animals, have been found to have reductions in both of these validated markers of MDMA-induced 5-HT neurotoxicity. In particular, squirrel monkeys treated with oral or systemic neurotoxic dosages of MDMA develop selective reductions in a variety of brain 5-HT neuronal markers including cerebrospinal fluid (CSF) 5-HIAA (Ricaurte et al., 1988a, 1988b; Insel et al., 1989; Ricaurte et al., 1992a; Fischer et al, 1995; Hatzidimitriou et al., 1999). Similar to MDMA-treated animals, abstinent MDMA users have been found to have selective reductions in CSF 5-HIAA (McCann et al., 1994, 1999a,b). While these findings are certainly consistent with the notion that MDMA leads to 5-HT neurotoxic injury, CSF measures are indirect. As such, reductions in CSF reductions in 5-HIAA are not conclusive evidence of MDMA-induced 5-HT damage.

The advent of positron emission tomography (PET) and single photon emission computed tomography (SPECT) methods has permitted direct visualization of neuron-specific markers in living humans, including markers of brain 5-HT axons and axon terminals (Huang et al., 2010). The first study validating molecular neuroimaging methods for detecting MDMA-induced 5-HT neurotoxicity strategies employed baboons treated with a known neurotoxic regimen of MDMA and PET scans with the first-generation PET SERT ligand, [^{11}C] McN5652. PET studies at 13, 19, and 40 days post-MDMA treatment revealed significant reductions in [^{11}C] McN 5652 binding ranging from 44% in the pons to 89% in the occipital cortex. Subsequent PET studies, 9 and 13 months later, revealed recovery in some regions, but persistent reductions in neocortical regions more than a year after MDMA treatment. This observation prompted parallel studies involving abstinent human MDMA users. These studies revealed significant reductions in SERT binding potential (McCann et al., 1998, 2005, 2008; Buchert et al., 2003, 2004, 2006, 2007; Thomasius et al., 2006). Subsequent studies using the second generation PET SERT ligand, [^{11}C] DASB, have revealed similar results in most (McCann et al., 2005, 2008; Kish et al., 2010b; Urban et al., 2012), but not all (Selvaraj et al., 2009) studies. In the one negative PET SERT study in MDMA users (Selvaraj et al., 2009), the absence of significant differences between groups may, in part, have been related to lack of power in the study design.

SPECT methods with the SERT ligand [^{123}I] β-CIT have also been used to study abstinent recreational MDMA users. Although reductions of SERT density in the hypothalamus and midbrain of monkeys with documented 5-HT lesions is associated with reductions in [^{123}I]β-CIT measured by SPECT (Reneman et al., 2002), studies in abstinent MDMA users by the same group and others find persistent reductions of [^{123}I] β-CIT in cortical, but not midbrain regions (Semple et al., 1999; Reneman et al., 2001). These regional differences in [^{123}I]β-CIT changes in human MDMA users (as compared to the validated animal model), combined with the fact that [^{123}I]β-CIT binds to the DAT, in addition to the SERT, may limit the utility of SPECT for detecting MDMA-induced SERT reductions in humans.

To our knowledge, there are only two studies in which markers of brain 5-HT integrity have been directly assessed in human MDMA users (Kish et al., 2000, 2010a). These postmortem case studies found reductions in 5-HT, SERT, and TPH in deceased MDMA users that had recently used the drug. Acknowledging its limitations (study size and recency of drug use), these observations are consistent with the notion that MDMA is neurotoxic toward brain 5-HT neurons in humans.

2.4. LASTING CONSEQUENCES OF MDMA EXPOSURE

Given the known important role of brain 5-HT in a variety of behavioral functions (Jacobs and Fornal, 2000; Henninger, 1995), and the considerable preclinical and clinical data demonstrating that MDMA is neurotoxic toward brain 5-HT neurons reviewed above, numerous research laboratories have made efforts to identify potential functional consequences of MDMA-induced 5-HT damage. In 2009, the National Institute for Health Research in the United Kingdom funded a Health Technology Assessment to investigate the harmful health effects of recreational MDMA use (Rogers et al., 2009). This technology involved a systematic review of over 4000 publications that had been published prior to that date, and addressed potential consequences of recreational MDMA use. Although, since that time, more than 250 articles have been published on the acute and long-term effects of MDMA in humans, conclusions by Rogers and colleagues are still valid.

MDMA has both acute and subacute effects on a variety of behaviors (presumably related to its acute and downstream pharmacological effects). However, in the present context, "consequences" of MDMA refers to lasting (weeks to years) behavioral changes seen in MDMA users, unless otherwise stated. In addition, it is important to note that most recreational MDMA users have also been exposed to other illicit and licit recreational substances, and these are used in uncontrolled circumstances. As such, the strength and composition of MDMA is not generally verified and the role of exposure to other substances in behavioral changes found in MDMA users cannot be excluded. Given these caveats, MDMA users, when compared to a variety of control groups, have been found to have lasting alterations in cognitive function, sleep, neuroendocrine function, and pain modulation. These differences will be reviewed, in turn.

2.4.1. Cognitive Function

Numerous researchers have explored the possibility that MDMA leads to altered cognitive processes, particularly in the area of memory function. A meta-analysis of findings from these studies (Rogers et al., 2009) indicates that, compared to controls, MDMA users have impaired memory on the six most commonly employed tests of immediate and delayed memory, including the ray auditory verbal learning test (immediate and delayed), the Rivermead behavioral test (prose recall, immediate, and delayed), and digit span (forward and backwards) (Bolla et al., 1998; Croft et al., 2001; Reneman et al., 2001; Reneman et al., 2006; Quednow et al., 2006; Lamers et al., 2006; de Win et al., 2008; Morgan, 1999; Morgan et al., 2002; Curran and Verheyden, 2003; Dafters et al., 2004; Gouzoulis-Mayfrank et al., 2000; Halpern et al., 2004; Wareing et al., 2004; McCardle et al., 2004). Subsequent studies questioning the validity of earlier results (e.g. Halpern et al., 2011), in fact, demonstrated significant effects despite a relative lack of power, as recently noted by Parrot (2011). Consistent with the notion that alterations in cognitive function in MDMA users are related to MDMA-induced neurotoxicity is the observation that deficits appear to be related to both lifetime exposure of MDMA and intensity of individual exposures (i.e. doses used) (Rogers et al., 2009; Parrott, 2011), both of which are known to influence the development of neurotoxicity in preclinical studies.

2.4.2. Sleep

Two studies (Allen et al., 1993; McCann et al., 2009) have used formal polysomnographic methods in abstinent MDMA users to assess the possibility that MDMA exposure leads to alterations in normal sleep neurophysiology. Both studies suggest that MDMA leads to chronic changes in sleep. In particular, MDMA users exhibit alterations in sleep architecture and circadian rhythms (Allen et al., 1993), and have an increased risk for obstructive sleep apnea (McCann et al., 2009). Given the importance of circadian rhythms and sleep quality in a variety of daytime functions (e.g. alertness, cognitive function, and endocrine function), these findings suggest that some of the behavioral differences that have been observed in MDMA users could be related, in part, to alterations in sleep timing, quality, or continuity.

2.4.3. Neuroendocrine Effects

Two studies have employed the neuroendocrine challenge method with 5-HT probes to assess the possibility that MDMA exposure leads to altered 5-HT-mediated endocrine function in abstinent MDMA users (McCann et al., 1999b; Gerra et al., 1998). In the first study, the mixed 5-HT agonist and 5-HT releaser, m-chlorophenylpiperazine (m-CPP), was administered to abstinent MDMA users and non–MDMA using controls. MDMA users were found to have reduced cortisol and prolactin responses to m-CPP, in addition to diminished behavioral responses. It was concluded that differential responses to m-CPP in the two subject groups could be related to MDMA-induced neurotoxicity, and alterations in 5-HT input on neuroendocrine function (McCann et al., 1999b).

In the second study, Gerra and colleagues conducted a longitudinal study in abstinent MDMA users at baseline (time 0), as well as 3 and 12 months later. MDMA users and controls were challenged with the 5-HT releaser, fenfluramine, which is known to induce increases in both prolactin and cortisol. MDMA users were found to have persistent reductions in fenfluramine-induced prolactin, but reversible reductions in cortisol (i.e. reductions did not persist for the full 12-month period). Further, the extent of alterations in prolactin responses at the 12-month time point was directly related to MDMA exposure. Taken together, these observations were taken as evidence that MDMA can lead to persistent alterations in 5-HT-mediated prolactin responses.

2.4.4. Pain

Several studies (McCann et al., 1994; O'Regan and Clow, 2004; McCann et al., 2011) have explored the possibility that pain perception is altered in abstinent MDMA users, given the known role of 5-HT in the modulation of pain (Yoshimura and Furue, 2006; Lopez-Garcia, 2006). One study, conducted in MDMA users who had been abstinent, on average, for 5 months noted no differences in MDMA users' response to an ischemic pain task (McCann et al., 1994) when compared to control subjects. The second study found reduced cold-pain tolerance in MDMA users who had been abstinent from MDMA for only 3–4 days (O'Regan and Clow, 2004). Most recently, MDMA users who had been abstinent for weeks to months and matched non-MDMA-exposed controls were compared on a variety of standardized pain measurements. Further, given increasing data that MDMA users have alterations in sleep and the known influence of sleep disruption on pain (Edwards et al., 2008; Tiede et al., 2010), the potential relationship between polysomnographic sleep measures and pain outcomes were explored. Abstinent MDMA users were found to have reduced pressure pain thresholds, increased cold-pain ratings, and increased pain ratings during testing of diffuse noxious inhibitory control. Although numerous significant relationships between sleep and pain measures were identified, meditational modeling indicted that altered pain perception in MDMA users was not related to changes in sleep.

2.4.5. Neuropsychiatric Effects/Complications

In addition to acute and subacute neuropsychiatric effects of MDMA, which are presumably related to its pharmacological properties, there have also been numerous reports of chronic psychiatric problems in MDMA users. Although beyond the scope of this chapter, the most common psychiatric problems seen in MDMA users include depression, anxiety, phobias, psychotic symptoms, and somatization disorders (Parrott et al., 2000, 2001, 2002; MacInness et al., 2001; Bobes et al., 2002; McCardle et al., 2004; Dafters et al., 2004; Lieb et al., 2002; Daumann et al., 2004; Sumnall and Cole, 2005; Karlsen et al., 2008). Although most of the authors of these reports draw attention to the fact that MDMA users tend to use "other drugs" in addition to MDMA, and that psychiatric problems may have drawn people to use MDMA (rather than having been caused by them), several studies have found a relationship between the extent of MDMA use and psychopathology (Parrott et al., 2000, 2002; de Win et al., 2004). While not conclusive, this observation is consistent with the notion that MDMA neurotoxicity could play a role in the psychiatric symptoms experienced by some MDMA users.

3. CONCLUSIONS

MDMA is a ring-substituted amphetamine analog with reinforcing psychoactive effects, in addition to neurotoxic potential toward central 5-HT neurons. Its major pharmacological effects appear to be secondary to actions at brain monoaminergic neurons, both as a potent monoamine releaser and as a reuptake inhibitor, although MDMA also binds directly to several brain 5-HT receptor subtypes. The mechanisms of MDMA-induced 5-HT neurotoxicity are not fully understood, but are known to be exacerbated by increased temperature, attenuated by hypothermia, and blocked by 5-HT reuptake inhibitors. A growing body of data indicates that humans, like animals treated with oral or systemic MDMA, are susceptible to MDMA-induced 5-HT neural injury. It is not clear whether 5-HT neural injury, which is best described as a distal axotomy of brain 5-HT neurons, is fully reversible in all brain regions. Studies in abstinent MDMA users have revealed deficits in a variety of behavioral functions known to be modulated by brain 5-HT, although it has not been definitively demonstrated that deficits are related to MDMA-induced 5-HT injury. Future research should be directed toward determining the mechanism of action of MDMA neurotoxicity, defining the parameters of MDMA use associated with neurotoxicity in humans, and determining the functional consequences (and reversibility) of MDMA neurotoxicity in both animals and humans.

REFERENCES

Aguirre, N., Barrionuevo, M., Ramírez, M.J., Del Río, J., Lasheras, B., 1999. Alpha-lipoic acid prevents 3,4-methylenedioxy-methamphetamine (MDMA)-induced neurotoxicity. Neuroreport 10 (17), 3675–3680.

Allen, R.P., McCann, U.D., Ricaurte, G.A., 1993. Persistent effects of (±)3,4-methylenedioxymethamphetamine (MDMA, "ecstasy") on human sleep. Sleep 16 (6), 560–564.

Axt, K., Mamounas, L., Molliver, M., 1994. Structural features of amphetamine neurotoxicity. In: Cho, A.K., Segal, D.S. (Eds.), Amphetamine and Its Analogs. Academic Press, San Diego, CA, pp. 315–370.

Battaglia, G., Yeh, S.Y., De Souza, E.B., 1988a. MDMA-induced neurotoxicity: parameters of degeneration and recovery of brain serotonin neurons. Pharmacol. Biochem. Behav. 29 (2), 269–274.

Battaglia, G., Brooks, B.P., Kulsakdinun, C., De Souza, E.B., 1988b. Pharmacologic profile of MDMA (3,4-methylenedioxymethamphetamine) at various brain recognition sites. Eur. J. Pharmacol. 149, 159–163.

Battaglia, G., Sharkey, J., Kuhar, M.J., De Souza, E.B., 1991. Neuroanatomic specificity and time course of alterations in rat brain serotonergic pathways induced by MDMA (3,4-methylenedioxymethamphetamine): assessment using quantitative autoradiography. Synapse 8 (4), 249–260.

Baumgarten, H.G., Lachenmayer, L., 2004. Serotonin neurotoxins–past and present. Neurotox. Res. 6 (7–8), 589–614.

Beck, J., 1978. The public implications of MDMA use. In: Stillman, R.C., Willette, R.E. (Eds.), The Psychopharmacology of Hallucinogens. Pergamon Press, New York, pp. 74–83.

Bendotti, C., Baldessari, S., Pende, M., Tarizzo, G., Miari, A., Presti, M.L., Mennini, T., Samanin, R., 1994. Does GFAP mRNA and mitochondrial benzodiazepine receptor binding detect serotonergic neuronal degeneration in rat? Brain Res. Bull. 34, 389–394.

Bhide, N.S., Lipton, J.W., Cunningham, J.I., Yamamoto, B.K., Gudelsky, G.A., 2009. Repeated exposure to MDMA provides neuroprotection against subsequent MDMA-induced serotonin depletion in brain. Brain Res. 1286, 32–41.

Biezonski, D.K., Meyer, J.S., 2010. Effects of 3,4-methylenedioxymethamphetamine (MDMA) on serotonin transporter and vesicular monoamine transporter 2 protein and gene expression in rats: implications for MDMA neurotoxicity. J. Neurochem. 112 (4), 951–962.

Blessing, W.W., Seaman, B., Pedersen, N.P., Ootsuka, Y., 2003. Clozapine reverses hyperthermia and sympathetically mediated cutaneous vasoconstriction induced by 3,4-methylenedioxymethamphetamine (ecstasy) in rabbits and rats. J. Neurosci. 23 (15), 6385–6391.

Bobes, J., Sáiz, P.A., González, M.P., Bascaran, M.T., Bousono, M., Ricaurte, G.A., McCann, U.D., 2002. Use of MDMA and other illicit drugs by young adult males in northern Spain. A five-year study. Eur. Addict. Res. 8, 147–154.

Bolla, K.I., McCann, U.D., Ricaurte, G.A., 1998. Memory impairment in abstinent MDMA ("Ecstasy") users. Neurology 51 (6), 1532–1537.

Buchert, R., Thomasius, R., Nebeling, B., Petersen, K., Obrocki, J., Jenicke, L., Wilke, F., Wartberg, L., Zapletalova, P., Clausen, M., 2003. Long-term effects of "ecstasy" use on serotonin transporters of the brain investigated by PET. J. Nucl. Med. 44 (3), 375–384.

Buchert, R., Thomasius, R., Wilke, F., Petersen, K., Nebeling, B., Obrocki, J., Schulze, O., Schmidt, U., Clausen, M., 2004. A voxel-based PET investigation of the long-term effects of "Ecstasy" consumption on brain serotonin transporters. Am. J. Psychiatry 161 (7), 1181–1189.

Buchert, R., Thomasius, R., Petersen, K., Wilke, F., Obrocki, J., Nebeling, B., Wartberg, L., Zapletalova, P., Clausen, M., 2006. Reversibility of ecstasy-induced reduction in serotonin transporter availability in polydrug ecstasy users. Eur. J. Nucl. Med. Mol. Imaging 33 (2), 188–199.

Buchert, R., Thiele, F., Thomasius, R., Wilke, F., Petersen, K., Brenner, W., Mester, J., Spies, L., Clausen, M., 2007. Ecstasy-induced reduction of the availability of the brain serotonin transporter as revealed by [11C](+)McN5652-PET and the multi-linear reference tissue model: loss of transporters or artifact of tracer kinetic modelling? J. Psychopharmacol. 21 (6), 628–634.

Callahan, B.T., Cord, B.J., Ricaurte, G.A., 2001. Long-term impairment of anterograde axonal transport along fiber projections originating in the rostral raphe nuclei after treatment with fenfluramine or methylenedioxy-methamphetamine. Synapse 40, 113–121.

Cami, J., Farré, M., Mas, M., Roset, P.N., Poudevida, S., Mas, A., San, L., de la Torre, R., 2000. Human pharmacology of 3,4-methylenedioxymethamphetamine ("ecstasy"): psychomotor performance and subjective effects. J. Clin. Psychopharmacol. 20 (4), 455–466.

Chu, T., Kumagai, Y., DiStefano, E.W., Cho, A.K., 1996. Disposition of methylenedioxymethamphetamine and three metabolites in the brains of different rat strains and their possible roles in acute serotonin depletion. Biochem. Pharmacol. 51 (6), 789–796.

Commins, D.L., Vosmer, G., Virus, R., Woolverton, W., Schuster, C., Seiden, L., 1987. Biochemical and histological evidence that methylenedioxymethylamphetamine (MDMA) is toxic to neurons in the rat brain. J. Pharmacol. Exp. Ther. 241, 338–345.

Croft, R.J., Mackay, A.J., Mills, A.T., Gruzelier, J.G., 2001. The relative contributions of ecstasy and cannabis to cognitive impairment. Psychopharmacology (Berl.) 153 (3), 373–379.

Cunningham, J.I., Raudensky, J., Tonkiss, J., Yamamoto, B.K., 2009. MDMA pretreatment leads to mild chronic unpredictable stress-induced impairments in spatial learning. Behav. Neurosci. 123 (5), 1076–1084.

Curran, H.V., Verheyden, S.L., 2003. Altered response to tryptophan supplementation after long-term abstention from MDMA (ecstasy) is highly correlated with human memory function. Psychopharmacology (Berl.) 169 (1), 91–103.

Dafters, R.I., Hoshi, R., Talbot, A.C., 2004. Contribution of cannabis and MDMA ("ecstasy") to cognitive changes in long-term polydrug users. Psychopharmacology (Berl.) 173 (3–4), 405–410.

Daumann, J., Hensen, G., Thimm, B., Rezk, M., Till, B., Gouzoulis-Mayfrank, E., 2004. Self-reported psychopathological symptoms in recreational ecstasy (MDMA) users are mainly associated with regular cannabis use: further evidence from a combined cross-sectional/longitudinal investigation. Psychopharmacology 173, 398–404.

de la Torre, R., Farré, M., Ortuño, J., Mas, M., Brenneisen, R., Roset, P.N., Segura, J., Camí, J., 2000. Nonlinear pharmacokinetics of MDMA ('ecstasy') in humans. Br. J. Clin. Pharmacol. 49 (2), 104–109.

de la Torre, R., Farre, M., Roset, P.N., Pizarro, N., Abanades, S., Segura, M., Segura, J., Cami, J., 2004. Human pharmacology of MDMA: pharmacokinetics, metabolism, and disposition. Ther. Drug Monit. 26, 137–144.

de Win, M.M., Jager, G., Booij, J., Reneman, L., Schilt, T., Lavini, C., Olabarriaga, S.D., den Heeten, G.J., van den Brink, W., 2008. Sustained effects of ecstasy on the human brain: a prospective neuroimaging study in novel users. Brain 131 (Pt 11), 2936–2945.

Downing, J., 1986. The psychological and physiological effects of MDMA on normal volunteers. J. Psychoactive Drugs 18, 335–340.

Easton, N., Marsden, C.A., 2006. Ecstasy: are animal data consistent between species and can they translate to humans? J. Psychopharmacol. 20 (2), 194–210.

Edwards, R.R., Almeida, D.M., Klick, B., Haythornthwaite, J.A., Smith, M.T., 2008. Duration of sleep contributes to next-day pain report in the general population. Pain 137, 202–207.

Fischer, C.A., Hatzidimitriou, G., Wlos, J., Katz, J.L., Ricaurte, G.A., 1995. Reorganization of ascending 5-HT axon projections in animals previously exposed to the recreational drug (+)3,4-methylenedioxymethamphetamine (MDMA, "ecstasy"). J. Neurosci. 15, 5476–5485.

Fox, M.A., Andrews, A.M., Wendland, J.R., Lesch, K.P., Holmes, A., Murphy, D.L., 2007. A pharmacological analysis of mice with a targeted disruption of the serotonin transporter. Psychopharmacology (Berl.) 195 (2), 147–166.

Frankfurt, M., Azmitia, E., 1984. Regeneration of serotonergic fibers in the rat hypothalamus following unilateral 5,7-dihydroxytyptamine injection. Brain Res. 298, 273–282.

Frankfurt, M., Beaudet, A., 1987. Ultrastructural organization of regenerated serotonin axons in the dorsomedial hypothalamus of the adult rat. J. Neurocytol. 16 (6), 799–809.

Gehlert, D.R., Schmidt, C.J., Wu, L., Lovenberg, W., 1985. Evidence for specific methylenedioxymethamphetamine (Ecstasy) binding sites in the rat brain. Eur. J. Pharmacol. 119, 135.

Gerra, G., Zaimovic, A., Giucastro, G., Maestri, D., Monica, C., Sartori, R., Caccavari, R., Delsignore, R., 1998. Serotonergic function after (±)3,4-methylene-dioxymethamphetamine ('Ecstasy') in humans. Int. Clin. Psychopharmacol. 13 (1), 1–9.

Gibb, J.W., Hanson, G.R., Johnson, M., 1994. Neurochemical mechanisms of toxicity. In: Cho, A.K., Segal, D.S. (Eds.), Amphetamine and Its Analogs. Academic Press, San Diego, CA, pp. 269–295.

Gouzoulis-Mayfrank, E., Daumann, J., Tuchtenhagen, F., Pelz, S., Becker, S., Kunert, H.J., Fimm, B., Sass, H., 2000. Impaired cognitive performance in drug free users of recreational ecstasy (MDMA). J. Neurol. Neurosurg. Psychiatry 68 (6), 719–725.

Green, A.R., Mechan, A.O., Elliott, J.M., O'Shea, E., Colado, M.I., 2003. The pharmacology and clinical pharmacology of 3,4-methylenedioxymethamphetamine (MDMA, "ecstasy"). Pharmacol. Rev. 55, 463–508.

Greer, G., 1985. Using MDMA in psychotherapy. Advances 2, 57–67.

Greer, G., Tolbert, R., 1986. Subjective reports of the effects of MDMA in a clinical setting. J. Psychoactive Drugs 18, 319–327.

Gudelsky, G.A., Yamamoto, B.K., 2003. Neuropharmacology and neurotoxicity of 3,4-methylenedioxymethamphetamine. Methods Mol. Med. 79, 55–73.

Gudelsky, G.A., Yamamoto, B.K., 2008. Actions of 3,4-methylenedioxymethamphetamine (MDMA) on cerebral dopaminergic, serotonergic and cholinergic neurons. Pharmacol. Biochem. Behav. 90 (2), 198–207.

Halpern, J.H., Pope Jr., H.G., Sherwood, A.R., Barry, S., Hudson, J.I., Yurgelun-Todd, D., 2004. Residual neuropsychological effects of illicit 3,4-methylenedioxymethamphetamine (MDMA) in individuals with minimal exposure to other drugs. Drug Alcohol Depend. 75 (2), 135–147.

Halpern, J.H., Sherwood, A.R., Hudson, J.I., Gruber, S., Kozin, D., Pope Jr, H.G., 2011. Residual neurocognitive features of long-term ecstasy users with minimal exposure to other drugs. Addiction 106 (4), 777–786.

Hardin, H., Bernard, A., Rajas, F., Fevre-Montange, M.F., Derrington, E., Belin, M.F., Didier-Bazes, M., 1994. Modifications of glial metabolism of glutamate after serotonergic neuron degeneration in the hippocampus of the rat. Mol. Brain Res. 26, 1–8.

Hardman, H.F., Haavik, C.O., Seevers, M.H., 1973. Relationship of the structure of mescaline and seven analogs to toxicity and behavior in five species of laboratory animals. Tox. Appl. Pharmacol. 25, 299–309.

Hatzidimitriou, G., McCann, U.D., Ricaurte, G.A., 1999. Altered serotonin innervation patterns in the forebrain of monkeys treated with (±)3,4-methylenedioxymethamphetamine seven years previously: factors influencing abnormal recovery. J. Neurosci. 19 (12), 5096–5107.

Hekmatpanah, C.R., Peroutka, S.J., 1990. 5-hydroxytryptamine uptake blockers attenuate the 5-ydroxytryptamine-releasing effect of 3,4-methylenedioxymethamphetamine and related agents. Eur. J. Pharmacol. 177 (1–2), 95–98.

Henninger, G.R., 1995. Indoleamines: the role of serotonin in clinical disorders. In: Bloom, F.E., Kupfer, D.J. (Eds.), Psychopharmacology: the Fourth Generation of Progress. Raven Press, New York.

Henry, J.A., Jeffreys, K.J., Dawling, S., 1992. Toxicity and deaths from 3, 4-methylenedioxymethamphetamine ("Ecstasy"). Lancet 340, 384–387.

Holland, J. (Ed.), 2001. Ecstasy, the Complete Guide: a Comprehensive Look at the Risks and Benefits of MDMA. Park Street Press, p. 11.

Hua, Y.S., Liang, R., Liang, L., Huang, G.Z., 2009. Contraction band necrosis in two ecstasy abusers: a latent lethal lesion associated with ecstasy. Am. J. Forensic Med. Pathol. 30 (3), 295–297.

Huang, Y., Zheng, M.Q., Gerdes, J.M., 2010. Development of effective PET and SPECT imaging agents for the serotonin transporter: has a twenty-year journey reached its destination? Curr. Top. Med. Chem. 10 (15), 1499–1526.

Hysek, C.M., Vollenweider, F.X., Liechti, M.E., 2010. Effects of a beta-blocker on the cardiovascular response to MDMA (Ecstasy). Emerg. Med. J. 27 (8), 586–589

Insel, T.R., Battaglia, G., Johannessen, J.N., Marra, S., De Souza, E.B., 1989. 3,4-methylenedioxymethamphetamine ("ecstasy") selectively destroys brain serotonin terminals in rhesus monkeys. J. Pharmacol. Exp. Ther. 249, 713–720.

Jacobs, B.L., Fornal, C.A., 1995. Serotonin and behavior. A general hypothesis. In: Bloom, F.E., Kupfer, D.J. (Eds.), Psychopharmacology: the Fourth Generation of Progress. Raven Press, New York.

Jonsson, G., 1980. Chemical neurotoxins as denervation tools in neurobiology. Annu. Rev. Neurosci. 3, 169–187.

Karlsen, S.N., Spigset, O., Slørdal, L., 2008. The dark side of ecstasy: neuropsychiatric symptoms after exposure to 3,4-methylenedioxymethamphetamine. Basic Clin. Pharmacol. Toxicol. 102 (1), 15–24.

Kish, S.J., Furukawa, Y., Ang, L., Vorce, S.P., Kalasinsky, K.S., 2000. Striatal serotonin is depleted in brain of a human MDMA (Ecstasy) user. Neurology 55 (2), 294–296.

Kish, S.J., Fitzmaurice, P.S., Chang, L.J., Furukawa, Y., Tong, J., 2010a. Low striatal serotonin transporter protein in a human polydrug MDMA (ecstasy) user: a case study. J. Psychopharmacol. 24 (2), 281–284.

Kish, S.J., Lerch, J., Furukawa, Y., Tong, J., McCluskey, T., Wilkins, D., Houle, S., Meyer, J., Mundo, E., Wilson, A.A., Rusjan, P.M., Saint-Cyr, J.A., Guttman, M., Collins, D.L., Shapiro, C., Warsh, J.J., Boileau, I., 2010b. Decreased cerebral cortical serotonin transporter binding in ecstasy users: a positron emission tomography/[(11)C]DASB and structural brain imaging study. Brain 133 (Pt 6), 1779–1797.

Kleven, M.S., Woolverton, W.L., Seiden, L.S., 1989. Evidence that both intragastric and subcutaneous administration of methylenedioxymethylamphetamine (MDMA) produce serotonin neurotoxicity in rhesus monkeys. Brain Res. 488 (1–2), 121–125.

Kolbrich, E.A., Goodwin, R.S., Gorelick, D.A., Hayes, R.J., Stein, E.A., Huestis, M.A., 2008a. Plasma pharmacokinetics of 3,4-methylenedioxymethamphetamine after controlled oral administration to young adults. Ther. Drug Monit. 30 (3), 320–332.

Kolbrich, E.A., Goodwin, R.S., Gorelick, D.A., Hayes, R.J., Stein, E.A., Huestis, M.A., 2008b. Physiological and subjective responses to controlled oral 3,4-methylenedioxymethamphetamine administration. J. Clin. Psychopharmacol. 28 (4), 432–440.

Lamers, C.T., Bechara, A., Rizzo, M., Ramaekers, J.G., 2006. Cognitive function and mood in MDMA/THC users, THC users and non-drug using controls. J. Psychopharmacol. 20 (2), 302–311.

Lawn, J.C., 1985. Schedules of controlled substances; temporary placement of 3,4-methylenedioxymethamphetamine (MDMA). Fed. Regist. 50 (105), 23118–23120.

Leonardi, E.T., Azmitia, E.C., 1994. MDMA (ecstasy) inhibition of MAO type A and type B: comparisons with fenfluramine and fluoxetine (Prozac). Neuropsychopharmacology 10 (4), 231–238.

Lew, R., Malberg, J.E., Ricaurte, G.A., Seiden, L.S., 1997. Evidence for and mechanism of action of neurotoxicity of amphetamine related compounds. In: Kostrzewa, R.M. (Ed.), Highly Selective Neurotoxins: Basic and Clinical Applications. Humana Press, Totowa, NJ, pp. 235–268.

Lieb, R., Schuetz, C.G., Pfister, H., von Sydow, K., Wittchen, H., 2002. Mental disorders in ecstasy users: a prospective-longitudinal investigation. Drug Alcohol Depend. 68, 195–207.

Liechti, M.E., Baumann, C., Gamma, A., Vollenweider, F.X., 2000. Acute psychological effects of 3,4-methylenedioxymethamphetamine (MDMA, "Ecstasy") are attenuated by the serotonin uptake inhibitor citalopram. Neuropsychopharmacology 22 (5), 513–521.

Liechti, M.E., Vollenweider, F.X., 2000. Acute psychological and physiological effects of MDMA ("Ecstasy") after haloperidol pretreatment in healthy humans. Eur. Neuropsychopharmacol. 10 (4), 289–295.

Liechti, M.E., Vollenweider, F.X., 2001. Which neuroreceptors mediate the subjective effects of MDMA in humans? A summary of mechanistic studies. Hum. Psychopharmacol. 16 (8), 589–598.

Logan, B.J., Laverty, R., Sanderson, W.D., Yee, Y.B., 1988. Differences between rats and mice in MDMA (methylenedioxymethylamphetamine) neurotoxicity. Eur. J. Pharmacol. 152 (3), 227–234.

Lopez-Garcia, J.A., 2006. Serotonergic modulation of spinal sensory circuits. Curr. Top. Med. Chem. 6, 1987–1996.

Lyon, R.A., Glennon, R.A., Titeler, M., 1986. 3,4-Methylenedioxymethamphetamine (MDMA): stereoselective interactions at brain 5-HT1 and 5-HT2 receptors. Psychopharmacology (Berl.) 88 (4), 525–526.

MacInnes, N., Handley, S.L., Harding, G.F.A., 2001. Former chronic methylenedioxymethamphetamine (MDMA or ecstasy) users report mild depressive symptoms. J. Psychopharmacol. 15, 181–186.

Mas, M., Farré, M., de la Torre, R., Roset, P.N., Ortuño, J., Segura, J., Camí, J., 1999. Cardiovascular and neuroendocrine effects and pharmacokinetics of 3,4-methylenedioxymethamphetamine in humans. J. Pharmacol. Exp. Ther. 290 (1), 136–145.

McCann, U.D., Ricaurte, G.A., 1993. Reinforcing subjective effects of (±)3,4-methylenedioxymethamphetamine ("ecstasy") may be separable from its neurotoxic actions: clinical evidence. J. Clin. Psychopharmacol. 13 (3), 214–217.

McCann, U.D., Ridenour, A., Shaham, Y., Ricaurte, G.A., 1994. Serotonin neurotoxicity after (±)3,4-methylenedioxymethamphetamine (MDMA; "Ecstasy"): a controlled study in humans. Neuropsychopharmacology 10 (2), 129–138.

McCann, U.D., Szabo, Z., Scheffel, U., Dannals, R.F., Ricaurte, G.A., 1998. Positron emission tomographic evidence of toxic effect of MDMA ("Ecstasy") on brain serotonin neurons in human beings. Lancet 352 (9138), 1433–1437.

McCann, U.D., Mertl, M., Eligulashvili, V., Ricaurte, G.A., 1999a. Cognitive performance in (±)3,4-methylenedioxymethamphetamine (MDMA, "Ecstasy") users: a controlled study. Psychopharmacology (Berl.) 143 (4), 417–425.

McCann, U.D., Eligulashvili, V., Mertl, M., Murphy, D.L., Ricaurte, G.A., 1999b. Altered neuroendocrine and behavioral responses to m-chlorophenylpiperazine in 3,4-methylenedioxymethamphetamine (MDMA) users. Psychopharmacology 147, 56–65.

McCann, U.D., Szabo, Z., Seckin, E., Rosenblatt, P., Mathews, W.B., Ravert, H.T., Dannals, R.F., Ricaurte, G.A., 2005. Quantitative PET studies of the serotonin transporter in MDMA users and controls using [11C]McN5652 and [11C]DASB. Neuropsychopharmacology 30 (9), 1741–1750.

McCann, U.D., Ricaurte, G.A., 2007. Effects of (±)3,4-methylenedioxymethamphetamine (MDMA) on sleep and circadian rhythms. Sci. World J. 7, 231–238.

McCann, U.D., Szabo, Z., Vranesic, M., Palermo, M., Mathews, W.B., Ravert, H.T., Dannals, R.F., Ricaurte, G.A., 2008. Positron emission tomographic studies of brain dopamine and serotonin transporters in abstinent (±)3,4-methylenedioxymethamphetamine ("ecstasy") users: relationship to cognitive performance. Psychopharmacology (Berl.) 200 (3), 439–450.

McCann, U.D., Sgambati, F.P., Schwartz, A.R., Ricaurte, G.A., 2009. Sleep apnea in young abstinent recreational MDMA ("ecstasy") consumers. Neurology 73 (23), 2011–2017.

McCann, U.D., Edwards, R.R., Smith, M.T., Kelley, K., Wilson, M., Sgambati, F., Ricaurte, G., 2011. Altered pain responses in abstinent (±)3,4-methylenedioxymethamphetamine (MDMA, "ecstasy") users. Psychopharmacology (Berl.) 217 (4), 475–484.

McCardle, K., Luebbers, S., Carter, J.D., Croft, R.J., Stough, C., 2004. Chronic MDMA (ecstasy) use, cognition and mood. Psychopharmacology (Berl.) 173 (3–4), 434–439.

McDowell, D.M., Kleber, H.D., 2004. MDMA: its history and pharmacology. Psychiatr. Ann. 24 (3), 127–130.

McLane, M.W., McCann, U., Ricaurte, G., 2011. Identifying the serotonin transporter signal in Western blot studies of the neurotoxic potential of MDMA and related drugs. Synapse 65 (12), 1368–1372.

Merck, 2012. Verfahren zur Darstellung von Alkyloxyaryl-, Dialkyloxyaryl–, und Alkylenedioxyarylamino-propanen bzw. derenam Stickstoff monoalkylierten Derivaten, 1914. German Patent 274,350, filed Dec. 24, 1912, issued May 16, 1914, and assigned to E. Merck in Darmstadt, Germany.

Meyer, M.R., Peters, F.T., Maurer, H.H., 2008. The role of human hepatic cytochrome P450 isozymes in the metabolism of racemic 3,4-methylenedioxymethamphetamine and its enantiomers. Drug Metab. Dispos. 36, 2345–2354.

Mills, E.M., Rusyniak, D.E., Sprague, J.E., 2004. The role of the sympathetic nervous system and uncoupling proteins in the thermogenesis induced by 3,4-methylenedioxymethamphetamine. J. Mol. Med. (Berl.) 82 (12), 787–799.

Molliver, M.E., Berger, U.V., Mamounas, L.A., Molliver, D.C., O'Hearn, E., Wilson, M.A., 1990. Neurotoxicity of MDMA and related compounds: anatomic studies. Ann. N.Y. Acad. Sci. 600, 649–661 Discussion 661–664.

Morgan, M.J., 1999. Memory deficits associated with recreational use of "ecstasy" (MDMA). Psychopharmacology (Berl.) 141 (1), 30–36.

Morgan, M.J., McFie, L., Fleetwood, H., Robinson, J.A., 2002. Ecstasy (MDMA): are the psychological problems associated with its use reversed by prolonged abstinence? Psychopharmacology (Berl.) 159 (3), 294–303.

Mueller, M., Kolbrich, E.A., Peters, F.T., Maurer, H.H., McCann, U.D., Huestis, M.A., Ricaurte, G.A., 2009. Direct comparison of (±)3,4-methylenedioxymethamphetamine ("ecstasy") disposition and metabolism in squirrel monkeys and humans. Ther. Drug Monit. 31 (3), 367–373.

Naranjo, C., Shulgin, A.T., Sargent, T., 1967. Evaluation of 3,4-methylenedioxyamphetamine (MDA) as an adjunct to psychotherapy. Med. Pharmacol. Exp. Int. J. Exp. Med. 17 (4), 359–364.

Nichols, D.E., Lloyd, D.H., Hoffman, A.J., Nichols, M.B., Yim, G.K., 1982. Effects of certain hallucinogenic amphetamine analogues on the release of [3H]serotonin from rat brain synaptosomes. J. Med. Chem. 25 (5), 530–535.

Nichols, D.E., Hoffman, A.J., Oberlender, R.A., Jacob 3rd, P., Shulgin, A.T., 1986. Derivatives of 1-(1,3-benzodioxol-5-yl)-2-butanamine: representatives of a novel therapeutic class. J. Med. Chem. 29 (10), 2009–2015.

O'Callaghan, J.P., Miller, D.B., 1994. Neurotoxicity profiles of substituted amphetamines in the C57BL/6J mouse. J. Pharmacol. Exp. Ther. 270, 741–751.

O'Hearn, E., Battaglia, G., De Souza, E.B., Kuhar, M.J., Molliver, M.E., 1988. Methylenedioxyamphetamine (MDA) and methylenedioxymethamphetamine (MDMA) cause selective ablation of serotonergic axon terminals in forebrain: immunocytochemical evidence for neurotoxicity. J. Neurosci. 8 (8), 2788–2803.

O'Regan, M.C., Clow, A., 2004. Decreased pain tolerance and mood in recreational users of MDMA. Psychopharmacology (Berl.) 173, 446–451.

Orio, L., O'Shea, E., Sanchez, V., Pradillo, J.M., Escobedo, I., Camarero, J., Moro, M.A., Green, A.R., Colado, M., 2004. 3,4-Methylenedioxymethamphetamine increases interleukin-1beta levels and activates microglia in rat brain: studies on the relationship with acute hyperthermia and 5-HT depletion. J. Neurochem. 89, 1445–1453.

Parrott, A.C., 2005. Chronic tolerance to recreational MDMA (3,4-methylenedioxymethamphetamine) or ecstasy. J. Psychopharmacol. 19 (1), 71–83.

Parrott, A.C., 2011. Residual neurocognitive features of ecstasy use: a re-interpretation of Halpern et al. (2011) consistent with serotonergic neurotoxicity. Addiction 106 (7), 1365–1368.

Parrott, A.C., 2012. MDMA and temperature: a review of the thermal effects of 'Ecstasy' in humans. Drug Alcohol Depend. 121 (1–2), 1–9.

Parrott, A.C., Sisk, E., Turner, J.J.D., 2000. Psychobiological problems in heavy 'ecstasy' (MDMA) polydrug users. Drug Alcohol Depend. 60, 105–110.

Parrott, A.C., Milani, R.M., Parmar, R., Turner, J.D., 2001. Recreational ecstasy/MDMA and other drug users from the UK and Italy: psychiatric symptoms and psychobiological problems. Psychopharmacology 159, 77–82.

Parrott, A.C., Buchanan, T., Scholey, A.B., Heffernan, T., Ling, J., Rodgers, J., 2002. Ecstasy/MDMA attributed problems reported by novice, moderate and heavy recreational users. Hum. Psychopharmacol. 17, 309–312.

Piper, B.J., Fraiman, J.B., Owens, C.B., Ali, S.F., Meyer, J.S., 2008. Dissociation of the neurochemical and behavioral toxicology of MDMA ('Ecstasy') by citalopram. Neuropsychopharmacology 33, 1192–1205.

Pubill, D., Canudas, A.M., Pallas, M., Camins, A., Camarasa, J., Escubedo, E., 2003. Different glial response to methamphetamine- and methylenedioxymethamphetamine-induced neurotoxicity. Naunyn–Schmiedebergs Arch. Pharmacol. 367, 490–499.

Quednow, B.B., Jessen, F., Kuhn, K.U., Maier, W., Daum, I., Wagner, M., 2006. Memory deficits in abstinent MDMA (ecstasy) users: neuropsychological evidence of frontal dysfunction. J. Psychopharmacol. 20 (3), 373–384.

Reneman, L., Booij, J., Majoie, C.B., van den Brink, W., Den Heeten, G.J., 2001. Investigating the potential neurotoxicity of ecstasy (MDMA): an imaging approach. Hum. Psychopharmacol. 16 (8), 579–588.

Reneman, L., Booij, J., Habraken, J.B., De Bruin, K., Hatzidimitriou, G., Den Heeten, G.J., Ricaurte, G.A., 2002. Validity of [123I]beta-CIT SPECT in detecting MDMA-induced serotonergic neurotoxicity. Synapse 46 (3), 199–205.

Reneman, L., de Win, M.M., van den Brink, W., Booij, J., den Heeten, G.J., 2006. Neuroimaging findings with MDMA/ecstasy: technical aspects, conceptual issues and future prospects. J. Psychopharmacol. 20 (2), 164–175.

Ricaurte, G.A., Bryan, G., Strauss, L., Seiden, L., Schuster, C., 1985. Hallucinogenic amphetamine selectively destroys brain serotonin nerve terminals. Science 229, 986–988.

Ricaurte, G.A., Forno, L.S., Wilson, M.A., DeLanney, L.E., Irwin, I., Molliver, M.E., Langston, J.W., 1988a. (±)3,4-Methylenedioxymethamphetamine (MDMA) selectively damages central serotonergic neurons in non-human primates. JAMA 260, 51–55.

Ricaurte, G.A., DeLanney, L.E., Irwin, I., Langston, J.W., 1988b. Toxic effects of MDMA on central serotonergic neurons in the primate: importance of route and frequency of drug administration. Brain Res. 446, 165–168.

Ricaurte, G.A., DeLanney, L.E., Wiener, S.G., Irwin, I., Langston, J.W., 1988c. 5-Hydroxyindoleacetic acid in cerebrospinal fluid reflects serotonergic damage induced by 3,4-methylenedioxymethamphetamine in CNS of non-human primates. Brain Res. 474 (2), 359–363.

Ricaurte, G.A., Martello, A.L., Katz, J.L., Martello, M.B., 1992a. Lasting effects of (±)3,4-methylenedioxymethamphetamine on central serotonergic neurons in non-human primates: neurochemical observations. J. Pharmacol. Exp. Ther. 261, 616–622.

Ricaurte, G.A., McCann, U.D., 1992b. Neurotoxic amphetamine analogues: effects in monkeys and implications for humans. Ann. N.Y. Acad. Sci. 648, 371–382.

Ricaurte, G.A., Yuan, J., McCann, U.D., 2000. (±)3,4-Methylenedioxymethamphetamine ('Ecstasy')-induced serotonin neurotoxicity: studies in animals. Neuropsychobiology 42 (1), 5–10.

Ricaurte, G.A., McCann, U.D., 2005. Recognition and management of complications of new recreational drug use. Lancet 365 (9477), 2137–2145.

Rogers, G., Elston, J., Garside, R., Roome, C., Taylor, R., Younger, P., Zawada, A., Somerville, M., 2009. The harmful health effects of recreational ecstasy: a systematic review of observational evidence. Health Technol. Assess. 13 (6), iii–iv, ix–xii, 1–315.

Rothman, R.B., Jayanthi, S., Wang, X., Dersch, C.M., Cadet, J.L., Prisinzano, T., Rice, K.C., Baumann, M.H., 2003. High-dose fenfluramine administration decreases serotonin transporter binding, but not serotonin transporter protein levels, in rat forebrain. Synapse 50 (3), 233–239.

Rothman, R.B., Jayanthi, S., Cadet, J.L., Wang, X., Dersch, C.M., Baumann, M.H., 2004. Substituted amphetamines that produce long-term serotonin depletion in rat brain amphetamines that produce long-term serotonin depletion in rat brain ("neurotoxicity") do not decrease serotonin transporter protein expression. Ann. N.Y. Acad. Sci. 1025, 151–161.

Rowland, N.E., Kalehua, A.N., Li, B.H., Semple-Rowland, S.L., Streit, W.J., 1993. Loss of serotonin uptake sites and immunoreactivity in rat cortex after dexfenfluramine occur without parallel glial cell reaction. Brain Res. 624, 35–43.

Rudnick, G., Wall, S.C., 1992. The molecular mechanism of "ecstasy" [3,4-methylenedioxy-methamphetamine (MDMA)]: serotonin transporters are targets for MDMA-induced serotonin release. Proc. Natl. Acad. Sci. U.S.A 89 (5), 1817–1821.

Rusyniak, D.E., Zaretskaia, M.V., Zaretsky, D.V., DiMicco, J.A., 2008. Microinjection of muscimol into the dorsomedial hypothalamus suppresses MDMA-evoked sympathetic and behavioral responses. Brain Res. 1226, 116–123.

Sadzot, B., Baraban, J.M., Glennon, R.A., Lyon, R.A., Leonhardt, S., Jan, C.R., Titeler, M., 1989. Hallucinogenic drug interactions at human brain 5-HT2 receptors: implications for treating LSD-induced hallucinogenesis. Psychopharmacology (Berl.) 98 (4), 495–499.

Sanchez, V., O'Shea, E., Saadat, K.S., Elliott, J.M., Colado, M.I., Green, A.R., 2004. Effect of repeated ('binge') dosing of MDMA to rats housed at normal and high temperature on neurotoxic damage to cerebral 5-HT and dopamine neurones. J. Psychopharmacol. 18 (3), 412–416.

Scanzello, C.R., Hatzidimitriou, G., Martello, A.L., Katz, J.L., Ricaurte, G.A., 1993. Serotonergic recovery after (±)3,4-(methylenedioxy) methamphetamine injury: observations in rats. J. Pharmacol. Exp. Ther. 264 (3), 1484–1491.

Schmidt, C.J., 1987. Neurotoxicity of the psychedelic amphetamine, methylenedioxymethamphetamine. J. Pharmacol. Exp. Ther. 240, 1–7.

Schmidt, C.J., Wu, L., Lovenberg, W., 1985. Methylenedioxymethamphetamine: a potentially neurotoxic amphetamine analogue. Eur. J. Pharmacol. 124, 175–178.

Schmidt, C.J., Levin, J.A., Lovenberg, W., 1987. In vitro and in vivo neurochemical effects of methylenedioxymethamphetamine on striatal monoaminergic systems in the rat brain. Biochem. Pharmacol. 36 (5), 747–755.

Selvaraj, S., Hoshi, R., Bhagwagar, Z., Murthy, N.V., Hinz, R., Cowen, P., Curran, H.V., Grasby, P., 2009. Brain serotonin transporter binding in former users of MDMA ('ecstasy'). Br. J. Psychiatry 194 (4), 355–359.

Semple, D.M., Ebmeier, K.P., Glabus, M.F., O'Carroll, R.E., Johnstone, E.C., 1999. Reduced in vivo binding to the serotonin transporter in the cerebral cortex of MDMA ('ecstasy') users. Br. J. Psychiatry 175, 63–69.

Shulgin, A., 1980. History of MDMA. In: Peroutka, S. (Ed.), Ecstasy: the Clinical, Psychological and Neurotoxicological Effects of the Drug MDMA. Kluwer Academic Press, Norwell, MA.

Shulgin, A.T., Nichols, D.E., 1978. Characterization of three new psychotomimetics. In: Stillman, R., Willette, R. (Eds.), The Psychopharmacology of Hallucinogens. Pergamon Press, New York, pp. 74–83.

Slikker, W., Al, i, S.F., Scallet, A.C., Frith, C.H., Newport, G.D., Bailey, J.R., 1988. Neurochemical and neurohistological alterations in the rat and monkey produced by orally administered methylenedioxymethamphetamine (MDMA). Toxicol. Appl. Pharmacol. 94, 448B57.

Slikker Jr., W., Holson, R.R., Ali, S.F., Kolta, M.G., Paule, M.G., Scallet, A.C., McMillan, D.E., Bailey, J.R., Hong, J.S., Scalzo, F.M., 1989. Behavioral and neurochemical effects of orally administered MDMA in the rodent and nonhuman primate. Neurotoxicology 10 (3), 529–542.

Steele, T.D., Nichols, D.E., Yim, G.K., 1987. Stereochemical effects of 3,4-methylenedioxymethamphetamine (MDMA) and related amphetamine derivatives on inhibition of uptake of [3H]monoamines into synaptosomes from different regions of rat brain. Biochem. Pharmacol. 36 (14), 2297–2303.

Steele, T.D., McCann, U.D., Ricaurte, G.A., 1994. 3,4-Methylenedioxymethamphetamine (MDMA, "Ecstasy"): pharmacology and toxicology in animals and humans. Br. J. Addict. 89 (5), 539–551.

Stone, D.M., Stahl, D.S., Hanson, G.L., Gibb, J.W., 1986. The effects of 3,4-methylenedioxymethamphetamine (MDMA) and 3,4-methylenedioxyamphetamine on monoaminergic systems in the rat brain. Eur. J. Pharmacol. 128, 41–48.

Stone, D.M., Hanson, G.R., Gibb, J.W., 1987. Differences in the central serotonergic effects of methylenedioxymethamphetamine (MDMA) in mice and rats. Neuropharmacology 26 (11), 1657–1661.

Straiko, M.M., Coolen, L.M., Zemlan, F.P., Gudelsky, G.A., 2007. The effect of amphetamine analogs on cleaved microtubule-associated protein-tau formation in the rat brain. Neuroscience 144 (1), 223–231.

Sumnall, H.R., Cole, J.C., 2005. Self-reported depressive symptomatology in community samples of polysubstance misusers who report ecstasy use: a meta-analysis. J. Psychopharmacol. 19, 89–97.

Teitler, M., Leonhardt, S., Appel, N.M., De Souza, E.B., Glennon, R.A., 1990. Receptor pharmacology of MDMA and related hallucinogens. Ann. N.Y. Acad. Sci. 600, 626–638.

Thomas, D.M., Dowgiert, J., Geddes, T.J., Francescutti-Verbeem, D., Liu, X., Kuhn, D.M., 2004. Microglial activation is a pharmacologically specific marker for the neurotoxic amphetamines. Neurosci. Lett. 367, 349–354.

Thomasius, R., Zapletalova, P., Petersen, K., Buchert, R., Andresen, B., Wartberg, L., Nebeling, B., Schmoldt, A., 2006. Mood, cognition and serotonin transporter availability in current and former ecstasy (MDMA) users: the longitudinal perspective. J. Psychopharmacol. 20 (2), 211–225.

Tiede, W., Magerl, W., Baumgartner, U., Durrer, B., Ehlert, U., Treede, R.D., 2010. Sleep restriction attenuates amplitudes and attentional modulation of pain-related evoked potentials, but augments pain ratings in healthy volunteers. Pain 148, 36–42.

Urban, N.B., Girgis, R.R., Talbot, P.S., Kegeles, L.S., Xu, X., Frankle, W.G., Hart, C.L., Slifstein, M., Abi-Dargham, A., Laruelle, M., 2012. Sustained recreational use of ecstasy is associated with altered pre- and postsynaptic markers of serotonin transmission in neocortical areas: a PET study with [(11)C]DASB and [(11)C]MDL 100907. Neuropsychopharmacology. http://dx.doi.org/10.1038/npp.2011.332 (Epub ahead of print).

Verrico, C.D., Miller, G.M., Madras, B.K., 2007. MDMA (ecstasy) and human dopamine, norepinephrine, and serotonin transporters: implications for MDMA-induced neurotoxicity and treatment. Psychopharmacology (Berl.) 189 (4), 489–503.

Vollenweider, F.X., Gamma, A., Liechti, M., Huber, T., 1998. Psychological and cardiovascular effects and short-term sequelae of MDMA ("ecstasy") in MDMA-naïve healthy volunteers. Neuropsychopharmacology 19 (4), 241–251.

Wang, S.S., Ricaurte, G.A., Peroutka, S.J., 1987. [3H]3,4-methylenedioxymethamphetamine (MDMA) interactions with brain membranes and glass fiber filter paper. Eur. J. Pharmacol. 138 (3), 439–443.

Wang, X., Baumann, M.H., Xu, H., Rothman, R.B., 2004. 3,4-methylenedioxymethamphetamine (MDMA) administration to rats decreases brain tissue serotonin but not serotonin transporter protein and glial fibrillary acidic protein. Synapse 53 (4), 240–248.

Wang, X., Baumann, M.H., Xu, H., Morales, M., Rothman, R.B., 2005. (±)-3,4-Methylenedioxymethamphetamine administration to rats does not decrease levels of the serotonin transporter protein or alter its distribution between endosomes and the plasma membrane. J. Pharmacol. Exp. Ther. 314 (3), 1002–1012.

Wareing, M., Fisk, J.E., Murphy, P., Montgomery, C., 2004. Verbal working memory deficits in current and previous users of MDMA. Hum. Psychopharmacol. 19 (4), 225–234.

Watson, L., Beck, J., 1991. New age seekers: MDMA use as an adjunct to spiritual pursuit. J. Psychoactive Drugs 23, 261–270.

Wilson, M.A., Ricaurte, G.A., Molliver, M.E., 1989. Distinct morphologic classes of serotonergic axons in primates exhibit differential vulnerability to the psychotropic drug 3,4-methylenedioxymethamphetamine. J. Neurosci. 28 (1), 121–137.

Wilson, M.A., Mamounas, L.A., Fasman, K.H., Axt, K.J., Molliver, M.E., 1993. Reactions of 5-HT neurons to drugs of abuse: neurotoxicity and plasticity. NIDA Res. Monogr. 136, 155–187.

Xie, T., Tong, L., McLane, M.W., Hatzidimitriou, G., Yuan, J., McCann, U., Ricaurte, G., 2006. Loss of serotonin transporter protein after MDMA and other ring-substituted amphetamines. Neuropsychopharmacology 12, 2639–2651 (Erratum in Neuropsychopharmacology (2008), 33 (3), pp. 712–713).

Yoshimura, M., Furue, H., 2006. Mechanisms for the anti-nociceptive actions of the descending noradren-ergic and serotonergic systems in the spinal cord. J. Pharmacol. Sci. 101, 107–117.

Yuan, J., Cord, B.J., McCann, U.D., Callahan, B.T., Ricaurte, G.A., 2002. Effect of depleting vesicular and cytoplasmic dopamine on methylenedioxymethamphetamine neurotoxicity. J. Neurochem. 80 (6), 960–969.

Zaczek, R., Hurt, S., Culp, S., De Souza, E.B., 1989. Characterization of brain interactions with methylene-dioxyamphetamine and methylenedioxymethamphetamine. NIDA Res. Monogr. 94, 223–239.

Zhou, F.C., Azmitia, E.C., 1986. Induced homotypic sprouting of serotonergic fibers in hippocampus. II. An immunocytochemistry study. Brain Res. 373 (1–2), 337–348.

CHAPTER SIXTEEN

Sedative Hypnotics

Domenic A. Ciraulo and Mark Oldham
Department of Psychiatry, Boston Medical Center, Boston University School of Medicine, Boston, MA, USA

The modern history of sedative hypnotics began with barbital, which was introduced in 1903, and later phenobarbital in 1912. In the mid-1930s, Sternbach synthesized several heptoxdiazines, although it wasn't until 1955 when one of these quinazolines was treated with methylamine that an active compound was developed. In 1957, the compound chlordiazepoxide was found to have hypnotic, sedative, and muscle-relaxing effects. The benzodiazepines provided advantages over the older barbiturates because they were less toxic in overdose and had fewer drug interactions. In addition, they had superior efficacy and safety compared to meprobamate, which was introduced as a tranquilizer in 1955. Nonbenzodiazepine hypnotics have been the most recent addition and may offer some advantages of lower risk tolerance and perhaps lower abuse liability although clinical experience is insufficient to make definitive judgments.

Almost all sedative-hypnotics in clinical use exert their primary pharmacodynamics effects in the GABA system, the major inhibitory system in the CNS. Recent studies have revealed important differences in the activity on GABA and the glutamate system among various agents. GABA-A receptors are pentameric ligand-gated ion channels, which upon activation lead to an influx of chloride ions. The subunit composition of these receptors, in conjunction with their precise location in the brain, determines their function and activity. There are now 21 different types of cloned GABA-A subunits, which are divided into eight distinct families. Subunit isoforms within a single class share approximately 70% of their sequence but between classes this drops to about 30–40%. The subunits are classified into α (alpha), β (beta), γ (gamma), δ (delta), ε (epsilon), π (pi), ω (omega), and ϱ (rho). In addition, spliced variants of several subunits have been reported. Clinically, the alpha, beta, delta, and gamma subunit families are the most important subunit combinations for drug action. The subunit composition of GABA-A receptors in the brain most frequently consists of two alpha-1 subunits and two beta-2 subunits that exist in association with a gamma or delta subunit. Receptors with the alpha-1, beta-2, and gamma-2 combination occur with the greatest frequency. GABA has a direct action on the receptor to open the chloride channel; GABA binds at the interface between the alpha and beta subunits. Benzodiazepines bind at the interface of alpha and gamma subunits. Whereas the binding of GABA to the receptor has a direct effect, benzodiazepine binding produces an allosteric modification in the GABA-binding site, making the receptor more sensitive to GABA. Zolpidem binds preferentially to

The Effects of Drug Abuse on the Human Nervous System
http://dx.doi.org/10.1016/B978-0-12-418679-8.00016-2

GABA-A receptors containing α_1 subunits, exhibiting over 20-times the binding affinity for alpha-1-containing GABA-A receptors than those containing alpha-2 or alpha-3 subunits and negligible affinity for those with α_5 subunits. Zaleplon may exhibit slightly less selectivity for alpha-1-containing receptors although murine studies do suggest relative binding preference for alpha-1-containing GABA-A receptors relative to alpha-2 or alpha-3 subtypess. Eszopiclone is the least selective of the Z-drugs with relative binding affinities to alpha-1-, alpha-2-, alpha-3-, and alpha-5-containing GABA-A receptors being 8:5:1:8.

Pentobarbital has greatest efficacy on GABA-A receptors containing the alpha-6 and beta-1 subunits. Data from molecular genetics have indicated that specific alpha subunits are associated with different clinical effects of benzodiazepines. Drug development that leads to more specific agents holds great promise for the discovery of antianxiety agents without such adverse effects as daytime sedation, memory impairment, and abuse potential.

GABA-A receptors are actively changing in response to the environment. Drugs acting on the receptor or other factors influence the activity of the GABA-A system by endocytosis from the plasma membrane, internalization which may lead to degradation, altered subunit composition, or posttranslational mechanisms such as phosphorylation.

In outpatient practice, benzodiazepines are approved for the short-term treatment of anxiety and insomnia. Of the currently marketed benzodiazepines, the high potency agents, such as alprazolam, clonazepam, and lorazepam, are probably more effective for panic disorder than other agents in the class. In conditions other than panic disorder, choice of benzodiazepine generally depends on pharmacokinetic characteristics of a particular agent. Clinicians should consider onset of effect, duration of action (including metabolites), and potential drug–drug interactions.

The short-term adverse effects of the benzodiazepines include excessive sedation, psychomotor impairment, learning, and memory disturbances. At high therapeutic doses that are often used to sedate agitated patients, slurred speech, ataxia, and impaired gag reflex may occur. The long-term adverse effects have generated more concern. Some relate to the development of tolerance, which can lead to escalating doses and a withdrawal syndrome upon abrupt discontinuation. Others relate to delayed and persistent effects of the drugs. For instance, late onset of cognitive decline has been suggested by epidemiological studies; however, it is unclear whether confounding by indication may account for a large portion of this finding. Short-term neurocognitive impairment related to benzodiazepine use may improve over time (i.e. tolerance may develop to a certain degree), but it likely never fully abates with chronic use. The risk of persistent psychomotor impairment has been associated with falls in the elderlyas well as traffic accidents—both of which are being increasingly recognized in relation to the Z-drugs.

Research evaluating abuse liability of sedative-hypnotics has used several predictive models. Animal models in rodents and primates include self-administration, reinstatement, place preference, brain stimulation, and reward. With the possible exception of flunitrazepam, benzodiazepines have lower abuse potential than barbiturates.

The use of benzodiazepines for their hedonic value is rare (however, flunitrazepam has been used surreptitiously in alcohol to impair memory and consciousness in sexual assaults). Data from the latest National Household Survey of Drug Use and Health show that nonmedical use of prescription pain relievers is more common than illicit use of "tranquilizers" or "sedatives" The Drug Abuse Warning Network (SAMHSA, 2011) reports drug-related problems from data reported by Emergency Departments (EDs) found that 33.6% of ED visits involving nonmedical use of pharmaceuticals in 2009 (the most recent report) involved sedative hypnotics, usually in combination with other substances.

In clinical practice, the use of medications to treat anxiety has increased, but benzodiazepine use has decreased as antidepressants have become first line agents in the treatment of anxiety. Similarly, the advent of the Z-drugs has reduced the number of prescriptions for benzodiazepine hypnotics. Several surveys of clinical practice indicate that most patients take benzodiazepines for less than one month with only about 1% taking them for longer than one year. Rates vary based on methodology and country studied. For example, 6-month use has been reported as 3% and 1-year use as 1.7%.

Since the advent of the benzodiazepines, barbiturate use has been limited in modern medicine. Phenobarbital is still prescribed as an anticonvulsant and as a sedative, especially for children. It is also a common component of many combination products, such as for headache and gastrointestinal distress, and reduces the stimulating effects of sympathomimetic agents. Butalbital is an intermediate-acting barbiturate found in a widely used combination product that also contains acetaminophen and caffeine (Fioricet) and is approved for the treatment of tension-type headaches. A major disadvantage of the use of the barbiturates is the development of pharmacokinetic and pharmacodynamic tolerance.

Abstinence syndromes associated with sedative hypnotics vary in severity depending on class of drug and pharmacokinetic characteristics. This chapter discusses common symptoms of the withdrawal syndrome and pharmacological management.

1. INTRODUCTION

1.1. Pharmacology of Sedative Hypnotics

The sedative hypnotic drugs represent a variety of chemical agents that depress central nervous system activity. They have many clinical uses in medicine including treatment of anxiety and insomnia, epilepsy, as muscle relaxants, and for induction of anesthesia. Some members of the class also possess significant abuse potential and a hazard of overdose fatalities, especially when combined with alcohol or other drugs. The focus of this chapter will be on benzodiazepines and nonbenzodiazepine hypnotics (colloquially called the

Z-drugs). The 1,4-benzodiazepine nucleus is the basic structure of the class and is shown in Figure 1. Substitutions to the ring and its constituents alter the potency, clinical efficacy, and adverse effects of individual agents. The most common benzodiazepine structures are shown in Figure 2. While there may be some variations in the classification schemes of this drug group, the following are commonly used: (1) the triazolo group: alprazolam, triazolam, estazolam; (2) the 2-keto group: diazepam, flurazepam, and clorazepate; (3) the 2-amino group: chlordiazepoxide; (4) the 3-hydroxy group: lorazepam, oxazepam, and temazepam; (5) the trifluoroethyl group: quazepam; (6) the imidazo group: midazolam, and (7) the 7-nitro group: nitrazepam and clonazepam.

A group of drugs, referred to as Z-drugs, are nonbenzodiazepine hypnotics: (1) zopiclone, a cyclopyrrolone; (2) eszopiclone, a steroselective isomer of zopiclone; (3) zaleplon, a pyrazolopyrimidine; and (4) zolpidem, an imidazopyridine. These will be discussed in greater detail later in the chapter.

The use of barbiturates has become more limited with the introduction of the benzodiazepines, which have lower risk of death in overdose, fewer drug–drug interactions, and less risk for addiction.

2.1. HISTORICAL PERSPECTIVE

The history of sedatives, hypnotics, and anxiolytics is replete with agents possessing low therapeutic indices, unacceptably high addiction potential, or both. From time immemorial, cultures have celebrated the anxiolytic, sedative, and hypnotic properties of

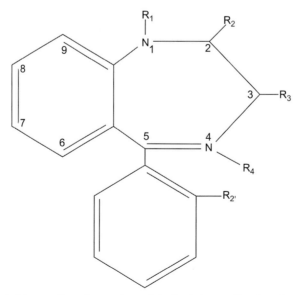

Figure 1 The 1, 4-benzodiazepine nucleus of the basic benzodizepine class of molecules.

ethanol. The dose-dependent nature of ethanol's effects is abundantly well-demonstrated in nearly any emergency room in that, with increasing blood alcohol levels, patients progress from alertness to somnolence, sleep, obtundation, cardiopulmonary suppression, and even potentially to the point of death. Despite its predictable ability to induce sleep, ethanol is a poor choice for a hypnotic agent given its relatively short duration of effects and zero-order kinetics, interference with sleep architecture, addiction potential, potential for causing life-threatening respiratory suppression and/or withdrawal (delirium tremens), hangover effects, and a host of medical (e.g. cancer, cirrhosis, dilated cardio-myopathy) and mental health (e.g. depression, suicide, dementia) complications (Arnedt et al., 2011; McKinney and Coyle, 2004).

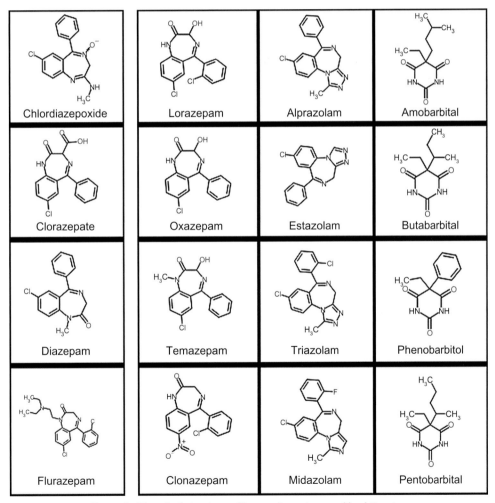

Figure 2 Chemical structure of benzodiazepines and barbiturates.

The twenty-first century has witnessed a series of medications aimed at inducing sleep without the unwanted (and potentially fatal) effects of ethanol (Matthew et al., 1972). Several analogs of ethanol (e.g. chloral hydrate, chlorobutanol, ethchlorvynol, and paraldehyde) as well as piperidinedione derivatives (e.g. glutethimide and methyprylon) had low therapeutic indices (McBay, 1973). Various nostrums containing bromide, most commonly potassium bromide, were prone to causing bromism, characterized by excessive sedation, seizures, and delirium (Noyes and Kolb, 1968). Barituric acid was first prepared in 1864 by von Baeyer Harvey (1975). Although not a central depressant itself, several of its derivatives have been used in medicine since the early 1900s. Barbital was introduced in 1903 and phenobarbital in 1912 (Harvey, 1975). Barbituric acid, named for the day on which it was discovered (Saint Barbara's Day), and its derivatives long held a place of prominence in the management of psychological and physical unrest, although the prevalence of their prescription combined with their potent ability to cause fatal respiratory suppression once made them the most common class of medications used by patients to commit suicide (Smith, 1972).

Reserpine, a derivative of *Rauwolfia*, gained limited traction in the 1950s as a weak antipsychotic and subsequently as an anxiolytic, but it was plagued with concerns of fatal cardiac arrhythmias and suicidality. In the 1950s, meprobamate, a carbamate derivative, with weak efficacy as an antianxiety agent was introduced. At clinical doses in humans, it has minimal muscle-relaxant effects, but it may have a mild analgesic effect in musculoskeletal pain and may potentiate analgesics. A discontinuance syndrome occurs after several weeks of treatment with a daily dose of 2,400 mg, and mild symptoms may be seen with long-term therapy at doses of 1,600 mg daily. Seizures may be common in withdrawal from meprobamate, and reports published in the 1970s suggest that serious withdrawal symptoms are more common with meprobamate than with barbiturates. Fatalities from overdose have been reported after doses as low as 12 g. Meprobamate induces microsomal enzymes and may exacerbate intermittent porphyria. Carisoprodol, currently available as a muscle relaxant, is metabolized to meprobamate, hydroxyl meprobamate, and hydroxyl carisoprodol and has been associated with abuse, dependence, and impairment of driving ability (Ciraulo and Sarid-Segal, 2009).

With respect to benzodiazepines, in the mid-1930s, Sternbach synthesized several heptoxdiazines, although it wasn't until 1955 when one of these quinazolines was treated with methylamine, that an active compound was developed (Greenblatt and Shader, 1974). In 1957, the compound chlordiazepoxide was found to have hypnotic, sedative, and muscle-relaxing effects, and it would become the first commercially available benzodiazepine (López-Muñoz, 2011). The benzodiazepines provided advantages over the older barbiturates because they were less toxic in overdose and had fewer drug interactions. In addition, they had superior efficacy and safety compared to meprobamate, which was introduced as a tranquilizer in 1955 (Harvey, 1975). Nonbenzodiazepine hypnotics have been the most recent addition and may offer some advantages of lower abuse liability, though clinical experience is insufficient to make definitive judgments.

In the 1990s, selective serotonin reuptake inhibitors (SSRIs) became first-line for most anxiety disorders; however, SSRIs lack sedative and hypnotic properties. Over the past several decades, other medications have vied with benzodiazepines for use as sedatives or hypnotics including those with antihistaminergic properties (e.g. diphenhydramine, hydroxyzine, doxylamine, mirtazapine), mixed-action agents (e.g. trazodone), melatonin agonists (e.g. melatonin, ramelteon), and several over-the-counter nutraceuticals (e.g. valerian root, kava, hops, passionflower, lemon balm, skullcap, catnip, lavender, linden flower, German chamomile, or rose hips) (Najib, 2006). Each of these agents, though, has properties that render them suboptimal in the management of insomnia—antihistamines often also possess anticholinergic side effects and are associated with weight gain; trazodone is ineffective for many people; melatonin agonists have limited hypnotic properties, and nutraceuticals are neither sufficiently studied nor currently regulated by the U.S. Food and Drug Administration (FDA).

In 2010, the U.S. FDA approved a novel low dose of doxepin, a tricyclic agent that has been labeled as an antidepressant for decades, for the management of maintenance insomnia. Also in 2010 the European Medicines Agency—the FDA analog for the European Union—approved an extended-release version of melatonin under the brand name Circadin® for insomnia for use for up to three months. In addition, evidence continues to grow for a class of investigational agents known as dual orexin receptor antagonists (DORAs) for the management of insomnia. Orexin (alternatively known as hypocretin) is secreted by the hypothalamus and plays a role in regulating sleep–wake cycles. Further, low levels of orexin in the cerebrospinal fluid have been implicated in both the pathogenesis and, hence, diagnosis of narcolepsy (APA, 2013). Notably, a U.S. FDA Advisory Committee recently found the DORA suvorexant efficacious for the management of insomnia at doses that do not affect cognition; however, a vote regarding formal approval by the U.S. FDA is expected in mid-2013 (Uslaner, 2013).

2.2. MECHANISM OF ACTION OF SEDATIVE HYPNOTICS

The GABA system is the predominant inhibitory neurotransmitter system in the CNS. Almost all sedative-hypnotics in clinical use exert their primary pharmacodynamics effects in the GABA system, although recent studies have revealed important differences in the activity on GABA and the glutamate system among various agents (Rudolph and Knoflach, 2011; Vithlani et al., 2011).

Most neurons of the system are interneurons producing widespread inhibitory actions in brain neurocircuits. Two receptor types, GABA-A and GABA-B, mediate this action, via two different mechanisms. Benzodiazepines and other sedative hypnotics exert their pharmacologic action at the GABA-A receptors, which are members of the superfamily of rapid acting Cys-loop ligand gated ion channels. Baclofen acts on the metabotropic GABA-B receptors linked to G proteins, which modulate either inhibit calcium or activate potassium channels to exert a slower, more prolonged response.

The structure of the GABA-A receptor is shown in Figure 3. While our knowledge of its basic structure and function remains much as described in our last review of the topic (Ciraulo and Knapp, 2009), recent evidence has expanded the concept of activity-related changes in the receptor. GABA-A receptors are pentameric ligand-gated ion channels, which upon activation, lead to an influx of chloride ions. The subunit composition of these receptors, in conjunction with their precise location in the brain determines their function and activity. There are now 21 different types of cloned GABA-A subunits, which are divided into eight distinct families. Subunit isoforms within a single class share approximately 70% of their sequence but between classes this drops to about 30–40%. The subunits are classified into alpha, beta, gamma, delta, epsilon, pi, omega, and rho. In addition, spliced variants several subunits have been reported, complicating interpretation of function. For the purposes of examining sedative hypnotic drugs actions, the alpha, beta, delta, and gamma families are the most important subunit combinations for drug action. The subunit composition of GABA-A receptors in the brain most frequently consists of two alpha-1 subunits and two beta-2 subunits that exist in association with a gamma or delta subunit. Receptors with the alpha-1, beta-2, and gamma-2 combination occur with the greatest frequency and the brain (Bergmann, 2013).

GABA has a direct action on the receptor to open the chloride channel; GABA binds at the interface between the alpha and beta subunits. Benzodiazepines bind at the interface of alpha and gamma subunits. While the binding of GABA to the receptor has a direct effect, benzodiazepine binding produces an allosteric modification in the

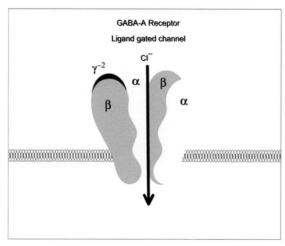

Figure 3 Diagram of the GABA receptor.

GABA-binding site, making the receptor more sensitive to GABA. Barbiturates at low doses produce allosteric modulation similar to benzodiazepines, however at higher doses, barbiturates have a direct action on enhancing channel opening. Barbiturates may also act by reducing excitatory neurotransmission through AMPA inhibition and blockage of voltage-gated calcium channels. GABA-A receptors that are sensitive to the benzo-diazepines contain alpha subunits 1, 2, 3, or 5, combined with two beta subunits and a gamma subunit. Receptors with alpha four and alpha six appear to be insensitive to the action of benzodiazepines. The alpha four and alpha six subunits, in combination with delta subunits are localized primarily in extrasynaptic regions of the neuron and most likely regulate tonic inhibitory, as opposed to the synaptic phasic inhibitory effects of GABA. Substantial evidence supports that tonic currents play a major role in controlling neuronal excitability.

Zolpidem binds preferentially to GABA-A receptors containing alpha-1 subunits (Pritchett and Seeburg, 1990; Hadingham et al., 1993; Atack, 2003), exhibiting over 20-times the binding affinity for alpha-1-containing GABA-A receptors than those con-taining alpha-2 or three subunits and negligible affinity for those with alpha-5 subunits (Nutt and Stahl, 2010). Zaleplon may exhibit slightly less selectivity for α_1-containing receptors (Noguchi et al., 2002) although murine studies do suggest relative binding preference for alpha-1-containing GABA-A receptors relative to alpha-2 or alpha-3 subtypes (Nutt and Stahl, 2010). Eszopiclone is the least selective of the Z-drugs with relative binding affinities to alpha-1, 2, 3, or 4-containing GABA-A receptors being 8:5:1:8 (Brunello et al., 2009). When accounting for its relative potency and efficacy at each of these four receptor subtypes, though, it appears to exert effects primarily via alpha-2 and 3-containing receptors (Nutt and Stahl, 2010), suggesting clinically relevant anxiolytic effects (An et al., 2008). Investigations on (S)-desmethylzopiclone, the prin-cipal metabolite of eszopiclone, have suggested that it demonstrates nonselective partial agonism at alpha-1, 2, and 3-containing GABA-A receptors with potential anxiolytic effects of its own (Fleck, 2002).

Specific subunit composition of GABA-A receptors has several intriguing clinical implications for pharmacology. First, subunit composition alters drug efficacy, so the presence of alpha four and alpha six subunits, or elimination of the gamma two subunit, render the receptor insensitive to benzodiazepines. On the other hand, pentobarbital has greatest efficacy in GABA-A receptors containing the alpha six and beta one sub-units. Second, data from molecular genetics have indicated that specific alpha subunits are associated with different clinical effects of benzodiazepines (Collinson et al., 2002; Kralic et al., 2002). Although considerable overlap exists (Rudolph and Knoflach, 2011; Vithlani et al., 2011), Table 1 below indicates the pharmacodynamics associated with specific alpha subunits. Drug development that leads to more specific agents holds great promise for the discovery of antianxiety agents without such adverse effects as daytime sedation, memory impairment, and abuse potential. It was suggested that the absence of

Table 1 Pharmacologic Effects

Alpha Subunit	Pharmacologic Effects
Alpha-1	Sedative, amnestic, anticonvulsant, addiction
Alpha-2	Anxiolytic, muscle relaxation, cognitive, antidepressant, analgesic
Alpha-3	Anxiolytic, muscle relaxation
Alpha-5	Learning and memory, muscle relaxation

abuse liability and dependence of TPA023 in baboons is due to lack of efficacy at alpha-1 and decreased efficacy at apha-2 and 3. This line of research may lead to novel analgesics, cognitive enhancers, and anticonvulsants (Greenfield, 2013).

Other clinical implications come from recent studies of the GABA-A receptor that indicate that mutations have been found in alpha-1 and gamma-2 subunits in epilepsy, SNP's in the gene encoding for the alpha-2 subunit in alcoholism and drug addiction, and other genes encoding for subunits of the GABA-A receptor linked to alcoholism, bipolar disorder and schizophrenia.

It is now known that GABA-A receptors are actively changing in response to the environment. Drugs acting on the receptor or other factors influence the activity of the GABA-A system by endocytosis from the plasma membrane, internalization which may lead to degradation, altered subunit composition, or posttranslational mechanisms such as phosphorylation. This has been proposed a mechanism of tolerance, although glutamatergic and serotonergic 5-HT 2A receptors have been implicated as contributing to the withdrawal syndrome (Benyamina et al., 2012). A recent primate study investigating midazolam and the GABA-A agonist pregnanolone (which acts outside benzodiazepine receptor binding sites) suggested that acute tolerance to benzodiazepines does not involve modification of the benzodiazepine binding site because flumazenil inhibited the effects of midazolam but not pregnanolone in acutely benzodiazepine-tolerant monkeys (Zanettini et al. 2013). Ongoing research continues to define the roles of GABA-A subunits to tolerance and clinic effect. For instance, chronic activation of alpha-2 or 3-containing GABA-A receptors appears not to produce pharmacodynamic tolerance, which is unlike the tolerance that occurs after persistent activation of alpha-1-containing GABA-A receptors (Vinkers et al., 2012).

2.3. CLINICAL USE OF BENZODIAZEPINES
2.3.1. Clinical Indications

Benzodiazepines are used as antianxiety agents, hypnotics, adjunctive medications in some seizure disorders, and off-label to potentiate antidepressant activity of SSRI. Table 2 lists the FDA approved indications of the most commonly used agents and drug characteristics that should be considered in addition to effects common to the class.

Table 2 FDA Approved Indications

Benzodiazepine	Indications	Comments
Alprazolam	Anxiety disorders, short-term management of anxiety symptoms, panic disorder, with or without agoraphobia	Use caution or avoid use with potent CYP3A inhibitors, such azole antifungals. Inducers include carbamazepine, propoxyphene, smoking
Clonazepam	Adjunct or monotherapy in Lennox–Gastaut syndrome, akinetic and myoclonic seizures, absence seizures. Panic disorder with or without agoraphobia	Myoclonic jerks common in withdrawal from clinical doses. Withdrawal seizures may appear later than expected based on elimination half-life. Affected by CYP3A4 induction or inhibition
Clorazepate	Anxiety disorders, short-term management of anxiety symptoms. Adjunctive management of partial seizures. Symptomatic relief of alcohol withdrawal (authors do not recommend use for alcohol withdrawal)	Prodrug for desmethyldiazepam. Metabolite rapidly absorbed in absence of altered gut acidity (as with co-administered antacids)
Chlordiazepoxide	Anxiety disorders, short-term management of anxiety symptoms, alcohol withdrawal, preoperative anxiety and apprehension	Limbitrol (combined with amitriptyline) for anxious depression, Librax (combined with clidinium bromide) for irritable bowel syndrome, acute enterocolitis, peptic ulcer
Diazepam	Oral preparation for short-term management of anxiety disorders, situational anxiety, alcohol withdrawal symptoms, adjunct for medical diagnostic/surgical procedures, skeletal muscle spasms, athetosis, stiff-man syndrome, tetanus. Adjunctive therapy in status epilepticus, severe recurrent seizures	Diazepam has a rapid onset of action and a euphoric effect in some individuals. As diazepam levels decline its metabolite desmethyldiazepam levels increase, providing long actin anxiolysis without a euphoric mood effect. Rectal preparation for seizure disorders
Lorazepam	Oral dosing used for anxiety, anxiety with depression, off-label as hypnotic and for panic disorder. IV formulation used for alcohol delirium tremens, seizures, status epilepticus, preanesthetic	Toxicity associated with vehicle with high doses of parenteral preparation. Major elimination via glucuronidation making it appropriate for patients with impaired hepatic function.

Continued

Table 2 FDA Approved Indications—cont'd

Benzodiazepine	Indications	Comments
Midazolam	IM or IV for preoperative sedation. Continuous IV for sedation of intubated or mechanically ventilated patients	Use only in closely monitored situations to avoid respiratory arrest or other adverse effects. Oral syrup formulation for pediatric use. Reported sex differences in plasma levels are not due to metabolic differences, but rather P-glycoprotein differences between men and women.
Oxazepam	Anxiety, anxiety with depression, anxiety or agitation in older patients, anxiety in alcoholism or alcohol withdrawal	Advantage of glucuronidation avoiding effects of hepatic disease. In alcoholism there is disadvantage in alcohol withdrawal because of lack of parenteral formulation.
Estazolam	Short-term management of insomnia, initial, middle and late	Complex nocturnal behavior disorders, angioedema, amnesia, paresthesias, paradoxical excitement
Flurazepam	Short-term management of insomnia, initial, middle and late	Active metabolite may lead to anti-anxiety effects or psychomotor impairment the following day
Temazepam	Short-term management of insomnia, initial, middle and late	Possibly the preferred agent of the older benzodiazepine hypnotics. Current formulations provide rapid onset that undergoes glucuronidation limiting effects on the following day. Expect adverse effects similar to other hypnotics
Triazolam	Short-term management of insomnia, initial, middle and late	Subject to CYP3A4 interactions. Cognitive disturbances may be more severe than other agents. Avoid use unless well justified
Quazepam	Short-term management of insomnia, initial, middle and late	Avoid use. Inhibits CYP2B6 and could lead to seizures with bupropion

In outpatient practice, benzodiazepines are approved for the short-term treatment of anxiety and insomnia. Of the currently marketed benzodiazepines, the high potency agents, such as alprazolam, clonazepam, and lorazepam, are probably more effective for panic disorder than other agents in the class. Nevertheless, anecdotal evidence supports the efficacy for other agents such as diazepam for panic disorder. It is our opinion that benzodiazepines effectively reduce anticipatory and generalized anxiety, which in turn reduces panic attacks. In conditions other than panic disorder, choice of benzodiazepine depends on pharmacokinetic characteristics of a particular agent. Clinicians should consider onset of effect, duration of action (including metabolites), and potential drug–drug interactions. For example, for treatment of early insomnia, a drug with rapid onset is of critical importance. Duration of action should be considered for middle and late insomnia, as well as the need for antianxiety effects the following day or the need to avoid daytime sedative effects. In the United States, temazepam is widely used because of its rapid onset and intermediate duration of action that provides coverage for the entire night. Occasionally lorazepam is used "off-label" for insomnia. Tolerance to hypnotic effects is common for all drugs of the class requiring short term or intermittent use. (Insomnia is discussed in greater detail below). Cognitive behavioral therapy (without medication) is the preferred long-term approach for treatment of insomnia, although this is not always acceptable to patients.

2.3.2. Risks and Adverse Effects of Benzodiazepines

2.3.2.1. Adverse Effects

The short-term adverse effects of the benzodiazepines include excessive sedation, psychomotor impairment, learning and memory disturbances. At high therapeutic doses that are often used to sedate agitated patients, slurred speech, ataxia, and impaired gag reflex may occur. The long-term adverse effects have generated more concern. For instance, late onset of cognitive decline has been suggested by epidemiological studies (Coporaso, 2013); however, it is unclear whether confounding by indication may account for a large portion of this finding (i.e. benzodiazepines may be prescribed in aging populations preferentially to those with early neuropsychiatric manifestations of incipient neurocognitive disorders, which may present with clinically significant anxiety). Short-term neurocognitive impairment related to benzodiazepine use may improve over time (i.e. tolerance may develop to a certain degree), but it likely never fully abates with chronic use. The risk of persistent psychomotor impairment has been associated with falls in the elderly as well as traffic accidents—both of which are being increasingly recognized in relation to the Z-drugs (Gunja, 2013).

2.3.2.2. Cognitive Function

Despite the number of studies that have examined cognitive function in long-term benzodiazepine users, results have been contradictory. In contrast to the finding that acute benzodiazepine use induces anterograde amnesia (Vermeeren and Coenen, 2011), there is not a consensus on deficits with long-term use. A meta-analysis of 13 studies published between 1980 and 2000 found persistent deficits in long-term users compared to controls, especially in the areas of sensory processing, verbal reasoning, verbal memory, attention and concentration (Barker et al., 2004). One study (Paterniti et al., 2002) reported that in a sample of individuals aged 6070, followed for

Table 3 Benzodiazepine and Z-Drugs, Their Active Metabolites, Half-life of Parent Compound ($T_{1/2}$), and Major P450 Cytochromes Implicated in Hepatic Metabolism

Drug	$T_{1/2}$	Active Metabolites	P450 Cytochrome
Benzodiazepines			
Alprazolam	12–15	α- Hydroxy-alprazolam	CYP2A4 & A5
Clonazepam	19–60	Inactive only	CYP3A4
Chlordiazepoxide	10–30	Desmethylchlordiazepoxide, demoxepam, nordazepam, oxazepam	
Clorazepate	Prodrug	Nordazepam, oxazepam	
Diazepam	20–70	Nordazepam, oxazepam	CYP2C19, CYP3A4
Estazolam	16	1-Oxo-estazolam	
Flurazepam	74	N-Hydroxyethyl-flurezapam N-Desalkyflurazepam	
Lorazepam	10–20	Inactive only	
Midazolam	1–4	α-Hydroxy-midazolam	CYP3A4 & A5
Oxazepam	5–10	Inactive only	
Quazepam	39	2-Oxo-quazepam N-Desalkyflurazepam	CYP2C19, CYP3A4
Temazepam	10–15	Oxazepam	
Triazolam	2–4	α-Hydroxy-triazolam	CYP3A4
Z-Drugs			
Eszopiclone	6	(S)-N-Desmenthyl zopiclone	CYP3A4, CYP2E1
Zaleplon	1	Inactive only	CYP3A4
Zolpidem	2.1	Inactive only	CYP3A4, CYP1A2, CYP2C9

4 years, those taking benzodiazepines had a more rapid decline in cognitive function than those who were not taking benzodiazepines. Puustinen et al. (2007) found no difference in cognitive function in elderly long-term users of zopiclone, temazepam, and oxazepam, and a group of nonusers. Supporting these findings, is the study of (Leufkens et al., 2009) that found no difference in cognitive performance or driving ability between chronic hypnotic users and controls. Most studies have found that discontinuation of benzodiazepines results in gradual improvement in cognitive function; however, the extent of recovery and duration of time required to recover is not consistent in the literature (Salzman et al., 1992; Curran et al., 2003; Pat McAndrews et al., 2003; Barker et al., 2004).

A number of studies have examined brain structure in chronic benzodiazepine users (Lader, 2011). Several computerized axial tomography studies have produced contradictory findings—from decreased ventricle to brain ratio (but still in normal range) in chronic benzodiazepine users (Lader et al., 1984; Schmauss et al, 1987; Uhde and Kellner, 1987) to no differences from control subjects (Perera et al., 1987; Moodley

Table 4 Signs and Symptoms of Benzodiazepine Discontinuation Syndrome

The following signs and symptoms may be seen when benzodiazepine therapy is discontinued; they may indicate
- Return of the original anxiety symptoms (recurrence),
- Worsening of the original anxiety symptoms (rebound),
- Emergence of new symptoms (true withdrawal):

> *Early symptoms (onset depends on elimination half-life of drug)* Anxiety, apprehension, insomnia, tinnitus, dysphoria, pessimism, irritability, blurred vision, hyperacusis, obsessive rumination,
> *Late symptoms* Tachycardia, elevated blood pressure, hyperreflexia, muscle tension, agitation/motor restlessness, tremor, myoclonus, muscle and joint pain, nausea, coryza, diaphoresis, ataxia, illusions, depersonalization, hallucinations, and grand mal seizures.

et al., 1993; Busto et al., 2000). It is very difficult to control for alcohol use, anxiety, dose, duration, and type of benzodiazepine taken, which contribute to inconsistent findings.

2.3.2.3. Behavioral Toxicity

Approximately 3–6% of patients taking benzodiazepines develop irritability, aggression, confusion, and impulsivity (Hall and Zisook, 1981; Honan, 1994; Ben-Porath and Taylor, 2002). It was proposed that changes in GABA-A receptor activity, induced by altered subunit composition of the receptor, may be one of several mechanisms underlying this paradoxical response.

2.3.2.4. Withdrawal Syndrome

Hollister and colleagues were the first to report a withdrawal syndrome after abrupt discontinuation of prolonged high-dose administration of chlordiazepoxide (Hollister et al., 1961) and diazepam (Hollister et al., 1963). Note that the American Psychiatric Association distinguished the terms withdrawal/abstinence and discontinuation syndrome: withdrawal represents the panoply of effects after abrupt cessation of substances of abuse where discontinuation syndrome refers to the signs and symptoms that occur when a prescribed pharmaceutical is abruptly discontinued (as often accompanies paroxetine or venlafaxine discontinuation). Smith and Wesson (1983) were the first to categorize withdrawal symptoms into *minor*—anxiety, insomnia, and nightmares—or *major*—seizures, psychosis, hyperpyrexia, and possibly death.

The signs and symptoms of benzodiazepine discontinuation are listed in Table 4. The time course of withdrawal is related to the elimination half-life of the benzodiazepine and its metabolites. Although these data provide a guide to treatment of withdrawal, every patient is different and clinicians are encouraged to know the progression of minor to major symptoms originally described by Smith and Wesson (1983). The appearance of minor symptoms should prompt reevaluation of the rate of drug taper and may require

Table 5 Guidelines for the Treatment of Benzodiazepine Discontinuation Syndrome

- Determine required dose of benzodiazepine or barbiturate for stabilization, guided by history, clinical presentation, and challenge dose.
- Some clinicians recommend switching to longer-acting benzodiazepine for withdrawal (e.g. diazepam (valium), clonazepam (Klonopin)), others recommend stabilizing on the drug that the patient was taking or on phenobarbital.
- After stabilization, reduce dosage by 10–25% percent on the second or third day and evaluate the response, keeping in mind that symptoms that occur after decreases in benzodiazepines with short elimination half-lives e.g. lorazepam (Ativan) appear sooner than with those with longer elimination half-lives (e.g. diazepam).
- Reduce dosage further by 10–25 percent every few days, if tolerated. Dose, duration of therapy, and severity of anxiety influence the rate of taper and the need of adjunctive medications.
- Hopkins phenobarbital protocol (Kawasaki and Freire, 2011) can be used. It is 3-day protocol that begins with one-time dose of 200 mg followed by 100 mg q 4 h for 5 doses, followed by 60 mg every 4 h for 5 doses, then 60 mg every 8 h for 3 doses.
- Use adjunctive medications if necessary- carbamazepine (Tegretol), pregabalin (Lyrica), gabapentin (Neurontin), β-adrenergic receptor antagonist, divalproex (Depakote), clonidine (Catapres, and sedative antidepressants have been used, but their efficacy in the treatment of the benzodiazepine abstinence syndrome has not been established. Flumazenil has been used intravenously in benzodiazepine withdrawal but it is not recommended for routine use.
- Cognitive behavior therapy and other psychotherapeutic interventions improve outcome.

Table 6 Approximate Equivalent Doses of Sedative-Hypnotics

Drug	Dose (mg)
Alprazolam	1
Chlordiazepoxide	25
Clonazepam	0.5–1.0
Clorazepate	15
Diazepam	10
Flurazepam	30
Lorazepam	2
Oxazepam	30
Temazepam	20–30
Triazolam	0.25
Zaleplon[1]	10
Zolpidem[1]	10
Eszopiclone[1]	2

[1]Cross tolerance with benzodiazepines or even within Z-drug class is not complete due to different GABA-A subunit potencies. If unable to taper using Z-drug patient is taking, use nonselective benzodiazepine (diazepam, lorazepam and others).
From Clinical Manual of Addiction Psychiatry.

Table 7 Pharmacokinetics of Zolpidem

Zolpidem: Immediate Release

Absorption: Under fasting conditions immediate-release zolpidem tartrate is rapidly absorbed, and 70% of the oral dose is bioavailable owing to limited first-pass metabolism (Holm, 2000; de Haas, 2010). In a premarketing, single-dose, cross-over study in 45 healthy subjects, maximum concentration (C_{max}) was achieved at 1.6 h (T_{max}) For both a 5- and 10-mg dose. Aftermarket studies of 10- and 20-mg doses have consistently demonstrated T_{max} to range from 0.5 to 2.5 h (Greenblatt et al., 1998; Drover et al., 2000; de Haas et al., 2010). Clinical effects of sedation occur within an hour of ingestion, and subjective sedation as measured by a visual analog scale does not last longer than 8 h (Drover et al., 2000). According to preclinical studies, ingesting zolpidem with food may prolong the T_{max} by 60% although this did not alter the elimination of half life (Holm and Goa, 2000). Notably, the threshold for sedation based on electroencephalographic findings generally occurs above 25 ng/ml (Patat et al., 1994).

Elimination: The elimination half life of zolpidem has been reported from 1 to 3 h, with an average of 2–2.5 h (Greenblatt et al., 1998; Drover, 2004; de Haas et al., 2010), and its pharmacokinetics are described using a one-compartment model with first-order absorption (Drover et al., 2000; de Haas et al., 2010). Metabolism of zolpidem occurs via hydroxylation at three separate sites, yielding three inactive metabolites: M-3, M-4, and M-11. M-3, the most abundant metabolite, accounts for more than 80% of the net clearance by liver enzymes based on in vitro studies. Three p450 enzymes are responsible for 97% of zolpidem's metabolism: 3A4 (61%), 2C9 (14%), and 1A2 (14%) (Von Moltke et al., 1999). Zolpidem elimination occurs primarily via urinary excretion (48−67%) with the remainder excreted in the bile (Darcourt et al., 1999).

Zolpidem: Extended Release

Formulation: Zolpidem extended release (zolpidem ER; also referred to as modified release or controlled release as in Ambien CR) was developed to improve sleep maintenance through the night (Hindmarch, 2006; Neubauer 2006). Zolpidem ER is formulated as a two-layered tablet with an outer layer that dissolves quickly for immediate release and a second layer that dissolves slowly to prolong its hypnotic effect.

Absorption: Like zolpidem, zolpidem ER is 70% bioavailable (Weinling et al, 2006) with a mean elimination half life of nearly 3 h. T_{max} occurs in 1.5 or 2 h with the 12.5- and 6.25-mg doses, respectively (Moen and Plosker, 2006). The optimal dose was established to be 12.5 mg based on the balance of improved sleep maintenance over 10 mg zolpidem (Stanley et al., 2005).

Zolpidem: Oral Spray

Zolpidem oral spray (Zolpimist®) is dispensed as a 60-actuation container (60 actuations after five priming pumps) designed to be sprayed onto the tongue (Neubauer, 2010). The ratio of metabolites of zolpidem oral spray to those of zolpidem tablet is 85%, suggesting that a portion of the oral spray is not exposed to first-pass metabolism and is absorbed through the oral mucosa (Dilone et al., 2008). Zolpidem levels of 20 ng/mL are generally sedating or "therapeutic" (Patat et al., 1994), and within 15 min 65% and 79% of subjects who ingested 5 and 10 mg zolpidem spray had reached this level compared with 19% and 26% of those taking equivalent doses of zolpidem tablets. Elimination half-life and T_{max} are nearly identical to that of zolpidem tablets.

Continued

516   Domenic A. Ciraulo and Mark Oldham

Table 7 Pharmacokinetics of Zolpidem—cont'd

Zolpidem: Sublingual

Formulations: Two separate sublingual (SL) tablets of zolpidem are FDA approved: Edluar®, which is designed to decrease SL by a larger degree than zolpidem, and Intermezzo®, which is designed for middle-of-the-night administration after awakening to facilitate return to sleep.

Sublingual zolpidem (Edluar®) is designed to be absorbed more rapidly than oral zolpidem and, as a result, is 30% faster at inducing sleep. It is bioequivalent in dosage to oral zolpidem, and, although it has a similar C_{max} (after single 10 mg dose, mean C_{max} = 106 ng/ml), its T_{max} is slightly shorter than oral zolpidem—80 min vs 100 min, respectively, which is thought to contribute to its more rapid onset of action (Lankford, 2009). Mean elimination half-life is 2.65 h at 10 mg dose and 2.85 h at 5 mg dose.

Intermezzo® at doses of 1.75 mg and 3.5 mg is designed to induce sleep for only a portion of the night after middle-of-the-night awakenings. In a study of patients with self-reported sleep fragmentation, both 1.75 and 3.5 mg low-dose sublingual zolpidem demonstrated a T_{max} of 40 min and an elimination half life of 2.5 h after a single 3.5 mg dose (Roth et al., 2008). Sedative effects occurred 20 min postdose and lasting between 2 and 4 h (Roth et al., 2008). It should not cause next-day impairments in healthy individuals (Lankford, 2009).

temporary dose increases. The goal of medication treatment of withdrawal is to keep the patient comfortable and prevent the progression of symptoms.

2.4. ABUSE AND MISUSE

2.4.1. Experimental Models

Research evaluating abuse liability of sedative-hypnotics has used several predictive models. Animal models in rodents and primates include self-administration, reinstatement, place preference, brain stimulation and reward. Although details of these studies are beyond the scope of this chapter, findings are consistent. With the possible exception of flunitrazepam, benzodiazepines have lower abuse potential than barbiturates.

There are two widely used predictive models of potential for abuse of drugs in humans: (1) assessment of single dose drug effects and (2) self-administration in a free choice paradigm with placebo control (Griffiths et al., 1980, 1983, 1985, 2003). The model of drug effects has the advantage of ability to test several doses of a drug against an active comparator, to examine reinforcing mood effects, and adverse events from a single dose. The disadvantage is that it relies heavily on self-report of mood effects, but this is mitigated somewhat by the use of standardized scales that are derived from the Addiction Research Clinical Inventory (Ciraulo et al., 2001). Using this model, Jaffe et al. (1983) found that diazepam produced euphoric mood effects but halazepam, a prodrug for desmethyldiazepam, did not. Carter et al. (2006) have developed a comprehensive battery that includes not only subjective mood and somatic effects (e.g. sedation)

Table 8 Pharmacokinetics of Zaleplon

Zaleplon is rapidly absorbed but undergoes extensive first-pass metabolism; absolute bioavailability of a 5-mg oral dose is 30% (Rosen et al., 1999). On average, T_{max} of a single dose is achieved 45 min after ingestion, and elimination half life is 1 h (Beer et al., 1994; Rosen et al., 1999).

Two studies that directly compared the pharmacokinetic profiles of zolpidem and zaleplon found zaleplon to have a significantly shorter half life than zolpidem—1 h vs 2 h, respectively (Greenblatt et al., 1998; Drover et al., 2000). Consistent with half life, duration of subjective sedation were half as long for zaleplon as for zolpidem—4 h vs 8 h (Drover et al., 2000), and in general the sedative effects of zolpidem, both self- and observer-rated, were nearly twice as potent as zolpidem at equal doses (Greenblatt et al., 1998).

Metabolism of zaleplon to three inactive metabolites occurs almost exclusively in the liver. The principal route of metabolism is via aldehyde oxidase, which produces 5-oxozaleplon (Lake et al., 2002). A secondary route of metabolism occurs by way of CYP3A4, which yields *N*-desethylzaleplon (Renwick et al., 1998). A third inactive metabolite, 5-oxo-*N*-desethylzaleplon, is generated via sequential metabolism by each of aldehyde oxidase and CYP3A4 (Mandrioli et al., 2010). The two 5-oxo metabolites undergo glucuronidation via UGT prior to excretion (Mandrioli et al., 2010).

Table 9 Pharmacokinetics of Eszopiclone

Racemic zopiclone is rapidly absorbed, with a bioavailability of 80% (Gaillot et al., 1983; Fernandez et al., 1995). In one study of 12 healthy adult Caucasian subjects that assayed the individual pharmacokinetic profiles of S(+)-zopiclone (eszopiclone, Lunesta®) and R(−)-zopiclone, T_{max} was found to be equivalent between the two enantiomers: 1.6 h (Fernandez et al., 1993). In that study, though, C_{max} and elimination half life of each enantiomer were found to differ significantly: C_{max} for eszopiclone was 87.3 ng/ml versus 44.0 ng/ml for R(−)-zopiclone whereas elimination half-life was 6 h 40 min versus 3 h 45 min, respectively. One study found that single-dose eszopiclone has a half-life of 6 h whereas multiple-dosing may yield a slightly longer half-life of 7 h (Maier et al., 2003), which is of unclear clinical import. Notably, the greater C_{max} of eszopiclone compared with R(−)-zopiclone has been attributed to its relatively slower clearance and smaller volume of distribution (Fernandez et al., 1995).

Eszopiclone is extensively metabolized via demethylation and oxidation. A study by Fernández et al., though, suggests that oxidation of zopiclone to *N*-oxide metabolites via the CYP 450 system accounts for only 30% of the initial dose (1995) with decarboxylation accounting for half of the metabolism. Oxidation yields the chiral metabolites (S)*N*-oxidezopiclone (inactive) and (S)*N*-desmethylzopiclone (moderately active) via CYP 3A4 and CYP 2E1 (Fernandez et al., 1993; Becquemont et al., 1999; Sanger, 2004).

but also measures of toxicity, such as psychomotor and cognitive performance, which is the Griffiths' model of abuse liability (Griffiths and Johnson, 2005). There is a consensus that the core of assessing abuse liability of sedative hypnotic agents is the drug property of inducing positive reinforcement due to direct action in the brain. While toxicity is important for any therapeutic agent, it does not predict the likelihood of continued use of the drug in nonmedical settings or continued use of a drug despite adverse consequences. The ability to produce tolerance and a withdrawal syndrome could be considered a somatic effect leading to dependence, but this effect must be associated with acute alterations in mood. For an interesting discussion of this topic readers are referred to Nielsen et al. (2012), Brady et al. (2012).

Other human models include drug choice in drug-dependent individuals. In this model, a placebo-controlled design is used and participants sample specific drugs and then are allowed to self-administer the drug that they prefer. This is a costly model that has the advantage of assessing direct ingestion of the preferred drug, but it loses the broad flexibility in dose ranging, pharmacokinetic and pharmacodynamics effects offered by the single dose model. Human laboratory studies indicate that benzodiazepines have lower abuse liability than barbiturates (for review see Ciraulo and Sarid-Segal (2009)).

Perhaps the least expensive method to determine abuse is by reviewing clinical records or by prospective studies. Examples of this approach include studies of the abuse liability of benzodiazepines in methadone clinics.

2.4.2. Survey Data

Assessing the public health impact of abuse and misuse of benzodiazepines requires differentiating use in medical settings from illicit use. With the exception of flunitrazepam (Rohypnol, "roofies"), the use of benzodiazepines for their hedonic value is rare (flunitrazepam has been used in combination surreptitiously with alcohol to impair memory and consciousness in sexual assaults). Data from the latest National Household Survey of Drug Use and Health show that nonmedical use of prescription pain relievers is more common than illicit use of "tranquilizers" or "sedatives" as shown in Figure 4.

Sedative-hypnotics are most often used in combination with other drugs, such as stimulants, opiates, and alcohol to enhance euphoric effects, to counteract symptoms of toxicity (e.g. cocaine) or to self-medicate withdrawal symptoms (e.g. alcohol). The Treatment Data Episode Data Set (Substance Abuse and Mental Health Services Administration, 2007), which provides information on individuals admitted to publicly funded treatment programs, has found that treatment for primary benzodiazepine dependence remains uncommon, but there is an increasing number of young adults (20–24 years old) admitted with benzodiazepines as part of a multiple drug abuse pattern.

The Drug Abuse Warning Network (SAMHSA, 2011) reports drug-related problems from data reported by EDs found that 33.6% of ED visits involving nonmedical use of pharmaceuticals in 2009 (the most recent report) involved sedative-hypnotics, usually in combination with other substances. From 2004 to 2009, there was an increase in the number of ED mentions of sedative-hypnotics from 177,394 to 363,270. Similar findings were reported for an Australian hospital, which found that benzodiazepines accounted for 3.9% of ED visits in 2007, which rose to 5.6% in 2011. Use of alprazolam

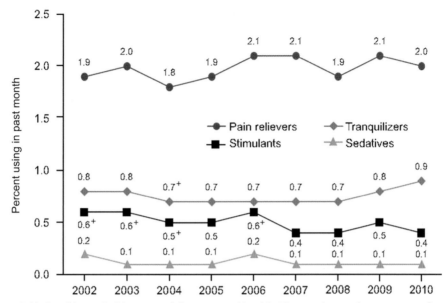

Figure 4 National household survey of drug use and health. The number and percentage of persons aged 12 or older who were current users of hallcinogens in 2010 (1.2 million or 0.5%) were similar to those in 2009 (1.3 million or 0.5%). These include similar number and percentages for current users of Ecstasy, with 695,000 (0.3%) current users in 2010 and 760,000 (0.3%) current users in 2009. Difference between this estimate and the 2010 estimate is statistically significant at the 0.05 level.

alone doubled over that five-year period (Reynolds et al., 2013). Clearly, benzodiazepine misuse continues to be a concern internationally.

In clinical practice, the use of medications to treat anxiety has increased, particularly as antidepressants have become first line agents in the treatment of anxiety. Similarly, the advent of the Z-drugs has reduced the number of prescriptions for benzodiazepine hypnotics (Ciraulo and Sarid-Segal, 2009). Several surveys of clinical practice indicate that most patients take benzodiazepines for less than one month with only about 1% taking them for longer than one year (Piper, 1995; Zandstra et al., 2002; Lagnaoui et al., 2004; Olfson et al., 2004; Veronese et al., 2007). Rates vary based on methodology and country studied. For example, 6-month use has been reported as 3% and 1-year use 1.7% by Zansdtra et al. (2002), which is similar to the United States rates. Long-term use in the elderly appears to remain a problem in many countries, with the main concern is increased risk of falls (Bradley et al., 2012). The survey data should be interpreted with caution. Surveys prior to the approval of SSRIs for anxiety disorders have higher rates of benzodiazepine prescriptions than recent studies. For example, a study of antidepressant and benzodiazepine use from 2005 to 2008 found anxiolytic prescriptions declined by 14% while antidepressants increased by 45% (Svab et al., 2011).

The class of hypnotic DORAs may exhibit a unique profile with regard to abuse and abuse liability. Orexin, as its name suggests, plays a role in appetite and feeding behavior, and lateral hypothalamic cells containing orexin project to reward-associated brain regions including the nucleus accumbens and ventral tegmental area. Studies in animal models have found orexin agonists to cause reinstatement of drug and food-seeking behavior (Cason and Aston-Jones, 2013); however, the role of orexin antagonists such as suvorexant has yet to be explored. Whether DORAs are associated with abuse liability or whether they may serve a protective role in addiction remains unknown.

Finally, emerging evidence suggests that cultural stereotypes of those with substance use disorders may lead one to overlook growing misuse of less traditionally abused sedative-hypnotics. For instance, (Earley and Finver, 2013) highlighted the increasing illicit use of propofol among HCPs particularly among anesthesiologists and certified registered nurse anesthetists.

2.5. INTRODUCTION TO THE Z-DRUGS: NONBENZODIAZEPINE GABA RECEPTOR AGONISTS

2.5.1. Terminology

The Z-drugs have garnered this more colloquial name as the generic names of two of the three currently approved agents in the U.S. begin with the letter Z—zolpidem and zaleplon—as does the racemate of the third—eszopiclone, the active stereoisomer of zopiclone. The Z-drugs are part of a larger category known as nonbenzodiazepine (in that they lack the defining fused benzene and diazepine rings of benzodiazepines) benzodiazepine receptor agonists. Although sometimes referred to simply as nonbenzodiazepines, this isolated term should be discouraged, given its imprecision (Figure 5).

2.5.2. Pharmacology

The Z-drugs—zolpidem, zaleplon, and eszopiclone—serve as positive modulators of GABA-A receptor–channel complex. Zolpidem and zaleplon preferentially bind to

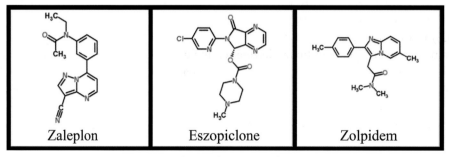

Figure 5 Chemical structure of Z drugs.

α_1-containing GABA-A receptors whereas eszopiclone exhibits a less selective binding profile. The time to maximal concentration (T_{max}) is about one and a half hours with zolpidem immediate or extended release or with eszopiclone. Zolpidem oral spray and sublingual formulations are expected to reach T_{max} in a little over an hour whereas zaleplon has the shortest T_{max} at 45 min postingestion. Clinical sedation occurs within an hour of ingestion, although zolpidem oral spray, zolpidem sublingual, and zaleplon are expected to cause sedation in less than 30 min. The elimination half life of zolpidem (including immediate release, oral spray, and sublingual formulations) is roughly 2–2.5 h. The half life of zolpidem extended release is 3 h. Zaleplon has the shortest half life at 1 h and eszopiclone the longest at 6 h. Metabolism of all three Z-drugs is primarily hepatic.

Zolpidem dosing requires special mention because the U.S. FDA recommended dosing changes as of January 2013. Because zolpidem may cause psychomotor impairment the morning following use—particularly in women—the FDA now cites recommended doses for zolpidem in women as 5 mg for immediate-release formulations (e.g. Ambien, Edluar, and Zolpimist) and 6.25 mg for extended release zolpidem (e.g. Ambien CR). These doses are consistent with recommended doses for the elderly. Doses for middle-of-the-night administration of low dose sublingual zolpidem (Intermezzo) are unchanged. The FDA further recommends that "health care professionals consider prescribing the lower doses" for men as well (U.S. FDA, 2013).

2.5.3. Applications

The Z-drugs have been studied extensively in the insomnias. Clinicians should note that sleep disturbances alone—difficulty falling or staying asleep—are insufficient to diagnose insomnia: insomnia must also include clinically significant distress or functional impairment per all three sleep nosologies (DSM-5, ICSD-R, and ICD-10). Also, over a half of patients with insomnia will have another psychiatric disorder; therefore, screening for co-morbid mental illness is *critical* in the assessment and management of all patients with insomnia.

Zolpidem immediate release (hereafter zolpidem) is effective on a wide range of sleep outcomes (sleep latency (SL), total sleep time (TST), number of awakenings [NAWs], wake after sleep onset (WASO), sleep efficiency (%SE), and sleep quality (SQ)) in the management of primary insomnia for up to a month and appears to have advantages over hypnotic benzodiazepines including limited rebound insomnia and onset of sedation nearly 30 min earlier than triazolam. Zolpidem extended release (zolpidem ER) has demonstrated similar findings to zolpidem on sleep outcomes for up to six months but may also improve sleep maintenance to a greater degree from 3 to 6 h postingestion. Sublingual zolpidem (zolpidem SL) appears to reduce latency to persistent sleep (LPS) by 10 min more than zolpidem although data on other sleep outcomes are limited. Zaleplon reliably shortens SL for up to a month but does not improve TST or measures of sleep maintenance such as WASO or NAW. Eszopiclone appears to improve the same range of sleep outcomes as zolpidem and zolpidem ER for up to six months.

The efficacy of zolpidem and eszopiclone in the management of various transient insomnias (lasting <1 month) is similar to their efficacy in primary insomnia. Zolpidem SL may shorten SL by up to 10 min more than zolpidem in transient insomnia although data are limited on other sleep measures.

Based on limited data, zolpidem may be efficacious as an adjunct to a SSRI in the management of insomnia related to major depressive disorder (MDD) or generalized anxiety disorder (GAD); however, one trial found no improvements on LPS for insomnia related to attention-deficit/hyperactivity disorder in children. Based on one trial, zolpidem ER may be effective as an adjunct to SSRI for insomnia related to MDD. Eszopiclone appears to be broadly efficacious as an adjunct to SSRI for the management of insomnia related to MDD, GAD, and—in one trial—posttraumatic stress disorder.

As-needed zolpidem for the management of primary insomnia consistently improved several global, subjective sleep outcomes; however, findings on measures such as SL, NAW, WASO, and TST were inconsistent relative to placebo. Low-dose zolpidem SL shortens LPS and enhances TST when taken after awakening in the middle of the night (MOTN) without causing residual sedation the following morning. Although not indicated for this purpose, zaleplon has been studied more extensively than low-dose zolpidem SL for MOTN administration and appears equally efficacious to low-dose zolpidem SL without next-morning sedation. Zolpidem does not appear to be effective for the management of phase advances or delays but may be useful for short-term administration for jet lag. At recommended doses, zolpidem does not adversely affect respiratory parameters in adults including the elderly or worsen hypopneas/apneas associated with obstructive sleep apnea. Eszopiclone administered during titration of continuous positive airway pressure Lettieri et al., 2009 may improve CPAP compliance.

2.5.4. Special Populations

The U.S. FDA's recent lowering of recommended doses of most zolpidem formulations in women identifies women as a special population for this purpose. Dosing of Z-drugs in the elderly is generally half that of recommended doses for adult men in most instances. The recommended dose of zolpidem and zaleplon in the elderly is 5 mg each and zolpidem ER 6.25 mg. Eszopiclone should be started at 1 mg but not exceed 2 mg. Zolpidem and zolpidem ER appear to shorten SL and enhance TST in elders with primary insomnia, but they have not been shown to improve sleep maintenance (WASO or NAW) at these recommended doses. Zaleplon improves SL and subjective SQ. Eszopiclone 1 mg (starting dose in elderly) improves only SL whereas 2 mg (maximum dose) improves SL, TST, WASO, and subjective SQ. Particularly in the elderly, given the concern for falls and psychomotor impairment during the night (as when getting up to use the restroom) or the following morning, the therapeutic index of any substance must be weighed very carefully. Clinicians are strongly cautioned not to exceed recommended doses of Z-drugs in the elderly.

Dose adjustments are not required for the Z-drugs in patients with kidney impairment. In mild-to-moderate liver impairment, zolpidem and zaleplon should be started at 5 mg although no changes are required for eszopiclone. In severe liver impairment, zolpidem may be cautiously initiated at 5 mg with close monitoring; zaleplon should be avoided; and eszopiclone should not exceed 2 mg. The Z-drugs are pregnancy category C and are not recommended during breast feeding. The Z-drugs are not indicated for use in children or adolescents. Zolpidem does not appear to worsen respiratory values including oxygenation in patients with patients with stable severe COPD.

Zolpidem was effective at improving SL and TST at simulated and in vivo high altitude conditions. Zolpidem is efficacious for sleep disturbances in patients with fibromyalgia, and eszopiclone is efficacious for sleep disturbances in patients with rheumatoid arthritis. One uncontrolled trial did not find zolpidem superior to haloperidol at enhancing overall sleep in pediatric burn patients. Zolpidem and eszopiclone have been shown to improve SL, TST, WASO, and NAW relative to placebo in peri/postmenopausal women with hot flashes.

2.5.5. Other Applications

Insufficient evidence suggests that zolpidem or zaleplon may enhance sustained wakefulness and alertness when limited time is available for sleep. Based on several reports, zolpidem appears efficacious in the management of retarded and excited catatonia, potentially even in patients whose catatonia is resistant to electroconvulsive therapy. Zolpidem may be associated with a dyskinetic effect in those with Parkinson disease and related neurodegenerative conditions. Zolpidem produces no more than mild improvements in enhanced awareness in patients in vegetative or minimally conscious states. Zolpidem, at higher than recommended doses, tends to causes hypnosis and amnesia that may avail its use as a preprocedural agent. Nightly zolpidem enhances sleep postoperatively although data on whether this correlates with improved pain control remain unclear. Zolpidem does not adversely affect EEG findings, and it appears to enhance polysomnography yield.

2.5.6. Tolerability

The Z-drugs are generally well tolerated. Common side effects due to zolpidem include drowsiness, dizziness, and diarrhea, and four large postmarketing surveys (greater than 23,000 patients total) support the overall safety and tolerability of zolpidem in the general population. Zolpidem ER most commonly causes headache, somnolence, dizziness, and nausea. The side effect profile of zaleplon is benign with most side effects broadly consistent with those related to placebo (e.g. headache, dizziness, abdominal pain, nausea, asthenia, and somnolence). Eszopiclone is often associated with an unpleasant taste in the mouth, and it may also cause headache, infections, somnolence, dizziness, dry mouth, and dyspepsia.

Zolpidem causes a dose- and time-dependent anterograde amnesia (for 4–5 h after a 10-mg dose) whereas zaleplon appears to be devoid of this effect. Although data on eszopiclone are limited, any anterograde amnesia is not expected to last longer than 8 h. No evidence exists for retrograde amnesia for the Z-drugs. Zolpidem causes psychomotor impairment for up to 6 h and sedation for up to 7 h on average. Zolpidem ER does not cause next-morning psychomotor impairment (8 h after administration). Zaleplon may cause impairment for one to 2 h. Limited evidence suggests that eszopiclone does not cause next-morning psychomotor impairment. Zolpidem may impair driving performance for up to 6 h, and zaleplon at 5 h was not shown to cause any driving impairment. Eszopiclone's effects on driving have not been studied systematically although limited evidence suggests that racemic zopiclone may impair driving for up to 10 h.

2.5.7. Tolerance

Tolerance to Z may be less of a problem than benzodiazepines and have lower abuse liability than benzodiazepines. Reports of Z-drug abuse and dependence have appeared with increasing frequency but zolpidem appears to be the least reinforcing of the three. Patients with a family or personal history of substance use disorder are particularly at risk of misusing Z-drugs. The Z-drugs do not cause any more than one night of rebound insomnia. Specifically, zaleplon appears to be devoid of rebound insomnia. The Z-drugs are remarkably safe in mono-drug overdose. Based on reported data, one fatality associated with zolpidem overdose may occur every 900 overdoses. Drug interactions with Z-drugs are rare. The benzodiazepine receptor antagonist flumazenil reverses sedation caused by zolpidem. Interactions may occur when metabolism of Z-drugs is inhibited, which may lead to excessive sedation.

2.5.8. Behavioral Toxicity

Sleep-related behaviors including cooking, eating, driving, intercourse, walking, and conversing have occurred after taking Z-drugs, particularly zolpidem. Such behaviors appear to be dose-dependent and almost universally resolve after discontinuation of the offending agent. Severe allergic reactions have rarely been reported due to the Z-drugs. Behavioral disinhibition, agitation, depersonalization, or depression with suicidal ideation may occur rarely on hypnotic agents. Abrupt withdrawal of chronic (generally higher-than-recommended-dose) treatment with Z-drugs may cause a benzodiazepine-responsive delirium. Delirium may occur with high-dose Z-drug therapy.

2.5.9. Other Adverse Effects

Large, retrospective studies have found increases in mortality and cancer among those prescribed Z-drugs; however, prospective, randomized trials would be needed to clarify any potential risk. Further, such studies do not account for the presence of insomnia or co-morbid mental illness. In premarketing trials, 12 newly diagnosed tumors were identified in those randomized to receive new hypnotics (Z-drugs plus ramelteon). Z-drugs (and ramelteon) may slightly predispose to the development of infections via an unclear mechanism. In premarketing trials, 2% of those randomized to the Z-drugs or ramelteon developed depression compared with 1% of those randomized to placebo. The significance of this finding is unclear given that Z-drugs have been used safely as an adjunct in the treatment of MDD.

2.6. BARBITURATES

2.6.1. Role in Modern Medicine

Since the advent of the benzodiazepines, barbiturate use has been limited in modern medicine. Phenobarbital is still prescribed as an anticonvulsant and as a sedative, especially for children. It is also a common component of many combination products and reduces the stimulating effects of sympathomimetic agents. Butalbital is an intermediate-acting barbiturate found in a widely used combination product that also contains acetaminophen and caffeine (Fioricet) and is approved for the treatment of muscle contraction headaches. Given the limited availability of barbiturates, it is uncommon, at least in the U.S., for addiction units to treat patients dependent on barbiturates other than butalbital and phenobarbital. The barbiturates marketed currently in the U.S. are listed in Table 10.

A major disadvantage of the use of the barbiturates is the development of pharmacokinetic and pharmacodynamic tolerance. Pharmacodynamic tolerance begins after acute

Table 10 Barbiturates and Other Sedative-Hypnotic Agents

Generic Name	Trade Name	Dose (mg)	Duration of Action	Therapeutic Use
Amobarbital	Amytal	100	Intermediate	Hypnotic, sedative, seizures
Aprobarbital	Alurate	40	Intermediate	Hypnotic, sedative, seizures
Butabarbital	Butisol	100	Intermediate	Hypnotic, sedative
Butalbital	Fiorinal	100	Intermediate	Combination products
Pentobarbital	Nembutal	100	Short	Sedation, seizures
Secobarbital	Seconal	100	Short	Sedation, seizures
Phenobarbital	Luminal	30	Long	Seizure, sedative hypnotic
Chloral hydrate	Generic	500		Not recommended
Ethchlorvynol	Placidyl	500		Not recommended
Glutethimide	Generic	250		Not recommended
Meprobamate	Miltown	400		Not recommended

Note: The substitution technique for sedative-hypnotic withdrawal requires calculation of equivalent doses of phenobarbital to replace the sedative-hypnotic agent that the patient is taking. The above are doses of various sedative-hypnotic agents for which a 30-mg dose of phenobarbital should provide adequate coverage of a withdrawal syndrome. Daily doses of phenobarbital should rarely exceed 600 mg using this protocol.
Sarid-Segal et al., 2009.

doses and continues to develop over weeks to months. Tolerance to the mood-altering and sedative effects develops to a greater extent than does tolerance to the lethal effects, increasing the risk of accidental overdose.

2.6.2. Intoxication

Intoxication from barbiturates begins with symptoms that resemble alcohol intoxication, and include sluggishness, incoordination, difficulty in thinking, poor memory, slowness of speech and comprehension, faulty judgment, disinhibition of sexual and aggressive impulses, a narrowed range of attention, emotional lability, and exaggeration of basic personality traits. The duration of intoxication depends primarily on the half-life of the barbiturate. At very high doses severe toxicity occurs, with nystagmus, diplopia, strabismus, ataxic gait, positive Romberg's sign, hypotonia, respiratory depression, and decreased superficial reflexes.

2.6.3. Dependence and Withdrawal

Physiological dependence may develop after a daily dose of 400 mg of pentobarbital (Nembutal) for 3 months at a daily dose of 600 mg of pentobarbital for one–2 months, a withdrawal syndrome characterized by anxiety, insomnia, anorexia, tremor, and EEG changes occurs in approximately one half of patients, and 10 percent may have a single seizure. At higher dosages of 800 to 2200 mg per day for several weeks to months, abrupt discontinuation begins with apprehension and uneasiness, insomnia, muscular weakness, twitches, coarse tremors, myoclonic jerks, postural faintness and orthostatic hypotension, anorexia, and vomiting. These may progress to more severe symptoms such as seizures and delirium, sometimes accompanied by hypothermia, which may be fatal. Disorientation, visual hallucinations, and frightening dreams may precede the onset of full delirium. Delirium may be exceedingly difficult to reverse, even with large doses of a barbiturate; thus, clinicians should never wait for the appearance of withdrawal symptoms before instituting therapy.

Table 11 Guidelines for Barbiturate Detoxification

Symptoms after Test Dose of 200 mg of Oral Pentobarbital	Estimated 24-h Oral Pentobarbital Dose (mg)	Estimated 24-h oral Phenobarbital Dose (mg)
Asleep, but can be aroused	0	0
Sedated, drowsy, slurred speech, nystagmus, ataxia, positive Romberg test result	500–600	150–200
Few signs of intoxification, patient is comfortable, may have lateral nystagmus	800	250
No drug effect	1000–1200	300–400

Note: Maximum phenobarbital dose is 600 mg.
Procedure modified from Ewing and Bakewell (1967).

2.6.4. Overdose

Barbiturates are lethal when taken in overdose, because they induce respiratory depression. In addition to intentional suicide attempts, accidental or unintentional overdoses are common. Barbiturates in home medicine cabinets are a common cause of fatal drug overdoses in children. As with benzodiazepines, the lethal effects of barbiturates are additive to those of other sedative-hypnotic drugs, including alcohol and benzodiazepines. Barbiturate overdose is characterized by induction of coma, respiratory arrest, cardiovascular failure, and death. For the most commonly abused barbiturates, the ratio of lethal to effective dose ranges between three to one and 30 to 1.

3. CONCLUSIONS

The history of the development of sedative hypnotic drugs demonstrates an evolution from pharmaceuticals with high-abuse liability, low therapeutic indices, questionable efficacy, and substantial risk of toxicity to therapeutic agents with strong evidence of efficacy and lower toxicity. Even though this chapter focused on benzodiazepines and Z-drugs, and much less with barbiturates, and not at all with chloral hydrate, glutethimide, methaqualone, meprobamate, or ethchlorvynol, readers should appreciate that the older agents caused severe dependencies and substantial risk of toxicity. It was not uncommon for individuals legitimately prescribed these agents to die from overdoses or during treatment of withdrawal syndromes. Patients suffered such severe memory and behavioral toxicity (disinhibition, aggression, hostility) that even therapeutic doses could be problematic.

The introduction of benzodiazepines and more recently the Z-drugs represent major advances in therapy of anxiety and insomnia, and the potential for a novel hypnotic class of agents centered on orexin antagonism seems promising. For clinicians lacking experience with older agents, the lower risk of the newer agents may not be obvious. There is a wealth of evidence indicating that benzodiazepines have lower risk of abuse, misuse, and toxicity than barbiturates and other older sedative hypnotics. It is not yet clear whether the Z-drugs offer substantial advantages over the nonselective benzodiazepines, but at least some data suggest that tolerance and rebound insomnia may be lower with Z-drugs. The Z-drugs are not without their own risks, most notably complex nocturnal behavioral disturbances. Still, the weight of the evidence supports that the benzodiazepines and Z-drugs are safer than older agents.

Our greater understanding of the mechanism of action of sedative hypnotics at the GABA-A receptor and new findings demonstrating that different alpha subunits have specific effects on sedation, anxiolysis, muscle relaxation, analgesia, and cognitive function holds great promise for the future. The benzodiazepines and Z-drugs have some unacceptable risks, especially in alcoholics, substance abusers, the elderly, and people with chronic pain syndromes. Addiction, persistent impairment of psychomotor and cognitive function, and discontinuation syndromes affect a substantial subgroup of individuals taking these medications. On the other hand, long-term use in the medical setting is low in the United States and other countries, and unauthorized dosage escalation is equivalent or lower than antidepressants. Survey data indicate that they are rarely the primary drug of abuse, and most often used in combination with alcohol or illicit drugs. Nonmedical use of the sedative hypnotics is substantially lower than pain medications. In conclusion, the benzodiazepines and Z-drugs offer advantages over older sedative hypnotics but are not without adverse effects. As more specific GABA-A positive modulators are developed improved risk to benefit ratios are expected.

REFERENCES

American Psychiatric Association, 2013. Diagnostic and Statistical Manual of Mental Disorders, Fifth ed. American Psychiatric Association, Arlington, VA.

An, J.Y., Kim, J.S., et al., 2008. Successful treatment of the Meige syndrome with oral zolpidem monotherapy. Mov. Disord. 23 (11), 1619–1621.

Arnedt, J.T., Rohsenow, D.J., et al., 2011. Sleep following alcohol intoxication in healthy, young adults: effects of sex and family history of alcoholism. Alcohol Clin. Exp. Res. 35 (5), 870–878.

Atack, J.R., 2003. Anxioselective compounds acting at the GABA(A) receptor benzodiazepine binding site. Curr. Drug Targets CNS Neurol. Disord. 2 (4), 213–232.

Barker, M.J., Greenwood, K.M., et al., 2004. Cognitive effects of long-term benzodiazepine use: a meta-analysis. CNS Drugs 18 (1), 37–48.

Barker, M.J., Greenwood, K.M., et al., 2004. Persistence of cognitive effects after withdrawal from long-term benzodiazepine use: a meta-analysis. Arch. Clin. Neuropsychol. 19 (3), 437–454.

Becquemont, L., Mouajja, S., et al., 1999. Cytochrome P-450 3A4 and 2C8 are involved in zopiclone metabolism. Drug Metab. Dispos. 27 (9), 1068–1073.

Beer, B., Ieni, J.R., et al., 1994. A placebo-controlled evaluation of single, escalating doses of CL 284,846, a non-benzodiazepine hypnotic. J. Clin. Pharmacol. 34 (4), 335–344.

Ben-Porath, D.D., Taylor, S.P., 2002. The effects of diazepam (valium) and aggressive disposition on human aggression: an experimental investigation. Addict. Behav. 27 (2), 167–177.

Benyamina, A., Naassila, M., et al., 2012. Potential role of cortical 5-HT(2A) receptors in the anxiolytic action of cyamemazine in benzodiazepine withdrawal. Psychiatry Res.

Bergmann, R., Kongsbak, K., et al., 2013. A unified model of the GABA$_A$ receptor comprising agonist and benzodiazepine binding sites. PLoS One 8 (1), 1–13.

Bradley, M.C., Fahey, T., et al., 2012. Potentially inappropriate prescribing and cost outcomes for older people: a cross-sectional study using the Northern Ireland Enhanced Prescribing Database. Eur. J. Clin. Pharmacol.

Brunello, N., Cooper, J., et al., 2009. Differential Pharmacological Profiles of the GABAA Receptor Modulators Zolpidem, Zopiclone, Eszopiclone, and (S)-desmethylzopiclone. World Psychiatric Association International Congress, Florence, WPA.

Busto, U.E., Bremner, K.E., et al., 2000. Long-term benzodiazepine therapy does not result in brain abnormalities. J. Clin. Psychopharmacol. 20 (1), 2–6.

Carter, L.P., Richards, B.D., et al., 2006. Relative abuse liability of GHB in humans: a comparison of psychomotor, subjective, and cognitive effects of supratherapeutic doses of triazolam, pentobarbital, and GHB. Neuropsychopharmacology 31 (11), 2537–2551.

Cason, A.M., Aston-Jones, G., 2013. Role of orexin/hypocretin in conditioned sucrose-seeking in rats. Psychopharmacology 226 (1), 155–165.

Ciraulo, D., Knapp, C., 2009. The pharmacology of nonalcohol sedative hypnotics. In: Ries, R.K., Fiellin, D.A., Miller, S.C., Saitz, R. (Eds.), Principles of Addiction Medicine. Wolters Kluwer/Lippencott Williams & Wilkens, Philadelphia, pp. 99–112.

Ciraulo, D.A., Sarid-Segal, O., 2009. Sedative, hypnotic, or anxiolytic-related disorders. In: Kaplan, B., Philadelphia, S.V. (Eds.), Comprehensive Textbook of Psychiatry. Williams & Wilkins, Lippincott, pp. 1397–1418.

Ciraulo, D.A., Knapp, C.M., et al., 2001. A benzodiazepine mood effect scale: reliability and validity determined for alcohol-dependent subjects and adults with a parental history of alcoholism. Am. J. Drug Alcohol Abuse 27 (2), 339–347.

Collinson, N., Kuenzi, F.M., et al., 2002. Enhanced learning and memory and altered GABAergic synaptic transmission in mice lacking the alpha 5 subunit of the GABAA receptor. J. Neurosci. 22 (13), 5572–5580.

Coporaso, G.L., 2013. Medications and Cognition in Older Adults. In: Ravdin, L.D., Katzen, H.L. (Eds.), Handbook on the Neuropsychology of Aging and Dementia. Springer Science, New York, pp. 89–107.

Curran, H.V., Collins, R., et al., 2003. Older adults and withdrawal from benzodiazepine hypnotics in general practice: effects on cognitive function, sleep, mood and quality of life. Psychol. Med. 33 (7), 1223–1237.

Darcourt, G., Pringuey, D., et al., 1999. The safety and tolerability of zolpidem–an update. J. Psychopharmacol. 13 (1), 81–93.

de Haas, S.L., Schoemaker, R.C., et al., 2010. Pharmacokinetics, pharmacodynamics and the pharmacokinetic/pharmacodynamic relationship of zolpidem in healthy subjects. J. Psychopharmacol. 24 (11), 1619–1629.

Dilone, E., Arumugam, U., et al., 2008. Evaluation of the effect of the route of administration on the human metabolite pattern of zolpidem: ZoloiMist oral spray compared to Ambien tablets. AAPSJ 10 (S12).

Drover, D.R., 2004. Comparative pharmacokinetics and pharmacodynamics of short-acting hypnosedatives: zaleplon, zolpidem and zopiclone. Clin. Pharmacokinet. 43 (4), 227–238.

Drover, D., Lemmens, H., et al., 2000. Pharmacokinetics, pharmacodynamics, and relative pharmacokinetic/pharmacodynamic profiles of zaleplon and zolpidem. Clin. Ther. 22 (12), 1443–1461.

Earley, P.H., Finver, T., 2013. Addiction to propofol: a study of 22 treatment cases. J. Addict Med. 7 (3), 169–176.

Fernandez, C., Maradeix, V., et al., 1993. Pharmacokinetics of zopiclone and its enantiomers in Caucasian young healthy volunteers. Drug Metab. Dispos. 21 (6), 1125–1128.

Fernandez, C., Martin, C., et al., 1995. Clinical pharmacokinetics of zopiclone. Clin. Pharmacokinet. 29 (6), 431–441.

Fleck, M.W., 2002. Molecular actions of (S)-desmethylzopiclone (SEP-174559), an anxiolytic metabolite of zopiclone. J. Pharmacol. Exp. Ther. 302 (2), 612–618.

Gaillot, J., Heusse, D., et al., 1983. Pharmacokinetics and metabolism of zopiclone. Pharmacology 27 (Suppl. 2), 76–91.

Greenblatt, D., Shader, R., 1974. Benzodiazepines in Clinical Practice. Raven Press, New York.

Greenblatt, D.J., Harmatz, J.S., et al., 1998. Comparative kinetics and dynamics of zaleplon, zolpidem, and placebo. Clin. Pharmacol. Ther. 64 (5), 553–561.

Greenfield L. J., in press. Molecular mechanisms of antiseizure drug activity at GABAA receptors. Eur. J. Epilepsy. http://dx.doi.org/10.1016/j.seizure.2013.04.015.

Griffiths, R.R., Johnson, M.W., 2005. Relative abuse liability of hypnotic drugs: a conceptual framework and algorithm for differentiating among compounds. J. Clin. Psychiatry 66 (Suppl. 9), 31–41.

Griffiths, R.R., Bigelow, G.E., et al., 1980. Drug preference in humans: double-blind choice comparison of pentobarbital, diazepam and placebo. J. Pharmacol. Exp. Ther. 215 (3), 649–661.

Griffiths, R.R., Bigelow, G.E., et al., 1983. Differential effects of diazepam and pentobarbital on mood and behavior. Arch. Gen. Psychiatry 40 (8), 865–873.

Griffiths, R.R., Lamb, R.J., et al., 1985. Relative abuse liability of triazolam: experimental assessment in animals and humans. Neurosci. Biobehav Rev. 9 (1), 133–151.

Griffiths, R.R., Bigelow, G.E., et al., 2003. Principles of initial experimental drug abuse liability assessment in humans. Drug Alcohol Depend. 70 (3 Suppl), S41–S54.

Gunja, N., 2013. In the zzz zone: the effects of z-drugs on human performance and driving. J. Med. Toxicol. 9 (2), 163–171.

Hadingham, K.L., Wingrove, P., et al., 1993. Cloning of cDNA sequences encoding human alpha 2 and alpha 3 gamma-aminobutyric acidA receptor subunits and characterization of the benzodiazepine pharmacology of recombinant alpha 1-, alpha 2-, alpha 3-, and alpha 5-containing human gamma-aminobutyric acidA receptors. Mol. Pharmacol. 43 (6), 970–975.

Hall, R.C., Zisook, S., 1981. Paradoxical reactions to benzodiazepines. Br. J. Clin. Pharmacol. 11 (Suppl. 1), 99S–104S.

Harvey, D., 1975. Hypnotics and Sedatives. The Pharmacological Basis of Therapeutics, fifth ed. Macmillan Publishing, New York 102–136.

Hindmarch, I., 2006. Zolpidem extended-release. CNS Drugs 20 (5), 427.

Hollister, L.E., Motzenbecker, F.P., et al., 1961. Withdrawal reactions from chlordiazepoxide ("Librium"). Psychopharmacologia 2, 63–68.

Hollister, L.E., Bennett, J.L., et al., 1963. Diazepam in newly admitted schizophrenics. Dis. Nerv Syst. 24, 746–750.

Holm, K.J., Goa, K.L., 2000. Zolpidem: an update of its pharmacology, therapeutic efficacy and tolerability in the treatment of insomnia. Drugs 59 (4), 865–889.

Honan, V.J., 1994. Paradoxical reaction to midazolam and control with flumazenil. Gastrointest. Endosc. 40 (1), 86–88.

Jaffe, J.H., Ciraulo, D.A., et al., 1983. Abuse potential of halazepam and of diazepam in patients recently treated for acute alcohol withdrawal. Clin. Pharmacol. Ther. 34 (5), 623–630.

Kawasaki, Y., Freire, E., 2011. Finding a better path to drug selectivity. Drug Discov. Today 16 (21–22), 985–990.

Kralic, J.E., Korpi, E.R., et al., 2002. Molecular and pharmacological characterization of GABA(A) receptor alpha1 subunit knockout mice. J. Pharmacol. Exp. Ther. 302 (3), 1037–1045.

Lader, M., 2011. Benzodiazepines revisited–will we ever learn? Addiction 106 (12), 2086–2109.

Lader, M.H., Ron, M., et al., 1984. Computed axial brain tomography in long-term benzodiazepine users. Psychol. Med. 14 (1), 203–206.

Lagnaoui, R., Depont, F., et al., 2004. Patterns and correlates of benzodiazepine use in the French general population. Eur. J. Clin. Pharmacol. 60 (7), 523–529.

Lake, B.G., Ball, S.E., et al., 2002. Metabolism of zaleplon by human liver: evidence for involvement of aldehyde oxidase. Xenobiotica 32 (10), 835–847.

Lankford, A., 2009. Sublingual zolpidem tartrate lozenge for the treatment of insomnia. Expert Rev. Clin. Pharmacol. 2 (4), 333–337.

Lettieri, C.J., Shah, A.A., et al., 2009. Effects of a short course of eszopiclone on continuous positive airway pressure adherence: a randomized trial. Ann. Intern. Med. 151 (10), 696–702.

Leufkens, T.R., Lund, J.S., et al., 2009. Highway driving performance and cognitive functioning the morning after bedtime and middle-of-the-night use of gaboxadol, zopiclone and zolpidem. J. Sleep Res. 18 (4), 387–396.

López-Muñoz, F., Alamo, C., et al., 2011. The discovery of chlordiazepoxide and the clinical introduction of benzodiazepines: half a century of anxiolytic drugs. J. Anxiety Disord. 25 (4), 554–562.

Maier, G., Koch, P., et al., 2003. Dose proportionality and time steady-state of eszopiclone in health adult volunteers following single and multiple dosing. AAPS Pharm. Sci. 5 (Suppl. 1), 1806.

Mandrioli, R., Mercolini, L., et al., 2010. Metabolism of benzodiazepine and non-benzodiazepine anxiolytic-hypnotic drugs: an analytical point of view. Curr. Drug Metab. 11 (9), 815–829.

Matthew, H., Roscoe, P., et al., 1972. Acute poisoning. A comparison of hypnotic drugs. Practitioner 208 (244), 254–258.

McBay, A.J., 1973. Toxicological findings in fatal poisonings. Clin. Chem. 19 (4), 361–365.

McKinney, A., Coyle, K., 2004. Next day effects of a normal night's drinking on memory and psychomotor performance. Alcohol Alcohol. 39 (6), 509–513.

Moen, M.D., Plosker, G.L., 2006. Zolpidem extended-release. CNS Drugs 20 (5), 419–426 discussion 427–418.

Moodley, P., Golombok, S., et al., 1993. Computed axial brain tomograms in long-term benzodiazepine users. Psychiatry Res. 48 (2), 135–144.

Najib, J., 2006. Eszopiclone, a nonbenzodiazepine sedative-hypnotic agent for the treatment of transient and chronic insomnia. Clin. Ther. 28 (4), 491–516.

Neubauer, D.N., 2006. Zolpidem extended release: a viewpoint by David N. Neubauer. CNS Drugs 20 (5), 427–428.

Neubauer, D.N., 2010. ZolpiMist™: a new formulation of zopidem tartrate for the short-term treatment of insomnia in the U.S. Nat. Sci. Sleep 2, 79–84.

Nielsen, M., Hansen, E.H., et al., 2012. What is the difference between dependence and withdrawal reactions? A comparison of benzodiazepines and selective serotonin re-uptake inhibitors. Addiction 107 (5), 900–908.

Noguchi, H., Kitazumi, K., et al., 2002. Binding and neuropharmacological profile of zaleplon, a novel non-benzodiazepine sedative/hypnotic. Eur. J. Pharmacol. 434 (1–2), 21–28.

Noyes, A.P., Kolb, L.C., 1968. Modern Clinical Psychiatry. Saunders, Philadelphia.

Nutt, D.J., Stahl, S.M., 2010. Searching for perfect sleep: the continuing evolution of GABAA receptor modulators as hypnotics. J. Psychopharmacol. 24 (11), 1601–1612.

Olfson, M., Marcus, S., et al., 2004. National trends in the outpatient treatment of anxiety disorders. J. Clin. Psychiatry 65 (9), 1166–1173.

Pat McAndrews, M., Weiss, R.T., et al., 2003. Cognitive effects of long-term benzodiazepine use in older adults. Hum. Psychopharmacol. 18 (1), 51–57.

Patat, A., Naef, M.M., et al., 1994. Flumazenil antagonizes the central effects of zolpidem, an imidazopyridine hypnotic. Clin. Pharmacol. Ther. 56 (4), 430–436.

Paterniti, S., Dufouil, C., et al., 2002. Long-term benzodiazepine use and cognitive decline in the elderly: the epidemiology of vascular aging study. J. Clin. Psychopharmacol. 22 (3), 285–293.

Perera, K.M., Powell, T., et al., 1987. Computerized axial tomographic studies following long-term use of benzodiazepines. Psychol. Med. 17 (3), 775–777.

Piper, A.J., 1995. Addiction to benzodiazepines-how common? Arch. Fam. Med. 4, 964–970.

Pritchett, D.B., Seeburg, P.H., 1990. Gamma-aminobutyric acidA receptor alpha 5-subunit creates novel type II benzodiazepine receptor pharmacology. J. Neurochem. 54 (5), 1802–1804.

Puustinen, J., Nurminen, J., et al., 2007. Associations between use of benzodiazepines or related drugs and health, physical abilities and cognitive function: a non-randomised clinical study in the elderly. Drugs Aging 24 (12), 1045–1059.

Renwick, A.B., Mistry, H., et al., 1998. Metabolism of Zaleplon by human hepatic microsomal cytochrome P450 isoforms. Xenobiotica 28 (4), 337–348.

Reynolds, M., Fulde, G., et al., 2013. Trends in benzodiazepine abuse: 2007–2011. Emerg. Med. Aust. 25 (2), 199–200.

Rosen, A.S., Fournie, P., et al., 1999. Zaleplon pharmacokinetics and absolute bioavailability. Biopharm. Drug Dispos. 20 (3), 171–175.

Roth, T., Hull, S.G., et al., 2008. Low-dose sublingual zolpidem tartrate is associated with dose-related improvement in sleep onset and duration in insomnia characterized by middle-of-the-night (MOTN) awakenings. Sleep 31 (9), 1277–1284.

Rudolph, U., Knoflach, F., 2011. Beyond classical benzodiazepines: novel therapeutic potential of GABAA receptor subtypes. Nat. Rev. Drug Discov. 10 (9), 685–697.

Salzman, C., Fisher, J.L., et al., 1992. Cognitive improvement following benzodiazepine discontinuation in elderly nursing home residents. Int. J. Geriatr. Psychiatry 7, 89–93.

Substance Abuse and Mental Health Services Administration, Office of Applied Studies, 2007. Treatment Episode Data Set (TEDS) Highlights–2006 National Admissions to Substance Abuse Treatment Services, OAS Series #s-40. DHHS, Rockville, MD (SMA) 08–4313.

SAMHSA, 2011. Drug Abuse Warning Network (DAWN): National Estimates of Drug-related Emergency Department Visits. DAWN, Rockville, MD, HHS.

Sanger, D.J., 2004. The pharmacology and mechanisms of action of new generation, non-benzodiazepine hypnotic agents. CNS Drugs 18 (Suppl. 1), 9–15 discussion 41, 43–15.

Sarid-Segal, O., Knapp, C.M., et al., 2009. The anticonvulsant zonisamide reduces ethanol self-administration by risky drinkers. Am. J. Drug Alcohol Abuse 35 (5), 316–319.

Schmauss, C., Apelt, S., et al., 1987. Characterization of benzodiazepine withdrawal in high- and low-dose dependent psychiatric inpatients. Brain Res. Bull. 19 (3), 393–400.

Smith, A.J., 1972. Self-poisoning with drugs: a worsening situation. Br. Med. J. 4 (5833), 157–159.

Smith, D.E., Wesson, D.R., 1983. Benzodiazepine dependency syndromes. J. Psychoactive Drugs 15 (1–2), 85–95.

Stanley, N., Hindmarch, I., et al., 2005. Zolpidem modified-release 12.5 mg improves measures of sleep disturbance (traffic noise) compared with standard zolpidem 10 mg. Pharmacotherapy 25 (10), 1504.

Svab, V., Subelj, M., et al., 2011. Prescribing changes in anxiolytics and antidepressants in Slovenia. Psychiatr. Danub 23 (2), 178–182.

U. S. Food and Drug Administration. Questions and Answers: risk of next-morning impairment after use of insomnia drugs; FDA requires lower recommended doses for certain drugs containing zolpidem (Ambien, Ambien CR, Edluar, and Zolpimist). Retrieved at: http://www.fda.gov/Drugs/DrugSafety/ucm334041.htm on June 15, 2013.

Uhde, T.W., Kellner, C.H., 1987. Cerebral ventricular size in panic disorder. J. Affect Disord. 12 (2), 175–178.

Uslaner, J.M., Tye, S.J., et al., 2013. Orexin receptor antagonists differ from standard sleep drugs by promoting sleep at doses that do not disrupt cognition. Sci. Transl. Med. 5 (179), 179ra44.

Vermeeren, A., Coenen, A.M., 2011. Effects of the use of hypnotics on cognition. Prog. Brain Res. 190, 89–103.

Veronese, A., Garatti, M., et al., 2007. Benzodiazepine use in the real world of psychiatric practice: low-dose, long-term drug taking and low rates of treatment discontinuation. Eur. J. Clin. Pharmacol. 63 (9), 867–873.

Vinkers, C.H., van Oorschot, R., et al., 2012. GABA(A) receptor α subunits differentially contribute to diazepam tolerance after chronic treatment. PLoS One 7 (8), e43054.

Vithlani, M., Terunuma, M., et al., 2011. The dynamic modulation of GABA(A) receptor trafficking and its role in regulating the plasticity of inhibitory synapses. Physiol. Rev. 91 (3), 1009–1022.

Von Moltke, L.L., Greenblatt, C.L., et al., 1999. Zolpidem metabolism in vitro: responsible cytochromes, chemical inhibitors, and in vivo correlations. Br. J. Clin. Pharmacol. 48 (1), 89–97.

Weinling, E., McDougall, S., et al., 2006. Pharmacokinetic profile of a new modified release formulation of zolpidem designed to improve sleep maintenance. Fundam. Clin. Pharmacol. 20 (4), 397–403.

Zandstra, S., Furer, J., et al., 2002. Different study criteria affect the prevalence of benzodiazepine use. Soc. Psychiatry Psychiatr. Epidemiol. 37 (3), 139–144.

Zanettini, C., Yoon, S.S., et al., 2013. Acute tolerance to chlordiazepoxide qualitatively changes the interaction between flumazenil and pregnanolone and not the interaction between flumazenil and midazolam in rhesus monkeys discriminating midazolam. Eur. J. Pharmacol. 700 (1–3), 159–1564.

CHAPTER SEVENTEEN

Hallucinogens

Ryan HA. Chan and John E. Mendelson

Addiction and Pharmacology Research Laboratory, California Pacific Medical Center Research Institute, CA, USA

1. INTRODUCTION

Hallucinogens are drugs that alter consciousness by inducing sensory and perceptual disturbances. Distorted perceptions of visual and auditory stimuli occur most commonly but tactile, gustatory, and olfactory distortions can also occur. A phenomenon called *synesthesia* is a combination of sensory distortions where sounds are "seen'" or colors are "heard" (Ries et al., 2009). Hallucinogens also induce intense emotional responses and thoughts that can influence the human psyche. However, the term *hallucinogen* can be misleading because it is not common for users to experience true hallucinations such as manifestations of something nonexistent. Instead, users experience illusions or a perceptual distortion of normal environmental stimuli (Lowinson et al., 2004). Although hallucinogens are relatively nontoxic, even in experienced users, they can produce dramatic effects that often have unpredictable variability in onset, duration, and intensity. Nondrug factors such as the "set and setting," where hallucinogens are used, can alter the experience and produce adverse effects in even highly experienced users (Lowinson et al., 2004).

Because hallucinogen is an umbrella term and almost any drug can produce sensory or perceptual distortions, this chapter is limited to drugs that are indolealkylamines. These drugs share structural similarities to serotonin and act through activation of specific serotonin receptors (Ries et al., 2009). Phenethylamines are structurally similar to the monoamines (dopamine, norepinephrine, and epinephrine) that have predominately stimulant effects. The phenylisopropylamines are unique because they share both stimulant and sensory distorting properties. They are sometimes referred to as *stimulant hallucinogens*. Because phenylisopropylamines are often synthesized to evade drug control laws or produce a particular form of intoxication, they are also called *designer drugs*. From 1970 to 2010, numerous analogs of substituted phenethylamines and phenylisopropylamines have been introduced to the public, reflecting both the ingenuity of underground medicinal chemists and the continuing appeal of hallucinogens to society.

In this chapter, we focus on the serotonergic hallucinogens. There are four major drugs in this group: psilocin, psilocybin, dimethyl tryptamine (DMT), and lysergic acid

The Effects of Drug Abuse on the Human Nervous System
http://dx.doi.org/10.1016/B978-0-12-418679-8.00017-4

diethylamide (LSD). As seen in Figures 1–5, all share an indolealkylamine structure that is similar to serotonin (5-hydroxytryptamine (5-HT)).

Of these four serotonergic hallucinogens, LSD is the most widely studied and can be considered the prototype for its class. Discovered by the Swiss chemist Albert Hoffman in 1938, LSD is a semisynthetic derivative of a lysergic acid, a natural product of the ergot fungus *Clavicus purpurea*. On April 16, 1943, Hoffman intentionally ingested 250 µg of LSD and experienced anxiety, paranoia, and fear as well as geometric visual

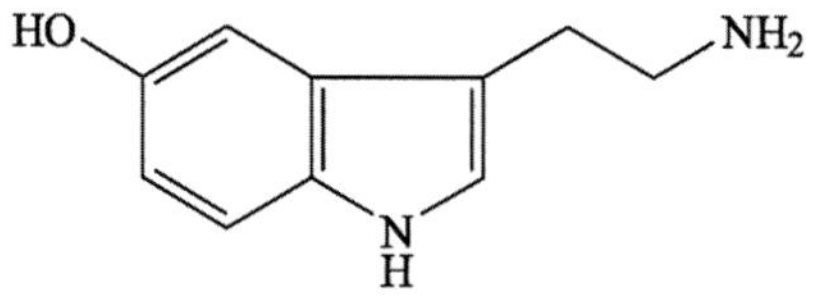

Figure 1 2-D Molecular structure of serotonin.

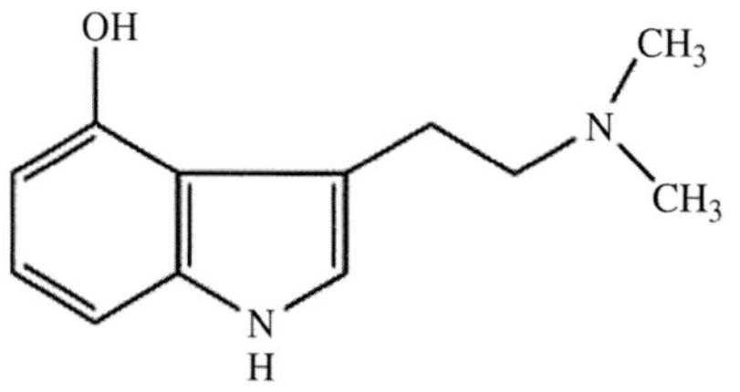

Figure 2 2-D Molecular structure of psilocin.

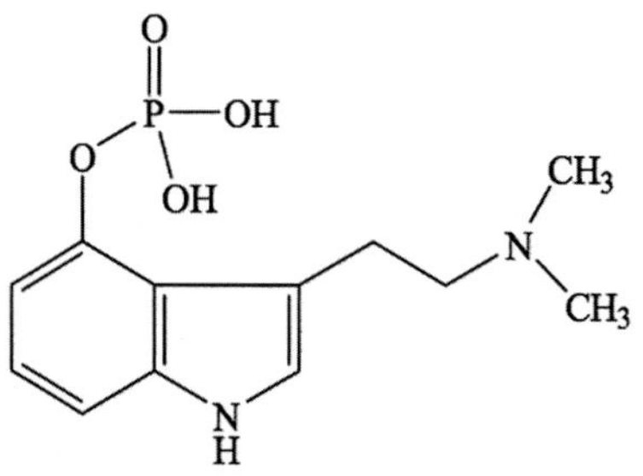

Figure 3 2-D Molecular structure of psilocybin.

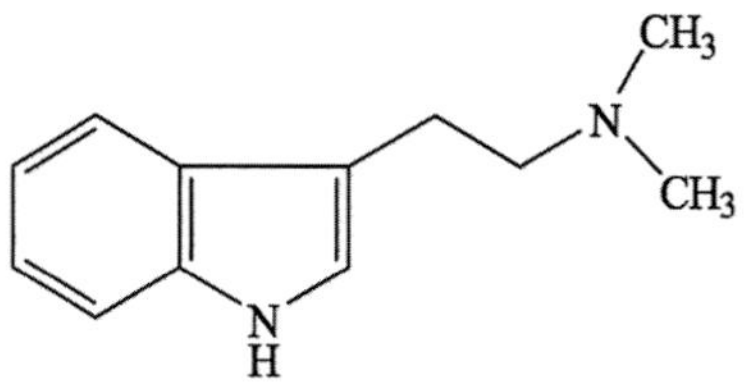

Figure 4 2-D Molecular structure of *N,N*-DMT.

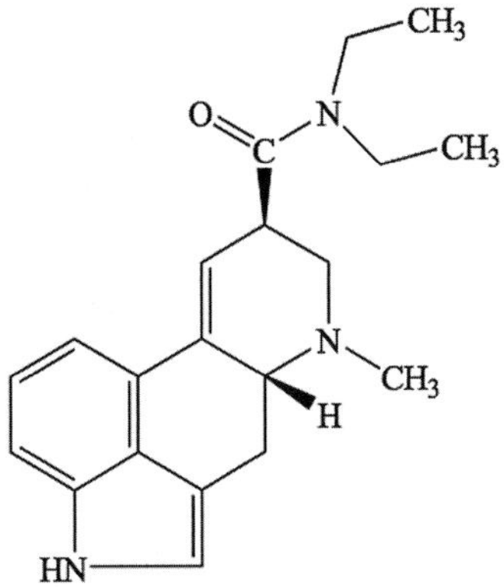

Figure 5 2-D Molecular structure of lysergic acid diethylamide.

alterations (Hoffman, 1979, 1994). This experience set in motion scientific discoveries and cultural upheavals as complex and contradictory as his first drug experience.

Hallucinogens are also known as *psychedelics*, a term Osmond coined to describe the more "mind-manifesting" aspects of the user experience (Osmond, 1957). Whatever the label, drugs that distort sensations have been associated with spiritual or religious enlightenment. This association has its roots in the shamanic practices of indigenous cultures (Winkelman and Roberts, 2007). Even in Western civilization of the nineteenth and twentieth centuries, there is evidence linking psychotomimetic mushroom use in Christian subcultures and the legal usage of the peyote cactus in the Native American church (Henrich, 2002). In many cultures, the experience of an altered state of consciousness facilitates a belief that a person is able to see beyond the boundaries of reality. These mind-manifesting experiences have deeply affected modern neuroscience with scientists questing for both mechanisms and meanings of consciousness. From the mid-twentieth century to the present, researchers have hoped that studying hallucinogens would provide a window into the genesis of severe psychiatric diseases such as the schizophrenias as well the neural basis of creativity, spirituality, and sensory processing. Coincident with scientific exploration, widespread public experimentation with hallucinogens caused an American government backlash that inhibited hallucinogenic research. We are just emerging from this period of scientific and cultural reticence to learn more about hallucinogens.

During the 1950s, hallucinogens were studied as a method for inducing a temporary psychosis in normal people and called *psychotomimetics* (producer of psychosis). Studies hoped that hallucinogens would be useful toward understanding the pharmacology of psychosis and the schizophrenias. By 1961, just prior to widespread abuse, there was a robust research effort with nearly 1000 articles published. However, scientific interest may have laid the foundation for hallucinogens' emergence in the 1960s "counter culture" movement. For example, Harvard psychologist Timothy Leary launched a movement built around the LSD experience where hallucinogens were used in a controlled manner that sought the enhancement of creativity and positive emotions (Lattin, 2010). Increasing publicity and widespread availability of LSD contributed to an epidemic of use in a predominately youth culture.

In tandem with the increasing popularity of hallucinogens, other abusable intoxicants emerged including marijuana, heroin, cocaine, methamphetamine, and methylenedioxymethamphetamine (MDMA). The combination of wide spread abuse and media sensationalism resulted in a growing public consensus that viewed these drugs as a danger to society. In the United States, the Comprehensive Drug Abuse Prevention and Control Act of 1970 was enacted to prosecute President Nixon's declaration of the War on Drugs. Under this legislation, hallucinogens were classified in Schedule 1, which is the highest level of control in the United States. By definition, Schedule 1 drugs have "no currently accepted medical use in the United States, a lack of accepted safety for use under medical supervision, and a high potential for abuse". Scheduling hallucinogens increased barriers to basic and clinical research. Publication rates decreased and American investigators moved to animal models for hallucinogenic research. With the exception of a very small cadre of researchers, one legacy of the 1970 law has been the lack of clinical studies for more than 40 years.

2.1. PHARMACOLOGY, ANTAGONISTS, AND NEUROANATOMY OF HALLUCINOGEN ACTION

2.1.1. Pharmacology: Serotonergic Mechanism

The family of serotonergic receptors has grown from two subtypes—5-HT_1 and 5-HT_2, discovered by Peroutka and Snyder in 1970s—to a total of 14 receptors categorized into 7 families. LSD has a high receptor affinity for many subcategories of 5-HT receptors including the 5-HT_{1B}, 5-HT_{1D}, 5-HT_7, and 5-HT_6 but there is a fairly clear consensus that agonism at the 5-HT_{2A} receptor mediates the effects of LSD (Ray, 2010; Nichols, 2004). In both animal and human studies, modulation of 5-HT_{2A} receptors appears to medicate the effects of LSD. For example, LSD behavioral effects in mice can be blocked by the highly selective 5-HT_{2A} receptor antagonist M100907 (Sorenson et al., 1993). A 2007 study with 5-HT_{2A} receptor knockout (KO) mice saw a restoration of LSD behavior (head twitch response) when the receptor was genetically restored (González-Maeso et al., 2007). In a rat study comparing chronic administration of either a nonhallucinogenic LSD analog or an LSD, the density of 5-HT_{2A} receptors decreased in LSD treated rats. This decrease suggests a molecular mechanism to explain the rapid development of tolerance to LSD and importance of this receptor (Buckholtz et al., 1990). In humans, Vollenweider et al. (1998) tested the effects of oral ketanserin and risperidone given before 0.25 mg of oral psilocybin. He found these relatively nonselective 5-HT_{2A} receptor blockers attenuated psilocybin effects. Although the effects of 5-HT_{2A} blockade on human LSD effects have not been assessed, similar findings are likely.

The postreceptor pharmacology of LSD is complex due to its actions as a G-protein-coupled receptor. Agonism at the 5-HT_{2A} receptor elevates intracellular phospholipase C that hydrolyzes phosphatidyl inositol 3,4–bisphosphate (PIP_2) into inositol 1,4,5–triphosphate (IP_3) and diacylglycerol (DAG). Both IP_3 and DAG act as secondary messengers to trigger protein phosphorylation and transcription of growth factors in the cell body. However, LSD stimulates PIP_2 hydrolysis weakly and there is speculation that additional activation of the independent phospholipase A_2 signaling pathway is necessary for hallucinogenic action (Fantegrossi et al., 2008).

The pharmacological mechanism of LSD's action and other hallucinogens is far from settled. For example, lisuride, a structural analog of LSD, is a dopamine and serotonin partial agonist that activates 5-HT_{2A} receptors. Yet, it lacks hallucinogenic properties (Nichols, 2004). Lisuride can be distinguished from LSD by using different receptor antagonists to stop their respective disruptions of prepulse inhibition, a measure of sensorimotor gating. While a selective 5-HT_{2A} antagonist attenuated only LSD's disruption, a dopamine D2/D3 antagonist attenuated only lisuride's disruption, suggesting two different receptor pathways (Halberstadt and Geyer, 2010). LSD also activates a pertussis-sensitive heterotrimeric G-protein and a tyrosine kinase protein SRC (often involved in cellular proliferation and growth) to stabilize a distinct conformation that is different from lisuride (Gonzalez-Maeso and Sealfon, 2009). The concept that G-protein receptors can have multiple agonist conformations explains these differences in pharmacologic effects.

Although LSD is primarily associated with the 5-HT_{2A} receptors, the drug's affinity for the receptor as a weak agonist is similar to other hallucinogens that are 20–30 times less potent (Kurrasch-Orbaugh et al, 2003). This weak agonism suggests that LSD must activate other receptors but whether it is other serotonin receptors or a different receptor class is not yet clear. LSD acts on pyramidal 5-HT neurons, which have a complex pharmacology with several 5-HT receptors likely modulating hallucinogenic activity. All agents that activate 5-HT_{2A} receptors also activate 5-HT_{2C} receptors, suggesting 5-HT_{2C} receptors may also be needed for hallucinogenic effects. In some rodent studies, compounds selective for the 5-HT_{2C} receptors appear more active than 5-HT_{2A} compounds in producing hallucinogenic effects (Nichols, 2004). Thus, although 5-HT_{2A} receptors are needed for a hallucinogenic stimulus to progress, they might not be sufficient and require additional contribution from the 5-HT_{2C} receptors. Adding complexity, the regulation of 5-HT receptors in pyramidal neurons is blocked when a 5-HT_{1A} antagonist is added (Arnt and Hyttel, 1989). When both 5-HT_{1A} and 2A receptor subtypes are coactivated, there is selective enhancement of response to the stronger excitatory stimuli in these pyramidal neurons (Araneda &Andrade, 1991). Lisuride has shown to be a strong 5-HT_{1A} agonist and exhibits no hallucinogenic effect. Its strong promotion of 5-HT_{1A} receptors probably leads to an inhibition of the hallucinogenic pathway over the excitatory hallucinogenic response of 5-HT_{2A} receptors (Marona-Lewicka et al., 2002).

In addition to serotonergic agonism, LSD also modulates the activity of glutamate and dopamine receptors. With glutamate, the metabotropic glutamate receptor 2 (mGluR2) dimerizes with 5-HT_{2A} to form a receptor complex that goes down two different G-protein pathways depending on the ligand (de Bartolomeis et al., 2013). In an mGluR2-KO study, mGluR2-KO mice did not exhibit a hallucinogenic response to injections of LSD compared to the control (Moreno et al., 2011). This result suggests that mGluR2 is necessary for hallucinosis by promoting a higher affinity conformation for hallucinogenic ligand binding to 5-HT_{2A} as a receptor complex. However, activation of only mGluR2 using a glutamate agonist, LY379268, causes an increase of glutamate in the prefrontal cortex (PFC) and a suppression of hallucinogen signaling (Wischhof and Koch, 2012; Gonzalez-Maeso et al., 2008). This G-protein signaling activates adenyl cyclase, cAMP, and then protein kinase A for an inhibitory response that has a higher prevalence in a normal basal metabolism. A decrease in mGluR2 receptors would maintain the prohallucinogenic pathway but limit the suppression. With schizophrenia and serotonergic hallucinosis having many similarities in pathways, a postmortem study of schizophrenic patients showed high cortical densities of 5-HT_{2A} receptors and low cortical mGluR2 to provide further evidence for this mGluR2 theory (González-Maeso et al., 2008). This theory of two competing G-protein pathways indicates that these receptors regulate each other and act as a signal gate for the cortex. It also suggests that these receptors have different responses as individual components than as a receptor complex (De Bartolomeis et al., 2013).

LSD also has dopaminergic effects with D2 receptor binding at behaviorally relevant doses (Nichols, 2004). In an animal study, risperidone, a mixture of 5-HT_2 and D2 antagonists, was more effective in blocking LSD discrimination in rats than ritanserin, a pure $5\text{-HT}_{2A}/5\text{-HT}_{2C}$ antagonist (Meert et al., 1989; Halberstadt et al., 2011). Even the efficacy of ritanserin against LSD increases substantially when given with haloperidol, an inverse dopamine agonist (Meert and Awouters, 1991). There is also evidence of a heteromer between D2 receptors and 5-HT_{2A} receptors in HEK293 cells (Albizu et al., 2011). Efficacy of hallucinogen binding to the 5-HT_{2A} receptor was increased when the D2 receptor was present and decreased after introducing quinpirole, a D2 agonist. Like the mGluR2 receptors, the D2 receptors promote a conformational change in the 5-HT_{2A} receptors. This conformational change allows increased coupling of the 5-HT_{2A} receptor to the G-protein with a concomitant increase in the pathway's activation. Introduction of the D2 agonist changes the D2 receptor's conformation, which alters the 5-HT_{2A} receptor's conformation and decreases its coupling to the G-protein for a suppressed hallucinogenic response.

Activation of dopamine receptors by LSD is also time-dependent (Marona-Lewicka et al., 2005, Marona-Lewicka and Nichols, 2007). In rodents, a 5-HT_{2A} receptor antagonist blocks behavior 30 min after LSD injection while D2 antagonists are only effective after 90 min. The study's results suggest LSD effects are separated into two phases. The first is dictated by serotonergic receptors with the second, later phase by dopaminergic receptors. In line with these laboratory findings, clinical observations of human LSD experiences indicate a temporal change to a more paranoid state at later time points (Freedman, 1984). These observations suggest a possible serotonergic etiology for the symptoms of paranoia and delusions seen in schizophrenia.

2.1.2. Antagonists

MAO inhibitors and selective serotonin reuptake inhibitors have been used to attenuate acute LSD hallucinations (Resnick et al., 1964; Strassman, 1992; Bonson and Murphy, 1996; Bonson et al., 1996). Similar to the development of LSD behavioral tolerance through continual usage, chronic administration of these inhibitors decreases the density of serotonin receptors (5-HT), which may mitigate behavioral effects of LSD. Pretreatment with fluoxetine reduced the effects of LSD. In contrast, lithium or tricyclic antidepressants exacerbate LSD-induced hallucinations (Bonson and Murphy, 1996, Bonson et al., 1996). Recent rodent studies suggest that clozapine attenuates LSD behavior by downregulating the 5-HT_{2A} receptors while olanzapine reduces hyperactivity in rats with chronic LSD injections (Moreno et al., 2012; Marona-Lewicka et al., 2011).

Trials addressing the human clinical pharmacology of LSD have only recently resumed. Unfortunately there are limited human serotonergic pharmacologic probes that are safe and well characterized. Vollenweider et al. (1998) used ketanserin and risperidone to attenuate the hallucinogenic effects of psilocybin. A selective serotonin antagonist, m100907 (Sorenson et al., 1993), that successfully inhibits LSD behavior in rats now has the name volinanserin. This drug is undergoing human safety trials for the treatment of insomnia and interaction studies between volinanserin and LSD may soon be possible.

2.2. NEUROANATOMY OF HALLUCINOGENS

The anatomical distribution of 5-HT_{2A} receptors is broad with systemic and central nervous system (CNS) expression of 5-HT2A receptors. CNS pyramidal neurons in cortical layer V have the highest concentrations of 5-HT_{2A} receptors. Activation of these neurons in the prefrontal and the anterior cingulate cortex is most likely responsible for the altered sensory perceptions induced by hallucinogens (Beique et al., 2007; Gresch et al, 2007). When the serotonin antagonist M100907 is infused directly into the anterior cingulate cortex of rats, the stimulus effects of systemic LSD are blocked, supporting the anatomic importance of this region in mediating LSD effects (Gresch et al, 2007).

In addition to pyramidal neurons, LSD induces more activation of C-fos in oligodendrocytes than pyramidal neurons in cortical layer V (Reissig et al., 2008). Oligodendrocytes are responsible for maintaining the myelin sheath. Some recent studies suggest that oligodendrocyte dysfunction is associated with psychiatric disorders including schizophrenia, adding some support for similarities between schizophrenia and serotonergic hallucinosis (Tkachev et al., 2003; Chew et al., 2013).

Human PET studies using either the 5-HT$_{2A}$ receptor ligand [^{18}F] altanserin or fluorodeoxyglucose suggest that psilocybin acts in the anterior cingulate cortex and medial prefrontal cortex (mPFC). These brain areas regulate emotional responses, interpreting visual memories and correlating emotions (Quednow et al., 2010; Vollenweider et al., 1997; Gouzoulis-Mayfrank et al., 1999). The fluorodeoxyglucose studies also showed a substantial decrease in brain activity in the thalamus. However, not all data are in agreement. One study showed decreased activity in the anterior cingulate, medial prefrontal, and posterior cingulate cortex and thalamus that was directly correlated with intensity of the user's altered experience (Carhart-Harris et al., 2012a, 2012b). This study suggests decoupling between the posterior cingulate cortex and mPFC because they had the most consistent decreased activity after psilocybin and the highest levels of glucose metabolism after placebo. The mPFC is important for its ability to inhibit the limbic activity in the brain, which defines our brain's processing of emotions and links to our autonomic nervous system. This study challenges the idea that hallucinogens cause only an influx of brain activation. Instead, 5-HT$_{2A}$ receptor activation leads to widespread deactivation and most importantly, the mPFC region that contributes to serotonergic hallucinosis.

2.3. HALLUCINOGENS IN MEDICINE

2.3.1. Acute Pharmacological Effects of LSD

LSD is highly potent and orally active. Albert Hoffman, the discoverer of LSD, took the first known dose of LSD in 1943 when the agent accidently came in contact with his skin. A day later, he deliberately repeated the experience with a 0.25 mg dose, which is now considered a large dose (Hoffman, 1979, 1994). Current estimates of illicit doses range from 0.025 to 0.08 mg compared to 0.10–0.20 mg range of the late 1960s (Nichols, 2004). Hallucinogenic experiences produce anxiety with lower doses less likely to result in adverse effects and require medical intervention (Nichols, 2004). LSD is taken orally with an onset of psychological and physiological symptoms within 60 min. Peak effects occur between 2 and 4 h after administration with effects persisting for 10–12 h (Lowinson et al., 2004). Most users suffer no long-term complications. Compared to LSD, other serotonergic hallucinogens tend to have lesser effects. Psilocybin effects last 4–6 h following 20–30 mg oral doses (Griffiths et al., 2011). DMT effects last 30 min after 60–100 mg smoked/injected doses (Ott, 2011).

From 1 to 4h after administration ("The Trip"), an LSD user experiences certain somatic, perceptual, and psychic symptoms. These symptoms recorded by Hollister (1984) are still pertinent to our current understanding of LSD interactions with the human body (Table 1).

Other specific autonomic effects include pupil dilation, hyperreflexia, diaphoresis, urinary retention, increase in blood pressure and body temperature, piloerection, and tachycardia. However, anxiety from undergoing this experience is a suspected contributor to these sympathetic effects (Ries et al., 2009).

As mentioned earlier, an LSD user often experiences distortions in his sensory perceptions and rarely manifestations of visual/auditory images (true hallucinations). Users often have intensified emotions and they can occur at the same time. Many users feel that the effects wear off after several hours but develop paranoid thoughts and ideas of reference (Lowinson et al., 2004). Although there is no immediate craving for the drug, there is a feeling of fatigue within the next 12h after abatement of LSD activity (Lowinson et al., 2004). The user has a very clear recollection of the drug experience and it is not uncommon for users to claim a change in their psyche and belief system.

In 1967, Cohen stated that death was not "directly caused by the toxicity of LSD" (Cohen, 1967). This statement remains true more than 40 years later (Jaffe, 1985; Nichols, 2004). The acute cardiovascular, renal, and hepatic toxicity of LSD is minimal, possibly because the drug has little affinity for the biological receptors that target these organs (Nichols, 2004). There is no recorded lethal dose in humans and the only reported acute overdose death occurred in an elephant administered a 300mg dose. Based on brain weight, this dose was ~1200 times stronger than the dose taken by Hoffman (West et al., 1962). There is a single 1971 case report of hemiplegia after LSD, suggesting that LSD can produce vasospasms (Sobel et al., 1971). A healthy patient experienced rhabdomyolysis after ingesting a "moderate" but unknown dose of LSD (Berrens et al., 2010). Another report described a fibrotic inflammatory mass in the mesentery of a chronic LSD user and multifocal cerebral demyelination for a user of psilocybin (Spengos et al., 2000). In general, there is "no accepted evidence of brain cell damage, chromosomal abnormalities, or teratogenic effects" when using LSD (Strassman, 1984; Li and Lin, 1998; Nichols, 2004).

Table 1 Typical Clinical Symptoms of LSD Users According to Hollister in 1984

Somatic symptoms	Dizziness, weakness, tremors, nausea, drowsiness, paresthesias, and blurred vision
Perceptual symptoms	Altered shapes and colors, difficulty in focusing on objects, sharpened sense of hearing, and rarely synesthesias
Psychic symptoms	Alterations in mood (happy, sad, or irritable at varying times), tension, distorted time sense, difficulty in expressing thoughts, depersonalization, dreamlike feelings, and visual hallucinations

Griffiths and colleagues have conducted several studies characterizing the human pharmacology of psilocybin. At 7 h, 20–30 mg/70 kg psilocybin produces anxiety or fear in ~40% of participants but in ~70%, the drug induces a "mystical" or spiritual experience. In addition to anxiety, psilocybin can induce headaches with a frequency, severity, and duration that increase with dose. In a pooled analysis of 110 people in 8 double-blind psilocybin studies, headaches were the most consistent acute side effect followed by a feeling of tiredness and exhaustion (Studerus et al., 2011). There were no serious complications in any of these studies suggesting that psilocybin, like LSD, is generally well tolerated. For more detailed descriptions of hallucinogenic experiences, there is extensive scientific documentation by Hollister and Griffiths with several personal accounts on Erowid.com.

However, suicidal behavior has occurred in intoxicated LSD users (Ungerleider et al., 1966). The development of paranoid thoughts, when combined with the user's alterations in perception, can create a "bad trip" with severe acute anxiety. Sober companions or "trip sitters" are used to safeguard hallucinogenic users from making poor decisions. Many users believe that there is a chance for dangerous behavior in a hallucinogenic user without assistance from a sober companion. The development of these symptoms was hypothesized as the mechanisms of schizophrenia in psychiatric patients.

2.3.2. Risk of Addiction

Repeated use of LSD results in tolerance and cross-tolerance with other serotonergic hallucinogens, which suggests that their mechanisms of tolerance development follow similar pathways (Lowinson et al., 2004). However, LSD is not considered addicting. Behavioral tolerance develops quickly after a few days of administration and this tolerance disappears if drug use is discontinued for some period of time. Therefore, users will take LSD 1–2 times spaced over a week (Ungerleider and Fisher, 1967). Although tolerance develops to the psychological effects, less tolerance seems to develop to autonomic effects. There are no withdrawal symptoms after terminating repeated usage, suggesting a low risk for addiction (Lowinson et al., 2004).

2.3.3. Long-Term Adverse Effects

Hallucinogen persisting perception disorder (HPPD) or "flashbacks" is the only recognized long-term complication of hallucinogen use. Under current DSM-IV guidelines, HPPD is the return of physiological and psychological symptoms from the original drug experience without use of the triggering drug. Chronic hallucinogen users may have an increased risk of developing HPPD. Although repeated use may increase the risk, long-term adverse effects can occur after only one exposure to LSD (Levi and Miller, 1990). A web questionnaire with 2455 respondents enrolled from Erowid website concluded that 60.6% had drug-free visual experiences that resembled hallucinogenic effects but only 4.2% found them distressing enough to seek help (Baggott et al., 2011). Although these data describe participants who had used several hallucinogenic drugs, the prevalence of visual experiences in non-LSD users was 18.1% compared to 34.5% who used LSD. Currently, there is no defined range of HPPD severity but this survey reveals the possibility of a higher rate of milder visual alterations with LSD use. Fortunately, the incidence of HPPD appears to be very small based on millions of hallucinogen users (Nichols, 2004). Overall, the understanding of HPPD triggers is limited because there is still no clear mechanism of LSD's influence on the brain.

2.3.4. Treating Complications of Hallucinogen Use

LSD often produces anxiety but severe complications requiring hospitalization are uncommon. Because no safe or specific antagonist is available, treatment of severe intoxication or an acute adverse reaction is symptomatic. Hallucinogen-associated anxiety can be treated by sedation with benzodiazepines. Despite the interest in LSD as a psychotomimetic, hallucinogens rarely produce more than transient episodes of thought disorder. Some recommend using phenothiazines such as haloperidol to treat acute LSD-induced psychosis (Dewhurst and Hatrick, 1972). However, use of antipsychotics should be approached with caution due to a potential paradoxical increase in psychotomimetic effects with illegal substances like phencyclidine and 2,5-Dimethoxy-4-methylamphetamine. Therefore, if the drug is unknown, benzodiazepines may be safer and just as effective (Vollenweider et al., 1998).

Patients occasionally seek treatment for HPPD but pharmacological management can be unsatisfying. Treatment with a serotonin reuptake inhibitor antidepressants or dopaminergic antagonists such as risperidone have reportedly induced or worsened symptoms (Lerner et al., 2002a; Markel et al., 1994; Alcántara, 1998). Mechanistically, increased synaptic serotonin may activate sensitized 5-HT_{2A} neurons for a heightened hallucinogenic response. Some benefit has been seen with the centrally acting alpha 2A agonist clonidine and the benzodiazepine clonazepam, or a combination of fluoxetine and olanzapine (Lerner et al., 2002b; Aldurra and Crayton, 2001). Most treatment options include sedatives, which are likely effective due to nonspecific sedation rather than impacting specific pharmacologic disruptions induced by LSD. Nonpharmacological avenues such as "talk down therapy," which works by openly addressing and reassuring a user's symptoms and anxiety, should be used in addition to these options for acute and chronic adverse events.

2.3.5. Epidemiology

In the United States, abuse of LSD decreased from peak levels in the 1960s (due to limited availability) but reversed in the 1990s as usage increased amongst young adults. In the most recent 2012 MTF study (The Monitoring the Future study, also known as the National High School Senior Survey), high-school students found that LSD was only "fairly accessible," with 27.6% of 12th graders, 14.9% of 10th graders, and 7.5% of 8th graders reporting that they can obtain LSD easily (Johnston et al., 2013). The combined 8th, 10th, and 12th grade annual LSD use prevalence rate has declined from 2.4% in 2002 to 1.6% in 2012 and this change in annual prevalence has fluctuated only marginally from 2003 to 2012 (0.1–0.3%). However, the use of other hallucinogens such as psilocybin has increased with annual prevalence rates of 3.1% in 2011 and 2.7% in 2012. Many 12th graders rate psilocybin as more available than LSD, which explains the increased prevalence of psilocybin use. The declining prevalence of LSD use has been accompanied by a fall in ratings of disapproval and perceived risk of LSD. Therefore, the decline in use is more a function of availability than perceived risk. The MTF study calls this "generational forgetting". Unfortunately, this could set the stage for a revival of abuse, with a new generation that seeks novel experiences.

As a much broader age population sampling of 12 years old and above, the 2011 National Survey on Drug Use and Health (NSDUH) data estimates newly initiated LSD use is rising but not as sharply as other drugs like MDMA. The NSDUH estimated that 358,000 people initiated use of LSD in 2011 as seen in Figure 6. In contrast, 922,000 are estimated to have initiated MDMA during this period. This number of LSD initiates is similar to 2009 but higher than in 2003–2007 when there were 200,000–270,000, respectively (SAMSHA, 2012). From 2002 to 2012, LSD use has remained relatively stable with only a small increase in use. Youth 12 and older perceive LSD as "fairly accessible" (12.9%); this perception has declined from 19.4% in 2002. In general, 972,000 (0.4%) people of the sampling ages 12 and older were using hallucinogens in the past month, which is lower than 1.2 million (0.5%) in 2010. The use of hallucinogens was most prevalent in young adults of 18–25 years of age (SAMSHA, 2012). Hallucinogen use was also highest in males and people of Native American descent or a mixed racial background (SAMSHA, 2012). Some suggest that hallucinogen use is most likely during life transition points such as graduating from high school, starting a career, going to college, or post secondary school (Wu et al., 2006).

2.3.6. Relation to Schizophrenia

The relationship between LSD use and development of schizophrenia is complex. It is unclear if LSD precipitates psychosis or merely exposes psychosis that was already present in the individual (Nichols, 2004). Hoffman noted psychotomimetic effects that were analogous to the behaviors seen in schizophrenia (Hoffman, 1979). From the positive experience of heightened senses and euphoria to the negative experiences such as paranoia and anxiety, the range of effects seen in serotonergic hallucinogenic users is more similar to earliest phases of schizophrenic psychoses than the chronic phases (Geyer and Vollenweider, 2008). Other similarities are seen on a neurological level with common hyperfrontal metabolic patterns between $5-HT_{2A}$ receptor activation and acute schizophrenia/psychosis (Vollenweider et al., 1997).

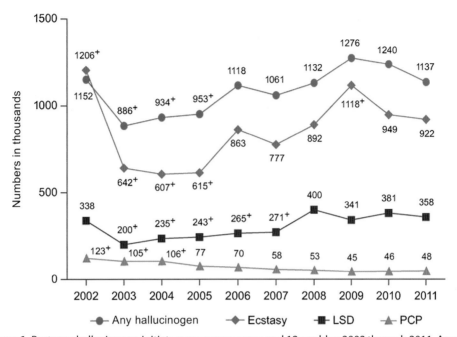

Figure 6 Past year hallucinogen initiates among persons aged 12 or older: 2002 through 2011. Among persons aged 12 or older, the number of past year initiates of LSD was 338,000 in 2002; 200,000 in 2003; 235,000 in 2004; 243,000 in 2005; 265,000 in 2006; 271,000 in 2007; 400,000 in 2008; 341,000 in 2009; 381,000 in 2010; and 358,000 in 2011. The differences between the 2011 estimate and the 2003 through 2007 estimates were statistically significant. *Substance Abuse and Mental Health Services Administration, Results from the 2011 National Survey on Drug Use and Health: Summary of National Findings, NSDUH Series H-44, HHS Publication No. (SMA) 12-4713.*

Vollenweider and Geyer (2001) hypothesized that serotonergic hallucinogens stimulate 5-HT_{2A} receptors in the corticostriatothalamocortical feedback loops to disrupt thalamocortical transmission. The thalamic gating of sensory and cognitive information is disrupted which leads to an overflow of stimuli in the cortex seen in early stages of schizophrenia.

Carhart-Harris et al. (2012c) proposes a model of early schizophrenia that involves competing brain networks. His model specifies two competing networks—the default mode network (DMN) for high-level, internal cognition like introspection and task-positive network (TPN) for processes like goal-oriented tasks (Carhart-Harris et al., 2012c). As competing networks, DMN and TPN have an inverse relationship with one being activated while the other is switched off during normal cognitive activity. Carhart's studies suggest that psilocybin activates both DPN and TPN. He believes that coactivation of these reciprocal networks blurs the boundary between a person's internal self and the environment leading to ego dissolution, a sign that may accompany early psychosis and schizophrenia.

However, there are differences between a serotonergic, hallucinogenic psychosis and psychosis in a schizophrenic patient. One is the life-long progression of schizophrenic disorders compared to the rapid tolerance of psychological effects in serotonergic hallucinogens. Another difference is that schizophrenic patients often experience auditory hallucinations instead of the visual hallucinations seen with serotonergic hallucinogens (Geyer and Vollenweider, 2008). Yet, there is evidence showing visual hallucinations are experienced in the acute and incipient stages of schizophrenic patients, which supports the claim that serotonergic hallucinations mimic early schizophrenia (Geyer and Vollenweider, 2008). It is suggested that schizophrenia is "dominated by neurobiological abnormalities" focusing on serotonin receptors in its earliest stages (Geyer and Vollenweider, 2008). As the schizophrenic disorder progresses, the brain compensates by involving dopaminergic and glutamatergic systems, which explains the striking differences between chronic schizophrenia and acute schizophrenia/serotonergic hallucinosis.

2.3.7. Potential Therapeutic Uses of Serotonergic Hallucinogens

With its potential to induce positive experiences through an altered perception of reality, serotonergic hallucinogens like naturally occurring psilocybin have been accepted into religious cultures. It is suggested that the ingestion of psilocybin in ceremonies for Aztecs and Mayans was a rite of passage or an opportunity to evoke visions in the spiritual realm and ultimately change the user's outlook on life (Stamets, 1996). Although initial studies with serotonergic hallucinogens did not seek to evoke spiritual experiences, recent studies have attempted to study the biological basis of spiritual experiences. This work started with studies of LSD in treatment of psychological disorders. For example, LSD can decrease anxiety and improve the emotional well-being of terminally ill patients (Kast, 1966, 1970). Studies conducted at the Spring Grove State hospital in Maryland found improvements in two-thirds of their terminal cancer patient, which allowed for a reduction in pain medication (Grof et al., 1973; Pahnke et al., 1969). Unfortunately, these studies were terminated after the Controlled Substance Act of 1970. Recently, studies in end of life care have resumed. Grob et al. (2011) showed that a moderate dose (0.2 mg/kg) of psilocybin decreased anxiety in 12 advanced stage cancer patients. After psilocybin, interviews suggest that patients "no longer considered themselves as being overly, anxious or worried people" (Grob et al., 2011).

Case reports suggest that hallucinogens may be useful in treating obsessive–compulsive disorder (OCD). In an early 1962 report, a patient experienced dramatic improvement after two unspecified doses LSD (Savage et al., 1962). Another early study used weekly LSD treatments for 15 months to produce a steady resolution of symptoms. Patients were symptom free for 3 years posttreatment (Brandup and Vanggaard, 1977). There are several other reports that provide case evidence of serotonergic hallucinogens impact on OCD (Hanes, 1996; Moreno and Delgado, 1997). Studies on the efficacy of psilocybin for anxiety are underway at John Hopkins and New York University.

Another focus of serotonergic hallucinogens on psychological disorders is substance abuse such as alcoholism. In a 1971 review of 31 studies from 1953 to 1969, with data of 1100 alcoholics, no definitive statements of LSD's impact on alcoholism could be made because of the wide range in study designs, outcomes, dosages, and definitions of alcoholism (Abuzzahab and Anderson, 1971). Nonetheless, at a 10-month follow up, a single dose of LSD produced improvement in 75% of treated patients vs 44% of the controls (Abuzzahab and Anderson, 1971). For subjects receiving multiple doses, 58% were improved at 20 months vs 54% of controls. Krebs and Johansen (2012) performed a retrospective analysis of data from six randomization controlled trials where participants were dosed once with up to 50 mcg of LSD; notably these trials excluded patients with severe psychiatric illness. They observed a significant beneficial effect of single LSD doses on alcohol misuse and saw a significant improvement between 2 and 6 months posttreatment. However, effects dissipated by 12 months, suggesting that repeated dosing of LSD may be necessary (Krebs and Johansen, 2012). When LSD is compared to treatment with approved medications like naltrexone, acamprosate, or disulfiram, comparable efficacy is seen. Although limited, these data suggest that single LSD doses are comparable to chronic treatment with these approved medications in reducing alcohol relapses.

Studies of hallucinogens to improve psychological well-being are being conducted on several centers. Notably, Griffiths and Johnson are using psilocybin to supplement cognitive behavioral therapy for smoking cessation (Griffiths and Grob, 2010). Overall, the proposed idea of addressing substance abuse with hallucinogens is becoming more rational as new neuroimaging and pharmacological techniques allow for a deeper analysis of drug effects.

3. CONCLUSIONS

Having minimal adverse effects in humans and a powerful influence on the human psyche, LSD was a fascinating tool for neuroscientists in the 1950s. Although the methods and designs of clinical trials have changed over the past 60 years, these early studies provided direction and inspiration for further research on serotonergic hallucinogens. More recent studies have focused on mechanisms of action, associations with disease states and possible therapeutic applications. Psilocybin has emerged as the drug with the most therapeutic potential but is now also the most widely abused member of this class of drugs. LSD is the prototype of the serotonergic hallucinogen class but its long duration of action and its powerful psychopharmacological effects may make LSD less attractive as a therapeutic when compared to shorter acting drugs like DMT or less powerful agents like psilocybin. Other widely abused drugs are also being "rehabilitated" and considered for therapeutic uses. There are also pilot studies into the effects of MDMA in treating posttraumatic stress disorder, ibogaine (psychoactive alkaloid) for opioid addiction, and ketamine for cocaine dependence and treatment–resistant depression (Ross, 2012). Hopefully, these studies will revitalize an area of research that has been stagnant for more than 40 years in the United States.

REFERENCES

Abuzzahab, F.S., Anderson, J., 1971. A review of LSD treatment in alcoholism. Int. Pharmacopsychiatry 6, 223–235.

Albizu, L., Holloway, T., González-Maeso, J., Sealfon, S.C., 2011. Functional crosstalk and heteromerization of serotonin 5-HT2A and dopamine D2 receptors. Neuropharmacology 61 (4), 770–777. http://dx.doi.org/10.1016/j.neuropharm.2011.05.023.

Alcántara, A.G., 1998. Is there a role for the alpha2 antagonism in the exacerbation of hallucinogen-persisting perception disorder with risperidone? J. Clin. Psychopharmacol. 18 (6), 487–488.

Aldurra, G., Crayton, J.W., 2001. Improvement of hallucinogen persisting perception disorder by treatment with a combination of fluoxetine and olanzapine: case report. J. Clin. Psychopharmacol. 21 (3), 343–344.

Araneda, R., Andrade, R., 1991. 5-hydroxytryptamine 2 and 5-hydroxy- tryptamine 1A receptors mediate opposing responses on membrane excitability in rat association cortex. Neuroscience 40, 399–412.

Arnt, J., Hyttel, J., 1989. Facilitation of 8-OHDPAT-induced forepaw treading of rats by the 5-HT2 agonist DOI. Eur. J. Pharmacol. 161, 45–51.

Baggott, M.J., Coyle, J.R., et al., 2011. Abnormal visual experiences in individuals with histories of hallucinogen use: a web-based questionnaire. Drug Alcohol Depend. 114 (1), 61–67.

Béïque, J.C., Imad, M., Mladenovic, L., Gingrich, J.A., Andrade, R., 2007. Mechanism of the 5-hydroxytryptamine 2A receptor-mediated facilitation of synaptic activity in prefrontal cortex. Proc. Natl. Acad. Sci. U.S.A. 104 (23), 9870–9875.

Berrens, Z., Lammers, J., et al., 2010. Rhabdomyolysis after LSD ingestion. Psychosomatics 51 (4) 356-356 e353.

Bonson, K.R., Murphy, D.L., 1996. Alterations in responses to LSD in humans associated with chronic administration of tricyclic antidepressants, monoamine oxidase inhibitors or lithium. Behav. Brain Res. 73 (1–2), 229–233.

Bonson, K.R., Buckholtz, J.W., Murphy, D.L., 1996. Chronic administration of serotonergic antidepressants attenuates the subjective effects of LSD in humans. Neuropsychopharmacology 14 (6), 425–436. http://dx.doi.org/10.1016/0893-133X(95)00145-4.

Brandrup, E., Vanggaard, T., 1977. LSD treatment in a severe case of compulsive neurosis. Acta Psychiatr. Scand. 55, 127–141.

Buckholtz, N.S., Zhou, D.F., Freedman, D.X., Potter, W.Z., 1990. Lysergic acid diethylamide (LSD) administration selectively downregulates serotonin2 receptors in rat brain. Neuropsychopharmacology 3 (2), 137–148.

Carhart-Harris, R.L., Leech, R., Williams, T.M., Erritzoe, D., Abbasi, N., Bargiotas, T., Nutt, D.J., 2012a. Implications for psychedelic-assisted psychotherapy: functional magnetic resonance imaging study with psilocybin. Br. J. Psychiatry 200 (3), 238–244. http://dx.doi.org/10.1192/bjp.bp.111.103309.

Carhart-Harris, R.L., Erritzoe, D., Williams, T., Stone, J.M., Reed, L.J., Colasanti, A., Nutt, D.J., 2012b. From the Cover: neural correlates of the psychedelic state as determined by fMRI studies with psilocybin. Proc. Natl. Acad. Sci. U.S.A. 109 (6), 2138–2143. http://dx.doi.org/10.1073/pnas.1119598109.

Carhart-Harris, R.L., Leech, R., Erritzoe, D., Williams, T.M., Stone, J.M., Evans, J., Nutt, D.J., 2012c. Functional connectivity measures after psilocybin inform a novel hypothesis of early psychosis. Schizophr. Bull. http://dx.doi.org/10.1093/schbul/sbs117.

Chew, L.J., Fusar-Poli, P., Schmitz, T., 2013. Oligodendroglial Alterations and the Role of Microglia in White Matter Injury: Relevance to Schizophrenia. Center for Neuroscience Research, Children's Research Institute, Children's National Medical Center, Washington D.C., U.S.A.

Cohen, S., 1967. Psychotomimetic agents. Annu. Rev. Pharmacol. 7, 301–318.

De Bartolomeis, A., Buonaguro, E.F., et al., 2013. Serotonin-glutamate and serotonin-dopamine reciprocal interactions as putative molecular targets for novel antipsychotic treatments: from receptor heterodimers to postsynaptic scaffolding and effector proteins. Psychopharmacology (Berl.) 225 (1), 1–19.

Dewhurst, K., Hatrick, J.A., 1972. Differential diagnosis and treatment of lysergic acid diethylamide induced psychosis. Practitioner 209 (251), 327–332.

Fantegrossi, W.E., Murnane, A.C., Reissig, C.J., 2008. The behavioral pharmacology of hallucinogens. Biochem. Pharmacol. 75 (1), 17.

Freedman, D.X., 1984. LSD: the bridge from human to animal. In: Jacobs, B.L. (Ed.), Hallucinogens: Neurochemical, Behavioral, and Clinical Perspectives. Raven Press, New York, pp. 203–e226.

Geyer, M.A., Vollenweider, F.X., 2008. Serotonin research: contributions to understanding psychoses. Trends Pharmacol. Sci. 29 (9), 445–453. http://dx.doi.org/10.1016/j.tips.2008.06.006.

González-Maeso, J., Weisstaub, N.V., Zhou, M., Chan, P., Ivic, L., Ang, R., Lira, A., Bradley-Moore, M., Ge, Y., Zhou, Q., Sealfon, S.C., Gingrich, J.A., 2007. Hallucinogens recruit specific cortical 5-HT2A receptor-mediated signaling pathways to affect behavior. Neuron 53, 439–452.

González-Maeso, J., Ang, R.L., Yuen, T., Chan, P., Weisstaub, N.V., López-Giménez, J.F., Sealfon, S.C., 2008. Identification of a serotonin/glutamate receptor complex implicated in psychosis. Nature 452 (7183), 93–97. http://dx.doi.org/10.1038/nature06612.

Gonzalez-Maeso, J., Sealfon, S.C., 2009. Agonist-trafficking and Hallucinogens. Curr. Med. Chem. 16 (8), 1017–1027.

Gouzoulis-Mayfrank, E., Schreckenberger, M., Sabri, O., Arning, C., Thelen, B., Spitzer, M., Kovar, K.A., Hermle, L., Büll, U., Sass, H., 1999. Neurometabolic effects of psilocybin, 3,4-methylenedioxyethylamphetamine (MDE) and d-methamphetamine in healthy volunteers. A double-blind, placebo-controlled PET study with [18F]FDG. Neuropsychopharmacology 20 (6), 565–581.

Gresch, P.J., Barrett, R.J., Sanders-Bush, E., Smith, R.L., 2007. 5-Hydroxytryptamine (serotonin) 2A receptors in rat anterior cingulate cortex mediate the discriminative stimulus properties of d-lysergic acid diethylamide. J. Pharmacol. Exp. Ther. 320 (2), 662–669.

Griffiths, R.R., Grob, C.S., 2010. Hallucinogens as medicine. Sci. Am. 303 (6), 76–79.

Griffiths, R.R., Johnson, M.W., et al., 2011. Psilocybin occasioned mystical-type experiences: immediate and persisting dose-related effects. Psychopharmacology (Berl.) 218 (4), 649–665.

Grob, C.S., Danforth, A.L., Chopra, G.S., Hagerty, M., McKay, C.R., Halberstadt, A.L., Greer, G.R., 2011. Pilot study of psilocybin treatment for anxiety in patients with advanced-stage cancer. Arch. Gen. Psychiatry 68 (1), 71–78. http://dx.doi.org/10.1001/archgenpsychiatry.2010.116.

Grof, S., Goodman, L.E., Richards, W.A., Kurland, A.A., 1973. LSD-assisted psychotherapy in patients with terminal cancer. Int. Pharmacopsychiatry 8, 129–144.

Halberstadt, A.L., Geyer, M.A., 2010. LSD but not lisuride disrupts prepulse inhibition in rats by activating the 5-HT(2A) receptor. Psychopharmacology (Berl.) 208 (2), 179–189.

Halberstadt, A.L., Geyer, M.A., 2011. Multiple receptors contribute to the behavioral effects of indoleamine hallucinogens. Neuropharmacology 61 (3), 364–381.

Hanes, K.R., 1996. Serotonin, psilocybin, and body dysmorphic disorder: a case report. J. Clin. Psychopharmacol. 16, 188–189.

Heinrich, C., 2002. Magic Mushrooms in Religion and Alchemy. Park Street Press, Rochester, VT.

Hofmann, A., 1979. How LSD originated. J. Psychedelic Drugs 11, 53–60.

Hofmann, A., 1994. Notes and documents concerning the discovery of LSD. 1970. Agents Actions 43, 79–81.

Hollister, L.E., 1984. Effects of hallucinogens in humans. In: Jacobs, B.L. (Ed.), Hallucinogens: Neurochemical, Behavioral, and Clinical Perspectives. Raven Press, New York, pp. 19–33.

Jaffe, J.H., 1985. Drug addiction and drug abuse. In: Gilman, A.G., Rall, L.S., Rall, T.W., Murad, F. (Eds.), Goodman and Gilman's the Pharmacological Basis of Therapeutics. Macmillan, New York, pp. 532–581.

Johnston, L.D., O'Malley, P.M., Bachman, J.G., Schulenberg, J.E., 2013. Monitoring the Future National Results on Adolescent Drug Use: Overview of Key Findings, 2012. Institute for Social Research, The University of Michigan, Ann Arbor.

Kast, E., 1966. LSD and the dying patient. Chicago Med. Sch. Q. 26, 80–87.

Kast, E.C., 1970. A concept of death. In: Aaronson, B., Osmond, H. (Eds.), Psychedelics: the Uses and Implications of Hallucinogenic Drugs. Anchor Books, Garden City, NY, pp. 366–381.

Krebs, T.S., Johansen, P.O., 2012. Lysergic acid diethylamide (LSD) for alcoholism: meta-analysis of randomized controlled trials. J. Psychopharmacol. 26 (7), 994–1002. http://dx.doi.org/10.1177/0269881112439253.

Kurrasch-Orbaugh, D.M., Watts, V.J., Barker, E.L., Nichols, D.E., 2003. Serotonin 5-hydroxytryptamine 2A receptor-coupled phospholipase C and phospholipase A2 signaling pathways have different receptor reserves. J. Pharmacol. Exp. Ther. 304, 229–237.

Lattin, Don, 2010. The Harvard Psychedelic Club: How Timothy Leary, Ram Dass, Huston Smith, and Andrew Weil Killed the Fifties and Ushered in a New Age for America. HarperCollins Publishers, New York.

Lerner, A.G., Shufman, E., Kodesh, A., Rudinski, D., Kretzmer, G., Sigal, M., 2002a. Risperidone-associated, benign transient visual disturbances in schizophrenic patients with a past history of LSD abuse. Isr. J. Psychiatry Relat. Sci. 39 (1), 57–60.

Lerner, A.G., Gelkopf, M., Skladman, I., Oyffe, I., Finkel, B., Sigal, M., Weizman, A., 2002b. Flashback and Hallucinogen Persisting Perception Disorder: clinical aspects and pharmacological treatment approach. Isr. J. Psychiatry Relat. Sci. 39 (2), 92–99.

Levi, L., Miller, N.R., 1990. Visual illusions associated with previous drug abuse. J. Clin. Neuroophthalmol. 10 (2), 103–110.

Li, J.H., Lin, L.F., 1998. Genetic toxicology of abused drugs: a brief review. Mutagenesis 13 (6), 557–565.

Lowinson, J.H., Ruiz, P., Langrod, J.G., Millman, R.B., 2004. Substance of Abuse: A Comprehensive Textbook, fourth ed. Lippincott Williams & Wilkins, Philadelphia.

Markel, H., Lee, A., Holmes, R.D., Domino, E.F., 1994. LSD flashback syndrome exacerbated by selective serotonin reuptake inhibitor antidepressants in adolescents. J. Pediatr. 125 (5 Pt 1), 817–819.

Marona-Lewicka, D., Kurrasch-Orbaugh, D.M., Selken, J.R., Cumbay, M.G., Lisnicchia, J.G., Nichols, D.E., 2002. Re-evaluation of lisuride pharmacology: 5-hydroxytryptamine1A receptor-mediated behavioral effects overlap its other properties in rats. Psychopharmacology (Berl.) 164, 93–107.

Marona-Lewicka, D., Thisted, R.A., Nichols, D.E., 2005. Distinct temporal phases in the behavioral pharmacology of LSD: dopamine D2 receptor-mediated effects in the rat and implications for psychosis. Psychopharmacology 180 (3), 427–435. http://dx.doi.org/10.1007/s00213-005-2183-9.

Marona-Lewicka, D., Nichols, D.E., 2007. Further evidence that the delayed temporal dopaminergic effects of LSD are mediated by a mechanism different than the first temporal phase of action. Pharmacol. Biochem. Behav. 87 (4), 453–461. http://dx.doi.org/10.1016/j.pbb.2007.06.001.

Marona-Lewicka, D., Nichols, C.D., et al., 2011. An animal model of schizophrenia based on chronic LSD administration: old idea, new results. Neuropharmacology 61 (3), 503–512.

Meert, T.F., de Haes, P., Janssen, P.A., 1989. Risperidone (R 64 766), a potent and complete LSD antagonist in drug discrimination by rats. Psychopharmacology (Berl.) 97, 206–212.

Meert, T.F., Awouters, F., 1991. Serotonin 5-HT2 antagonists: a preclinical evaluation of possible therapeutic effects. In: Idzikowski, C., Cowen, P.J. (Eds.), Serotonin, Sleep and Mental Disorders. Biomedical Publishing, Washington, pp. 65–76.

Moreno, F.A., Delgado, P.L., 1997. Hallucinogen-induced relief of obsessions and compulsions. Am. J. Psychiatry 154, 1037–1038.

Moreno, J.L., Holloway, T., Albizu, L., Sealfon, S.C., González-Maeso, J., 2011. Metabotropic glutamate mGlu2 receptor is necessary for the pharmacological and behavioral effects induced by hallucinogenic 5-HT2A receptor agonists. Neurosci. Lett. 493 (3), 76–79. http://dx.doi.org/10.1016/j.neulet.2011.01.046.

Moreno, J.L., Holloway, T., Umali, A., Rayannavar, V., Sealfon, S.C., González-Maeso, J., 2012. Persistent effects of chronic clozapine on the cellular and behavioral responses to LSD in mice. Psychopharmacology 225 (1), 217–226. http://dx.doi.org/10.1007/s00213-012-2809-7.

(N. D.) Controlled Substance Schedules. US Department of Justice Drug Enforcement Administration. Retrieved on March 19, 2013 from http://www.deadiversion.usdoj.gov/schedules/index.html#define.

Nichols, D.E., 2004. Hallucinogens. Pharmacol. Ther. 101 (2), 131–181. http://dx.doi.org/10.1016/j.pharmthera.2003.11.002.

Osmond, H., 1957. A review of the clinical effects of psychotomimetic agents. Ann. N. Y. Acad. Sci. 66, 418–434.

Ott, J., 2001. Pharmepena-psychonautics: human intranasal, sublingual and oral pharmacology of 5-methoxy-N, N-dimethyl-tryptamine. J. Psychoactive Drugs 33, 403–407.

Pahnke, W.N., Kurland, A.A., Goodman, L.E., Richards, W.A., 1969. LSD-assisted psychotherapy with terminal cancer patients. Curr. Psychiatry Ther. 9, 144–152.

Quednow, B.B., Geyer, M.A., Halberstadt, A.L., 2010. Serotonin and schizophrenia. In: Muller, C.P., Jacobs, B. (Eds.), Handbook of the Behavioral Neurobiology of Serotonin. Academic Press, London, pp. 585–620.

Ray, T.S., 2010. Psychedelics and the human receptorome. PLoS One 5 (2), e9019.

Reissig, C.J., Rabin, R.A., et al., 2008. d-LSD-induced c-Fos expression occurs in a population of oligodendrocytes in rat prefrontal cortex. Eur. J. Pharmacol. 583 (1), 40–47.

Resnick, O., Krus, D.M., Raskin, M., 1964. LSD-25 action in normal subjects treated with a monoamine oxidase inhibitor. Life Sci. 3, 1207–1214.

Ries, R.K., Fiellin, D.A., Miller, S.C., Saitz, R., 2009. Principles of Addiction Medicine, fourth ed. Lippencott Williams & Wilkins, Philadelphia, PA.

Ross, S., 2012. Serotonergic hallucinogens and emerging targets for addiction pharmacotherapies. Psychiatry Clin. North Am. 35 (2), 357–374. http://dx.doi.org/10.1016/j.psc.2012.04.002.

Savage, C., Jackson, D., Terrill, J., 1962. LSD, transcendence, and the new beginning. J. Nerv. Ment. Dis. 135, 425–439.

Sobel, J., Espinas, O., Friedman, S., 1971. Carotid artery obstruction following LSD capsule ingestion. Arch. Int. Med. 127, 290–291.

Sorensen, S.M., Kehne, J.H., Fadayel, G.M., Humphreys, T.M., Ketteler, H.J., Sullivan, C.K., Schmidt, C.J., 1993. Characterization of the 5-HT2 receptor antagonist MDL 100907 as a putative atypical antipsychotic: behavioral, electrophysiological and neurochemical studies. J. Pharmacol. Exp. Ther. 266 (2), 684–691.

Spengos, K., Schwartz, A., Hennerici, M., 2000. Multifocal cerebral demyelination after magic mushroom abuse. J. Neurol. 247 (3), 224–225.

Stamets, P., 1996. Psilocybin Mushrooms of the World: An Identification Guide. Ten Speed Press, Berkeley, California.

Strassman, R.J., 1984. Adverse reactions to psychedelic drugs. A review of the literature. J. Nerv. Ment. Dis. 172 (10), 577–595.

Strassman, R.J., 1992. Human hallucinogen interactions with drugs affecting serotonergic neurotransmission. Neuropsychopharmacology 7 (3), 241–243.

Studerus, E., Kometer, M., et al., 2011. Acute, subacute and long-term subjective effects of psilocybin in healthy humans: a pooled analysis of experimental studies. J. Psychopharmacol. 25 (11), 1434–1452.

Substance Abuse and Mental Health Services Administration, 2012. Results from the 2011 National Survey on Drug Use and Health: Summary of National Findings. NSDUH Series H-44, HHS Publication No. (SMA) 12-4713. Substance Abuse and Mental Health Services Administration, Rockville, MD.

Tkachev, D., Mimmack, M.L., Ryan, M.M., Wayland, M., Freeman, T., Jones, P.B., Starkey, M., Webster, M.J., Yolken, R.H., Bahn, S., 2003. Oligodendrocyte dysfunction in schizophrenia and bipolar disorder. Lancet 362, 798–805.

Ungerleider, J.T., Fisher, D.D., Fuller, M., 1966. The dangers of LSD. Analysis of seven months' experience in a university hospital's psychiatric service. JAMA 197 (6), 389–392.

Ungerleider, J.T., Fisher, D.D., 1967. The problems of LSD-25 and emotional disorder. Calif. Med. 106 (1), 49–55.

Vollenweider, F.X., Leenders, K.L., Scharfetter, C., Maguire, P., Stadelmann, O., Angst, J., 1997. Positron emission tomography and fluorodeoxyglucose studies of metabolic hyperfrontality and psychopathology in the psilocybin model of psychosis. Neuropsychopharmacology 16, 357–372.

Vollenweider, F.X., Vollenweider-Scherpenhuyzen, M.F., Bäbler, A., Vogel, H., Hell, D., 1998. Psilocybin induces schizophrenia-like psychosis in humans via a serotonin-2 agonist action. Neuroreport 9 (17), 3897–3902.

Vollenweider, F.X., Geyer, M.A., 2001. A systems model of altered consciousness: integrating natural and drug-induced psychoses. Brain Res. Bull. 56, 495–507.

West, L.J., Pierce, C.M., Thomas, W.D., 1962. Lysergic acid diethylamide: its effects on a male asiatic elephant. Science (New York, N.Y.) 138 (3545), 1100–1103.

Winkelman, M.J., Roberts, T.B., 2007. Psychedelic Medicine: New Evidence for Hallucinogenic Substances as Treatments. Praeger Publishers, Westport.

Wischhof, L., Koch, M., 2012. Pre-treatment with the mGlu2/3 receptor agonist LY379268 attenuates DOI-induced impulsive responding and regional c-Fos protein expression. Psychopharmacology (Berl.) 219 (2), 387–400.

Wu, L.T., Schlenger, W.E., Galvin, D.M., 2006. Concurrent use of methamphetamine, MDMA, LSD, ketamine, GHB, and flunitrazepam among American youths. Drug Alcohol Depend. 84 (1), 102–113.

CHAPTER EIGHTEEN

Inhalants: Addiction and Toxic Effects in the Human

Scott Bowen[1],* and **Silvia L. Cruz[2],***
[1]Department of Psychology, Wayne State University, Detroit, MI, USA
[2]Departamento de Farmacobiología, Cinvestav, Sede Sur, Mexico, Federal District, Mexico

1. INTRODUCTION

1.1. What are Inhalants?

1.1.1. Overview

"Bagging", "glue sniffing", "dusting", and "huffing". These are all terms that have been used to describe inhalant abuse. But what exactly are inhalants? Inhalants are a special group of misused substances that are classified based on a common administration route rather than on similar pharmacological properties, toxicological profile, or shared mechanism of action. As such, this category comprises a wide variety of chemicals with some physicochemical properties in common: they are volatile, flammable, and with high affinity for lipids. Some are gases, others become vapors at room temperature, and all of them are intentionally self-administered through inhalation to achieve intoxication. Other features that make inhalants unique are that their possession is legal, they are easily available in many commercial products, and their psychoactive effects are almost immediate. This, together with the wrong perception that breathing vapors is rather harmless, contributes to a low-risk perception of inhalant misuse among the general population (Cruz and Balster, 2013).

1.1.2. Classification of Inhalants

Currently, classifying inhalants can be a difficult task because (1) there is little information on the behavioral and pharmacological effects of this large group of compounds; (2) a number of compounds have several different names; (3) many commercial product labels give little information about components of the product; and (4) multiple formulations can be utilized to manufacture similar products (e.g. paint thinners) (Balster et al., 2009). As an example, gasoline is frequently listed as a solvent because of its physicochemical properties. However, gasoline is a combination of a number of substances, several of these being solvents as well. Nitrous oxide is another difficult compound to classify because it is a propellant in a number of commercial products as well as an

---

* Dr Bowen and Dr Cruz contributed equally to this Chapter.

The Effects of Drug Abuse on the Human Nervous System
http://dx.doi.org/10.1016/B978-0-12-418679-8.00018-6

anesthetic gas used in dentistry. Other examples are butane and propane, which are both propellants and fuels. Regardless, there are classifications that can be differentiated based on structure (e.g. ethers, hydrocarbons, etc.) or the physical state (aerosols, gases, or liquids). There is not enough information to establish a pharmacological classification based on mechanisms of action of the many misused inhalants, but a few more or less defined groups can be distinguished based on similar characteristics (Table 1).

Solvents and fuels comprise the larger group of inhaled substances. Industry, along with many households, uses large quantities of solvents for cleaning purposes (i.e. dissolving grease) and these compounds are typically in the form of liquids with low boiling points and high vapor pressures (Table 2). Among solvents, toluene is the most commonly misused inhalant throughout the world, not only because it is available in many commercial products (glues, paints, and degreasers, to mention a few) but also because it is a potent psychoactive substance. Gasoline is a mixture of hydrocarbons including toluene itself, hexane, and benzene. In some countries, gasoline still has lead, a component to be considered which contributes to gasoline's toxicological effects. Paint thinners are composed with variable mixtures of aromatic hydrocarbons (e.g. toluene, xylene, ethylbenzene, etc.), of which toluene is usually predominant. Other specific solvents are xylene (ortho-, meta-, and para-xylene), used in leather manufacturing, rubber

Table 1 Some Commonly Misused Inhalants

Group	Characteristics	Effects	Commercial Products
Alkyl nitrites	They have a nitrite group (–O–N|O)	Vasodilation, smooth muscle relaxation.	"Poppers", room odorizers
Nitrous oxide	Inorganic anesthetic gas also used as propellant.	CNS depression, hilarity, analgesia, and anesthesia.	Whipped dairy cream chargers, nitrous oxide tanks.
Other anesthetics	Gases or volatile liquids.	Relaxation, numbness analgesia, anesthesia, and respiratory depression.	Anesthesia bottles (ether, halothane, isoflurane, and desflurane).
Propellants-fuels	Flammable gases that are liquefied under pressure.	Relaxation, hallucinations, and illusions.	Cigarette lighters, cooking fuels, hairsprays, and central air conditioning units.
Propellant			Dust spray.
Solvents	Liquids that dissolve other substances (solutes) without any change in chemical composition.	Initial disinhibition followed by a more prolonged depression. Illusions, hallucinations.	Paints, paint thinners, glues, inks, lacquers, varnish removers, degreaser agents, spot removers, and motor fuels.

factories, printing, and histology; and benzene, which is a constituent of crude oil, an ingredient of gasoline and a chemical precursor for organic synthesis. 1,1,1-trichloroethane was extensively used a few years ago in correction fluids, and as a degreasing agent and propellant, but its use is being phased out because it is an ozone-depleting substance, now regulated by the Montreal Protocol. Another relevant chlorinated organic solvent is trichloroethylene, a spot remover and metal degreaser used in dry cleaning and industry.

Ether, chloroform, and nitrous oxide are anesthetics that can also be misused as inhalants. Ether and chloroform are highly volatile flammable liquids. Safer compounds have replaced them, but voluntary intoxication with these chemicals is well documented in the literature (Alt et al., 2011; Flanagan and Ives, 1994; Hutchens and Kung, 1985). Nitrous oxide, frequently identified as "laughing gas", can be inhaled from balloons filled with gas for this specific purpose or from whipped cream chargers ("whippets"). Misuse of other halogenated anesthetics has also been reported, including halothane, enflurane, and isoflurane (Krajcovic et al., 2011; Pavlic et al., 2002). All these substances act as central nervous system (CNS) depressants and their effects are perceived as relaxant at subanesthetic doses.

Butane and propane are other inhalant gases used as fuels, either alone or in combination. Butane is the gas commonly used in cigarette lighters, portable camping stoves and in smaller canisters for plumbing torches. Propane has similar uses and together they are employed in common house stoves as liquefied under pressure gas. Both gases are used for residential central heating. Depending on how extreme the temperature conditions are, propane can be preferred over butane because it has a lower melting point and it is more difficult to freeze. These alkane gases replaced chlorinated fluorocarbons (CFC) as propellants in many commercial products. CFCs, also known as freons, are organic compounds analogous to hydrocarbons in which the hydrogen carbons have been replaced by chlorine or fluoride atoms. They have many physicochemical advantages that made them easy to handle but were phased out because they damage the ozone atmospheric layer. A related agent is 1,1-difluoroethane, which is still in use as propellant in PC cleaners (air duster). This gas is considered "safe" for the ozone layer and marketed as such, but there is no reason to extend this label to human use. On the contrary, there is enough evidence that proves that accidental death can result from its inhalation (Avella et al., 2006; Hahn et al., 2006).

Amyl nitrite is a potent vasodilator that was introduced for the treatment of angina pectoris in the late nineteenth century. More reliable compounds such as nitroglycerine replaced it and its medical use was restricted to conditions when a rapid absorption was needed for specific cardiac problems. Amyl nitrite is also used to treat cyanide poisoning. Apart from that, amyl, butyl, and isobutyl nitrites are largely inhaled for recreational purposes. They can be sold in glass ampules that pop when opened, hence the generic term "poppers". Some room odorizers contain isobutyl nitrites that are inhaled to achieve intoxication. This practice has been associated with the desire to enhance sexual performance among homosexual men (McManus et al., 1982; Mullens et al., 2009).

Table 2 Physicochemical Properties and Chemical Classes of Some Compounds that Can be Used as Inhalants

Use	Chemical Class	Compound/Physical State	Molecular Formula	Structure	MW	VP (mm Hg)	BP (°C)	MP (°C)	Specific Gravity (water=1)	Vapor Density (air=1)	Partition Coeff. Log Kow
Organic solvents	Hydrocarbon, acyclic	n-Hexane/Liquid	C_6H_{14}		86.17	153	68.7	−94.3	0.655	2.97	3.9
	Hydrocarbon, alicyclic	Cyclohexane/Liquid	C_6H_{12}		84.2	96.9	80.7	6.5	0.778	2.98	3.4
	Hydrocarbons, aromatic	Benzene/Liquid	C_6H_6		78.1	94.8	80.1	5.5	0.878	2.8	2.13
		Toluene/Liquid	C_7H_8		92.14	28.4	110.6	−94.9	0.863	3.1	2.73
		o-Xylene m-xylene p-xylene/Liquid	C_8H_{10}		106.2	6.6 8.29 8.84	144 139 138	−25.2 −47.4 13.25	0.88 0.87 0.86	3.7 3.66 3.7	3.12 3.2 3.15
		Ethyl benzene/liquid	C_8H_{10}		106.2	9.6	136.1	−94.9	0.867	3.66	3.15
		Propyl benzene/liquid	C_9H_{12}		120.2	3.4	159.2	−99.5	0.862	4.14	3.69

Class	Name/State	Formula	Structure	MW	VP	BP	MP			
Hydrocarbons, halogenated	1,1,1-TCE/liquid	$C_2H_3Cl_3$		133.4	124	74	−30.1	1.34	4.63	2.49
Anesthetics	Trichloroethylene/liquid	C_2HCl_3		131.4	69	87.2	−84.7	1.46	4.53	2.61
	Diethyl ether/liquid	$C_4H_{10}O$		74.12	538	34.6	−116.3	0.713	2.55	0.89
	Chloroform/liquid	$CHCl_3$		119.37	197	61.2	−63.4	1.48	4.12	1.97
	Halothane/liquid	$C_2HBrClF_3$		197.38	302	50.2	−118	1.87	6.9	2.3
Inorganic	Nitrous oxide/gas	N_2O		44.01	42,900	−88.5	−90.8	1.23	1.53	0.36
Propellants, Hydrocarbon, acyclic	Propane/liquified gas	C_3H_8		44.09	7150	−42.1	−187.6	0.49	1.56	2.36
	Butane/liquified gas	C_4H_{10}		58.12	1820	−0.5	−138.2	0.6	2.07	2.89
Hydrocarbon, halogenated	1,1-difluoroethane	$C_2H_4F_2$		66.05	4550	−24.7	−117	0.896	2.3	0.75

MW, molecular weight; VP, vapor pressure; BP, boiling point; MP, melting point; log Kow, octanol/water partition coefficient.
Source: Data taken from http://pubchem.ncbi.nlm.nih.gov

2.1. HOW AND WHY ARE THEY USED?

It is clear that the misuse of inhalants is prevalent around the world (Dell et al., 2011). There are a number of reasons for the continued interest in inhalants. Many of these compounds (Table 1) are relatively inexpensive as compared to other abused drugs (e.g. alcohol and nicotine) and in many instances, individuals do not have to even purchase them. In fact, many of the products containing abused solvents are widely available in the home, grocery stores, gas stations, hardware stores, and schools, among others. Another reason for the continued interest in inhalants is that there are several administration methods, simple and uncomplicated with each delivering a rapidly high blood concentration of the volatile product being used. Among these various methods are, inhaling fumes from products located in a paper or plastic bag, or placed in a balloon ("bagging" or "ballooning"), pouring the product onto clothing or cuffs and placing the clothing near the nose or mouth to inhale ("cuffing"), placing a rag or cloth soaked in solvent near or on the nose and/or mouth ("huffing"), painting the product onto the fingernails and raising the fingernails to the nose and/or mouth ("nailing"), inhaling the product directly from the container ("sniffing" or "snorting"), soaking cotton balls, cotton swabs, or cloth and placing them into the nose or mouth ("soaking"), and turning a spray can upside down and spraying the aerosolized compounds directly into the nose and/or mouth ("spraying"). With each of these methods, deep breaths are taken and effects occur very rapidly (usually within only a couple of seconds) with the intoxicating effects continuing from 15 to 60 min. The intoxication or "high" can be maintained for several hours by repeatedly inhaling the vapors.

Although misuse of inhalants has been traditionally associated with marginalized groups, recent epidemiological data indicate that it is being generalized to mainstream populations (Dell et al., 2011; Hynes-Dowell et al., 2011; Medina-Mora and Real, 2008). As it occurs with other psychoactive substances, inhalant experimentation could be driven by curiosity, peer pressure, personal expectations, and the desire to get high or escape boredom. Depending on the population, reasons to continue using inhalants might vary. Among marginalized groups such as "street children" (a broad category with different connotations), inhalation may have to do with harsh lifestyle conditions such as coping with hunger, cold weather, and violence. On the other hand, school children may report that they want to have fun or that inhalants help them deal with stressful situations at school and home. Regardless of their condition, some solvent users mention the desire to experience hallucinations as a reason to inhale (Cruz and Dominguez, 2011; MacLean, 2007, 2008).

2.2. EPIDEMIOLOGY DATA

The use of inhalants for intoxicating purposes is considered a public health problem in high- and low-income countries from every continent. In some parts of Australia and Canada, inhalant misuse is more prevalent among marginalized and indigenous populations, with variations in the substances of choice: paint sprays in Australia (MacLean, 2007; Takagi et al., 2011), gasoline, propane, and spray paint in Canada (Dell and Hopkins, 2011). In Egypt, the problem is highly prevalent among street children that exposed themselves to gasoline vapors or glue (Elkoussi and Bakheet, 2011), whereas in India, the preferred products are whitening fluid, glues, and gasoline products (Sharma and Lal, 2011). Inhalants of choice in New Zealand are butane and propane (Beasley et al., 2006); while paint thinner and a toluene-based liquid called "activo" are preferred in Mexico. This particular product is sold on the black market mixed with odorizers to offer different "flavors" such as cinnamon, orange, or chocolate, to conceal the strong solvent odor and make the product more appealing (Villatoro et al., 2011).

On the other hand, according to the First Report on drug use in the Americas, inhalant misuse is also prevalent in the Americas and the Caribe. This report compiles data from 34 countries gathered during 2002–2009, which helps identify trends and commonalities. Among the most commonly used substances in the region are toluene-based products such as paint thinner and adhesives. In general, the lowest rates of inhalant use are found in Central American countries and the highest in the Caribbean region and Brazil. Past-year prevalence of inhalant use among secondary school students ranges from 0.5% in Dominican Republic to 14.34% in Brazil. Of interest is that inhalant use is higher among boys than girls in Latin American countries, but in other countries in the Caribbean, inhalant prevalence is higher among girls than boys (OAS & Hynes-Dowell, 2011).

Taking into account the available epidemiological evidence, at least three trends can be identified in countries where inhalant misuse is prevalent: (1) it is increasing among high school students; (2) the sex gap among male and female users is narrowing; and (3) the use of inhalants is no longer associated almost exclusively with children and adolescents (Medina-Mora and Real, 2008). In fact, according to a recent study by the Substance Abuse and Health Service Administration (SAMHSA), 54% of treatment admissions related to inhalants abuse in 2008 in the United States involved adults ages 18 or older. Of these adult admissions, 32% involved people aged 30–44, and 16%, people aged 45 or older (SAMHSA, 2011).

2.3. WHAT ARE THE MEDICAL CONSEQUENCES OF ABUSE?

2.3.1. Acute

The acute symptoms of someone who has been abusing an inhalant are dependent on the solvent being abused and the concentrations achieved. Generally, these symptoms include initial excitation (i.e. disinhibition) followed by a more prolonged inhibition, emotional volatility (i.e. irritability), dizziness, loss of coordination, cognitive impairment, disorientation, distorted perceptions, slurred speech, nausea, wheezing, and irritation of the eyes and respiratory system. It should be noted that because these acute effects are transitory, they may be missed by both medical and law-enforcement personnel. Use of higher concentrations may result in hallucinations, seizures, coma, and eventually death.

A number of medical consequences can result from abuse of solvents and other volatile compounds. The most observable of these are most likely the different types of skin damage that can occur outside the mouth and nose or inside the mouth on the tongue or cheeks. These can include burns, frostbite, sores, dermatitis, and eczema. Frostbite can occur with products containing butane, propane, fluorocarbons, or nitrous oxide and are the result of the lower temperatures that occur when the liquefied gas volatilizes when leaving the container and contacts soft tissue (the skin or the tongue). It should be noted that inhalation of many gases does not leave marks on the skin and as such, may be more difficult to diagnose. Additionally, inhalation of vapors that are irritating to the mucosa may result in allergic or asthmatic reactions, which can produce coughing, sneezing, wheezing, and an asthmatic response. Breathing vapors can also produce hypoxia, which is thought to contribute to its psychological effects. Short-term use of inhalants may result in CNS dysfunction without pronounced changes in anatomical structure. However, memory impairment (Huerta-Rivas et al., 2012; Win-Shwe and Fujimaki, 2012), sleep disturbances (Alfaro-Rodriguez et al., 2011; Fakier and Wild, 2011), visual, and optical changes (Lomax et al., 2004) have also been reported following acute inhalant exposure.

Inhalation of high concentrations of amyl nitrite favors the ferric (Fe^{3+}) form of hemoglobin, termed methemoglobin, which is incapable of transporting oxygen. As a consequence, nitrite misuse results in poor oxygenation, cyanosis (blue color of the skin), and dyspnea. Very high levels of amyl nitrite can produce lethargy, coma, and death (Edwards and Ujma, 1995).

One of the biggest concerns following acute inhalant use is "sudden sniffing death". While rare among inhalant abusers, death can occur anytime an inhalant is abused, even after a single use. Death can occur from accidental trauma (falling, etc.), asphyxiation by the bag covering the face, aspiration of the compound or vomit, anoxia, cardiac dysfunction, CNS depression, severe allergic reaction, or extreme injury to the lungs (Adgey et al., 1995; Bowen, 2011; Bowen et al., 1999a,b; Garriott and Petty, 1980; Garriott, 1992; Hahn et al.,

2006; Kringsholm, 1980; Maxwell, 2001; Novosel et al., 2011; Potocka–Banas et al., 2011). Death can also occur because of fire as some canisters are heated to enhance the amount of solvent available which may result in combustion. It is important to stress that death can happen with any abused volatile substance and it can occur after only a single use.

Sudden sniffing death was first documented in the 1980s in the United Kingdom (Anderson et al., 1985). Recent estimates have placed the number of inhalant fatalities per year in just the United States between 100 and 125 (NIPC, 2010). However, as was stated previously, many feel that the deaths attributed to the misuse of volatile solvents is probably underreported as a result of the fatality not being recognized as a volatile solvent death. From the information that has been gathered from previous reported cases of deaths in the United Kingdom and in the United States resulting from volatile substance misuse, we know that the characteristic case of lethality involves a young Caucasian male (Bowen, 2011). Similarly, males make up a large proportion (95%) of the 282 inhalant deaths reported between 1971 and 1983 in the United Kingdom (Anderson et al., 1985).

Similar patterns emerge from the deaths reported in most of the U.S. surveys of lethalities from volatile substance misuse with male deaths making up 79–95% of all reported deaths (Bowen et al., 1999a,b; Garriott and Petty, 1980; Garriott, 1992). For example, in the State of Virginia, Bowen et al. (1999a,b) reported that 95% of fatalities reported from volatile substance misuse were males ($N = 37$), accounting for 0.3% of all deaths in males aged 13–22. These findings are somewhat surprising in light of the most current surveys suggesting that, at least in the United States, rates of misusing volatile substances among females at certain ages are equal or greater than males (SAMHSA, 2007). Identifying if there are factors that are causing males to have such a greater risk for lethality as compared to females remains to be determined.

2.3.2. Chronic

Long-term, repeated abuse of inhalants can result in toxic effects to many parts of the body especially the brain, bones, heart, kidneys, liver, and lungs (Hormes et al., 1986; Rosenberg et al., 1988, 2002; Rosenberg and Sharp, 1997). Damage to the brain and nervous system are frequently reported with common diagnoses including ataxia and atrophy of the cerebellum, encephalopathy, convulsions and tremors, degeneration of myelin, peripheral, and sensorimotor neuropathy. The severe impact on CNS myelin can result in toluene leukoencephalopathy and neurological disorders in abusers and may explain the forms of neuropathological destruction observed in brains of individuals who chronically abuse toluene (Filley, 2013; Filley et al., 2004). Myeloneuropathy, a neurological condition where neuron's myelin sheath is damaged, can result from nitrous oxide misuse, secondary to vitamin B_{12} deficiency (Hathout and El-Saden, 2011). This condition can lead to ascending lower extremity numbness and weakness (Alt et al., 2011; Lin et al., 2011). Although rare, fatal cases of death have also been associated with nitrous oxide inhalation (Potocka–Banas et al., 2011).

Long-term misuse of inhalants has also been reported to result in cardiovascular damage including myocardial edema and ischemia, fibrous, and congestive heart failure (Avella et al., 2006; Flanagan et al., 1990; Novosel et al., 2011). Inhalant abuse may also produce cardiovascular sensitivity to endogenous catecholamines, so much so that any increase in excitement or physical activity may cause death (Adgey et al., 1995). Both the aromatic and chlorinated hydrocarbons are known to be damaging to the kidneys and the liver resulting in distal renal tubular acidosis and centrilobular tumors and necrosis (Meadows and Verghese, 1996; Wick et al., 2007). Chronic inhalation to several inhalants (isobutyl nitrite, naphthalene, etc.) has resulted in suppression of bone marrow production, which can lead to anemia and leukemia. Another complication associated with chronic isopropyl nitrite inhalation is progressive bilateral vision loss due to retinal damage (Audo et al., 2011; Fledelius, 1999).

Long-term exposure to air dusters, containing 1,1-difluoroethane, produces a rare condition known as *heterotrophic ossification*, i.e. the formation of bone in soft tissues other than the periosteum. This has been documented in chronic users that recovered from fractures caused by motor accidents and continued misusing 1,1-difluoroethane during their convalescence (Little et al., 2008).

Damage to the pulmonary system is also reported with repeated abuse resulting in coughing, wheezing, shortness of breath and pneumonia. This has been described in chronic abusers of keyboard cleaners (e.g. 1,1-difluoroethane) leading to alveolar hemorrhage and eventually death (Schloneger et al., 2009). Inhaling fuel gases (e.g. butane) has also been reported to increase the incidence of laryngeal edema and laryngospasm (i.e. seizing of the vocal chords) (Adgey et al., 1995).

Lastly, chronic inhalant abusers have been documented with increased incidences of psychiatric comorbidity. Severe emotional and social deprivation, anxiety, major depression, as well as delirium, dementia, and psychosis have been documented in these individuals. Antisocial and delinquent behaviors are also frequently reported in inhalant abusers stressing the importance of intervention and treatment (Howard et al., 2011; Ridenour, 2005).

2.3.3. Gestational Effects

Volatile substance misuse is an extremely harmful behavior among pregnant women with a number of medical reports reporting that misuse of inhalants during pregnancy results in a myriad of developmental problems and physical malformations to the fetus (Arai et al., 1997; Arnold et al., 1994; Goodwin et al., 1981; Goodwin, 1988; Hersh, 1989; Hersh et al., 1985; Pearson et al., 1994; Toutant and Lippmann, 1979). Death of the fetus has also been documented in females with histories of intentional inhalant abuse involving very high levels of solvent exposure (Arnold et al., 1994; Bukowski, 2001; Wilkins-Haug and Gabow, 1991). Indeed, Arnold et al. (1994) documented that prenatal death occurred in 3 out of 35 confirmed antenatal exposures to toluene. While there are a number of uncertainties in the literature, including use of multiple solvents, questions of dosing, recall problems, retrospective subject selection, etc., these toxic effects have been recently and extensively reviewed and the reader is directed to these articles and reviews for more detailed information (Bowen, 2011; Bowen et al., 2005; Bowen and Hannigan, 2006, 2013; Hannigan and Bowen, 2010; Jones and Balster, 1998).

Briefly, the phrase *fetal solvent syndrome* (FSS), which has been used to describe the pattern of behavioral and morphological effects resulting from inhalant misuse during pregnancy, builds on the fetal alcohol syndrome model. When comparisons are made between the two syndromes, a number of similarities arise. Most notable of these are the alterations in birth weight, deep-set eyes, craniofacial features, such as undersized palpebral fissures, thinning of the upper lip, and hypoplasia of the mid-face that can occur after abuse of either alcohol or solvents like toluene. Arnold et al. (1994) reported on 31 cases for which delivery records were available, and found that 13/31 (42%) of the infants were classified as premature with 16/31 (52%) of the infants having birth weights below the 10th percentile. A smaller group of these babies, 7/22 (32%), were also microencephalic (i.e. small head and underdeveloped brain) (Arnold et al., 1994). Other reported FSS anomalies include ear malformations, irregular scalp hair patterning, down-turned corners of the mouth, micrognathia, enlargement of the fontanelle, and spatulate fingertips and small fingernails.

Follow-up examinations of these children at an older age (3 years) demonstrated that these development delays and impairments persisted. Reports include retardation of growth, hyperactivity, impairments in language as well as neurological dysfunction, including balance and walking problems, and movements of the arms and hands (Arnold et al., 1994; Goodwin, 1988; Hersh, 1989; Hersh et al., 1985; Hunter et al., 1979; Toutant and Lippmann, 1979). It is unclear whether it is possible to reverse these delays and impairments or whether these delays and impairments last into adulthood. We are not aware of any reports of clinical remedies for fetal solvent effects, nor were we able to find any follow-up studies of these affected children, as they became adults.

2.4. PHARMACOLOGICAL PROPERTIES/EFFECTS

2.4.1. Solvents

Characterization of inhalants' pharmacological effects is far from complete partly due to the great variability of substances involved; however, a significant amount of work has been made in the past several years to elucidate how solvents produce their effects. Basic research has provided evidence that several misused solvents have cellular mechanisms in common with other CNS depressants, specifically with ethanol. Nonetheless, animal research has shown that one of these misused solvents, toluene, shares discriminative stimulus effects not only with alcohol (Bowen, 2009; Evans and Balster, 1991) but also with amphetamines (Bowen, 2006) and phencyclidine (Bowen et al., 1999a,b), which alludes to the complexity of its actions.

Inhaled solvents are depressant drugs with peculiar characteristics. Toluene produces a generalized CNS depression through a complex mechanism that involves both enhanced inhibitory neurotransmission and attenuated excitatory signaling. The first evidence of the existence of molecular targets affected by this solvent came from studies in oocytes expressing recombinant N-methyl-D-aspartate (NMDA) receptors assembled from the NR1 and NR2 subunits. Toluene inhibited NMDA receptors at micromolar concentrations that did not alter cellular membranes. This inhibition was noncompetitive, reversible, and almost complete. The most sensitive NMDA receptor composition was that of NR1/NR2B subunits followed by NR1/NR2A and NR1/NR2C NMDA receptors (Cruz et al., 1998). Interestingly, toluene had no significant effects on the closely related AMPA and kainate receptors at concentrations that were effective to block NMDA receptors. Similar results were found for benzene, m-xylene, and ethylbenzene, propylbenzene, 1,1,1-trichloroethane (TCE) (Cruz et al., 2000), and a number of other volatile benzene analogs (Raines et al., 2004).

As occurs with other CNS depressants having antianxiety effects, solvents act as positive allosteric modulators of $GABA_A$ and glycine receptors. This has been demonstrated for toluene, trichloroethylene, and 1,1,1-TCE (Beckstead et al., 2000). The combination of NMDA receptor antagonism and $GABA_A$ receptor potentiation has been documented in brain slices of the medial prefrontal cortex (Beckley and Woodward, 2011) and cultured hippocampal cells (Bale et al., 2005a,b). These actions might be responsible, at least in part, for the cognitive function impairment associated with toluene misuse.

Toluene, trichloroethylene, and 1,1,1-TCE significantly potentiate submaximal $5\text{-}HT_3$ receptor-mediated current responses evoked by 5-HT (Lopreato et al., 2003). $5\text{-}HT_3$ receptors are thought to play important roles in regulating the cardiovascular and nervous systems. Presynaptic $5\text{-}HT_3$ receptors modulate GABA release (Koyama et al., 2000; Turner et al., 2004), an effect that may contribute to behavioral solvents' effects.

Neuronal nicotinic acetylcholine receptors are likewise targets for solvents. Toluene and perchloroethylene inhibit $\alpha_4\beta_2$, $\alpha_3\beta_2$ and homomeric α_7 nicotinic receptors with different potencies; i.e. these ligand-gated ionotropic receptors, like NMDA receptors, show different subunit-dependent sensitivity to solvents (Bale et al., 2005a,b, Bale et al., 2002).

Solvents not only affect the function of neuronal ionotropic receptors, they also act on voltage-gated ion channels. Toluene concentration dependently inhibits sodium channels, which are responsible for the initial depolarizing phase of cardiac action potential (Cruz et al., 2003), an effect that could be related with toluene's arrhythmogenic actions. Interestingly, a single mutation on the local anesthetics binding site is enough to abolish the inhibitory effects of this solvent, suggesting that toluene has affinity for this site (Gauthereau et al., 2005; Scior et al., 2009). Toluene inhibits dihydropyridine-sensitive and -insensitive calcium channels (Tillar et al., 2002), which is also observed with other halogenated hydrocarbons (trichloroethylene and perchloroethylene) using different experimental preparations (Shafer et al., 2005).

Not surprisingly toluene, like other misused drugs, increases dopamine release in the nucleus accumbens by a direct activation of neurons in ventral tegmental area (Beckley et al., 2013; Riegel et al., 2007). This pharmacological effect is consistent with toluene's reported rewarding effects (Gerasimov et al., 2002) with several animal testing paradigms providing support consistent with such a hypothesis (self-administration, intracranial self-administration (ICSS), and conditioned place preference (CPP)). For example, several earlier studies reported that monkeys would work to gain short access to both toluene vapor (Weiss et al., 1979) and nitrous oxide (Wood et al., 1977). A more recent report demonstrated that mice will self-administer intravenously administered toluene and TCE (Blokhina et al., 2004). Glue vapors (toluene 25%, benzene fraction 37%, ethyl acetate 31%, and methylene chloride 7%) at a concentration of 7200 ppm have been shown to increase response rates of ICSS in the lateral hypothalamus and enhance the threshold current of self-stimulation (contrary to what other drugs of abuse to ICSS) (Yavich et al., 1994; Yavich and Zvartau, 1994). When these glue vapors were increased to concentrations of 14,400 ppm and higher, the response rate of ICSS was completely eliminated. Along with ICSS changes, there have been several recent reports that the CPP model can be used to demonstrate the reinforcing effects of toluene. Animals have been shown to spend more time in a chamber that was previously paired with toluene than they did prior to the drug/chamber pairings (Funada et al., 2002; Gerasimov et al., 2003; Lee et al., 2006) with combinations of commonly abused inhalants (including toluene, benzene, ethyl acetate, and methyl chloride) demonstrating CPP as well (Yavich et al., 1994). Taken together, these studies suggest that toluene has reinforcing properties, similar to other commonly abused drugs.

The multiple actions of solvents' actions are somewhat surprising but this is also true for ethanol, another misused solvent with particular characteristics (Alfonso-Loeches and Guerri, 2011). The prevalence of one or several mechanisms of action over others can be related with differences in affinities of the molecular targets involved in a determined behavioral response and the relative density of receptor subtypes found in different tissues.

Chronic exposure to solvents can lead to molecular adaptations. This has been documented with repeated exposure to toluene that results in changes in $GABA_A$ and NMDA receptor subunit brain levels as well as dopamine levels (Alfaro-Rodriguez et al., 2011; Williams et al., 2005). More research is warranted in this particular area.

2.4.2. Anesthetics

Anesthetics, like toluene, also act on the so-called anesthetic-sensitive superfamily of Cys-loop ligand-gated ion channels that include the $GABA_A$, glycine, nicotinic acetylcholine, and serotonin receptors (Forman and Miller, 2011). Most if not all general anesthetics at relevant concentrations enhance the function of $GABA_A$ and glycine receptors (Mihic et al., 1997; Raines et al., 2003). Physically small, halogenated volatile anesthetics such as halothane and chloroform significantly potentiate submaximal $5-HT_3$ receptor function (Stevens et al., 2005). Of interest is nitrous oxide because it acts as partial opioid agonist (Gillman and Lichtigfeld, 1994) and inhibits NMDA and non-NMDA ionotropic receptors (Akk et al., 2008; Mennerick et al., 1998).

2.4.3. Nitrites

Nitrites differ from other inhalants because they are used for their vascular effects—they are potent vasodilators, rather than by their CNS effects. Nitrites produce a direct peripheral arteriolar dilation, which results in decreased systolic tension and oxygen utilization of the heart (Mason et al., 1971). It has been proposed that the vasodilation produced by nitrites is produced, at least in part, because the functional chemical group nitrite (–O–N]O) is transformed into nitric oxide (N]O) (Pinder et al., 2009). Nitric oxide (NO)—not to be confused with nitrous oxide (N_2O), the "laughing gas"—is an important cellular signaling molecule that is continuously synthesized in the body and has multiple functions. Under normal conditions, NO is formed by vascular endothelium and it rapidly diffuses into the blood where it binds to hemoglobin. It also diffuses into the vascular smooth muscle cells where it activates the enzyme guanylyl cyclase. This enzyme catalyzes the dephosphorylation of guanidine triphosphate to cyclic guanidine monophosphate (cGMP), which, in turn induces smooth muscle relaxation. The combination of nitrites and drugs used for the treatment of erectile dysfunction is dangerous because these drugs block the action of phosphodiesterase PDE5, the enzyme that lowers the cGMP levels. The result is unusually elevated NO levels that could lead to serious hypotension and heart failure (Krenzelok, 2000).

2.5. SCREENING

Despite being one of the most common forms of drug abuse, inhalants are not routinely included in drug screenings. The difficulty may lie in identifying inhalant misuse because clinicians may not be knowledgeable of inhalant use problems and the behavioral and physiological effects can be subtle (Anderson and Loomis, 2003). Suspicions should be raised if (1) chemical odors are noticed on breath; (2) paint or other stains on face, hands, or clothes; (3) hidden empty spray paint or solvent containers, chemical-soaked rags or clothes, small cylinders (nitrous oxide), or small broken (amyl and butyl nitrites); (4) frostbite around mouth or nose; (5) drunk or disoriented appearance, slurred speech, nausea or loss of appetite, inattentiveness, lack of coordination, irritability, and depression.

While early assessment and screening of inhalant misuse can provide essential opportunities for prevention and treatment, there are only a limited number of pharmacological/toxicological laboratory tests that exist which can successfully detect inhalant misuse.

In detecting occupational toluene exposure, the most frequently used method is determination of hippuric acid concentrations in the urine. Because this method uses urine instead of blood, it is nonevasive, allows for easy sample collection, and could be used for screening of inhalant misuse (Thiesen et al., 2007). Based on additional occupational toxicology research, there are endeavors to advance the laboratory diagnostic abilities of other potential urinary tests for detecting toluene metabolites including o-cresol and benzylmercapturic acid in inhalant abusers (Chakroun et al., 2008; Cok et al., 2003; Inoue et al., 2004; Thiesen et al., 2007). Other more established and current clinical tests include determination of blood counts as well as levels of electrolytes, calcium, and phosphorus as well as liver and kidney profiles and analysis of cardiac/muscle enzymes (Broussard, 2000).

Similar to the progression reported with the pharmacological/toxicological tests, few questionnaires exist which are specifically designed to successfully detect inhalant misuse. It was reported on the development of the Yeniden Inhalant Use Severity Scale which is aimed at determining the severity of substance use, especially in adolescents. Recently, Howard et al. (2011) developed the volatile solvent screening inventory (VSSI) and the comprehensive solvent assessment interview (CSAI). The VSSI assesses the past-year and lifetime frequency of use of more than 50 volatile compounds and requires approximately 20 min to complete. The individual's medical history, demographic characteristics, current psychiatric symptoms, suicidal thoughts and attempts, trauma history, temperamental traits such as impulsivity and frequency, and nature of antisocial behavior in the prior year are recorded. The second assessment, the CSAI, evaluates reasons for beginning the misuse of inhalants and takes 20–90 min to complete. This assessment inquires about locations, contexts, and the subjective effects of misuse; assesses any adverse acute problems of inhalant misuse as well as perceived risks of inhalant misuse; a projected estimate of future misuse; and uses DSM-IV inhalant abuse and dependence criteria. Both instruments are available without charge and can be accessed at the following site: dx.doi.org/10.1016/j.drugalcdep.2007.08.023 (Howard et al., 2011).

2.6. RECOVERY POTENTIAL AND TREATMENT

Inhalant misuse is prevalent in many countries and the demand for treatment is growing. The negative consequences of inhalant exposure are such that it is tempting to assume that they will be relatively permanent; however, this perception is an oversimplified generalization that hampers the development of effective treatment programs such as some already at work (e.g. Dell and Hopkins, 2011). As Garland and Howard (2010) pointed out, there is evidence that agents with different physicochemical properties have distinct cognitive, affective, and somatic effects, and that the consequences of inhalant misuse might vary considerably in sporadic versus chronic users.

Despite the impact of the negative effects of inhalant misuse, there are few studies on the recovery potential of cognitive and neurological deleterious effects. A recent review suggests that recovery occurs following abstinence from solvent misuse, and it depends on the extension and duration of inhalant misuse (Dingwall and Cairney, 2011). In this regard, it has been reported that the myeloneuropathy associated with chronic use of nitrous oxide improves with inhalant discontinuation and vitamin B_{12} supplementation (Alt et al., 2011). Similarly, retinal damage produced by chronic nitrite inhalation can recover after cessation (Audo et al., 2011). Some negative sequelae seen in inhalant users seem to be more devastating, an example of which is benzene-induced leukemia or liver toxicity produced by halogenated compounds. Nonetheless, there is no doubt that discontinuation of inhalant misuse has clear beneficial effects and efforts in this direction should be encouraged (Cairney et al., 2013). Support with behavioral cognitive therapy, attention to organic damage (hearing or sight loss, for example), and treatment of psychiatric comorbid disorders when needed would be fundamental to successful treatment programs.

To date, there is no available pharmacological therapy for treating inhalant misuse. However, there are a few case reports in the literature reporting limited success, including one in which risperidone was used to control the paranoid psychosis in a male who had been inhaling gasoline and carburetor cleaner daily for 5 years (Misra et al., 1999). Hernandez-Avila et al. (1998) also reported that about half of the men with inhalant induced psychotic disorder histories demonstrated a reduction in the severity of their symptoms when treated with either carbamazepine or haloperidol (Hernandez-Avila et al., 1998). Finally, Shen (2007) reported that daily administration of lamotrigine decreased the cravings for misusing inhalants in a 21-year-old male with a 4 year history of inhalant misuse (Shen, 2007). It is hoped that these studies, combined with the growing knowledge of the molecular targets involved in inhalant actions, would help developing this area in the near future.

3. CONCLUSIONS

Not all inhalant groups are similar. There are important differences among nitrites, nitrous oxide, anesthetic gases, and solvents, but these groups are rarely identified as different whenever inhalant abuse is documented. Efforts are needed to include standardized items in international surveys to identify distinctive groups or individual substances.

Although the mechanisms of action of several inhalants, mainly anesthetic gases, have been well documented for many years now, this specialized literature is hardly consulted by researchers working on drug addiction. On the other hand, knowledge of how solvents act has substantially improved in the last decade, but inhalants still remain the least studied drugs of abuse. At this point, the challenge is not only to generate sound scientific evidence on how the different substances exert their pharmacological and toxic effects but also to make this information available for teachers, parents, policy makers, and the general public.

Inhalants constitute a unique cluster of misused drugs that are grouped based on a common administration route rather than on distinctive pharmacological effects. This condition provides a potential source of confusion and oversimplification for users and the general public, which can also extend to researchers and policy makers. The particular noninvasive route of administration, along with the legal status of inhalants, favors a general low-risk perception toward their use (e.g. why should inhaling "air" (vapors) be dangerous?). This perception prevails despite the evidence that inhalants are particularly toxic and potentially fatal. One of the main challenges in this field is to change this perception using the scientific knowledge generated in the last years.

REFERENCES

Adgey, A.A., Johnston, P.W., McMechan, S., 1995. Sudden cardiac death and substance abuse. Resuscitation 29 (3), 219–221.

Akk, G., Mennerick, S., Steinbach, J.H., 2008. Actions of anesthetics on excitatory transmitter-gated channels. Handb. Exp. Pharmacol. 182, 53–84.

Alfaro-Rodriguez, A., Bueno-Nava, A., Gonzalez-Pina, R., Arch-Tirado, E., Vargas-Sanchez, J., Avila-Luna, A., 2011. Chronic exposure to toluene changes the sleep-wake pattern and brain monoamine content in rats. Acta Neurobiol. Exp. (Wars) 71 (2), 183–192.

Alfonso-Loeches, S., Guerri, C., 2011. Molecular and behavioral aspects of the actions of alcohol on the adult and developing brain. Crit. Rev. Clin. Lab. Sci. 48 (1), 19–47.

Alt, R.S., Morrissey, R.P., Gang, M.A., Hoffman, R.S., Schaumburg, H.H., 2011. Severe myeloneuropathy from acute high-dose nitrous oxide (N_2O) abuse. J. Emerg. Med. 41 (4), 378–380.

Anderson, C.E., Loomis, G.A., 2003. Recognition and prevention of inhalant abuse. Am. Fam. Physician 68 (5), 869–874.

Anderson, H.R., Macnair, R.S., Ramsey, J.D., 1985. Deaths from abuse of volatile substances: a national epidemiological study. Br. Med. J. (Clin. Res. Ed.), 290 (6464), 304–307.

Arai, H., Yamada, M., Miyake, S., Yamashita, S., Iwamoto, H., Aida, N., Hara, M., 1997. [Two cases of toluene embryopathy with severe motor and intellectual disabilities syndrome]. No To Hattatsu 29 (5), 361–366.

Arnold, G.L., Kirby, R.S., Langendoerfer, S., Wilkins-Haug, L., 1994. Toluene embryopathy: clinical delineation and developmental follow-up. Pediatrics 93 (2), 216–220.

Audo, I., El Sanharawi, M., Vignal-Clermont, C., Villa, A., Morin, A., Conrath, J., Fompeydie, D., Sahel, J.A., Gocho-Nakashima, K., Goureau, O., Paques, M., 2011. Foveal damage in habitual poppers users. Arch. Ophthalmol. 129 (6), 703–708.

Avella, J., Wilson, J.C., Lehrer, M., 2006. Fatal cardiac arrhythmia after repeated exposure to 1,1-difluoroethane (DFE). Am. J. Forensic Med. Pathol. 27 (1), 58–60.

Bale, A.S., Meacham, C.A., Benignus, V.A., Bushnell, P.J., Shafer, T.J., 2005a. Volatile organic compounds inhibit human and rat neuronal nicotinic acetylcholine receptors expressed in *Xenopus* oocytes. Toxicol. Appl. Pharmacol. 205 (1), 77–88.

Bale, A.S., Smothers, C.T., Woodward, J.J., 2002. Inhibition of neuronal nicotinic acetylcholine receptors by the abused solvent, toluene. Br. J. Pharmacol. 137 (3), 375–383.

Bale, A.S., Tu, Y., Carpenter-Hyland, E.P., Chandler, L.J., Woodward, J.J., 2005b. Alterations in glutamatergic and gabaergic ion channel activity in hippocampal neurons following exposure to the abused inhalant toluene. Neuroscience 130 (1), 197–206.

Balster, R.L., Cruz, S.L., Howard, M.O., Dell, C.A., Cottler, L.B., 2009. Classification of abused inhalants. Addiction 104 (6), 878–882.

Beasley, M., Frampton, L., Fountain, J., 2006. Inhalant abuse in New Zealand. N. Z. Med. J. 119 (1233), U1952.

Beckley, J.T., Evins, C.E., Fedarovich, H., Gilstrap, M.J., Woodward, J.J., 2013. Medial prefrontal cortex inversely regulates toluene-induced changes in markers of synaptic plasticity of mesolimbic dopamine neurons. J. Neurosci. 33 (2), 804–813.

Beckley, J.T., Woodward, J.J., 2011. The abused inhalant toluene differentially modulates excitatory and inhibitory synaptic transmission in deep-layer neurons of the medial prefrontal cortex. Neuropsychopharmacology 36 (7), 1531–1542.

Beckstead, M.J., Weiner, J.L., Eger 2nd, E.I., Gong, D.H., Mihic, S.J., 2000. Glycine and gamma-aminobutyric acid(A) receptor function is enhanced by inhaled drugs of abuse. Mol. Pharmacol. 57 (6), 1199–1205.

Blokhina, E.A., Dravolina, O.A., Bespalov, A.Y., Balster, R.L., Zvartau, E.E., 2004. Intravenous self-administration of abused solvents and anesthetics in mice. Eur. J. Pharmacol. 485 (1–3), 211–218.

Bowen, S.E., 2006. Increases in amphetamine-like discriminative stimulus effects of the abused inhalant toluene in mice. Psychopharmacology (Berl) 186 (4), 517–524.

Bowen, S.E., 2009. Time course of the ethanol-like discriminative stimulus effects of abused inhalants in mice. Pharmacol. Biochem. Behav. 91 (3), 345–350.

Bowen, S.E., 2011. Two serious and challenging medical complications associated with volatile substance misuse: sudden sniffing death and fetal solvent syndrome. Subst. Use Misuse 46 (Suppl. 1), 68–72.

Bowen, S.E., Batis, J.C., Mohammadi, M.H., Hannigan, J.H., 2005. Abuse pattern of gestational toluene exposure and early postnatal development in rats. Neurotoxicol. Teratol. 27 (1), 105–116.

Bowen, S.E., Daniel, J., Balster, R.L., 1999a. Deaths associated with inhalant abuse in Virginia from 1987 to 1996. Drug Alcohol Depend. 53 (3), 239–245.

Bowen, S.E., Hannigan, J.H., 2006. Developmental toxicity of prenatal exposure to toluene. AAPS J. 8 (2), E419–E424.

Bowen, S.E., Hannigan, J.H., 2013. Binge toluene exposure in pregnancy and pre-weaning developmental consequences in rats. Neurotoxicol. Teratol. 38C, 29–35.

Bowen, S.E., Wiley, J.L., Jones, H.E., Balster, R.L., 1999b. Phencyclidine- and diazepam-like discriminative stimulus effects of inhalants in mice. Exp. Clin. Psychopharmacol. 7 (1), 28–37.

Broussard, L.A., 2000. The role of the laboratory in detecting inhalant abuse. Clin. Lab. Sci. 13 (4), 205–209.

Bukowski, J.A., 2001. Review of the epidemiological evidence relating toluene to reproductive outcomes. Regul. Toxicol. Pharmacol. 33 (2), 147–156.

Cairney, S., O' Connor, N., Dingwall, K.M., Maruff, P., Shafiq-Antonacci, R., Currie, J., Currie, B.J., 2013. A prospective study of neurocognitive changes 15 years after chronic inhalant abuse. Addiction 108 (6), 1107–1114.

Chakroun, R., Faidi, F., Hedhili, A., Charbaji, K., Nouaigui, H., Laiba, M.B., 2008. Inhalant abuse detection and evaluation in young Tunisians. J. Forensic Sci. 53 (1), 232–237.

Cok, I., Dagdelen, A., Gokce, E., 2003. Determination of urinary hippuric acid and o-cresol levels as biological indicators of toluene exposure in shoe-workers and glue sniffers. Biomarkers 8 (2), 119–127.

Cruz, S.L., Balster, R., 2013. Neuropharmacology of inhalants. In: Miller, P.M. (Ed.), Biological Research on Addiction: Comprehensive Addictive Behaviors and Disorders. Academic Press, San Diego, Elsevier Inc., pp. 637–645.

Cruz, S.L., Balster, R.L., Woodward, J.J., 2000. Effects of volatile solvents on recombinant N-methyl-D-aspartate receptors expressed in Xenopus oocytes. Br. J. Pharmacol. 131 (7), 1303–1308.

Cruz, S.L., Dominguez, M., 2011. Misusing volatile substances for their hallucinatory effects: a qualitative pilot study with Mexican teenagers and a pharmacological discussion of their hallucinations. Subst. Use Misuse 46 (Suppl. 1), 84–94.

Cruz, S.L., Mirshahi, T., Thomas, B., Balster, R.L., Woodward, J.J., 1998. Effects of the abused solvent toluene on recombinant N-methyl-D-aspartate and non-N-methyl-D-aspartate receptors expressed in Xenopus oocytes. J. Pharmacol. Exp. Ther. 286 (1), 334–340.

Cruz, S.L., Orta-Salazar, G., Gauthereau, M.Y., Millan-Perez Pena, L., Salinas-Stefanon, E.M., 2003. Inhibition of cardiac sodium currents by toluene exposure. Br. J. Pharmacol. 140 (4), 653–660.

Dell, C.A., Gust, S.W., MacLean, S., 2011. Global issues in volatile substance misuse. Subst. Use Misuse 46 (Suppl. 1), 1–7.

Dell, D., Hopkins, C., 2011. Residential volatile substance misuse treatment for indigenous youth in Canada. Subst. Use Misuse 46 (Suppl. 1), 107–113.

Dingwall, K.M., Cairney, S., 2011. Recovery from central nervous system changes following volatile substance misuse. Subst. Use Misuse 46 (Suppl. 1), 73–83.

Edwards, R.J., Ujma, J., 1995. Extreme methaemoglobinaemia secondary to recreational use of amyl nitrite. J. Accid. Emerg. Med. 12 (2), 138–142.

Elkoussi, A., Bakheet, S., 2011. Volatile substance misuse among street children in Upper Egypt. Subst. Use Misuse 46 (Suppl. 1), 35–39.

Evans, E.B., Balster, R.L., 1991. CNS depressant effects of volatile organic solvents. Neurosci. Biobehav. Rev. 15 (2), 233–241.

Fakier, N., Wild, L.G., 2011. Associations among sleep problems, learning difficulties and substance use in adolescence. J. Adolesc. 34 (4), 717–726.

Filley, C.M., 2013. Toluene abuse and white matter: a model of toxic leukoencephalopathy. Psychiatr. Clin. North Am. 36 (2), 293–302.

Filley, C.M., Halliday, W., Kleinschmidt-DeMasters, B.K., 2004. The effects of toluene on the central nervous system. J. Neuropathol. Exp. Neurol. 63 (1), 1–12.

Flanagan, R.J., Ives, R.J., 1994. Volatile substance abuse. Bull. Narc. 46 (2), 49–78.

Flanagan, R.J., Ruprah, M., Meredith, T.J., Ramsey, J.D., 1990. An introduction to the clinical toxicology of volatile substances. Drug Saf. 5 (5), 359–383.

Fledelius, H.C., 1999. Irreversible blindness after amyl nitrite inhalation. Acta Ophthalmol. Scand. 77 (6), 719–721.

Forman, S.A., Miller, K.W., 2011. Anesthetic sites and allosteric mechanisms of action on Cys-loop ligand-gated ion channels. Can. J. Anaesth. 58 (2), 191–205.

Funada, M., Sato, M., Makino, Y., Wada, K., 2002. Evaluation of rewarding effect of toluene by the conditioned place preference procedure in mice. Brain Res. Protoc. 10 (1), 47–54.

Garland, E.L., Howard, M.O., 2010. Phenomenology of adolescent inhalant intoxication. Exp. Clin. Psychopharmacol. 18 (6), 498–509.

Garriott, J., Petty, C.S., 1980. Death from inhalant abuse: toxicological and pathological evaluation of 34 cases. Clin. Toxicol. 16 (3), 305–315.

Garriott, J.C., 1992. Death among Inhalant Abusers. National Institute on Drug Abuse Research Monograph 129. DHHS Pub. No. (93-3475). National Institute on Drug Abuse, Rockville, MD, pp. 181–191.

Gauthereau, M.Y., Salinas-Stefanon, E.M., Cruz, S.L., 2005. A mutation in the local anaesthetic binding site abolishes toluene effects in sodium channels. Eur. J. Pharmacol. 528 (1–3), 17–26.

Gerasimov, M.R., Collier, L., Ferrieri, A., Alexoff, D., Lee, D., Gifford, A.N., Balster, R.L., 2003. Toluene inhalation produces a conditioned place preference in rats. Eur. J. Pharmacol. 477 (1), 45–52.

Gerasimov, M.R., Schiffer, W.K., Marstellar, D., Ferrieri, R., Alexoff, D., Dewey, S.L., 2002. Toluene inhalation produces regionally specific changes in extracellular dopamine. Drug Alcohol Depend. 65 (3), 243–251.

Gillman, M.A., Lichtigfeld, F.J., 1994. Pharmacology of psychotropic analgesic nitrous oxide as a multipotent opioid agonist. Int. J. Neurosci. 76 (1–2), 5–12.

Goodwin, J.M., Geil, C., Grodner, B., Metrick, S., 1981. Inhalant abuse, pregnancy, and neglected children. Am. J. Psychiatry 138 (8), 1126.

Goodwin, T.M., 1988. Toluene abuse and renal tubular acidosis in pregnancy. Obstet. Gynecol. 71 (5), 715–718.

Hahn, T., Avella, J., Lehrer, M., 2006. A motor vehicle accident fatality involving the inhalation of 1,1-difluoroethane. J. Anal. Toxicol. 30 (8), 638–642.

Hannigan, J.H., Bowen, S.E., 2010. Reproductive toxicology and teratology of abused toluene. Syst. Biol. Reprod. Med. 56 (2), 184–200.

Hathout, L., El-Saden, S., 2011. Nitrous oxide-induced B_{12} deficiency myelopathy: perspectives on the clinical biochemistry of vitamin B_{12}. J. Neurol. Sci. 301 (1–2), 1–8.

Hernandez-Avila, C.A., Ortega-Soto, H.A., Jasso, A., Hasfura-Buenaga, C.A., Kranzler, H.R., 1998. Treatment of inhalant-induced psychotic disorder with carbamazepine versus haloperidol. Psychiatr. Serv. 49 (6), 812–815.

Hersh, J.H., 1989. Toluene embryopathy: two new cases. J. Med. Genet. 26 (5), 333–337.

Hersh, J.H., Podruch, P.E., Rogers, G., Weisskopf, B., 1985. Toluene embryopathy. J. Pediatr. 106 (6), 922–927.

Hormes, J.T., Filley, C.M., Rosenberg, N.L., 1986. Neurologic sequelae of chronic solvent vapor abuse. Neurology 36 (5), 698–702.

Howard, M.O., Bowen, S.E., Garland, E.L., Perron, B.E., Vaughn, M.G., 2011. Inhalant use and inhalant use disorders in the United States. Addict. Sci. Clin. Pract. 6 (1), 18–31.

Huerta-Rivas, A., Lopez-Rubalcava, C., Sanchez-Serrano, S.L., Valdez-Tapia, M., Lamas, M., Cruz, S.L., 2012. Toluene impairs learning and memory, has antinociceptive effects, and modifies histone acetylation in the dentate gyrus of adolescent and adult rats. Pharmacol. Biochem. Behav. 102 (1), 48–57.

Hunter, A.G., Thompson, D., Evans, J.A., 1979. Is there a fetal gasoline syndrome? Teratology 20 (1), 75–79.

Hutchens, K.S., Kung, M., 1985. "Experimentation" with chloroform. Am. J. Med. 78 (4), 715–718.

Hynes-Dowell, M., Mateu-Gelabert, P., Barros, H.M., Delva, J., 2011. Volatile substance misuse among high school students in South America. Subst. Use Misuse 46 (Suppl. 1), 27–34.

Inoue, O., Kanno, E., Kasai, K., Ukai, H., Okamoto, S., Ikeda, M., 2004. Benzylmercapturic acid is superior to hippuric acid and o-cresol as a urinary marker of occupational exposure to toluene. Toxicol. Lett. 147 (2), 177–186.

Jones, H.E., Balster, R.L., 1998. Inhalant abuse in pregnancy. Obstet. Gynecol. Clin. North Am. 25 (1), 153–167.

Koyama, S., Matsumoto, N., Kubo, C., Akaike, N., 2000. Presynaptic 5-HT$_3$ receptor-mediated modulation of synaptic GABA release in the mechanically dissociated rat amygdala neurons. J. Physiol. 529 (Pt 2), 373–383.

Krajcovic, J., Novomesky, F., Stuller, F., Straka, L., Mokry, J., 2011. An unusual case of anesthetic abuse by a full-face gas mask. Am. J. Forensic Med. Pathol.

Krenzelok, E.P., 2000. Sildenafil: clinical toxicology profile. J. Toxicol. Clin. Toxicol. 38 (6), 645–651.

Kringsholm, B., 1980. Sniffing-associated deaths in Denmark. Forensic Sci. Int. 15 (3), 215–225.

Lee, D.E., Gerasimov, M.R., Schiffer, W.K., Gifford, A.N., 2006. Concentration-dependent conditioned place preference to inhaled toluene vapors in rats. Drug Alcohol Depend. 85 (1), 87–90.

Lin, R.J., Chen, H.F., Chang, Y.C., Su, J.J., 2011. Subacute combined degeneration caused by nitrous oxide intoxication: case reports. Acta Neurol. Taiwan 20 (2), 129–137.

Little, J., Hileman, B., Ziran, B.H., 2008. Inhalant abuse of 1,1-difluoroethane (DFE) leading to heterotopic ossification: a case report. Patient Saf. Surg. 2 (1), 28.

Lomax, R.B., Ridgway, P., Meldrum, M., 2004. Does occupational exposure to organic solvents affect colour discrimination? Toxicol. Rev. 23 (2), 91–121.

Lopreato, G.F., Phelan, R., Borghese, C.M., Beckstead, M.J., Mihic, S.J., 2003. Inhaled drugs of abuse enhance serotonin-3 receptor function. Drug Alcohol Depend. 70 (1), 11–15.

MacLean, S., 2007. Global selves: marginalised young people and aesthetic reflexivity in inhalant drug use. J. Youth Stud. 10 (4), 399–418.

MacLean, S., 2008. Volatile bodies: stories of corporeal pleasure and damage in marginalised young people's drug use. Int. J. Drug Policy 19 (5), 375–383.

Mason, D.T., Zelis, R., Amsterdam, E.A., 1971. Actions of the nitrites on the peripheral circulation and myocardial oxygen consumption: significance in the relief of angina pectoris. Chest 59 (3), 296–305.

Maxwell, J.C., 2001. Deaths related to the inhalation of volatile substances in Texas: 1988–1998. Am. J. Drug Alcohol Abuse 27 (4), 689–697.

McManus, T.J., Starrett, L.A., Harris, J.R., 1982. Amyl nitrite use by homosexuals. Lancet 1 (8270), 503.

Meadows, R., Verghese, A., 1996. Medical complications of glue sniffing. South. Med. J. 89 (5), 455–462.

Medina-Mora, M.E., Real, T., 2008. Epidemiology of inhalant use. Curr. Opin. Psychiatry 21 (3), 247–251.

Mennerick, S., Jevtovic-Todorovic, V., Todorovic, S.M., Shen, W., Olney, J.W., Zorumski, C.F., 1998. Effect of nitrous oxide on excitatory and inhibitory synaptic transmission in hippocampal cultures. J. Neurosci. 18 (23), 9716–9726.

Mihic, S.J., Ye, Q., Wick, M.J., Koltchine, V.V., Krasowski, M.D., Finn, S.E., Mascia, M.P., Valenzuela, C.F., Hanson, K.K., Greenblatt, E.P., Harris, R.A., Harrison, N.L., 1997. Sites of alcohol and volatile anaesthetic action on GABA(A) and glycine receptors. Nature 389 (6649), 385–389.

Misra, L.K., Kofoed, L., Fuller, W., 1999. Treatment of inhalant abuse with risperidone. J. Clin. Psychiatry 60 (9), 620.

Mullens, A.B., Young, R.M., Hamernik, E., Dunne, M., 2009. The consequences of substance use among gay and bisexual men: a Consensual Qualitative Research analysis. Sex. Health 6 (2), 139–152.

NIPC, 2010. Guidelines for Medical Examiners, Coroners and Pathologists: Determining Inhalant Deaths. . Retrieved 26 July, 2010, from: www.inhalants.org.

Novosel, I., Kovacic, Z., Gusic, S., Batelja, L., Nestic, M., Seiwerth, S., Skavic, J., 2011. Immunohistochemical detection of early myocardial damage in two sudden deaths due to intentional butane inhalation. Two case reports with review of literature. J. Forensic Leg. Med. 18 (3), 125–131.

OAS, Hynes-Dowell, M., 2011. Report on Drug Use in the Americas 2011. Organization of American States, Inter-American Drug Abuse Control Commission, Washington, D.C.

Pavlic, M., Haidekker, A., Grubwieser, P., Rabl, W., 2002. Fatal accident caused by isoflurane abuse. Int. J. Legal Med. 116 (6), 357–360.

Pearson, M.A., Hoyme, H.E., Seaver, L.H., Rimsza, M.E., 1994. Toluene embryopathy: delineation of the phenotype and comparison with fetal alcohol syndrome. Pediatrics 93 (2), 211–215.

Pinder, A.G., Pittaway, E., Morris, K., James, P.E., 2009. Nitrite directly vasodilates hypoxic vasculature via nitric oxide-dependent and -independent pathways. Br. J. Pharmacol. 157 (8), 1523–1530.

Potocka-Banas, B., Majdanik, S., Dutkiewicz, G., Borowiak, K., Janus, T., 2011. Death caused by addictive inhalation of nitrous oxide. Hum. Exp. Toxicol. 30 (11), 1875–1877.

Raines, D.E., Claycomb, R.J., Forman, S.A., 2003. Modulation of GABA(A) receptor function by nonhalogenated alkane anesthetics: the effects on agonist enhancement, direct activation, and inhibition. Anesth. Analg. 96 (1), 112–118 table of contents.

Raines, D.E., Gioia, F., Claycomb, R.J., Stevens, R.J., 2004. The N-methyl-D-aspartate receptor inhibitory potencies of aromatic inhaled drugs of abuse: evidence for modulation by cation-pi interactions. J. Pharmacol. Exp. Ther. 311 (1), 14–21.

Ridenour, T.A., 2005. Inhalants: not to be taken lightly anymore. Curr. Opin. Psychiatry 18 (3), 243–247.

Riegel, A.C., Zapata, A., Shippenberg, T.S., French, E.D., 2007. The abused inhalant toluene increases dopamine release in the nucleus accumbens by directly stimulating ventral tegmental area neurons. Neuropsychopharmacology 32 (7), 1558–1569.

Rosenberg, N.L., Grigsby, J., Dreisbach, J., Busenbark, D., Grigsby, P., 2002. Neuropsychologic impairment and MRI abnormalities associated with chronic solvent abuse. J. Toxicol. Clin. Toxicol. 40 (1), 21–34.

Rosenberg, N.L., Kleinschmidt-DeMasters, B.K., Davis, K.A., Dreisbach, J.N., Hormes, J.T., Filley, C.M., 1988. Toluene abuse causes diffuse central nervous system white matter changes. Ann. Neurol. 23 (6), 611–614.

Rosenberg, N.L., Sharp, C.W., 1997. Solvent toxicity: a neurological focus. Subst. Use Misuse 32 (12&13), 1859–1864.

SAMHSA, March 17, 2011. Data Spotlight: Adults Represent the Majority of Inhalant Treatment Admissions. Substance Abuse and Mental Health Services Administration, Office of Applied Studies, Rockville, MD.

SAMHSA, 2007. The NSDUH Report March 15, 2007 Patterns and Trends in Inhalant Use by Adolescent Males and Females: 2002–2005. Substance Abuse and Mental Health Services Administration, U.S. Department of Health and Human Services. No. NSDUH07-0315.

Schloneger, M., Stull, A., Singer, J.I., 2009. Inhalant abuse: a case of hemoptysis associated with halogenated hydrocarbons abuse. Pediatr. Emerg. Care 25 (11), 754–757.

Scior, T.R.F., Martinez-Morales, E., Cruz, S.L., Salinas-Stefanon, E.M., 2009. In silico modeling of toluene binding site in the pore of voltage gate sodium channel. J. Receptor Ligand Channel Res. 2, 1–9.

Shafer, T.J., Bushnell, P.J., Benignus, V.A., Woodward, J.J., 2005. Perturbation of voltage-sensitive Ca^{2+} channel function by volatile organic solvents. J. Pharmacol. Exp. Ther. 315 (3), 1109–1118.

Sharma, S., Lal, R., 2011. Volatile substance misuse among street children in India: a preliminary report. Subst. Use Misuse 46 (Suppl. 1), 46–49.

Shen, Y.C., 2007. Treatment of inhalant dependence with lamotrigine. Prog. Neuropsychopharmacol. Biol. Psychiatry 31 (3), 769–771.

Stevens, R.J., Rusch, D., Davies, P.A., Raines, D.E., 2005. Molecular properties important for inhaled anesthetic action on human 5-HT_{3A} receptors. Anesth. Analg. 100 (6), 1696–1703.

Takagi, M., Lubman, D.I., Cotton, S., Fornito, A., Baliz, Y., Tucker, A., Yucel, M., 2011. Executive control among adolescent inhalant and cannabis users. Drug Alcohol Rev. 30 (6), 629–637.

Thiesen, F.V., Noto, A.R., Barros, H.M., 2007. Laboratory diagnosis of toluene-based inhalants abuse. Clin. Toxicol. (Phila) 45 (5), 557–562.

Tillar, R., Shafer, T.J., Woodward, J.J., 2002. Toluene inhibits voltage-sensitive calcium channels expressed in pheochromocytoma cells. Neurochem. Int. 41 (6), 391–397.

Toutant, C., Lippmann, S., 1979. Fetal solvents syndrome. Lancet 1 (8130), 1356.

Turner, T.J., Mokler, D.J., Luebke, J.I., 2004. Calcium influx through presynaptic 5-HT$_3$ receptors facilitates GABA release in the hippocampus: in vitro slice and synaptosome studies. Neuroscience 129 (3), 703–718.

Villatoro, J.A., Cruz, S.L., Ortiz, A., Medina-Mora, M.E., 2011. Volatile substance misuse in Mexico: correlates and trends. Subst. Use Misuse 46 (Suppl. 1), 40–45.

Weiss, B., Wood, R.W., Macys, D.A., 1979. Behavioral toxicology of carbon disulfide and toluene. Environ. Health Perspect. 30, 39–45.

Wick, R., Gilbert, J.D., Felgate, P., Byard, R.W., 2007. Inhalant deaths in South Australia: a 20-year retrospective autopsy study. Am. J. Forensic Med. Pathol. 28 (4), 319–322.

Wilkins-Haug, L., Gabow, P.A., 1991. Toluene abuse during pregnancy: obstetric complications and perinatal outcomes. Obstet. Gynecol. 77 (4), 504–509.

Williams, J.M., Stafford, D., Steketee, J.D., 2005. Effects of repeated inhalation of toluene on ionotropic GABA$_A$ and glutamate receptor subunit levels in rat brain. Neurochem. Int. 46 (1), 1–10.

Win-Shwe, T.T., Fujimaki, H., 2012. Acute administration of toluene affects memory retention in novel object recognition test and memory function-related gene expression in mice. J. Appl. Toxicol. 32 (4), 300–304.

Wood, R.W., Grubman, J., Weiss, B., 1977. Nitrous oxide self-administration by the squirrel monkey. J. Pharmacol. Exp. Ther. 202 (3), 491–499.

Yavich, L., Patkina, N., Zvartau, E., 1994. Experimental estimation of addictive potential of a mixture of organic solvents. Eur. Neuropsychopharmacol. 4 (2), 111–118.

Yavich, L., Zvartau, E., 1994. A comparison of the effects of individual organic solvents and their mixture on brain stimulation reward. Pharmacol. Biochem. Behav. 48 (3), 661–664.

CHAPTER NINETEEN

Emerging Designer Drugs

David E. Nichols[1] and William E. Fantegrossi[2]
[1]Division of Chemical Biology and Medicinal Chemistry, Eshelman School of Pharmacy, University of North Carolina, Chapel Hill, NC, USA
[2]Department of Pharmacology and Toxicology, University of Arkansas for Medical Sciences, Little Rock, AR, USA

1. INTRODUCTION

The consideration of Emerging Designer Drugs seems very topical today. In the past few years, a plethora of relatively obscure compounds have appeared on the illicit drug market, and in some cases, have proven to be quite popular. In many, but not all cases, these new drugs of abuse have turned out to be established research chemicals that have diffused out of laboratories and scientific journals and onto the streets. As novel pharmacological entities, the legal ramifications for selling and possessing these drugs are initially unclear, and enterprising individuals typically exploit the novelty of these substances to make rapid and substantial profits selling them over-the-counter and online. Indeed, emerging drugs of abuse occupy a legal grey area until emergency scheduling powers are invoked, typically first at the municipal and state level, then nationally. Although the proliferation of designer drugs toward the end of the 1990s was likely fueled by the desire to profit off the sale of "legal ecstasy alternatives," the market has since expanded tremendously, and now includes not only potential ecstasy replacements, but a variety of psychedelic phenethylamines and tryptamines, cathinone derivatives with bizarre psychostimulant properties, and synthetic cannabinoids. A few modified ketamine derivatives also have recently appeared, and one might even expect eventually to see some modified salvinorin analogues for sale.

Some years ago, a graduate student in one of our laboratories speculated, "make one drug illegal, and a more dangerous one will take its place." Today, this seems almost axiomatic. Certainly, no knowledgeable person could present a reasonable argument that marijuana is less safe than the new synthetic cannabinoids that are now appearing. It may be a case of unintended consequences of current drug policy that new, untested drugs are proliferating on the black market at an unprecedented rate. One might reasonably believe that if marijuana, LSD, and ecstasy had remained legal, or at least decriminalized, many of these new designer drugs would never have caught on. For example, it is not clear that any of the potential replacements for ecstasy is as satisfactory to users as is the original molecule, MDMA (3,4-methylenedioxy-*N*-methylamphetamine), at least from a psychopharmacology point of view. Similarly, it seems unlikely that the products containing potent synthetic

The Effects of Drug Abuse on the Human Nervous System
http://dx.doi.org/10.1016/B978-0-12-418679-8.00019-8

cannabinoids would ever have appeared if marijuana and/or its many preparations were legal and readily available. It seems evident that much of the demand for new drugs that is the motivation for illegal manufacture is based on finding new "legal highs" that can be quickly marketed before they are identified and restricted by the various drug control agencies throughout the world.

Similarly, it may be the case that strict controls on the majority of recreational drugs with pronounced reinforcing effects have driven those searching for new compounds to explore the pharmacological "back bench." In that regard, although it is the case that essentially all preclinical evaluations of the abuse potential of psychostimulants (like cocaine or methamphetamine) or opioids (like heroin or alfentanil) detect strong reinforcing effects, many emerging drugs of abuse tend to function as relatively weak reinforcers. In what may be the only case of a compound being released from Schedule I controls after emergency scheduling powers were invoked, 1-[3-(trifluoromethyl) phenyl]piperazine (TFMPP) failed to engender self-administration in rhesus monkeys (Fantegrossi et al., 2005) despite eliciting a hallucinogen-like head twitch response in the mouse, and substituting for (MDMA) in a drug discrimination paradigm (Yarosh et al., 2007). Thus, it may be the case that many emerging drugs of abuse deliver less acceptable rewarding effects than do the drugs for which they are marketed as "legal alternatives."

What are other factors that are driving the explosion of new designer drugs? First of all, the Internet has provided a venue for rapid dissemination of information about drugs. Not only can users search public databases for keywords such as "psychostimulant," "hallucinogen," or "cannabinoid," but numerous blogs now exist where users of new chemicals can describe in detail not only their experiences with these chemicals but also where they can be obtained. Previously, availability of such chemicals was largely limited to aficionados who had contacts with small organic chemistry laboratories. When the desired substance is not actually proscribed, it can be manufactured in laboratories in other countries, particularly China. A recent Google search of "research chemicals" and "China" revealed that most of the popular research chemicals could be shipped "from multiple warehouses: China, EU, USA, Russia, India, and Kazakhstan." "Best Seller" chemicals listed on one site included, for example, the synthetic cannabinoid JWH-018 in 10 g or 50 g lots, the "bath salt" psychostimulant/entactogen methylenedioxypyrovalerone (MDPV) in 5 g lots, and the psychedelic 2C-I in 50 g lots.

Internet marketing sites couple ready access to information about the doses and effects of research chemicals with easy accessibility. Furthermore, the economics of synthetic research chemicals can be very favorable. For example, 10 g of JWH-018 was listed on one Internet site at a cost of only $100! With a smoked dose of 2–5 mg it is about twice the potency of THC, meaning that the intoxicating potential of 10 g of JWH-018 is approximately comparable to one-quarter pound of very high grade (16–18% THC) cannabis. From the user's perspective, such a synthetic cannabinoid is much less expensive than cannabis itself. Amusingly, many internet distributors will include purity analysis of their drugs, including results obtained using thin layer chromatography or mass spectrometry, and the purity of products from illicit sources often (but not always) rivals that of established research chemical companies. It is often quite difficult to justify the cost of purchasing these drugs from legitimate distributors of research chemicals instead of from overseas clandestine laboratories, but the requirement to disclose the source of all materials in scientific manuscripts and grant applications probably keeps the behavior of most scientists in check.

New research chemicals, in general, do not present significant challenges to competent synthetic chemists. Synthetic cannabinoids require relatively few simple steps. Synthetic hallucinogens typically require three or four synthetic steps. Although the syntheses are not formidable for the typical synthetic chemist, when we consider that small pilot plants may be dedicated to their synthesis in countries such as China, the number of synthetic steps is irrelevant. The cost of starting materials such as substituted benzaldehydes, indoles, and reducing agents is relatively trivial compared with the marketed prices of the final products.

Given that we presently do not have ways to control the appearance of these substances on illicit markets, we shall attempt to identify the emerging trends that seem to be appearing on the current drug scene. Our intention here is to give a survey of the major classes of substances that are appearing, and in some cases, at least, to predict future trends. The reader should be aware, however, that prediction in this game cannot be reliable because someone may serendipitously discover a new molecule that finds tremendous acceptance with potential users, much as what happened with MDMA. It is impossible to predict such random occurrences, and despite our best efforts to offer what we believe are likely trends, the discovery of such a drug would completely derail all of our reasoned arguments.

Finally, it must be kept in mind that it is not only the research chemicals, per se, that present the problems, but also the absence of quality controls used in their manufacture. Unlike the pharmaceutical industry, there is no standard of purity required for research chemicals. Savvy buyers may seek sellers who offer the equivalent of material safety data sheets, or documentation of purity by NMR or IR, but one must first of all trust the integrity of the seller, and that the documentation is actually for the specific lot of chemical that was purchased, when in fact there is no guarantee of that. Indeed, enormous variability in the identity and dose of active constituents has recently been documented both within and between brands of internet-available "legal high" products (Baron et al., 2011; Zuba and Byrska, 2013). Second, manufacturing contaminants may be included and unless it has been definitively established to be safe, a minor contaminant may lead to an adverse reaction. Recall that MPTP, the potent dopamine neurotoxin, was a contaminant generated during the crystallization of a meperidine analogue. Had the final meperidine analogue been pure, and free of the MPTP contaminant, none of the heroin addicts who used it would have been afflicted with severe parkinsonism.

2.1. TYPES OF DESIGNER DRUGS

2.1.1. Phenethylamines

From a chemistry perspective, phenethylamines represent the largest category of current designer drugs, as well as potentially a vast reservoir for new and untested substances. Phenethylamines are probably the largest category because they are the easiest to synthesize, and the possible ring modifications that can be carried out on them are almost uncountable. In addition, depending on what substituents are attached to the aromatic ring of the phenethylamine template, phenethylamines can have psychopharmacological effects ranging from classic hallucinogenic action, to psychostimulant effects, to various molecules that could possess mixtures of direct effects on GPCRs, as well as being either inhibitors or substrates for the three monoamine uptake carrier proteins (See Figure 1). Effects on monoamine reuptake carriers can range from molecules that possess amphetamine- or methamphetamine-like psychostimulant effects, to other compounds that may have more of an MDMA-like or entactogenic effect. The extent to which activity at each of these targets can be "mixed and matched" is unknown, and it is likely that new designer drugs will continually appear that are built upon the phenethylamine scaffold.

2.1.2. Hallucinogens

There are two categories of hallucinogens that have appeared on the street. The first are the phenethylamines, originally inspired by the structure of mescaline (1). A large compendium of active hallucinogenic phenethylamines has been presented in the book PIHKAL, by Shulgin and Shulgin (1991).

Mescaline 1

Despite its historic importance and being the only naturally occurring phenethylamine hallucinogen, mescaline itself is perhaps one of the least potent such substances. Replacing the 4-methoxy with larger alkoxy groups (ethoxy, propoxy, alloxy, methallyloxy) or with alkylthio groups leads to compounds that are significantly more potent than mescaline itself (Shulgin and Shulgin, 1991; Nichols, 2004). Although relatively potent, these molecules have not appeared on the illicit market, and reasoning to explain that can perhaps be found in their more difficult synthesis, compared to other ring substitution patterns, as well as the unfavorable economics of manufacturing and distributing compounds that require relatively large doses (tens of milligrams vs a few

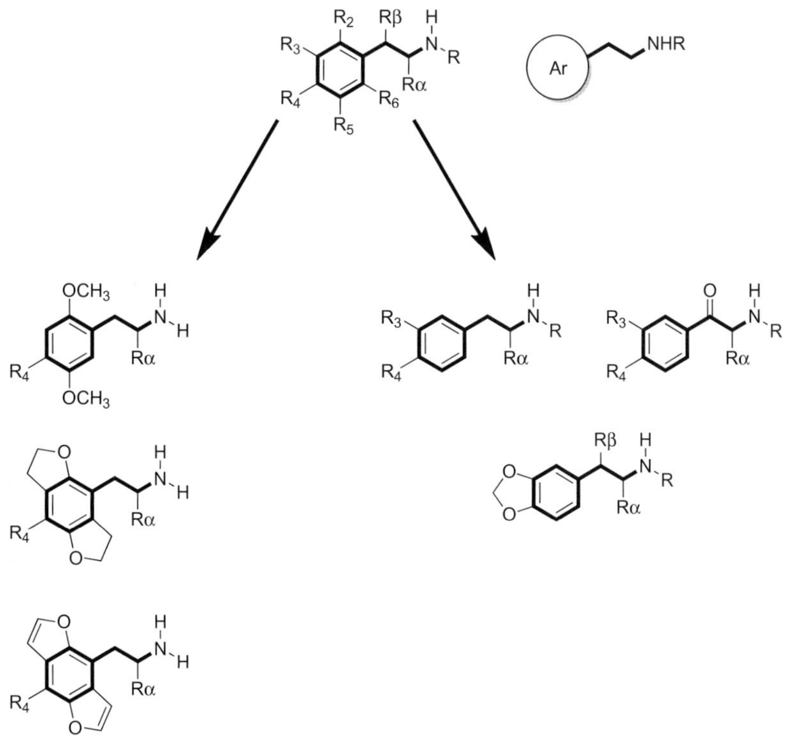

Figure 1 General examples of how the simple phenethylamine template can be modified to produce hallucinogens, illustrated on the left, or molecules that have effects at the monoamine transporters, exemplified on the right.

milligrams or less). It is the author's opinion that molecules closely resembling mescaline (i.e. 3,4,5-trisubstituted phenethylamines) are unlikely to appear on the street in significant quantities.

Transposing the 3- and 5-methoxy groups to the 2,6-positions leads to compounds that also are active, but again the relatively low potency and particularly the significant synthetic challenges in making 2,6-dimethoxy-4-substituted phenethylamines probably mean that these also will not emerge as new problems.

By contrast, transposition of the 3-methoxy of mescaline to the 2-position, leading to a 2,5-dimethoxy substitution pattern, is the most commonly seen type of hallucinogen. Although several of these have the simple two-carbon side chain, such as 2C-D, 2C-B, or 2CT-2 (Figure 2), the most potent have an alpha-methyl attached to the side chain (e.g. DOM, DOB, and Aleph-2). These types of molecules are generally referred to as substituted "amphetamine" hallucinogens, because unsubstituted alpha-methylphenethylamine itself is amphetamine. Only a small set of these compounds is illustrated in Figure 2, because a very large library of similar compounds can be generated

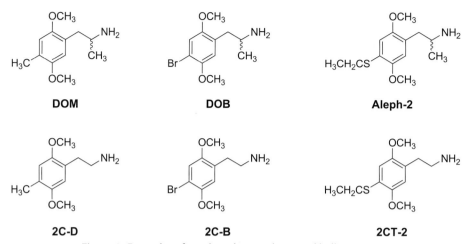

Figure 2 Examples of 2,5-dimethoxy-substituted hallucinogens.

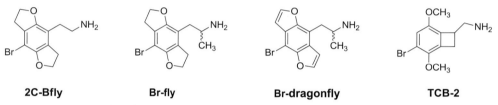

Figure 3 Examples of rigid analogues of hallucinogenic amphetamines with high potency.

simply by changing the 4–substituent. Halogens (other than fluorine), short unbranched alkyl groups, and a variety of alkylthio substituents can be introduced into 4–position of the molecule to give potent compounds, both in the phenethylamine and in the amphetamine series. Interestingly, extending the alpha-methyl in the side chain to the two carbon alpha-ethyl completely abolishes hallucinogenic activity.

Further increases in potency can be obtained by constraining the 2,5-dimethoxy substituents into a dihydrofuran ring (e.g. 2C-Bfly, Br-Fly, and Br-Dragonfly; Figure 3). Indeed, Br-dragonfly approaches the potency of LSD using in vitro or rodent models, and overdose deaths have resulted from its use (e.g. Andreasen et al., 2009). Although the synthesis of these furan-type compounds is slightly more complex than for simple phenethylamines, their higher potency holds economic incentives for illicit laboratories. In principle, a variety of new designer drugs could emerge from this template. For example, 2,5-dimethoxy-4-ethylphenethylamine (2C-E) has been cited as an important hallucinogen with unusual properties (Shulgin and Shulgin, 1991). The corresponding 4-ethyl could be readily prepared for any of the rigid molecules, and indeed 4-propyl, and a variety of 4-alkylthio compounds could safely be predicted to be quite psycho-active. A potent molecule that was developed by constraint of the side chain is TCB-2

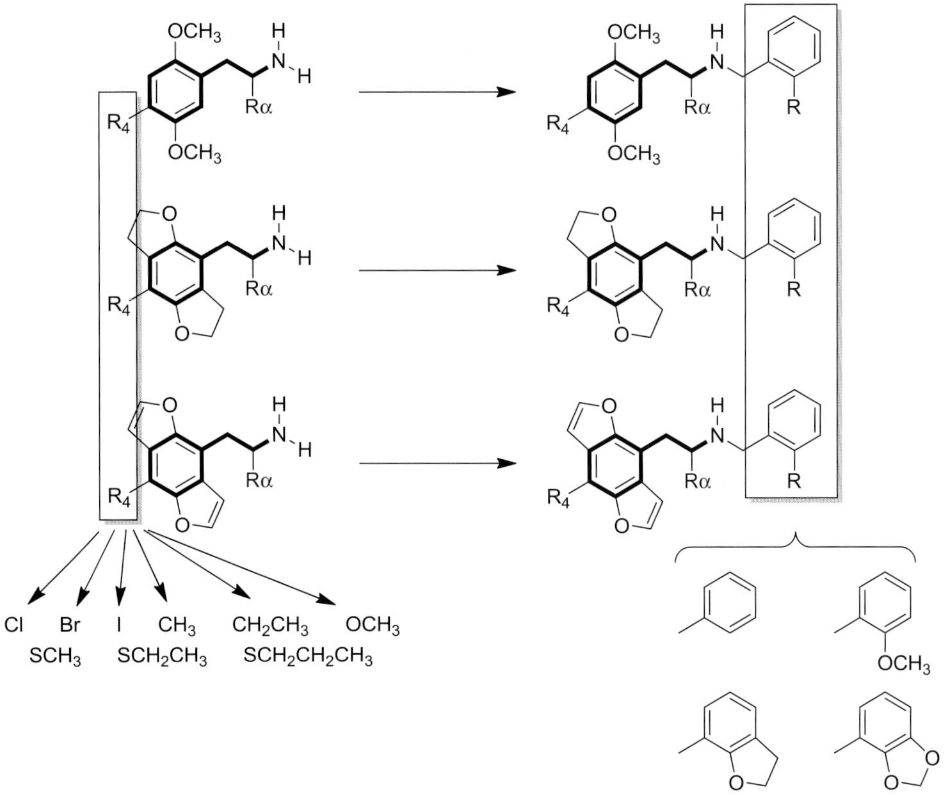

Figure 4 Examples of possible permutations and combinations of *N*-benzylphenethylamines that could appear as new designer drugs.

(McLean et al., 2006), now commercially available as a 5-HT$_{2A/2C}$ agonist for experimental laboratory studies. Although its synthesis is tedious enough to prevent its manufacture from being economical, it does exemplify the fact that relatively modest structural changes can lead to active compounds.

The toxicity of phenethylamine hallucinogens has generally been considered to be low. Nevertheless, deaths are occasionally reported, usually following overdose, and in some cases associated with polydrug abuse. The stimulation of 5-HT$_{2A}$ receptors in the vasculature also can lead to severe vasoconstriction resulting in limb amputation or death (Bowen et al., 1983; Winek et al., 1981).

N-Benzylphenethylamines. Following earlier studies by Elz (2002) and Heim (2003) *N*-benzylphenethylamines are now recognized to be highly potent 5-HT$_{2A}$ receptor agonists, with potential as hallucinogens (Braden et al., 2006). Although to date, only the 2,5-dimethoxy-4-iodo and 2,5-dimethoxy-4-chlorophenethylamines with *N*-(o-methoxybenzyl) groups seem to have gained any popularity as recreational drugs, it is possible to construct a large library of *N*-benzylphenethylamines that might be expected to have activity, as illustrated in Figure 4. Once the phenethylamine is in hand, it is a trivial matter to add the *N*-benzyl substituent from readily available benzaldehydes. These compounds do not appear to have oral activity, and the few blogs describing them generally report administration either rectally or by buccal absorption. These compounds are the most potent 5-$HT_{2A/2C}$ agonists known, with picomolar receptor affinities. Unfortunately, it also appears that their high potency can easily lead to overdose, and in some cases death (Geller, 2012; Araiza, 2012).

The second category of hallucinogens is comprised of the tryptamines, which would include LSD, a semisynthetic ergoline derivative. LSD is an extremely potent compound, with a typical minimum active human dose of about 0.05 mg. The subjective effects resulting from LSD ingestion can last up to 12 h and include alterations of mood, perceptual changes, and cognitive impairment. Thus, any novel analogues maintaining this high potency and relatively long duration could be problematic to new users not familiar with appropriate dosing. As it happens, however, no structural analogues have been developed that retain the unique psychopharmacological characteristics of LSD. Although it is relatively easy to produce lysergic acid amides other than the diethylamide seen in LSD, none of the ones that have been tested show potencies comparable to LSD. The methyl group on the basic nitrogen of LSD can be replaced with an ethyl to afford a quite potent analogue of LSD, but the economics of producing LSD, and then transforming it to the (N6)-ethyl compound are unfavorable.

The core structure of tryptamines is comprised of a bicyclic indole ring system with an aminoethyl moiety attached at the 3-position. The basic structure for all the tryptamines is derived from tryptophan, which serves as an essential amino acid in some animals. Enzymatic decarboxylation of tryptophan then leads to tryptamine. The biosynthesis of various endogenous tryptamines proceeds through differential modification of the tryptophan structure. For example, serotonin biosynthesis commences with hydroxylation at the 5-position of tryptophan by tryptophan hydroxylase and then proceeds through decarboxylation of the side chain by aromatic amino acid decarboxylase. The production of other endogenous tryptamines such as melatonin proceeds through different biosynthetic sequences, but all tryptamines contain the basic indole ring system, and one can consider them all to be structurally similar to serotonin. The bicyclic indole ring system contains six positions (not counting the site where the tryptamine side chain is attached) that are available for chemical modification; however, the majority of medicinal chemistry efforts have thus far focused on modification of the 4- and 5-positions. One reason for that is because it has been shown that modification of either the 6- or 7-positions significantly reduces or abolishes the psychoactive effects of the resulting compound. The addition of untested functional groups could potentially change this view, however, perhaps one day giving rise to novel 6- or 7-position substituted tryptamines that retain pharmacological activity.

2.2. PSYCHOSTIMULANTS

2.2.1. MDMA and Its Replacements

The unique psychoactive properties of MDMA have so far not been discovered in any other molecule, although various substances are claimed to have "similar" effects. Research chemicals that have been marketed as possible MDMA-replacements include several structures illustrated in Figure 5, including 4-fluoroamphetamine and 4-fluoromethamphetamine, as well as 4,5-methylenedioxy-2-aminoindan (MDAI), 5-iodo-2-aminoindan (5AI), 5,6-methylenedioxy-2-aminotetralin (MDAT), and the dihydrofuran compound 6-(2-aminopropyl)-2,3-dihydrobenzofuran (6-APB) (Figures 6, 7).

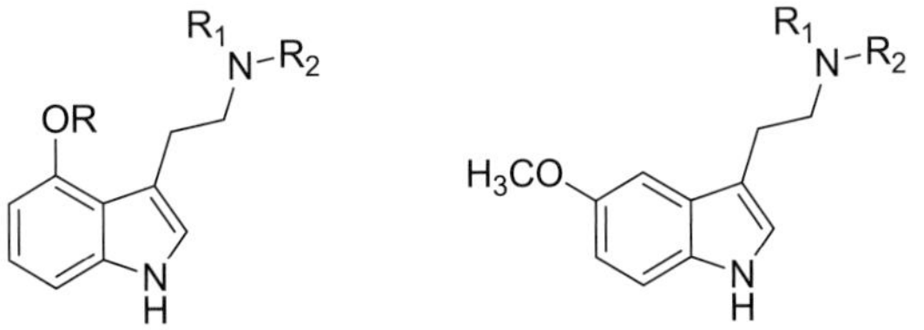

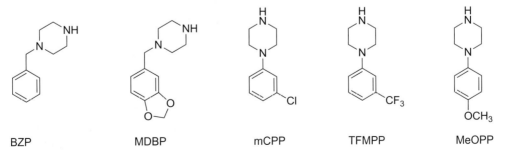

Figure 5 General structural features of simple hallucinogenic tryptamines.

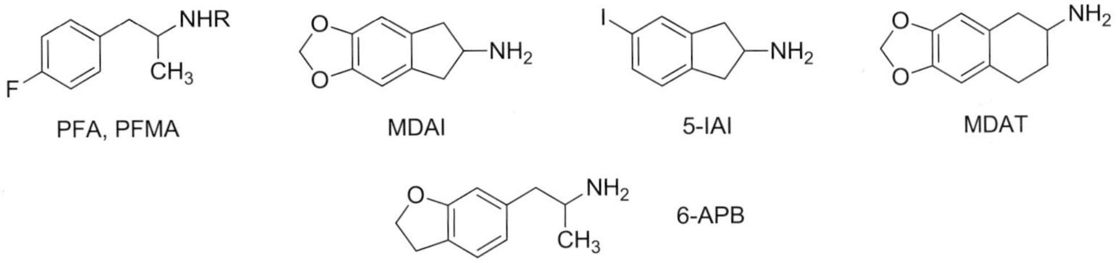

Figure 6 MDMA-related "research chemicals."

Figure 7 Substituted piperazine compounds with abuse potential.

It is perhaps important to note that MDMA itself, as well as a few of its analogues, has a chiral carbon, giving rise to stereoisomers. That is not particularly uncommon among drugs of abuse: for example, the psychostimulant methamphetamine and the psychedelic DOI are both phenethylamine derivatives with chirality in their side chains, although the stereochemistry is reversed for psychostimulants and hallucinogens. In most cases, however, one enantiomer is more biologically active, whereas the other will be either inactive, or "functionally inactive" due to a markedly decreased potency. That is not the case with MDA, MDMA, and perhaps several other analogues. Indeed, the S-(+)- and R-(−)- enantiomers of MDA and MDMA are both active at approximately the same dose, although their biological effects appear to be qualitatively distinct from one another. In that regard, we observe stimulant-like reinforcing and discriminative stimulus effects with S-(+)-MDMA that we do not see with the R-(−)- enantiomer (Fantegrossi et al., 2005). Similarly, in animal models, hallucinogen-like effects of R-(−)-MDA and R(−)-MDMA are not induced by their S-(+)- enantiomers. In humans, S-(+)-MDMA appears to be responsible for the unique psychopharmacology of racemic MDMA, but the pure S enantiomer did not completely reproduce the effects of the racemate, suggesting some contribution from the R enantiomer (Anderson III et al., 1978).

Although some reasonably efficient stereoselective syntheses for substituted amphetamine isomers are known, thus far, no chiral products have been seen on the illicit market (other than S-(+)-methamphetamine). The economics of a stereoselective synthesis are very unfavorable, as is postsynthesis resolution of individual enantiomers, but it may be the case that some novel pharmacological entity may appear where one enantiomer has desirable properties, but the racemate has undesirable effects or has some toxic effect. In such a case, we might expect one day to see the appearance of two new drugs of abuse, with perhaps very different pharmacological effects, that turn out to be enantiomers of one another. Current drug-scheduling regulations, at least in the United States, cover all stereoisomers of a specific controlled substance.

2.2.2. Benzyl- and Phenylpiperazines

Following the appearance of N-benzylpiperazine (BZP) in the United States in 1996 (Austin and Monasterio, 2004), a number of substituted N-substituted piperazines appeared as drugs of abuse (Arbo et al., 2012). Although these molecules have not been as popular as some other types of drugs, they still represent a distinct class of designer drugs, and one might surmise that a variety of aromatic ring substituents can be introduced to provide new substances that might have abuse potential. Molecules that have so far appeared on the illicit market include the following.

Substituted piperazines often have been sold as ecstasy or ecstasy replacements with names such as A2, Bliss, Charge, Frenzy, Herbal Ecstasy, and Rapture, among others. These preparations often consist of 1-benzylpiperazine (BZP) and 1-(3-trifluoromethylphenyl) piperazine (TFMPP) in a 2:1 ratio, as estimated by the DEA System to Retrieve Information From Drug Evidence (STRIDE) program. It was have reported that a mixture of

BZP and TFMPP can mimic the effect of MDMA in humans. Similarly, drug users have posted experiences with meta-chlorophenylpiperazine (mCPP) to internet sites specializing in the dissemination of drug information, such as erowid.org and lycaeum.org, and the drug has been used as a positive control for MDMA in human studies (Tancer and Johanson, 2001, 2003). Interestingly, some cocaine (Buydens-Branch et al., 1997), alcohol (Benkelfat et al., 1991), and MDMA abusers (McCann et al., 1999) have reported "euphoric" responses to mCPP, perhaps explaining its recreational use.

Perhaps not surprisingly, based on their illicit use, piperazines are reported to have substrate activity at the dopamine and serotonin reuptake transporters, DAT and SERT, respectively, a pharmacology that is shared with MDMA and other psychostimulants. We have previously investigated the reinforcing and discriminative stimulus effects of BZP and TFMPP in rhesus monkeys (Fantegrossi et al., 2005). In these studies, BZP was self-administered and amphetamine-like in drug discrimination, whereas TFMPP was not self-administered and did not have amphetamine-like interoceptive effects. An extensive review on patterns and motivation of BZP use, target populations, legal status around the world, pharmacology, toxicology, kinetics and new developments in analytical and detection techniques has recently been published (Monteiro et al., 2013). In accordance with these studies, stimulant-like effects of BZP have been demonstrated in humans Campbell et al., 1972 (Campbell et al., 1972), rats (Baumann et al., 2005), and mice (Yarosh et al., 2007). The binding profile of TFMPP at various serotonin receptors is complex, as similar potencies have been reported for TFMPP at 5-HT_{1A}, 5-HT_{1B}, and 5-HT_{2C} receptors (Schoeffter and Hoyer, 1989). Additional studies have suggested that TFMPP may be either an antagonist (Conn and Sanders-Bush, 1987) or a weak partial agonist (Grotewiel et al., 1994) at 5-HT_{2A} receptors as well. This promiscuous pattern of binding to 5-HT receptors and monoamine transporters likely provides quite a bit of room for optimization of novel pharmacological entities built upon the piperazine scaffold. It seems likely that piperazine-like designer stimulants, psychedelics, and perhaps drugs of mixed action will appear as new drugs of abuse in the future.

2.2.3. Substituted Cathinone Derivatives

A variety of ring- and sidechain-substituted cathinones (β-ketophenethylamines) have appeared over the years. Initially, only cathinone and methcathinone were seen on the illicit market. More recently, mephedrone has been widely used, and its effects have been compared to those of MDMA. Mixtures of mephedrone and pyrovalerone have been marketed as "bath salts," although that is nothing but a marketing ploy, because they have no value in bathing or cleaning. One frequent constituent of these products is 3,4-methylenedioxypyrovalerone (MDPV), which is structurally similar to both MDMA and methamphetamine. MDPV surprisingly acts as a cocaine-like reuptake inhibitor at dopamine transporters (Baumann et al., 2013), although it has MDMA- and methamphetamine-like actions in mice (Fantegrossi et al., 2013). By inspection of the structures in Figure 8 it can be surmised that a variety of side chain lengths and amines

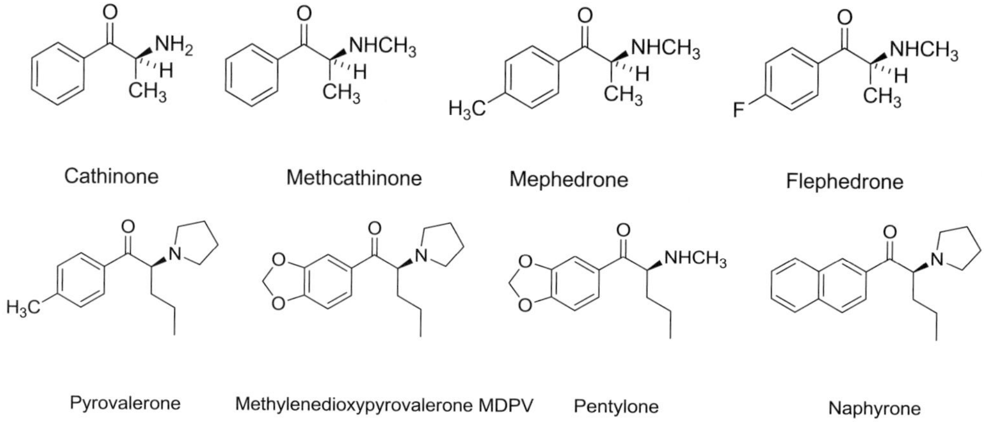

Cathinone   Methcathinone   Mephedrone   Flephedrone

Pyrovalerone   Methylenedioxypyrovalerone MDPV   Pentylone   Naphyrone

Figure 8 Examples of substituted cathinone derivatives.

can be used to create new compounds that likely will have similar pharmacology. And, as was noted for some phenethylamines, the presence of one or more chiral carbon atoms allows for stereoisomerism among the cathinones. Of all the designer drugs to have appeared recently, these may have some of the most serious reported adverse effects, primarily affecting the heart and cardiovascular system (see e.g. Warrick et al. (2012)). The reader should consult the chapter in this volume by Richard Glennon for a more detailed discussion of substituted cathinone derivatives.

2.3. SYNTHETIC CANNABINOIDS

2.3.1. Overview

Although structure-activity studies of the psychoactive component of marijuana (THC) have been carried out for decades, it is only recently that synthetic cannabinoids have become popular as recreational drugs. These compounds clearly are the result of mining the literature on cannabimetics.

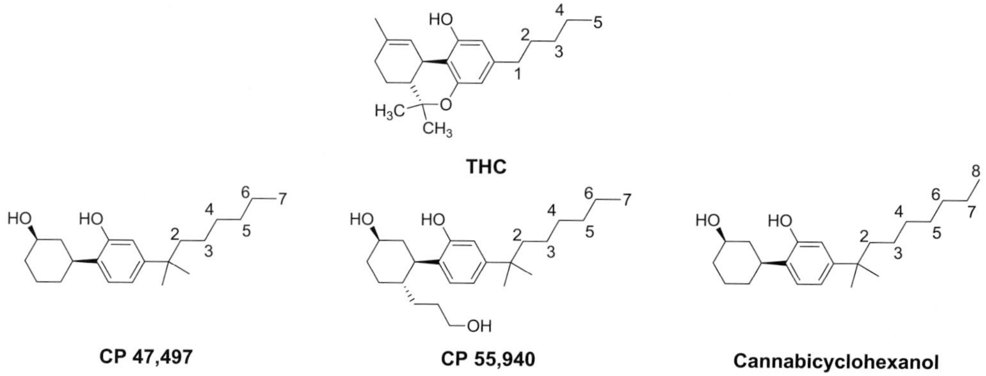

THC

CP 47,497   CP 55,940   Cannabicyclohexanol

The structurally dissected synthetic cannabinoids CP 47,497 and CP 55,940 were originally developed by Pfizer, and took advantage of earlier work showing that the 1,1-dimethylhelptyl alkyl chain provided optimum activity in THC congeners.

JWH-007   JWH-018   JWH-019   JWH-073

JWH-081   JWH-098   JWH-122   JWH-164

JWH-182   JWH-198   JWH-200   JWH-210

In recent years, products sold as incense in "head shops" have commonly been referred to as "K2" or "Spice" and have been shown to contain one or more of these synthetic cannabinoids. Although marketed as "natural" herbal blends, K2 products are usually comprised of nonpsychotropic plant matter adulterated with various mixtures of these chemicals, most of which are aminoalkylindoles (AAIs) of the JWH family (a series of WIN-55,212-2 analogues created in 1994 by Dr John W. Huffman for structure-activity relationship studies of the cannabinoid receptors). They, along with other synthetic cannabinoids, such as CP-47,497 and HU-210, were first discovered to be in "natural" herbal smoking blends in 2008. One particular AAI, JWH-018, is quite prevalent across many different brands and batches of K2 products. JWH-018 and other cannabinoids, such as $\Delta 9$-tetrahydrocannabinol ($\Delta 9$-THC), the major active constituent in marijuana, produce their psychoactivity by binding and activating, to varying degrees, cannabinoid 1 receptors (CB1Rs) in the CNS, which are Gi/o-protein coupled receptors (GPCRs).

Although the desired effects of K2 products are probably those that are generally similar to marijuana, the frequency and severity of adverse effects caused by synthetic cannabinoids certainly seems to be much greater than that of marijuana, which has been used for millennia and is the most commonly abused illegal drug in the U.S. Although smoking or oral consumption of marijuana acutely produces relatively mild and tolerable side effects in most users, such as appetite stimulation and orthostatic hypotension, it very rarely causes the adverse effects observed rather commonly with similar use of K2 products, such as hypertension, agitation, hallucinations, psychoses, seizures, and panic attacks. In one case, seizures and supraventricular tachycardia were characterized after ingestion of pure JWH-018; it should be noted, however, that the afflicted user dissolved "a spoonful" of compound in a mug of warm ethanol in order to swallow the drug in a presumably very large bolus dose. In extreme THC overdose cases, similar symptoms can be observed but they are not generally associated with marijuana use. In addition to acute adverse effects produced by K2, one case report indicates that chronic abuse may also result in a severe withdrawal and dependence syndrome. The use of K2 has even been causally linked to at least one death by overdose and has been implicated for likely involvement in several other fatalities, resulting in over 2500 calls to poison control centers in 2010 alone and numerous visits to emergency departments across the United States and in Europe.

The rapid increase in recreational use of synthetic cannabinoids, their current inability to be detected by standard drug urine tests, and the constant introduction of new structurally similar products of unknown content pose a significant risk to public health. Most importantly, the pharmacological and toxicological profiles of these products are virtually unknown, and the mechanisms underlying the discrepancies in the frequency and severity of K2 adverse effects relative to the well-established cannabis have yet to be elucidated.

Internet sites may sell either "herbal" blends (plant material impregnated with active compound) or the pure drugs themselves. In one case, interested potential users were offered instructions in how to make their own smoking blend. After purchasing 1 g of JWH-018 ($50–$70 on various websites), they were told to obtain mullein or marshmallow leaves as a substrate, acetone to dissolve the pure compound, and an acetone-proof spray bottle to distribute the drug solution. Users were instructed to dissolve 1 g JWH-018 in 4 ml acetone and place the resulting solution into the spray bottle. The instructions continued, "Now this is the most dangerous part. You must spray the leaves as evenly as possible, or you can get "hotspots" or localized areas in your mixture that have much higher concentrations of JWH-018 that can be dangerous." Indeed, we have detected these "hot spots" even in commercial preparations, where different extractions from a single divided sample can contain 2–3 times the amount of active compound (data presented by Cindy Moran, Arkansas State Crime Laboratory, at the 2011 College on Problems of Drug Dependence.) Similarly, there is no consistency within a given "brand" of these commercial preparations, as the amount of compound varies from lot to lot, and even the identity of the active compounds themselves can change over time. All of these factors could easily lead to overdose, even by the most cautious user.

UR-144: R = CH$_3$

XLR-11: R = CH$_2$F

Recently, a few novel cannabinoid compounds have emerged with structures that differ from those of Δ^9-THC or the AAIs. These compounds would not likely be captured by current analogue scheduling laws that require "substantial chemical similarity" between the novel compound and a previously scheduled drug of abuse, which may explain their sudden emergence onto the drug scene. Examples of such compounds include (1-pentylindol-3-yl)-(2,2,3,3-tetramethylcyclopropyl)methanone (UR-144), and its 5"-fluoro analogue (1-(5-fluoropentyl)-1H-indol-3-yl)(2,2,3,3-tetramethylcyclopropyl)methanone (XLR-11). Both UR-144 and XLR-11 have recently been detected in herbal smoking blends, first in New Zealand, but now in the US as well. Interestingly, UR-144 (and presumably XLR-11 as well) acts as a selective and highly efficacious agonist at cannabinoid CB2 receptors, and has substantially lower affinity for the CB1 receptor (Frost et al., 2010). This finding challenges the notion that CB1 receptors are the primary site of action for psychoactive cannabinoids, and perhaps implies that new compounds of this class may specifically target CB2. Indeed, users describing their experiences with these compounds on various internet forums report that UR-144 produces a "less freaky, wired high than JWH-081" or "a nice high similar to THC/Cannabis."

Health problems associated with the use of "Spice" products are reported to be similar to those after cannabis use. For some particular products, however, e.g. "Lava Red", increasing numbers of users have been hospitalized with severe intoxications. A potential problem to be aware of is the unknown cumulative toxic effects these compounds or their metabolites may have. In this regard, we have recently reported that several phase I hydroxylated metabolites of JWH-018 and JWH-073 retain high affinity for cannabinoid CB1 receptors, and display a range of efficacies from neutral antagonism, to partial agonism, to full agonism. Similar results have been published with the fluorinated analogue of JWH-018, which is also a prevalent constituent of K2/'Spice' products, AM-2201 (Chimalakonda et al., 2013). More recently, other phase I metabolites of these same synthetic cannabinoids were demonstrated to act as agonists at cannabinoid CB2 receptors, where they induce qualitatively and quantitively distinct signaling events, as compared to traditional cannabinoids (Rajasekaran et al., 2013). Finally, a phase II glucuronidated metabolite of JWH-018 exhibiting antagonist affinity at CB1 receptors has been described (Seely et al., 2012). Liver metabolism of xenobiotics is highly variable across the population, so the effects of these active metabolites might be expected to blunt (in the case of antagonist metabolites) or enhance (in the case of high-efficacy metabolites) the effects of the parent drug, depending on one's individual liver enzyme profile, perhaps resulting in wildly unpredictable effects across individuals using the same drug supply.

2.3.2. Toxicity of Synthetic Cannabinoids

Recent reports of acute kidney injury following the use of synthetic cannabinoids (Thornton et al., 2013), particularly XLR-11 (CDC, 2013), as well as reports of seizures following administration of AM-2201 (McQuade et al., 2013) or various compounds from the JWH series (Hermanns-Clausen et al., 2013) raise concerns about the potential toxicity associated with use of these substances. One may speculate that the compounds containing the naphthyl moiety could have carcinogenic potential but without broad screening across a large library of receptors and channels one cannot know what potential for severe or even life-threatening intoxications might exist, particularly in overdose. Further, a pattern of chronic use for these compounds increases concerns about toxicity. It also might be noted that some of these compounds, such as HU-210, CP-55,940 and WIN-55,212-2 are full agonists at the CB1 receptor, whereas THC acts only as a partial agonist. Recently, an increase in the number and severity of symptoms observed in hospitalized persons after consumption of herbal mixtures containing JWH-122, e.g. "Lava Red" and "OMG", has been reported in Germany and Italy. Some of these patients suffered from generalized muscular spasms and/or loss of consciousness that required artificial ventilation. Such severe symptoms had not been reported with JWH-018, emphasizing the fact that slight changes in molecular structure can lead to a dramatic increase in toxicity.

2.4. SALVINORIN

Salvinorin A (SVA) is one of several diterpenes isolated from the Mexican mint Ska Maria Pastora (*Salvia divinorum*). This plant has been used by indigenous peoples in the Oaxaca region of Mexico for hundreds of years (Valdes III et al., 1983; Sheffler and

Roth, 2003), presumably for its psychoactive effects. Both the plant and SVA extracts are now widely available via the internet, where they are marketed as legal short-acting hallucinogens. In this regard, the psychoactive potency of SVA rivals that of lysergic acid diethylamide, although the intoxication induced by SVA is reported to be qualitatively different from that produced by the classical serotonergic hallucinogens (Siebert, 1994). Interestingly, the mechanism of action for SVA was unknown until Roth and colleagues (2002) demonstrated that this compound binds as a potent and selective κ-opioid receptor agonist. The agonist effects of SVA at κ-opioid receptors were further elaborated when in vitro in studies demonstrated that this compound functions as a full agonist at this receptor (Chavkin et al., 2004). SVA is thus the first naturally occurring exogenous κ-opioid receptor agonist to be discovered. Similarly, SVA is the only non-nitrogenous compound known to bind to opioid receptors. The structure of SVA is lipid-like, completely distinct from those of all previously identified opioid ligands, and thus defines a new structural class of κ-opioid receptor selective drugs. The action of SVA is extremely brief, yet a number of analogues of SVA are now known that have a longer duration of action. There are synthetic challenges to making longer-acting SVA analogues, and the only economical approach is to start with SVA itself, so the likelihood that any of them might appear on the street seems low. Nevertheless, given the high potency of SVA, and the fact that *S. divinorum* can be grown on a relatively large scale, it is not inconceivable that a long-acting analogue of SVA could appear on the street. Given the tendency of SVA to cause disorientation and loss of insight, a longer acting version of SVA could be very problematic.

2.5. DISSOCIATIVES

Although this category has so far not expanded like the others, the potential still exists for the creation of a number of new designer drugs to emerge. To date, the only member of this class to gain popularity is methoxetamine (Mket; MXE) (Corazza et al., 2012). It is a structural analogue of PCP (phencyclidine) and ketamine, which are non-competitive NMDA antagonists, and high affinity binding of methoxetamine to this same receptor has recently been demonstrated in vitro (Roth et al., 2013).

Methoxetamine   Ketamine   Phencyclidine

Methoxetamine has noticeable effects after a 20 mg sublingual dose, and higher doses lead to a disconnection from reality, with loss of motor coordination and sensory distortions. A dose of 60 mg and higher can produce feelings of floating or falling into another place that is different from normal reality. The hallucinations can be realistic and frightening, although they may not be well remembered after the drug effect wears off. The drug can cause bizarre and reckless behavior. Adverse effects seen in emergency room situations can include hypertension, tachycardia, and agitation. More extreme effects can also be observed after high doses or in combination with other drugs, including apparent reversible cerebellar toxicity (Shields et al., 2012) or even death (Wikström et al., 2013). The psychoactive effects of methoxetamine are longer-lasting than for ketamine, and can be unpredictable. Ketamine can lead to addiction, and there are some suggestions that methoxetamine also may present risk of addiction.

There are at least two locations in the molecule where structural modifications could be made relatively easily by clandestine chemists. In particular, the aromatic ring could be substituted by a variety of substituents, and the starting materials, substituted benzonitriles, are generally available commercially. The method of synthesis precludes the production of anything other than a secondary amine, but one would certainly expect that propylamine could be employed to give an active congener. In addition, although phencyclidine (PCP) is not a popular substance, substitutions on the aromatic right might lead to more acceptable materials that would technically be "legal."

Legal Control Issues

The first attempt to regulate new "designer drugs" arose as a result of illicit production in the early 1980s of 4-methylfentanyl, an opiate that was about 30 times more potent than fentanyl itself. Marketed as "China White," it had resulted in a number of overdose deaths. At that time, any substance that was controlled had to be explicitly named, and the chemical structure had to be known. Realizing that there were a number of similar modifications of fentanyl that could be produced with equally serious properties, but unable to predict which of them might arise, Congress enacted the Controlled Substance Analogue Enforcement Act of 1986. This Act sets out the following definition of a "controlled substance analogue" as:

1. the chemical structure of which is substantially similar to the chemical structure of a controlled substance in schedule I or II;

2. which has a stimulant, depressant, or hallucinogenic effect on the central nervous system that is substantially similar to or greater than the stimulant, depressant, or hallucinogenic effect on the central nervous system of a controlled substance in schedule I or II;

3. or with respect to a particular person, which such person represents to have a stimulant, depressant, or hallucinogenic effect on the central nervous system that is substantially similar to or greater than the stimulant, depressant, or hallucinogenic effect on the central nervous system of a controlled substance in schedule I or II.

The first point (1) is considered necessary, but either (2) or (3) may serve to define the activity and relate to point (1). Thus, the act has two "prongs," each of which must be satisfied. It served for about two decades to give enforcement agencies authority to arrest and prosecute manufacturers and distributors of new "designer drugs" as quickly as the structures could be identified. This act was particularly useful in controlling a number of simple chemical analogues of MDMA. With the plethora of new structures that began to appear in the 1990s, however, this law did not prove to be comprehensive enough. In particular, the synthetic cannabinoids did not fulfill the first prong of the act (1), in that they had no similarity to THC or to the chemical structures of other controlled substances. Furthermore, legal issues also had frequently arisen in interpreting what was meant by "substantially similar."

When the illicit market started to overflow with new and unknown drug molecules, additional regulation was clearly warranted. This action took the form of H.R. 1254, the Synthetic Drug Control Act of 2011. This new law amended the earlier Controlled Substances Act by adding a specific provision to schedule "cannabimimetic agents," defined as "any substance that is a cannabinoid receptor type 1 (CB1 receptor) agonist as demonstrated by binding studies and functional assays within any of the following structural classes." and then listing five very broad chemical types, and explicitly naming 15 distinct compounds within those classes. The specification that the molecules should have cannabimimetic pharmacology offers the possibility of including new molecules that might arise that were not included within the five broad chemical types specified.

The new law goes on to include a number of "Other Drugs," and includes all of the substituted cathinones that have made an appearance or seem likely to appear on the market, as well as one MDMA-like compound (e.g. MDAI). The new law also then schedules a series of hallucinogenic phenethylamines (not amphetamines, which were covered by the earlier act) that includes 2C-E, 2C-D, 2C-C, 2C-I, 2C-T-2, 2C-T-4, 2C-H, 2C-N, and 2C-P. Although some of these have never had a significant presence on the black market, the DEA is perhaps attempting to anticipate possible problems. Importantly, with these compounds now explicitly described in the amended law, it broadens the range of substances covered in the original Controlled Substance Analogue Enforcement Act that will be considered "substantially similar." Interestingly, for all practical purposes 2C-H is inactive, but it is the precursor to several of the newly restricted compounds.

3. CONCLUSIONS

In summary, although underground chemists mining the scientific literature have a vast database from which to identify potential new "research chemicals," it will be increasingly difficult for them to market them as "legal highs." In the meantime, one hopes that some new chemical will not emerge that proves to have unexpected toxicity, with disastrous consequences for the adolescent and young adult population who are the main consumers of these "research chemicals."

REFERENCES

Anderson III, G.M., Braun, G., Braun, U., Nichols, D.E., Shulgin, A.T., 1978. Absolute configuration and psychotomimetic activity. NIDA Res. Monogr. 8–15.

Andreasen, M.F., Telving, R., Birkler, R.I., Schumacher, B., Johannsen, M., 2009. A fatal poisoning involving Bromo-Dragonfly. Forensic Sci. Int. 183, 91–96.

Araiza, V., 2012. New Street Drug Causing Concern among Medics. http://www.wsfa.com/story/16977573/new-street-drug-causng-concern-among-medics.

Arbo, M.D., Bastos, M.L., Carmo, H.F., 2012. Piperazine compounds as drugs of abuse. Drug Alcohol Depend. 122, 174–185.

Austin, H., Monasterio, E., 2004. Acute psychosis following ingestion of 'Rapture'. Australas Psychiatry 12, 406–408.

Baron, M., Elie, M., Elie, L., 2011. An analysis of legal highs: do they contain what it says on the tin? Drug Test Anal. 3 (9), 576–581.

Baumann, M.H., Clark, R.D., Budzynski, A.G., Partilla, J.S., Blough, B.E., Rothman, R.B., 2005. N-substituted piperazines abused by humans mimic the molecular mechanism of 3,4-methylenedioxy-methamphetamine (MDMA, or 'Ecstasy'). Neuropsychopharmacology 30, 550–560.

Baumann, M.H., Partilla, J.S., Lehner, K.R., Thorndike, E.B., Hoffman, A.F., Holy, M., Rothman, R.B., Goldberg, S.R., Lupica, C.R., Sitte, H.H., Brandt, S.D., Tella, S.R., Cozzi, N.V., Schindler, C.W., 2013. Powerful cocaine-like actions of 3,4-Methylenedioxypyrovalerone (MDPV), a principal constituent of psychoactive 'bath salts' products. Neuropsychopharmacology 38 (4), 552–562.

Benkelfat, C., Murphy, D.L., Hill, J.L., George, D.T., Nutt, D., Linnoila, M., 1991. Ethanollike properties of the serotonergic partial agonist m-chlorophenylpiperazine in chronic alcoholic patients. Arch. Gen. Psychiatry 48, 383.

Bowen, J.S., Davis, G.B., Kearney, T.E., Bardin, J., 1983. Diffuse vascular spasm associated with 4-bromo-2,5-dimethoxyamphetamine ingestion. JAMA 249, 1477–1479.

Braden, M.R., Parrish, J.C., Naylor, J.C., Nichols, D.E., 2006. Molecular interaction of serotonin 5-HT2A receptor residues Phe339(6.51) and Phe340(6.52) with superpotent N-benzyl phenethylamine agonists. Mol. Pharmacol. 70, 1956–1964.

Buydens-Branch, L., Branchey, M., Fergeson, P., Hudson, J., McKernin, C., 1997. The meta-chlorophenylpiperazine challenge test in cocaine addicts: hormonal and psychological responses. Biol. Psychiatry 41, 1071–1086.

Campbell, H., Cline, W., Evans, M., Lloyd, J., Peck, A.W., 1972. Proceedings: comparison of the effects of dexamphetamine and 1-benzylpiperazine in former addicts. Br. J. Pharmacol. 44, pp. 369–370.

Centers for Disease Control and Prevention (CDC), 2013. Acute kidney injury associated with synthetic cannabinoid use–multiple states, 2012. MMWR Morb. Mortal Wkly Rep. 62 (6), 93–98.

Chavkin, C., Sud, S., Jin, W., Stewart, J., Zjawiony, J.K., Siebert, D.J., Toth, B.A., Hufeisen, S.J., Roth, B.L., 2004. Salvinorin A, an active component of the hallucinogenic sage *salvia divinorum* is a highly efficacious kappa-opioid receptor agonist: structural and functional considerations. J. Pharmacol. Exp. Ther. 308, 1197–1203.

Chimalakonda, K.C., Seely, K.A., Bratton, S.M., Brents, L.K., Moran, C.L., Endres, G.W., James, L.P., Hollenberg, P.F., Prather, P.L., Radominska-Pandya, A., Moran, J.H., 2013. Cytochrome P450-mediated oxidative metabolism of abused synthetic cannabinoids found in K2/Spice: identification of novel cannabinoid receptor ligands. Drug Metab. Dispos. 40 (11), 2174–2184.

Conn, P.J., Sanders-Bush, E., 1987. Relative efficacies of piperazines at the phosphoinositide hydrolysis-linked serotonergic (5-HT-2 and 5-HT-1c) receptors. J. Pharmacol. Exp. Ther. 242, 552–557.

Corazza, O., Schifano, F., Simonato, P., Fergus, S., Assi, S., Stair, J., Corkery, J., Trincas, G., Deluca, P., Davey, Z., Blaszko, U., Demetrovics, Z., Moskalewicz, J., Enea, A., di Melchiorre, G., Mervo, B., di Furia, L., Farre, M., Flesland, L., Pasinetti, M., Pezzolesi, C., Pisarska, A., Shapiro, H., Siemann, H., Skutle, A., Enea, A., di Melchiorre, G., Sferrazza, E., Torrens, M., van der Kreeft, P., Zummo, D., Scherbaum, N., 2012. Phenomenon of new drugs on the Internet: the case of ketamine derivative methoxetamine. Hum. Psychopharmacol. 27 (2), 145–149.

Elz, S., Klass, T.H.R., Wamke, U., Pertz, H.H., 2002. Development of highly potent partial agonists and chiral antagonists as tools fo the study of 5-HT2A-receptor mediated functions. Nauyn Schmiedebergs Arch. Pharmacol. 365 (Suppl. 1), R29.

Fantegrossi, W.E., Winger, G., Woods, J.H., Woolverton, W.L., Coop, A., 2005. Reinforcing and discriminative stimulus effects of 1-benzylpiperazine and trifluoromethylphenylpiperazine in rhesus monkeys. Drug Alcohol Depend. 77, 161–168.

Fantegrossi, W.E., Gannon, B.M., Zimmerman, S.M., Rice, K.C., 2013. In vivo effects of abused 'bath salt' constituent 3,4-methylenedioxypyrovalerone (MDPV) in mice: drug discrimination, thermoregulation, and locomotor activity. Neuropsychopharmacology 38 (4), 563–573.

Frost, J.M., Dart, M.J., Tietje, K.R., Garrison, T.R., Grayson, G.K., Daza, A.V., El-Kouhen, O.F., Yao, B.B., Hsieh, G.C., Pai, M., Zhu, C.Z., Chandran, P., Meyer, M.D., 2010. Indol-3-ylcycloalkyl ketones: effects of N1 substituted indole side chain variations on CB(2) cannabinoid receptor activity. J. Med. Chem. 53, 295–315.

Geller, L., 2012. Kids Overdosing on New Drug. WWBT NBC12. http://www.nbc12.com/story/16964534/kids-overdosing-on-new-drug.

Grotewiel, M.S., Chu, H., Sanders-Bush, E., 1994. m-chlorophenylpiperazine and m-trifluoromethylphenylpiperazine are partial agonists at cloned 5-HT2A receptors expressed in fibroblasts. J. Pharmacol. Exp. Ther. 271, 1122–1126.

Heim, R., 2003. Synthese und Pharmakologie potenter 5-HT2A-Rezeptoragonisten mit N-2-Methoxybenzyl-Partialstruktur Entwicklung eines neen Struktur-Wirkungskonzepts (Ph.D. thesis). Freien Universität Berlin, Berlin.

Hermanns-Clausen, M., Kneisel, S., Szabo, B., Auwärter, V., 2013. Acute toxicity due to the confirmed consumption of synthetic cannabinoids: clinical and laboratory findings. Addiction 108 (3), 534–544.

McCann, U.D., Eligulashvili, V., Mertl, M., Murphy, D.L., Ricaurte, G.A., 1999. Altered neuroendocrine and behavioral responses to m-chlorophenylpiperazine in 3,4-methylenedioxymethamphetamine (MDMA) users. Psychopharmacology (Berl.) 147, 56–65.

McLean, T.H., Parrish, J.C., Braden, M.R., Marona-Lewicka, D., Gallardo-Godoy, A., Nichols, D.E., 2006. 1-Aminomethylbenzocycloalkanes: conformationally restricted hallucinogenic phenethylamine analogues as functionally selective 5-HT2A receptor agonists. J. Med. Chem. 49, 5794–5803.

McQuade, D., Hudson, S., Dargan, P.I., Wood, D.M., 2013. First European case of convulsions related to analytically confirmed use of the synthetic cannabinoid receptor agonist AM-2201. Eur. J. Clin. Pharmacol. 69 (3), 373–376.

Monteiro, M.S., Bastos Mde, L., Guedes de Pinho, P., Carvalho, M., 2013. Update on 1-benzylpiperazine (BZP) party pills. Arch. Toxicol. 87 (6), 929–947.

Nichols, D.E., 2004. Hallucinogens. Pharmacol. Ther. 101, 131–181.

Rajasekaran, M., Brents, L.K., Franks, L.N., Moran, J.H., Prather, P.L., 2013. Human metabolites of synthetic cannabinoids JWH-018 and JWH-073 bind with high affinity and act as potent agonists at cannabinoid type-2 receptors. Toxicol. Appl. Pharmacol. 269 (2), 100–108.

Roth, B.L., Baner, K., Westkaemper, R., Siebert, D., Rice, K.C., Steinberg, S., Ernsberger, P., Rothman, R.B., 2002. Salvinorin A: a potent naturally occurring nonnitrogenous kappa opioid selective agonist. Proc. Natl. Acad. Sci. U.S.A. 99, 11934–11939.

Roth, B.L., Gibbons, S., Arunotayanun, W., Huang, X.P., Setola, V., Treble, R., Iversen, L., 2013. The ketamine analogue methoxetamine and 3- and 4-methoxy analogues of phencyclidine are high affinity and selective ligands for the glutamate NMDA receptor. PLoS One 8 (3), e59334.

Seely, K.A., Brents, L.K., Radominska-Pandya, A., Endres, G.W., Keyes, G.S., Moran, J.H., Prather, P.L., 2012. A major glucuronidated metabolite of JWH-018 is a neutral antagonist at CB1 receptors. Chem. Res. Toxicol. 25 (4), 825–827.

Schoeffter, P., Hoyer, D., 1989. Interaction of arylpiperazines with 5-HT1A, 5-HT1B, 5-HT1C and 5-HT1D receptors: do discriminatory 5-HT1B receptor ligands exist? Naunyn Schmiedebergs Arch. Pharmacol. 339, 675–683.

Sheffler, D.J., Roth, B.L., 2003. Salvinorin A: the "magic mint" hallucinogen finds a molecular target in the kappa opioid receptor. Trends Pharmacol. Sci. 24, 107–109.

Shields, J.E., Dargan, P.I., Wood, D.M., Puchnarewicz, M., Davies, S., Waring, W.S., 2012. Methoxetamine associated reversible cerebellar toxicity: three cases with analytical confirmation. Clin. Toxicol. 50 (5), 438–440.

Shulgin, A., Shulgin, A., 1991. PIHKAL: A Chemical Love Story. Transform Press, Berkeley, CA 94701.

Siebert, D.J., 1994. *Salvia divinorum* and salvinorin A: new pharmacologic findings. J. Ethnopharmacol. 43, 53–56.

Tancer, M., Johanson, C.E., 2003. Reinforcing, subjective, and physiological effects of MDMA in humans: a comparison with d-amphetamine and mCPP. Drug Alcohol Depend. 72, 33–44.

Tancer, M.E., Johanson, C.E., 2001. The subjective effects of MDMA and mCPP in moderate MDMA users. Drug Alcohol Depend. 65, 97–101.

Thornton, S.L., Wood, C., Friesen, M.W., Gerona, R.R., 2013. Synthetic cannabinoid use associated with acute kidney injury. Clin. Toxicol. 51 (3), 189–190.

Valdes III, L.J., Diaz, J.L., Paul, A.G., 1983. Ethnopharmacology of ska Maria Pastora (*Salvia divinorum*, Epling and Jativa-M.). J. Ethnopharmacol. 7, 287–312.

Warrick, B.J., Wilson, J., Hedge, M., Freeman, S., Leonard, K., Aaron, C., 2012. Lethal serotonin syndrome after methylone and butylone ingestion. J. Med. Toxicol. 8, 65–68.

Wikström, M., Thelander, G., Dahlgren, M., Kronstrand, R., 2013. An accidental fatal intoxication with methoxetamine. J. Anal. Toxicol. 37 (1), 43–46.

Winek, C.L., Collom, W.D., Bricker, J.D., 1981. A death due to 4-bromo-2,5-dimethoxyamphetamine. Clin. Toxicol. 18, 267–271.

Yarosh, H.L., Katz, E.B., Coop, A., Fantegrossi, W.E., 2007. MDMA-like behavioral effects of N-substituted piperazines in the mouse. Pharmacol. Biochem. Behav. 88, 18–27.

Zuba, D., Byrska, B., 2013. Prevalence and co-existence of active components of 'legal highs'. Drug Test Anal. http://dx.doi.org/10.1002/dta. 1365 (Epub ahead of print).

INDEX

Note: Page numbers followed by "f" denote figures; "t" tables.